NANOTECHNOLOGY FOR BIOMEDICAL IMAGING AND DIAGNOSTICS

NANOTECHNOLOGY FOR BIOMEDICAL IMAGING AND DIAGNOSTICS

From Nanoparticle Design to Clinical Applications

Edited by

MIKHAIL Y. BEREZIN

WILEY

Published by John Wiley & Sons, Inc., Hoboken, New Jersey
Published simultaneously in Canada

Library of Congress Cataloging-in-Publication Data

Nanotechnology for biomedical imaging and diagnostics : from nanoparticle design to clinical applications / [edited by] Mikhail Y. Berezin.
p. ; cm.
Includes bibliographical references and index.
ISBN 978-1-118-12118-4 (cloth : alk. paper)
I. Berezin, Mikhail Y., editor.
[DNLM: 1. Diagnostic Imaging. 2. Nanotechnology. 3. Nanoparticles. WN 180]
RC78.7.D53
616.07′54–dc23

2014029620

Printed in the United States of America

oBook ISBN: 9781118873151
ePDF ISBN: 9781118873175
ePub ISBN: 9781118873144

10 9 8 7 6 5 4 3 2 1

This book is dedicated to the memory of my parents Yuliy and Esfir and my twin sister Yelena. Their courageous battle with cancer motivated me to start and complete this book.

CONTENTS

CONTRIBUTORS ix

PREFACE xiii

ACKNOWLEDGMENTS xix

1 **Historical Perspective on Nanoparticles in Imaging from 1895 to 2000** 1
Mikhail Y. Berezin

PART I NANOPARTICLE DESIGN, SYNTHESIS AND CHARACTERIZATION 25

2 **Iron Oxide-Based Magnetic Nanoparticles Synthesized from the Organic Solution Phase for Advanced Biological Imaging** 27
Sen Zhang and Shouheng Sun

3 **Lipid-Based Pharmaceutical Nanocarriers for Imaging Applications** 49
Tamer Elbayoumi and Vladimir Torchilin

4 **Hollow Nanocapsules in Biomedical Imaging Applications** 83
Sergey A. Dergunov and Eugene Pinkhassik

5 **Nanoparticles as Contrast Agents for Optoacoustic Imaging** 111
Anton V. Liopo and Alexander A. Oraevsky

6 **Nanoparticles for Bioimaging: Analytical Characterization and Measurements** 151
Kate Nelson, Patrick Winter, Monica Shokeen, Steven Wang, and Mikhail Y. Berezin

PART II IMAGING MODALITIES: FROM CONCEPTS TO APPLICATIONS 193

7 **Radio-Labeled Nanoparticles for Biomedical Imaging** 195
Tolulope Aweda, Deborah Sultan, and Yongjian Liu

8 **MRI with Gadolinium-Based Nanoparticles** 223
François Guérard, Geoffrey L. Ray, and Martin W. Brechbiel

9 ***In Vivo* Molecular Fluorescence Imaging** 263
Yasaman Ardeshirpour, Victor Chernomordik, Moinuddin Hassan, Dan Sackett, and Amir H. Gandjbakhche

10 **Photoacoustic and Ultrasound Imaging with Nanosized Contrast Agents** 293
Mansik Jeon and Chulhong Kim

11 **Surface-Enhanced Raman Scattering-Based Bioimaging** 325
Limei Tian and Srikanth Singamaneni

PART III NANOTECHNOLOGY IN BIOMEDICAL IMAGING AND BEYOND 347

12 **Pandia®: Gold Nanorods and their Applications in Cancer Therapy and *In Vivo* Imaging in Companion Animals and their Potential Application to Humans** 349
Christian Schoen and Cheryl London

13 **Imaging Genetic Information** 373
John-Stephen Taylor

14 **The Application of Plant Viral Nanoparticles in Tissue-Specific Imaging** 401
Amy M. Wen, Choi-Fong Cho, John D. Lewis, and Nicole F. Steinmetz

15 **Design and Development of Theranostic Nanomedicines** 429
Jelena M. Janjic and Mingfeng Bai

16 **Animal Models for Preclinical Imaging** 467
Grayson Talcott and Walter J. Akers

INDEX 487

CONTRIBUTORS

Walter J. Akers Department of Radiology, Washington University School of Medicine, St. Louis, MO, USA

Yasaman Ardeshirpour Section on Functional and Analytical Biophotonics, Program of Pediatrics Imaging and Tissue Sciences, Eunice Kennedy Shriver National Institutes of Child Health and Human Development, National Institutes of Health, Bethesda, MD, USA

Tolulope Aweda Department of Radiology, Washington University School of Medicine, St. Louis, MO, USA

Mingfeng Bai Molecular Imaging Lab, Department of Radiology, University of Pittsburgh, Pittsburgh, PA, USA

Mikhail Y. Berezin Department of Radiology, Washington University School of Medicine, St. Louis, MO, USA

Martin W. Brechbiel Radioimmune & Inorganic Chemistry Section, Radiation Oncology Branch, NCI, National Institutes of Health, Bethesda, MD, USA

Victor Chernomordik Section on Functional and Analytical Biophotonics, Program of Pediatrics Imaging and Tissue Sciences, Eunice Kennedy Shriver National Institutes of Child Health and Human Development, National Institutes of Health, Bethesda, MD, USA

Choi-Fong Cho Harvey Cushing Neuro-Oncology Laboratories, Brigham and Women's Hospital, Harvard Medical School, Boston, MA, USA

Sergey A. Dergunov Department of Chemistry, Saint Louis University, St. Louis, MO, USA

Tamer Elbayoumi Department of Pharmaceutical Sciences, Midwestern University, Glendale, AZ, USA

Amir H. Gandjbakhche Section on Functional and Analytical Biophotonics, Program of Pediatrics Imaging and Tissue Sciences, Eunice Kennedy Shriver National Institutes of Child Health and Human Development, National Institutes of Health, Bethesda, MD, USA

François Guérard Radioimmune & Inorganic Chemistry Section, Radiation Oncology Branch, NCI, National Institutes of Health, Bethesda, MD, USA

Moinuddin Hassan Section on Functional and Analytical Biophotonics, Program of Pediatrics Imaging and Tissue Sciences, Eunice Kennedy Shriver National Institutes of Child Health and Human Development, National Institutes of Health, Bethesda, MD, USA

Jelena M. Janjic Graduate School of Pharmaceutical Sciences, Mylan School of Pharmacy, Duquesne University, Pittsburgh, PA, USA

Mansik Jeon Departments of Electrical Engineering and Creative IT Engineering, Pohang University of Science and Technology, Pohang, Republic of Korea

Chulhong Kim Departments of Electrical Engineering and Creative IT Engineering, Pohang University of Science and Technology, Pohang, Republic of Korea

John D. Lewis Translational Prostate Cancer Research Group, Department of Oncology, University of Alberta, Edmonton, Alberta, Canada

Anton V. Liopo TomoWave Laboratories, Inc., Houston, TX, USA

Yongjian Liu Department of Radiology, Washington University School of Medicine, St. Louis, MO, USA

Cheryl London Department of Veterinary Biosciences, Ohio State University, Columbus, OH, USA

Kate Nelson Nano Research Facility, Washington University School of Medicine, St. Louis, MO, USA

Alexander A. Oraevsky TomoWave Laboratories, Inc., Houston, TX, USA

Eugene Pinkhassik Department of Chemistry, Saint Louis University, St. Louis, MO, USA

Geoffrey L. Ray Radioimmune & Inorganic Chemistry Section, Radiation Oncology Branch, NCI, National Institutes of Health, Bethesda, MD, USA

Dan Sackett Section of Cell Biophysics, Program in Physical Biology, Eunice Kennedy Shriver National Institute of Child Health and Human Development, National Institutes of Health, Bethesda, MD, USA

Christian Schoen Nanopartz Inc., Loveland, CO, USA

Monica Shokeen Department of Radiology, Washington University School of Medicine, St. Louis, MO, USA

Srikanth Singamaneni Department of Mechanical Engineering and Materials Science, Washington University, St. Louis, MO, USA

Nicole F. Steinmetz Departments of Biomedical Engineering, Radiology, Materials Science and Engineering, and Macromolecular Science and Engineering, Case Western Reserve University, Cleveland, OH, USA

Department of Macromolecular Science and Engineering, Case Western Reserve University, Cleveland, OH, USA

Deborah Sultan Department of Radiology, Washington University School of Medicine, St. Louis, MO, USA

Shouheng Sun Department of Chemistry, Brown University, Providence, RI, USA

Grayson Talcott Department of Radiology, Washington University School of Medicine, St. Louis, MO, USA

John-Stephen Taylor Department of Chemistry, Washington University, St. Louis, MO, USA

Limei Tian Department of Mechanical Engineering and Materials Science, Washington University, St. Louis, MO, USA

Vladimir Torchilin Department of Pharmaceutical Sciences and Center for Pharmaceutical Biotechnology and Nanomedicine, Northeastern University, Boston, MA, USA

Steven Wang Department of Radiology, Washington University School of Medicine, St. Louis, MO, USA

Amy M. Wen Department of Biomedical Engineering, Case Western Reserve University, Cleveland, OH, USA

Patrick Winter Cincinnati Children's Hospital, Imaging Research Center, Cincinnati, OH, USA

Sen Zhang Department of Chemistry, Brown University, Providence, RI, USA

PREFACE

SCOPE OF THE BOOK: IMAGING AND NANOPARTICLES

From the first X-rays at the beginning of the twentieth century to the ultrasonic, computed tomography (CT), magnetic resonance imaging (MRI), optical, and nuclear modalities of the twenty-first century, medical imaging has transformed the practice of diagnosis. Today, more than 300 million imaging services are provided to patients in the United States every year. The ability to noninvasively scan for pathologies in a relatively painless and facile way has dramatically increased the effectiveness of medicine, leading to more efficient treatments and a number of benefits including the reduction of hospital length stay to an increase in human lifespan. Nanotechnology promises to advance medical imaging to the next level by increasing the *resolution* of current techniques. High resolution is especially important for early diagnostics before complications occur since a number of serious illnesses can be successfully treated if detected early. This book also describes how the unique designs of nanoconstructs are expected to enhance the *specificity* of targeted imaging. The book reflects upon the increasing role of nanomaterials in biological and medical imaging research (from <0.1% in 1976 to >3% in 2013), presents the state-of-the-art current research, and delves into future research directions.

So, why are nanoparticles combined with imaging? The simplest answer is to improve the contrast. Image contrast can be poor and, hence, contrast agents need to be utilized. These contrast agents could be represented by any entity that provides a strong imaging signal, is biologically harmless, and has at least some biological specificity. Traditionally, small molecule contrast agents such as ^{18}F-FDG, Magnevist, and indocyanine green have dominated the research and markets due to the ease of their synthesis and straightforward formulation. In the past decade, the situation started shifting toward nanotechnology. The advantages of nanoparticles compared to small molecules lie in (i) the ability to pack more reporting/targeting functionalities to increase signal strength, (ii) the increased retention time and reroute clearance to their target hard-to-image organs, and (iii) their enhancement of sensitivity and specificity of the imaging agent by including targeting groups. To enhance the contrast, nanoparticle architecture can be varied based upon the application (disease) and route of administration (e.g., intravenous, oral, and intratracheal). Nanoparticle size can be controlled and spanned from a few nanometers for metal colloids to hundreds of nanometers for fully assembled liposomes and microbubbles. Imaging nanoparticles can also be made from a variety of materials (e.g., polymers, metals, lipids, and sugars) and can mimic or be made of naturally-occurring nanoconstructs (e.g., viruses and exosomes). They might carry several reporters (multimodal nanoparticles) or even drugs for therapy (theranostic nanoparticles). All these properties make nanoparticles an incredibly versatile platform for designing new imaging agents that often exceed the potential of small molecules.

MY INTEREST IN THIS FIELD

Like most of my peers who began their research carriers in the early and mid-1990s, my interest in nanoparticles for imaging started from something remote. As I was working as a postdoc at Monsanto, I was developing metal colloids as catalysts for the chemical degradation of formic acid. Commonly used Pt and Pd colloids worked fine, but I needed, what is called in biology, a negative control or something that should have zero reactivity. Naturally, I looked at a colloid made from gold, the most noble of all the elements. The result was quite unexpected. The gold turned out to be the most reactive catalyst of all the metals that I tested. After a year of work, I figured that the remarkable activity of this gold colloid was size dependent and could be further tuned by the coating of nanoparticles through a process that we called at that time "self-assembled monolayers." Although this process has never been commercialized, this study helped me recognize the hidden power, unpredictability, and potential of nanoparticles. It also taught me to appreciate work with nanoparticles that reached reproducibility—the ultimate nanoparticle challenge.

My interest to imaging came in the late 2003 when I joined Washington University in St. Louis and started working in the newly formed Optical Radiology Laboratory of Sam Achilefu at the Mallinckrodt Institute of Radiology. The institute is one of the oldest radiology departments in the world and is probably one of the largest centers, if not the largest center of imaging. Many of the seminal discoveries of imaging have started here. Naturally, many researchers from radiology, including myself, turned to nanoparticles as alternative vehicles for the delivery of imaging contrast agents when small molecules were not satisfactory. Simultaneously, a growing group of nanoparticle developers—chemists and materials scientists—became aware of opportunities in radiology. The merging of the two fields proved to be synergistic in creating several new directions and new types of expertise. Today, nanoparticles dominate research in some areas such as contrast agents for MRI and Raman spectroscopy, and they are a big part of imaging in optical, photoacoustic, nuclear, and other modalities. I am glad to see that my interest in combining nanotechnologies with imaging is shared by thousands of my colleagues.

PURPOSE OF THE BOOK AND WHAT THE READER WILL GAIN

The purpose of this book is to cover recent trends in nanotechnology in imaging and *in vivo* diagnostics. The reader should gain not just the knowledge of different approaches in nanoparticle design and get the breath of chemistry used in synthesis and imaging nanoparticles, but also the limitations of what biological media impose on the design of the nanoparticles. The reader will also learn of the interactions between modern branches of radiology and contrast agents, the rationale for selecting nanoparticles for animal testing, and the limitations of the nanoparticle approach in medical imaging.

AUTHORS

Imaging with nanoparticles is based on two large posts: imaging instrumentation that are developed mostly by radiologists, physicists, and biomedical engineers, and nanotechnology that is broadly presented by chemists, material scientists, and biochemists. Hence, the book presents a combined effort of experts in nanotechnology and imaging from academia, industry, and healthcare from different specialties (nanoparticles synthesis, analytical instrumentation, physics, engineering, biology, and medicine) who are actively working to bring nanotechnology to clinical imaging. There are a total of 40 authors representing universities, companies, and governmental agencies. The firsthand knowledge, experience, and, foremost, the future vision in this field of the authors in nanotechnology, medical instrumentation, and medicine are expected to be of high interest to a broad audience of scientists, medical engineers, and health care professionals.

WHO SHOULD READ THIS BOOK

The book is written for a research-oriented audience with a general knowledge in chemistry. Some chapters require a minimum knowledge of mathematics, physics, and biology. Topics are introduced in an order that is typical in nanoparticle research: nanoparticle synthesis, their characterization, imaging instrumentation, and biological applications for imaging. Although many examples presented are cell studies, the main focus of the book is on *in vivo* imaging. The chapters introduce the readers to terminology in medical imaging and nanoparticles, the typical train of thought behind nanoparticle design, to rules of thumb, challenges, imaging modalities, and animal models. The major aspects of nanotechnology and medical imaging are covered from the design and synthesis of nanoparticles to imaging instrumentation and modalities and to potential clinical applications. This book will be invaluable for senior undergraduates, graduate students, and researchers arriving from different backgrounds including those working in areas of chemistry, materials science, biomedical engineering, biology, and medicine. With a multidisciplinary approach and a balance of research and diagnostic topics, this book is an essential resource for a broad range of scientists interested in emerging medical technologies.

BOOK STRUCTURE

The chapters provide a comprehensive coverage of the field, ranging from the architectural design of nanomaterials to their broad imaging applications in medicine. They are grouped into three parts: Part I, "Nanoparticle Design, Synthesis and Characterization" describes the fundamental principles of nanoparticle design relevant to imaging including fundamental concepts that establish nanoparticles as contrast agents and a detailed explanation of their classes and distinguished properties; Part II, "Imaging Modalities" describes established and novel imaging modalities and the design of nanoparticles tailored for specific imaging techniques; and Part III, "Nanotechnology in Biomedical Imaging and Beyond" describes the emerging role of nanotechnology in diagnostics, image–guided therapies, and other critical areas of radiology.

We will start this book with a historical account of the discovery of X-ray imaging and the first contrast agents to show the evolution of nanoparticles from anecdotal usage at the end of the 1940s to the developed concept formed in the 1980s and 1990s. Chapter 1 discusses nanoparticles in imaging research with in the context of stages marked by significant milestones, new directions, and redirections. This short historical account allows us to acknowledge and appreciate the seminal contributions of great chemists, engineers, biologists, and physicians to the development of the presented field.

In Part I, Chapter 2 introduces magnetic nanoparticle magnetism and its contrast effect in MRI. The chapter also reviews the organic–phase synthesis of iron oxide magnetic nanoparticles and hybrid nanoconstructs for applications in multimodality biological imaging. Chapter 3 describes the design and application of lipid-based vesicles, mostly liposomes and micelles, as pharmaceutical carriers for biomedical and diagnostic imaging agents. Chapter 4 discusses the synthesis and characterization of hollow nanocapsules, strategies for entrapment of molecules, functional performance of nanocapsules, and examples of their potential applications in biomedical imaging. Chapter 5 illustrates the growing role of metal-based nanoparticles as optoacoustic (photoacoustic) contrast agents. Chapter 6 describes the current analytical methods, state-of-the-art instrumentation, and emerging approaches for the characterization of nanoparticles that are relevant to imaging.

In Part II, Chapter 7 describes the advantages and applications of radio-labeled nanoparticles of different origins for SPECT and PET imaging. Chapter 8 describes the mechanisms involved in the production of contrasted images with Gd^{3+}-based nanoparticles for MRI. Chapter 9 describes the fundamentals of optical imaging and focuses on *in vivo* optical imaging techniques for fluorescent nanoparticles as contrast agents. Chapter 10 discusses principles of photoacoustic imaging with nano-sized photoacoustic contrast agents. Chapter 11 presents recent advances in the application of surface-enhanced Raman spectroscopy in bioimaging, highlighting several recent results as key examples that demonstrate the breadth of applications in noninvasive probing inside living tissue.

In Part III, Chapter 12 describes the synthesis and application of imaging gold nanorods for image-guided thermal treatment of solid tumors. Chapter 13 presents the challenges and nanoparticle-based approaches in noninvasive imaging of DNA and RNA to monitor disease states that are associated with the expression of a unique gene. Chapter 14 presents the emerging field of viral-type nanoparticles in tissue-specific imaging and considers the advantages and versatility of the viral nanoparticle platform compared to conventional nanoparticles. Chapter 15 focuses on examples of theranostic nanomedicine evaluated for imaging and drug delivery in animal disease models with a discussion of future theranostic designs from a pharmaceutical development view point. Chapter 16 presents a survey of animal models that broadly reflects the biology of human disease sufficient for molecular imaging with nano-material contrast agents.

Given the increasing number of publications on the use of nanoparticles in the basic and medical sciences, it is nearly impossible and impractical to cover all aspects of such a broad topic in one book. Some of the missing topics are covered in the recently published book *Nanoplatform-Based Molecular Imaging* (Wiley, 2011, ed. X. Chen), and some of the emerging subjects such as CT nanoparticles, nanoparticles for second harmonic generation imaging, upconverted nanoparticles, and temperature sensitive nanoconstructs are only briefly mentioned. These will be covered in future editions.

ACKNOWLEDGMENTS

I voice a special note of gratitude to my wife Tatiana for her strong moral and editorial support, my brother Oleg—my first teacher in science—and my daughter Sophia, who was born during the preparation of this book, for her way of organizing my time.

I would like to express my gratitude to the authors and my colleagues who contributed to the chapters and discussion. I would also like to acknowledge many people whom I contacted that helped me outline and shape the book, and my students who were the first readers and critics of the chapters.

Finally, I would particularly like to thank the readers, for taking the time to explore the content of the book. I am positive that the chapters of this book will prove interesting and useful to you.

1

HISTORICAL PERSPECTIVE ON NANOPARTICLES IN IMAGING FROM 1895 TO 2000

Mikhail Y. Berezin

Department of Radiology, Washington University School of Medicine, St. Louis, MO, USA

1.1 INTRODUCTION

Out of the two main subjects covered in this book—imaging and technology—imaging, or more commonly referred to as radiology, "the eye of medicine," is certainly the oldest. Prior to the appearance of nanoscience, radiology had already been well established through several generations of physicians who themselves processed thousands of images every year. Still, the persistent quest to "see the invisible" to better diagnose patients forced radiologists to pay close attention to the research and development of new imaging technologies. In the past two decades, nanoparticle contrast agents, stemming from the earliest contrast agents discovered soon after the discovery of X-rays over a hundred years ago, have become the holy grail of imaging. Today, an impressive number of radiological procedures that routinely utilize nanoparticles in clinics with even more impressive number are under preclinical testing and medical research.

The National Institutes of Health (NIH) in 2002 prioritized the most pressing problems facing medical science and identified three key areas in need of research: biological pathways, molecular imaging, and nanotechnology. The focus on these three critical components, backed by substantial investments from the NIH, transformed classic radiology and early disorchestrated attempts with nanoparticles

Nanotechnology for Biomedical Imaging and Diagnostics: From Nanoparticle Design to Clinical Applications, First Edition. Edited by Mikhail Y. Berezin.

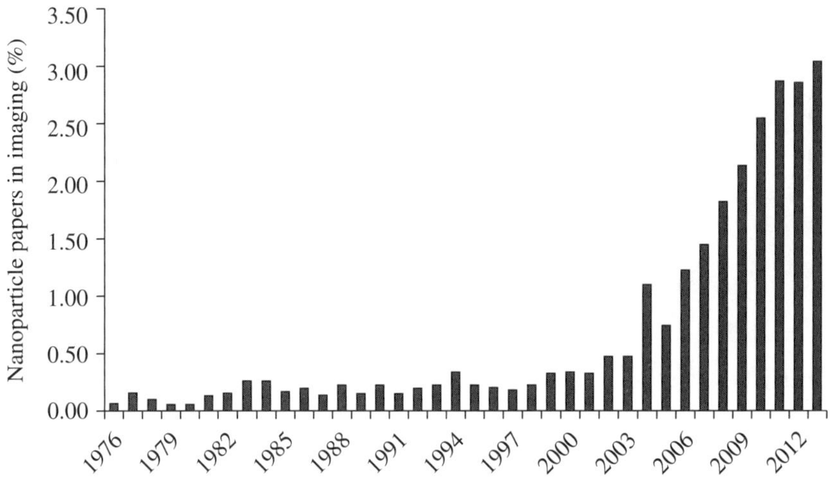

FIGURE 1.1 Growth of the nanoparticle research in biomedical imaging. Solid arrows show the appearance of imaging techniques, and dotted arrows show the emergence of nanoparticles. A number of citations are given from PubMed database.

into a mature field known today as molecular imaging. Figure 1.1 reflects a remarkable tenfold increase in nanoparticle-related medical imaging research from a relatively modest approximately 0.25–0.3% in the twentieth century to the current 3%. This growth resulted in more than 1500 nanoparticle imaging-related publications in 2012 alone.

From the onset of radiology and the first contrast agents to the end of the twentieth century, imaging techniques such as X-ray, PET, SPECT, ultrasound, MRI, optical, and photoacoustics have emerged. The first imaging nanoparticles appeared only in the middle of the twentieth century. The progress and the application of imaging nanoparticles followed the advent of new imaging modalities and diverged into two equally important directions. In one direction, de novo nanoparticle designs were developed for specific imaging modalities. Some examples include magnetic particles for MRI, quantum dots (QDs) for optical, and nanobubbles for ultrasound. The other direction adopted previously established designs of nanoparticles (for instance, for drug delivery) and modified them for imaging applications. Some examples include liposomes, virions, cross-linked nanoparticles, and surface modification to increase the nanoparticles' imaging specificity. Regardless of direction, many nanoparticles applications often began as unexpected discoveries. Many steps to refine their design were necessary to turn them from a mere curiosity to a clinically acceptable tool. Today, the continued improvement in nanoparticle synthesis, conjugation technique, and novel biomarkers made the nanoparticle approach a unique and well-differentiated scientific direction that blends seamlessly with clinical imaging. The historical trend illustrated in Figure 1.2 highlights the most important milestones toward this direction and is discussed in this chapter.

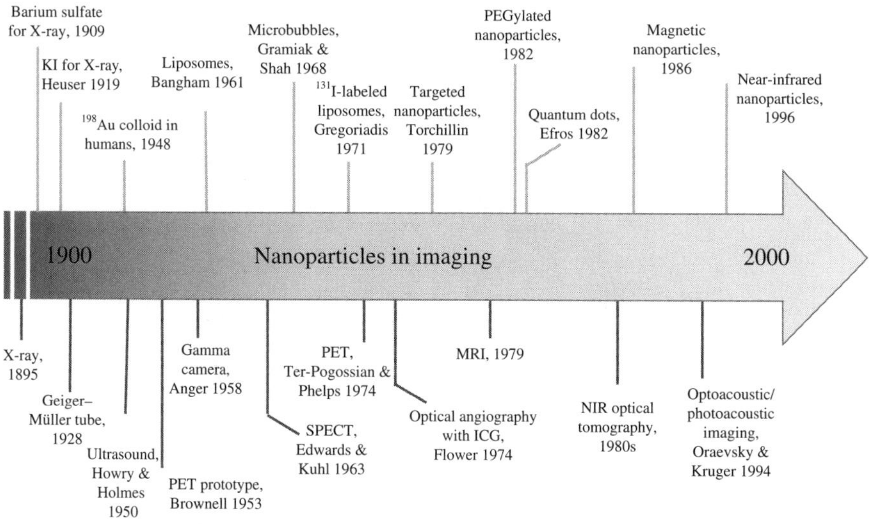

FIGURE 1.2 Timeline of the most important events in the development of nanoparticles for imaging and diagnostics covering the period from the twentieth century. The upper part corresponds to nanoparticles, and the lower part to the development of imaging modalities. (*See insert for color representation of the figure.*)

1.2 X-RAY AND FIRST CONTRAST AGENTS (1895–1930s)

The history of medical imaging started on November 8, 1895, when a 50-year-old Wilhelm Conrad Röntgen—a physicist from the University of Würzburg in Germany—observed a greenish glow from a recently invented Crookes tube. A new form of radiation, which Röntgen called an "X-ray," freely penetrated through biological tissue but was absorbed by dense material such as bones. Recorded on radiation-sensitive photographic plates, a well-recognized X-ray image was made. This entirely noninvasive imaging technique quickly spread across the world after its demonstration to the public in 1896. A review of major medical colleges across the United States conducted by the *American X-Ray Journal* (Fig. 1.3 shows the cover of this journal) in 1899 revealed more than 80 institutions where X-ray machines were available for patients [1], a remarkable rate given that it was just 4 years after X-ray discovery. With X-ray imaging, bone fractures, kidney stones, and metallic objects such as bullets and needles could be reliably located. With further refinement, physicians could even recognize and visualize certain organs. However, imaging inside the organs was impossible since the low and uniform density of soft tissue composed of transparent to X-rays water and organic media provided little contrast within the tissue.

To address this shortcoming, W. Cannon from Harvard Medical School began developing "contrast agents," biocompatible compounds that could absorb X-rays. In 1905, he discovered that high-density metal salts such as bismuth-based compounds provided the desired contrast in the intestines: "The animals thus fed with food mixed with bismuth subnitrate were exposed to the X-rays and, without disturbing the

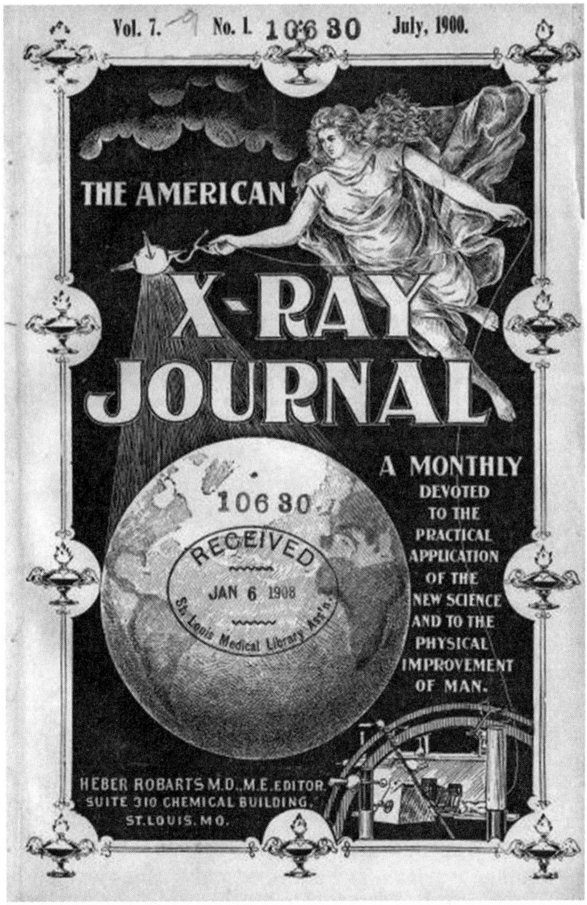

FIGURE 1.3 The *American X-Ray Journal* established in May 1897 was one of the first imaging journals. Launched by Dr. H. Robarts, a prominent radiologist from St. Louis, his biography is described in Ref. [2]. The journal existed until 1905. (Courtesy of Becker Library, Washington University School of Medicine.)

processes of digestion, the movements of the food in the stomach and small intestine were observed by means of the shadows cast on a fluorescent screen" [3]. A few years later, a less toxic barium sulfate mixed with foodstuffs became the first broadly used contrast agent in X-ray imaging of the digestive tract [4]. This water-insoluble salt (to prevent barium toxicity) was swallowed with food prior to the imaging procedure to outline the esophagus, stomach, and small intestines. The contrast could also be inserted via enemas to visualize the colon. This practice allowed the visualization of tumors, strictures, blockages, and ulcers and has been so simple and successful that it is still in use today.

The next advancement in the development of contrast agents came from Argentina, where in 1919 the radiologist Dr. C. Heuser intravenously injected a water-soluble potassium iodide to image the circulatory system. High-density iodide provided significant attenuation to X-ray radiation, causing the blood vessels to appear lighter on film. A few years later, Heuser utilized another iodinated compound called Lipiodol synthesized in 1901 by the French chemist M. Guerbet. Lipiodol is a low-viscosity radio-opaque diagnostic agent formed by the iodination of the fatty acids in poppy-seed oil and was applied to investigate the uterine cavity and fallopian tubes. Due to its high density and low toxicity, many iodinated compounds are commonly used today in X-ray and computer tomography (CT) imaging—a successor of the X-ray technique. (One of the leading companies of X-ray contrast agents is the Guerbet Group established by the son of Lipiodol's inventor in 1926.) However, despite several decades of continuous efforts to improve X-ray instrumentation and expand X-ray imaging to soft tissue with contrast agents, diagnosing diseases of internal organs suffered from unacceptably low contrast. New technologies were desperately needed.

1.3 RISE OF THE NUCLEAR IMAGING TECHNIQUES (1940s–1950s)

Shortly after World War II in 1946, the U.S. Congress passed the Atomic Energy Act that transferred nuclear weapon development and nuclear power management to civilian, rather than military control. The Oak Ridge Laboratory in Tennessee was directed to provide radioisotopes for peaceful purposes, especially for medical applications. One of the first isotopes made available was ^{198}Au colloid. It was produced by bombarding gold foil with slow neutrons in a uranium pile and was immediately (1947) utilized for cancer therapy in patients [5]. Since gold cations are extremely reactive due to their high reduction potential ($Au^{3+}(aq) + 3e^- \rightarrow Au(s)$, +1.50 V vs. NHE), they are incompatible with biological tissues. In contrast, gold colloid is chemically stable for storage, and the author recalls seeing bottles of colloidal gold that were several decades old. In addition, gold colloid is biologically inert and has been known in medicine since the time of Paracelsus [6].

^{198}Au emits radiation consisting of 0.97 MeV beta (β^-)- and 0.411 MeV gamma (γ)-rays with a half-life of 2.7 days [7]. The beta radiation from this isotope is absorbed under several millimeters of tissue rendering its importance for cancer treatment. The gamma emission that penetrated freely through the body became important for imaging. Produced colloidal gold nanoparticles were small enough (3–7 nm) [8] to pass through the pulmonary capillaries (<7 μm) but were accumulating mostly in the liver and spleen [9]. At higher dosages, even bone marrow could be visualized. The problem with ^{198}Au was its high radiation dosage of 50–100 rad/μCi that limited its clinical utility. In the search for compounds offering better imaging properties, ^{99m}Tc–sulfur colloid has been explored. Subsequently, other radioactive colloids such as ^{68}Ga ferric oxide and ^{113}In ferric hydroxide have been employed. With the help of these nanoparticles, untreated leukemia with grossly expanded marrow compartments was shown to be distinguished from aplastic anemia or myelofibrosis with less than normal activity of marrow [10].

Following the acceptance of isotopes in imaging, the 1940s and 1950s witnessed a rapid development of imaging instrumentation. The diagnostics with radioactive metals were generally conducted by "external counting" or "scintillation scanning." For that, a handheld Geiger–Muller counter introduced in 1928 capable of measuring gamma-rays and its mica-window modification for simultaneous detection of energetic beta-rays from *in vivo* sources was utilized [11]. By applying a Geiger–Muller counter to the surface of the skin at the site of interest, the distribution of the isotopes in the blood and extracellular tissue fluids could be followed. This method was a widely accepted standard in clinics until in 1958 when H. Anger from Berkeley Lab described a new scintillation camera (Anger camera), where gamma-rays were detected by a scintillating crystal. Upon contact with a gamma photon, a scintillator such as NaI crystal emits a photon at much lower energy, approximately 430 nm, thus converting ionizing radiation into light energy that could be detected by a photomultiplier tube (PMT). With many of the PMT tubes attached to the same crystal, many points could be imaged simultaneously. One of the first applications of the Anger camera was in a knee injected with ^{198}Au to diagnose an acute knee diffusion [12], a pathology that describes an excessive amount of fluid that accumulates around the joint and causes swelling.

Positron emission tomography (PET) and single-photon emission computed tomography (SPECT) have made their appearance in the 1950s. At the beginning of this decade, a team from MIT led by G. Brownell and physician W. Sweet from Massachusetts General Hospital [13] and independently F. Wrenn *et al.* [14] constructed the first PET detector to exploit the positron–electron annihilation effect for use as an imaging tool. D. Kuhl at the University of Pennsylvania and his colleagues at the University of Pennsylvania built the Mark II scanner, an ancestor of today's CT and SPECT scanners. The historical reviews on the development of imaging techniques written by the pioneers of this field describe these early efforts in great detail [15–17]. One of the first human scanners Mark III is shown in Figure 1.4.

Although the period of the 1940s–1950s has demonstrated the potential of imaging with nanoparticles in diagnostics and treatment monitoring, the use of nanoparticles was accidental. The majority of the efforts were directed toward the discovery of less expensive and more available sources of radioisotopes (cyclotrons, nuclear reactors), the development of imaging instrumentation, and the medical assessment of the techniques. Nanoparticles were produced mostly in the form of colloids, their chemistry has more or less been established, and their formulations were straightforward. Minimum efforts have been made to modify the nanoparticles for specific medical applications. These efforts started and went into full swing throughout the next decades.

1.4 IMAGING WITH LIPOSOMES (1960s–1970s)

1.4.1 Discovery of Liposomes

In the beginning of the 1960s, A. Bangham and his colleagues from the University of Cambridge (London) visualized the dispersion of lecithin-type phospholipids under an electron microscope and discovered their unusual multilamellar architecture (Fig. 1.5). "Toward the end of 1962, we had persuaded ourselves that we were seeing minute sacs of approximately 50 nm diameter, the first 'lipid somes' as we have come to know them." Intensive studies of the liposomes led to the discovery of aqueous

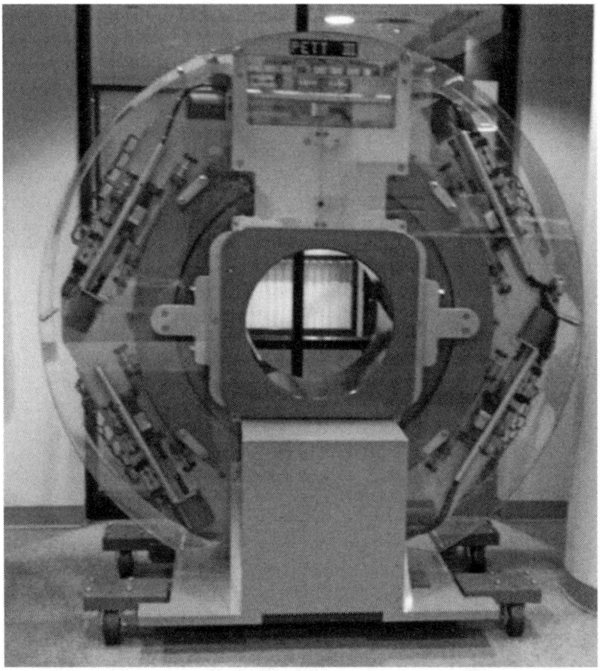

FIGURE 1.4 First human PEN scanner PETT III (1974) located in the hall of the Department of Radiology Washington University School of Medicine in St. Louis, where this scanner had been invented. The inventors had given the name "positron emission transaxial tomography" (PETT). The name was reduced to PET because transaxial was no longer the only plane used for image reconstruction. (*See insert for color representation of the figure.*)

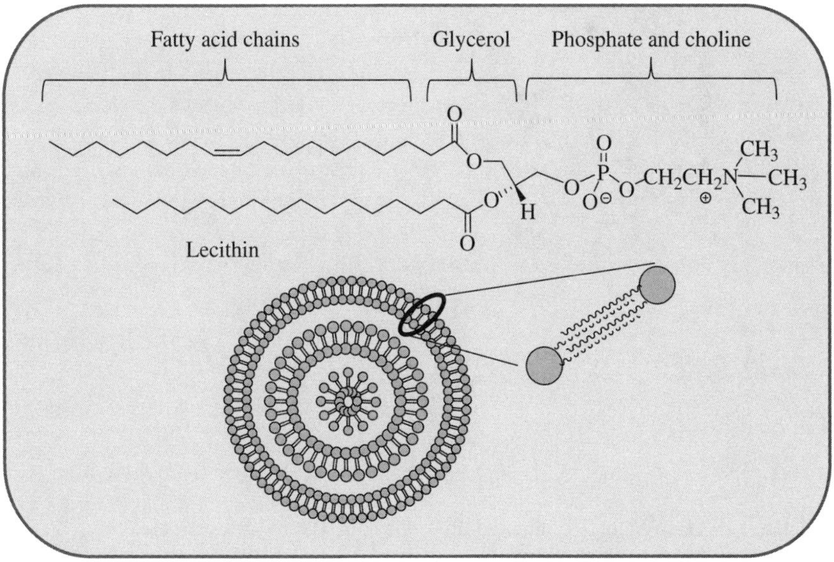

FIGURE 1.5 Structure of a multilamellar liposome and of a typical lecithin component phosphatidylcholine. The latter is composed from choline and phosphate group, glycerol, and long-chain fatty acid. Lecithin was first isolated in 1846 by the French chemist and pharmacist Theodore Gobley.

channels between lamellar structures that could be widened by the introduction of charged molecules into previously uncharged lecithin layers [18, 19]. These multilamellar liposomes were found to capture a variety of cationic species from tiny Li^+ ions to relatively large cholines and, as soon to be shown, imaging reporters that were dissolved in the aqueous phase at the time of liposome formation.

Following the discovery and characterization of multilamellar liposomes, D. Papahadjopoulos and N. Miller in 1967 described the structure of small unilamellar vesicles (SUVs) [20, 21]. This was an important development, since SUVs could be formed with better reproducibility and could serve as a technological platform for molecular imaging.

1.4.2 Visualization of Liposomes *in Vivo*

The majority of liposome clinical applications were historically centered in drug delivery. However, the visualization of the liposome distribution *in vivo* was critical for their clinical success and was the driving force behind the labeling of the liposomes with imaging reporters. In the beginning of the 1970s, G. Gregoriadis with colleagues from the Royal Free Hospital School of Medicine in London prepared liposomes labeled with entrapped ^{131}I-labeled albumin [22, 23] (Fig. 1.6). Upon *in vivo* administration, these liposomes were primarily deposited into the liver (major)

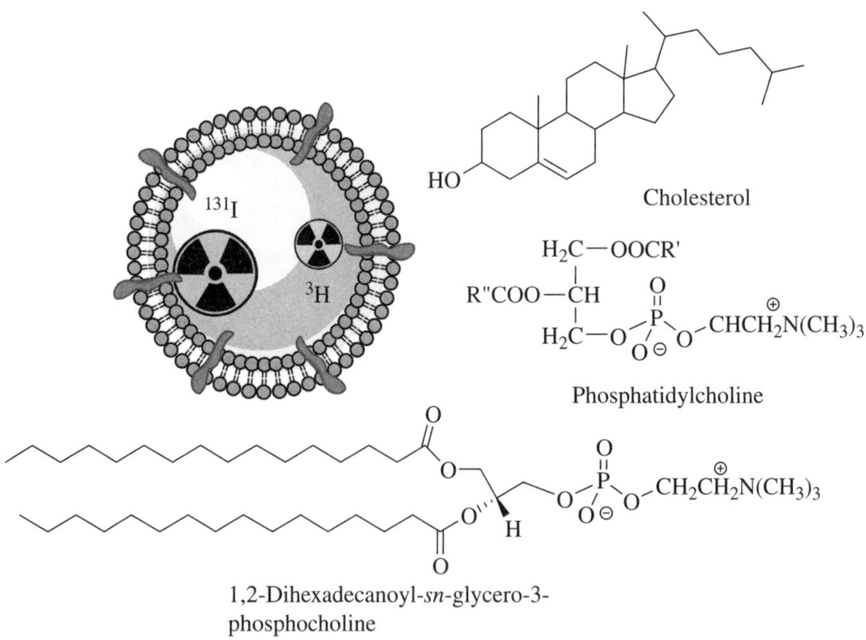

1,2-Dihexadecanoyl-*sn*-glycero-3-phosphocholine

FIGURE 1.6 Design of ^{131}I-albumin liposomes. [^{3}H]Amyloglucosidase and ^{131}I-labeled albumin were entrapped into liposomes composed of phosphatidyl choline, cholesterol, and dicetyl phosphate. ^{131}I-labeled albumin was also entrapped in [^{3}H]cholesterol liposomes. (Based on Refs. [22] and [23].)

and spleen (minor), targeting Kupffer cells as well as parenchymal cells. A few years later (1974), this contrast agent was applied to cancer patients [24] to monitor the deposition of a liposome carrying radiopharmaceutical to the area of treatment. It was, in fact, the first time liposomes were administered into humans.

Realizing the potential of radiolabeled liposomes in imaging, groups of V. Caride from Yale and V. Richardson from University of London in the middle of 1970s developed a ^{99m}Tc-labeled liposome for tumor detection [25, 26]. By that time, the medical community recognized ^{99m}Tc as an excellent radioisotope for diagnostics because of its short half-life of approximately 6 h allowing for the complete clearance of radioactivity after 24 h and, also, readily detectable 140 keV γ-rays by conventional X-ray diagnostic equipment. ^{99m}Tc labels were attached to a variety of preformed liposomes made from phosphatidylcholine and cholesterol via a so-called tin chloride method shown in Figure 1.7. This method allowed ^{99m}Tc originally generated in the form of highly hydrophilic pertechnetate anion TcO_4^-, to be reduced into more hydrophobic Tc^{3+} compound and be integrated into organic environments. Upon administration of this radiolabeled liposome to mice, a significant increase of the clearance rate of radioactivity was achieved compared to free pertechnetate. More importantly, the preferential uptake of the contrast agent by tumors was observed. However, the liposomes with Tc^{3+} had low stability *in vivo* resulting in the dissociation of the radionuclide from the liposome. Hence, the tin chloride reduction method was later enhanced by the addition of ^{99m}Tc-specific chelators.

Around the same time in the 1970s, the group of J. Baldeschwieler from Stanford University came up with the visualization of liposomes from another angle. The original intention of the group was to develop a technique for monitoring the release of a therapeutic cargo from the liposomes. Certain isotopes, such as ^{111}In known well today, sequentially emit two gamma-rays of different energy. These two sequential emissions can be measured by a technique known as gamma-ray perturbed angular correlation (PAC) coincidence spectroscopy. The result of the measurement is a parameter that is related to the rotational correlation time of the label. Molecules trapped inside the liposomes rotate faster than the released ones and bound by high-molecular-weight serum proteins. Hence, the change of the environment affects (perturbs) the angular correlation and alters its rotational time. In the series of publications from 1972 to 1980, the group refined this technique, synthesized a variety of ^{111}In carrying liposomes, measured the liposomes' structural integrity *in vivo*, and

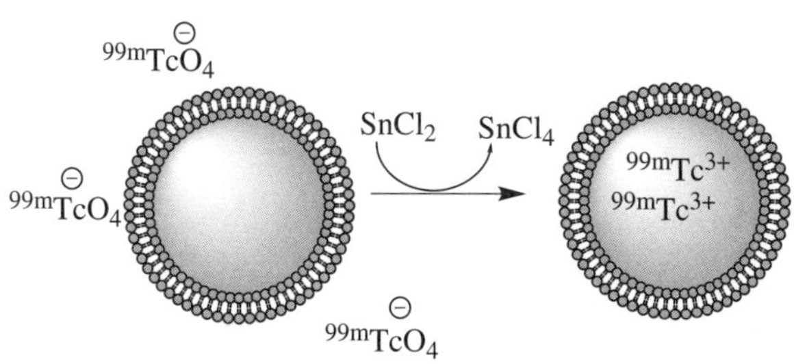

FIGURE 1.7 Formation of ^{99m}Tc liposomes via tin chloride method. (Based on Ref. [25].)

finally demonstrated tissue specificity for targeting liposomes [27–29]. One of the outcomes of this study was the recognition and the rise of a variety of ^{111}In-labeled nanoparticles for SPECT imaging [30–33].

1.5 MAGNETIC IMAGING WITH NANOPARTICLES (1980s–2000)

1.5.1 MRI Nanoparticles with Paramagnetic Ions

Clinical applications of MRI in the beginning of the 1980s, initiated by the first commercially available MRI scanners from General Electric, opened a new era in imaging. The fascinating early history of MRI contrast agents has been well documented, and the reader is encouraged to review the Reference [34]. In MRI, a strong magnetic field pulse is applied to a body causing certain atoms, such as protons, to be excited. The rate of the following relaxation is recorded and transferred into an image. This concept of relaxation came from nuclear magnetic resonance (NMR), the predecessor of today's MRI, as a method for characterizing organic compounds. By the middle of the 1970s, chemists recognized that beside three major NMR parameters, chemical shifts, coupling constants, and integrated areas, there were two more parameters, namely, spin relaxation time (T_1) and lattice relaxation time (T_2), that could be used to characterize the structure of the organic molecule [35]. Hence, there were also a number of ways whereby spin–lattice relaxation times could be chemically manipulated. Among them was the use of paramagnetic metal ions including gadolinium (Gd) that affected the relaxation times of associated ligands and nearby solvent molecules. Not surprisingly, the first MRI contrast agents described in the early 1980s were based on Gd complexes.

To enhance the MR signal, metal atoms should be freely exposed to biological water. This requirement demands the location of the metal at the exterior of the nanoparticle. To address this problem, G. Kabalka from the University of Tennessee in 1987 prepared Gd complexes with DTPA carrying two lipophilic long chains that were integrated into the lamellar phase of liposome particles [36] (Fig. 1.8). This proof-of-concept approach showed that the relaxation rate in the liver, post *ex vivo*, increased by 180%. Other metal complexes such as Mn^{2+} complexes have been also explored. The early attempts to entrap Mn^{2+}-DTPA in multilamellar liposomes [37] were not entirely successful; the complex from nanoparticles leaked out although the image showed the difference in contrast agent biodistribution compared to the free complex. In a parallel effort, the group of G. Navon from Tel-Aviv University added serum albumin to stabilize the complex inside the liposome and enhance their effect on water proton relaxation rates [38]. However, these promising *in vitro* results were not followed by imaging *in vivo*.

1.5.2 Supermagnetic Nanoparticles

The breakthrough in the development of MRI contrast agents began from the introduction of supermagnetic nanoparticles in the mid-1980s. In contrast to by that time popular T_1 Gd paramagnetic agents, this class of contrast agents gave rise to a dramatic

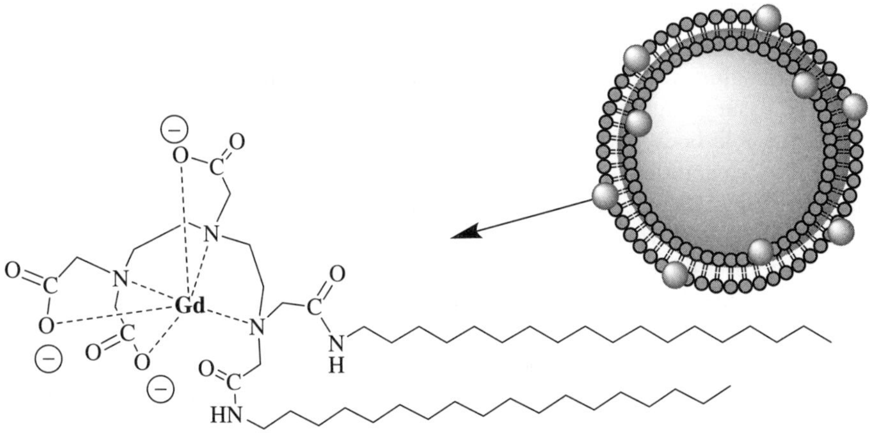

FIGURE 1.8 Amphipathic Gd complex with DTPA featuring hydrophobic tails. DTPA anhydride was reacted with stearyl amines and integrated into the lamellar phase of liposome particles. The T_1 in the liver increased by 180%. (Based on Ref. [36].)

shortening of the T_2 or T_2* relaxation times. A group led by J. Leigh Jr., a radiology scientist at the University of Pennsylvania, introduced ferromagnetic, albumin-coated magnetite particles of 1–50 nm in diameter [39]. These nanoparticles, prepared from ferric and ferrous chlorides, were covered with cross-linked albumin to decrease potential metal toxicity and showed significant reduction of both T_1 and T_2 at relatively low particle concentration in rats. At the same time, P. Lauterbur's group from the State University of New York, Stony Brook, reported enhanced imaging of the abdomen in dogs using 50 nm magnetite nanoparticles [40]. (The contribution of P. Lauterbur in MRI was acknowledged with a Nobel Prize in 2003.) Both Leigh's and Lauterbur's groups predicted the success of ferromagnetic nanoparticles in future biomedical research and clinical diagnosis.

In the upcoming years, the potential of many types and sizes of superparamagnetic particles (SMP), micrometer-sized paramagnetic iron oxide (MPIO, micron size), superparamagnetic iron oxide (SPIO, submicron size), and ultrasmall superparamagnetic iron oxide (USPIO, <50 nm) has been evaluated in many preclinical and clinical trials. The acceptance of these particles in research and clinical applications was mainly due to the following advantages: (i) a strong change in the signal per iron atom, in particular on T_2-weighted images (given thousands of iron atoms in the nanoparticle, the strength of the signal overcomes the typical low contrast agent sensitivity of MRI); (ii) the biocompatibility of relatively low toxicity iron—free iron can be metabolized via normal biochemical pathways; and (iii) coating and surface modifications, allowing the active targeting of the nanoparticles [41]. Many of the SMP nanoparticles have been commercialized and approved by healthcare systems in Europe and Asia under the trade names of Endorem®, Guerbet SA, or Ferridex®, Berlex Laboratories, Inc. By the turn of the twenty-first century, contrast agents were used in nearly 25% of all MRI procedures.

1.6 OPTICAL IMAGING WITH NANOPARTICLES (1980s–2000)

Optical imaging was one of the newest imaging modalities that marked its debut for deep tissue imaging less than two decades ago. Although observation of visually accessible body parts and some organs (such as the colon, throat) has been accessed for centuries, the use of light to visualize internal organs did not begin until the mid-1970s. Diaphanography [42] (a method for the evaluation of the female breast) and eye angiography (see Section 1.6.1 "NIR cyanine dyes") were one of the first clinical applications of imaging with light in the 1970s. Typical visible photons do not penetrate through biological tissues deeper than several millimeters due to extensive scattering of photons by the tissue and significant levels of endogenous chromophores such as melanin and hemoglobin that absorb visible photons. In contrast, near-infrared (NIR) light (700–900 nm) can penetrate the tissue much deeper: the scattering is lower and the endogenous absorption in this spectral range is weak. With the development of NIR instrumentation such as detectors, light sources, and especially NIR contrast agents in the middle of the 1990s, optical imaging quickly advanced from a preclinical stage to use in clinical applications.

1.6.1 NIR Cyanine Dyes

A large number of today's nanoparticle research involves decorating nanoparticles with fluorescent NIR dyes from a class of cyanines. The history of cyanine dyes for imaging applications started in the middle of the 1950s when an executive of Eastman Kodak Company, who was pleased with the care he received from Dr. I. Fox at the Mayo Clinic, offered to help him in his research. Dr. Fox needed a biocompatible dye that could be detected in blood. The executive sent an array of dyes from the Eastman Kodak laboratories for evaluation. Included among them was indocyanine green (ICG) (Fig. 1.9) developed as a sensitizer for photography by D. Heseltine and L. Brooker [43]. The testing for measuring hepatic function began in 1957 [44], and due to the low toxicity, the compound was rapidly approved by the FDA for this application in the beginning of 1959. Several years later, this dye was clinically used for measuring cardio function from ICG blood clearance curves.

The first attempt at using ICG for angiography was conducted by K. Kogure and coworkers from the University of Miami in 1968 when they demonstrated

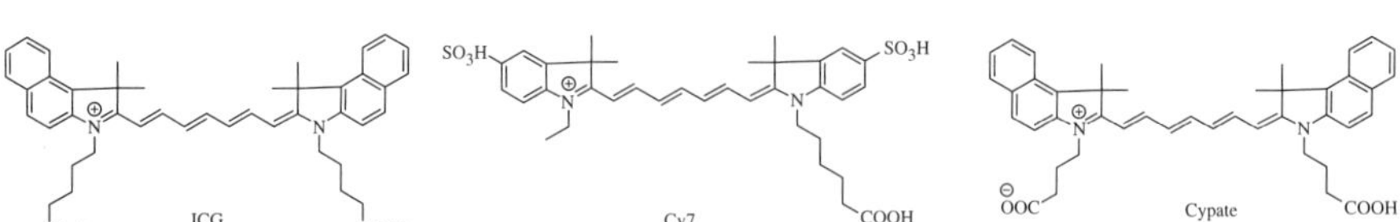

FIGURE 1.9 Structures of NIR dyes: indocyanine green, first approved near-infrared dye developed by Kodak chemists Heseltine and Brooker in 1957 [44], Cy7 developed by Waggoner (1987) and later commercialized by Amersham [47, 48], and cypate (2000) synthesized by Achilefu [49].

infrared absorption angiography of a monkey's brain vasculature [45]. Following intra-arterial ICG injection, the brain was illuminated by a flash from a bright lamp, and images were recorded on a Kodak Ektachrome film with sensitivity up to 900 nm. However, the method did not work with intravenous injections. In the early 1970s, R. Flower and B. Hochheimer at Johns Hopkins discovered the NIR fluorescence of ICG that was especially strong in blood samples and that could be visualized with night vision scopes or sensitive NIR films. In 1972, the first intravenously injected ICG fluorescence angiograms were produced. The early history of these events and their following development are described in References 43 and 46.

Following its success in ophthalmology, ICG was exploited in cancer diagnostics in 1994 to delineate tumor margins [50]. To overcome the disadvantage of fast clearance (half-life is 3.4 min in humans) and for better confinement in blood vessels, Devoisselle with colleagues prepared ICG incorporated in liposomes [51]. Expectedly, a substantially longer residence time (>60 min) was achieved when administered to rodents compared to the free dye.

The use of ICG was restricted to mostly simple formulations or direct injections of the dye. Lack of active functionalities precluded the use of ICG for labeling. A. Waggoner from Carnegie Mellon University addressed this problem by developing a set of cyanine dyes including NIR dye Cy7 with reactive functionalities such as NHS esters and isothiocyanate [47, 48]. Cy7 commercialized by Amersham (now GE Healthcare) became one of the most utilized NIR dyes in the 1990s. In parallel, S. Achilefu (then at Mallinckrodt, Inc.) developed cypate, an activatable analog of ICG, with almost identical optical properties to ICG (molar absorptivity, quantum yield) [49]. Other reactive NIR dyes followed [52], and the imaging using the dyes of conjugated to antibodies and small peptides has been demonstrated *in vivo* in mouse models [53]. These functionalizable NIR dyes have later become the dyes of choice for labeling many types of nanoparticles.

1.6.2 QDs

The term "quantum dots" was coined by Mark Reed (then at Texas Instruments) in his paper published in 1988 [54]. The introduction of QDs started with the realization that the optical properties of semiconductor particles were strongly dependent on particle size due to the quantum confinement of the charge carriers in small spaces. Thus, at a certain small size, the semiconductors could turn into bright emitters. A theoretical framework for these size-dependent properties was first described by brothers Alexander and Aleksei Efros from Ioffe Institute in the USSR in 1982 [55]. Shortly after their publication, publications describing synthesis of QDs emerged from several researchers. Among them, L. Brus (from Bell Labs) was awarded with the first Kavli Prize in Nanoscience in 2008 for his pioneering efforts in this field [56]. (The prize was actually shared with Sumio Iijima, of Meijo University in Japan, for his discovery of carbon nanotubes—another prominent imaging agent—of which its potential in imaging became recognized in the 2000s.)

For almost a decade, QDs with their low quantum yield and instability remained a subject of interest chiefly among physicists and material scientists. Three steps toward

FIGURE 1.10 High-temperature synthesis of colloidal CdSe quantum dots. A Cd precursor is dissolved in the tri-n-octylphosphine oxide (TOPO)-coordinating solvent. Under continuous stirring at high temperature (~320°C), a Se precursor dissolved in tri-n-octylphosphine (TOP) is injected into the flask, initiating nucleation of CdSe nanoparticles. The growth phase occurs at lower temperature (to ~290°C). When the desired size has been obtained, the solution is cooled (~220°C) to prevent further particle growth. A passivating layer of ZnS is deposited by injection of zinc and sulfur precursors dissolved in TOP. Upon cooling to room temperature, these core–shell QDs are isolated via precipitation. (Based on Ref. [57].)

bioimaging applications have been made in following years. The first step was to increase the brightness and stability of QDs with a protective layer. The synthesis of such protected CdSe/ZnS QTs with tunable emission was developed by the M. Bawendi group [57] and is shown in Figure 1.10. The produced QDs were highly fluorescent but only soluble in organic solvents such as toluene. Hence, another critical step toward biomedical applications was making the QDs hydrophilic. In 1998, two groups led by P. Alivisatos at UC, Berkeley, and S. Nie at Indiana University, Bloomington, independently demonstrated that QDs could be made water soluble and could be conjugated with biological molecules [58, 59]. The final step toward *in vivo* imaging was the development of QDs emitting above 700 nm. This step was introduced only in 2004 in a joint publication from Bawendi and Frangioni [60] and is part of the new history of imaging nanoparticles.

1.6.3 Nanoparticles as Photoacoustics Contrast Agents

One of the promising applications of optical nanoparticles is their utility as contrast agents in photoacoustics. Although the effect of photoacoustics or (optoacoustics) has been demonstrated long time ago in 1880 with the pioneering work of Alexander Graham Bell, photoacoustics became an imaging modality only recently. The early works by Oraevsky *et al.* [61] and Kruger *et al.* [62] demonstrated the value of photoacoustic signal in imaging of deep tissue in the early 1990s. Several years later, prolated metal nanoparticles absorbing at 1064 nm were suggested as contrast agents by Oraevsky's group [63, 64].

1.7 ULTRASOUND MICROBUBBLE CONTRAST AGENTS (1970s–2000)

Technically, microbubbles are vehicles that exceed the size of nanoparticles with a diameter in the 1–10 μm range. However, the recent trend of minimizing their size to nanosized scales prompts us to cover microbubbles in this chapter.

Ultrasound was primarily developed for military applications for detecting submarines in World War I. In the late 1940s, several enthusiasts recognized the potential of this technique in medical imaging. Among them was a radiologist D. Howry. In his basement in Denver, Howry, later together with the nephrologist J. Holmes, built his first ultrasound scanner to image soft tissue [65]. The principle of imaging was based on the reflection of an ultrasound signal generated by a piezoelectric transducer at some fundamental frequency. This ultrasound signal travels through the biological tissue, gets reflected off a structure, and travels back to the transducer to produce an image. After a decade of development, the technique underwent significant modifications, and by the mid-1960s, the first commercial system, Vidoson® by Siemens, Inc., for ultrasound imaging was released.

However, due to poor reflection of signal against biological tissue, measured signal intensity was low, producing images of poor quality. By serendipity, Raymond Gramiak, a radiologist at the University of Rochester, while examining a patient for cardiac output with indocyanine green—a standard method at that time—observed an intense contrast improvement at the site of injection [66]. In the following paper, the authors identified a small amount of foam formed upon dissolution of ICG with water responsible for the intracardiac contrast effect produced by ICG. "It is our belief that the contrast effect represents the ultrasonic detection of miniature bubbles within the heart produced by gaseous cavitation, which occurs when the contrast agent is injected rapidly, or by miniature bubbles injected in the foam of indocyanine-green solutions [67]." This discovery paved the way to an era of numerous designs of micro- and, later, nanobubbles in the following decades that are covered in many reviews and books [68–70].

The clinical use of early contrast agents that were made on site by agitation of saline solutions filled with air were unsuccessful due to the low stability of air-filled microbubbles and rapid disappearance of the contrast signal *in vivo*. The early agents also suffered from large and heterogeneous sizes and low permeability through the pulmonary system. Subsequent efforts showed that the stability of the microbubbles could be enhanced via a sonication technique instead of agitation and could be further improved by stabilizing agents such as sorbitol and dextrose [71]. Among the new techniques, a patent filed by Feinstein in 1985 [72] introduced a new class of stabilized microbubbles. These microbubbles were formed by sonication of an albumin solution that led to the formation of a protective coating of serum albumin on the surface of the bubbles [68] (Fig. 1.11). This discovery headed to the development of the first generation of the commercial contrast agent Albunex® (Mallinckrodt, Inc.) with a thin layer of cross-linked albumin (<15 nm) around an air bubble. The advantage of this contrast agent was that Albunex could be purchased in a standardized prepackaged form, minimizing handling complications and errors with dosage.

Despite the presence of protective layers, the microbubbles were still suffering from short lifetimes in circulation. Due to their relatively large size, they were unable to cross the lung barrier, thus limiting the imaging to the right side of the heart. Addressing these problems, the second-generation contrast agents were filled with a heavy gas such as perfluoropropane. The use of the high-density gas as a filling material was a breakthrough since it drastically reduced the diffusion of the gas from

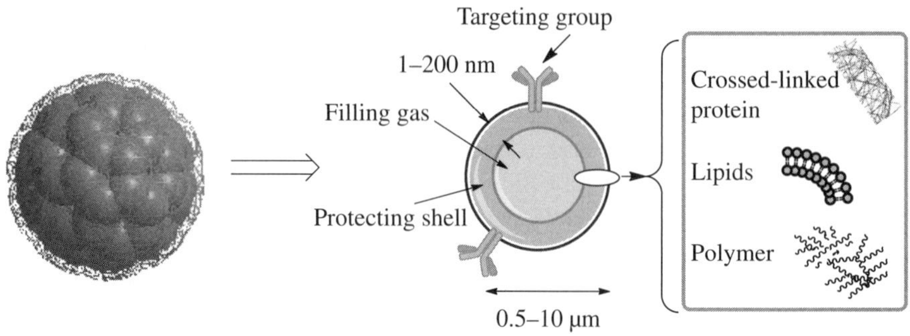

FIGURE 1.11 Schematic of a microbubble with a protecting shell and targeting groups. (Based on Ref. [68].)

the bubble due to the low solubility of perfluorinated molecules in the blood. The high stability of the microbubbles allowed imaging at much lower dosages and with longer imaging time. In addition, these new-generation microbubbles such as Optison (Mallinckrodt, Inc.) were approved by the FDA in 1997 and were much smaller with diameters less than 4 μm, allowing for the imaging of the whole heart.

1.8 MATURITY: NANOPARTICLES SURFACE MODIFICATIONS (LATE 1980s–EARLY 2000s)

By the middle of 1980s, several imaging techniques became suitable for the identification of internal organ pathologies. The ability to perform diagnostics noninvasively had a profound effect on the practice of medicine. New contrast agents became commercialized and rapidly adopted by clinics, which further accelerated the search for new contrast agents including nanoparticles. However, the available methods and contrast agents to highlight pathological changes could not be utilized for early diagnostics, where the change in signal was often below the resolution of the instrumentation. Furthermore, practically all types of nanoparticles were removed rapidly from circulation by phagocytic cells of the reticuloendothelial system (RES). As a result, nanoparticles mostly resided in the liver and the spleen. The effect of the RES could have been hypothetically evaded by using large doses of nanoparticles or pharmacological interventions to inactivate macrophages, but such strategies could compromise the immune system of patients [73]. Thus, a realization emerged that nanoparticles should be modified to alter their clearance pathway and directed (targeted) to specific tissues.

1.8.1 Targeting of Nanoparticles

By the middle of the 1970s, liposomes have been the most advanced type of nanoparticles and presented an excellent scaffold for the incorporation of different types of molecules including multiple targeting moieties from small molecules to antibodies.

Multiple ligands were sought to provide additional benefits increasing the specificity of the probe due to many interactions with cell receptors overexpressed at the site of interest. The first effort to modify the surface of the nanoparticles with antibodies appeared in the mid-1970s with the pioneering work of Gregoriadis. He demonstrated that ^{111}In-labeled liposomes carrying "homing probes" such as IgG and desialylated fetuin improved the selectivity of internalization in cell studies [33]. Further progress was made by V. Torchillin's group that decorated liposomes with covalently conjugated anticanine cardiac myosin antibodies. These targeted liposomes carrying ^{111}In radiolabels, named "immunoliposomes," were administered to dogs and were shown to localize in acute canine myocardial infarctions, thus providing one of the first examples of target imaging *in vivo* [42]. In addition to antibodies, the surfaces of the nanoparticles were also modified with carbohydrates. Exposed 6-aminomannose moieties, through covalent attachments to cholesterol imbedded in the lipid bilayers (see Fig. 1.12), produced a dramatic change in the biodistribution of nanoparticles labeled with radiolabels and significantly reduced the uptake by the RES [74].

Following liposomes, the development of targeted microbubbles was initiated during the late 1990s by F. Villanueva *et al.* from the University of Pittsburgh. A 40-fold increase in the extent of monoclonal antibody-labeled bubble adhesions to activated coronary artery endothelium cells compared to nontargeted contrast agents was observed [75].

Similar to other nanoparticles of that time, early magnetite nanoparticles as a result of RES sequestration were mostly utilized for imaging of the liver, spleen, and bone narrow system. Because of RES, these nanoparticles also suffered a short blood lifetime. Hence, a significant effort over the following several years had been made to develop new synthetic routes and surface modification techniques to increase the lifetime and alter biodistribution and pharmacokinetics of iron oxide nanoparticles [41, 76].

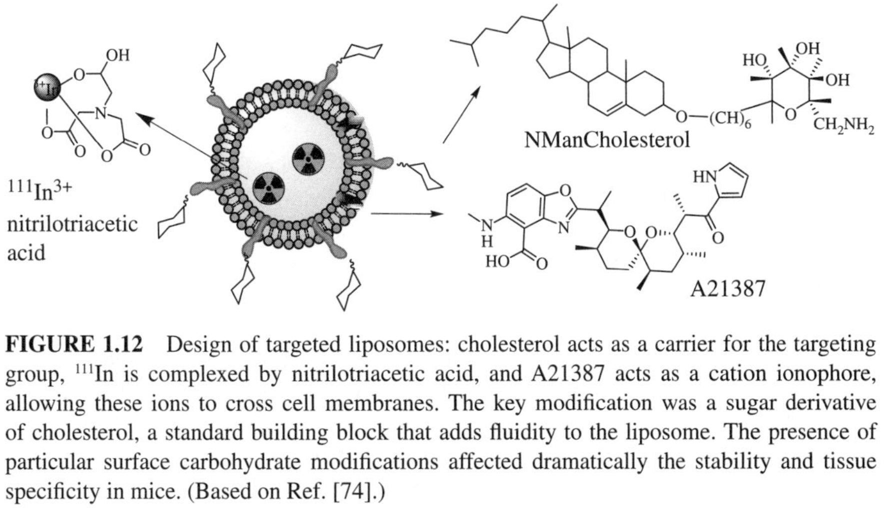

FIGURE 1.12 Design of targeted liposomes: cholesterol acts as a carrier for the targeting group, ^{111}In is complexed by nitrilotriacetic acid, and A21387 acts as a cation ionophore, allowing these ions to cross cell membranes. The key modification was a sugar derivative of cholesterol, a standard building block that adds fluidity to the liposome. The presence of particular surface carbohydrate modifications affected dramatically the stability and tissue specificity in mice. (Based on Ref. [74].)

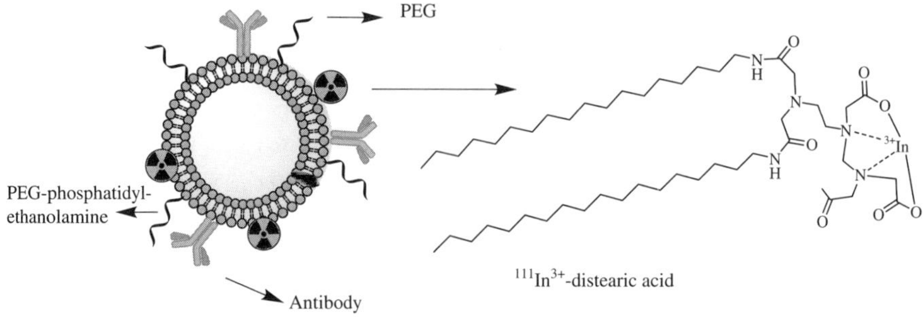

FIGURE 1.13 Immunoliposomes labeled with ^{111}In-EDTA-SA complex incorporated into the lipidic bilayer were modified with a PEG5000 derivative of phosphatidylethanolamine in combination with a monoclonal antibody to target myosin heavy chain. (Based on Ref. [79].)

1.8.2 PEGylation of Nanoparticles

Besides targeting, another important development in nanoparticle design was the introduction of polyethylene glycol (PEG). This inert hydrophilic polymer was introduced around 1970 by F. Davis, at Rutgers University, to modify bioactive proteins for medical applications to extend blood life and control immunogenicity of the proteins [77]. In 1988, A. Gabizon and D. Papahadjopoulos published a seminal paper describing a concept of PEGylated liposomes with increased retention time [78]. In this publication, a shell of PEG of up to 5–10kDa around liposomes largely prevented the uptake by RES rendering liposomes more accessible to other organs including implanted tumors. The potential of the PEGylated nanoparticles in imaging was further revealed in 1991 by a publication from Torchillin's group [79]. In this report, liposomes labeled with ^{111}In complex were modified with a PEG5000 derivative of phosphatidylethanolamine (Fig. 1.13). Because of the presence of PEG, the liposomes cleared slowly from the blood after intravenous injection and showed up to 6–18-fold at the specific localization of the target.

1.9 CONCLUDING REMARKS

By the beginning of the twenty-first century, many diagnostic modalities, including PET, SPECT, CT, MRI, optical angiography, and ultrasound, became routine clinical tools. Emerging modalities such as optical tomography, photoacoustics, and Raman and fluorescence imaging have also begun their way into clinics. Hardware development, novel and faster algorithms, sophisticated image acquisition, and better processing software to further improve the resolution and accelerate the diagnostic procedures also emerged. After several decades of research, facing problems such as batch-to-batch variation, low *in vivo* stability, liver clogging, and legal issues that required testing of each nanoparticle component, large clinical trials with nanoparticles have finally been completed. Many imaging agents were approved by the FDA and other regulatory agencies abroad. The advances in chemistry, material science, and analytical chemistry along with progress in molecular and cell biology, genome decoding, and the unraveling of the molecular pathways with the discovery of many disease-specific biomarkers gave rise to a new generation of high affinity contrast agents. These targeted agents increasingly in a form of nanoparticles opened up a new and exciting field of *molecular imaging*. This concept defined by R. Weisleder in 2001 as "*in vivo* characterization and measurement of biologic processes at the cellular and molecular level" [80] in contrast to the classic imaging techniques relying on changes in gross anatomy became the guide for the development of many modern imaging modalities. Many of the new advances are further discussed in this book.

REFERENCES

[1] Robarts H, editor. *X-Ray in Medical Colleges*. Volume **5**, St. Louis, MO: American X-Ray Publishing Company; 1899. p 679–686.

[2] Brown P. American martyrs to radiology. Heber Robarts (1852–1922). 1936. Am J Roentgenol 1995;**165**:473–476.

[3] Cannon WB. The passage of different food-stuffs from the stomach and through the small intestine. Am J Physiol 1904;**12**:387–418.

[4] Cannon WB. *The Mechanical Factors of Digestion*. New York: Longmans, Green & Co.; 1911.

[5] Sheppard C, Goodell J, Hahn P. Colloidal gold containing the radioactive isotope Au-198 in the selective internal radiation therapy of diseases of the lymphoid system. J Lab Clin Med 1947;**32**:1437–1441.

[6] Dykman LA, Khlebtsov NG. Gold nanoparticles in biology and medicine: recent advances and prospects. Acta Naturae 2011;**3**:34–55.

[7] Levy PW, Greuling E. The radiations from 2.7-day Au^{198}. Phys Rev 1949;**75**:819.

[8] Wheeler HB, Jaques WE, Botsford TW. Experiences with the use of radioactive colloidal gold in the treatment of cancer. Ann Surg 1955;**141**:208.

[9] Mellgren JAN, Knutsson F, Norin T. Localization and effect of injected radioactive gold (Au198) in the treatment of malignant tumors. Acta Pathol Microbiol Scand 1954;**34**: 393–409.

[10] Wagner HN. Current research in nuclear medicine. Am J Roentgenol 1970;**109**: 668–675.

[11] Strajman E. Small mica window Geiger-Müller counter for measurements of radioactive isotopes *in vivo*. Rev Sci Instrum 1946;**17**:232–234.

[12] Mallard JR, Peachey CJ. A quantitative automatic body scanner for the localisation of radioisotopes *in vivo*. Br J Radiol 1959;**32**:652–657.

[13] Brownell GL, Sweet WH. Localization of brain tumors with positron emitters. Nucleonics 1953;**11**:40–45.

[14] Wrenn FR, Good ML, Handler P. The use of positron-emitting radioisotopes for the localization of brain tumors. Science 1951;**113**:525–527.

[15] Nutt R. The history of positron emission tomography. Mol Imaging Biol 2002;**4**:11–26.

[16] Brownell GL. A history of positron imaging. Physics Research Laboratory; October 15, 1999; Massachusetts General Hospital, MIT; 1999.

[17] Thomas A, Banerjee AK, Busch U. *Classic Papers in Modern Diagnostic Radiology*. Berlin/New York: Springer; 2005.

[18] Bangham AD, Horne RW. Negative staining of phospholipids and their structural modification by surface-active agents as observed in the electron microscope. J Mol Biol 1964;**8**:660–668.

[19] Bangham AD, Standish MM, Watkins JC. Diffusion of univalent ions across the lamellae of swollen phospholipids. J Mol Biol 1965;**13**:238–252.

[20] Papahadjopoulos D, Miller N. Phospholipid model membranes. I. Structural characteristics of hydrated liquid crystals. Biochim Biophys Acta 1967;**135**:624–638.

[21] Papahadjopoulos D, Watkins JC. Phospholipid model membranes. II. Permeability properties of hydrated liquid crystals. Biochim Biophys Acta 1967;**135**:639–652.

[22] Gregoriadis G, Ryman BE. Liposomes as carriers of enzymes or drugs: a new approach to the treatment of storage diseases. Biochem J 1971;**124**:58P.

[23] Gregoriadis G, Ryman BE. Fate of protein-containing liposomes injected into rats. Eur J Biochem 1972;**24**:485–491.

[24] Gregoriadis G, Wills EJ, Swain CP, Tavill AS. Drug-carrier potential of liposomes in cancer chemotherapy. Lancet 1974;**1**:1313–1316.

[25] Richardson VJ, Jeyasingh K, Jewkes RF, Ryman BE, Tattersall MH. Properties of [99mTc] technetium-labelled liposomes in normal and tumour-bearing rats. Biochem Soc Trans 1977;**5**:290–291.

[26] Caride VJ, Taylor W, Cramer JA, Gottschalk A. Evaluation of liposome-entrapped radioactive tracers as scanning agents. Part 1: organ distribution of liposome [99mTc-DTPA] in mice. J Nucl Med 1976;**17**:1067–1072.

[27] Hwang KJ, Mauk MR. Fate of lipid vesicles *in vivo*: a gamma-ray perturbed angular correlation study. Proc Natl Acad Sci U S A 1977;**74**:4991–4995.

[28] Mauk MR, Gamble RC. Stability of lipid vesicles in tissues of the mouse: a gamma-ray perturbed angular correlation study. Proc Natl Acad Sci U S A 1979;**76**:765–769.

[29] Meares CF, Sundberg MW, Baldeschwieler JD. Perturbed angular correlation study of a haptenic molecule. Proc Natl Acad Sci U S A 1972;**69**:3718–3722.

[30] Sundberg MW, Meares CF, Goodwin DA, Diamanti CI. Selective binding of metal ions to macromolecules using bifunctional analogs of EDTA. J Med Chem 1974;**17**:1304–1307.

[31] Grove RB, Eckelman WC, Reba RC. Distribution of labeled bleomycin in normal and tumor-bearing mice. J Nucl Med 1973;**14**:917–919.

[32] Thakur ML. The preparation of Indium-111 labelled bleomycin for tumour localisation. Int J Appl Radiat Isot 1973;**24**:357–359.

[33] Gregoriadis G, Neerunjun ED. Homing of liposomes to target cells. Biochem Biophys Res Commun 1975;**65**:537–544.

[34] de Haën C. Conception of the first magnetic resonance imaging contrast agents: a brief history. Top Magn Reson Imaging 2001;**12**:221–230.

[35] Hall LD. Tate and Lyle Lecture. Spin-lattice relaxation: a fourth dimension for proton n. m. r. spectroscopy. Chem Soc Rev 1975;**4**:401–420.

[36] Kabalka G, Buonocore E, Hubner K, Moss T, Norley N, Huang L. Gadolinium-labeled liposomes: targeted MR contrast agents for the liver and spleen. Radiology 1987;**163**:255–258.

[37] Caride VJ, Sostman HD, Winchell RJ, Gore JC. Relaxation enhancement using liposomes carrying paramagnetic species. Magn Reson Imaging 1984;**2**:107–112.

[38] Navon G, Panigel R, Valensin G. Liposomes containing paramagnetic macromolecules as MRI contrast agents. Magn Reson Med 1986;**3**:876–880.

[39] Renshaw PF, Owen CS, McLaughlin AC, Frey TG, Leigh JS. Ferromagnetic contrast agents: a new approach. Magn Reson Med 1986;**3**:217–225.

[40] Dias MHM, Lauterbur PC. Ferromagnetic particles as contrast agents for magnetic resonance imaging of liver and spleen. Magn Reson Med 1986;**3**:328–330.

[41] Tiefenauer LX, Kuehne G, Andres RY. Antibody-magnetite nanoparticles: in vitro characterization of a potential tumor-specific contrast agent for magnetic resonance imaging. Bioconjug Chem 1993;**4**:347–352.

[42] Ohlsson B, Gundersen J, Nilsson D-M. Diaphanography: a method for evaluation of the female breast. World J Surg 1980;**4**:701–705.

[43] Flower RW. Evolution of indocyanine green-dye choroidal angiography. Opt Eng 1995; **34**:727–736.

[44] Fox I, Brooker L, Heseltine D, Essex H, Wood E. A tricarbocyanine dye for continuous recording of dilution curves in whole blood independent of variations in blood oxygen saturation. Proc Staff Meet Mayo Clin 1957;**32**:478–484.

[45] Choromokos EA, David NJ, Justice J Jr. The beginnings of indocyanine green angiography. J Ophthalmic Photogr 1994;**16**:70.

[46] Yannuzzi LA, Flower RW, Slakter JS. *Indocyanine Green: Angiography*. St. Louis (MO): Mosby; 1997.

[47] Ernst LA, Gupta RK, Mujumdar RB, Waggoner AS. Cyanine dye labeling reagents for sulfhydryl groups. Cytometry 1989;**10**:3–10.

[48] DeBiasio R, Bright GR, Ernst LA, Waggoner AS, Taylor DL. Five-parameter fluorescence imaging: wound healing of living Swiss 3T3 cells. J Cell Biol 1987;**105**:1613–1622.

[49] Achilefu S, Dorshow RB, Bugaj JE, Rajagopalan R. Novel receptor-targeted fluorescent contrast agents for *in vivo* tumor imaging. Invest Radiol 2000;**35**:479–485.

[50] Haglund MM, Hochman DW, Spence AM, Berger MS. Enhanced optical imaging of rat gliomas and tumor margins. Neurosurgery 1994;**35**:930–940; discussion 940–941.

[51] Devoisselle J-M, Soulie-Begu S, Mordon SR, Mestres G, Desmettre T, Maillols H. Effect of indocyanine green formulation on blood clearance and *in vivo* fluorescence kinetic profile of skin. Proc SPIE 1995;**2627**:100–108.

[52] Becker A, Riefke B, Ebert B, Sukowski U, Rinneberg H, Semmler W, Licha K. Macromolecular contrast agents for optical imaging of tumors: comparison of indotricarbocyanine-labeled human serum albumin and transferrin. Photochem Photobiol 2000;**72**: 234–241.

[53] Mariani G, Lasku A, Balza E, Gaggero B, Motta C, Luca LD, Dorcaratto A, Viale GA, Neri D, Zardi L. Tumor targeting potential of the monoclonal antibody BC-1 against oncofetal fibronectin in nude mice bearing human tumor implants. Cancer 1997;**80**: 2378–2384.

[54] Reed MA, Randall JN, Aggarwal RJ, Matyi RJ, Moore TM, Wetsel AE. Observation of discrete electronic states in a zero-dimensional semiconductor nanostructure. Phys Rev Lett 1988;**60**:535–537.

[55] Efros AL, Efros AL. Interband absorption of light in a semiconductor sphere. Sov Phys Semicond 1982;**16**:772–775.

[56] Editorial. The many aspects of quantum dots. Nat Nanotechnol 2010;**5**:381.

[57] Murray CB, Norris DJ, Bawendi MG. Synthesis and characterization of nearly monodisperse CdE (E = sulfur, selenium, tellurium) semiconductor nanocrystallites. J Am Chem Soc 1993;**115**:8706–8715.

[58] Bruchez M, Moronne M, Gin P, Weiss S, Alivisatos AP. Semiconductor nanocrystals as fluorescent biological labels. Science 1998;**281**:2013–2016.

[59] Chan WCW, Nie SM. Quantum dot bioconjugates for ultrasensitive nonisotopic detection. Science 1998;**281**:2016–2018.

[60] Kim S, Lim YT, Soltesz EG, De Grand AM, Lee J, Nakayama A, Parker JA, Mihaljevic T, Laurence RG, Dor DM, Cohn LH, Bawendi MG, Frangioni JV. Near-infrared fluorescent type II quantum dots for sentinel lymph node mapping. Nat Biotechnol 2004;**22**:93–97.

[61] Oraevsky AA, Jacques SL, Esenaliev RO, Tittel FK. Laser-based optoacoustic imaging in biological tissues. Proc SPIE 1994;**2134**:122–128.

[62] Kruger RA, Liu P. Photoacoustic ultrasound: pulse production and detection in 0.5% Liposyn. Med Phys 1994;**21**:1179–1184.

[63] Oraevsky A, Oraevsky AN. On a plasmon resonance in ellipsoidal nanoparticles. Quantum Electron 2002;**32**:79.

[64] Oraevsky AA, Karabutov AA, Savateeva EV. Enhancement of optoacoustic tissue contrast with absorbing nanoparticles. Proc SPIE 2001;**4434**:60–69.

[65] Goldberg BB, Gramiak R, Freimanis AK. Early history of diagnostic ultrasound: the role of American radiologists. Am J Roentgenol 1993;**160**:189–194.

[66] Gramiak R, Shah PM. Echocardiography of the aortic root. Invest Radiol 1968;**3**:356–366.

[67] Gramiak R, Shah PM, Kramer DH. Ultrasound cardiography: contrast studies in anatomy and function. Radiology 1969;**92**:939–948.

[68] Stride E, Edirisinghe M. Novel microbubble preparation technologies. Soft Matter 2008;**4**:2350–2359.

[69] Dijkmans PA, Juffermans LJ, Musters RJ, van Wamel A, ten Cate FJ, van Gilst W, Visser CA, de Jong N, Kamp O. Microbubbles and ultrasound: from diagnosis to therapy. Eur J Echocardiogr 2004;**5**:245–256.

[70] Deshpande N, Willmann JK. Microparticle-and nanoparticle-based contrast-enhanced ultrasound imaging. In: Chen X, editor. *Nanoplatform-Based Molecular Imaging*. Hoboken (NJ): John Wiley & Sons; 2011. p 233.

[71] Feinstein SB, ten Cate FJ, Zwehl W, Ong K, Maurer G, Tei C, Shah PM, Meerbaum S, Corday E. Two-dimensional contrast echocardiography. I. In vitro development and quantitative analysis of echo contrast agents. J Am Coll Cardiol 1984;**3**:14–20.

[72] Feinstein SB. Contrast agents for ultrasonic imaging. US patent 4,718,433. 1988.

[73] Kim S. Liposomes as carriers of cancer chemotherapy. Drugs 1993;**46**:618–638.

[74] Mauk MR, Gamble RC, Baldeschwieler JD. Targeting of lipid vesicles: specificity of carbohydrate receptor analogues for leukocytes in mice. Proc Natl Acad Sci U S A 1980;**77**:4430–4434.

[75] Villanueva FS, Jankowski RJ, Klibanov S, Pina ML, Alber SM, Watkins SC, Brandenburger GH, Wagner WR. Microbubbles targeted to intercellular adhesion molecule-1 bind to activated coronary artery endothelial cells. Circulation 1998;**98**:1–5.

[76] Weissleder R, Bogdanov A, Neuwelt EA, Papisov M. Long-circulating iron oxides for MR imaging. Adv Drug Deliv Rev 1995;**16**:321–334.

[77] Davis FF. The origin of pegnology. Adv Drug Deliv Rev 2002;**54**:457–458.

[78] Gabizon A, Papahadjopoulos D. Liposome formulations with prolonged circulation time in blood and enhanced uptake by tumors. Proc Natl Acad Sci U S A 1988;**85**:6949–6953.

[79] Klibanov AL, Khaw BA, Nossiff N, O'Donnell SM, Huang L, Slinkin MA, Torchilin VP. Targeting of macromolecular carriers and liposomes by antibodies to myosin heavy chain. Am J Physiol 1991;**261**:60–65.

[80] Weissleder R, Mahmood U. Molecular imaging. Radiology 2001;**219**:316–333.

PART I

NANOPARTICLE DESIGN, SYNTHESIS AND CHARACTERIZATION

2

IRON OXIDE-BASED MAGNETIC NANOPARTICLES SYNTHESIZED FROM THE ORGANIC SOLUTION PHASE FOR ADVANCED BIOLOGICAL IMAGING

SEN ZHANG AND SHOUHENG SUN

Department of Chemistry, Brown University, Providence, RI, USA

2.1 INTRODUCTION

Magnetic nanoparticles (MNPs) based on magnetite (Fe_3O_4) and maghemite (γ-Fe_2O_3) have been extensively explored in various biological applications such as magnetic resonance imaging (MRI), magnetic fluid hyperthermia (MFH), smart drug delivery, bioseparation, and biosensor [1–4]. Especially in MRI, iron oxide MNPs (IOMNPs) serve as an important class of contrast agent and provides an appealing nanoplatform for enhancing the sensitivity and accuracy of MRI in biomedical diagnostics. Unlike the conventional paramagnetic gadolinium complex contrast agents, IOMNPs at the size smaller than 20 nm are normally superparamagnetic and reach ferromagnetic saturation magnetization under a MRI magnetic field, leading to drastic contrast enhancement in MRI. This, plus their comparable size to biomacromolecules and low toxicity, makes the IOMNPs extremely promising in advanced MRI detection and diagnosis [5, 6].

The typical paradigm of clinically acceptable IOMNPs has been built up by the Ferridex. It is composed of Fe_3O_4 NPs and biocompatible polymer dextran or its

Nanotechnology for Biomedical Imaging and Diagnostics: From Nanoparticle Design to Clinical Applications, First Edition. Edited by Mikhail Y. Berezin.

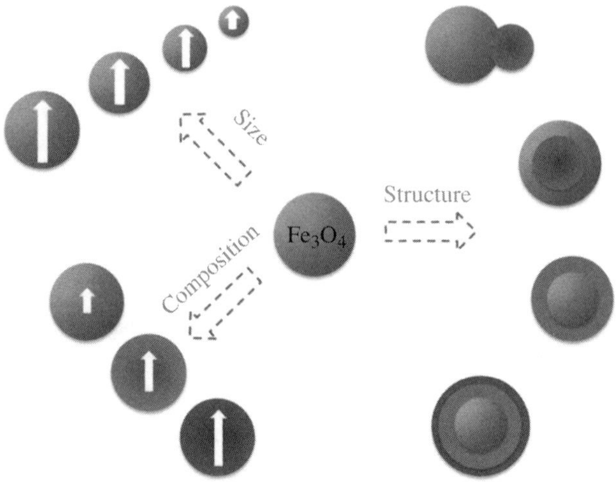

FIGURE 2.1 The schematic illustration of representative IOMNPs summarized in this chapter: size, composition, and structure-controllable hybrid IOMNPs including dumbbell, core–shell, and multilayer core–shell NPs.

derivative [7]. The polymer passivates and stabilizes the IOMNPs against aggregation in the aqueous media. However, these IOMNPs produced *via* the coprecipitation of Fe(II) and Fe(III) in a basic solution at low temperature show a low contrast effect in clinical MRI applications due to their large size distribution and poor magnetic properties. Recently, IOMNPs were successfully synthesized in organic solution phases with great control in size, crystallinity, composition, structure, and magnetic properties [8–10]. They become promising new-generation contrast agents for biomedical imaging applications. In this chapter, we focus on the controlled synthesis of IOMNPs and their hybrid NPs *via* this organic solution phase method and their use as advanced MRI contrast agents. The main types of IOMNPs described in this chapter are shown in Figure 2.1.

2.2 NANOMAGNETISM AND MRI

MNPs have a unique size-dependent magnetism. With the decrease of physical size from bulk form to nanoregime, an MNP shows a transition from multidomain structure to single-domain structure within which all free spins are ferromagnetic coupled into one direction [11, 12]. The critical size for the formation of single-domain particles is determined by the innate magnetic anisotropy energy of the material and usually falls in the range of tens of nanometers [12–14]. An IOMNP of less than 100 nm contains one single-domain structure. For a group of single-domain IOMNPs, their overall magnetization (M) reaches saturation (M_s) when their magnetization directions are aligned parallel under a large enough external magnetic field (H). The resistance to the magnetization alignment is measured by susceptibility (M/H). Similar to bulk ferromagnets, these magnetized NPs can retain some collective remnant magnetization

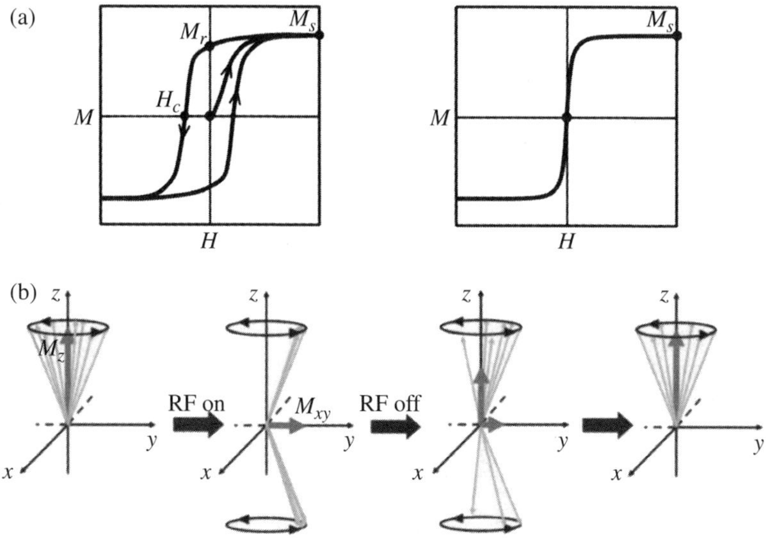

FIGURE 2.2 Schematic illustrations of (a) the hysteresis loops of a group of ferromagnetic (left) and superparamagnetic NPs (right). (b) Schematic illustration of spin orientation of the proton under a permanent magnetic field (in *z*-direction) and a resonant RF perturbation field that is perpendicular to the permanent field direction. The RF field perturbs the magnetization of spins into *xy*-directions. Subsequent removal of this RF irradiation causes spin to realign along the *z*-direction—a process called spin relaxation. The process leading to moment increase in *z*-direction is called T_1 relaxation, while that leading to moment decrease in *xy*-directions is called T_2 relaxation. (Reproduced with permission from Ref. [6]. © Wiley.)

(M_r) after the removal of the external field. To fully demagnetize these MNPs requires an external field in the opposite direction and in certain strength, called coercive field or coercivity (H_c) [11, 14]. When the size of a single-domain MNP decreases to a certain degree, superparamagnetism occurs and the remnant M_r drops to zero in the absence of the external magnetic field. Figure 2.2a displays the representative hysteresis loops of a group of ferromagnetic NPs and superparamagnetic NPs. The superparamagnetism originates from the thermal-induced magnetization reversal and is defined by

$$\tau = \tau_0 e^{K_u V/2kT} \tag{2.1}$$

in which τ is the magnetization relaxation time, K_u is the particle's magnetic anisotropy constant, V is the particle volume, k is Boltzmann's constant, and T is temperature [3, 15]. In the critical size where the thermal energy (kT) is comparable to magnetization reversal energy ($K_u V$), the single-domain MNPs can be demagnetized by thermal fluctuation without the external field reaching the superparamagnetic state. The superparamagnetic NPs have negligible magnetic dipole interactions and therefore can be dispersed and stabilized in the liquid media for biological applications [3].

MRI is based on the principal of nuclear magnetic relaxation of water protons [3, 5]. The protons align with an applied magnetic field in two ways in an MRI scanner: parallel (low energy state) and antiparallel (high energy state). The population difference between these two states is very small with more protons (in ppm level) being in low energy state but is proportionally increased with the field strength. This is the reason why an MRI scanner with a higher field has a better sensitivity. If a resonance electromagnetic radio-frequency (RF) pulse is applied transversely, the protons absorb the energy and then jump to an antiparallel state. Correspondingly, the longitudinal magnetization vector paralleling to external magnetic field is decreased (M_z in Fig. 2.2b) while giving rise to transverse magnetization (M_{xy} in Fig. 2.2b) [3, 6, 16]. The subsequent relaxation occurs when the RF is turned off, allowing protons to return to the origin state. Two relaxation processes are monitored: longitudinal magnetization recovery (T_1-recovery) and transverse magnetization decay (T_2-decay), to generate a bright (T_1-weighted) and a dark (T_2-weighted) image, respectively. The image contrast depends on local variation of relaxation time, resulting from the proton density and the physiological environment of specimen [3]. Here, T_2 represents the time it takes to drop the transverse magnetization to its 36.8% original magnitude. In the presence of MNPs, the T_2 of the surrounding protons is shortened due to the impact of secondary magnetic field produced by MNPs. As a result, MNPs can act as a T_2 contrast agent to give out a "darker" image. The sensitivity of a T_2 contrast agent is indicated by the relaxation rate, $R_2 = 1/T_2$ (s^{-1}), and the relaxivity, $r_2 = R_2/$concentration ($mM^{-1} \cdot s^{-1}$) [5]. Based on a quantum mechanical outer-sphere theory, the relaxation rate of the MNPs contrast agent in solution is defined by

$$\frac{1}{T_2} = \frac{\left(256\pi^2\gamma^2 / 405\right) V^* M_s^2 r^2}{D\left(1 + L / r\right)} \tag{2.2}$$

where γ is the proton gyromagnetic ratio, V^* is the volume fraction, M_s is saturation magnetization, r is the radius of MNP core, D is the diffusivity of water molecules, and L is the thickness of an impermeable surface coating [17, 18]. Therefore, an ideal MNP contrast agent is expected to possess high saturation magnetization (high M_s), large size (large r), and thin coating (small L). It is also favorable that these superparamagnetic NPs are monodisperse (size distribution in diameter with standard deviation <10%) with an overall hydrodynamic size letter than 50 nm to avoid the nonspecific uptake and have a long circulation time [19].

2.3 ORGANIC SOLUTION PHASE SYNTHESES OF IOMNPs

Organic solution phase reactions have been proven to be ideal for synthesizing IOMNPs in high quality. In a typical synthesis, an iron precursor is thermally decomposed in an organic solution and then generates the IOMNPs through the nucleation and growth steps (Fig. 2.3a) [3, 8, 20–22]. Iron acetylacetonate ($Fe(acac)_3$/$Fe(acac)_2$) or iron oleate complex is the common precursor for the synthesis. An organic solvent

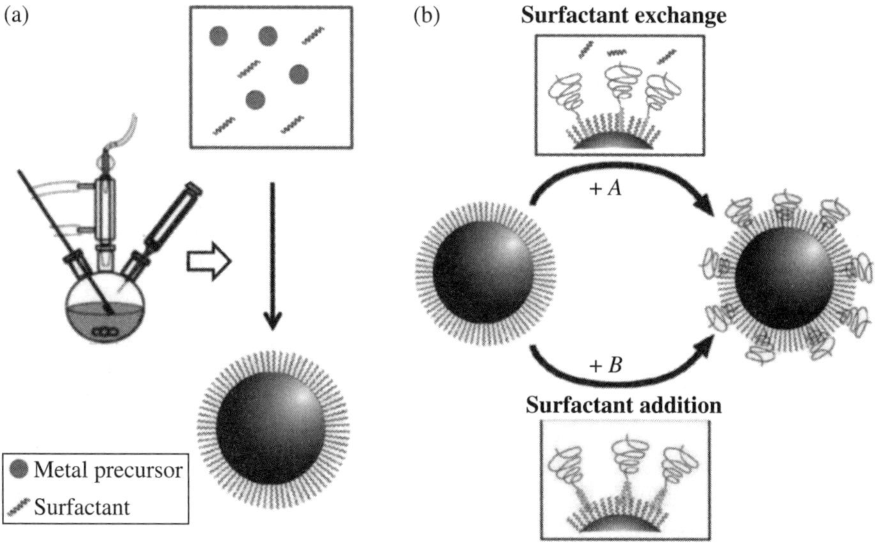

FIGURE 2.3 Schematic illustrations of (a) a typical organic solution phase synthesis of IOMNPs (b) two strategies used for converting hydrophobic NPs into hydrophilic ones. (Reprinted with permission from Ref. [10]. © Wiley.)

with high boiling point is also necessary to provide the desired decomposition condition. To control NP nucleation and growth, the iron precursor is either "hot injected" into the preheated solvent or premixed with other surfactants in the organic solvent followed by heating. The nucleation happens when the rapid decomposition of iron precursors at high temperature yields an over-supersaturated monomer solution in a short interval, which is called "burst nucleation." This burst nucleation consumes most monomers and prevents any further nucleation [21, 22]. The nuclei then grow up to monodisperse NPs at the same rate, depleting the rest of monomers.

Another avenue to IOMNPs is *via* oxidation of the metallic Fe NPs, which are preformed by the thermal decomposition of iron pentacarbonyl ($Fe(CO)_5$) [23, 24]. The Fe NPs can also be oxidized in a more controlled manner, giving core–shell Fe/iron oxide NPs, iron oxide NPs, or hollow iron oxide NPs [25].

In organic solution phase synthesis, the presence of a lipid-type surfactant is critical to control the NP growth into different sizes and shapes and to stabilize the produced MNPs [26]. The bipolar surfactant always has a polar group binding with the surface atoms and a hydrocarbon chain to provide steric protection of the NPs. Oleic acid and oleylamine are the common surfactant combination used in NP synthesis and stabilization. These surfactant-capped MNPs can form very stable dispersion in a nonpolar solvent. However, to make them suitable for biological applications, these MNPs need to be further modified. As shown in Figure 2.3b, one well-established method is to do ligand exchange, replacing the initial hydrophobic surfactant with a hydrophilic capping agent that has a stronger binding power to the NP surface [10]. As a typical example, 2, 3-dimercaptosuccinic acid (DMSA) can replace the hydrophobic capping via metal–S bond and chelate bonding of carboxyl group to MNPs [27]. Polyethylene glycol (PEG)–dopamine is also a common surfactant for stabilizing IOMNPs in physiological solutions [28–30]. In this stabilization strategy, catechol unit in dopamine is used to bind to iron oxide surface *via* the strong bidentate coordination, and PEG is responsible for long-term NP stabilization. The physiological stability of the NPs can be further enhanced by charges carried on the PEG units.

The second strategy used for NP functionalization is "ligand addition" of amphiphilic surfactant outside the initial hydrophobic capping (Fig. 2.3b) [10]. The amphiphilic surfactant inserts its hydrophobic segment into the initial capping layer while exposing its hydrophilic segment to aqueous solution, forming a liposome-type structure with NPs embedded in the alternate hydrophobic and hydrophilic layers. In this scheme, the amphiphilic PEG–phospholipids and poly(maleic anhydride-alt-1-octadecene)–PEG block copolymer are often applied to stabilize IOMNPs in aqueous solutions [7, 31]. The hydrophilic layer around each IOMNP also provides a hotbed for conjugating other functional molecules, such as biological targeting agents and fluorescence organic dyes, making the MNPs both target specific and capable of MRI–fluorescence dual-modality imaging [7, 10, 31].

2.4 DESIGNS AND FABRICATIONS OF IOMNPs AS MRI CONTRAST AGENT

2.4.1 NP Size Control

Size of MNPs is crucial to control MRI contrast effect. First, according to Equation 2.2, the relaxivity is proportional to the square of radius of MNPs. Thus, higher relaxivity can be predicted for the larger MNPs if the volume fraction V^* is constant. Second, the saturation magnetization M_s of a certain weight of iron oxide NP is strongly dependent on the size—small NPs often have the reduced moment due to surface-canting effect as illustrated in Figure 2.4a [32]. It is known that the surface spins of MNPs are disorderedly aligned due to the reduced spin–spin exchange coupling energy at the surface. These disordered spins reduce the magnetization m_s of smaller MNPs as in these smaller MNPs the surface atom ratio is dramatically increased, as described in

$$m_s = M_s \left[\frac{r-d}{r} \right]^3 \tag{2.3}$$

with r, M_s, d being the size, the saturation magnetization of bulk materials, and the thickness of disordered surface layer, respectively [33].

The size-controlled synthesis of IOMNPs was first realized *via* the thermal decomposition of $Fe(acac)_3$ [33, 34]. The $Fe(acac)_3$ was dissolved in the diphenyl ether solution of 1,2-hexadecanediol, oleic acid, and oleylamine, where the oleic acid and oleylamine act as surfactants and 1,2-hexadecanediol as a mild reducing agent to reduce Fe(III) to Fe(II) at high temperature. When the solution was heated at 265°C, 4 nm Fe_3O_4 NPs were generated (Fig. 2.4b). By replacing the diphenyl

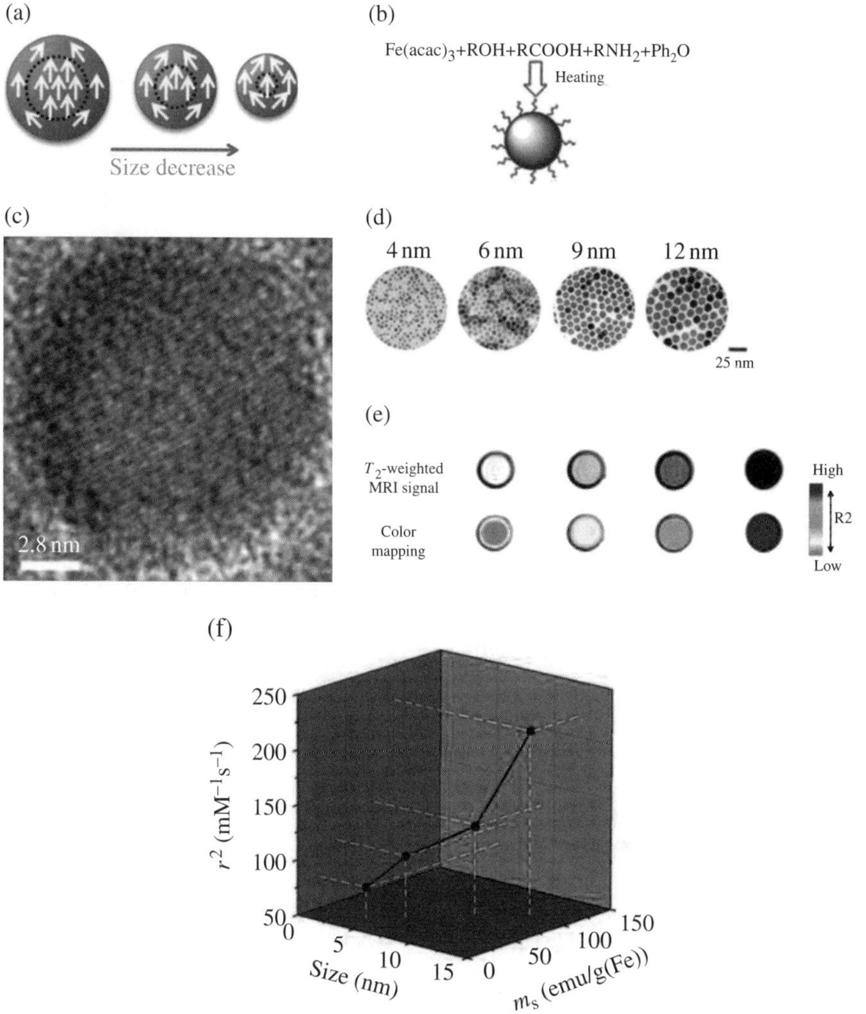

FIGURE 2.4 (a) Schematic illustration of size-induced surface spin-canting effect in MNPs. (b) The synthesis of Fe_3O_4 NPs *via* thermal decomposition of $Fe(acac)_3$ and (c) the HRTEM image of a 16 nm Fe_3O_4 NP. (d)–(f) Size effect on MRI contrast: (d) TEM images of Fe_3O_4 NPs with different sizes, (e) the relevant T_2-weighted MR images and color-coded maps, and (f) relation of size, saturation magnetization, and relaxivity r_2. (Reprinted with permission from Refs. [32] and [33]. © American Chemical Society). (*See insert for color representation of the figure.*)

ether with benzyl ether, a solvent with higher boiling point (300°C), 6 nm Fe_3O_4 NPs could be produced. Larger Fe_3O_4 NPs were also made *via* the seed-mediated growth. For instance, 8 nm Fe_3O_4 NPs were produced by thermally decomposing $Fe(acac)_3$ in the presence of 6 nm Fe_3O_4 NP seeds. Varying the seeds with different sizes and the ratio of seeds/Fe(acac)$_3$ could further generate IOMNPs with sizes up to 20 nm. Transmission electron microscopy (TEM) and high-resolution TEM (HRTEM) indicate that the size-controllable IOMNPs were monodisperse and single crystalline (Fig. 2.4c). The 16 nm Fe_3O_4 NPs produced through this route display the saturation magnetization M_s at 82 emu·g^{-1} (Fe_3O_4), which is very close to the value of bulk Fe_3O_4 (90 emu·g^{-1} (Fe_3O_4)). Fe_3O_4 NPs > 20 nm were also reported *via* the decomposition of Fe(acac)$_3$ in the presence of only oleic acid in benzyl ether [35]. Their sizes in the range of 20–160 nm were controlled by the reaction time and concentration of Fe(acac)$_3$. But the Fe_3O_4 NPs > 20 nm show ferrimagnetism at room temperature and are difficult to stabilize under the synthetic conditions.

Apart from Fe(acac)$_3$, iron oleate is also a common precursor used to synthesize iron oxide NPs with controllable sizes [36]. It is prepared from iron chloride ($FeCl_3 \cdot 6H_2O$) and sodium oleate in a mixture solvent of ethanol, water, and hexane at 70°C. Thermal decomposition of iron oleate in the reflux of different organic solvents produced IOMNPs with different sizes, for example, 5, 9, 12, 16, and 22 nm IOMNPs were yielded in the solvent of 1-hexadecene (b.p. 274°C), octyl ether (b.p. 287°C), 1-octadecene (b.p. 317°C), 1-eicosene (b.p. 330°C), and trioctylamine (b.p. 365°C), respectively. However, these IONPs are not the pure Fe_3O_4 or γ-Fe_2O_3, but contain their mixture in the form of $(\gamma\text{-}Fe_2O_3)_{1-x}(Fe_3O_4)_x$ where x varies with the size.

The IOMNP's size effect on MRI contrast was first demonstrated by Cheon's group [32, 37]. They synthesized 4, 6, 9, and 12 nm Fe_3O_4 NPs *via* the decomposition of Fe(acac)$_3$ (Fig. 2.4d). The as-synthesized NPs were made water soluble after ligand exchange with DMSA. The DMSA-protected Fe_3O_4 NPs were dispersed in phosphate-buffered saline (PBS). The magnetizations of the DMSA-capped Fe_3O_4 NPs at 1.5 T were measured to be 25 (4 nm) to 43 (6 nm), 80 (9 nm), and 102 emu·g^{-1} (Fe) (12 nm). Consistent with the magnetization data, their T_2-weighted MR images (same iron concentration) show significant change from white to black with the size increasing from 4 to 12 nm, suggesting the much enhanced MRI sensitivity in larger NPs (Fig. 2.4e). Using a 1.5 T MRI scanner, the relaxivity r_2 values were calculated to be 78, 106, 130, and 218 mM^{-1}·s^{-1} for 4, 6, 9, and 12 nm DMSA-protected Fe_3O_4 NPs, respectively, as shown in Figure 2.4f.

2.4.2 Composition Control

The magnetism of Fe_3O_4 is based on the inverse spinel structure in the form of $(Fe^{3+})_T(Fe^{2+}Fe^{3+})_OO_4$. Fe^{2+} and half Fe^{3+} occupy the octahedral (O) sites, while the other half Fe^{3+} take the tetrahedral (T) sites. As the spins from the two classes of Fe^{3+} cancel each other, the overall magnetic moment of each unit of $(Fe^{3+})_T(Fe^{2+}Fe^{3+})_OO_4$ is from the d^6 configuration of Fe^{2+}, which is 4 μ_B in high spin state. Considering that the dopant substitution of Fe^{2+} with M^{2+} (M = Mn, Co, Ni, Zn, etc.) is capable of changing the overall magnetic moment, synthesis of ferrites provides another method to tune the performance of IOMNP contrast agent [32, 34].

The controlled synthesis of ferrite MFe_2O_4 MNPs can be directly modified from the abovementioned methods used to produce Fe_3O_4 NPs. High-quality 3–20 nm

$CoFe_2O_4$ and 3–18 nm $MnFe_2O_4$ NPs were made by the decomposition of $Fe(acac)_3$ with $Co(acac)_2$ and $Mn(acac)_2$ in stoichiometric compositions [34]. The seed-mediated growth was also applied to control the size of these ferrite NPs. $M(acac)_{2/3}$ (M = Fe, Co, Mn, Zn) can be replaced by the $MCl_{2/3}$ [38], and the thermal decomposition of metal oleate complex can be extended to synthesize MFe_2O_4 NPs, which requires the MFe_2–oleate complex to be prepared in homogeneous fashion, not the simple mixture of M–oleate and Fe–oleate [39].

A representative composition effect on MRI contrast was demonstrated on 12 nm $MnFe_2O_4$, Fe_3O_4, $CoFe_2O_4$, and $NiFe_2O_4$ NPs [27]. The NPs were transferred to aqueous medium by DMSA (Fig. 2.5a). The $MnFe_2O_4$, Fe_3O_4, $CoFe_2O_4$, and $NiFe_2O_4$ NPs show magnetizations of 110, 101, 99, and 85 emu·g^{-1} (M + Fe atoms), respectively (Fig. 2.5b). The relevant relaxivities r_2 were measured to be 358, 218, 172, and 152 mM^{-1}·s^{-1} at a 1.5 *T* MRI scanner, accordingly (Fig. 2.5d). All these NPs outperformed the commercial cross-linked iron oxide (CLIO), which show a r_2 at 62 mM^{-1}·s^{-1} under the same condition (Fig. 2.5d). Most importantly, the Mn-doped ferrite NPs exhibits the highest magnetization and best contrast sensitivity in T_2-weighted MR image and color-coded MR image. The doping-induced magnetic modulation is proposed in Figure 2.5c. In $MnFe_2O_4$, *T* sites are occupied by $Mn^{2+}{}_{1-x}Fe^{3+}{}_x$ $(0 < x < 1)$ and *O* sites are taken by $Mn^{2+}{}_xFe^{3+}{}_{2-x}$, leading to a magnetic moment of approximate 5 μ_B in each unit of $MnFe_2O_4$. This value is estimated to be 4, 3, 2 μ_B in Fe_3O_4, $CoFe_2O_4$, and $NiFe_2O_4$ because of the complete occupations of M^{2+} in their *O* sites. In the image demonstration, the 12 nm $MnFe_2O_4$ with the best magnetic properties was modified with the antibody Herceptin, which could specifically target the HER2/neu marker overexpressed in breast and ovarian cancers. The Herceptin-conjugated $MnFe_2O_4$ (MnMEIO–Herceptin) NPs were tested *in vitro* and compared with CLIO–Herceptin in various cell lines with different levels of HER2/neu overexpressions: Bx-PC-3, MDA-MB-231, MCF-7, and NIH3T6.7 (Fig. 2.5e). The much more conspicuous MRI detection sensitivity was found for MnMEIO–Herceptin conjugates in all cell lines, even in Bx-PC-3 cell line (low HER2/neu overexpression), whereas CLIO–Herceptin conjugates show negligible signals in most cell lines.

To further enhance the magnetic property of ferrite, the nonmagnetic Zn^{2+} was doped to form $(Zn_xFe_{1-x})Fe_2O_4$ with x being controlled from 0 to 0.1, 0.2, 0.3, 0.4, and 0.8 [38]. With the addition of Zn^{2+}, *T* sites were occupied and induced the partial cancelation of antiferromagnetic coupling between Fe^{3+} in *T* and *O* sites, giving the highest magnetic moment of 161 emu·g^{-1} (Zn + Fe atoms) in $(Zn_{0.4}Fe_{0.6})Fe_2O_4$ (Fig. 2.6a and b). The higher level of Zn doping caused the drop of magnetic moment as antiferromagnetic coupling interactions between Fe^{3+} ions in each *O* site would be dominant. According to this theory, the maximal M_s 175 emu·g^{-1} (Zn + Mn + Fe atoms) was achieved in the 15 nm $(Zn_{0.4}Mn_{0.6})Fe_2O_4$ NPs (Fig. 2.6b). In MRI test under a 4.5 *T* field, DMSA-stabilized $(Zn_{0.4}Mn_{0.6})Fe_2O_4$ NPs and $(Zn_{0.4}Fe_{0.6})Fe_2O_4$ NPs exhibit the r_2 value of 860 and 687 mM^{-1}·s^{-1} (Fig. 2.6c), a significant enhancement compared to CLIO (62 mM^{-1}·s^{-1}), Ferridex (110 mM^{-1}·s^{-1}), undoped Fe_3O_4 (276 mM^{-1}·s^{-1}), and $MnFe_2O_4$ (422 mM^{-1}·s^{-1}) under the identical condition (Fig. 2.6d) [38].

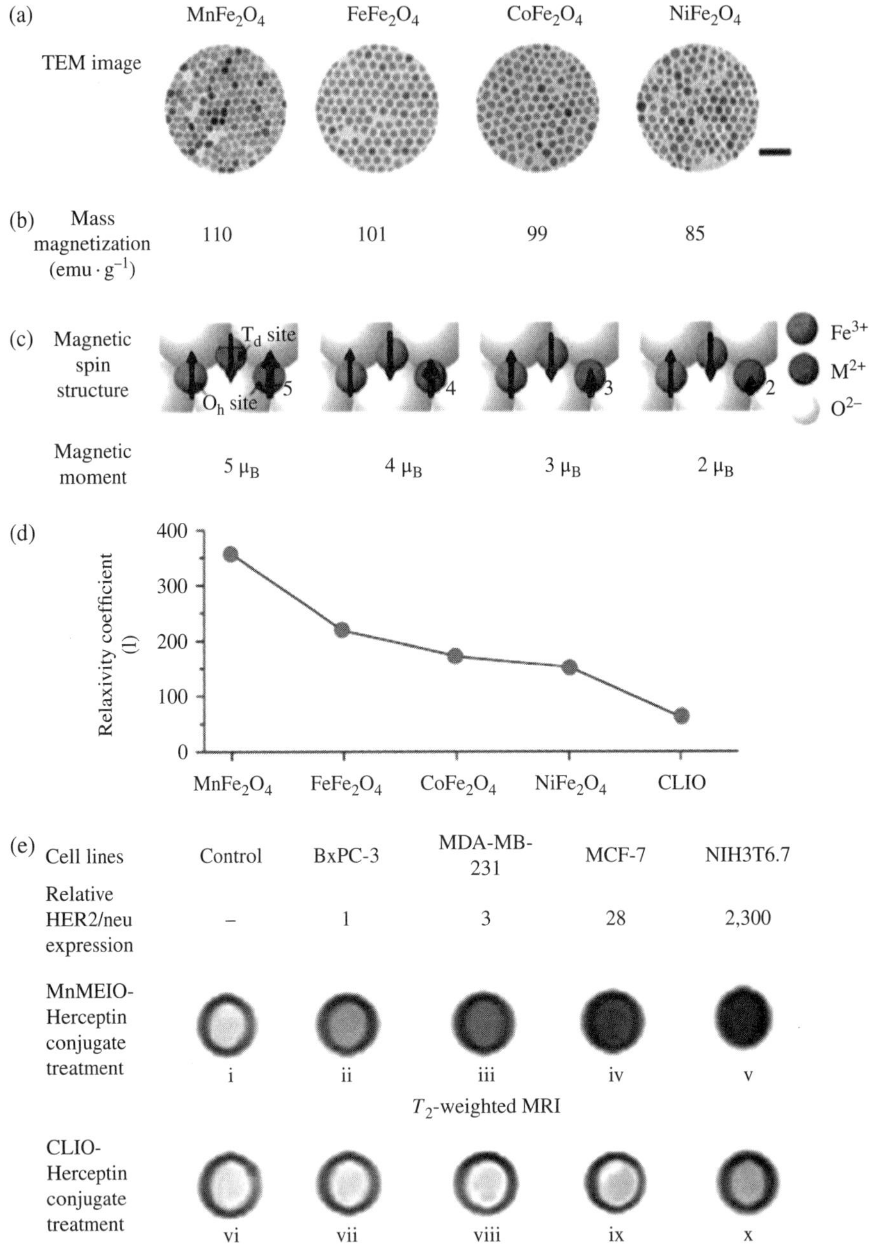

FIGURE 2.5 (a) TEM images of 12 nm $MnFe_2O_4$ NPs (M = Mn, Fe, Co, Ni) and (b) their mass magnetizations and (c) their schematic spin moments. (d) Relaxivity r_2 values of 12 nm $MnFe_2O_4$ NPs (M = Mn, Fe, Co, Ni) and CLIO at 1.5 *T*. (e) T_2-weighted MR images of cell lines treated with MnMEIO–Herceptin conjugate and CLIO–Herceptin conjugates. (Reprinted with permission from Ref. [27]. © Macmillan Publishers Ltd.)

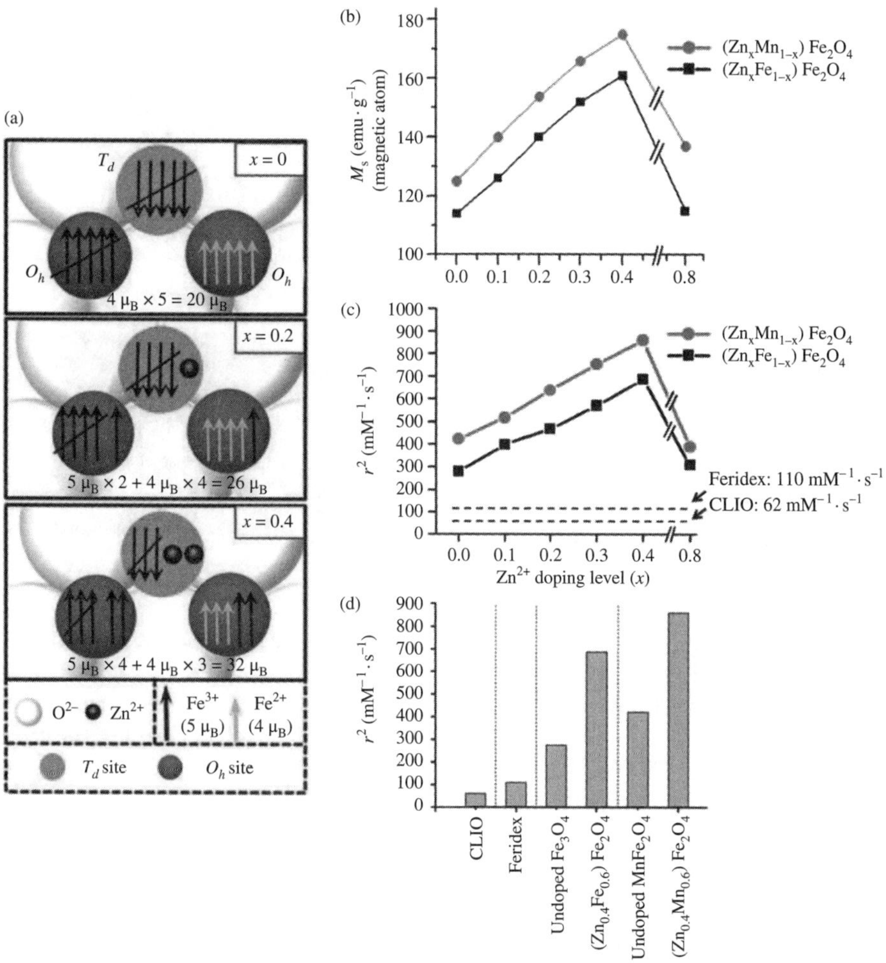

FIGURE 2.6 (a) Schematic illustration of spin moment of $(Zn^xFe^{1-x})Fe_2O_4$ NPs ($x=0$, 0.2, and 0.4). (b) Plot of mass magnetization versus Zn^{2+} doping level of $(Zn^xMn^{1-x})Fe_2O_4$ NPs and $(Zn^xFe^{1-x})Fe_2O_4$ NPs. (c) Plot of relaxivity r_2 value versus Zn^{2+} doping level of $(Zn^xMn^{1-x})Fe_2O_4$ NPs and $(Zn^xFe^{1-x})Fe_2O_4$ NPs at 4.5 T. (d) Comparison of relaxivity r_2 values of different NP contrast agents. (Reprinted with permission from Ref. [38]. © Wiley.)

2.4.3 Structure Control

The iron oxide and ferrite are just magnetic materials with medium magnetizations. But their magnetic properties are very stable due to the chemical stability of oxides. In contrast, metallic Fe, Co, and FeCo alloy can display high moment (Fe: Ms = 218 emu·g^{-1}; Co: 162 emg·g^{-1}; $Fe_{65}Co_{35}$: 245 emu·g^{-1}) but are chemically and magnetically unstable due to their fast oxidation. For biological application, they must be protected to ensure their stability and biocompatibility.

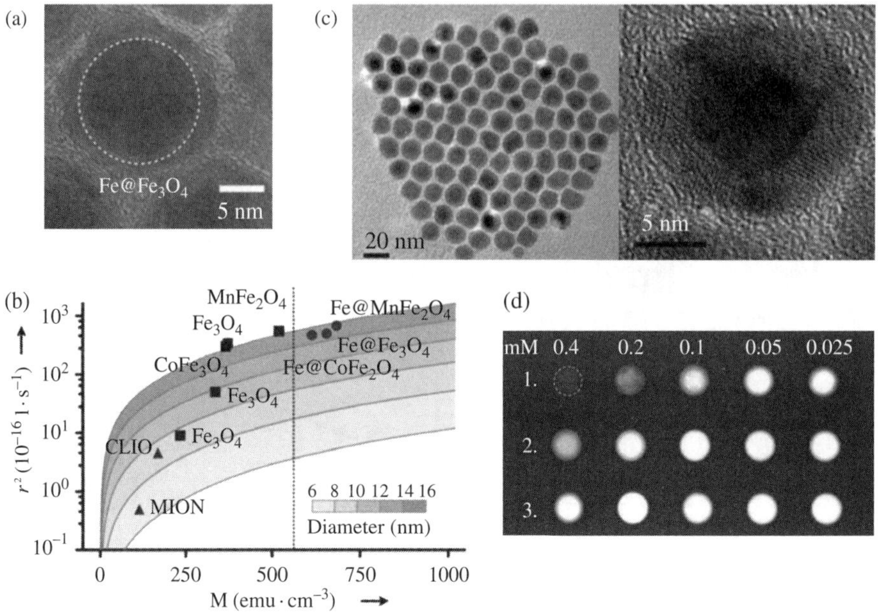

FIGURE 2.7 (a) HRTEM image of an individual amorphous Fe/*fcc*-Fe_3O_4 core–shell NPs produced by thermal decomposition of Fe–oleate complex with Fe NPs. (b) Relaxivity r_2 of various ferrite NPs and amorphous Fe/ferrite core–shell NPs, indicating that the relaxivity of NPs increases with size and magnetization and amorphous Fe/*fcc*-$MnFe_2O_4$ core–shell NPs show the highest r_2 value. (c) TEM and HRTEM images of *bcc*-Fe/*fcc*-Fe_3O_4 core–shell NPs. (d) T_2-weighted images of the 16 nm *bcc*-Fe/*fcc*-Fe_3O_4 NPs (1), amorphous Fe/*fcc*-Fe_3O_4 NPs (2), and Fe_3O_4 NPs (3). (Reprinted with permission from Ref. [41]. © Wiley and Reprinted with permission from Ref. [43]. © American Chemical Society.)

Size-controllable monodisperse Fe NPs were synthesized by the thermal decomposition of $Fe(CO)_5$ in octadecene solvent with oleylamine as the surfactant [24]. The burst decomposition of $Fe(CO)_5$ at 180°C, yielding Fe atoms, CO gas and initiates the nucleation and consequent growth [40]. The ratio of $Fe(CO)_5$/oleylamine was reported to be the critical factor to control the size of Fe NPs [24, 41]. The Fe NPs produced in this method was amorphous and subject to air corrosion to form amorphous iron oxide on the surface. To protect the Fe NPs, the oxygen-transferring agent $(CH_3)_3NO$ was used to react with freshly made amorphous NPs at 250°C to generate amorphous Fe/*fcc*-Fe_3O_4 core–shell NPs, which had enhanced chemical and magnetic stability [24]. Another method was to thermally decompose the metal oleate complex to form polycrystalline MFe_2O_4 (M = Mn, Fe, Co) on the surface of amorphous Fe NPs, as labeled in Figure 2.7a [41, 42]. The produced 11 nm core/2.5 nm shell Fe/$MnFe_2O_4$ NPs, Fe/Fe_3O_4 NPs, and Fe/$CoFe_2O_4$ NPs show the saturation magnetizations of 149, 142, and 133 emu·g^{-1} (metal atoms), which were larger than one of $MnFe_2O_4$ of the same size (110 emu·g^{-1} (metal)). These amorphous Fe/MFe_2O_4 NPs were also made hydrophilic by DMSA. Measured at 0.47 *T* field, the 16 nm

$Fe/MnFe_2O_4$ NPs with a relaxivity of 356 mM^{-1}·s^{-1} produced the same contrast effect at 10 times lower dose compared to CLIO and displayed better MRI sensitivity than single component ferrite NPs (Fig. 2.7b) [41].

The single-crystalline *bcc*-Fe/*fcc*-Fe_3O_4 NPs with further enhanced magnetization were also synthesized by the decomposition of $Fe(CO)_5$ in the presence of oleylamine and hexadecylamine chloride [43]. The use of hexadecylamine chloride was critical for the formation of single-crystalline *bcc*-Fe NPs. After exposure to air, a shell of crystalline Fe_3O_4 was formed (Fig. 2.7c). The 10.5 nm core/2.5 nm shell bcc-Fe/*fcc*-Fe_3O_4 NPs displayed a very stable magnetization of 164 emu·g^{-1} (Fe), much higher than 16 nm amorphous Fe/*fcc*-Fe_3O_4 NPs and Fe_3O_4 NPs. Therefore, a darker T_2-weighted MR image was visualized with bcc-Fe/*fcc*-Fe_3O_4 NPs (Fig. 2.7d). In biological test, the 15 nm *bcc*-Fe/*fcc*-Fe_3O_4 NPs were stabilized by oleylamine–PEG and show a relaxivity r_2 of 220 mM^{-1}·s^{-1} at a 3 *T* field, better than Ferridex (110 mM^{-1}·s^{-1}) and comparable with the $MnFe_2O_4$ NPs stabilized by DMSA with 2 nm coating thickness. It is noteworthy that the r_2 value of *bcc*-Fe/*fcc*-Fe_3O_4 is measured with the influence of thick coating (around 15 nm).

2.5 DESIGNS AND FABRICATIONS OF MULTIMODALITY IMAGING AGENTS BASED ON IOMNPs

The intrinsic "dark" T_2-weighted contrast image given by IOMNPs sometimes causes ambiguity with hypointense area in the practically clinical detection, such as bleeding, calcification, and metal deposition [6]. To address this issue, the multimodality imaging combining MRI with other imaging modality has been studied to enhance the diagnostic accuracy. For instance, the conjugate of $MnFe_2O_4$ with ^{124}I through the reaction between ^{124}I and a surface functional group of NPs can be used as MRI–positron emission tomography (PET) dual-modality contrast agent [44]. The surface of IOMNPs can be modified by some organic dye to enable the MRI–fluorescence dual imaging [45–48]. But the multistep surface conjugating reaction may lead to degradation/leaking of surface conjugates, decrease in efficiency, and increase in toxicity. In this section, we focus on fabrications of multimodality imaging agents based on hybrid inorganic IOMNPs where other inorganic components are crystallographically attached to iron oxide in the structure such as core–shell and dumbbell.

2.5.1 MRI + Optical Imaging

A variety of optically active inorganic materials have been successfully loaded onto the surface of IOMNPs, forming a core–shell structure. The Fe_3O_4/ZnO core–shell NPs were made by a two-step thermal decomposition of $Fe(acac)_3$ and of $Zn(acac)_2$ in octyl ether [49]. The Fe_3O_4/ZnO core–shell NPs show dual superparamagnetic and fluorescent properties and could be detected *in vitro* by confocal microscopy and *in vivo* by MRI [49]. Another example is Fe_3O_4/Au core–shell NPs in which the Au shell is a plasmonically active component [50]. The Fe_3O_4 NP seeds were first prepared through thermal decomposition and dispersed in chloroform with oleylamine.

The Au shell was formed on the surface of Fe_3O_4 NPs after the addition of $HAuCl_4$ to the chloroform solution and the slow reduction of $HAuCl_4$ by oleylamine. The produced Fe_3O_4/Au core–shell NPs were dispersed in hexane and also in aqueous medium after ligand exchange with hexadecyltrimethylammonium bromide and sodium citrate. The Fe_3O_4/Au core–shell NPs could further act as seeds to obtain Fe_3O_4/Au (thicker shell) and Fe_3O_4/Au/Ag NPs, where the shell thickness was tuned by seed/precursor ratio [50].

Dumbbell-like Fe_3O_4/Au NPs represent another class of hybrid dual-functional probe. The growth toward dumbbell NPs requires exquisite control in the solvent choice, precursor concentration, and reaction temperature in favor of heterogeneous nucleation. As shown in Figure 2.8a, the decomposition of $Fe(CO)_5$ initiates the nucleation over the surface of Au seeds and air oxidizes Fe into Fe_3O_4 [51]. The size of the Fe_3O_4 part is controlled by the ratio of $Fe(CO)_5$/Au seeds. The dumbbell NPs have two different types of surface, making the surface modification with different targeting and drug molecules much easier in comparison to single component NPs (Fig. 2.8b and c) [52, 53]. As an example, Fe_3O_4 surface was linked to dopamine–PEG–NH_2 through the chelating catechol unit in dopamine to Fe_3O_4. The Au surface was modified by SH-PEG-NH_2 *via* the formation of Au/S bonding. The water-soluble Au/Fe_3O_4 NPs with amine group was further conjugated with epidermal growth factor receptor antibody (EGFRA) for their targeting A431 human epithelial carcinoma cells that overexpress the epidermal growth factor receptor (Fig. 2.8d) [52]. *In vitro* experiments show that the EGFRA-labeled 8–20 nm Au/Fe_3O_4 dumbbell NPs had high specificity to A431 cells with a r_2 of 80.4 $mM^{-1}\cdot s^{-1}$ at a 3 *T* field (Fig. 2.8e). Meanwhile, the A431 cells labeled with 8–20 nm Au/Fe_3O_4 NPs could also be monitored with a scanning confocal microscopy (Fig. 2.8f). Using the wavelength of 594 nm to detect the strong reflectance from Au, the morphology of A431 cells was clearly visualized. More importantly, there was no image intensity change even after 3 days, providing a long-term optical tracking of the cells in addition to MRI detection.

2.5.2 MRI + Computed Tomography

X-ray computed tomography (CT) is a powerful and widely used clinical imaging modality, which can offer high-resolution 3D tomography information of the anatomic structure. But it has relatively low sensitivity, which is disadvantageous in diagnosing soft tissue [54, 55]. MRI with high sensitivity can be combined with CT to achieve more accurate detection. The first reported NP-based MRI–CT dual-functional contrast agent is FePt NPs, where FePt NPs are superparamagnetic and Pt has a strong X-ray absorption [54]. Relevant *in vivo* tests have demonstrated that the FePt NPs have potential to enhance the MRI T_2 contrast and CT contrast simultaneously. Recently, core–shell Fe_3O_4/TaO_x NPs were tested for simultaneous MRI–CT dual detection [55]. Figure 2.9a summarizes the synthesis of this class of core–shell NPs. A Fe_3O_4 core is premade and stabilized with oleic acid and then further transferred into reverse micelles after mixing with cyclohexane, ethanol, IGEPAL CO-520, and NaOH. The TaO_x shell is formed through sol–gel reaction of tantalum ethoxide over the surface of Fe_3O_4 NP core. Subsequently, the sol–gel reaction can also be utilized to modify the core–shell NPs with PEG–silane and rhodamine B isothiocyanate

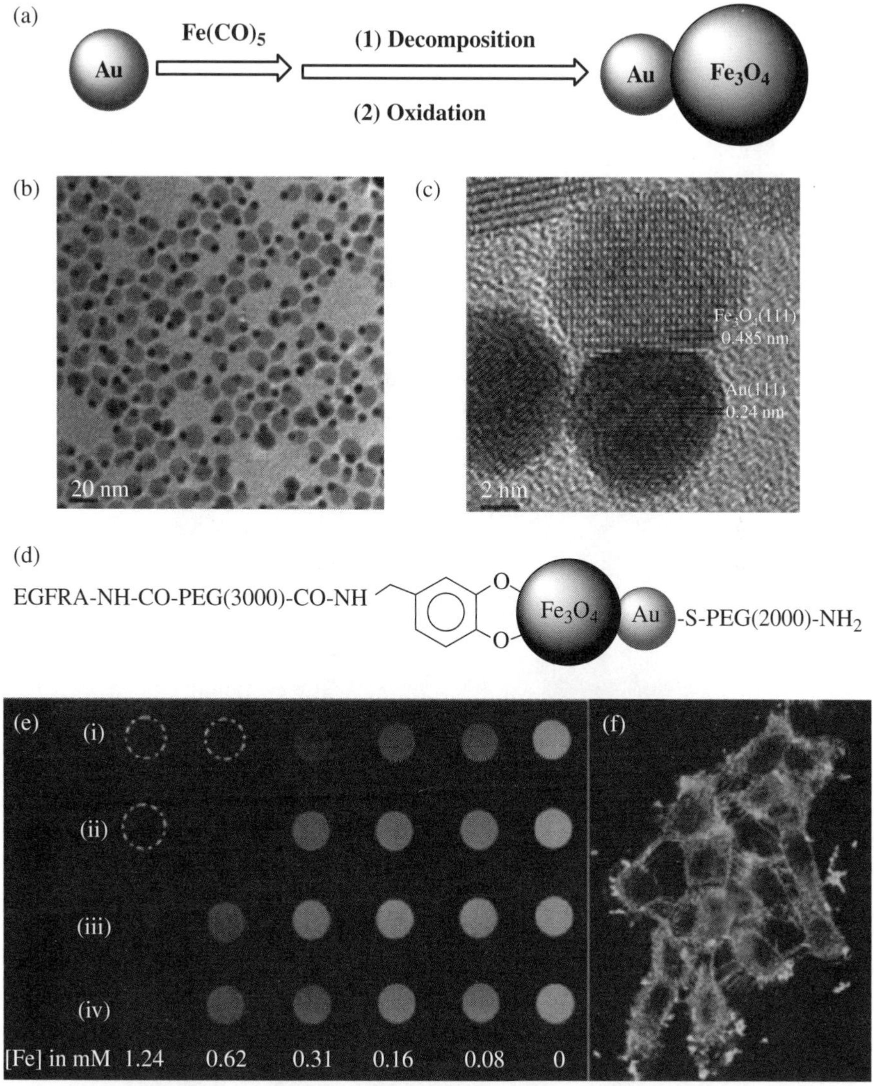

FIGURE 2.8 (a) Scheme for the formation of dumbbell-like Fe_3O_4/Au NPs. (b) TEM and (c) HRTEM images of the dumbbell Fe_3O_4/Au NPs. (d) Surface modification of the dumbbell Fe_3O_4/Au NPs. (e) T_2-weighted MR images of (i) 20 nm Fe_3O_4, (ii) 3–20 nm Au/Fe_3O_4 dumbbell NPs, (iii) 8–20 nm Au/Fe_3O_4 dumbbell NPs, and (iv) A431 cells labeled with 8–20 nm Au/Fe_3O_4 dumbbell NPs. (f) Reflection image of A431 cells labeled with 8–20 nm Au/Fe_3O_4 dumbbell NPs. (Reprinted with permission from Ref. [51]. © American Chemical Society and Reprinted with permission from Ref. [52]. © Wiley.)

(RITC)–silane, which make the NPs water soluble and potentially useful for fluorescence imaging. Dispersed in deionized water, the core–shell Fe_3O_4/TaO_x NPs show a r_2 value of 81.2 $mM^{-1}\cdot s^{-1}$ at a 3 T field. The CT signal intensity constantly increases with the dose increase. *In vivo* tests on rats bearing mammary adenocarcinoma cells

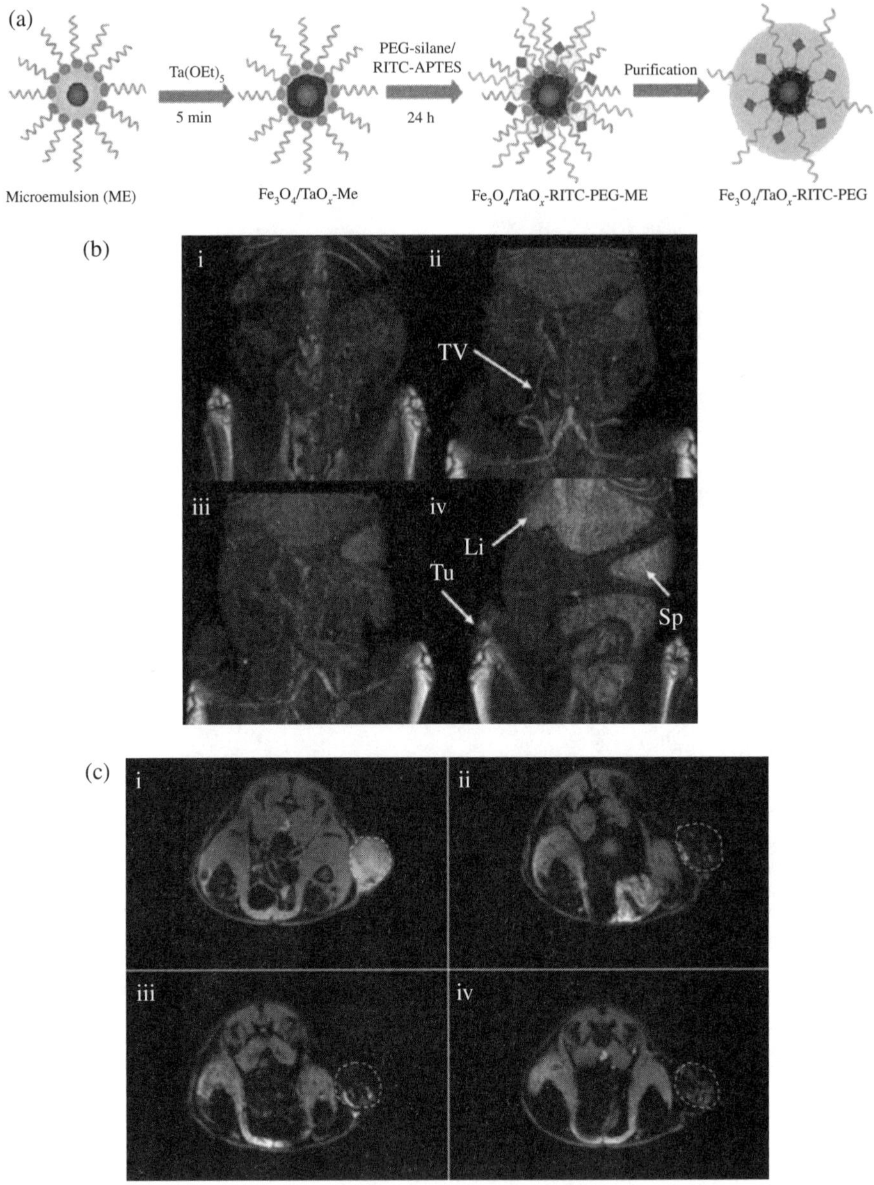

FIGURE 2.9 (a) Scheme for the synthesis and surface modification of Fe_3O_4/TaO_x core–shell NPs. (b) *In vivo* X-ray CT images and (c) *in vivo* T_2-weighted MR images of a rat bearing a MAT B III tumor (i) before and (ii) 1 h, (iii) 2 h, and (iv) 24 h after the injection of Fe_3O_4/TaO_x core–shell NPs. TV, Li, Tu, and Sp in CT images indicate the tumor-associated vessel, liver, tumor, and spleen, respectively. (Reprinted with permission from Ref. [55]. © American Chemical Society.)

(MAT B III) indicated the advantages of bimodal imaging [55]. Since the injected NPs tended to accumulate in tumor due to enhanced permeability and retention (EPR) effect, 3D CT images gave the enhanced signals from tumor-associated areas, especially those contrast from newly formed blood vessel to tumors, which are the main targets in practical clinical treatment to tumors (Fig. 2.9b) [55]. Simultaneously, MRI with higher sensitivity than CT was used to reveal the more precise shape and structure of the tumor (Fig. 2.9c). As the multimodality imaging probe, the Fe_3O_4/TaO_x NPs show no obvious toxicity to rats.

2.5.3 Simultaneous T_1–T_2 Images in MRI

Considering requirements in toxicity, sensitivity, and spatial/time resolutions, MRI is a unique multimodality imaging technique as it provides two independent imaging modalities, T_1-weighted image and T_2-weighted image, to generate "brighter" and "darker" images. Therefore, the preferential combination of T_1 and T_2 contrast effect in a single NP will enable the so-called "self-confirming" imaging under a single instrument [56]. As Figure 2.10a demonstrates, the fringe field created by superparamagnetic NP T_2 contrast agent perturbs the relaxation process of paramagnetic T_1 contrast agent when two probes are in proximity, dramatically quenching the T_1 signal. To magnetically decouple the T_1 and T_2 contrast effect in a single agent, a dual-mode nanoparticle contrast agent (DMCA) was designed and synthesized through fabricating multilayer core–shell NPs. Figure 2.10b shows the structure and TEM image of this hybrid NPs [56]. The inner core is 15 nm $MnFe_2O_4$ NPs acting as T_2 contrast agent, while the outer shell of 1.5 nm $Gd_2O(CO_3)_2$ is used to enhance T_1 contrast. A separating SiO_2 layer is inserted between these two layers and is controlled from 4 to 20 nm to study the decoupling effect (Fig. 2.10c and d). With the 16 nm layer of SiO_2 employed, the DMSA shows a three times higher r_1 (33.1 $mM^{-1}\cdot s^{-1}$) than the conventional Gd-DTPA T_1 contrast agent (9.6 $mM^{-1}\cdot s^{-1}$) and a two times higher r_2 (274 $mM^{-1}\cdot s^{-1}$) than the conventional Ferridex T_2 contrast agent (108 $mM^{-1}\cdot s^{-1}$) (Fig. 2.10c and d) [56]. *In vivo* tests among Gd-DTPA, Ferridex, and DMCA implanted in mice, only DMCA displayed both brighter T_1 and darker T_2 responses (Fig. 2.10e). On the contrary, Gd-DTPA and Ferridex were only effective in one specific imaging mode. This study provides a new pathway to the multimodality imaging contrast agent with a higher level of accuracy and self-confirming capability in one instrumental system.

2.6 CONCLUSION AND PERSPECTIVE

Recent advance in synthetic methodology in organic solution phase allows for the desirable control of IOMNPs' size, composition, crystallinity, and structures, leading to the optimal magnetic property of IOMNPs. Superparamagnetic Fe_3O_4 and ferrite NPs with high magnetizations were used as advanced MRI contrast agent to enhance

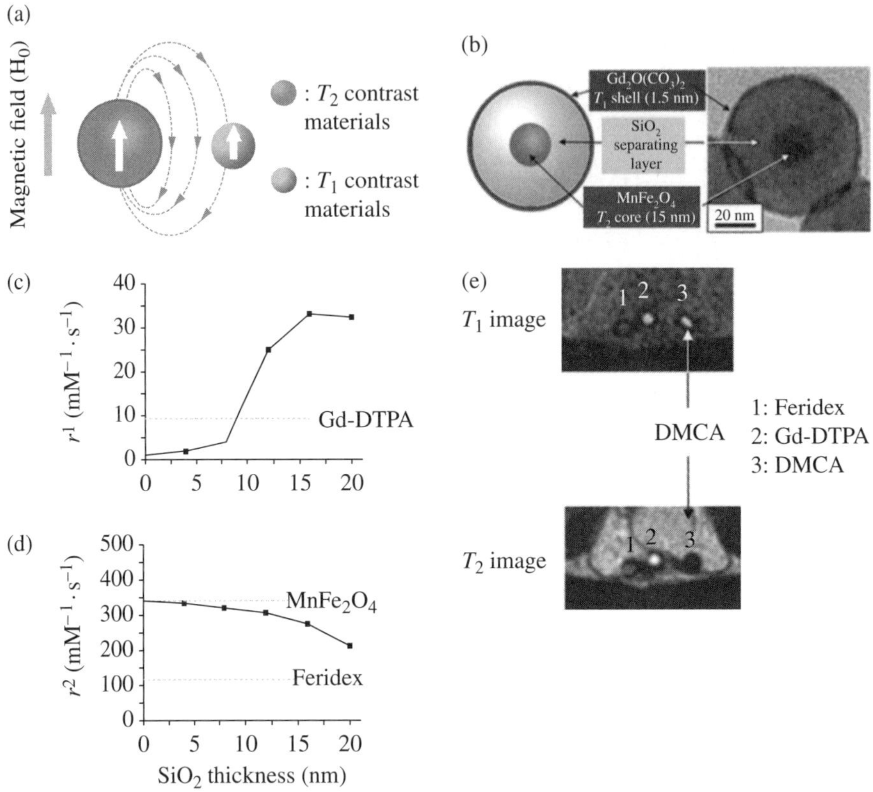

FIGURE 2.10 (a) Schematic illustration of the magnetic coupling of a T_1 contrast agent and a T_2 contrast agent in proximity. (b) Structure and HRTEM image of "magnetic-decoupled" DMCA. (c) Plot of relaxivity r_1 value and (d) relaxivity r_2 value *versus* the thickness of separating SiO_2 layer in DMCA. (e) *In vivo* T_1-weighted MR image and T_2-weighted MR image of mice implanted by DMCA, Ferridex, and Gd-DTPA in separated tubing. (Reprinted with permission from Ref. [56]. © American Chemical Society.)

the sensitivity and efficiency of biomedical imaging. The high magnetization Fe NPs with Fe_3O_4 and ferrite protective shell provided another appealing class of probe for sensitive MRI. To optimize the efficiency of biomedical imaging, IOMNPs have been synthetically combined with other NP components in different hybrid formats such as core–shell and dumbbell, enabling superior multimodality imaging. The as-fabricated MRI–optical, MRI–CT, and MRI T_1–T_2 multimodality nanoprobes have significantly enhanced the biological imaging in accuracy, sensitivity, and special/time resolution.

Despite the exciting progress in this field, the IOMNPs are still in lab demonstration stage. The advanced MRI and multimodality imaging require the formation of IOMNPs with high magnetic moments, high stability, and biocompatibility, which should be further improved. The surface engineering of IOMNPs has not been well established to avoid the unspecific uptake. Besides, the IOMNP-related biocirculation, toxicological effect, and pharmacokinetics should be further studied and properties should be optimized. Once these are better understood, more advanced IOMNPs will be designed and synthesized for clinical multimodality imaging, and the much expected early diagnosis will then be realized.

REFERENCES

[1] Laurent S, Forge D, Port M, Roch A, Robic C, Vander Elst L, Muller RN. Magnetic iron oxide nanoparticles: synthesis, stabilization, vectorization, physicochemical characterizations, and biological applications. Chem Rev 2008;**108**:2064–2110.

[2] Yoo D, Lee J-H, Shin T-H, Cheon J. Theranostic magnetic nanoparticles. Acc Chem Res 2011;**44**:863–874.

[3] Frey NA, Peng S, Cheng K, Sun S. Magnetic nanoparticles: synthesis, functionalization, and applications in bioimaging and magnetic energy storage. Chem Soc Rev 2009;**38**: 2532–2542.

[4] Sun X, Don H, Lacroix LM, Xiao JQ, Sun S. Magnetic nanoparticles for magnetoresistance-based biodetection. IEEE Trans Nanobioscience 2012;**11**:46–53.

[5] Ho D, Sun X, Sun S. Monodisperse magnetic nanoparticles for theranostic applications. Acc Chem Res 2011;**44**:875–882.

[6] Lee N, Hyeon T. Designed synthesis of uniformly sized iron oxide nanoparticles for efficient magnetic resonance imaging contrast agents. Chem Soc Rev 2012;**41**:2575–2589.

[7] Xie J, Liu G, Eden HS, Ai H, Chen X. Surface-engineered magnetic nanoparticle platforms for cancer imaging and therapy. Acc Chem Res 2011;**44**:883–892.

[8] Park J, Joo J, Kwon SG, Jang Y, Hyeon T. Synthesis of monodisperse spherical nanocrystals. Angew Chem Int Ed Engl 2007;**46**:4630–4660.

[9] Jeong U, Teng X, Wang Y, Yang H, Xia Y. Superparamagnetic colloids: controlled synthesis and niche applications. Adv Mater 2007;**19**:33–60.

[10] Hao R, Xing R, Xu Z, Hou Y, Gao S, Sun S. Synthesis, functionalization, and biomedical applications of multifunctional magnetic nanoparticles. Adv Mater 2010;**22**:2729–2742.

[11] Morrish AH. *The Physical Principles of Magnetism*. Volume **1**, New York: Wiley-VCH; 2001.

[12] Skomski R, Coey J. *Permanent Magnetism*. Bristol: Institute of Physics Publishing; 1999.

[13] Lu A-H, Salabas EL, Schüth F. Magnetic nanoparticles: synthesis, protection, functionalization, and application. Angew Chem Int Ed Engl 2007;**46**:1222–1244.

[14] Cullity BD. *Introduction to Magnetic Materials*. Reading (MA): Addison-Wesley Publishing Company; 1972.

[15] Sun S. Recent advances in chemical synthesis, self-assembly, and applications of FePt nanoparticles. Adv Mater 2006;**18**:393–403.

[16] Na HB, Song IC, Hyeon T. Inorganic nanoparticles for MRI contrast agents. Adv Mater 2009;**21**:2133–2148.

[17] Tong S, Hou S, Zheng Z, Zhou J, Bao G. Coating optimization of superparamagnetic iron oxide nanoparticles for high T2 relaxivity. Nano Lett 2010;**10**:4607–4613.

[18] Kellar KE, Fujii DK, Gunther WH, Briley-Sæbø K, Bjørnerud A, Spiller M, Koenig SH. NC100150 injection, a preparation of optimized iron oxide nanoparticles for positive-contrast MR angiography. J Magn Reson Imaging 2000;**11**:488–494.

[19] Moghimi SM, Hunter AC, Murray JC. Long-circulating and target-specific nanoparticles: theory to practice. Pharmacol Rev 2001;**53**:283–318.

[20] Hyeon T. Chemical synthesis of magnetic nanoparticles. Chem Commun (Camb) 2003;**21**:927–934.

[21] Murray CB, Kagan C, Bawendi M. Synthesis and characterization of monodisperse nanocrystals and close-packed nanocrystal assemblies. Ann Rev Mater Sci 2000;**30**: 545–610.

[22] LaMer VK, Dinegar RH. Theory, production and mechanism of formation of monodispersed hydrosols. J Am Chem Soc 1950;**72**:4847–4854.

[23] Pei W, Kakibe S, Ohta I, Takahashi M. Controlled monodisperse Fe nanoparticles synthesized by chemical method. IEEE Trans Magn 2005;**41**:3391–3393.

[24] Peng S, Wang C, Xie J, Sun S. Synthesis and stabilization of monodisperse Fe nanoparticles. J Am Chem Soc 2006;**128**:10676–10677.

[25] Peng S, Sun S. Synthesis and characterization of monodisperse hollow Fe_3O_4 nanoparticles. Angew Chem Int Ed Engl 2007;**46**:4155–4158.

[26] Jun Y-W, Choi J-S, Cheon J. Shape control of semiconductor and metal oxide nanocrystals through nonhydrolytic colloidal routes. Angew Chem Int Ed Engl 2006;**45**: 3414–3439.

[27] Lee J-H, Huh Y-M, Jun Y-W, Seo J-W, Jang J-T, Song H-T, Kim S, Cho E-J, Yoon H-G, Suh J-S, Cheon J. Artificially engineered magnetic nanoparticles for ultra-sensitive molecular imaging. Nat Med 2007;**13**:95–99.

[28] Xu C, Xu K, Gu H, Zheng R, Liu H, Zhang X, Guo Z, Xu B. Dopamine as a robust anchor to immobilize functional molecules on the iron oxide shell of magnetic nanoparticles. J Am Chem Soc 2004;**126**:9938–9939.

[29] Xie J, Xu C, Kohler N, Hou Y, Sun S. Controlled PEGylation of monodisperse Fe_3O_4 nanoparticles for reduced non-specific uptake by macrophage cells. Adv Mater 2007;**19**: 3163–3166.

[30] Wang B, Xu C, Xie J, Yang Z, Sun S. pH controlled release of chromone from chromone-Fe_3O_4 nanoparticles. J Am Chem Soc 2008;**130**:14436–14437.

[31] Medintz IL, Uyeda HT, Goldman ER, Mattoussi H. Quantum dot bioconjugates for imaging, labelling and sensing. Nat Mater 2005;**4**:435–446.

[32] Jun Y-W, Seo J-W, Cheon J. Nanoscaling laws of magnetic nanoparticles and their applicabilities in biomedical sciences. Acc Chem Res 2008;**41**:179–189.

[33] Sun S, Zeng H. Size-controlled synthesis of magnetite nanoparticles. J Am Chem Soc 2002;**124**:8204–8205.

[34] Sun S, Zeng H, Robinson DB, Raoux S, Rice PM, Wang SX, Li G. Monodisperse MFe_2O_4 (M=Fe, Co, Mn) nanoparticles. J Am Chem Soc 2004;**126**:273–279.

[35] Kim D, Lee N, Park M, Kim BH, An K, Hyeon T. Synthesis of uniform ferrimagnetic magnetite nanocubes. J Am Chem Soc 2008;**131**:454–455.

[36] Park J, An K, Hwang Y, Park J-G, Noh H-J, Kim J-Y, Park J-H, Hwang N-M, Hyeon T. Ultra-large-scale syntheses of monodisperse nanocrystals. Nat Mater 2004;**3**:891–895.

[37] Jun Y-W, Huh Y-M, Choi J-S, Lee J-H, Song H-T, Kim S, Yoon S, Kim K-S, Shin J-S, Suh J-S, Cheon J. Nanoscale size effect of magnetic nanocrystals and their utilization for cancer diagnosis via magnetic resonance imaging. J Am Chem Soc 2005;**127**: 5732–5733.

[38] Jang J-T, Nah H, Lee J-H, Moon SH, Kim MG, Cheon J. Critical enhancements of MRI contrast and hyperthermic effects by dopant-controlled magnetic nanoparticles. Angew Chem Int Ed Engl 2009;**48**:1234–1238.

[39] Bao N, Shen L, Wang Y, Padhan P, Gupta A. A facile thermolysis route to monodisperse ferrite nanocrystals. J Am Chem Soc 2007;**129**:12374–12375.

[40] Huber DL. Synthesis, properties, and applications of iron nanoparticles. Small 2005;**1**: 482–501.

[41] Yoon T-J, Lee H, Shao H, Weissleder R. Highly magnetic core–shell nanoparticles with a unique magnetization mechanism. Angew Chem Int Ed Engl 2011;**50**:4663–4666.

[42] Lee H, Yoon T-J, Weissleder R. Ultrasensitive detection of bacteria using core–shell nanoparticles and an NMR-filter system. Angew Chem Int Ed Engl 2009;**48**: 5657–5660.

[43] Lacroix L-M, Frey Huls N, Ho D, Sun X, Cheng K, Sun S. Stable single-crystalline body centered cubic Fe nanoparticles. Nano Lett 2011;**11**:1641–1645.

[44] Choi J-S, Park JC, Nah H, Woo S, Oh J, Kim KM, Cheon GJ, Chang Y, Yoo J, Cheon J. A hybrid nanoparticle probe for dual-modality positron emission tomography and magnetic resonance imaging. Angew Chem Int Ed Engl 2008;**47**:6259–6262.

[45] Veiseh O, Sun C, Gunn J, Kohler N, Gabikian P, Lee D, Bhattarai N, Ellenbogen R, Sze R, Hallahan A, Olson J, Zhang M. Optical and MRI multifunctional nanoprobe for targeting gliomas. Nano Lett 2005;**5**:1003–1008.

[46] Josephson L, Tung C-H, Moore A, Weissleder R. High-efficiency intracellular magnetic labeling with novel superparamagnetic-Tat peptide conjugates. Bioconjug Chem 1999; **10**:186–191.

[47] Xie J, Chen K, Huang J, Lee S, Wang J, Gao J, Li X, Chen X. PET/NIRF/MRI triple functional iron oxide nanoparticles. Biomaterials 2010;**31**:3016–3022.

[48] Lee J-H, Jun Y-W, Yeon S-I, Shin J-S, Cheon J. Dual-mode nanoparticle probes for high-performance magnetic resonance and fluorescence imaging of neuroblastoma. Angew Chem Int Ed Engl 2006;**45**:8160–8162.

[49] Cho N-H, Cheong T-C, Min JH, Wu JH, Lee SJ, Kim D, Yang J-S, Kim S, Kim YK, Seong S-Y. A multifunctional core-shell nanoparticle for dendritic cell-based cancer immunotherapy. Nat Nanotechnol 2011;**6**:675–682.

[50] Xu Z, Hou Y, Sun S. Magnetic core/shell Fe_3O_4/Au and Fe_3O_4/Au/Ag nanoparticles with tunable plasmonic properties. J Am Chem Soc 2007;**129**:8698–8699.

[51] Yu H, Chen M, Rice PM, Wang SX, White RL, Sun S. Dumbbell-like bifunctional Au–Fe_3O_4 nanoparticles. Nano Lett 2005;**5**:379–382.

[52] Xu C, Xie J, Ho D, Wang C, Kohler N, Walsh EG, Morgan JR, Chin YE, Sun S. Au–Fe_3O_4 dumbbell nanoparticles as dual-functional probes. Angew Chem Int Ed Engl 2008;**47**:173–176.

[53] Xu C, Wang B, Sun S. Dumbbell-like Au–Fe_3O_4 nanoparticles for target-specific platin delivery. J Am Chem Soc 2009;**131**:4216–4217.

[54] Chou S-W, Shau Y-H, Wu P-C, Yang Y-S, Shieh D-B, Chen C-C. In vitro and *in vivo* studies of FePt nanoparticles for dual modal CT/MRI molecular imaging. J Am Chem Soc 2010;**132**:13270–13278.

[55] Lee N, Cho HR, Oh MH, Lee SH, Kim K, Kim BH, Shin K, Ahn T-Y, Choi JW, Kim Y-W, Choi SH, Hyeon T. Multifunctional Fe_3O_4/TaOx core/shell nanoparticles for simultaneous magnetic resonance imaging and x-ray computed tomography. J Am Chem Soc 2012;**134**:10309–10312.

[56] Choi J-S, Lee J-H, Shin T-H, Song H-T, Kim EY, Cheon J. Self-confirming "AND" logic nanoparticles for fault-free MRI. J Am Chem Soc 2010;**132**:11015–11017.

3

LIPID-BASED PHARMACEUTICAL NANOCARRIERS FOR IMAGING APPLICATIONS

TAMER ELBAYOUMI

Department of Pharmaceutical Sciences, Midwestern University, Glendale, AZ, USA

VLADIMIR TORCHILIN

Department of Pharmaceutical Sciences and Center for Pharmaceutical Biotechnology and Nanomedicine, Northeastern University, Boston, MA, USA

3.1 INTRODUCTION

A quite actively growing area in current experimental and clinical medicine is the application of lipid-based vesicles, mostly liposomes and micelles, as pharmaceutical carriers for biomedical and diagnostic imaging agents.

Conditional on the contrast agent, sensitivity, and resolution limit of the scanning method, specific minimum amounts of the contrast agent have to be delivered into the area of interest. While conventional nonenhanced imaging techniques are useful only when relatively large tissue areas are involved in the pathological process, in the majority of cases, this may not be feasible due to the variable ability of different tissues to attenuate contrast signal, like X-rays, sound waves, radiation, or radio frequency. Hence, in order to achieve sufficient attenuation, contrast agents are utilized to discriminate target tissue, owing to their capability to absorb signal and irradiation, much stronger than surrounding tissue.

Nanotechnology for Biomedical Imaging and Diagnostics: From Nanoparticle Design to Clinical Applications, First Edition. Edited by Mikhail Y. Berezin.

Consequently, chief objectives for successful biomedical imaging become to accumulate a sufficient quantity of the contrast moiety in the area of interest and keep its presence in normal tissues on the most minimal level possible.

A very attractive approach is the application of contrast-loaded pharmaceutical nanocarriers as contrast materials providing an enhanced accumulation of reporter moieties in pathological areas. For almost three decades now, liposomes have been identified as promising vehicles to deliver a wide range of encapsulated and/or membrane-associated therapeutic or diagnostic agents into the required areas [1–3], and in more recent years, micelle-loaded imaging agents have also shown great clinical promise. In particular, these nanocarriers have been utilized for carrying the diagnostic moieties used with all imaging modalities: gamma-scintigraphy, magnetic resonance imaging (MRI), computed tomography (CT) imaging, and sonography [4, 5]. In the last case, acoustic characterization of echogenic systems, such as liposomes, becomes especially important as well as the choice of the optimal frequency to obtain a required signal [6]. In several important recent review papers, the use of magnetoliposomes for molecular imaging and cancer theranostics is described [7], as well as the development of liposomes from drug delivery vehicles to the leading platform in diagnostic and theranostic nanomedicine [8], and general problems of using liposomes in molecular imaging [9] are analyzed. Some of those reviews concentrate on specific problems, such as liposomal contrast agents for brain tumor imaging [10] or for cardiovascular diseases [11] or the use of liposomal contrast agents in specific imaging modalities, such as positron emission tomography (PET) [12].

Now, the combination of advanced imaging methods and equipments, providing high sensitivity and spatial resolution, and the use of nanoscale devices to deliver diagnostic agents with high target specificity allow for more accurate detection and staging, leading to more accurate therapy planning of the disease state.

Among the current pharmaceutical nanoscale drug vehicles, lipid-based nanovesicles, such as liposomes (predominantly, for water-soluble drugs) and micelles (principally, for water-insoluble drugs), are considered the most extensively studied, demonstrating marked advantages over other delivery systems, and possess the most suitable characteristics for encapsulation of many drugs, genes, and diagnostic (imaging) agents [13, 14]. Simply, possibility to easily control composition, size, and *in vivo* stability of a nanoreservoir is a major advantage. More importantly, these nanovesicular systems retain remarkable loading capacity of various imaging and contrast agents, of either hydrophilic or lipophilic nature or both [15].

Liposomes are bilayered phospholipid vesicles of size ranging from 50 to 1000 nm in most cases, which can be loaded with a variety of water-soluble cargos (into their inner aqueous compartment) and water-insoluble molecules (into the hydrophobic compartment of their phospholipid membrane), and are considered as promising drug carriers for well over three decades [15]. The use of targeted liposomes, that is, liposomal carriers specifically accumulating inside the affected organ or tissue, would increase the efficacy of the liposome-incorporated drug and decrease the loss of their contents in the reticuloendothelial system (RES) [15]. Several successful protocols have been employed to actively and passively target liposomal nanocarriers to various pathological tissues and into target cells in particular, via various targeting ligands and homing moieties [16].

Analogously, micellar assemblies, including lipopolymeric micelles, represent another promising type of pharmaceutical carriers. Micelles are amphiphilic compound-formed colloidal particles with hydrophobic core and hydrophilic corona with size range between 5 and 100 nm [17], with exceptional capability to enhance the solubility and bioavailability of poorly soluble pharmaceuticals [18]. The use of certain amphiphilic molecules, such as polyethylene glycol (PEG)–phosphatidylethanolamine (PE), as micelle-building blocks can extend the blood half-life of the micellar carrier. In addition, the micelles could be combined with different targeting ligands, strategically incorporated into the micelle structure to make micelles targeted. There are several main micelle parameters (size, critical micelle concentration (CMC), and loading capacity of the hydrophobic core of the micelle) essential for successful preparation of micelle-incorporated pharmaceuticals and imaging agents [19]. As a rule, these parameters are as follows: the size of a pharmaceutical micelle is between 10 and 100 nm, an optimal CMC value is below micromolar concentration, and the loading efficiency toward a hydrophobic drug is between 5 and 25 wt.%. In the case of targeted micelles, the release of a free drug from micelles limited to the target organ should lead to the increased efficacy of the drug, while maintaining stability of the micelles in the blood should contribute to drug solubility and toxicity reduction due to minimal interaction with nontarget organs [14, 18].

3.2 GENERAL APPROACHES FOR LOADING LIPOSOMES AND MICELLES WITH CONTRAST AGENTS

Regardless of the imaging modality to be employed, lipid-based nanoformulations, namely, liposomes and micelles, have been long studied as vehicles for contrast agents to facilitate sufficient accumulation of contrast agent within the required area of interest, in order to differentiate this area from surrounding tissue. Table 3.1 shows the reporter groups and their minimal required concentration at area of interest, which have to be attained, for each particular case.

One can easily recognize that different chemical nature of reporter moieties utilized in different imaging modalities requires different protocols to load liposomes and micelles with a certain contrast agent. In addition, all the imaging modalities exemplified here differ not only in their sensitivity and resolution, but for appropriate

TABLE 3.1 Concentration of a Contrast Agent Required for Diagnostically Significant Tissue Attenuation in Various Imaging Modalities

Modality	Reporter group	Required tissue concentration (M)
Gamma-scintigraphy	Radionuclide (^{111}In, ^{99m}Tc)	10^{-10}
MRI	Paramagnetic metal (Gd, Mn)	10^{-4}
CT imaging	Heavy element (iodine, bromine, barium)	10^{-2}

tissue attenuation, they also require different levels of the corresponding diagnostic label to be delivered in the target region.

A large variety of techniques have been developed to load liposomes and lipid–polymer micelles with different reporter labels. With the exception of gas-loaded vesicles intended for ultrasound imaging, the following criteria were established to be the main characteristics of a successful lipid nanovesicle labeling method: (i) sufficient shelf stability of the formulation to be labeled; (ii) high labeling efficiency; (iii) convenient and efficient labeling at room temperature; (iv) use of readily available isotopes and metals that have good physical imaging characteristics, dosimetry, and half-life; (v) reasonable *in vitro* and *in vivo* stability of label association with the carriers; and (vi) universal applicability to all types of liposomes and micelles, regardless of size, surface charge, or lipid composition [20, 21].

A number of general approaches developed for the labeling of liposomes—successfully applied for lipid–polymer micelles also—are summarized here in order of complexity and development from the poorer to higher *in vivo* stability [21].

3.2.1 Label Encapsulation during Nanocarrier Preparation

Lipid-based nanocarriers can be labeled by the incorporation of the gamma-imaging agent into the lipid phase [22] or encapsulation in the aqueous interior of the vesicle during formulation [23, 24]. This method has many disadvantages including low encapsulation efficiency, high radiation exposure, difficulty in maintaining nanocarrier sterility, and the necessity of making nanocarriers within a very short period of the intended clinical use, due to the overall instability of the formulation [20]. This is particularly true with radioisotopes with half-lives under 24 h, which includes ^{99m}Tc and ^{18}F. The only method reported to date for labeling nanocarriers with a PET radionuclide is the encapsulation of ^{18}F-fluorodeoxyglucose (^{18}F-FDG) (110 min half-life) within vesicles during liposome and micelle manufacture [25]. The development of methods to label nanocarriers with PET radionuclides after their manufacture will be very valuable. A better approach is the labeling of nanocarriers after their manufacture and just prior to use [26].

3.2.2 Surface Labeling after Nanocarrier Preparation

Nanocarriers can be labeled by attaching the radioisotope to the surface of preformed phospholipid vesicles. This method in fact offers an improvement over the encapsulation method as the preformed nanocarriers can be made in a different location than the labeling site, where the labeling process can occur just prior to the clinical use. This surface labeling employs the basic electrostatic interactions between the charged isotope and the nanocarrier surface. Labeling using this method has the problems of inconsistent labeling efficiencies and insufficient *in vivo* stability due to exposure of the external surface of nanocarriers to plasma proteins in the circulation [20].

Initial methods described for labeling postmanufactured liposomes with ^{99m}Tc relied on the labeling of the lipid membrane with reduced ^{99m}Tc. This method proved to produce ^{99m}Tc-labeled nanocarriers that were not stable when injected *in vivo*, with frequent dissociation of the ^{99m}Tc from the nanocarrier after administration [27].

3.2.3 Label Incorporation into the Lipid Bilayer of Preformed Nanocarriers

Preformed nanocarriers can be also labeled by incubation with a lipophilic chelator into which the radioisotope has been chelated, prior to the incubation with the lipid suspension. Although this method provides high labeling efficiency, its *in vivo* stability was questionable, because most of these lipophilic labels can readily exchange with lipoprotein components in blood [28].

Morgan *et al.* [29] were among the first to label liposomes and micelles (MLVs) with ^{99m}Tc by reducing pertechnetate with tin (stannous) chloride in the presence of the nanovesicle. In animal studies, the liposomal preparation labeled according to this "stannous chloride method" showed relatively high uptake of the radiolabel in kidneys and bladder, suggesting extensive *in vivo* release of the radiolabel from liposomes. This in fact was confirmed by other groups [30, 31]. Several methods to label nanocarriers with ^{99m}Tc have been developed. Alafandy *et al.* developed the oxidoreduction method [32], as an improved version of the stannous chloride method. Nanocarriers were prepared with tin(II) dioxinate complexes inserted in the bilayers. However, this method required a series of extensive washing in order to remove the free pertechnetate from the nanocarriers [33].

3.2.4 Surface Labeling of Preformed Nanocarrier by Incorporating the Lipid–Chelator Conjugate During Preparation

Another approach is to incorporate a lipid chelator into the vesicles during the formulation stage, particularly during the preparation of the lipid film [20]. These amphiphilic chelators are composed of diethylene triamine pentaacetic acid (DTPA) conjugated to the head group region of a long-chain fatty acid or phospholipid molecule. The chelator is incorporated into the lipid components of the bilayer prior to the actual preparation of the nanocarriers. After hydration of the phospholipids, the chelator is exposed on the outer surface of the nanocarriers, thus being easily accessible for the radionuclide. The nanocarriers are then incubated with the radioisotope, which becomes bound to the chelator present on the external surface of the nanocarrier. Still, lipid–chelator systems described to date suffer from decreased *in vivo* stability, mainly due to the exposure of the radiolabel on the surface of nanocarrier [33].

Historically, Hnatowich *et al.* described the synthesis of a long hydrocarbon chain covalently coupled to DTPA [22]. These nanocarriers could be labeled with ^{67}Ga or ^{99m}Tc. However, labeling efficiencies did not exceed 60%. During the incubation in 50% human serum, approximately 30% of the ^{99m}Tc activity was released from the nanocarriers, indicating the instability of these radiolabeled formulations. Later, Goto *et al.* developed nanocarriers containing stearylamine (SA)–DTPA conjugates to label nanocarriers with ^{99m}Tc [34]. These preparations showed good *in vitro* stability in 50% human serum.

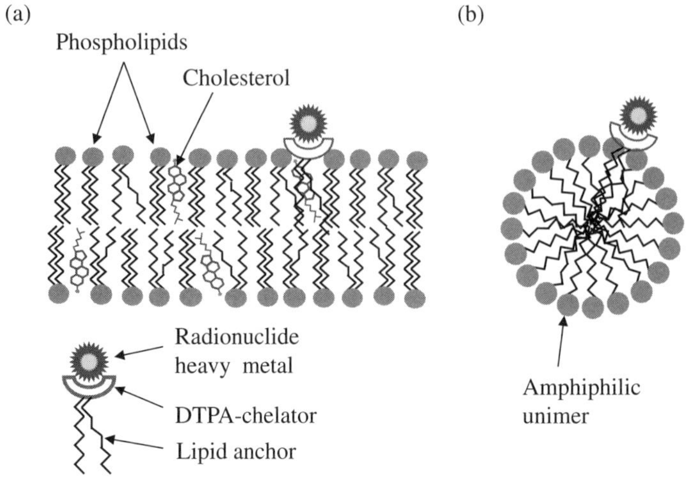

FIGURE 3.1 Schematics of radionuclide or MR contrast metal association with liposomes or micelles via the amphiphilic chelate.

In the case of MRI imaging agents, to achieve a better MR signal, all reporter metal atoms should be freely exposed for interaction with water. This requirement makes metal encapsulation into the nanocarrier less attractive than metal coupling with polymeric chelators exposed into the outer water space. Nanocarriers loaded with chelated paramagnetic ions (Gd, Dy, Mn, Fe) have been demonstrated to be useful as MRI contrast agents mostly for the visualization of the macrophage-rich tissues such as RES organs [35].

The low-molecular-weight water-soluble paramagnetic probes may leak from nanocarriers upon contact with body fluids, which destabilizes most liposomal membranes. Moreover, it has been shown that when too high concentrations of Gd-DTPA are encapsulated inside nanocarriers for the better enhancement, the relaxivity of the compound might be even lower than for nonencapsulated Gd-DTPA complex, probably because of decreased residence lifetime of water molecules inside vesicles [36]. Membranotropic chelating agents—DTPA-SA [37] and DTPA-PE [38]—consist of the polar head containing chelated paramagnetic atom and the lipid moiety, which anchors the metal–chelate complex in the nanocarrier membrane. This approach has been shown to be far more superior in terms of the relaxivity of the final preparation when compared with nanocarrier-encapsulated paramagnetic ions due to the decrease in the rotational correlation times of the paramagnetic moiety rigidly connected with relatively large particle [36]. Nanocarriers with membrane-bound paramagnetic ion demonstrate also the reduced risk of the leakage in the body. Membranotropic chelates are suitable for micelle incorporation (they anchor in the hydrophobic micelle core) and may also serve to load liposomes and micelles with heavy radiometals (see Fig. 3.1).

A second method has been recently developed by Laverman *et al.* [39]. With this method, the highly stable ^{99m}Tc chelator, hydrazino nicotinamide (HYNIC), is attached to distearoyl phosphatidylethanolamine (DSPE), which is added to the liposomal nanocarrier formulation during the manufacture. Reduced ^{99m}Tc is then incubated with the HYNIC nanocarriers where it becomes chelated by the HYNIC on the nanocarrier surface.

Another approach aimed at the amplification of the diagnostic signal using polychelating polymers, for higher loading capacity of reporter molecules, was suggested [40–42]. Since it is rather difficult to enhance the signal intensity from a given reporter metal in nanovesicles, one may attempt to increase the quantity of carrier-associated reporter metal (such as Gd for MRI or ^{111}In for scintigraphy) [43]. In this design, a series of single terminus-activated polychelating polymers were synthesized using poly-L-lysine (PLL) as a main chain and chelating moieties (such as DTPA) as side groups. These polychelating polymers were additionally modified with a hydrophobic residue to assure their firm incorporation into the nanocarrier membrane. As a result, polychelating amphiphilic polymers (PAP) have been obtained, a family of copolymers containing a hydrophilic fragment with multiple chelating groups and relatively short but highly hydrophobic phospholipid fragment on one end of a polymeric chain suitable for incorporation into hydrophobic nanocarrier membrane or micelle core [44]. The chemistry was based on the use of carbobenzoxy (CBZ)-protected PLL with free terminal amino group, which was used to modify the PLL chain with the phospholipid anchor, with subsequent CBZ deprotection and incorporation of DTPA residues at lysine amino groups in PLL. This polychelator was found to easily incorporate into the liposomal membrane in the process of nanocarrier preparation and sharply increases the number of chelated heavy metal atoms attached to a single lipid anchor (Fig. 3.2, panel A).

As noted by the authors of [45], one of the major challenges in MRI is the quantification of the local concentration of contrast agents. To solve the problem, they have proposed a multimodal radiolabeled paramagnetic liposomal agent, which allows for the simultaneous imaging with SPECT and MRI. SPECT-based quantification permits to determine the concentration of Gd and thus deconvolute the MRI signal.

The amphiphilic polychelator *N,N*-(DTPA-polylysyl)glutaryl phosphatidylethanolamine (DTPA-PLL-NGPE) easily incorporates into the liposome membrane or micelle core in the process of liposome of micelle preparation and sharply increases the number of chelated Gd atoms attached to a single lipid anchor. After the saturation of polychelator with Gd^{3+} ions, the elemental analysis has revealed that Gd-DTPA-PLL-NGPE contains about 40% (w/w) Gd, which corresponds to 8–10 metal atoms per single lipid-modified polymer molecule assuming its molecular weight to be 3500–4000 Da. Higher Gd content leads to better relaxivity and, consequently, to greater MR signal intensity (assuming the Gd tissue concentration does not exceed millimolar range). Additionally, upon the incorporation into the liposome membrane or micelle core, the NGPE anchor grafted with PAP keeps chelating groups above the liposomal membrane or within the micelle's corona. In both cases, metal atoms chelated into PAP are directly exposed to the exterior water environment and have better access to the adjacent tissue water protons. This might lead to the corresponding enhancement of the vesicle contrast properties. As a result, amphiphilic polychelator-containing nanovesicles have higher relaxation influence on water protons compared

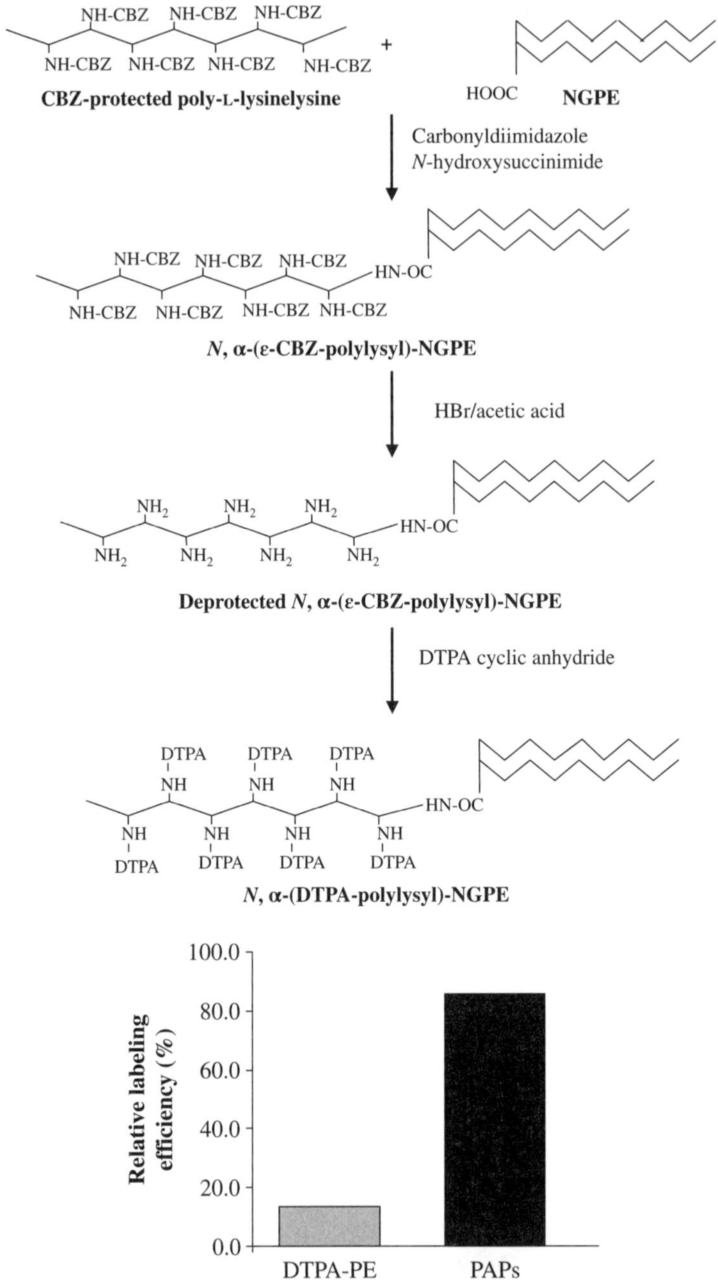

FIGURE 3.2 Synthesis of amphiphilic polychelate DTPA-PLL-NGPE consisting of hydrophilic DTPA-polylysyl (PLL) (DTPA-PLL) moiety and hydrophobic N-glutaryl phosphatidylethanolamine (NGPE) moiety (**top**); labeling efficiency of liposomes containing the same molar fraction of PAP or DTPA-PE (**bottom**).

with conventional preparations at the same phospholipid content. In addition to the enhanced relaxivity, the PEG polymer attached on liposomal or micellar carriers can help in avoiding the contrast agent (both for gamma-imaging and MRI) uptake in the site of injection by resident phagocytic cells.

Nanocarriers loaded with ^{111}In via the membrane-incorporated PAP demonstrated significantly higher labeling efficiency than nanocarriers loaded with ^{111}In via the monomeric DTPA-PE (Fig. 3.2, panel B). This technique allowed to increase the number of bound reporter metal atoms per vesicle and to decrease the dosage of an administered total liposomal lipid without compromising the image signal intensity [46].

3.2.5 Aqueous Phase Loading of Preformed Nanocarrier

Several methods have been developed for loading the radioisotope into the aqueous phase of preformed nanocarriers. This approach generally uses lipophilic chelator capable of carrying the radioisotope through the lipid membrane into the aqueous phase, where the isotope is rechelated into a more firm complex with an entrapped compound and remains inside the nanocarrier. The success of this method depends mainly on the high affinity of the radioisotope for the agent entrapped in the aqueous phase of the nanocarrier. Otherwise, the radioisotope will shuttle back and forth across the lipid membrane. This method has shown high labeling efficiency (more than 90%). Moreover, it is likely to have the greatest *in vivo* stability, because of the protected location of the radioisotope inside the nanocarrier [26].

Initially, the approach was demonstrated with ^{67}Ga transported through the liposomal bilayer as the lipophilic ^{67}Ga–oxine complex and then irreversibly trapped inside nanocarriers via the presence of aqueous interior-loaded hydrophilic chelator in preformed nanocarriers [47]. The same procedure was suggested for labeling nanocarriers with ^{111}In as well [48]. In the last decade, two reproducible methods of labeling postmanufactured nanocarriers with ^{99m}Tc have been reported. In the first method, ^{99m}Tc is chelated into the lipophilic carrier hexamethylpropyleneamine oxime (HMPAO) [49]. The ^{99m}Tc-HMPAO complex is then added to nanocarriers with encapsulated glutathione (GSH) and incubated with the GSH-containing nanocarriers for 30 min, during which the ^{99m}Tc-HMPAO migrates across the lipid membrane inside nanocarriers, where it is converted to a hydrophilic complex by reductive partial decomposition via the interaction with GSH. The hydrophilic ^{99m}Tc-HMPAO complex remains trapped within the nanocarrier. This method proved reliable, yielding *in vivo* stable radiolabeled nanocarriers with high labeling efficiency (~85%).

In a similar way, Suresh *et al.* proposed the use of the hepatobiliary imaging agent ^{99m}Tc–diisopropyl iminodiacetic acid (^{99m}Tc-DISIDA) to label preformed nanocarriers with encapsulated GSH [50]. As in the previously mentioned study, the ^{99m}Tc-DISIDA complex undergoes the reductive decomposition to more hydrophilic species via the reaction with encapsulated GSH, yielding the incorporation of the ^{99m}Tc-DISIDA into the nanocarriers with 90% efficiency. The labeled nanocarriers were stable in saline and 90% FBS for up to 24 h. Biodistribution studies in mice showed that ^{99m}Tc-labeled nanocarriers accumulated in liver and spleen at 24 h postinjection. The use of ^{99m}Tc-DISIDA was preferred compared to HMPAO, since ^{99m}Tc-DISIDA is much cheaper and has a longer shelf life when used for nanocarrier labeling.

A few years ago, Phillips *et al.* [51] introduced a novel method for radiolabeling preformed nanocarriers with ^{99m}Tc–N-ethyl-substituted bis(2-mercaptoethyl)amine and phenylthiolate coligand (S) complexes (SNS/S), which can be applied to nanocarrier radiolabeling with therapeutic radionuclides, ^{186}Re and ^{188}Re. In his technique, GSH-loaded nanocarriers were incubated with ^{99m}TC-SNS/S complexes having a neutral core coordination structure. These ^{99m}TC-SNS/S complexes can change into hydrophilic complexes following their entry into the inner space of nanocarriers and reaction with pre-encapsulated GSH, cysteine, or other hydrophilic chemicals containing a thiol group. It was also found that the higher the lipophilicity of the ^{99m}TC-SNS/S complex, the easier it can pass through the double layer of the nanocarrier membrane. This method resulted in good labeling efficiency (about 50–70%) and high *in vitro* stability of ^{99m}Tc-BMEDA nanocarriers. Moreover, these ^{99m}Tc-GSH nanocarriers also showed the slow blood pool clearance and low spleen accumulation typical for PEG-coated nanocarriers.

In interpreting and analyzing scintigraphic images with radiolabeled nanocarriers, one has to recognize that these radionuclide labels track the biodistribution of nanocarriers and do not always track the disposition of encapsulated therapeutic agents [52]. The particular chelate with which the radionuclide is bound may also influence the behavior of the radionuclide after the nanocarrier leaves the blood and becomes metabolized. Thus, Ogihara-Umeda *et al.* [53] have shown that when nanocarrier-encapsulated ^{67}Ga is in the form of ^{67}Ga–nitrilotriacetic acid (NTA) complex, the ^{67}Ga accumulation in a tumor is twice as high as when ^{67}Ga is used as ^{67}Ga–deferoxamine even at similar blood level of ^{67}Ga-nanocarrier activities at 24 h. It was speculated that this difference was due to the high affinity of the unbound ^{67}Ga for the tumor tissue, which resulted in ^{67}Ga tumor accumulation after metabolism of the nanocarrier in the region of the tumor and release of ^{67}Ga from the low-affinity NTA chelate. Hence, the characteristics of the radionuclide and the radionuclide–ligand complex used in nanocarrier labeling must be always carefully considered when analyzing the scintigraphic images, particularly after the nanocarrier metabolism starts [26].

In a recent comparative analysis, the relative efficacy of entrapment of radiopaque materials into different nanocarriers and advantages and disadvantages of various nanocarrier types were reviewed [35]. The maximum entrapment into the inner nanovesicles interior might be achieved for the vesicles prepared by reverse-phase evaporation, dehydration/rehydration, and interdigitation/fusion methods. However, all these methods are difficult to control and scale up. So, from the practical point of view, the optimum method for nanovesicle loading with contrast material by entrapment is still to be developed.

CT contrast agents (primarily heavily iodinated organic compounds) were included in the inner water compartment of nanocarriers or incorporated into the nanocarrier membrane during the nanocarrier preparation [54]. Thus, iopromide-containing nanocarriers were prepared by ethanol evaporation method with encapsulation efficiency up to 40% [55–57].

3.2.6 Key Challenges in Using Liposomes and Micelles as Carriers for Imaging Agents

To facilitate the accumulation of contrast in the required zone, various nanovesicles, such as liposomes and micelles, draw a special attention because of their easily controlled properties and good pharmacological characteristics. Pursuing different *in vivo* delivery purposes, one can easily change the size, charge, and surface properties of these nanocarriers simply by adding new ingredients to the lipid or amphiphilic mixture before liposome or micelle preparation and/or by variation of preparation methods.

The use of lipidic nanovesicles for the delivery of imaging agents has quite a long history [14, 58]. The ability of these nanocarriers to entrap different substances into both the aqueous phase and the nanocarrier membrane compartment made them suitable for carrying the diagnostic moieties used with all imaging modalities: gamma-scintigraphy, MRI, CT imaging, and even sonography. The different chemical nature of reporter moieties used in different modalities requires different protocols to load nanocarriers with the given contrast agent. All the imaging modalities, as listed in Table 3.1, not only differ in their sensitivity and resolution but also require different amounts of a diagnostic label to be delivered into the area of interest. These general considerations, taken together, led to the development of the whole family of lipid-based contrast agents for various purposes. Since the micelle-based delivery of contrast agents is a relatively new approach [18, 43], most of the analysis of current trends in carrier-mediated transport of imaging agents will, predominantly, involve the data obtained with the use of liposomal nanocarriers.

Phospholipid nanocarriers, if introduced into the circulation, are very rapidly (usual half-clearance time is within 30 min) sequestered by the cells of RES. Liver cells are primarily responsible [59], and the sequestration is almost independent on their size, charge, and composition of the nanovesicles. From the pharmacokinetic point of view, it is important that circulating peripheral blood monocytes can also endocytose nanovesicles and later infiltrate tissues and deliver endocytosed nanovesicles to certain pathological areas in the body [44]. The decrease in RES uptake can be achieved by decreasing nanovesicle size, increasing nanovesicle dose, presaturating RES with "empty" nanolipidic vesicles or other particles, or modifying nanovesicle surface with certain "protective" polymers. To increase nanovesicle accumulation in the "required" areas, the use of targeted nanocarriers has been suggested [60]. Nanocarriers with a specific affinity for an affected organ or tissue were believed to increase the efficacy of liposomal pharmaceutical agents, including the imaging ones. Immunoglobulins, primarily of the IgG class, are the most promising and widely used targeting moieties for various drugs and drug carriers including nanolipidic vesicles.

However, pharmacokinetic properties of immunoliposomes (as well as immunomicelles, in a similar fusion) in many cases were far from desirable. Despite evident success in the development of antibody-to-nanovesicle coupling technique and improvements in the targeting efficacy, the majority of antibody-coupled nanovesicles still ended in the liver, which was usually a consequence of insufficient time for the interaction between the target and targeted nanocarrier. This is especially true in cases when a target of choice has diminished blood supply (ischemic or necrotic areas), and only a small number of nanovesicles passes through the target with the blood during the time period when nanovesicles are present in the circulation. The same lack of targeting occurs if the concentration of the target antigen is very low and the number of "productive collisions" between target antigens and immuno-nanocarrier is too small. In both cases, much better accumulation can be achieved if the nanovesicles can stay in the circulation long enough. This will increase the total quantity of immunocarrier passing through the target and the number of collisions between the immunocarrier and target antigen [61].

Different methods have been suggested to achieve long circulation, including coating the nanocarrier lipid surface with inert, biocompatible polymers, such as PEG, which form a protective layer over the nanocarrier surface and slow down the nanocarrier recognition by opsonins and subsequent clearance. Long-circulating nanocarriers are now widely used in biomedical *in vitro* and *in vivo* studies and even found their way into clinical practice [62]. The use of PEGylated liposomes and micelles as carriers for contrast agents also forms now an important area of research [5, 43, 63]. Very special pharmacokinetic properties of long-circulating (usually PEGylated) drugs and drug carriers deserve at least a short separate discussion.

3.2.7 Pharmacokinetics and Clearance of Long-Circulating Drugs and Drug Carriers

The most important pathways for the clearance of long-circulating drugs and drug carriers—extravasation, renal clearance, and uptake by cells from the blood—are independent processes. However, namely, extravasation determines interstitial and lymphatic transport and interstitial and lymphatic uptake of long-circulating substances. With this in mind, it was suggested to divide all long-circulating drugs/carriers into extravasating and nonextravasating ones [64]. The main difference between these two groups is their size, with a border zone being at about 5–10 nm. The nonextravasating group includes cells, cell ghosts, particles, and large nanovesicles, while the extravasating group includes small nanovesicles, proteins and their derivatives (including antibodies and PEG-modified enzymes), and various polymers that are large enough (more than 40 kDa) to avoid fast renal clearance [65]. In certain pathological areas with increased endothelial permeability, the pattern can change completely. However, such conditional division permitted to build idealized models describing the biological behavior of long-circulating substances [64].

Interactions of nonextravasating or very slowly extravasating drugs/carriers are limited to blood components and cells exposed into the blood. Usually, such substances are removed from the circulation via opsonization-mediated phagocytosis, and surface coating with PEG or PEG-like polymers may prevent opsonization or, at least, significantly slow it down.

An important analysis of pharmacokinetics of long-circulating (PEG-coated) nanocarriers was performed by Allen [66]. It clearly follows from this analysis that association of drugs with carriers such as nanocarriers has pronounced effects of pharmacokinetic profiles of both the drug and the carrier. The most important consequences of such association for the drug are delayed drug absorption, restricted drug biodistribution, decreased volume of drug biodistribution, delayed drug clearance, and retarded drug metabolism [66–68]. All these effects are determined by hindered interstitial penetration of a drug and lesser drug accessibility for the biological milieu because of entrapment into nanolipidic vesicle (or other drug carriers). The presence of protective polymer on the carrier surface changes all these parameters still further [69]. Thus, while "plain" nanocarriers have nonlinear, saturable kinetics, long-circulating nanocarriers demonstrate dose-independent, nonsaturable, and log-linear kinetics [66–68]. The appearance of log-linear kinetics for long-circulating nanocarriers (and, evidently, other sterically protected carriers) is explained by a significant decrease in the first phase of particulate carrier clearance into a high-affinity and relatively low-capacity system, such as RES [70]. All pharmacokinetic effects depend on the route of nanocarrier administration and their size and composition and always are less expressed for sterically protected PEG carriers [71]. Naturally, many of these features are common for all long-circulating drugs and carriers including imaging agents.

3.3 SPECIAL CONSIDERATIONS FOR TARGET VISUALIZATION VIA CONTRAST-LOADED LIPID NANOCARRIERS

3.3.1 Effect of Nanocarrier Properties on the *In Vivo* Behavior of the Radiolabeled Formulation

It was demonstrated earlier that the mean vesicle size is a major factor affecting the circulation time of "plain" nanocarriers [72]. A nonlinear pharmacokinetics of MLV was observed, and the variability in the MLV sizes was assumed to be the underlying reason. The effect of particle size was later studied [73], and the circulation times of the carriers of identical lipid composition but varying in mean size (200 vs. 400 nm) were compared in mice. The larger carriers cleared much faster ($t_{1/2}$ = 0.2 h) than the smaller vesicles ($t_{1/2}$ = 1.5 h). The vesicle *in vitro* and *in vivo* properties depend also on their composition. The lipid composition and the preparation technique determine the characteristics of the nanocarriers. Overall, the use of phospholipids with longer saturated acyl chains and the inclusion of cholesterol (Chol) yield more stable vesicles. Therefore, for application as a radiopharmaceutical (when the radionuclide is incorporated in the aqueous phase), the nanocarriers should preferably contain such components to prevent leakage of the radionuclide from the nanovesicle [59]. The incorporation of Chol into the liposomal bilayers and the use of sphingomyelin (SM) and/or phospholipids with long and saturated acyl chains prolong also the vascular residence of nanocarriers [74–76]. Moreover, nanocarriers are cleared from the circulation by RES in a dose-dependent manner [77, 78]. It was shown in mice that small unilamellar vesicles (SUVs, 80 nm, SM/Chol = 2:1) administered at the dose of 6 mg lipid/kg body weight show high liver uptake (about 60% of injected dose 23 h postinjection), while at doses exceeding 120 mg/kg, liver uptake is much lower and reaches plateau at 20% of injected dose [79].

The surface charge and the charge density of nanocarriers can be controlled by introducing certain components in the lipid bilayers. In general, the presence of a high electrostatic surface charge promotes the interaction of nanovesicles with biomolecules (e.g., serum proteins) and cells. It has been shown that neutral and positively charged nanocarriers acquire a negative charge in the circulation, probably due to the adsorption of plasma components (e.g., α2-macroglobulin) [80]. However, regardless of charge acquired on contact with plasma, the tissue uptake of negatively charged nanocarriers after intravenous (IV) injection is more rapid than of neutral or positively charged vesicles as indicated by their faster clearance from the circulation [1, 59].

3.3.2 Background Clearance

To clearly visualize the tumor, it has to accumulate a sufficient amount of radioactivity above the background level of activity demonstrated by normal tissues. Generally, a target-to-background ratio of at least 1.5–2 is necessary for detecting a pathological site and providing diagnostically significant image [81]. As the radioactivity of agents remaining in the blood form the main source for the background, proper identification and selection of the radiopharmaceutical plasma half-life is essential. Long circulation is important for higher tumor accumulation via the enhanced permeability and retention (EPR) effect [82, 83]. At the same time, the remaining blood activity decreases the target-to-background ratio. On the other hand, too short plasma half-life of a contrast agent results in great reduction in tumor accumulation, although providing low background activity. Thus, a half-life of about 6–12 h appears to be ideal to allow for sufficient accumulation of nanocarriers in inflamed or cancerous targets yet provides enough background clearance to permit early identification of the target [20, 84, 85].

3.3.3 Active Background Removal

The target-to-background ratio can also be increased by developing special protocols for decreasing the background signal. Thus, the method was described for the rapid removal of nanocarriers from the circulation prior to the imaging [86]. In this method, biotin-coated nanocarriers are injected intravenously and allowed to accumulate in the tumor. Then, after 4 h, avidin is injected to induce the aggregation of the biotin-coated nanocarriers due to strong avidin–biotin interaction. This resulted in rapid removal of aggregated nanocarriers from the circulation by liver and spleen. Using this technique, the tumor-to-blood ratio increased to 13.8 compared to about 0.9 in the control. This method has great potential for rapid imaging using nanolipidic vesicles.

3.4 DIAGNOSTIC APPLICATIONS OF LIPOSOMAL IMAGING AGENTS

The main body of experimental data on lipidic nanocarrier contrast agents relates to liposomal formulations. Some key examples of *in vivo* imaging utilizing contrast liposomal vehicles are presented here (additional information can be found in Reference [19]).

3.4.1 Liver and Spleen Imaging

The imaging of the most macrophage-rich organs of RES, liver and spleen, is the easiest to perform, as RES organs are the natural targets for intravenously administered nanovesicles. The diagnostic imaging of liver and spleen is usually aimed at discovering tumors and metastases in those organs, as well as certain blood flow irregularities and inflammatory processes. The use of at least three different imaging modalities for this purpose was described, namely, gamma-imaging, MRI, and CT imaging.

Gamma-scintigraphy of the liver and spleen with liposomal contrast is used rather rarely. However, SUVs loaded with ^{99m}Tc–glucoheptonate were found to accumulate well in the liver of mice and rabbits and provide good scintigraphic images 90 min upon IV administration of the liposomal contrast [87].

Currently, MRI and CT imaging of the liver and spleen with diagnostic nanocarriers are well developed. Nanocarriers containing a soluble Gd-DTPA complex were demonstrated to be effective contrast agents for MRI of liver, spleen, and hepatic metastases [88–90]. Besides, Gd-DTPA in 70 nm vesicles showed better enhancement of rat liver and major blood vessels than the equivalent dose of free Gd-DTPA [35, 91].

Liposomes containing X-ray contrast agents (such as verografin, ioxaglate, iohexol, iopromide, different nonionic contrast media, etc.) were used for CT imaging [81]. Iopromide-containing nanocarriers were even obtained in the lyophilized state and then used for effective detection of liver tumors in mice and demonstrated good storage longevity [55, 56]. Iodine-containing interdigitation/fusion vesicles [92] with high iodine content and a size of more than 1 μm were shown to provide sufficient parenchyma enhancement at a dose of 0.1 g of iodine/kg. Recently, the application of serine- and mannose-modified liposomes loaded with a nonionic X-ray contrast media, iohexol, for the parenchymal enhancement in rabbit liver VX-2 tumor model was described, since Kupffer cell imaging is important for the detection of the liver cancer [93]. Similar model was also used for tumor CT imaging with iopromide liposomes [94].

3.4.2 Tumor Imaging with Contrast Nanocarriers

The main mechanism of nanocarrier accumulation in tumors is via extravasation through leaky tumor capillaries into the interstitial space [95]. As in many other cases, the efficacy of such accumulation can be sharply increased by using long-circulating PEG-coated nanocarriers [96]. Nanocarrier-based imaging agents were already

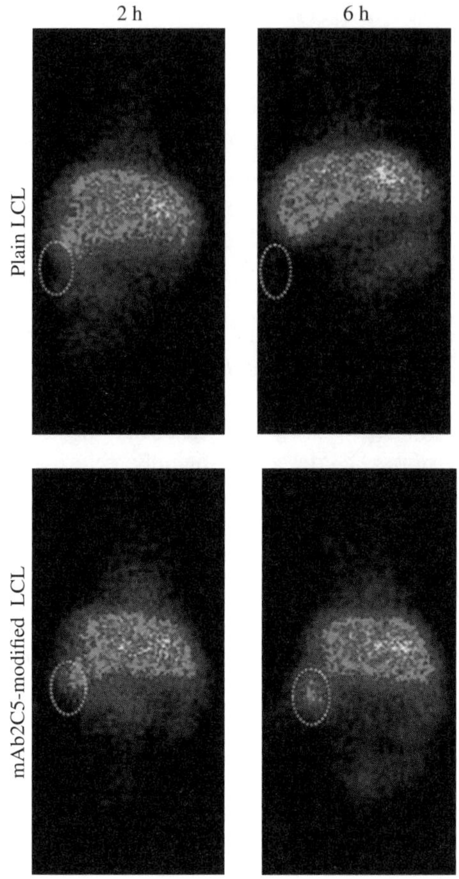

FIGURE 3.3 Whole-body gamma-images of LLC tumor-bearing mice receiving various ^{111}In-labeled liposomal preparations (nontargeted long-circulating liposomes (LCLs), plain LCL; tumor-targeted LCLs modified with tumor-specific monoclonal antibody 2C5 (mAb2C5), mAb2C5-modified LCL), captured at different times post IV tail vein injection. Tumor is encircled. (Reprinted with kind permission from Ref. [101]. © Springer Science and Business Media.) (*See insert for color representation of the figure.*)

successfully used for gamma-imaging, MRI, CT imaging, and sonographic imaging of tumors. ^{111}In-labeled nanocarriers for tumor imaging (VesCan®, Vestar, Inc.) are still progressing in phase II–III clinical trials. The results obtained via the utilization of small unilamellar liposomal nanocarriers (with or without PEG coating) are quite encouraging; sensitivity of this preparation is up to 82%, and specificity 98% in the detection of various carcinomas as well as for melanoma, sarcoma, and lymphoma [97, 98]. ^{111}In nanocarriers were also used for gamma-visualization of human tumor xenografts in mice [85, 99–101]. It was shown that additional targeting of radiolabeled liposomes with tumor-specific antibodies results in significantly accelerated tumor imaging [101] (see Fig. 3.3). Clinical data with ^{111}In-labeled nanocarriers are already available on the visualization of primary lung cancer, Kaposi sarcoma, and lymphoma in AIDS patients [102] and various other tumors [21, 58]. The use of ^{99m}Tc-DTPA nanocarriers for experimental tumor imaging is also described [103], as is the use of nanocarriers labeled with ^{67}Ga [104].

PEG-coated, pH-sensitive, and folate-modified liposomes labeled with ^{159}Gd were used for simultaneous antitumor treatment, biodistribution studies, and scintigraphic imaging in mice with Ehrlich tumor [105]. Pharmacokinetics and dosimetry were studies using PEGylated liposomes loaded with ^{111}In-labeled vinorelbin and with ^{188}Re-labeled doxorubicin in two different colon carcinoma models in mice [106]. The imaging was able to prove that both agents have comparable antitumor activity.

Interestingly, liposomal contrast agent begin to find their application in veterinary, and safety and biodistribution study of ^{99m}Tc-labeled liposomes has been performed in horses [107].

Both animal experiments and clinical studies revealed that in certain cases contrast nanocarriers accumulate in tumors indirectly, via loading resident macrophages [98, 108]. The sensitivity of the method reaches 85%, far exceeding that for traditional contrast agents.

The use of lipid-coated stable gas-containing microbubbles for sonographic imaging of experimental brain glyomas in rats is also described earlier [109].

The imaging of lymph nodes plays a major role in the early detection of neoplastic involvement in cancer patients. Lymph nodes with malignancies contain less normal macrophage-rich tissue that actively absorbs particulates, and so, on the image (no matter which imaging modality is used), an abnormal lymph node should have certain “filling defect” [27]. Over 30 years ago, conventional nanovesicles were shown to accumulate in macrophages of regional lymph nodes after subcutaneous injection. It takes 6–24 h in rat to get the nodes visualized with gamma- or paramagnetic-labeled nanocarriers [110].

Liposomes modified with already mentioned polychelating polymers (PAP) loaded with Gd together with PEG and tumor-specific antibody demonstrated significantly enhanced tumor MRI with good tumor visualization in mice already after 4 h (see Fig. 3.4) [111].

Folate-targeted liposomes loaded with SPION demonstrated strong association with cancer cells and good signal from these cells *in vitro* [112]. Ferriliposomes (liposomes loaded with magnetic nanoparticle clusters) have been successfully used to monitor drug delivery into tumors and tumor microenvironment [113]. Similarly, liposomes coloading with chelated Gd and prednisolone phosphate allowed following the antitumor activity of the preparation in murine tumor model [114].

More and more often liposomes are used simultaneously bearing contrast moieties for different imaging modalities. Thus, fluorescent paramagnetic liposomes (calcein + Gd chelate) provided multimodal imaging of the cellular uptake of folate-modified liposomes [115]. Allen’s group has described an interesting liposomal system for simultaneous CT imaging and MRI for various image-guided applications [116]. Another interesting example is multicolor molecular MRI using diamagnetic chemical exchange saturation transfer liposomes [117], which becomes possible since the diamagnetic saturation transfer contrast is frequency dependent.

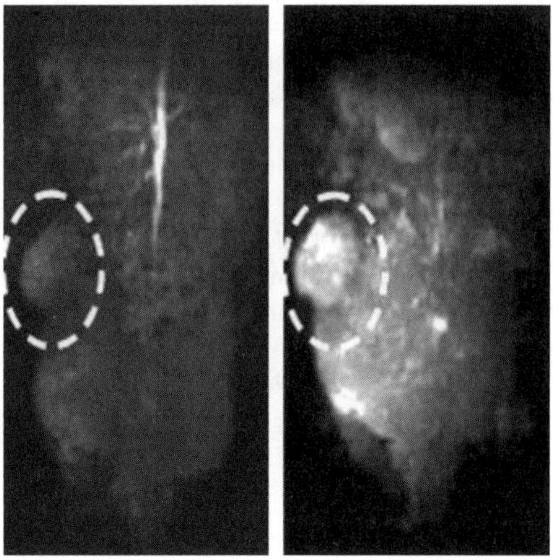

FIGURE 3.4 Three-dimensional angiography of mice injected with antitumor mAb2C5-modified Gd-PAP-containing PEGylated liposomes. Dotted ovals show tumor location. Notice good tumor visualization with 2C5-modified liposomes already after 4 h. (With kind permission from Ref. [111]. © John Wiley & Sons, Inc.)

PEGylated liposomes labeled with positron emitter (^{18}F) have been used for PET imaging of brain cancer (glioma) in rats [118]. ^{64}Cu-labeled liposomes represent another example of PET imaging agent and have been successfully used for tumor imaging in mice [119]. Similar liposomes have also been used for PET imaging of the human neuroendocrine carcinoma in mice [120].

3.4.3 Contrast-Loaded Nanocarriers in Cardiovascular System

Blood pool imaging is of special interest for the evaluation of the current state of blood flow and for the discovery of its irregularities caused by atherosclerotic lesions, thrombi, or tumors. Blood pool imaging (experimental so far) requires the prolonged circulation of diagnostic nanocarriers and is usually based on the utilization of sterically protected nanocarriers loaded with ^{111}In, ^{67}Ga, Gd, heavily iodinated organic compounds, and gas bubbles. Blood perfusion and various cardiac parameters (ventricular ejection fraction, cardiac output, and wall motion being among them) can be evaluated in all imaging modalities [92, 121]. Therefore contrast-loaded nanocarriers may serve as a good alternative for currently used ^{99m}Tc-labeled red blood cells for the same purposes, which do not allow acquisition of desirable quality images because of the fast accumulation of the latter in the liver.

Plain, nonmodified nanocarriers loaded with ^{99m}Tc-DTPA accumulate to some extent in infarcted areas (experimental myocardial infarction) via the impaired filtration mechanism (extravasation through ischemically damaged blood vessels) [122]. However, the degree of this accumulation is insufficient to obtain diagnostically significant images. Therefore, immunoliposomes were used for targeted delivery of drugs and imaging agents [123].

The first visualization of experimental myocardial infarction with nanocarriers was achieved in dogs using $^{111}InCl_3$-loaded immunoliposomes modified with infarct-specific, antimyosin antibody [124]. The further attempts to use nanocarriers as imaging agents for myocardial infarction involved long-circulating immunoliposomes (including those modified with antimyosin antibody). ^{111}In-labeled PEG-modified antimyosin nanocarriers have been shown to accumulate successfully in the areas of experimental myocardial infarction in rabbits and dogs (via both specific and nonspecific mechanisms) and provide good gamma-immunoscintigraphic images of the infarct within several hours upon IV administration [125].

Myocardial infarct was also visualized by sonography, when gas-filled nanocarriers were used for infarct imaging in rabbits [126]. Still, infarct imaging with nanocarriers remains at the level of animal experimentation.

Gd-containing phosphatidylserine (PS) liposomes have been used for molecular imaging of atherosclerosis, since PS residues trigger rapid phagocytosis by macrophages know to be present in atherosclerotic lesions [127].

Four-dimensional MR microscopy of the mouse heart using Gd-liposomes was used for spatial and temporal resolution as well as for high-resolution fast isotropic imaging [128].

Liposomes loaded with PET/MR contrast agents were used to follow the therapeutic action of anti-inflammation drugs in atherosclerosis model in rabbits [129].

3.4.4 Other Organ and Tissue Imaging with Contrast-Loaded Liposomes

The use of microparticulate imaging agents for the visualization of infection and inflammation sites is based on the ability of microparticulates to extravasate from the circulation and accumulate in those sites similar to what we already described for tumors and infarcted tissues. ^{99m}Tc nanocarriers were shown to accumulate within 30 min in experimental staphylococcal abscess in rats [29]. The same approach was successfully used in preliminary clinical studies, where deep-seated infections were visualized with labeled nanocarriers in several patients [130]. Later on, more advanced ^{99m}Tc nanocarriers labeled using lipophilic chelator and reduced GSH were shown to accumulate in rats in sites injected with *Staphylococcus aureus* [131]. Abscesses were visualized within 2 h after the injection of ^{99m}Tc nanocarriers.

Despite the fact that tumors also accumulate nanocarriers via the impaired filtration mechanism, there exists a possibility to discriminate between infection and tumor using positively charged ^{67}Ga nanocarriers, which do accumulate in tumors and do not accumulate in infection sites [104].

Inflammation sites have also been shown to accumulate contrast-loaded lipid nanocarrier. Thus, gamma-imaging demonstrated specific accumulation of ^{99m}Tc-loaded negatively charged liposomes in the inflamed paws of rats with adjuvant-induced arthritis [132]. Liposomes labeled with ^{67}Ga well enough detected experimentally induced inflammations in rat, the inflammation-to-muscle accumulation ratio reaching maximum (32:1) in the case of neutral nanocarriers [104].

MRI of rabbit brain was achieved after intracarotid administration of large multivesicular nanocarriers loaded with paramagnetic metals chelated by DTPA [133]. Certain data are available on the accumulation of liposomal nanocarriers in sites of experimentally induced arthritis in a rat model [132]. Phagocytes within the inflammatory lesion may also play a role in the enhanced accumulation of liposomes.

Nebulized liposomal gadobenate dimeglumine contrast agent was also used for MRI of larynx and trachea [134] and could be utilized for improved mucosal surface visualization and early diagnosis of diseases originating in these organs. Cationic Gd-DTPA liposomes have been used also for highly efficient labeling of mesenchymal stem cells and cell tracking with MRI [135]. Paramagnetic probes incorporated into drug-loaded pH-sensitive liposomes have also been used to follow the process of the pH-mediated drug release [136].

3.5 DIAGNOSTIC APPLICATIONS OF MICELLAR IMAGING AGENTS

While the delivery of therapeutics by micelles is relatively well developed and elaborated [18], the application of micellar formulation as vehicles for contrast agents is still under development. Apart from the practical value for diagnosis itself, this area has significant importance for delivery of drugs as it allows visualization of the exact sites of the drug carrier deposition within the body with extraordinary anatomical resolution [43, 137].

3.5.1 *In Vivo* Gamma and Magnetic Resonance Visualization with Micellar Imaging Agents

In experiments on lymphatic imaging [138], amphiphilic chelating probes, Gd(^{111}In)-DTPA-PE and ^{111}In-DTPA-SA, have been successfully incorporated into PEG (5 kDa)-PE micelles (20 nm in diameter), and these nanoparticulate agents were used in experimental percutaneous lymphography using γ-scintigraphy and MRI in rabbits. The localization of ^{111}In-labeled DTPA-SA/PEG (5 kDa)-PE micelles in local lymphatics after subcutaneous administration of a 20 μCi dose into the dorsum of a rabbit hind paw has been demonstrated. The popliteal lymph node can be visualized within seconds after injection. It was shown that the micelles as smaller particles exhibit higher accumulation in the primary lymph node.

As the aforementioned data demonstrates, the micellar particulates due to their size and surface properties can be moved with ease from the injection site along the lymphatics to the systemic circulation with the lymph flow. In rabbits, T1-weighted transverse MR images of the axillary/subscapular lymph node area could be obtained by very fast after subcutaneous injection of small doses of Gd-DTPA-PE/PEG (5 kDa)-PE micelles. The collecting lymph vessel and axillary lymph node become visible only 4 min after administration of Gd-containing micelles [138]. Taking into account the ease and speed with which the primary lymph nodes may be visualized, one can suggest that unlike the other lymphotropic contrast media, polymeric micelles are lymphangiographic in nature. Their action is based on the visualization of lymph flowing through different elements of the lymphatics. The action of other lymphotropic contrast media is based primarily on their active uptake by the nodal macrophages (i.e., lymph nodes become visible only after the concentration of the medium in the lymph node has achieved certain threshold value, which usually requires prolonged periods of time). Similar experiments with PEG-PE mixed micelles with a core-incorporated amphiphilic ^{111}In- or Gd-loaded PAP have also demonstrated fast and efficient gamma and MR visualization of different compartments of the lymphatic system. Micelles mostly stay within lymph fluid rather than accumulate in the nodal macrophages (because of protective effect of surface PEG fragments) and rapidly move via the lymphatic pathway. They can serve as fast and efficient lymphangiographic agents for scintigraphy or MRI.

Blood pool imaging is of special interest for the evaluation of the current state of blood flow and for the discovery of irregularities caused by atherosclerotic lesions, thrombi, or tumors. Blood pool imaging requires prolonged circulation of contrast agents and is usually based on the utilization of sterically protected polymer-modified microparticulates with contrast properties.

CT represents an imaging modality with high spatial and temporal resolution. The diagnostic value of CT might be further significantly increased when contrast agents (i.e., substances containing X-ray absorbing heavy elements, such as iodine) are used to attenuate tissues and organs of interest. Since, as was already told, providing diagnostically acceptable imaging requires the iodine concentration on the order of millimoles per milliliter of tissue [139], large doses of low-molecular-weight CT contrast agent, such as iodine-containing organic molecules, are normally administered to patients. In order to more specifically target contrast agents, attempts have been made to use contrast agent-loaded microparticulate carriers for CT imaging [57, 140]. However, currently suggested particulate contrast agents possess relatively large particle size (between 0.25 and 3.5 μm) and are actively cleared by phagocytosis. Thus, in order to prepare a material whose distribution is limited to the blood pool, certain simple properties seem necessary to be met: a size larger than fenestrated capillaries (>10 nm), resistance to phagocytosis, and the radiopaque moiety structurally incorporated within the particulate.

Amphiphilic polymers which are able to spontaneously form stable and long-circulating 10–80 nm micelles in an aqueous media seem to be carriers of choice that meet all the aforementioned requirements [141]. The synthesis and *in vivo* properties of a block copolymer of methoxy-poly(ethylene glycol) (MPEG) and iodine-substituted poly-L-lysine have been described [142, 143]. This copolymer easily micellizes in the solution forming stable and heavily iodine-loaded particles (up to 35% of iodine by weight) with a size well below 100 nm. The iodine content in these micelles is about 30–35% of total weight. Other amphiphilic polymeric micelles formed by block copolymers have sizes around 10–50 nm [144, 145]. The slightly increased particle size of MPEG–iodolysine micelles may be explained by noncovalent binding of triiodobenzoic acid to the core of the micelle.

In vivo, the iodine-containing polymeric micelles demonstrated prolonged circulation time [138], where IV injections into rats and rabbits and the X-ray signal were monitored from several organs using a CT scanner. A significant enhancement

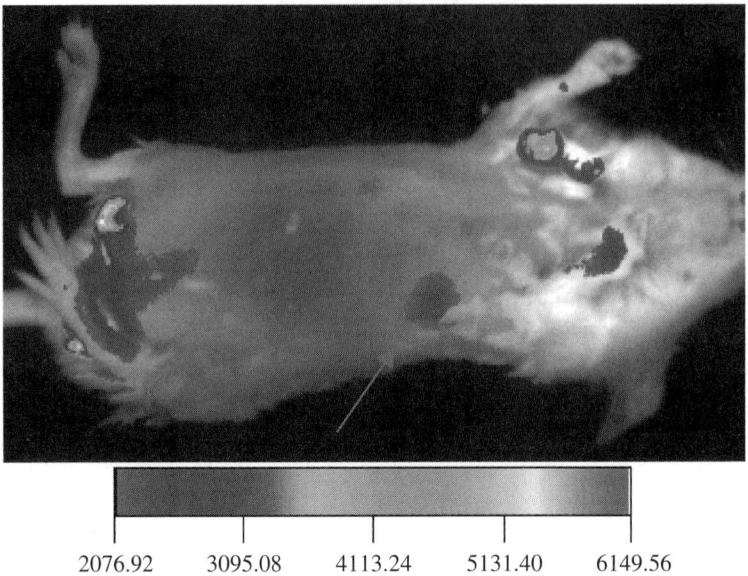

FIGURE 3.5 NIR imaging. Side view of mouse injected with Alexa-labeled micelles, at 2 h after the injection. Arrow points the tumor. (With permission from Ref. [153]. © DovePress.) (*See insert for color representation of the figure.*)

of three- to four-fold in the blood pool (aorta and heart) was visually observed in rats and rabbits for at least a period of 2 h following the injection. The quantification of some of these images with the help of image-processing software showed no sign of a decrease in the blood opacification during at least a 3 h period following the injection. A similar pattern was also observed in rats. Micelles with such properties would certainly offer the potential for clinical use in the diagnostic CT imaging of the blood pool. The clinical utility of selective blood pool contrast agents may be variable: minimally invasive angiography, image guidance of minimally invasive procedures, oncologic imaging of angiogenesis, ascertaining organ blood volume tomographically, and identifying hemorrhage are applications that could benefit from a long-lived intravascular agent.

Diacyllipid micelles coloaded with the photosensitizer drug and iron oxide nanoparticles have been suggested for simultaneous photodynamic therapy and MRI of tumors [146]. Similar drug-/iron oxide-coloaded micelles as the possible agents for combined chemotherapy and imaging of cancer were also suggested in [147]. General platform of target-specific micelles for molecular MRI is discussed in [148]. In particular, CB2 receptor- and NGAL-targeted paramagnetic micelles were shown to nicely image atherosclerotic plaques [149].

Of special interest is the use of PEG–lipid micelles for encapsulation of quantum dots (QD) [150]. QD encapsulated in lipid-based micelles have been successfully used for visualization of tumors and metastases *in vivo* in mice with tumors [151–153], and QD-loaded micelles specifically targeted to tumors with antitumor antibodies provided the best and fastest optical near-infrared images (see Fig. 3.5).

3.5.2 Some Recent Trends

Recent developments with diagnostic liposomes and lipid-core micelles deal mainly with utilizing new imaging modalities (such as optical imaging) and with making more and more advanced theranostic agents.

In addition to already mentioned nanocarriers loaded with QD, cetuximab-labeled liposomes have been colabeled with fluorescent indocyanine green, showed good binding with A431 colon carcinoma cells, demonstrated high quantum yield, and were suggested as nanosystems combining targetability and imaging properties [154]. Near-infrared quantitative imaging of lymphatic functions (lymphatic flow) was performed using the liposomal formulation of indocyanine green, which in free form is very unstable in solution [155]. PEGylated DSPC-liposomes with incorporated fluorescence-labeled dodecaborate lipid were shown to accumulate well in tumors except hypoxic regions [156]. These are just few examples of fast developing near-infrared optical imaging (especially useful for animal experiments) using lipid-based nanocarrier systems.

Developing theranostic nanosystems, liposomes have been prepared coated with D-alpha-tocopheryl PEG 1000 succinate monoester and containing both docetaxel and QD [157]. In cell culture experiments, these liposomes showed good uptake by MCF-7 cancer cells and are suggested to improve cancer imaging and therapy. Very interesting theranostic liposomes have been described in [158], where the authors combined in a single preparation doxorubicin (as a drug postloaded into contrast-labeled liposomes) and fluorescent tracer (for near-infrared imaging), ^{64}Cu (for PET imaging) or ^{99m}Tc (for SPECT imaging), and Gd in amphiphilic chelate (for MRI). A novel class of dual 1H-MRI probes—Gd-doped LipoCEST agents—was described, where liposomes contain a shift reagent in the aqueous core and Gd complex on the external surface [159]. Such probe can serve as T1 contrast, but after the cleavage and removal of the Gd coating, it works as a CEST contrast. Combined magneto-photo-acoustic imaging was achieved by using liposomes coloaded with gold nanorods and iron oxide nanoparticles [160]. Similar studies are too numerous to be listed all, but the examples presented demonstrate well the general direction in which the approach is developing.

3.6 CONCLUDING REMARKS

Perhaps, the key benefit of liposome- or micelle-based diagnostic is the ability to create high local concentrations of imaging/contrast agent and thus better "image" (i.e., evaluate) functional characteristic of a tissue or other region of interest. As long as an understanding of the characteristics of the lipid-based nanocarrier and how it interacts with the biological milieu exists, it should be possible to design diagnostics that make these functional measurements possible. The data obtained so far leads to a definite conclusion that contrast-loaded liposomes and micelles may be and will be successfully used for visualization of numerous organs, tissues, and disease sites in all imaging modalities. The advantages of gamma-imaging and MRI with liposomes (and, more recently, with micelles) have been already demonstrated in many experimental and clinical studies. Gamma-imaging with radiometal-loaded liposomes and micelles seems to be the most attractive because of relatively small quantities of both carrier and reporter metal needed for successful imaging, which greatly minimizes the problems of side effects. By using polymeric chelates, liposomes and micelles can be easily loaded with sufficient quantities of radioactive (for gamma-imaging) or MR-active (for MRI) agents. Additionally, grafted PEG can serve as a capture-avoiding agent permitting effective accumulation of the diagnostic label in the target and also for improvement of MR contrast properties.

Still, before liposomes and micelles become imaging agents of choice, several important questions have to be answered. We have to know much more about the pharmacokinetics of liposomal and micellar imaging agents; we have to find a way to still further increase carrier load with reporter moieties in order to minimize overloading patient with a carrier itself and to decrease cost of the preparation; we have to learn more about possible toxicity of metal-loaded liposomes and micelles owing to the intracellular liberation of metal; we have to achieve cost-effective large-scale production of homogenous preparations of diagnostic liposomes and micelles. These are the aims for the nearest future.

REFERENCES

[1] Gregoriadis G, Putman D, Louis L, Neerunjun D. Comparative effect and fate of non-entrapped and liposome-entrapped neuraminidase injected into rats. Biochem J 1974;**140**: 323–330.

[2] Papahadjopoulos D. Liposome formation and properties: an evolutionary profile. Biochem Soc Trans 1988;**16**:910–912.

[3] Lasic DD. Novel applications of liposomes. Trends Biotechnol 1998;**16**:307–321.

[4] Torchilin VP, editor. *Handbook of Targeted Delivery of Imaging Agents*. 1st ed. Volume **1**, Boca Raton: CRC Press; 1995.

[5] Torchilin VP. Liposomes as delivery agents for medical imaging. Mol Med Today 1996;**2**:242–249.

[6] Kopechek JA, Haworth KJ, Raymond JL, Douglas Mast T, Perrin SR, Klegerman ME, Huang S, Porter TM, McPherson DD, Holland CK. Acoustic characterization of echogenic liposomes: frequency-dependent attenuation and backscatter. J Acoust Soc Am 2011;**130**:3472–3481.

[7] Fattahi H, Laurent S, Liu F, Arsalani N, Vander Elst L, Muller RN. Magnetoliposomes as multimodal contrast agents for molecular imaging and cancer nanotheragnostics. Nanomedicine (Lond) 2011;**6**:529–544.

[8] Al-Jamal WT, Kostarelos K. Liposomes: from a clinically established drug delivery system to a nanoparticle platform for theranostic nanomedicine. Acc Chem Res 2011;**44**: 1094–1104.

[9] Silindir M, Erdogan S, Ozer AY, Maia S. Liposomes and their applications in molecular imaging. J Drug Target 2012;**20**:401–415.

[10] Ghaghada KB, Colen RR, Hawley CR, Patel N, Mukundan S Jr. Liposomal contrast agents in brain tumor imaging. Neuroimaging Clin N Am 2010;**20**:367–378.

[11] Bowey K, Tanguay JF, Tabrizian M. Liposome technology for cardiovascular disease treatment and diagnosis. Expert Opin Drug Deliv 2012;**9**:249–265.

[12] Silindir M, Ozer AY, Erdogan S. The use and importance of liposomes in Positron Emission Tomography. Drug Deliv 2012;**19**:68–80.

[13] Torchilin VP. Nanocarriers. Pharm Res 2007;**24**:2333–2334.

[14] Torchilin VP. Targeted pharmaceutical nanocarriers for cancer therapy and imaging. AAPS J 2007;**9**:E128–E147.

[15] Elbayoumi TA, Torchilin VP. Current trends in liposome research. In: Weissig V, editor. *Liposomes*. Totowa: Springer & Humana Press; 2009.

[16] Kirpotin DB, Drummond DC, Shao Y, Shalaby MR, Hong K, Nielsen UB, Marks JD, Benz CC, Park JW. Antibody targeting of long-circulating lipidic nanoparticles does not increase tumor localization but does increase internalization in animal models. Cancer Res 2006;**66**:6732–6740.

[17] Kakizawa Y, Kataoka K. Block copolymer micelles for delivery of gene and related compounds. Adv Drug Deliv Rev 2002;**54**:203–222.

[18] Torchilin VP. Micellar nanocarriers: pharmaceutical perspectives. Pharm Res 2007;**24**: 1–16.

[19] Torchilin VP. Lipid-core micelles for targeted drug delivery. Curr Drug Deliv 2005;**2**: 319–327.

[20] Phillips WT. Delivery of gamma-imaging agents by liposomes. Adv Drug Deliv Rev 1999;**37**:13–32.

[21] Goins B, Phillips T. Radiolabeled liposomes for imaging and biodistribution studies. In: Torchilin V, Weissig V, editors. *Liposomes: A Practical Approach*. 2nd ed. London: Oxford University Press; 2003. p 319–336.

[22] Hnatowich DJ, Friedman B, Clancy B, Novak M. Labeling of preformed liposomes with Ga-67 and Tc-99m by chelation. J Nucl Med 1981;**22**:810–814.

[23] Caride VJ, Taylor W, Cramer JA, Gottschalk A. Evaluation of liposome-entrapped radioactive tracers as scanning agents. Part 1: organ distribution of liposome (99mTc-DTPA) in mice. J Nucl Med 1976;**17**:1067–1072.

[24] Torchilin VP, Trubetskoy VS. *In vivo* visualizing of organs and tissues with liposomes. J Liposome Res 1995;**5**:795–812.

[25] Oku N. Delivery of contrast agents for positron emission tomography imaging by liposomes. Adv Drug Deliv Rev 1999;**37**:53–61.

[26] Goins BA, Phillips WT. The use of scintigraphic imaging as a tool in the development of liposome formulations. Prog Lipid Res 2001;**40**:95–123.

[27] Patel HM, Boodle KM, Vaughan-Jones R. Assessment of the potential uses of liposomes for lymphoscintigraphy and lymphatic drug delivery. Failure of 99m-technetium marker to represent intact liposomes in lymph nodes. Biochim Biophys Acta 1984;**801**:76–86.

[28] Damen J, Regts J, Scherphof G. Transfer and exchange of phospholipid between small unilamellar liposomes and rat plasma high density lipoproteins. Dependence on cholesterol content and phospholipid composition. Biochim Biophys Acta 1981;**665**: 538–545.

[29] Morgan JR, Williams KE, Davies RL, Leach K, Thomson M, Williams LA. Localisation of experimental staphylococcal abscesses by 99MTC-technetium-labelled liposomes. J Med Microbiol 1981;**14**:213–217.

[30] Love WG, Amos N, Williams BD, Kellaway IW. Effect of liposome surface charge on the stability of technetium (99mTc) radiolabelled liposomes. J Microencapsul 1989;**6**: 105–113.

[31] Ahkong QF, Tilcock C. Attachment of 99mTc to lipid vesicles containing the lipophilic chelate dipalmitoylphosphatidylethanolamine-DTTA. Int J Rad Appl Instrum B 1992;**19**: 831–840.

[32] Alafandy M, Goffinet G, Umbrain V, D'Haese J, Camu F, Legros FJ. 99mTechnetium-stannous oxinate as marker of liposome formulations. Nucl Med Biol 1996;**23**:881–887.

[33] Boerman OC, Laverman P, Oyen WJ, Corstens FH, Storm G. Radiolabeled liposomes for scintigraphic imaging. Prog Lipid Res 2000;**39**:461–475.

[34] Goto R, Kubo H, Okada S. Liposomes prepared from synthetic amphiphiles. I. Their technetium labeling and stability. Chem Pharm Bull (Tokyo) 1989;**37**:1351–1354.

[35] Tilcock C, Unger E, Cullis P, MacDougall P. Liposomal Gd-DTPA: preparation and characterization of relaxivity. Radiology 1989;**171**:77–80.

[36] Tilcock C. Liposomal paramagnetic MR contrast agents. In: Gregoriadis G, editor. *Liposome Technology*. 2nd ed. Boca Raton: CRC Press; 1993. p 65–87.

[37] Kabalka GW, Davis MA, Buonocore E, Hubner K, Holmberg E, Huang L. Gd-labeled liposomes containing amphipathic agents for magnetic resonance imaging. Invest Radiol 1990;**25** (Suppl 1):S63–S64.

[38] Grant CW, Karlik S, Florio E. A liposomal MRI contrast agent: phosphatidylethanolamine-DTPA. Magn Reson Med 1989;**11**:236–243.

[39] Laverman P, Dams ET, Oyen WJ, Storm G, Koenders EB, Prevost R, van der Meer JW, Corstens FH, Boerman OC. A novel method to label liposomes with 99mTc by the hydrazino nicotinyl derivative. J Nucl Med 1999;**40**:192–197.

[40] Khaw BA, Klibanov A, O'Donnell SM, Saito T, Nossiff N, Slinkin MA, Newell JB, Strauss HW, Torchilin VP. Gamma imaging with negatively charge-modified monoclonal antibody: modification with synthetic polymers. J Nucl Med 1991;**32**:1742–1751.

[41] Slinkin MA, Klibanov AL, Torchilin VP. Terminal-modified polylysine-based chelating polymers: highly efficient coupling to antibody with minimal loss in immunoreactivity. Bioconjug Chem 1991;**2**:342–348.

[42] Torchilin VP, Klibanov AL. The antibody-linked chelating polymers for nuclear therapy and diagnostics. Crit Rev Ther Drug Carrier Syst 1991;**7**:275–308.

[43] Torchilin VP. Polymeric contrast agents for medical imaging. Curr Pharm Biotechnol 2000;**1**:183–215.

[44] Torchilin VP. Pharmacokinetic considerations in the development of labeled liposomes and micelles for diagnostic imaging. Q J Nucl Med 1997;**41**:141–153.

[45] de Vries A, Kok MB, Sanders HM, Nicolay K, Strijkers GJ, Grull H. Multimodal liposomes for SPECT/MR imaging as a tool for in situ relaxivity measurements. Contrast Media Mol Imaging 2012;**7**:68–75.

[46] Erdogan S, Roby A, Sawant R, Hurley J, Torchilin VP. Gadolinium-loaded polychelating polymer-containing cancer cell-specific immunoliposomes. J Liposome Res 2006;**16**: 45–55.

[47] Hwang KJ, Merriam JE, Beaumier PL, Luk KF. Encapsulation, with high efficiency, of radioactive metal ions in liposomes. Biochim Biophys Acta 1982;**716**:101–109.

[48] Gabizon A, Huberty J, Straubinger RM, Price DC, Papahadjopoulos D. An improved method for *in vivo* tracking and imaging of liposomes using a gallium-67-deferoxamine complex. J Liposome Res 1988;**1**:123–135.

[49] Phillips WT, Rudolph AS, Goins B, Timmons JH, Klipper R, Blumhardt R. A simple method for producing a technetium-99m-labeled liposome which is stable *in vivo*. Int J Rad Appl Instrum B 1992;**19**:539–547.

[50] Suresh M, Cao Y. A simple and efficient method for radiolabeling of preformed liposomes. J Pharm Pharm Sci 1998;**1**:31–37.

[51] Bao A, Goins B, Klipper R, Negrete G, Mahindaratne M, Phillips WT. A novel liposome radiolabeling method using 99mTc-"SNS/S" complexes: in vitro and *in vivo* evaluation. J Pharm Sci 2003;**92**:1893–1904.

[52] Barenholz Y, Amselem S, Goren D, Cohen R, Gelvan D, Samuni A, Golden EB, Gabizon A. Stability of liposomal doxorubicin formulations: problems and prospects. Med Res Rev 1993;**13**:449–491.

[53] Ogihara-Umeda I, Sasaki T, Nishigori H. Development of a liposome-encapsulated radionuclide with preferential tumor accumulation—the choice of radionuclide and chelating ligand. Int J Rad Appl Instrum B 1992;**19**:753–757.

[54] Seltzer SE. The role of liposomes in diagnostic imaging. Radiology 1989;**171**:19–21.

[55] Sachse A, Leike JU, Rossling GL, Wagner SE, Krause W. Preparation and evaluation of lyophilized iopromide-carrying liposomes for liver tumor detection. Invest Radiol 1993;**28**:838–844.

[56] Leike JU, Sachse A, Rupp K. Characterization of continuously extruded iopromide-carrying liposomes for computed tomography blood-pool imaging. Invest Radiol 2001; **36**:303–308.

[57] Krause W, Leike J, Sachse A, Schuhmann-Giampieri G. Characterization of iopromide liposomes. Invest Radiol 1993;**28**:1028–1032.

[58] Elbayoumi TA, Torchilin VP. Use of radiolabeled liposomes for tumor imaging. In: Bulte JWM, Modo M, editors. *Nanoparticles in Biomedical Imaging*. 1st ed. New York: Springer; 2007. p 211–236.

[59] Senior JH. Fate and behavior of liposomes *in vivo*: a review of controlling factors. Crit Rev Ther Drug Carrier Syst 1987;**3**:123–193.

[60] Torchilin VP. Drug targeting. Eur J Pharm Sci 2000;**11** (Suppl 2):S81–S91.

[61] Torchilin V. Antibody-modified liposomes for cancer chemotherapy. Expert Opin Drug Deliv 2008;**5**:1003–1025.

[62] Torchilin VP. Recent advances with liposomes as pharmaceutical carriers. Nat Rev Drug Discov 2005;**4**:145–160.

[63] Weissig VV, Babich J, Torchilin VV. Long-circulating gadolinium-loaded liposomes: potential use for magnetic resonance imaging of the blood pool. Colloid Surf B 2000;**18**:293–299.

[64] Papisov MI. Modeling *in vivo* transfer of long-circulating polymers (two classes of long-circulating polymers and factors affecting their transfer *in vivo*). Adv Drug Deliv Rev 1995;**16**:127–139.

[65] Torchilin VP, Omelyanenko VG, Papisov MI, Bogdanov AA Jr, Trubetskoy VS, Herron JN, Gentry CA. Poly(ethylene glycol) on the liposome surface: on the mechanism of polymer-coated liposome longevity. Biochim Biophys Acta 1994;**1195**:11–20.

[66] Allen TM. The use of glycolipids and hydrophilic polymers in avoiding rapid uptake of liposomes by the mononuclear phagocyte system. Adv Drug Deliv Rev 1994;**13**: 285–309.

[67] Allen TM. Long-circulating (sterically stabilized) liposomes for targeted drug delivery. Trends Pharmacol Sci 1994;**15**:215–220.

[68] Allen TM, Haanen C, Menenez DEL. Pharmacokinetics of long-circulating liposomes. Adv Drug Deliv Rev 1995;**16**:141–155.

[69] Torchilin VT. How do polymers prolong circulation time of liposomes. J Liposome Res 1996;**6**:99–116.

[70] Klibanov AL, Maruyama K, Torchilin VP, Huang L. Amphipathic polyethyleneglycols effectively prolong the circulation time of liposomes. FEBS Lett 1990;**268**:235–237.

[71] Torchilin VP. Structure and design of polymeric surfactant-based drug delivery systems. J Control Release 2001;**73**:137–172.

[72] Gregoriadis G, Neerunjun DE. Control of the rate of hepatic uptake and catabolism of liposome-entrapped proteins injected into rats. Possible therapeutic applications. Eur J Biochem 1974;**47**:179–185.

[73] Senior J, Crawley JC, Gregoriadis G. Tissue distribution of liposomes exhibiting long half-lives in the circulation after intravenous injection. Biochim Biophys Acta 1985; **839**:1–8.

[74] Kirby C, Clarke J, Gregoriadis G. Effect of the cholesterol content of small unilamellar liposomes on their stability *in vivo* and in vitro. Biochem J 1980;**186**:591–598.

[75] Kirby C, Clarke J, Gregoriadis G. Cholesterol content of small unilamellar liposomes controls phospholipid loss to high density lipoproteins in the presence of serum. FEBS Lett 1980;**111**:324–328.

[76] Senior J, Gregoriadis G. Stability of small unilamellar liposomes in serum and clearance from the circulation: the effect of the phospholipid and cholesterol components. Life Sci 1982;**30**:2123–2136.

[77] Bosworth ME, Hunt CA. Liposome disposition *in vivo* II: dose dependency. J Pharm Sci 1982;**71**:100–104.

[78] Abra RM, Hunt CA. Liposome disposition *in vivo* IV: the interaction of sequential doses of liposomes having different diameters. Res Commun Chem Pathol Pharmacol 1982;**36**:17–31.

[79] Beaumier PL, Hwang KJ, Slattery JT. Effect of liposome dose on the elimination of small unilamellar sphingomyelin/cholesterol vesicles from the circulation. Res Commun Chem Pathol Pharmacol 1983;**39**:277–289.

[80] Goins B, Klipper R, Rudolph AS, Phillips WT. Use of technetium-99m-liposomes in tumor imaging. J Nucl Med 1994;**35**:1491–1498.

[81] Phillips TW, Goins B. Targeted delivery of imaging agents by liposomes. In: Torchilin VP, editor. *Handbook of Targeted Delivery of Imaging Agents*. Boca Raton: CRC Press; 1995. p 149–176.

[82] Maeda H, Wu J, Sawa T, Matsumura Y, Hori K. Tumor vascular permeability and the EPR effect in macromolecular therapeutics: a review. J Control Release 2000;**65**:271–284.

[83] Maeda H. The enhanced permeability and retention (EPR) effect in tumor vasculature: the key role of tumor-selective macromolecular drug targeting. Adv Enzyme Regul 2001;**41**:189–207.

[84] Gabizon A, Price DC, Huberty J, Bresalier RS, Papahadjopoulos D. Effect of liposome composition and other factors on the targeting of liposomes to experimental tumors: biodistribution and imaging studies. Cancer Res 1990;**50**:6371–6378.

[85] Harrington KJ, Rowlinson-Busza G, Syrigos KN, Uster PS, Abra RM, Stewart JS. Biodistribution and pharmacokinetics of 111In-DTPA-labelled pegylated liposomes in a human tumour xenograft model: implications for novel targeting strategies. Br J Cancer 2000;**83**:232–238.

[86] Ogihara-Umeda I, Sasaki T, Nishigori H. Active removal of radioactivity in the blood circulation using biotin-bearing liposomes and avidin for rapid tumour imaging. Eur J Nucl Med 1993;**20**:170–172.

[87] Jaggi M, Khar R, Chauhan U, Gangal S. Liposomes as carriers of technetium-99m glucoheptonate for liver imaging. Int J Pharm 1991;**69**:77–79.

[88] Unger EC, Winokur T, MacDougall P, Rosenblum J, Clair M, Gatenby R, Tilcock C. Hepatic metastases: liposomal Gd-DTPA-enhanced MR imaging. Radiology 1989;**171**:81–85.

[89] Kabalka GW, Buonocore E, Hubner K, Davis M, Huang L. Gadolinium-labeled liposomes containing paramagnetic amphipathic agents: targeted MRI contrast agents for the liver. Magn Reson Med 1988;**8**:89–95.

[90] Kabalka GW, Davis MA, Moss TH, Buonocore E, Hubner K, Holmberg E, Maruyama K, Huang L. Gadolinium-labeled liposomes containing various amphiphilic Gd-DTPA derivatives: targeted MRI contrast enhancement agents for the liver. Magn Reson Med 1991;**19**:406–415.

[91] Unger C, Shen DK, Fritz J. Status of liposomes as MR contrast agents. J Magn Reson Imaging 1993;**3**:195–198.

[92] Tilcock C. Imaging tools: liposomal agents for nuclear medicine, computed tomography, magnetic resonance, and ultrasound. In: Philippot JR, Schuber F, editors. *Liposomes as Tools in Basic Research and industry*. Boca Raton: CRC Press; 1995. p 225–240.

[93] Okada M, Isoda T, Kumano S, Kagawa Y, Araki T, Onishi H, Hori M, Kim T, Motokui Y, Wada T, Murakami T. Serine- and mannose-modified liposomal contrast agent for computed tomography: evaluation of the enhancement in rabbit liver VX-2 tumor model. Contrast Media Mol Imaging 2010;**5**:140–146.

[94] Krause W, Schonborn A, Rupp K. CT imaging with iopromide liposomes in a rabbit model. J Liposome Res 2011;**21**:229–236.

[95] Yuan F, Leunig M, Huang SK, Berk DA, Papahadjopoulos D, Jain RK. Microvascular permeability and interstitial penetration of sterically stabilized (stealth) liposomes in a human tumor xenograft. Cancer Res 1994;**54**:3352–3356.

[96] Gabizon AA. Liposome circulation time and tumor targeting: implications for cancer therapy. Adv Drug Deliv Rev 1995;**16**:285–294.

[97] Harrington KJ. Liposomal cancer chemotherapy: current clinical applications and future prospects. Expert Opin Investig Drugs 2001;**10**:1045–1061.

[98] Presant CA, Blayney D, Proffitt RT, Turner AF, Williams LE, Nadel HI, Kennedy P, Wiseman C, Gala K, Crossley RJ, *et al.* Preliminary report: imaging of Kaposi sarcoma and lymphoma in AIDS with indium-111-labelled liposomes. Lancet 1990;**335**:1307–1309.

[99] Nakamura K, Kuba A, Katayama M, Hashimoto S, Takatoku K, Wakui Y. Indium-111-labelled liposomes: quality and biodistribution in mice bearing human tumor xenografts. Drug Deliv 1993;**1**:69–74.

[100] Kubo A, Nakamura K, Sammiya T, Katayama M, Hashimoto T, Hashimoto S, Kobayashi H, Teramoto T. Indium-111-labelled liposomes: dosimetry and tumour detection in patients with cancer. Eur J Nucl Med 1993;**20**:107–113.

[101] Elbayoumi TA, Torchilin VP. Enhanced accumulation of long-circulating liposomes modified with the nucleosome-specific monoclonal antibody 2C5 in various tumours in mice: gamma-imaging studies. Eur J Nucl Med Mol I 2006;**33**:1196–1205.

[102] Harrington KJ, Mohammadtaghi S, Uster PS, Glass D, Peters AM, Vile RG, Stewart JS. Effective targeting of solid tumors in patients with locally advanced cancers by radiolabeled pegylated liposomes. Clin Cancer Res 2001;**7**:243–254.

[103] Goto R, Kubo H, Okada S. Effect of reticuloendothelial blockade on tissue distribution of technetium-99-labeled synthetic liposomes in ehrlich solid tumor-bearing mice. Chem Pharm Bull (Tokyo) 1991;**39**:230–232.

[104] Ogihara I, Kojima S, Jay M. Differential uptake of gallium-67-labeled liposomes between tumors and inflammatory lesions in rats. J Nucl Med 1986;**27**:1300–1307.

[105] Soares DC, de Oliveira MC, de Barros AL, Cardoso VN, Ramaldes GA. Liposomes radiolabeled with (159)Gd: in vitro antitumoral activity, biodistribution study and scintigraphic image in Ehrlich tumor bearing mice. Eur J Pharm Sci 2011;**43**: 290–296.

[106] Lin YY, Chang CH, Li JJ, Stabin MG, Chang YJ, Chen LC, Lin MH, Tseng YL, Lin WJ, Lee TW, Ting G, Chang CA, Chen FD, Wang HE. Pharmacokinetics and dosimetry of (111)In/(188)Re-labeled PEGylated liposomal drugs in two colon carcinoma-bearing mouse models. Cancer Biother Radiopharm 2011;**26**:373–380.

[107] Underwood C, van Eps AW, Ross MW, Laverman P, van Bloois L, Storm G, Schaer TP. Intravenous technetium-99m labelled PEG-liposomes in horses: a safety and biodistribution study. Equine Vet J 2012;**44**:196–202.

[108] Briele B, Hotze A, Oehr P, Biersack HJ, Rosanowski F, Gorgulla W, Hertberhold C, Harlapp JP. Tumour imaging with labelled liposomes. Lancet 1990;**336**:875–876.

[109] Simon RH, Ho SY, Perkins CR, D'Arrigo JS. Quantitative assessment of tumor enhancement by ultrastable lipid-coated microbubbles as a sonographic contrast agent. Invest Radiol 1992;**27**:29–34.

[110] Osborne MP, Richardson VJ, Jeyasingh K, Ryman BE. Radionuclide-labelled liposomes—a new lymph node imaging agent. Int J Nucl Med Biol 1979;**6**:75–83.

[111] Erdogan S, Medarova ZO, Roby A, Moore A, Torchilin VP. Enhanced tumor MR imaging with gadolinium-loaded polychelating polymer-containing tumor-targeted liposomes. J Magn Reson Imaging 2008;**27**:574–580.

[112] Liao Z, Wang H, Lv R, Zhao P, Sun X, Wang S, Su W, Niu R, Chang J. Polymeric liposomes-coated superparamagnetic Iron oxide nanoparticles as contrast agent for targeted magnetic resonance imaging of cancer cells. Langmuir 2011;**27**:3100–3105.

[113] Mikhaylov G, Mikac U, Magaeva AA, Itin VI, Naiden EP, Psakhye I, Babes L, Reinheckel T, Peters C, Zeiser R, Bogyo M, Turk V, Psakhye SG, Turk B, Vasiljeva O. Ferri-liposomes as an MRI-visible drug-delivery system for targeting tumours and their microenvironment. Nat Nanotechnol 2011;**6**:594–602.

[114] Cittadino E, Ferraretto M, Torres E, Maiocchi A, Crielaard BJ, Lammers T, Storm G, Aime S, Terreno E. MRI evaluation of the antitumor activity of paramagnetic liposomes loaded with prednisolone phosphate. Eur J Pharm Sci 2012;**45**:436–441.

[115] Ding N, Lu Y, Lee RJ, Yang C, Huang L, Liu J, Xiang G. Folate receptor-targeted fluorescent paramagnetic bimodal liposomes for tumor imaging. Int J Nanomed 2011;**6**:2513–2520.

[116] Zheng J, Liu J, Dunne M, Jaffray DA, Allen C. *In vivo* performance of a liposomal vascular contrast agent for CT and MR-based image guidance applications. Pharm Res 2007;**24**:1193–1201.

[117] Liu G, Moake M, Har-El YE, Long CM, Chan KW, Cardona A, Jamil M, Walczak P, Gilad AA, Sgouros G, van Zijl PC, Bulte JW, McMahon MT. *In vivo* multicolor molecular MR imaging using diamagnetic chemical exchange saturation transfer liposomes. Magn Reson Med 2012;**67**:1106–1113.

[118] Oku N, Yamashita M, Katayama Y, Urakami T, Hatanaka K, Shimizu K, Asai T, Tsukada H, Akai S, Kanazawa H. PET imaging of brain cancer with positron emitter-labeled liposomes. Int J Pharm 2011;**403**:170–177.

[119] Petersen AL, Binderup T, Rasmussen P, Henriksen JR, Elema DR, Kjaer A, Andresen TL. 64Cu loaded liposomes as positron emission tomography imaging agents. Biomaterials 2011;**32**:2334–2341.

[120] Petersen AL, Binderup T, Jolck RI, Rasmussen P, Henriksen JR, Pfeifer AK, Kjaer A, Andresen TL. Positron emission tomography evaluation of somatostatin receptor targeted (64)Cu-TATE-liposomes in a human neuroendocrine carcinoma mouse model. J Control Release 2012;**160**:254–263.

[121] Tilcock C, Ahkong QF, Fisher D. Polymer-derivatized technetium 99mTc-labeled liposomal blood pool agents for nuclear medicine applications. Biochim Biophys Acta 1993;**1148**:77–84.

[122] Palmer TN, Caride VJ, Caldecourt MA, Twickler J, Abdullah V. The mechanism of liposome accumulation in infarction. Biochim Biophys Acta 1984;**797**:363–368.

[123] Torchilin VP. Liposomes as targetable drug carriers. Crit Rev Ther Drug Carrier Syst 1985;**2**:65–115.

[124] Torchilin VP, Khaw BA, Smirnov VN, Haber E. Preservation of antimyosin antibody activity after covalent coupling to liposomes. Biochem Biophys Res Commun 1989;**89**: 1114–1119.

[125] Torchilin VP, Klibanov AL, Huang L, O'Donnell S, Nossiff ND, Khaw BA. Targeted accumulation of polyethylene glycol-coated immunoliposomes in infarcted rabbit myocardium. FASEB J 1992;**6**:2716–2719.

[126] Unger E, Shen DK, Fritz T, Lund P, Wu GL, Kulik B, DeYoung D, Standen J, Ovitt T, Matsunaga T. Gas-filled liposomes as echocardiographic contrast agents in rabbits with myocardial infarcts. Invest Radiol 1993;**28**:1155–1159.

[127] Maiseyeu A, Mihai G, Kampfrath T, Simonetti OP, Sen CK, Roy S, Rajagopalan S, Parthasarathy S. Gadolinium-containing phosphatidylserine liposomes for molecular imaging of atherosclerosis. J Lipid Res 2009;**50**:2157–2163.

[128] Bucholz E, Ghaghada K, Qi Y, Mukundan S, Johnson GA. Four-dimensional MR microscopy of the mouse heart using radial acquisition and liposomal gadolinium contrast agent. Magn Reson Med 2008;**60**:111–118.

[129] Lobatto ME, Calcagno C, Metselaar JM, Storm G, Stroes ES, Fayad ZA, Mulder WJ. Imaging the efficacy of anti-inflammatory liposomes in a rabbit model of atherosclerosis by non-invasive imaging. Methods Enzymol 2012;**508**:211–228.

[130] Morgan JR, Williams LA, Howard CB. Technetium-labelled liposome imaging for deep-seated infection. Br J Radiol 1985;**58**:35–39.

[131] Goins B, Klipper R, Rudolph AS, Cliff RO, Blumhardt R, Phillips WT. Biodistribution and imaging studies of technetium-99m-labeled liposomes in rats with focal infection. J Nucl Med 1993;**34**:2160–2168.

[132] Love WG, Amos N, Kellaway IW, Williams BD. Specific accumulation of technetium-99m radiolabelled, negative liposomes in the inflamed paws of rats with adjuvant induced arthritis: effect of liposome size. Ann Rheum Dis 1989;**48**:143–148.

[133] Turski P, Kalinke T, Strother L, Perman W, Scott G, Kornguth S. Magnetic resonance imaging of rabbit brain after intracarotid injection of large multivesicular liposomes containing paramagnetic metals and DTPA. Magn Reson Med 1988;**7**:184–196.

[134] Wei X, Wu H, Lu Q, Xu J, Xu Y. Nebulized liposomal gadobenate dimeglumine contrast formulation for magnetic resonance imaging of larynx and trachea. Int J Nanomed 2011;**6**:3383–3391.

[135] Guenoun J, Koning GA, Doeswijk G, Bosman L, Wielopolski PA, Krestin GP, Bernsen MR. Cationic Gd-DTPA liposomes for highly efficient labeling of mesenchymal stem cells and cell tracking with MRI. Cell Transplant 2012;**21**:191–205.

[136] Torres E, Mainini F, Napolitano R, Fedeli F, Cavalli R, Aime S, Terreno E. Improved paramagnetic liposomes for MRI visualization of pH triggered release. J Control Release 2011;**154**:196–202.

[137] Torchilin VP. PEG-based micelles as carriers of contrast agents for different imaging modalities. Adv Drug Deliv Rev 2002;**54**:235–252.

[138] Trubetskoy VS, Frank-Kamenetsky MD, Whiteman KR, Wolf GL, Torchilin VP. Stable polymeric micelles: lymphangiographic contrast media for gamma scintigraphy and magnetic resonance imaging. Acad Radiol 1996;**3**:232–238.

[139] Wolf GL. Targeted delivery of imaging agents: an overview. In: Torchilin VP, editor. *Handbook of Targeted Delivery of Imaging Agents*. Boca Raton: CRC Press; 1995. p 3–22.

[140] Seltzer SE, Blau M, Herman LW, Hooshmand RL, Herman LA, Adams DF, Minchey SR, Janoff AS. Contrast material-carrying liposomes: biodistribution, clearance, and imaging characteristics. Radiology 1995;**194**:775–781.

[141] Trubetskoy VS, Torchilin VP. Use of polyoxyethylene-lipid conjugates as long-circulating carriers for delivery of therapeutic and diagnostic agents. Adv Drug Deliv Rev 1995;**16**:311–320.

[142] Trubetskoy VS, Gazelle GS, Wolf GL, Torchilin VP. Block-copolymer of polyethylene glycol and polylysine as a carrier of organic iodine: design of long-circulating particulate contrast medium for X-ray computed tomography. J Drug Target 1997;**4**:381–388.

[143] Torchilin VP, Frank-Kamenetsky MD, Wolf GL. CT visualization of blood pool in rats by using long-circulating, iodine-containing micelles. Acad Radiol 1999;**6**:61–65.

[144] Kwon GS, Kataoka K. Block copolymer micelles as long-circulating drug vehicles. Adv Drug Deliv Rev 1995;**16**:295–309.

[145] Kwon GS. Diblock copolymer nanoparticles for drug delivery. Crit Rev Ther Drug Carrier Syst 1998;**15**:481–512.

[146] Cinteza LO, Ohulchanskyy TY, Sahoo Y, Bergey EJ, Pandey RK, Prasad PN. Diacyllipid micelle-based nanocarrier for magnetically guided delivery of drugs in photodynamic therapy. Mol Pharm 2006;**3**:415–423.

[147] Sawant R, Sawant R, Gultepe E, Nagesha D, Papahadjopoulos-Sternberg B, Sridhar S, Torchilin VP. Nanosized cancer cell-targeted polymeric immunomicelles loaded with superparamagnetic iron oxide nanoparticles. J Nanopart Res 2009;**11**:1777–1785.

[148] Straathof R, Strijkers GJ, Nicolay K. Target-specific paramagnetic and superparamagnetic micelles for molecular MR imaging. Methods Mol Biol 2011;**771**:691–715.

[149] te Boekhorst BC, Bovens SM, Rodrigues-Feo J, Sanders HM, van de Kolk CW, de Kroon AI, Cramer MJ, Doevendans PA, ten Hove M, Pasterkamp G, van Echteld CJ. Characterization and in vitro and *in vivo* testing of CB2-receptor- and NGAL-targeted paramagnetic micelles for molecular MRI of vulnerable atherosclerotic plaque. Mol Imaging Biol 2010;**12**:635–651.

[150] Kim YH, Subramanyam E, Im JH, Huh KM, Choi H, Choi JS, Lee YK, Park SW. A new PEG-lipid conjugate micelle for encapsulation of CdSe/ZnS quantum dots. J Nanosci Lett 2010;**10**:3275–3279.

[151] Papagiannaros A, Levchenko T, Hartner W, Mongayt D, Torchilin V. Quantum dots encapsulated in phospholipid micelles for imaging and quantification of tumors in the near-infrared region. Nanomedicine 2009;**5**:216–224.

[152] Papagiannaros A, Upponi J, Hartner W, Mongayt D, Levchenko T, Torchilin V. Quantum dot loaded immunomicelles for tumor imaging. BMC Med Imaging 2010;**10**:22.

[153] Papagiannaros A, Kale A, Levchenko TS, Mongayt D, Hartner WC, Torchilin VP. Near infrared planar tumor imaging and quantification using nanosized Alexa 750-labeled phospholipid micelles. Int J Nanomed 2009;**4**:123–131.

[154] Portnoy E, Lecht S, Lazarovici P, Danino D, Magdassi S. Cetuximab-labeled liposomes containing near-infrared probe for *in vivo* imaging. Nanomedicine 2011;**7**:480–488.

[155] Proulx ST, Luciani P, Derzsi S, Rinderknecht M, Mumprecht V, Leroux JC, Detmar M. Quantitative imaging of lymphatic function with liposomal indocyanine green. Cancer Res 2010;**70**:7053–7062.

[156] Nakamura H, Ueda N, Ban HS, Ueno M, Tachikawa S. Design and synthesis of fluorescence-labeled closo-dodecaborate lipid: its liposome formation and *in vivo* imaging targeting of tumors for boron neutron capture therapy. Org Biomol Chem 2012;**10**: 1374–1380.

[157] Muthu MS, Kulkarni SA, Raju A, Feng SS. Theranostic liposomes of TPGS coating for targeted co-delivery of docetaxel and quantum dots. Biomaterials 2012;**33**:3494–3501.

[158] Li S, Goins B, Zhang L, Bao A. A novel multifunctional theranostic liposome drug delivery system: construction, characterization, and multimodality MR, near-infrared fluorescent and nuclear imaging. Bioconjug Chem 2012;**23**:1322–1332.

[159] Terreno E, Boffa C, Menchise V, Fedeli F, Carrera C, Castelli DD, Digilio G, Aime S. Gadolinium-doped LipoCEST agents: a potential novel class of dual 1H-MRI probes. Chem Commun (Camb) 2011;**47**:4667–4669.

[160] Qu M, Mallidi S, Mehrmohammadi M, Truby R, Homan K, Joshi P, Chen YS, Sokolov K, Emelianov S. Magneto-photo-acoustic imaging. Biomed Opt Express 2011;**2**: 385–396.

4

HOLLOW NANOCAPSULES IN BIOMEDICAL IMAGING APPLICATIONS

Sergey A. Dergunov and Eugene Pinkhassik
Department of Chemistry, Saint Louis University, St. Louis, MO, USA

4.1 INTRODUCTION

This chapter will focus on the application of hollow polymer nanocapsules (NCs) in biomedical imaging. The main emphasis will be on vesicle-templated NCs. Other NCs, for example, those produced by layer-by-layer (LBL) assembly, have been discussed in several excellent reviews. We have focused on developments in the past decade, and research discussed in this chapter concerns synthesis and characterization of polymer NCs or applications in a biomedical setting. In addition, we have attempted to highlight relevant research and review articles from earlier work that had led to the foundation of this field.

A critical characteristic of hollow NCs is the interior core, suitable for the entrapment of contrast agents or molecules acting as chemosensors. Functional performance of NCs is determined by the chemical composition of the capsule walls. With this in mind, this chapter will discuss the synthesis and characterization of hollow NCs, strategies for entrapment of molecules, functional performance of NCs, and examples of potential applications of NCs in biomedical imaging (Fig. 4.1). The synthesis of NCs with different architectures as well as structural and functional characterization is presented in Section "Synthesis and characterization of nanocapsules." Section "Compartmentalization of molecules in hollow nanocapsules" deals with

Nanotechnology for Biomedical Imaging and Diagnostics: From Nanoparticle Design to Clinical Applications, First Edition. Edited by Mikhail Y. Berezin.

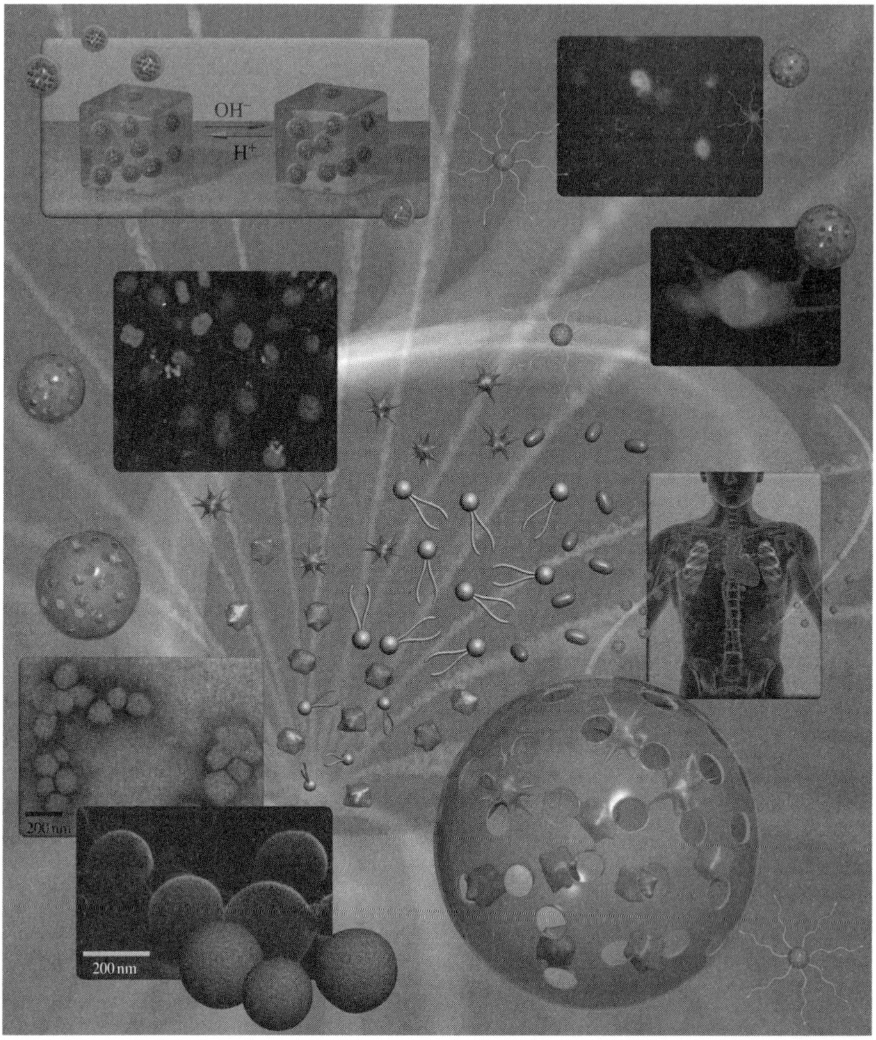

FIGURE 4.1 Scope of this chapter—from synthesis and characterization of hollow NCs to building nanodevices and biomedical applications. (*See insert for color representation of the figure.*)

compartmentalization of small molecules within polymer shells. Finally, Section "Biomedical applications of polymer nanocapsules" describes biomedical applications of polymer NCs, that is, *in vivo* studies and application of capsules as diagnostic devices or as delivery vehicles.

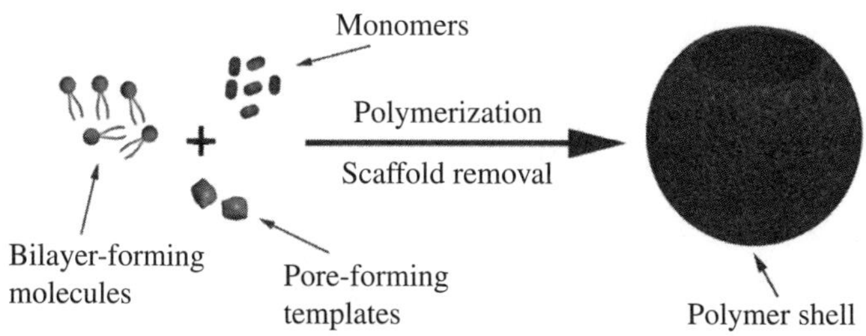

FIGURE 4.2 Overview of vesicle-templated synthesis of NCs.

4.2 SYNTHESIS AND CHARACTERIZATION OF NANOCAPSULES

Functionally and structurally, NCs can be considered as simple models of cells and their organelles. Not surprisingly, bioinspired synthetic strategies proved successful for the fabrication of NCs [1–5]. Vesicle-templated assembly is based on the polymerization of hydrophobic monomers distributed in the hydrophobic interior of bilayers formed by lipids [6–8] or other surfactants (Fig. 4.2) [9–12].

In this approach, the bilayer acts as a spontaneously formed two-dimensional solvent. The polymer network propagates laterally within the bilayer but does not grow in the dimension orthogonal to the plane of the bilayer. The polymerization creates a cross-linked network within the hydrophobic interior of the bilayer. After the polymerization, the lipid or surfactant scaffold can be removed to yield polymer NCs. While outside the scope of this review, other structures, such as hollow silica microspheres, can be prepared using bilayers as templates.

The permeability of the NC shell can be controlled by creating nanopores [13–18]. To achieve this goal, pore-forming templates are loaded into the bilayer interior together with monomers and cross-linkers (Fig. 4.3). After the polymerization, these templates become embedded in the cross-linked polymer network within the bilayer. These templates can be removed by chemical degradation or by washing NCs with an organic solvent to yield uniform nanopores. This method of the pore formation essentially constitutes molecular imprinting. When inert hydrophobic molecules are used as pore-forming templates, such as glucose pentaesters, the chemical composition of the pore orifice is the same as the composition of the NC shell, that is, a hydrophobic cross-linked polymer. An extension of the pore-imprinting method uses pore-forming templates containing a polymerizable moiety connected to a bulky group with a chemically degradable linker (Fig. 4.3). This template copolymerizes with the monomers and cross-linkers. Cleavage of the linker and removal of the bulky group create nanopores with functional group(s) in the pore orifice. Embedded functional groups can be modified further using click chemistry. This method enables controlling both size and chemical environment of nanopores, permitting control of permeability in a broad range. Structural characterization of vesicle-templated NCs

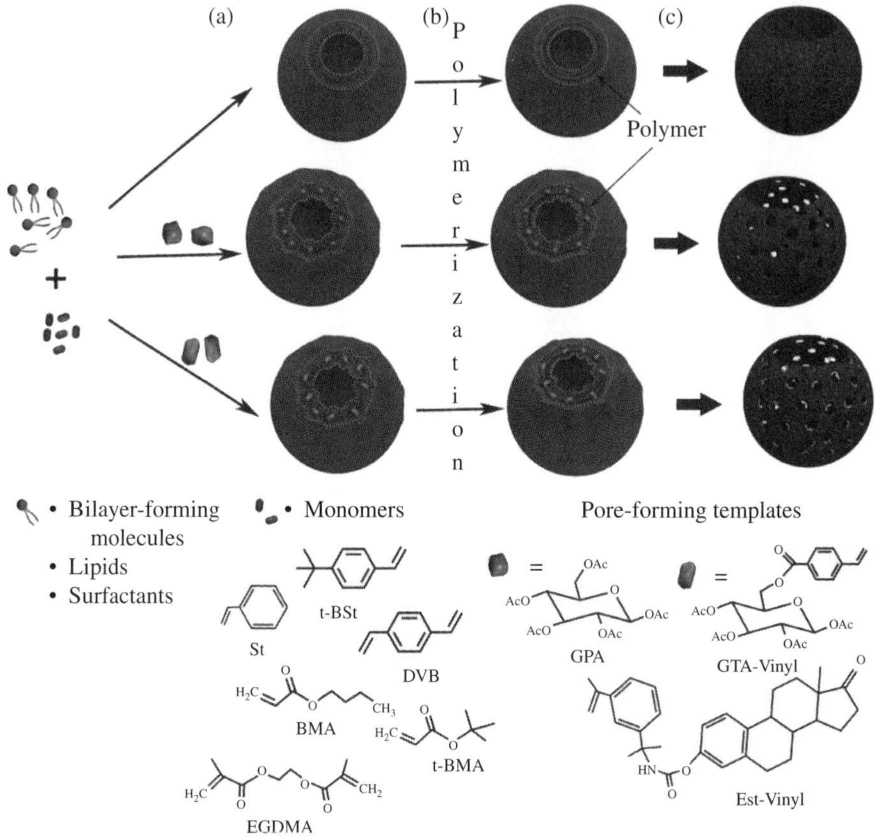

FIGURE 4.3 Bilayer-templated assembly of NCs. (a) Formation of bilayers with loaded monomers with or without pore-forming templates; (b) polymerization (thermally or UV induced) leading to the cross-linked polymer shell in the bilayer interior; (c) scaffold removal and formation of hollow NCs with nanometer-thin walls. (Reprinted and adapted with permission from Ref. [17]. © American Chemical Society.)

showed that the capsules prepared with a high concentration of cross-linkers (1:1 monomer-to-cross-linker ratio) [13] retain the shape and size of vesicles used for the capsule synthesis (Fig. 4.4). A combination of transmission electron microscopy (TEM) and dynamic light scattering (DLS) showed the same average size and size distribution throughout the NC synthesis (Fig. 4.4).

An alternative route to the synthesis of polyelectrolyte hollow spheres is spontaneous assembly of oppositely charged surfactants [19, 20]. Direct templating without morphological changes and by use of hydrophobic monomers was achieved with the help of equilibrium vesicles [21–24]. The hollow polymer capsules are stable enough to withstand dialysis, vacuum drying, and resuspension in water aided by nonionic surfactants. The stability of these polymerizable vesicles results from the high cross-link density of polystyrene, from the compatibility of the aromatic

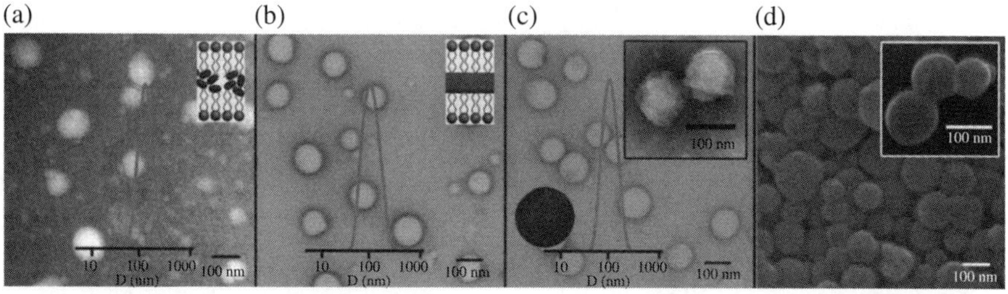

FIGURE 4.4 Electron microscopy images and DLS data of NCs: (a) liposomes loaded with monomers; (b) liposomes containing polymer NCs, after polymerization; (c) polymer NCs, after lipid removal. (d) SEM image of polymer NCs, after lipid removal, hydrolysis, precipitation, multiple washings, resuspension, and freeze-drying. (Adapted with permission from Ref. [13]. © Wiley.)

surfactants and polymer, and perhaps from the ability of the bilayer to rearrange to accommodate the growing polymer.

The monomer-to-surfactant and monomer-to-cross-linker ratios play a large role in determining the structural rigidity and permeability of NCs [17]. Figure 4.5 shows TEM and DLS data for experiments varying these ratios. Higher monomer-to-lipid ratio resulted in stable spherical NCs (Fig. 4.5a). Lowering the amount of monomers and cross-linkers in the bilayer resulted in capsules that appeared collapsed upon drying (Fig. 4.5b–d). Lowering the fraction of the cross-linker produced structures with larger diameter, consistent with the adoption of a "pita bread" structure upon drying (Fig. 4.5e). In the absence of cross-linkers, no NCs formed (Fig. 4.5f).

Permeability studies showed expected structure/performance relationship—lowering the fraction of the cross-linker rendered capsule shells more permeable. Variations in the amount of monomers and/or cross-linkers in the bilayer offer a likely explanation of previously reported variations in the structure of vesicle-templated architectures. For example, the polymerization of hydrophobic monomers within the bilayers of the surfactant dioctadecyldimethylammonium bromide (DODAB) was described in a series of papers. In several previous reports, polymer product did not resemble the vesicle structure, and usually, different hybrid morphologies were obtained (Fig. 4.6) [11, 25, 26]. The polymerization of styrene (or styrene and butyl methacrylate) with or without cross-linked produced the so-called parachute architecture or "necklace" morphology, which was characterized by dense polymer beads that are attached to a pure surfactant vesicle bilayer. The copolymerization of styrene with a small amount of polymerizable surfactant gave a "wrapped parachute" or "matrioshka" structure, wherein each parachute vesicle–polymer hybrid is encapsulated in a unilamellar vesicle. These types of architectures were formed probably because of phase separation of the polymer matrix from the bilayers and additionally because of insufficient amount of monomers in bilayer as discussed earlier [17].

Other methods for the synthesis of hollow NCs include the formation of polymersomes [27–31], LBL assembly [32–39], and cross-linking of supramolecular spherical assemblies (Fig. 4.7) [40–46]. By using different types of synthetic amphiphiles

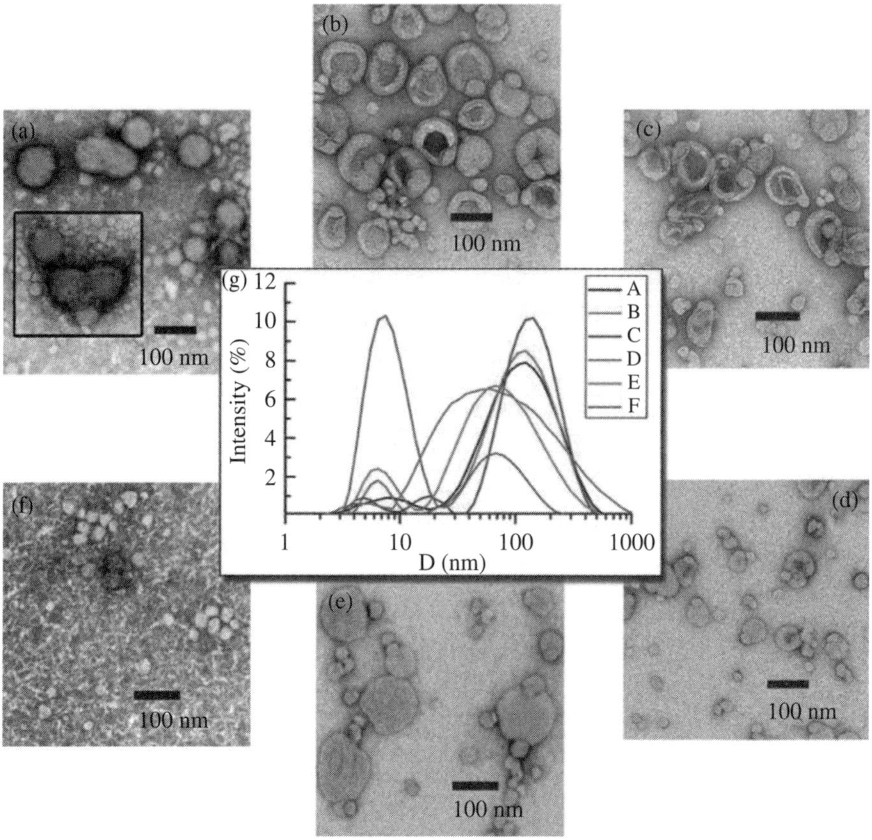

FIGURE 4.5 DLS and TEM images of NCs prepared by varying monomer-to-lipid (M:L) and monomer-to-cross-linker (M:C) ratios. In these experiments, liposomes loaded with monomers were irradiated with UV light for 90 min and treated with Triton X-100. Initial ratios used for the preparation of monomer-loaded liposomes were as follows: (a) M:L = 3:1, M:C = 1:1; (b) M:L = 2:1, M:C = 1:1; (c) M:L = 1:1, M:C = 1:1; (d) M:L = 0.5:1, M:C = 1:1; (e) M:L = 3:1, M:C = 10:1; (f) M:L = 3:1, no cross-linker; (g) DLS data of all NCs. (Reprinted and adapted with permission from Ref. [17]. © American Chemical Society.)

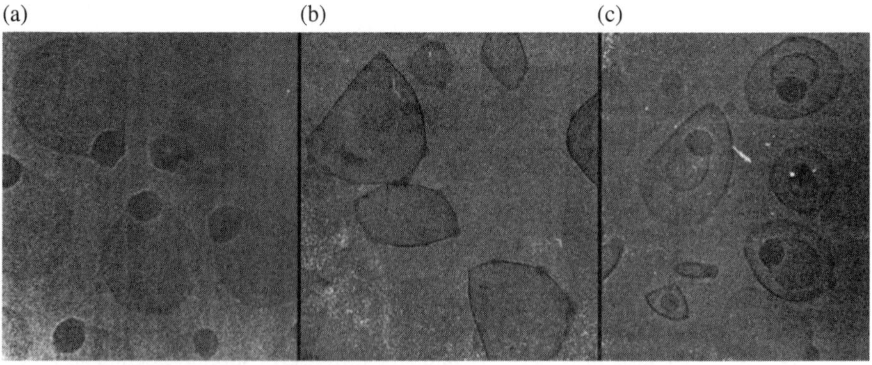

FIGURE 4.6 Different polymer morphologies from vesicle templating. (a) Parachutes; (b) necklaces; and (c) matrioshka structures. (Reprinted with permission from Ref. [12]. © Pergamon.)

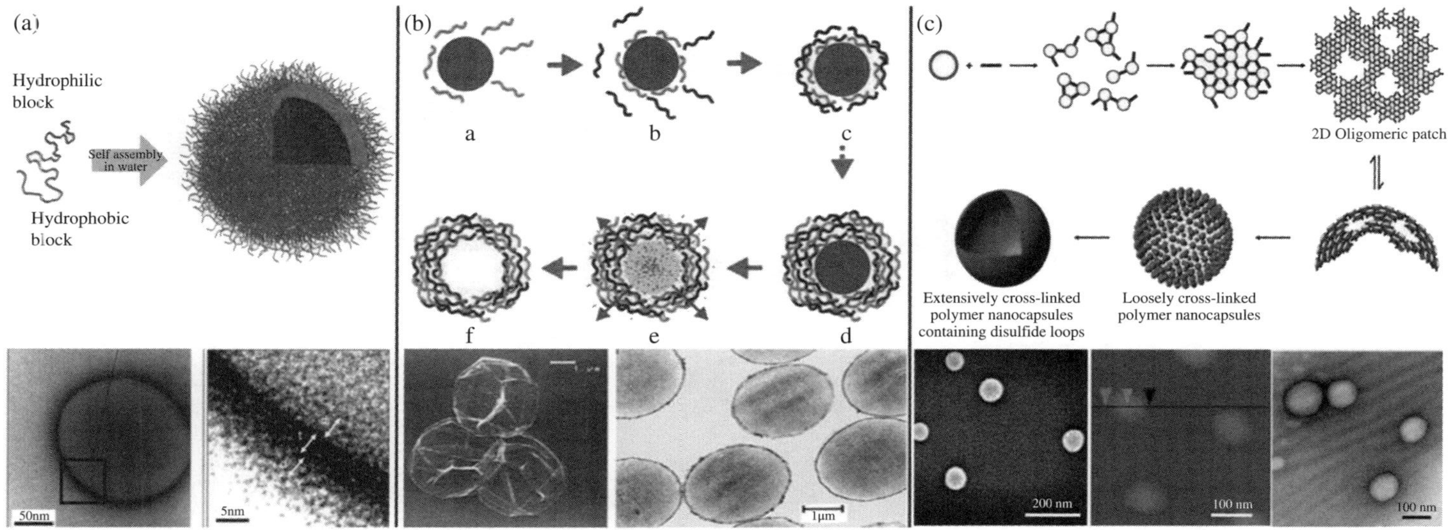

FIGURE 4.7 (A) Schematic of a polymersome formation and positive staining electron micrographs showing the membrane conformation of PEO50-PBO70 polymersomes. (B) Schematic illustration of the polyelectrolyte deposition process and of subsequent core decomposition and SEM and TEM images of polyelectrolyte shells. The initial steps (a–d) involve stepwise film formation by repeated exposure of the colloids to polyelectrolytes of alternating charge. The excess polyelectrolyte is removed by cycles of centrifugation and washing before the next layer is deposited. After the desired number of polyelectrolyte layers is deposited, the coated particles are exposed to 100 mM HCl (e). The core immediately decomposes, as evidenced by the fact that the initially turbid solution becomes essentially transparent within a few seconds. Three additional washings with 100 mM HCl ensure removal of the dissolved melamine formaldehyde (MF) oligomers. Finally, a suspension of free polyelectrolyte hollow shells is obtained (f). (Reprinted with permission from Ref. [32]. © Wiley.) (C) Proposed mechanism of the self-assembled polymer NC formation and SEM, AFM, and HRTEM images of polymer NCs. (Reprinted and adapted with permission from Ref. [40]. © American Chemical Society.)

[47, 48] to form polymersomes, it is possible to prepare vesicles with fairly high control of the size distribution and sizes ranging from tens of nanometers to tens of micrometers. Polymersome membranes comprise a highly entangled hydrophobic layer and two hydrophilic polymer brushes that stabilize the membrane. Both the hydrophilic brush and the hydrophobic membrane usually have similar thickness (Fig. 4.7a). Polymersomes form spontaneously by self-assembly of diblock or triblock copolymers. Typically, a diblock copolymer containing a hydrophilic and a hydrophobic block assembles spontaneously in water to form a vesicle with 10–15 nm-thick shells (Fig. 4.7a).

LBL assembly starts with a sacrificial core that is coated with sequential layers of polyelectrolytes (Fig. 4.7b). After depositing the desired number of layers, the core is dissolved in acidic medium. Repeated washings yield hollow microcapsules. These microcapsules have been studied extensively for diverse applications, including bioimaging. They have been reviewed in detail in recent reviews [49–54].

Another example of formation of capsules by self-assembly of building blocks (Fig. 4.7c) with narrow size distribution not requiring the use of templates was recently reported [40]. Different dyes (fluoresceins, rhodamines) can be easily entrapped inside these capsules. Another unique property of self-assembled polymer capsules is formation of stable host–guest complexes with polyamines with recognition properties of the accessible molecular cavities exposed on the capsule periphery. A different approach for self-assembly of hollow polymer NCs was used for cholesterol-modified dextran (Chol-Dex) and polylactic acid (PLA) by dialysis of their DMSO solution against water. The hollow capsules were formed initially by anchoring of most amphiphilic polysaccharide onto the surface of swelled aggregates because of the phase separation of amphiphilic Chol-Dex and hydrophobic PLA. Followed by the deposition of PLA on the inner interface of polysaccharide layer due to the "deswelling" of the aggregates, the remaining Chol-Dex inside the aggregates formed the innermost hydrophilic layer. The resulting hollow capsules may have potential application as drug and imaging agent carriers or biomembrane models [55].

The defining functional features of a membrane are the selectivity of permeability and the rate of mass transfer. The size-selective permeability is determined by the size of the pores. Ideally, only molecules that are smaller than the pore size should be able to diffuse across the polymer shell. The distinguishing characteristics of bilayer-templated membranes are uniform molecular size pores that span the polymer membranes. These pores offer very short path for molecules crossing the membrane. In contrast, traditional polymer membranes, such as dialysis membranes, contain a large network of interconnecting tunnels resulting in long diffusion times.

Using different pore-forming templates [14], pores with different sizes were created in vesicle-templated NCs. The size of the pores was determined by a size-selective retention assay. A mixture of colored size probes was encapsulated in vesicles prior to the polymerization. After the formation of the polymer shell and removal of surfactant scaffold and pore-forming templates, NCs were separated from the released size probes by size-exclusion chromatography. The color of the capsules with retained size probes indicated the pore size (Fig. 4.8).

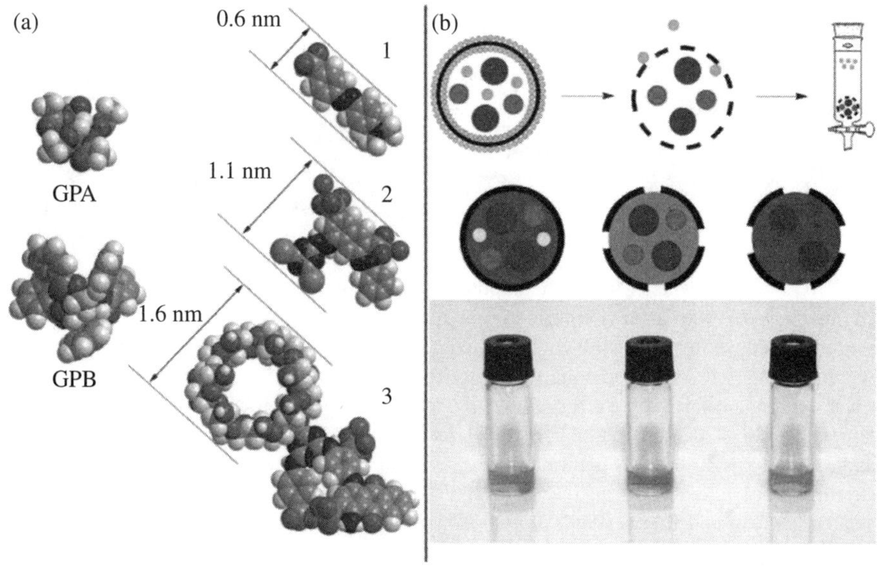

FIGURE 4.8 Determination of pore size in vesicle-templated NCs using a selective permeability assay. (a) Pore-forming templates. GPA, glucose pentaacetate, and GPB, glucose pentabenzoate. Size probes: 1, methyl orange (0.6 nm); 2, Procion Red (1.1 nm); 3, Reactive Blue 2/b-cyclodextrin conjugate (1.6 nm). The smallest cross section corresponds to the smallest pore the probe can cross in its most tightly packed conformation. (b) Selective permeability demonstrated by the size probe retention assay. Liposomes were loaded with the mixture of colored size probes: 0.6 nm yellow (1), 1.1 nm red (2), and 1.6 nm blue (3). Monomers and cross-linkers with or without pore-forming templates were loaded into the liposomal membrane, and polymerization was initiated, after which the phospholipids and pore-forming templates were removed and capsules were separated from released size probes on a size-exclusion column. The NC fraction is shown. In the absence of pore-forming templates, no probes escaped, and the capsules remained brown (left sample). When GPA was used as pore-forming template, 0.6 nm yellow probe was released, and the capsules were colored purple (middle sample). With GPB as a pore-forming template, 0.6 nm yellow and 1.1 nm red probes were released, and capsules were colored blue (right sample). (Reprinted with permission from Ref. [14]. © Wiley.)

Functional groups in the pore orifice can be modified further using click chemistry [13]. For example, carboxylic groups in the nanopores can be activated and converted to amide groups with a quantitative yield (Fig. 4.9). Functionalization of nanopores did not compromise the ability to control the pore size (Fig. 4.9).

The NCs can retain entrapped molecules for extended periods of time. In a previous study [17], we showed that polystyrene capsules did not release Procion Red, a dye used as 1.1 nm size probe, for 240 days. A follow-up study on the same samples showed no measurable release after 40 months. In these experiments, release of as little as 1% of encapsulated dye could be detected with UV–Vis analysis of the supernatant from the suspension of NCs. This long-term stability is likely the result of high degree of cross-linking in the NC shells.

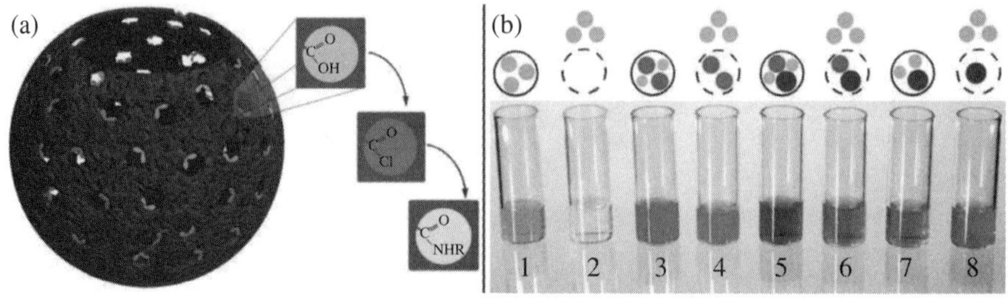

FIGURE 4.9 (a) The as-formed nanopores have a single carboxy functional group, which can be converted into an acyl chloride group and then into an amide group. (b) Colored size probe retention assay. NCs with encapsulated colored size probe mixtures were prepared and separated on a size-exclusion column to remove released probes. Photograph of the NC fractions: with encapsulated 0.6 nm (yellow) probes, (1) before template removal and (2) after template removal, demonstrating complete release of the 0.6 nm probes; with encapsulated 0.6 and 1.1 nm (red) probes, (3) before template removal and (4) after template removal, demonstrating complete release of the 0.6 nm probes and retention of the 1.1 nm probes; with encapsulated 0.6, 1.1, and 1.6 nm (blue) probes, (5) before template removal and (6) after template removal, demonstrating release of the 0.6 nm probes and retention of the 1.1 and 1.6 nm probes; and with encapsulated 0.6 and 1.6 nm probes, (7) before template removal and (8) after template removal, demonstrating release of the 0.6 nm probes and retention of the 1.6 nm probes. (Adapted with permission from Ref. [13]. © Wiley.)

To gain insights into the kinetics of transport through the shells of bilayer-templated NCs, the rates of protonation and deprotonation of encapsulated pH-sensitive dyes were investigated using stopped flow technique coupled with absorbance and fluorescence spectroscopy [56]. In these experiments, an aqueous suspension of surfactant-stabilized NCs containing pyranine or bromophenol blue was rapidly mixed with a base or an acid followed by measuring fluorescence increase for the deprotonated form of pyranine or absorbance decrease for the protonated form of bromophenol blue. Remarkably, the rate of absorbance was the same for the encapsulated and free dyes (Fig. 4.10). These observations suggest that the transport through the shell of the NCs is not the rate-limiting step in the overall process. Since protonation/deprotonation reactions have very low energy barrier and are diffusion controlled, we conclude that the mixing of two solutions is the rate-limiting step in the overall process involving encapsulated pH-sensitive dyes. These observations suggest that NCs can be used for real-time measurements of exterior ions or small molecules. In addition, NCs with negligible mass transfer resistance may lead to fast-acting devices for delivery of drugs or molecular imaging agents.

4.3 COMPARTMENTALIZATION OF MOLECULES IN HOLLOW NANOCAPSULES

Molecules can be entrapped in hollow NCs during the synthesis stage. Alternatively, larger molecules can be assembled from smaller components through a ship-in-a-bottle assembly within prefabricated NCs (Fig. 4.11a) [57]. Another type of loading

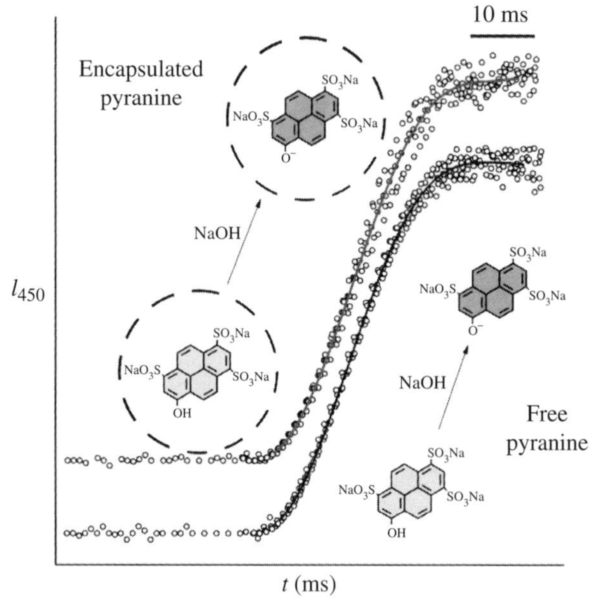

FIGURE 4.10 Change of fluorescence intensity as function of time in stopped flow experiments with free and NC-encapsulated dyes. Response of free pyranine (bottom curve) and pyranine entrapped within NCs (top curve) to NaOH. (Adapted with permission from Ref. [56]. © Royal Society of Chemistry.)

of molecules into hollow NCs is based on reversible coentrapment of components that form a noncovalent complex (Fig. 4.11b) [34, 57, 58]. Here, components can enter the capsule individually through the nanopores. The complex is too big to pass through the pores, and it remains entrapped [59, 60]. The dissociation of the complex upon external stimulus triggers the release of individual components. In one implementation [58], we used a conformationally flexible crown compound. It is likely that this concept will be applicable to other systems, where the minimal cross section of the complex is bigger than the effective minimal cross sections of individual components. One can envision further extension of this concept into metal-mediated release of biologically active molecules or into noncovalent assembly of functional devices, such as nanoreactors or nanosensors. This approach may be further extended to other materials with uniform nanopores.

Several examples of stimuli-responsive encapsulation of low- or high-molecular-weight species were reported for LBL microcapsules (Fig. 4.11c). For example, a pH change can induce electrostatic repulsion between the two types of polyelectrolyte, while osmotic pressure can build up when ions diffuse to the interior of the vesicles to compensate the charges. The resulting forces form holes in the polyion shells. Under standard aqueous conditions, the capsule shell is permeable to low-molecular-weight compounds such as ions and small drug molecules but impermeable to macromolecular components ($M_w > 5$ kDa) [61]. The loading of larger species

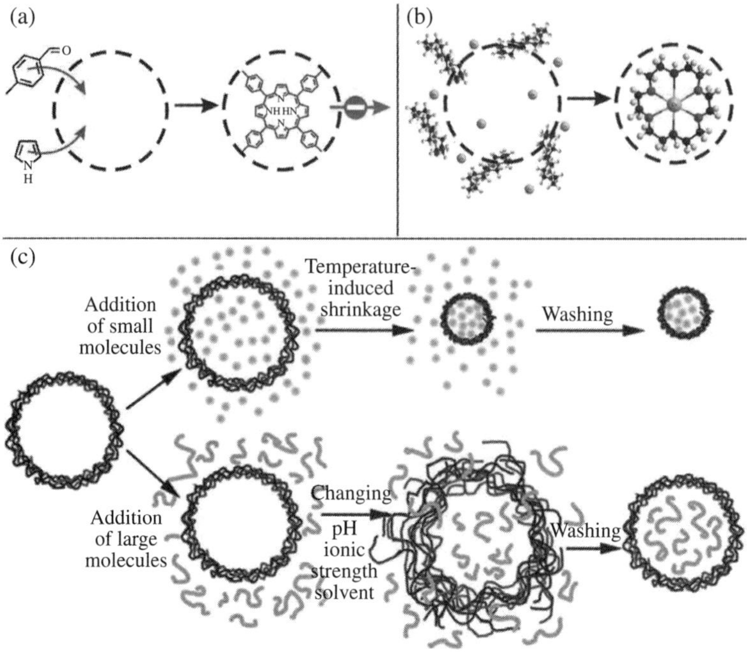

FIGURE 4.11 (a) Combined assembly and entrapment of 5,10,15,20-tetra-p-tolyl-21H,23H-porphine (H_2TTP) in a porous NC. The synthesis of H_2TTP was carried out by a 30 min reflux in acetic acid. (b) Synergistic coentrapment in porous hollow NCs: individual components are smaller than the pore size and can diffuse into the capsule. Their complex is bigger than the pores and cannot escape from the capsule. (Adapted with permission from Ref. [58]. © American Chemical Society.) (c) Schematic representation of the different strategies to generate postload polymeric multilayer capsules (PMLCs) by changing their physicochemical environment. (Adapted with permission from Ref. [34]. © Wiley.)

is possible by altering either their own solubility or by altering the permeable PMLC shell. Reversibly changing the permeability of the PMLC shell to macromolecules is achieved by changing the pH value, ionic strength, or solvent polarity, which leads to segregation of the polyelectrolyte network and defects in the shell [60, 62–64]. After macromolecules have passed through the capsule wall, the PMLCs are transferred to their original medium, thereby entrapping the species in their hollow void.

NCs can play a dual protective role in biomedical applications (Figs. 4.12 and 4.13). On the one hand, NCs can protect encapsulated molecules from biological environment, for example, from enzymatic degradation. On the other hand, NCs can protect the biological environment from potential adverse actions of encapsulated molecules. Molecules otherwise attractive as imaging contrasts may prove to be toxic to the cells. The combined effect of extremely rapid mass transfer and protective features of NCs makes them particularly attractive for the biomedical applications as smart containers for sensors or imaging contrasts as well as therapeutic agents [65].

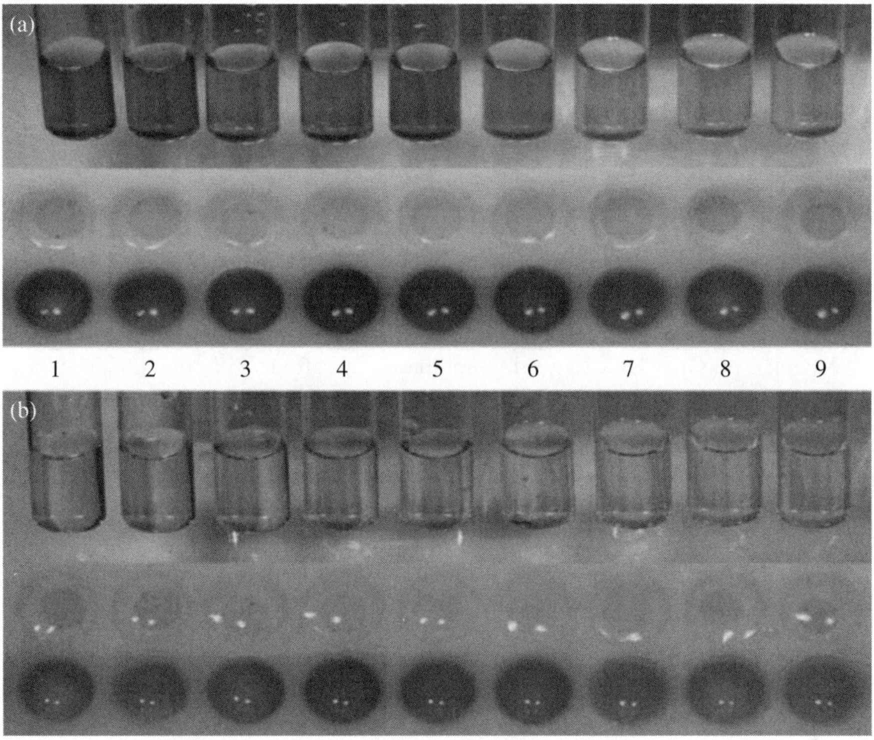

FIGURE 4.12 Photostability of Nile Blue (NB) A free and when encapsulated in porous NCs in buffers solutions with pH values ranging between pH 2.5 and 12.5. (a) Photos of freshly prepared solutions, microplate wells filled with PVA-based hydrogel specimens in which the NB A-loaded NCs are dispensed, and microplate wells filled with buffer solutions of different pH values in which the NB A-loaded NCs are dispensed. (b) Photographs of the same solutions and wells after 2 weeks of storage under regular laboratory conditions. The pH in the test tubes and wells labeled 1 through 9 was set to pH 2.5, 3.5, 5.5, 6.5, 9.5, 10.0, 10.5, 11.5, and 12.5 using citrate- (1, 2), MES-(2-(*N*-morpholino)ethanesulfonic acid) (3, 4), borate- (5, 6, 7), and phosphate-based (8, 9) buffers, respectively. (Adapted with permission from Ref. [65]. © American Chemical Society.)

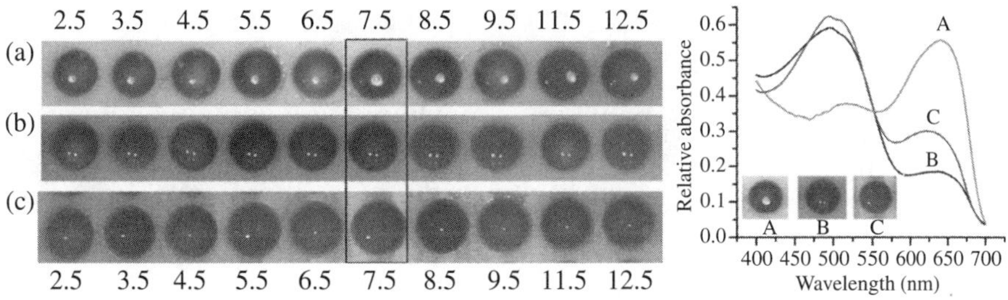

FIGURE 4.13 Aqueous suspensions and UV–Vis spectra (at pH 7.5) of NCs containing NB with β-CD sulfonate (a) NB and (b) NB with tetraphenyl borate (c) at different pH. (Adapted with permission from Ref. [65]. © American Chemical Society.)

4.4 BIOMEDICAL APPLICATIONS OF POLYMER NANOCAPSULES

The influence of capsules onto living cells have been studied recently addressing different aspects of interactions with living tissue [66], such as toxicity of capsules, uptake kinetics and mechanisms, intracellular degradation, as well as strategies to enhance or block the capsule uptake.

To enable biomedical applications of polymer NCs, it is important to know their biocompatibility with living cell. NCs representing different scenarios of preparation and handling can be taken up by human cells (Fig. 4.14) [67]. Among them, NCs with an outer shell of lipids have the best cell penetration ability. Dried NCs may need assistance with dispersions in aqueous media, for example, with mild surfactants. All investigated polymer NCs showed no inhibition of cell proliferation at a dose up to 200 mg/ml or up to 0.2% by volume. At this dose, no generation of reactive oxygen species (ROS), induction of apoptosis, or perturbation of cell cycle is detected.

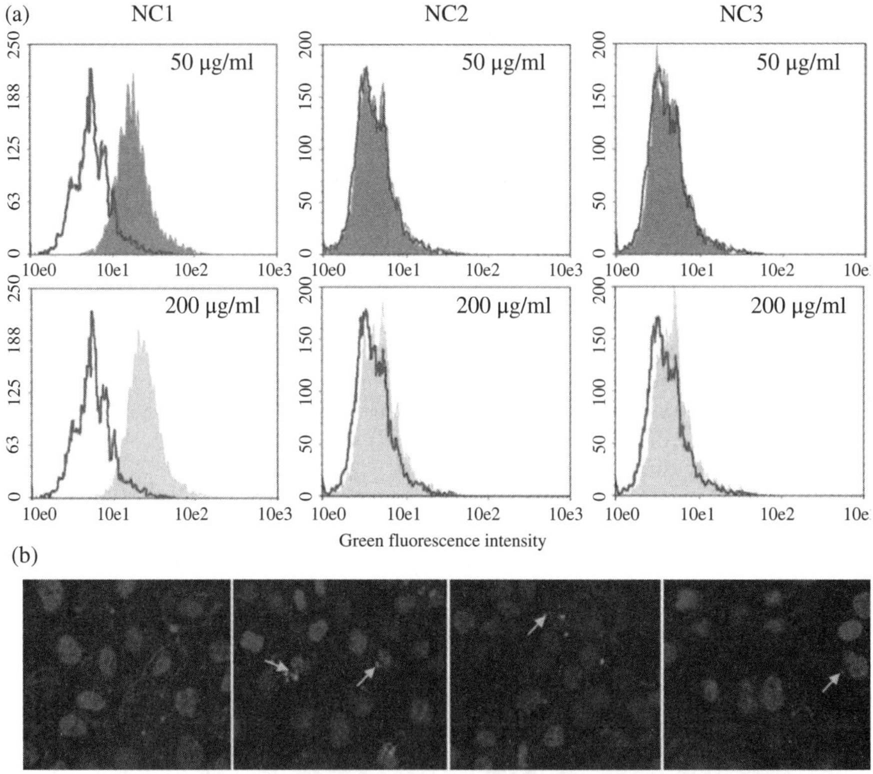

FIGURE 4.14 Cell uptake of NCs in A549 cell. Cell internalization was measured by flow cytometry (a) and confocal laser scanning microscopy (CLSM) (excitation at 488 nm) (b). In CLSM experiments, cell culture medium (a) or NC1 (b), NC2 (c), or NC3 (d) was incubated with cells for 24 h before measurements. (Adapted with permission from Ref. [67]. © Royal Society of Chemistry.)

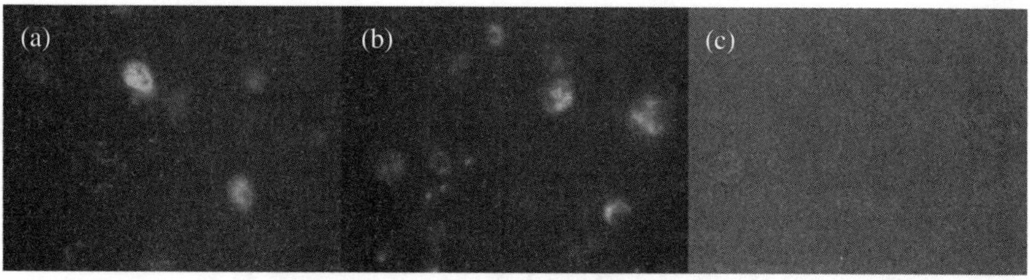

FIGURE 4.15 NCs (50 μg/ml) added to DRG neurons for 24 h, washed, and imaged. (a) Single slice fluorescence only. (b) 3D slice fluorescence only. (c) 3D slice DIC only.

In the case of PMLCs, several research groups have assessed their toxicity by performing *in vitro* cell viability assays such as the MTT test [68, 69]. Generally, no acute toxicity was observed at moderate capsule concentrations, and an outermost polyanionic layer appeared to further decrease the toxicity (cationic PMLCs exhibit a pronounced tendency to adhere to the cellular surface) [66]. Some toxicity was observed at elevated capsule concentrations; this is commonly attributed to sedimentation of the capsules on top of the cells as a result of competition for space between the capsules and cells.

Picart *et al.* have shown that polyelectrolyte multilayers can be digested by enzymatic action when placed in the peritoneal [70] and oral environments [71]. The biocompatibility and *in vivo* fate of dextran sulfate–poly-L-arginine polyelectrolyte capsules after subcutaneous injection were recently assessed by De Koker *et al.* [68].

In our recent experiments, hollow polymer NCs loaded with calcein were taken up by dorsal root ganglion (DRG) neurons (Fig. 4.15). Preliminary data indicate no adverse effect of NCs on function of DRG neurons, consistent with data described earlier.

In therapeutic applications, specific targeting of cells or receptors is desirable because many side effects of current drugs emerge from the undesired impact of a drug molecule on physiological pathways in a cell type or organ that is not involved in the disease process [72, 73]. Targeting specifically functionalized polymersomes to a desired cell type and location inside the body is of essential importance for biomedical imaging (Fig. 4.16). Nanocontainer-treated macrophages showed characteristic intracellular accumulations of vesicular structures with sizes in the range of the nanocontainers. Together with the confocal microscopy results, this finding is compatible with intracellular transport and accumulation of nanocontainers to certain locations in the cells, presumably linked to the endocytotic pathway.

A variety of porphyrin-based or other near-infrared (NIR) dyes loaded to the hydrophobic leaflet of polymersomes can be used for imaging *in vivo*. The total fluorescence emanating from the polymersomes gave rise to a localized signal of sufficient intensity to penetrate the dense tumor tissue of this living animal (Fig. 4.17), giving a signal-to-background ratio of at least 10 to 1 [74]. Conjugation of Tat [75], a cell permeable peptide, to NIR-emissive polymersomes allows tracking dendritic cells. An improved and selective intracellular delivery of Tat-functionalized vesicles was found in dendritic cells in comparison with nonfunctionalized polymersomes.

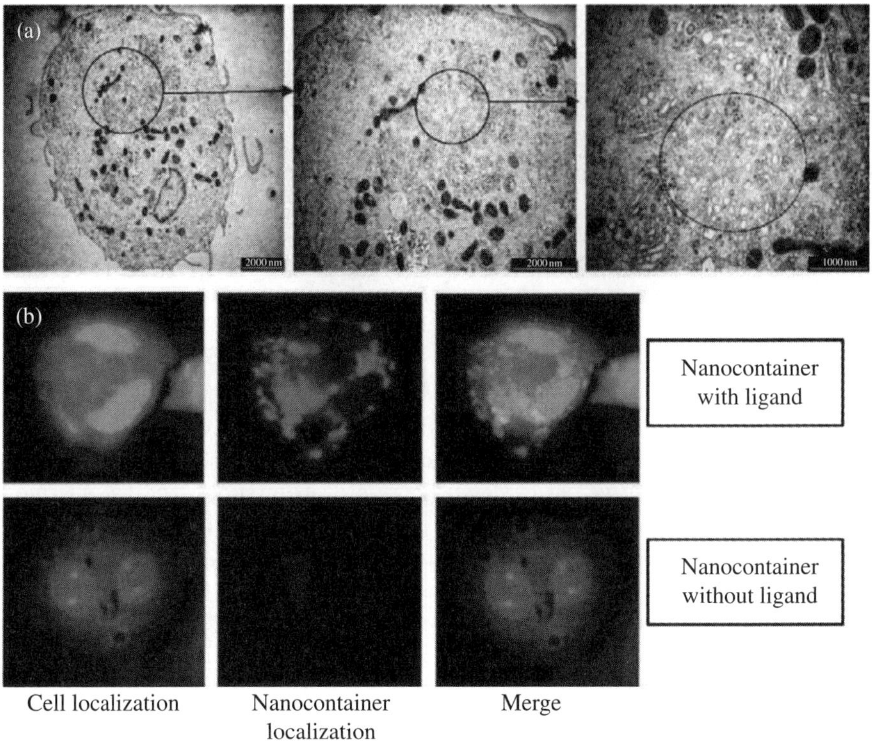

FIGURE 4.16 (a) TEM shows functionalized nanocontainers in THP-1 cells. (b) Fluorescently labeled nanocontainer found at the surface and inside macrophages. (Reprinted with permission from Ref. [73]. © Elsevier.)

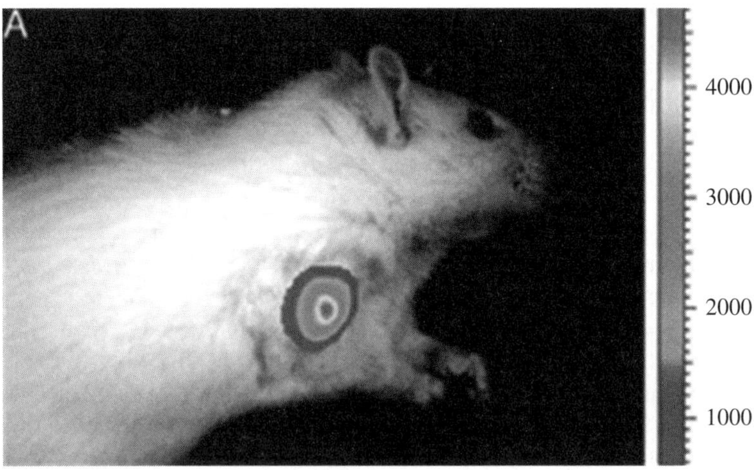

FIGURE 4.17 *In vivo* fluorescence image of 300 nm-sized NIR-emissive polymersomes taken 10 min after direct tumor injection of a glioma-bearing rat. The fluorescence signal intensity corresponded to only half the integrated emission from the polymersomes, and it remained constant between successive images taken during a 20 min interval postinjection. (Reprinted from Ref. [74]. © National Academy of Sciences.)

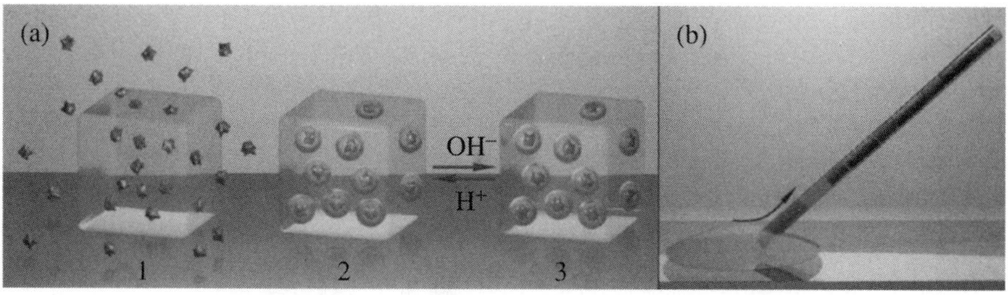

FIGURE 4.18 (a) The underlying concept of the sensor platform based on dye-loaded NCs embedded in a highly permeable matrix: (1) low-molecular-weight dye molecules leach out from the PVA matrix; (2 and 3) the same dye molecules encapsulated in porous NCs remain entrapped in the PVA matrix. The porous nanothin wall of the capsules allows unhindered transport of H^+ and OH^- ions into the capsules. (b) A multiband capillary can sample a small amount of fluid. (Adapted with permission from Ref. [65]. © American Chemical Society.)

An alternative imaging modality that can be used to image polymer vesicles is diagnostic ultrasound. Zhou *et al.* prepared air-encapsulated polymersomes via lyophilization and rehydration of previously formed polymer vesicles [76]. The polymer bubbles were imaged using a Pie Medical Scanner 350 and were visualized as bright spots, validating the acoustic activity of air-encapsulated polymersomes.

Ghoroghchian *et al.* [77] recently showed how to apply simple diblock copolymer system such as poly(ethylene oxide)-block-poly(ε-caprolactone) (PEO-*b*-PCL) and poly(ethylene oxide)-block-poly(γ-methyl-ε-caprolactone) (PEO-*b*-PMCL) to create polymersomes with special emissive properties. This can be achieved by incorporating porphyrin-based fluorophores within the polymeric membrane. The final structure shows optical properties similar to quantum dots from the visible to the infrared. In another work, fluorescent amphiphilic molecules have been incorporated to polymersome membrane and delivered within cells in order to generate a revolutionary noncytotoxic, nonimmunogenic cellular tracking system [78]. Polymersomes loaded with such fluorophores are therefore promising candidates for use as "nanoscale imaging agents" [79]. In addition to the incorporation of fluorophores within polymersomes [77, 78], the potential of those carriers to the field of imaging has been further exemplified by the encapsulation of magnetic resonance imaging (MRI) contrast agents within porous PEO-PBD polymersomes, in research work carried out by Cheng and Tsourkas [80].

Additionally to fluorescent markers, quantum dots embedded into the capsules can potentially be used as ion probes and luminescence sensors. The effect of pH conditions and various ions on the luminescence intensity of CdTe nanocrystals capped by thioglycolic acid has been studied [81]. Luminescence of both bare CdTe nanocrystals and CdTe nanocrystals embedded into polymer microcapsules was found to be pH sensitive within the range of pH 4–6. Another imaging application of PEO-PBD copolymers has been found by the Maskos research group, which encapsulates quantum dots within polymersome membranes [82].

NCs loaded with imaging contrasts or stimuli-responsive molecules can be embedded into a highly permeable matrix. This method can lead to devices for medical diagnostics (Fig. 4.18) that may use a broad range of optical chemosensors, such as pH-sensitive molecules [65]. Embedding dye-loaded NCs into a permeable matrix addresses a long-standing challenge of reconciling fast response of chemosensors with long-term stable performance of practical optical sensors. This concept can be expanded into biocompatible and/or biodegradable matrices to create implantable devices.

Cargo release from capsules can be controlled remotely and on demand by using an external stimulus. For biomedical applications, one of the extensively used external release triggers is NIR radiation because this stimulus can be applied in a focused area that allows localized release from individual capsules (Fig. 4.19). For therapeutic applications, the release of capsule cargo through an inherent biological stimulus is of immense interest. Changes in the environmental pH, salinity, and temperature are commonly exploited environmental triggers that have been used to induce release in polymer carriers. In addition, significant advances were reported in the field of triggered release using other intrinsic stimuli, including changes in redox potential (thiol–disulfide exchange), enzymatic degradation, change of solvent composition, and degradation in response to the presence of a specific metabolite.

Recently, we created NCs capable of burst release of encapsulated content upon contact with blood [84]. Rapid release is achieved by the expansion of intrinsic nanopores in the NC shell caused by the conformational change due to interactions with albumin. *In vivo* imaging showed rapid release of the model compound indocyanine green (ICG), an FDA-approved NIR fluorescent dye (Fig. 4.20).

To treat some forms of cancer and viral infections, a photosensitizer (hypocrellin B (HB)) was loaded into capsules applying a solvent exchange step, where ethanol was used as a solvent for HB [85]. Upon exposure to light, HB generates singlet oxygen, which is cytotoxic and induces cell death. However, HB does not display cytotoxicity in the absence of the light irradiation. Upon incubation with the living cells, the HB-loaded capsules were efficiently internalized and appeared to be noncytotoxic. However, upon irradiation with 488 nm light, the cell viability dropped by 70%. In this *in vitro* experiment, the light addressable microcapsules emerge as promising candidates for the site-specific delivery of drug molecules that need to be placed into a protective container on the way to the biological target.

Cancer therapy is a rapidly emerging application, due to the ability of polymersomes to load both hydrophilic and hydrophobic anticancer drugs within their lumen and membrane cores, respectively. Both hydrophilic (doxorubicin (DOX)) and hydrophobic (paclitaxel) anticancer drugs have been loaded simultaneously within PEO-PLA and PEO-PCL polymersomes (generated by mixing these hydrolytically degradable block copolymers (1:3 mol/mol) with inert PEO-PBD) by Ahmed, Discher, and coworkers [86, 87]. These polymersomes were not only able to deliver these drugs into a human breast tumor implanted into mice; growth retardation and a reduction of the tumor area were also observed. At a time point of 5 days after IV tail injection of the polymersomes into the mice, the tumor shrank to less than 50% of its original size. Due to the biodegradability of the ester blocks comprising these polymersomes and the blending of a degradable block copolymer with a chemically inactive block copolymer, the drug release rate could be controlled, with the mildly acidic environment within cellular endosomes triggering faster ester hydrolysis and

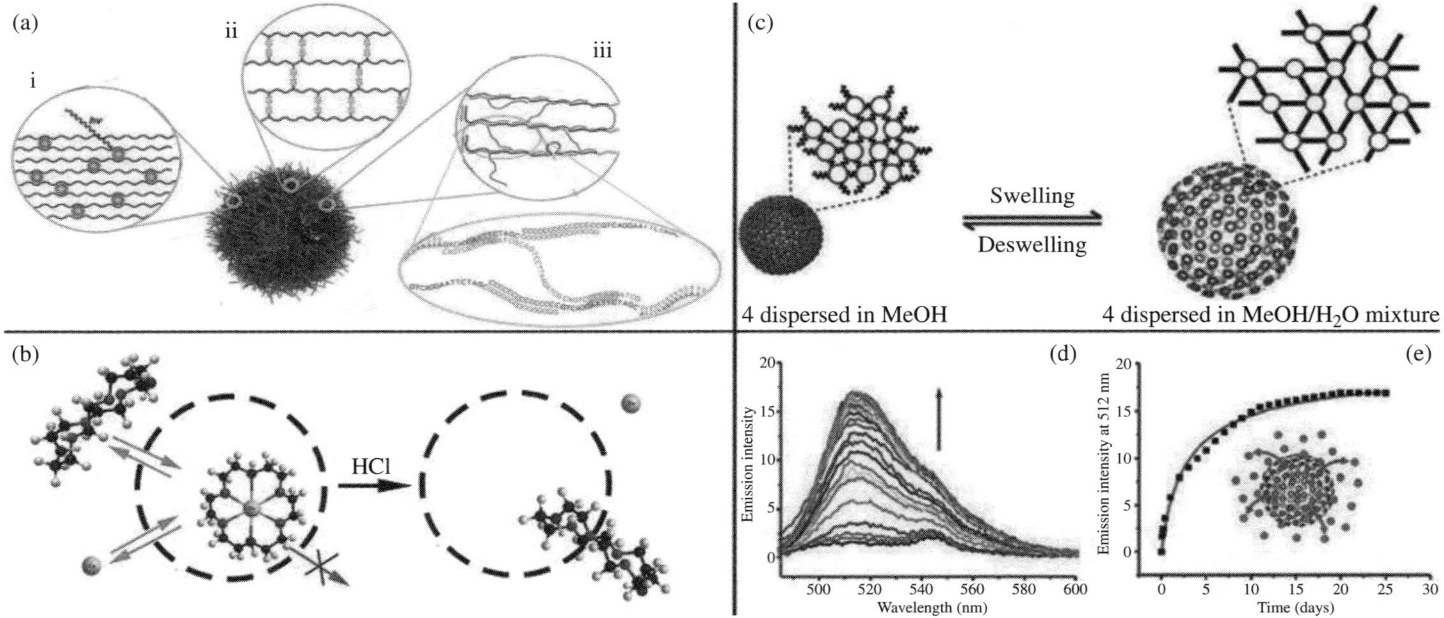

FIGURE 4.19 (a) Examples of triggers to release encapsulants from polymer carriers: (i) light (NIR light-absorbing nanoparticles), (ii) redox potential (cleavage of disulfide bonds), and (iii) enzymes (enzymatic cleavage). (b) Triggered release of entrapped molecules from NCs. The complex between DA18C6 and metal cations is dissociated upon lowering pH of the solution. The cation and DA18C6 are now free to leave the NC. (c) Reversible swelling and deswelling of polymer NC. (d) Changes in the emission spectra of carboxyfluorescein (CF) upon release from the interior of polymer NC by swelling in 10% methanol–water (l_{ex} = 458 nm, l_{em} = 512 nm). (e) Release profile of the encapsulated CF from the interior of the polymer NC in 10% methanol–water as a function of time. (Reprinted with permission from Ref. [83]. © Wiley.)

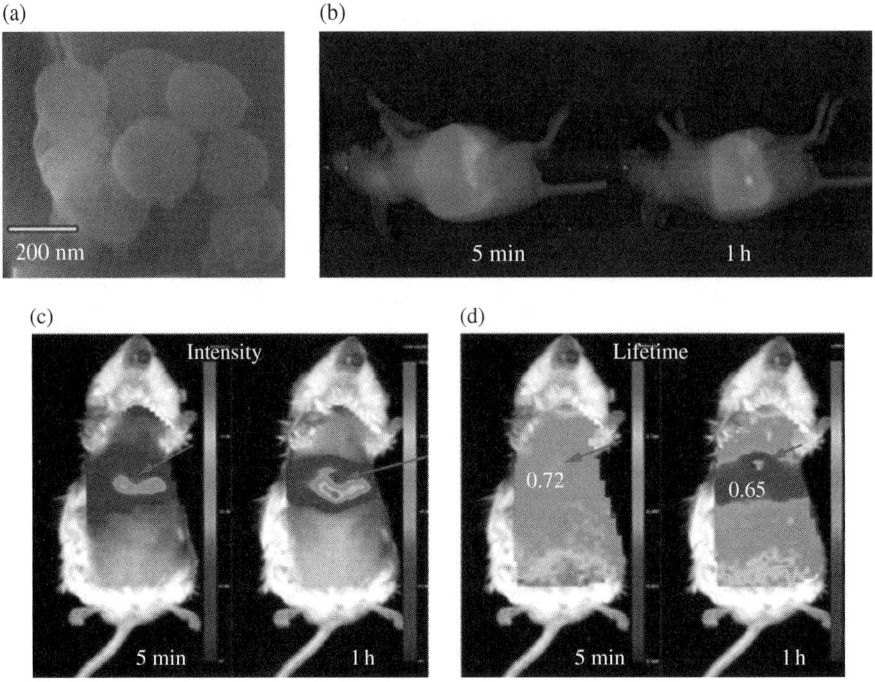

FIGURE 4.20 (a) SEM images of NCs with ICG after treatment with 4% w/v BSA. (b) Fluorescence intensity after administration of ICG-NCs. Red (ICG) channel emission (800 nm) and green channel emission (700 nm) correspond to endogenous fluorescence (from the diet). Red fluorescence is primarily localized in the liver. (c) Intravenous (IV) injection of ICG-NC shows an accumulation of the fluorescent dye in the liver. Left, fluorescence intensity; right, fluorescence lifetime, ex/em = 775/820 nm. *(See insert for color representation of the figure.)*

polymersome poration compared with the extracellular environment. Temperature was also found to have an effect on the rate of polymersome poration.

A near-infrared fluorophore (NIRF)-labeled peptosome was prepared by adding ICG-conjugated polypeptide. Diameter of the peptosome prepared from GA-PSar-PMLGs was in the range from 100 to 200 nm, which is the appropriate size for the passive targeting into the solid tumor by the EPR effect. *In vivo* retention in the blood of ICG-labeled peptosome was assayed by using rats as experimental animals, and the retention time was proven to be as long as the PEGylated liposome. ICG-labeled peptosome was also administered to the tumor-bearing mouse and successfully detected the tumor tissue due to the EPR effect [88].

The anticancer drug DOX and hydrophilic superparamagnetic iron oxide nanoparticles (SPIONs) were encapsulated inside the inner aqueous core of the vesicles for cancer therapy and MRI, respectively. It was shown that consequently effective suppression of cancer cell growth was detected [89].

Novel functional hollow chitosan–poly(acrylic acid)–gold (CS-PAA-Au) hybrid nanospheres, prepared in an aqueous solution via a one-pot route, can not only act as drug carriers for intracellular and intranuclear drug delivery but also act as a contrast agent in cancer cell imaging and in lighting up the nucleus, which results in a type of polymer hybrid nanosphere that can execute tumor cell imaging and anticancer drug delivery at the same time [90].

Polymersomes based on polyanionic block copolymers labeled with a NIRF were obtained by the Discher group [91] to quantify biodistribution in live mice and excised organs. It was shown that NIRF allows imaging of polymersomes in freshly excised organs using wavelengths available on any epifluorescence microscope. And additionally, authors mentioned that method allowed for the observation of the interaction between NIRF-labeled carriers and live cells in their native tissue microenvironment.

Novel polymer–lipid hybrid nanovesicles were produced from the aqueous coassembly of the diblock copolymer, poly(ethylene oxide)-block-polybutadiene (PEO-PBD), and the phospholipid, hydrogenated soy phosphatidylcholine (HSPC) [92]. The PEG-based vesicles were functionalized with either folic acid or anti-HER2/neu affibodies as targeting ligands to confer specificity for cancer cells. Authors showed that nanovesicles prepared from polymer–lipid blends led to significant improvement in cell binding compared to nanovesicles prepared from block copolymer alone in both *in vitro* cell studies and murine tumor models.

Hydrophobically modified maghemite (γ-Fe_2O_3) nanoparticles were encapsulated within the membrane of poly(trimethylene carbonate)-b-poly(L-glutamic acid) (PTMC-b-PGA) block copolymer vesicles using a nanoprecipitation process [93]. These superparamagnetic hybrid self-assemblies display enhanced contrast properties that open potential applications for MRI. They can also be guided in a magnetic field gradient. The feasibility of controlled drug release by radio-frequency magnetic hyperthermia was demonstrated in the case of encapsulated DOX molecules, showing the viability of the concept of magneto-chemotherapy. These magnetic polymersomes can be used as efficient multifunctional nanocarriers for combined therapy and imaging.

Authors [94] demonstrated how a membrane-impermeable Ru^{II}-based light-switch complex that usually binds to nuclear DNA can be successfully rerouted to specifically target mitochondria in live cells by employing a biocompatible, pH-sensitive polymersome vector.

4.5 CONCLUSIONS

The main focus of this chapter discussion was polymer capsules with high aspect ratio of their diameters to thickness of the walls. The small thickness of the walls is critical for increasing the payload capacity of NCs and for rapid mass transfer of molecules through the capsule shell. Vesicle-templated synthesis is particularly suitable for creation of NCs with nanometer-thin walls and controlled permeability characteristics.

The LBL approach is an attractive approach to the fabrication of well-defined microcapsules, and starting materials are typically commercially available. One challenge is to bring the size of LBL capsules to sub-100 nm scale while maintaining adequate payload capacity. The chemical toolbox available for constructing polymer micelles and polymersomes is diverse, and natural motifs have been actively incorporated into their designs.

Different types of polymeric capsules can be used for a broad range of biomedical applications, particularly for imaging or combination of imaging with therapeutic function. The field of hollow NCs is rapidly growing, and recent years saw significant advances in controlled encapsulation, triggered release, and programmed permeability of NC shells. Future developments are likely to focus on targeted delivery, expanded *in vivo* applications, and construction of practical devices for medical diagnostics.

ACKNOWLEDGMENTS

We thank Amy Harkins and Jason Papke for imaging NCs in neurons (Fig. 4.15). This work was supported by the National Science Foundation (CHE-1316680), National Institutes of Health (1 R21 AR060408-01A1), and Saint Louis University Presidential Research Fund.

REFERENCES

[1] Douglas T. A bright bio-inspired future. *Science* 2003;**299**:1192–1193.

[2] Elemans JAAW, Rowan AE, Nolte RJM. Mastering molecular matter. Supramolecular architectures by hierarchical self-assembly. *J Mater Chem* 2003;**13**:2661–2670.

[3] Hamley IW. Nanotechnology with soft materials. *Angew Chem Int Ed* 2003;**42**: 1692–1712.

[4] Capito RM, Azevedo HS, Velichko YS, Mata A, Stupp SI. Self-assembly of large and small molecules into hierarchically ordered sacs and membranes. *Science* 2008;**319**:1812–1816.

[5] Mann S. Life as a nanoscale phenomenon. *Angew Chem Int Ed* 2008;**47**:5306–5320.

[6] Ruysschaert T, Germain M, da Silva Gomes JFP, Fournier D, Sukhorukov GB, Meier W, Winterhalter M. Liposome-based nanocapsules. *IEEE Trans Nanobiosci* 2004;**3**:49–55.

[7] Meier W. Polymer nanocapsules. *Chem Soc Rev* 2000;**29**:295–303.

[8] Gomes JFPS, Sonnen AFP, Kronenberger A, Fritz J, Coelho MÁN, Fournier D, Fournier-Nöel C, Mauzac M, Winterhalter M. Stable polymethacrylate nanocapsules from ultraviolet light-induced template radical polymerization of unilamellar liposomes. *Langmuir* 2006;**22**:7755–7759.

[9] Hotz J, Meier W. Vesicle-templated polymer hollow spheres. *Langmuir* 1998;**14**: 1031–1036.

[10] Ali SI, Heuts JPA, van Herk AM. Vesicle-templated pH-responsive polymeric nanocapsules. *Soft Matter* 2011;**7**:5382–5390.

[11] Jung M, Hubert DHW, van Veldhoven E, Frederik P, van Herk AM, German AL. Vesicle–polymer hybrid architectures: a full account of the parachute architecture. *Langmuir* 2000;**16**:3165–3174.

[12] Hentze H-P, Kaler EW. Polymerization of and within self-organized media. *Curr Opin Colloid Interf Sci* 2003;**8**:164–178.

[13] Dergunov SA, Pinkhassik E. Functionalization of imprinted nanopores in nanometer-thin organic materials. *Angew Chem Int Ed* 2008;**47**:8264–8267.

[14] Danila DC, Banner LT, Karimova EJ, Tsurkan L, Wang XY, Pinkhassik E. Directed assembly of sub-nanometer thin organic materials with programmed-size nanopores. *Angew Chem Int Ed* 2008;**47**:7036–7039.

[15] Savariar EN, Krishnamoorthy K, Thayumanavan S. Molecular discrimination inside polymer nanotubules. *Nat Nanotechnol* 2008;**3**:112–117.

[16] Cockroft SL, Chu J, Amorin M, Ghadiri MR. A single-molecule nanopore device detects DNA polymerase activity with single-nucleotide resolution. *J Am Chem Soc* 2008;**130**:818–820.

[17] Dergunov SA, Kesterson K, Li W, Wang Z, Pinkhassik E. Synthesis, characterization, and long-term stability of hollow polymer nanocapsules with nanometer-thin walls. *Macromolecules* 2010;**43**:7785–7792.

[18] Ki CD, Chang JY. Preparation of a molecularly imprinted polymeric nanocapsule with potential use in delivery applications. *Macromolecules* 2006;**39**:3415–3419.

[19] McKelvey CA, Kaler EW, Zasadzinski JA, Coldren B, Jung HT. Templating hollow polymeric spheres from catanionic equilibrium vesicles: synthesis and characterization. *Langmuir* 2000;**16**:8285–8290.

[20] McKelvey CA, Kaler EW. Characterization of nanostructured hollow polymer spheres with small-angle neutron scattering (SANS). *J Colloid Interf Sci* 2002;**245**:68–74.

[21] Dergunov SA, Schaub SC, Richter A, Pinkhassik E. Time-resolved loading of monomers into bilayers with different curvature. *Langmuir* 2009;**26**:6276–6280.

[22] Richter AG, Dergunov SA, Ganus B, Thomas Z, Pingali SV, Urban V, Liu Y, Porcar L, Pinkhassik E. Scattering studies of hydrophobic monomers in liposomal bilayers: an expanding shell model of monomer distribution. *Langmuir* 2011;**27**:3792–3797.

[23] Banner LT, Danila DC, Sharpe K, Durkin M, Clayton B, Anderson B, Richter A, Pinkhassik E. Controlled loading of building blocks into temporary self-assembled scaffolds for directed assembly of organic nanostructures. *Langmuir* 2008;**24**: 11464–11473.

[24] Brückner E, Rehage H. Solubilization of toluene in phospholipid vesicles studies by video-enhanced contrast microscopy. In: Lagaly G, editor. *Horizons 2000 – Aspects of Colloid and Interface Science at the Turn of the Millennium*. Darmstadt: Steinkopff; 1998. p 21–28.

[25] Jung M, Hubert DHW, Bomans PHH, Frederik P, van Herk AM, German AL. A topology map for novel vesicle-polymer hybrid architectures. *Adv Mater* 2000;**12**:210–213.

[26] Jung M, Robinson BH, Steytler DC, German AL, Heenan RK. Polymerization of styrene in DODAB vesicles: a small-angle neutron scattering study. *Langmuir* 2002;**18**:2873–2879.

[27] Discher BM, Won Y-Y, Ege DS, Lee JC-M, Bates FS, Discher DE, Hammer DA. Polymersomes: tough vesicles made from diblock copolymers. *Science* 1999;**284**: 1143–1146.

[28] Vriezema DM, Comellas Aragonès M, Elemans JAAW, Cornelissen JJLM, Rowan AE, Nolte RJM. Self-assembled nanoreactors. *Chem Rev* 2005;**105**:1445–1490.

[29] Lazzari M, Liu G, Lecommandoux S. *Block Copolymers in Nanoscience*. Wiley-VCH: Weinheim; 2006.

[30] LoPresti C, Lomas H, Massignani M, Smart T, Battaglia G. Polymersomes: nature inspired nanometer sized compartments. *J Mater Chem* 2009;**19**:3576–3590.

[31] Discher DE, Eisenberg A. Polymer vesicles. *Science* 2002;**297**:967–973.

[32] Donath E, Sukhorukov GB, Caruso F, Davis SA, Möhwald H. Novel hollow polymer shells by colloid-templated assembly of polyelectrolytes. *Angew Chem Int Ed* 1998;**37**:2201–2205.

[33] Caruso F, Caruso RA, Möhwald H. Nanoengineering of inorganic and hybrid hollow spheres by colloidal templating. *Science* 1998;**282**:1111–1114.

[34] De Cock LJ, De Koker S, De Geest BG, Grooten J, Vervaet C, Remon JP, Sukhorukov GB, Antipina MN. Polymeric multilayer capsules in drug delivery. *Angew Chem Int Ed* 2010;**49**:6954–6973.

[35] Caruso F. Nanoengineering of particle surfaces. *Adv Mater* 2001;**13**:11–22.

[36] Kharlampieva E, Kozlovskaya V, Sukhishvili SA. Layer-by-layer hydrogen-bonded polymer films: from fundamentals to applications. *Adv Mater* 2009;**21**:3053–3065.

[37] Kozlovskaya V, Kharlampieva E, Mansfield ML, Sukhishvili SA. Poly(methacrylic acid) hydrogel films and capsules: response to pH and ionic strength, and encapsulation of macromolecules. *Chem Mater* 2005;**18**:328–336.

[38] Shenoy DB, Antipov AA, Sukhorukov GB, Möhwald H. Layer-by-layer engineering of biocompatible, decomposable core–shell structures. *Biomacromolecules* 2003;**4**: 265–272.

[39] Yi Q, Wen D, Sukhorukov GB. UV-cross-linkable multilayer microcapsules made of weak polyelectrolytes. *Langmuir* 2012;**28**:10822–10829.

[40] Kim D, Kim E, Lee J, Hong S, Sung W, Lim N, Park CG, Kim K. Direct synthesis of polymer nanocapsules: self-assembly of polymer hollow spheres through irreversible covalent bond formation. *J Am Chem Soc* 2010;**132**:9908–9919.

[41] Dong R, Liu W, Hao J. Soft vesicles in the synthesis of hard materials. *Acc Chem Res* 2012;**45**:504–513.

[42] Kim D, Kim E, Kim J, Park KM, Baek K, Jung M, Ko YH, Sung W, Kim HS, Suh JH, Park CG, Na OS, Lee D-k, Lee KE, Han SS, Kim K. Direct synthesis of polymer nanocapsules with a noncovalently tailorable surface. *Angew Chem Int Ed* 2007;**46**: 3471–3474.

[43] Harada A, Ichimura S-I, Yuba E, Kono K. Hollow nanocapsules prepared through stabilization of polymer vesicles formed from head-tail type polycations by introducing cross-linkages. *Soft Matter* 2011;**7**:4629–4635.

[44] Kim E, Kim D, Jung H, Lee J, Paul S, Selvapalam N, Yang Y, Lim N, Park CG, Kim K. Facile, template-free synthesis of stimuli-responsive polymer nanocapsules for targeted drug delivery. *Angew Chem Int Ed* 2010;**49**:4405–4408.

[45] Kim E, Lee J, Kim D, Lee KE, Han SS, Lim N, Kang J, Park CG, Kim K. Solvent-responsive polymer nanocapsules with controlled permeability: encapsulation and release of a fluorescent dye by swelling and deswelling. *Chem Commun (Camb)* 2009;**12**: 1472–1474.

[46] Kim H-J, Kim T, Lee M. Responsive nanostructures from aqueous assembly of rigid–flexible block molecules. *Acc Chem Res* 2010;**44**:72–82.

[47] Matyjaszewski K, Xia J. Atom transfer radical polymerization. *Chem Rev* 2001;**101**: 2921–2990.

[48] Matyjaszewski K, Spanswick J. Controlled/living radical polymerization. *Mater Today* 2005;**8**:26–33.

[49] De KS, Hoogenboom R, De GBG. Polymeric multilayer capsules for drug delivery. *Chem Soc Rev* 2012;**41**:2867–2884.

[50] Tong W, Song X, Gao C. Layer-by-layer assembly of microcapsules and their biomedical applications. *Chem Soc Rev* 2012;**41**:6103–6124.

[51] Volodkin D, Skirtach A, Moehwald H. Bioapplications of light-sensitive polymer films and capsules assembled using the layer-by-layer technique. *Polym Int* 2012;**61**:673–679.

[52] De Geest BG, De Koker S, Sukhorukov GB, Kreft O, Parak WJ, Skirtach AG, Demeester J, De Smedt SC, Hennink WE. Polyelectrolyte microcapsules for biomedical applications. *Soft Matter* 2009;**5**:282–291.

[53] Decher G, Schlenoff JB. *Multilayer Thin Films: Sequential Assembly of Nanocomposite Materials*. 2nd ed. Wiley-VCH Verlag GmbH & Co. KGaA: Weinheim; 2012.

[54] Ariga K, Ji Q, Richards GJ, Hill JP. Soft capsules, hard capsules, and hybrid capsules. *Soft Matter* 2012;**10**:387–412.

[55] Long L-x, Yuan X-b, Chang J, Zhang Z-h, Gu M-q, Song T-t, Xing Y, Yuan X-y, Jiang S-c, Sheng J. Self-assembly of polylactic acid and cholesterol-modified dextran into hollow nanocapsules. *Carbohydr Polym* 2012;**87**:2630–2637.

[56] Dergunov SA, Miksa B, Ganus B, Lindner E, Pinkhassik E. Nanocapsules with “invisible” walls. *Chem Commun* 2010;**46**:1485–1487.

[57] Shmakov SN, Dergunov SA, Pinkhassik E. Ship-in-a-bottle entrapment of molecules in porous nanocapsules. *Chem Commun (Camb)* 2011;**47**:8223–8225.

[58] Dergunov SA, Pinkhassik E. Synergistic co-entrapment and triggered release in hollow nanocapsules with uniform nanopores. *J Am Chem Soc* 2011;**133**:19656–19659.

[59] Sukhorukov GB, Fery A, Brumen M, Mohwald H. Physical chemistry of encapsulation and release. *Phys Chem Chem Phys* 2004;**6**:4078–4089.

[60] Sukhorukov GB, Antipov AA, Voigt A, Donath E, Moehwald H. pH-controlled macromolecule encapsulation in and release from polyelectrolyte multilayer nanocapsules. *Macromol Rapid Commun* 2001;**22**:44–46.

[61] Sukhorukov GB, Brumen M, Donath E, Möhwald H. Hollow polyelectrolyte shells: exclusion of polymers and donnan equilibrium. *J Phys Chem A* 1999;**103**:6434–6440.

[62] Ibarz G, Dähne L, Donath E, Möhwald H. Smart micro- and nanocontainers for storage, transport, and release. *Adv Mater* 2001;**13**:1324–1327.

[63] Lvov Y, Antipov AA, Mamedov A, Möhwald H, Sukhorukov GB. Urease encapsulation in nanoorganized microshells. *Nano Lett* 2001;**1**:125–128.

[64] Shutava T, Prouty M, Kommireddy D, Lvov Y. pH responsive decomposable layer-by-layer nanofilms and capsules on the basis of tannic acid. *Macromolecules* 2005;**38**: 2850–2858.

[65] Kim MD, Dergunov SA, Lindner E, Pinkhassik E. Dye-loaded porous nanocapsules immobilized in a permeable polyvinyl alcohol matrix: a versatile optical sensor platform. *Anal Meth* 2012;**84**:2695–2701.

[66] Muñoz-Javier A, Kreft O, Piera Alberola A, Kirchner C, Zebli B, Susha AS, Horn E, Kempter S, Skirtach AG, Rogach A, Rädler J, Sukhorukov GB, Benoit M, Parak WJ. Combined atomic force microscopy and optical microscopy measurements as a method to investigate particle uptake by cells. *Small* 2006;**2**:394–400.

[67] Zhou Q, Dergunov SA, Zhang Y, Li X, Mu Q, Zhang Q, Jiang G, Pinkhassik E, Yan B. Safety profile and cellular uptake of biotemplated nanocapsules with nanometre-thin walls. *Nanoscale* 2011;**3**:2576–2582.

[68] De Koker S, De Geest BG, Cuvelier C, Ferdinande L, Deckers W, Hennink WE, De Smedt SC, Mertens N. *In vivo* cellular uptake, degradation, and biocompatibility of polyelectrolyte microcapsules. *Adv Funct Mater* 2007;**17**:3754–3763.

[69] An Z, Kavanoor K, Choy ML, Kaufman LJ. Polyelectrolyte microcapsule interactions with cells in two- and three-dimensional culture. *Colloid Surf B* 2009;**70**:114–123.

[70] Picart C, Schneider A, Etienne O, Mutterer J, Schaaf P, Egles C, Jessel N, Voegel JC. Controlled degradability of polysaccharide multilayer films in vitro and *in vivo*. *Adv Funct Mater* 2005;**15**:1771–1780.

[71] Etienne O, Picart C, Taddei C, Keller P, Hubsch E, Schaaf P, Voegel JC, Haikel Y, Ogier JA, Egles C. Polyelectrolyte multilayer film coating and stability at the surfaces of oral prosthesis base polymers: an in vitro and *in vivo* study. *J Dent Res* 2006;**85**:44–48.

[72] Tanner P, Baumann P, Enea R, Onaca O, Palivan C, Meier W. Polymeric vesicles: from drug carriers to nanoreactors and artificial organelles. *Acc Chem Res* 2011;**44**:1039–1049.

[73] Brož P, Benito SM, Saw C, Burger P, Heider H, Pfisterer M, Marsch S, Meier W, Hunziker P. Cell targeting by a generic receptor-targeted polymer nanocontainer platform. *J Control Release* 2005;**102**:475–488.

[74] Ghoroghchian PP, Frail PR, Susumu K, Blessington D, Brannan AK, Bates FS, Chance B, Hammer DA, Therien MJ. Near-infrared-emissive polymersomes: self-assembled soft matter for *in vivo* optical imaging. *Proc Natl Acad Sci U S A* 2005;**102**:2922–2927.

[75] Christian NA, Milone MC, Ranka SS, Li G, Frail PR, Davis KP, Bates FS, Therien MJ, Ghoroghchian PP, June CH, Hammer DA. Tat-functionalized near-infrared emissive polymersomes for dendritic cell labeling. *Bioconjug Chem* 2006;**18**:31–40.

[76] Zhou W, Meng F, Engbers GHM, Feijen J. Biodegradable polymersomes for targeted ultrasound imaging. *J Control Release* 2006;**116**:e62–e64.

[77] Ghoroghchian PP, Frail PR, Li G, Zupancich JA, Bates FS, Hammer DA, Therien MJ. Controlling bulk optical properties of emissive polymersomes through intramembranous polymer–fluorophore interactions. *Chem Mater* 2007;**19**:1309–1318.

[78] Massignani M, Canton I, Sun T, Hearnden V, MacNeil S, Blanazs A, Armes SP, Lewis A, Battaglia G. Enhanced fluorescence imaging of live cells by effective cytosolic delivery of probes. *PLoS One* 2010;**5**:e10459.

[79] Duncan TV, Ghoroghchian PP, Rubtsov IV, Hammer DA, Therien MJ. Ultrafast excited-state dynamics of nanoscale near-infrared emissive polymersomes. *J Am Chem Soc* 2008;**130**:9773–9784.

[80] Cheng Z, Tsourkas A. Paramagnetic porous polymersomes. *Langmuir* 2008;**24**:8169–8173.

[81] Susha AS, Javier AM, Parak WJ, Rogach AL. Luminescent CdTe nanocrystals as ion probes and pH sensors in aqueous solutions. *Colloid Surf A* 2006;**281**:40–43.

[82] Mueller W, Koynov K, Fischer K, Hartmann S, Pierrat S, Basché T, Maskos M. Hydrophobic shell loading of PB-b-PEO vesicles. *Macromolecules* 2008;**42**:357–361.

[83] Johnston AP, Such GK, Caruso F. Triggering release of encapsulated cargo. *Angew Chem Int Ed* 2010;**49**:2664–2666.

[84] Gustafson TP, Dergunov SA, Akers WJ, Cao Q, Magalotti S, Achilefu S, Pinkhassik E, Berezin MY. Blood triggered rapid release porous nanocapsules. *RSC Adv* 2013;**3**:5547–5555.

[85] Wang K, He Q, Yan X, Cui Y, Qi W, Duan L, Li J. Encapsulated photosensitive drugs by biodegradable microcapsules to incapacitate cancer cells. *J Mater Chem* 2007;**17**: 4018–4021.

[86] Ahmed F, Pakunlu RI, Srinivas G, Brannan A, Bates F, Klein ML, Minko T, Discher DE. Shrinkage of a rapidly growing tumor by drug-loaded polymersomes: ph-triggered release through copolymer degradation. *Mol Pharm* 2006;**3**:340–350.

[87] Ahmed F, Pakunlu RI, Brannan A, Bates F, Minko T, Discher DE. Biodegradable polymersomes loaded with both paclitaxel and doxorubicin permeate and shrink tumors, inducing apoptosis in proportion to accumulated drug. *J Control Release* 2006;**116**: 150–158.

[88] Tanisaka H, Kizaka-Kondoh S, Makino A, Tanaka S, Hiraoka M, Kimura S. Near-infrared fluorescent labeled peptosome for application to cancer imaging. *Bioconjug Chem* 2007;**19**:109–117.

[89] Sun Q, Cheng D, Yu X, Zhang Z, Dai J, Li H, Liang B, Shuai X. A pH-sensitive polymeric nanovesicle based on biodegradable poly(ethylene glycol)-b-poly(2-(diisopropylamino) ethyl aspartate) as a MRI-visible drug delivery system. *J Mater Chem* 2011;**21**: 15316–15326.

[90] Hu Y, Chen Q, Ding Y, Li R, Jiang X, Liu B. Entering and lighting up nuclei using hollow chitosan–gold hybrid nanospheres. *Adv Mater* 2009;**21**:3639–3643.

[91] Christian DA, Garbuzenko OB, Minko T, Discher DE. Polymer vesicles with a red cell-like surface charge: microvascular imaging and *in vivo* tracking with near-infrared fluorescence. *Macromol Rapid Commun* 2010;**31**:135–141.

[92] Cheng Z, Elias DR, Kamat NP, Johnston ED, Poloukhtine A, Popik V, Hammer DA, Tsourkas A. Improved tumor targeting of polymer-based nanovesicles using polymer–lipid blends. *Bioconjug Chem* 2011;**22**:2021–2029.

[93] Sanson C, Diou O, Thévenot J, Ibarboure E, Soum A, Brûlet A, Miraux S, Thiaudière E, Tan S, Brisson A, Dupuis V, Sandre O, Lecommandoux S. Doxorubicin loaded magnetic polymersomes: theranostic nanocarriers for mr imaging and magneto-chemotherapy. *ACS Nano* 2011;**5**:1122–1140.

[94] Tian X, Gill MR, Cantón I, Thomas JA, Battaglia G. Live cell luminescence imaging as a function of delivery mechanism. *Chembiochem* 2011;**12**:548–551.

5

NANOPARTICLES AS CONTRAST AGENTS FOR OPTOACOUSTIC IMAGING

ANTON V. LIOPO AND ALEXANDER A. ORAEVSKY
TomoWave Laboratories, Inc., Houston, TX, USA

5.1 INTRODUCTION

Optoacoustic (OA) imaging is a noninvasive biomedical imaging modality that combines spectral selectivity and high optical contrast based on strong absorption of laser pulses by specific molecules and nanoparticles with the high resolution of ultrasound imaging [1–3]. OA imaging is *in vivo* imaging technology [4, 5], where short light energy pulses are absorbed by tissue and converted to thermal energy. The thermal energy is subsequently converted to ultrasound signals that allow high-resolution whole-body imaging. The term "OA imaging" used in this chapter may be replaced with the broader terms of either photoacoustic (PA) or thermoacoustic imaging, meaning that optical photons can be replaced, in principle, with photons having any electromagnetic energy. Only in the optical spectral range from red to near infrared (NIR), also known as the window of relative tissue transparency [6, 7], can one achieve a useful compromise between strong resonance absorption and deep penetration of electromagnetic radiation into tissues. Hence, lasers emitting in the spectral range from approximately 650 to approximately 1700 nm are normally used for generation of OA signals (and images) due to the relatively weak absorption of biological tissues in this spectral range.

Nanotechnology for Biomedical Imaging and Diagnostics: From Nanoparticle Design to Clinical Applications, First Edition. Edited by Mikhail Y. Berezin.

OA imaging utilizes absorption of short laser pulses in tissue and effective conversion of the deposited energy into heat. Rapid deposition of thermal energy in tissues results in an immediate pressure increase, which propagates out of the heated voxels in the form of transient ultrasonic waves. OA imaging combines the most compelling properties of light and sound in a hybrid modality, simultaneously avoiding limitations of light and sound when utilized separately. While scattering of optical photons is an obstacle for resolution and sensitivity of pure optical imaging in the depth of tissue beyond 1 mm, it only helps to deliver homogenized distribution of the optical energy. Hence, every photon within the tissue is utilized in OAs. OA imaging system detects acoustic waves generated at the location of absorbed photons. Acoustic waves propagate in a well-defined manner as expanding spheres, so that the distance between acoustic sources and ultrasonic transducers (detectors) can be accurately determined as the product of propagation time and the speed of sound. Therefore, OA imaging has much higher spatial resolution and deeper imaging depth when compared with other optical imaging methods (such as those based on fluorescence or optical scattering) limited by scattering in tissues.

Currently, two OA imaging techniques are being actively developed: optoacoustic tomography (OAT) in deep tissues [8–10] and OA microscopy of thin tissue layers [11, 12]. Typically, tomography utilizes either an array of ultrasound transducers scanned around/along the tissue surface and an inverse algorithm to reconstruct cross-sectional or three-dimensional (3D) images of biological tissue [13–15]. In contrast, microscopy uses a raster-scanned single-focused ultrasonic detector coupled with confocal optical illumination and requires no reconstruction algorithm. The advantage of tomography techniques is rapid acquisition of large-volume tissue images with relatively low resolution measured in hundreds of microns, while microscopy permits high-resolution (1–50 μm) imaging of thin tissue slices with relatively small total volume. Both imaging modalities demonstrate an excellent ratio of depth to resolution (about 100–200), high sensitivity resulting in their capability to resolve the optical absorption contrast of about $\Delta\mu_a = 0.5$/cm, and the ability to quantify the signal in terms of the optical absorption coefficient, which is proportional to the concentration of chromophores.

Ultrasound imaging contrast is defined by the mechanical properties (density and speed of sound), which have no resonances in biological tissues. Optical contrast of OA imaging is defined by the spectral resonances in molecules, which can be exploited by tuning the laser illumination wavelength to a desirable resonance. Administration of exogenous contrast agents with strong resonance absorption in the spectral range of weak endogenous optical absorption creates a unique opportunity to visualize biological tissues with molecular specificity and greater contrast compared with ultrasonic imaging. Nanotechnology-based methods that further enhance the realization of this opportunity are the subject of the present review. We demonstrate that OA imaging techniques with contrast agents based on strongly absorbing nanoparticles have the capability to bring greater value and safety to biomedical diagnostics and angiography, compared with methods based on more traditional small molecule-based imaging.

All nanoparticle-based OA contrast agents can be divided in two groups: (1) nanoparticles based on endogenous chromophores and (2) nanoparticles based on exogenous chromophores. OA imaging takes advantage of the high optical contrast of hemoglobin over a broad optical spectrum in the NIR. There is no other biological molecule that possesses strong NIR absorption close to that of hemoglobin, with the exception of melanin. Melanin is a protein that is naturally present only within the very thin basal layer of the epidermis. Hemoglobin, on the other hand, is distributed throughout the entire volume of the body. Both forms of hemoglobin, oxygenated and deoxygenated, have much stronger optical absorption coefficients in the NIR ($\mu_a = 1–10\,cm^{-1}$) compared with the NIR optical absorption of other tissues. Therefore, OA imaging represents, possibly, the most powerful modality for imaging blood distribution in organs, the circulation system, and micro blood vessels including angiogenesis-related microvasculature of tumors [2, 3, 16, 17].

Administration of exogenous contrast agents *in vivo* opens another opportunity for enhancement of the OA imaging contrast in organs and tissues of interest. Standard molecular contrast agents such as NIR organic dyes have been extensively used in OA imaging of small animal models [18–21]. A variety of chromophores are absorbed in the wavelength range of NIR. However, only a few types of strongly absorbing nanoparticles can compete with endogenous contrast agents such as hemoglobin and melanin. Nanoparticles are especially attractive as OA contrast agents compared with molecules, since the signal is proportional to the total volume of the optically absorbing chromophores. The volume of a nanoparticle is significantly larger than that of a single chromophore molecule.

The two most effective types of exogenous contrast agents designed for optical and OA imaging modalities are dye-containing nanoparticles (such as indocyanine green (ICG)-doped nanoparticles) and nanoparticles that possess strong surface plasmon resonance (SPR) absorption (such as gold nanorods, hollow gold nanospheres, and nanocages). NIR exogenous nanoparticles may be more effective as contrast agents when being targeted and selectively delivered and bound with specific molecules in cells and tissues with minimal NIR absorption. However, nanoparticles passively accumulating in organs with higher metabolic activity have the advantage of a relatively simple and low-cost design. Both types, therefore, have merit and will be discussed in the following text. Advances in nanotechnology have produced significant contributions to the nanoparticle-based optical and OA contrast agents [22–30]. In this review, we will describe the OA applications of nanoparticles based on both endogenous and exogenous chromophores.

5.2 OPTOACOUSTIC NANOPARTICLES BASED ON ENDOGENOUS CHROMOPHORES

Two of the key advantages of using endogenous contrast agents for imaging applications are safety and the possibility of revealing the physiological environment, because the physiological parameters of the tissue do not change during OA image acquisition [9]. OA imaging technique is suitable for monitoring endogenous optical contrast agents such as hemoglobin and melanin [18]. These and some other endogenous molecules have much stronger light absorption properties in the NIR region than the surrounding tissue; hence, the intensity of an OA signal in biological tissue is proportional to optical energy absorption of the contrast agent [31]. Endogenous OA contrast agents are usually not targeted when used, but they can be incorporated into the matrix of nanoparticles with specified binding of molecules of interest. Examples include hemoglobin and melanin, the two most important naturally occurring contrast agents for enhanced OA imaging.

5.2.1 Hemoglobin

As an endogenous contrast agent, hemoglobin has been most widely explored for OA imaging and tomography. OA imaging with hemoglobin was shown to greatly facilitate blood dynamics related to brain research [32–34] and help with visualizing brain structure and lesions [32], delineating tumor vasculature [35], monitoring hemodynamics, measuring microvascular blood flow [36], and imaging whole body of animals *in vivo* and *ex vivo* [37, 38].

OA noninvasive visualization of the brain's vascular system could be of great importance for studying the function of the brain, diagnosing possible disorders, and providing clinically translatable insights into the progression of various human neurological diseases. The structures of the surface of a rat brain with hemoglobin as contrast agent were mapped with OA imaging (Fig. 5.1) by Wang's group from the University of Michigan, School of Medicine. The report [32] describes the functional cerebral hemodynamic changes in cortical blood vessels around the whisker-barrel cortex evaluated in response to their stimulation, as well as hyper- and hypoxia-induced cerebral hemodynamic changes. Other reports from the same research group

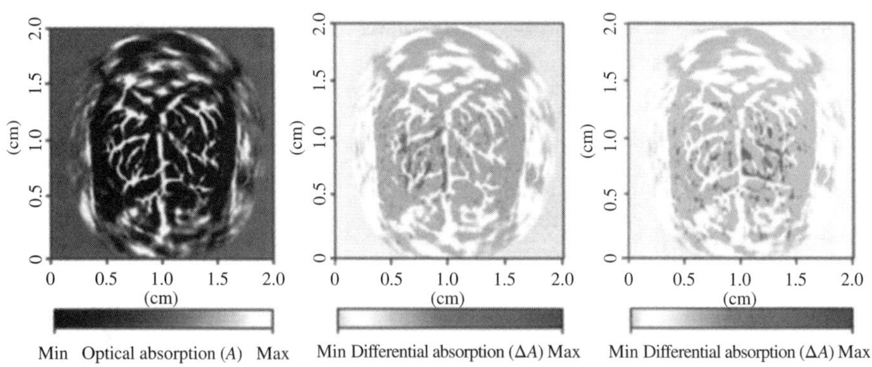

FIGURE 5.1 Nontargeted functional OA imaging of cerebral hemodynamic changes in response to whisker stimulation. Left: OA image of the vascular pattern in the superficial layer of the rat cortex acquired with the skin and the skull intact. Middle: Noninvasive functional OAT images corresponding to left-side whisker stimulation. Right: Noninvasive functional PAT images corresponding to right-side whisker stimulation. (Adapted with permission from Ref. [32]. © Macmillan Publishers Ltd.)

discussed using OA imaging techniques to observe the pattern of optical absorption in mouse brain regions [33]. These studies represented an important step for OA imaging, suggesting that this imaging modality showed promise for applications in neurophysiology, neuropathology, and neurotherapy [39]. The central neuronal system tissue contrast in addition to blood vessels also revealed other brain structures such as the cerebellum, hippocampus, and ventriculi lateralis [32, 39]. Recently, non-invasive, high-resolution imaging of mouse brain activity has been reported [40, 41]. With endogenous hemoglobin as the contrast agent, a contrast-to-noise ratio of 25 dB (decibels) was achieved [40].

Hemoglobin-based PA imaging has been shown to facilitate the monitoring of many biological processes of clinical importance. In burn recovery studies, for example, a two-wavelength OA imaging technique was used to discriminate coagulated and noncoagulated blood in a dermal burn phantom [42]. Coagulation and deoxygenation of blood could be related to OA changes, as demonstrated by TomoWave Labs [38] in the OAT visualization of the whole body of a mouse from life to postmortem (Fig. 5.2. Different optical absorption spectra of coagulated and noncoagulated blood produce OA signals of varied amplitudes, which could be utilized to differentiate blood samples. Multiwavelength OA measurements were performed to monitor the wound healing process of extensive and deep dermal burns in rats [43] suggesting that the multiwavelength approach could be useful for monitoring the changes in local hemodynamics including angiogenesis (the formation of new blood vessels in the wound) during wound healing.

The imaging of angiogenesis in cancer is another important application for hemoglobin-based OAT [13, 40]. The tumor growth is generally dependent on

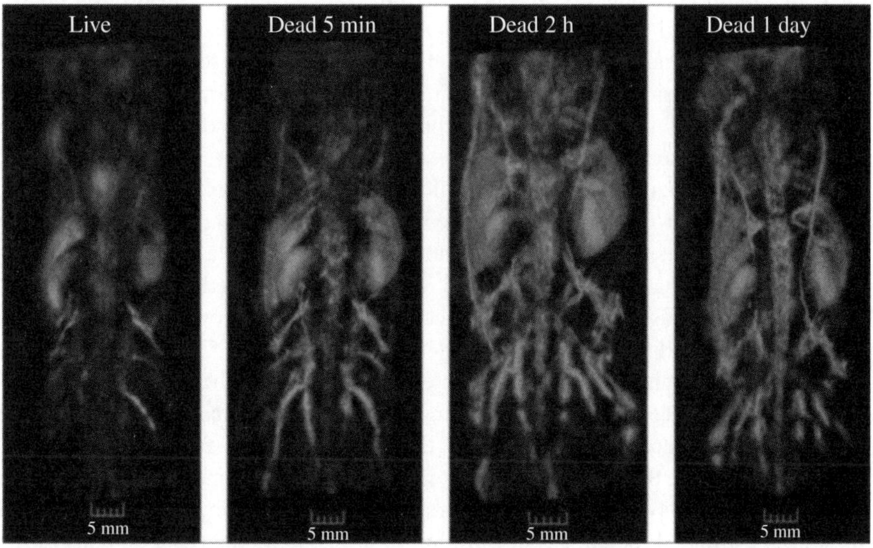

FIGURE 5.2 Postmortem changes of optoacoustic images of the mice. (Adapted with permission from Ref. [38]. © SPIE-International Society for Optical Engineers.)

the formation of new blood vessels, and noninvasive OAT of tumor angiogenesis could play a fundamental role in many aspects of preclinical research and for cancer patient management [44–46]. OA imaging capability for imaging blood vessels in highly scattering samples was demonstrated in 1998 [47]. The authors have shown that the sensitivity of this technique could reach single red blood cell detection on a glass plate. Later OA images of tumor neovascularization were obtained in dynamic state after subcutaneous inoculation of pancreatic tumor cells in rats [35]. 3D datasets were acquired to visualize the development of the blood vessels as well as to quantify the extent of individual blood vessels around the growing tumor and measure the blood concentration changes inside the tumor [35].

5.2.2 Melanin

The melanin layer serves as an optical gradient filter for sunlight, strongly absorbing ultraviolet and allowing sufficient transmission of visible light while modestly absorbing NIR due to its small thickness. In addition to hemoglobin, melanin is another primary absorber in tissue featuring a broad absorption spectrum [48, 49] that provides a strong absorption contrast for OA imaging [11]. OA imaging with melanin is mainly used for the diagnosis, prognosis, and treatment planning of melanotic melanoma (>90% of all melanomas) [50]. The combination of hemoglobin and melanin can be used for more accurate detection of melanomas, because more parameters can be measured. Dual-wavelength reflection-mode OA imaging has been developed to noninvasively obtain 3D images of subcutaneous melanomas and their surrounding vasculature in living nude mice [50]. At the two wavelengths used for hemoglobin and melanin, 584 and 764 nm, respectively, the absorption coefficients of blood and melanin-pigmented melanomas vary greatly relative to each other. *In vivo* OA imaging with a 764 nm light source was able to image the 3D melanin distribution inside the skin to a maximum thickness of 0.5 mm of melanoma. In another study, OAT was reported to be capable of detecting melanoma cells within the human circulation system [51, 52].

Several OA imaging applications are focused on tyrosinase, an enzyme responsible for melanin production in melanogenic cells, such as melanocytes. In addition to melanogenic cells, tyrosinase can be introduced into nonmelanogenic cells resulting in their pigmentation [53]. This effect was successfully used in the development of OA imaging focused on melanin-induced contrast [54]. Tyrosinase was used as a reporter gene, yielding melanin contrast for OAT *in vivo* [55]. In this study, the expression of melanin in cells was variable and depended on the efficiency of transfection, with some cells producing little melanin. Despite the fact that the extinction coefficient of melanin was lower than that of hemoglobin for the wavelengths studied, and despite the variable expression, there was enough melanin production in the transfected cells to increase concentration to detectable levels [55]. Figure 5.3 shows the estimated melanin concentration in a mouse model (ear) demonstrating that tyrosinase-derived melanin sufficiently increases the contrast for visualization of normally nonmelanogenic *in vivo* [55].

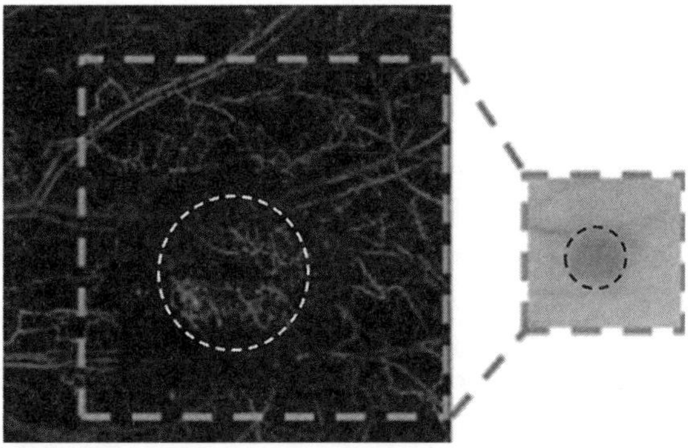

FIGURE 5.3 Optoacoustic image of a mouse ear showing post injection formation of tyrosinase-derived melanin on the background of hemoglobin in the tissue microvasculture. Inset shows a photograph of the OA image area outlined with a gray dashed square. Melanin production area is outlined with dashed circles on both, the OA image and the photograph. (Adapted with permission from Ref. [55] SPIE-International Society for Optical Engineers).

In addition to imaging and diagnostics, melanin was shown to be a suitable target for laser-induced thermotherapy and enhanced oncolytic viral therapy. Regarding the latter, Stritzker *et al.* [54] provided evidence that gene-evoked production of melanin could be used as a therapeutic and diagnostic mediator, as exemplified by insertion of only one or two genes into the genome of an oncolytic vaccinia virus strain. Such modified viruses resulted in a new class of theranostic agents. Oncolytic viruses are characterized by their largely tumor cell-specific replication, resulting in tumor cell lysis and efficient tumor regression. Tumor selectivity is an inherent property of viruses such as Newcastle disease virus, parvovirus, and vaccinia virus [54]. These could be further enhanced, for example, by deletion of viral genes, which are dispensable in malignant cells, or by production of surface receptors that bind ligands that are highly expressed on tumor cells [56]. The oncolytic vaccinia virus strain GLV-1h68 serves as the backbone for the insertion of key genes in melanogenesis genes into tumor cells. Being inserted, the genes might be rendered "visible" as a result of melanin overproduction by optical as well as OA imaging and might facilitate therapy and use the produced melanin as an additional theranostic mediator [55].

The oncolytic virus-mediated production of melanin and its optical absorption in the NIR enable the optical (and OA) imaging of tumors and metastases. It was found [54] that pellets of cells infected with various melanin-rVACV strains showed OA signals in the NIR range that were equivalent to optical absorption up to $2\,cm^{-1}$. A significant increase in melanin production in a cell culture is related to the darker color of eumelanin in cells and tissues and, as a result, serves as a better OA contrast. These values are an order of magnitude above the $0.3–0.5\,cm^{-1}$ tissue optical absorption in NIR and are comparable with the $2\,cm^{-1}$ absorption of whole blood, thus ensuring sufficient signal-to-background ratio. In combination with multispectral optoacoustic tomographic (MSOT) imaging of melanin-containing tumors, the authors of Reference [54] were able to generate an image-processing algorithm that allowed to visualize the presence and distribution of melanin not only on the surface but also deep within the mouse body. With this technique, near-surface A549 and PC-3 tumors, as well as PC-3-RFP lymph node metastases (Fig. 5.4), which, in a recent study, have been shown to be preferentially colonized versus the primary tumor [57], could be detected in live mice injected with melanin-rVACV.

Compared with fluorescent proteins (see following text), melanin can certainly not be used for subcellular protein distribution and interaction analysis, but it has the advantage of being visible via noninvasive deep tissue imaging with OAT imaging. The detection of fluorescent proteins is probably more sensitive than melanin in terms of generated signal per molecule, but melanin is produced enzymatically, which allows relatively fast detection (about 18 h post viral cell infection) with increasing intensity over time [54]. Another advantage of melanin versus fluorescent proteins is the extreme stability of melanin [58]. These data have shown that gene-evoked melanin production has a strong potential to become widely used in surgery and endoscopy with OA imaging. Finally, the ubiquitous presence of melanin in all kingdoms of life suggests that introduction of melanin synthesis as a diagnostic/ theranostic marker will be possible in most species [59] and should also translate into clinical settings [54]. However, the process of melanin production in transfected cells is thought to be toxic [60]. Future studies must investigate different ways of melanin delivery and focus on *in vivo* applications of this technology.

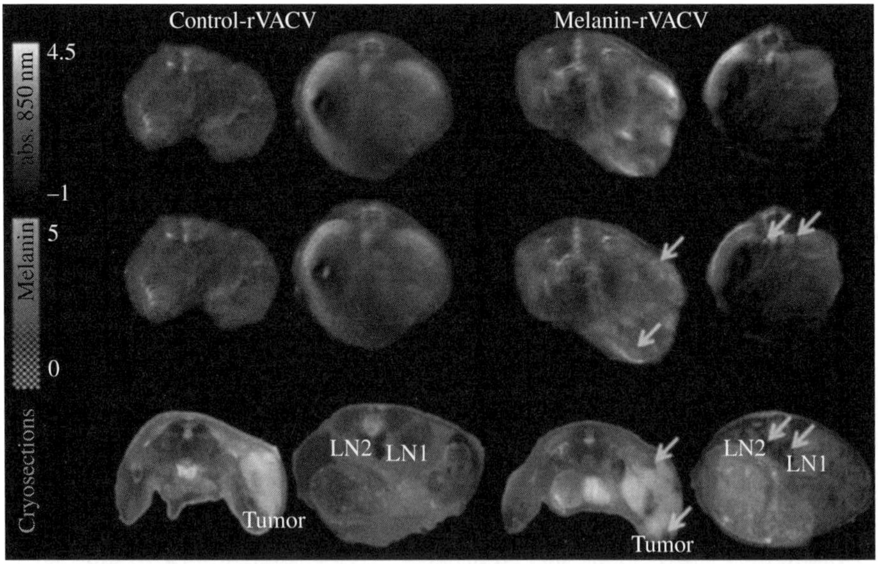

FIGURE 5.4 Optoacoustic imaging of melanin-rVACV in live mice. Multispectral optoacoustic tomography (MSOT) of PC-3-RFP tumor and lymph node metastasis (LN1 and LN2) bearing live mice 14 days after rVACV injection. (Top) Background optoacoustic images taken at 850 nm excitation wavelength. (Middle) Melanin spatial distribution (pseudocolor) is overlaid on the background images using a variable transparencies function (Bottom) As indicated by the *arrows*. (With permission from Ref. [54]. © PNAS.) (*See insert for color representation of the figure.*)

5.2.3 Fluorescent Proteins and Enzymatic Approach

Fluorescent proteins are widely used in cell biology and in the study of transgenic animals, because they can be genetically targeted to a specific molecule of interest [61]. One major advantage of using fluorescent proteins as the contrast agent for OA imaging is that they overcome the limitations of conventional fluorescence microscopy and allow imaging of gene expression in much deeper tissues [62]. It has been reported that MSOT can achieve tissue penetration of several millimeters (potentially centimeters) with a resolution of 20–100 μm, which remains constant as a function of depth and depends only on the ultrasonic detector characteristics [20]. This technique was capable of visualizing fluorescent proteins in small living organisms such as the zebra fish, *Drosophila melanogaster* pupae, and adult zebra fish for high-resolution morphological and functional observations (Fig. 5.5).

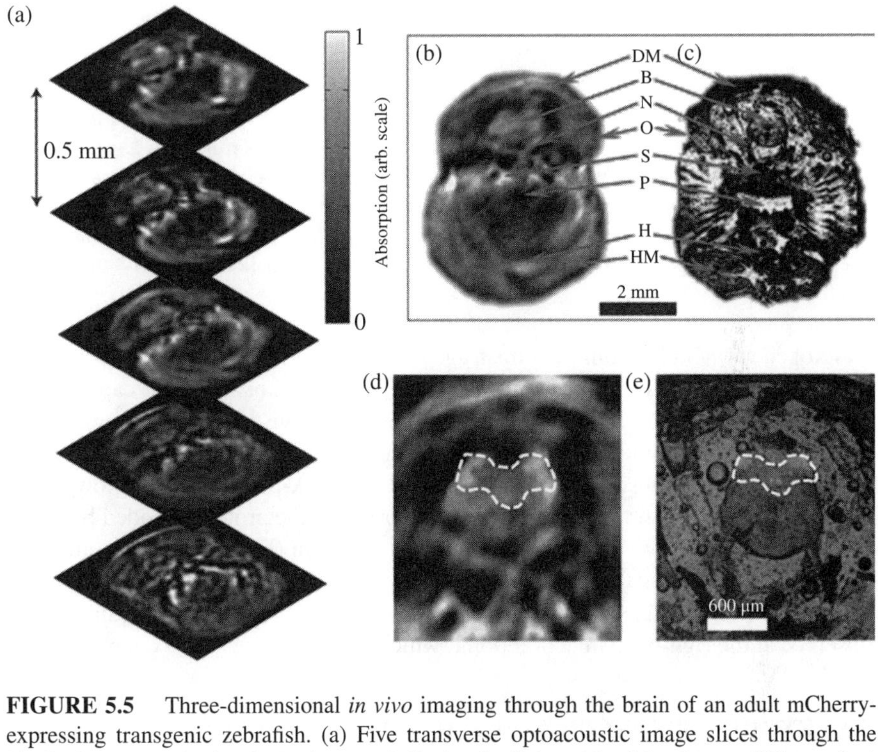

FIGURE 5.5 Three-dimensional *in vivo* imaging through the brain of an adult mCherry-expressing transgenic zebrafish. (a) Five transverse optoacoustic image slices through the hindbrain area at the level of crista cerebellaris of a living zebrafish taken at 585 nm. (b,c) Example of an imaged slice (b) and its corresponding histological section (c). DM, dorsal fin musculature; B, hindbrain; N, lateral line nerve; O, operculum; S, skull bones; P, pharynx; H, heart; HM, hypobranchial musculature. (d) MSOT image of the brain with mCherry expression shown in the area outlined with a dashed line; and (e) Corresponding epi-fluorescent histology of the fish showing the same area of mCherry expression. (Adapted with permission from Ref.[20]. Nature Publishing Group).

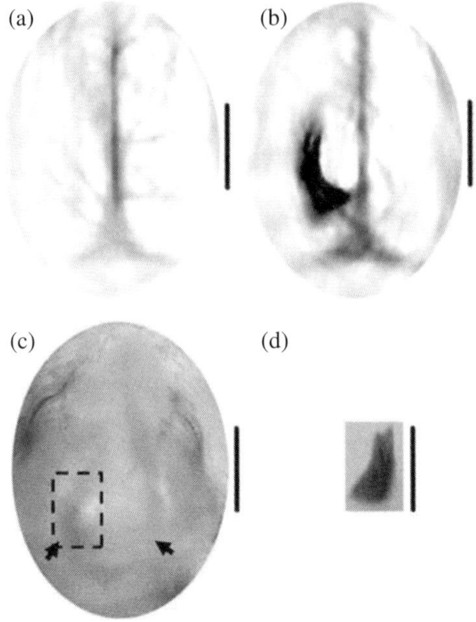

FIGURE 5.6 Optoacoustic images at 650 nm before and after the injection of X-gal (planes *a* and *b*, respectively). Plane *c* shows the photograph of the rat head following X-gal injection and just prior to imaging. The two arrows indicate the position of injection of X-gal. Plane *d* shows the underside of the dissected scalp in the rectangular region indicated in plane *c*, showing the X-gal stained tumor, which is dark blue in color (*d*). The line scale represents 5 mm. (Reproduced with permission from Ref. [63]. © SPIE-International Society for Optical Engineers.)

Exogenous nonplasmonic contrast agents can be used for enhancement of OA imaging with gene expression. The study [63] first demonstrated the capability of OAT for reporter gene imaging. Rats inoculated with 9L/lacZ gliosarcoma tumor cells were imaged with OAT before and after injection of X-gal, an optically transparent lactose-like substrate, which yields a stable dark-blue product following the cleavage of the glycosidic linkage by β-galactosidase (the protein encoded by lacZ). Figure 5.6 shows the OA image prior to X-gal injection (Fig. 5.6), and the tumor is clearly seen after X-gal administration (Fig. 5.6). A spatial resolution of around 400 μm, as well as a sensitivity of 0.5 μM, was achieved. No signal upregulation was observed at the right-side injection point, which served as a negative control.

5.3 EXOGENOUS NONPLASMONIC CONTRAST AGENTS FOR OPTOACOUSTIC IMAGING

Although the progress made to date by OAT and imaging with endogenous markers such as hemoglobin and melanin has been significant, the use of exogenous contrast agents has further increased the benefits of this imaging modality by enhancing the image contrast and quality [39], particularly in cancer research. Exogenous contrast agents are classified as nonplasmonic and as nanoparticles with SPR. Both classes could be used as nontargeted and targeted for a variety of OA techniques including microscopy, tomography, single wavelength, and multispectral imaging. Commonly used nonplasmonic contrast agents for PA imaging include dyes, such as ICG, various carbon nanotubes (CNTs), quantum dots, and noble metal nanoconstructs, generally gold nanoparticles.

5.3.1 Dye-Encapsulated Nanoparticles: ICG, Viruses, QDs, etc.

ICG, the cyanine dye, with peak absorption at approximately 800 nm has been successfully used as a diagnostic aid for measuring blood volume, cardiac output, and hepatic function. It is the only FDA-approved NIR chromophore. ICG (molecular weight of its sodium salt, $M_W \sim 775$ Da) is an excellent NIR light absorber with reported values of its molar extinction coefficient in aqueous media ranging between 8×10^4 and $1.2 \times 10^5\,M^{-1}\,cm^{-1}$ at 780 nm [64], which corresponds to absorption cross section for 780 nm, of about 3×10^{-16} to $4.6 \times 10^{-16}\,cm^2$ [65, 66].

Noninvasive OA angiography with ICG has been reported in animal brains [67]. When the dye is injected into the circulation system of a rat, ICG significantly enhances the absorption contrast between the blood vessels and the background tissues. Because NIR light can penetrate deep into brain tissues through the skin and the skull, the vascular distribution in the rat brain can be reconstructed from the OA signals with high spatial resolution and low background. It has been reported that OAT could image objects embedded at depths of up to several centimeters at a resolution of less than 1000 μm with a sensitivity of less than 10 pM of ICG in the blood [68]. Resolution of OA imaging has deteriorated slowly with increasing imaging depth for this contrast agent.

A free dye ICG has a half-life of only a few minutes in the blood stream. Dye encapsulation in nanoparticles significantly increases its circulation time [69]. Nanoparticles (~100 nm in diameter) containing ICG have been developed as a contrast agent for OA imaging using organically modified silicate (ormosil) as a matrix [69]. Encapsulation of the dye in the silicate matrix provided a number of benefits for OA. First, the nanoparticle surface can be engineered for incorporating a targeting moiety for selective contrast enhancement. Second, superior contrast is achieved because of the high dye concentration in the nanoparticle. For example, efficient encapsulation of ICG with up to 23,000 dye molecules entrapped in a single nanoparticle has been reported [69]. Finally, encapsulation in a nanoparticle stabilizes ICG dye molecules against the effects of an aqueous medium and other destabilizing effects from the biological environment.

In addition to the silica (and polymeric shells), viruses have also been used to encapsulate ICG [66]. In comparison to animal viruses, plant viruses are easier to produce and purify, and they are more chemically and structurally stable [70–72]. The surface of the viruses is amenable to modification, for example, surface conjugation of the cowpea mosaic virus (CPMV) with a folic acid–polyethylene glycol (PEG) moiety has been achieved for *in vitro* recognition of tumor cells bearing a folate receptor [73].

OA imaging using a contrast agent composed of a plant virus protein shell that encapsulates ICG has been recently demonstrated [66]. This nanoconstruct is comprised of a protein shell, purified from the plant-infecting brome mosaic virus (BMV) and encapsulated dye. These viruse-based nanoconstructs as optical viral ghosts (OVGs) since the genomic content of the wild-type BMV is eliminated and replaced with ICG, while the capsid protein (CP) subunits remain as the encapsulating shell [66]. These nanoconstructs were shown to provide higher PA signals than blood in tissue phantoms and display superior photostability compared to the nonencapsulated dye.

Besides ICG-based nanoparticles, other NIR-active compounds such quantum dots and dyes have been investigated. Quantum dots can convert the absorbed energy (up to 60%) into heat and then at certain condition into sound [74], which is sufficient for detection of quantum dots using thermal or ultrasound transducers. Gao *et al.* [75] summarized recent developments in the field of stem cell tracking with optically active nanoparticles including quantum dots for the OA and fluorescence imaging. Coregistration of the OAT and fluorescence images enabled simultaneous visualization of the tumor location, angiogenesis, and brain structure. The fluorescent properties of quantum dots-based contrast agents were used for OA imaging [74, 76]. Recently, Prussian blue nanoparticles have been described as a novel contrast agent for enhancing OAT *in vitro* and *in vivo* [77]. These Prussian blue nanoparticles of approximately 40 nm are composed of iron-based pigment $Fe^{III}_4[Fe^{II}(CN)_6]_3 \cdot nH_2O$ capped with citric acid and feature strong absorption NIR photons.

5.3.2 Carbon Nanotubes

Noninvasive detection of various molecular markers of diseases can allow for much earlier diagnosis, control for treatment, and better prognosis that will eventually lead to personalized medicine [78–80]. Numerous biomedical applications of single-walled and multiwalled carbon nanotubes (SWNTs and MWNTs, respectively, or CNTs in general) have been demonstrated over the years in this regard [26, 79, 81–83]. Many modifications of SWNTs have raised significant interest over the past decades for their unique physical and optical properties for OAT and imaging. Owing to their ability to absorb light at a wide range of wavelengths spanning the ultraviolet, visible, and NIR spectral ranges, CNTs are natural contrast agents for OA and photothermal (PT) technique. When compared with blood in phantom studies, SWNTs were found to show significant signal enhancement for OAT at 1064 nm excitation [26]. Recently, noninvasive SWNT-enhanced OA identification of sentinel lymph nodes (SLNs) in a rat model was reported [84]. In this study, SLNs were successfully imaged *in vivo* with high contrast-to-noise ratio (>80) and good resolution (~500 μm). Because SWNTs have optical absorption over a wide excitation wavelength range, the imaging depth could be maximized by varying the incident light wavelength to the NIR, where biological tissues (e.g., hemoglobin, tissue pigments, lipids, and water) have low light absorption. Targeted SWNTs have frequently been used for molecular OA imaging. Integrin αvβ3 and arginine–glycine–aspartic acid (or RGD peptide), a potent integrin αvβ3 antagonist, are among the most extensively studied and validated receptor-ligand pairs used in conjunctions with SWNTs [39, 85].

The first demonstration of CNTs as OA contrast agents was performed by Zharov *et al.* in 2007 [79] through *in vitro* and *in vivo* detection of circulating *Staphylococcus aureus* and *Escherichia coli* labeled with CNTs. The detection was performed in blood flow using *in vivo* PA and PT flow cytometry (PAFC/PTFC) system [86]. In 2008, de la Zerda *et al.* [87] demonstrated the first *in vivo* PA imaging of CNTs by molecularly targeting SWNTs to tumor neovasculature using arginine–glycine–aspartic acid (RGD) peptide in living mice. Since then, multiple groups have proposed various chemical modifications to CNTs that enhance their PA signals, as well as demonstrating new applications of the PA technique using CNTs. Chemical modifications included coating of CNTs with organic optical dyes [88], gold with folates, antibodies for molecular targeting of circulating tumor cells (CTCs), and primary tumors [89] as well as endothelial lymphatic LYVE-1 receptors [83, 90]. CNTs were also suggested as potential contrast agents for SLN imaging [87, 91].

SWNTs conjugated with cyclic RGD peptides were used as a contrast agent for OA imaging of tumors in living mice [87]. In this study, mice were injected subcutaneously with SWNTs at increasing concentrations from 50 to 600 nM (Fig. 5.7. The sensitivity in living mice was found to be 50 nM (i.e., 50 nM of SWNT-RGD gave the equivalent PA signal as the background tissue signal).

Mice bearing U87 glioblastoma xenograft tumors injected with SWNT-RGD showed significantly higher PA signal in the tumor as compared with untargeted SWNTs (Fig. 5.8. The high OA signal in the mouse injected with plain SWNTs (indicated by the white arrow) is not seen in the subtraction image, suggesting that it is due to a large blood vessel and not SWNTs. However, the minimum detectable

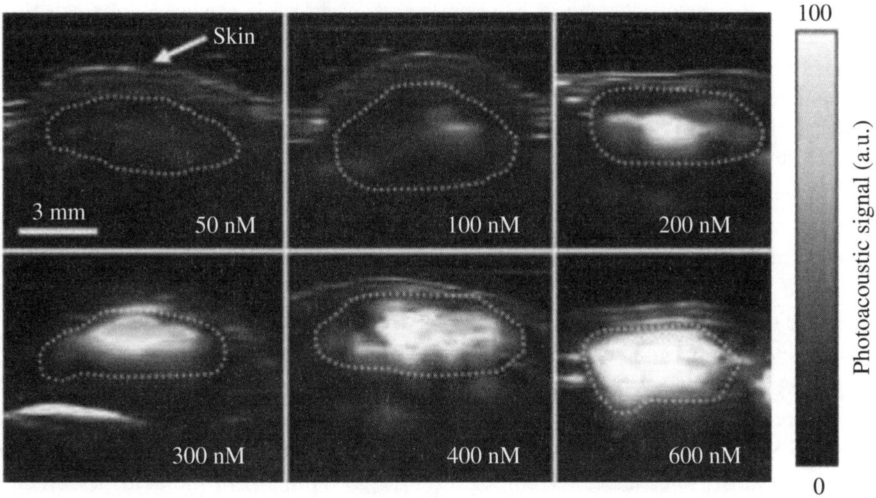

FIGURE 5.7 Optoacoustic detection of single-wall carbon nanotubes (SWNTs) in living mice. Mice were injected subcutaneously with single-wall carbon nanotubes at concentrations of 50–600 nM. The dotted lines on the images illustrate the edges of each inclusion. A gray scale ultrasound B-mode image showing the skin level was overlaid with the OA image (*white*). (Reproduced with permission from Ref. [87]. © Nature Publishing Group.)

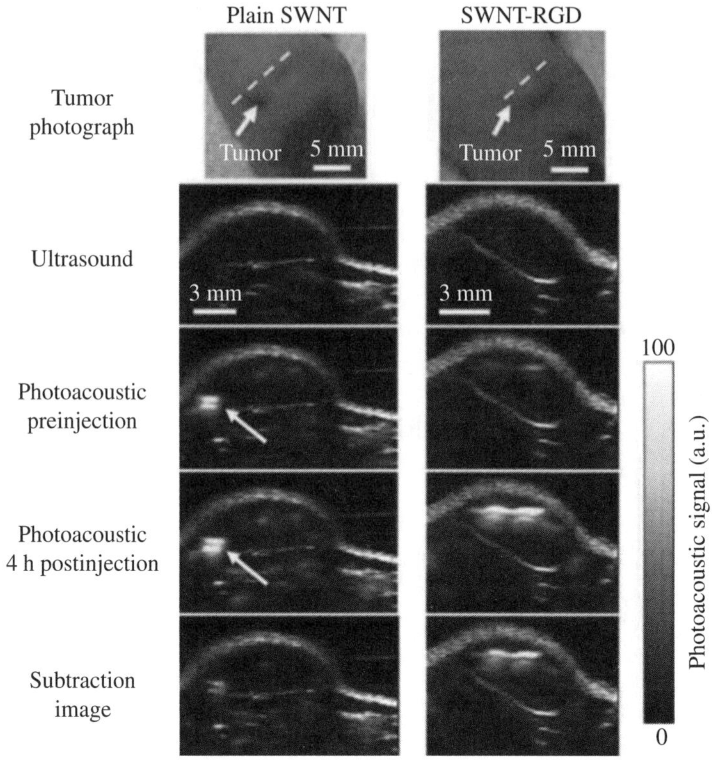

FIGURE 5.8 Tumor targeting in living mice with nanoparticle conjugates of single-wall carbon nanotube and arginine–glycine–aspartic acid (SWNT-RGD). The dotted lines on the images illustrate the edges of each inclusion. A gray scale ultrasound B-mode image showing the skin level was overlaid with the OA image (*white*). The high optoacoustic signal in the mouse injected with plain SWNTs (white arrows) is not seen in the subtraction image, suggesting that it is due to a large blood vessel and not SWNT. (Reproduced with permission from Ref. [87]. © Nature Publishing Group.)

concentration of SWNTs was likely less than that because the PA images were acquired before and after the administration of the contrast agent, which makes it possible to separate the signal of the SWNT from the background.

Targeted SWNTs can be used for both OAT imaging and PA therapy. Recently, the large OA effect of SWNTs was explored for targeting and selective destruction of cancer cells [92]. Under the irradiation of a 1064 nm Q-switched millisecond pulsed laser, SWNTs showed a large PA effect in suspension, which could trigger an explosion at the nanoscale. By conjugating the SWNTs with folic acid, which can bind to cancer cells overexpressing folate receptors on the cell membrane, the laser power used for cancer cell killing could be reduced 150–1500 times. This discovery has new perspectives for exploring the PA properties of SWNTs in cancer therapy.

de la Zerda *et al.* summarized the vast applications of advanced nanoparticles with a focus on CNT contrast agents for OA imaging, cytometry, and theranostics applications with a PT effect [81]. The authors also stated that incubation of SWNTs with ICG, a strong and stable binding of ICG molecules to the surface of the SWNT occurs, leads to a approximately 20-fold increase in the optical absorption of the SWNT [81]. While small optical dyes such as ICG have lower optical absorption absorptivity than SWNTs, they are also significantly smaller in size. The absorption coefficient of ICG divided by its molecular weight, however, is higher than that of SWNTs and significantly greater than that of gold-based nanoparticles. Thus, while ICG exhibits 30 times and 5×10^3 times lower optical absorbance than SWNTs and gold nanorods, respectively, it is also 220 times and 45×10^6 times lighter than SWNTs and nanorods. Hence, on an absorbance-per-weight measure, ICG is approximately 7 times more efficient than SWNT and approximately 9000 times more efficient than nanorods at absorbing light [81]. It should be noted that if volume of nanoparticle is considered instead of molecular weight, gold nanorods will be more efficient in absorbing than SWNTs.

SWNTs conjugated with RGD peptide through phospholipid–polyethylene glycol–amine (PL-PEG-NH_2) linker with long (5400 Da) PEG chains were also demonstrated to target $\alpha v\beta 3$ in tumors, a vascular integrin associated with tumor angiogenesis [82]. This included testing the PA signal stability when introduced to serum, cell uptake and photobleaching, and characterizing the signal–concentration curve. In another report [88], the sensitivity of ICG-modified SWNT-RGD in living mice has been estimated to be 170 pM. This represents a significant improvement in sensitivity of over 300-fold as compared with plain SWNT-RGD. This improvement in sensitivity was primarily due to the higher optical absorption induced by ICG, as well as the ability to shift the laser excitation wavelength from 690 to 780 nm (the absorption spectrum maximum of SWNT-ICG-RGD), at which the background tissue absorption is significantly reduced.

The combination of CNTs and their conjugates is well suited for theranostic applications and can be exploited for diagnostic purposes with OA imaging and drug delivery in cancer treatment [93, 94]. Control for local concentration and biodistribution of SWNT in these conditions is important, and the nanoparticle concentrations could be monitored with a 3D OAT *in vivo*, as demonstrated in Figure 5.9 by the

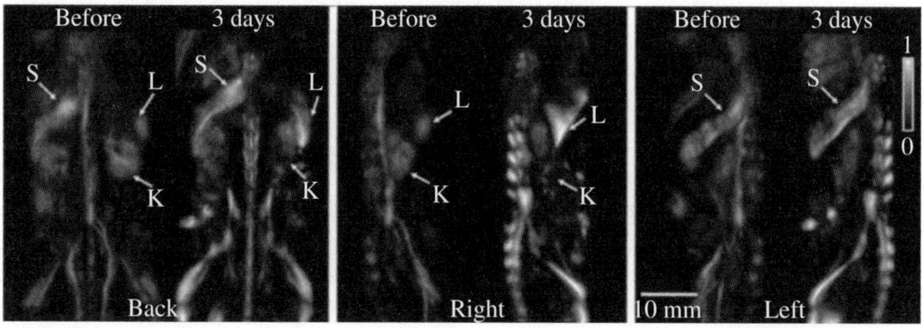

FIGURE 5.9 Three-dimensional OA images of a mouse before and 3 days after SWNT injection. The grayscale intensity is the same for all images. Dorsoventral, left mediolateral, and right mediolateral views are shown. S, L, and K show locations of the spleen, liver, and kidney, respectively. (Adopted with permission from Ref. [95]. © SPIE-International Society for Optical Engineers.)

TomoWave Labs research group [14, 95]. This figure shows OA images of a mouse before and after SWNT injection. A localized increase in OA image brightness generally implies an increase in the value at that location. OA signals from the organs were correlated with the measured concentrations of SWNTs accumulated in these organs as measured spectroscopically *ex vivo* in the harvested organs using the intrinsic fluorescence of SWNTs. The methodology presented in this report can further be extended to calibrate the sensitivity of an OA imaging system for a range of changes in optical absorption coefficient values at specific locations in a mouse body to enable noninvasive measurements of nanoparticle concentrations *in vivo*.

Recently, a novel class of hybrid CNTs (multicomponent nanoconstructs with several types of nanoparticles, such as nanotubes and gold nanoparticles) was shown to be more biocompatible and less toxic with better biodistribution and pharmacokinetic properties than individual nanocomponents. These hybrid constructs were shown to permit more efficient and versatile bioconjugation with targeting agents thus making these systems attractive for nanomedicine-related theranostics [81].

5.4 PLASMONIC NANOPARTICLES AS OPTOACOUSTIC CONTRAST AGENTS

5.4.1 Types and Properties of Plasmonic Nanoparticles

Surface plasmon nanoparticles (SPR), or plasmonic nanoparticles, with many different shapes have been designed, including nanospheres, gold nanorods, nanotubule, bipyramids, ribbons, nanorice, nanocubes, triangular gold nanoplates, hollow nanoshells, nanobowl multipods, gold nanostars, lumps, gold nanoshuttles, gold tetrahedra, octahedral, cubooctahedra, and polyhedral, branched, and hollow/porous nanowires and nanodisks [96–98]. Figure 5.10 presents typical examples of plasmonic nanoparticles. Recently, in relation to OA imaging, different alloys, plasmonic nanoparticles, and/or their compositions with other materials for manufacturing multilayered and multimaterial nanostructures, such as gold nanobeacons [29] and other composites [99], have been applied.

Despite the large number of shapes of plasmonic nanoparticles, little is published on the correlation of shapes and their utility in OA imaging. Plasmonic nanoparticles for *in vivo* applications need to be engineered with absorption peaks ranging from 600 to 900 nm [19]. As mentioned earlier, this spectral region is also known as the "optical window," where absorption is minimal and optical penetration is maximal [100]. The main plasmonic nanoparticles for OAT and imaging are gold nanoshells, gold nanorods, gold nanocubes, and hollow gold nanoshells [19, 101–104]. In an interesting example, antibody-conjugated, epidermal growth factor receptor (EGFR)-targeted spherical gold nanoparticles undergo molecular-specific aggregation when they bind to EGFR on the cell surface, thereby leading to a red shift in their plasmon resonance frequency leading to an enhanced PA imaging [80]. Many of these gold nanoparticles are targeted and used mainly for molecular imaging of cancer [105–108] due to high sensitivity of gold contrast in OA imaging.

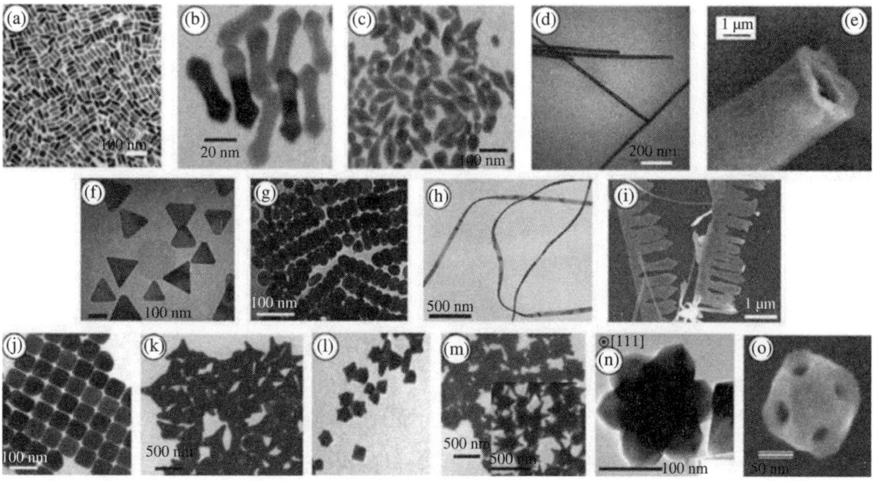

FIGURE 5.10 Representative examples of nonspherical 1D, 2D, and 3D nanoparticles (NPs) synthesized by the colloid-chemical synthetic approach. The first row shows the TEM (a–d) or SEM (e) images of NPs, where major growth has taken place in 1D: (a) nanorods, (b) nanoshuttles, (c) nanobipyramids of Au/Ag, (d) silver nanowires, and (e) Au nanotubule. The middle row shows some examples of 2D NPs: TEM images of (f) triangular Au nanoplate, (g) silver nanodisk, and (h) Au ribbons; (i) SEM image of Au belts. The bottom row shows TEM or SEM images of some typical 3D (such as polyhedral, branched, and hollow/porous) NPs: (j) nanocubes, (k) tetrapods, and (l, m) star-shaped Au particles; (n) Ag nanohexapods and (o) Au–Ag nanocage. (Reproduced with permission from Ref. [98]. © Wiley-VCH Verlag GmbH & Co. KGaA.)

For example, in the monitoring of antitumor necrosis factor (anti-TNF) *ex vivo*, gold nanorods with a concentration down to 1 pM in phantoms and 10 pM in biological tissues have been imaged with good signal-to-noise ratio and high spatial resolution [109].

In OA imaging techniques, the signal from a PA effect is proportional to optical absorption [110]. Thus, high optical absorption of the contrast agent is a very critical requirement for OA imaging. Hu *et al.* compared absorption and scattering spectra calculated for a variety of nanoparticles [103]. The presented review data revealed that gold nanocages and nanorods have much larger absorption and scattering cross sections than gold nanoshells: that is, $\boldsymbol{C}_{\mathbf{rod}} \approx \boldsymbol{C}_{\mathbf{cage}} > \boldsymbol{C}_{\mathbf{shell}}$ (Fig. 5.11 upper part), making gold nanocages and nanorods better candidates for OA imaging and PT therapy. It is also well known that by changing the shape of nanoparticles to that of elongated rods, the optical characteristics of the gold nanoparticles can be significantly changed leading to the absorption at longer wavelengths [111]. The calculated absorption and scattering of the surface plasmon band of gold nanorods is shown in Figure 5.11 (middle and bottom parts, respectively). The spectra demonstrate that the plasmon maximum of the nanorods (corresponding to the mode with the electric field parallel to the long axis of the nanorod) lies in the desirable NIR region, thus making gold nanorods potentially useful for *in vivo* applications.

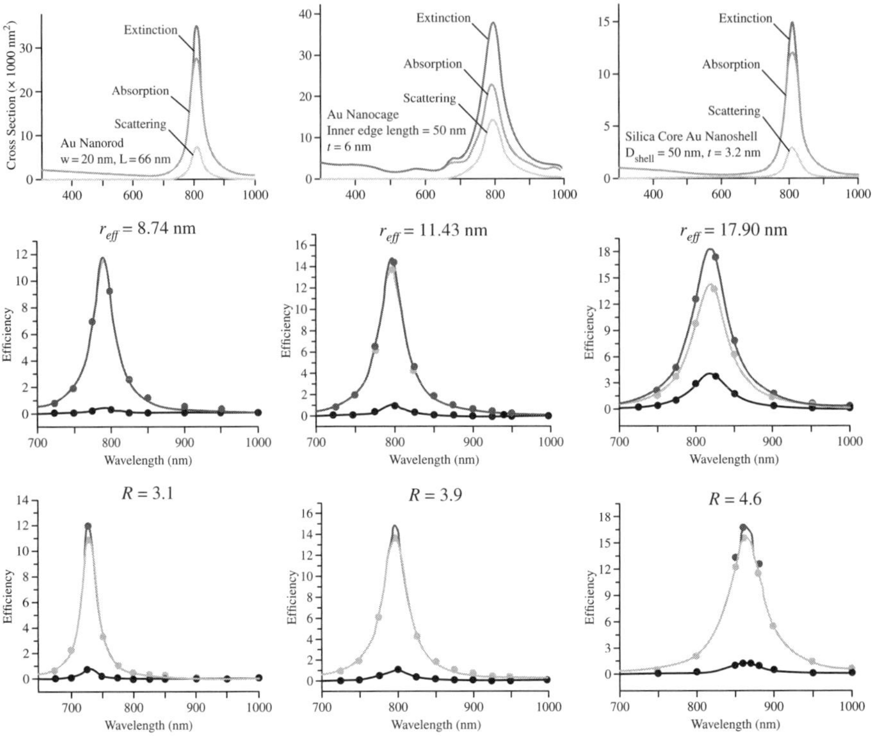

FIGURE 5.11 Upper Row: Comparison of the optical spectra calculated using discrete dipole approximation (DDA) for gold nanorods (GNR), nanocages (GNC), and nanoshells (GNS). For all the structures, the surface plasmon resonance peak was tuned to exactly 800 nm. The GNR has dimensions 20 nm × 66 nm. The GNC has an inner edge length of 50 nm and wall thickness of 6 nm. The GNS has a silica core of 50 nm in diameter and shell thickness of 3.2 nm. (Reproduced with permission from Ref. [27]. © John Wiley & Sons, Inc.) (Middle and Bottom Rows): Calculated spectra of the optical absorption efficiency, Q_{abs} (*light gray*), scattering Q_{sca} (*black*), and extinction Q_{ext} (*dark gray*) for GNR with fixed aspect ratio (R) = 3.9 and different effective radius (r_{eff}) 8.74, 11.43, and 17.90 nm (*middle*), and with fixed r_{eff} = 11.43 nm and different (R) 3.1, 3.9, and 4.6 (bottom). (Reproduced from [111] .with permission from American Chemical Society).

5.4.2 Gold Nanoshells

Nanoshells for biomedical applications usually consist of a dielectric core coated by a conductive, nanometer-thick metallic shell, typically gold [112]. Core size and/or shell thickness can shift the SPR that allows nanoparticle optical properties to be tuned over the visible and NIR spectral regions [113, 114]. Nanoshells are optically robust, exhibiting resistance photobleaching and chemical/thermal stability to denaturation [115]. Nanoshells comprised of gold offer a straightforward means to chemically modify surface characteristics by grafting on functional groups ranging from PEGs to integrins [27, 116]. Gold nanoshells were first adopted for OA imaging around a decade ago [117, 118]. The initial application of nanoshells as an OA contrast agent was to increase the NIR contrast of cortical blood vessels in rat brain

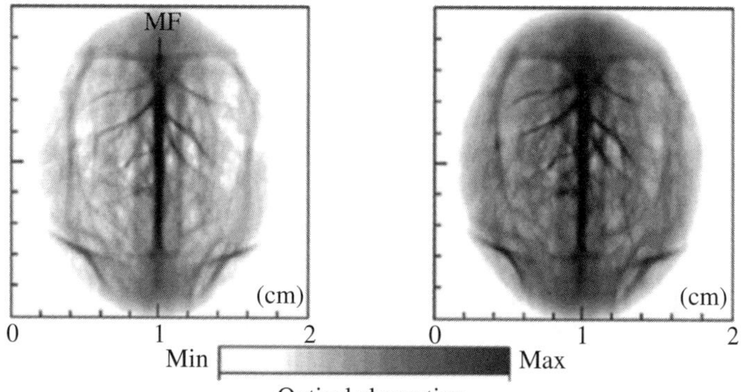

FIGURE 5.12 Noninvasive *in vivo* OA image of a rat cortical vasculature acquired using 800 nm irradiation before (left) administration of nanoshells (MF: median fissure) and after (right) approximately 20 min postinjection of the third dose of nanoshells. (Reproduced with permission from Ref. [117]. © American Chemical Society.)

in a living animal [33]. For that, PEGylated gold nanoshells (half-life is nearly 4 h in blood) with a diameter of approximately 135 nm and peak absorption of approximately 800 nm were injected into the tail vein of Sprague–Dawley rats. Cortical vasculature was noninvasively scanned using PA tomography following nanoshell injections (three in total). Successive OA image acquisition was used to quantify circulation dynamics, including clearance, in the cortical blood stream (Fig. 5.12. This work illustrated the feasibility of using nanoshells *in vivo* as an NIR contrast agent in OA imaging, with particular implications in enhancing the sensitivity of cortical blood volume measurements in noninvasive studies of small animal brain function [33].

5.4.3 Gold Nanorods

Gold nanorods represent another type of gold nanoparticles whose absorption mechanism is based on localized SPR. Unlike nanoshells, nanorods exhibit cylindrical symmetry where small changes in particle symmetry can significantly alter SPR characteristics [27, 104, 119–121]. The first proposal to use the relation between ellipsoidal and cylindrical plasmonic nanorods and OA imaging of different modalities was described by Oraevsky *et al.* in the beginning of the 2000s [119, 120, 122, 123]. Shortly after, nanorods were proposed for OA technique for flow measurements with the use of laser-induced shape transitions [124]. A series of studies were performed to investigate the shape dependence of the optical absorption of gold nanorods, as well as the shape transition induced by pulsed laser irradiation [125]. It was found that photon-induced shape transition of nanorods involves mainly a rod-to-sphere conversion and a shift in the peak optical absorption wavelength. The application of laser pulses induces shape changes in gold nanorods as they flow through a region of interest, and the quantitative flow information can be derived based on the PA signals (measured as a function of time) from the irradiated nanorods.

Since then, OA imaging experiments have been conducted using gold nanorods in a variety of phantoms, cells, and *in vivo* models of disease. Agarwal *et al.* [126] imaged gel beads dispersed with gold nanorods implanted in the hind leg of a mouse using a bimodal ultrasound—a PA system where a clinical ultrasonography system was synchronized to a laser. The groups of Li and Wang, who were active in many aspects of synthesis and applications of gold nanorods, published one of the first clear demonstrations of molecular imaging using antibody-coupled nanoparticles [127].

In the following publications, Song *et al.* used gold nanorods as tracers in OA mapping of lymph drainage to identify SLNs in rat models [128]. The system used was a deeply penetrating PA microscopy system. Cui *et al.* [129] also applied PA microscopy to visualize gold nanorods accumulated at a tumor site in a mouse model, to guide high-intensity focused ultrasound (HIFU) for therapy. Taruttis *et al.* [130] showed real-time imaging of circulating gold nanorods in blood vessels in the neck region of a mouse injected with the particles. Gold nanorods were also shown to be capable of targeting multiple molecules [96, 121, 131].

A number of studies show that antibody-conjugated gold nanorods as a molecular-specific contrast agent can enhance the sensitivity of OAT for imaging of tumors [106, 107]. In the studies [127, 132], two corresponding monoclonal antibodies were conjugated to nanorods with different aspect ratios to target HER2 and CXCR4 molecules. By simply switching the wavelength of the excitation laser, multiple molecular signatures could be obtained both *in vitro* and *in vivo*. This approach allowed monitoring of the expression levels of different oncogenes within cancer cells simultaneously. The feasibility of OA imaging of inflammatory responses using bioconjugated nanorods was demonstrated *ex vivo* [133]. Gold nanorods were conjugated to anti-intercellular adhesion molecule-1, and stimulated endothelial cells labeled with bioconjugated nanorods were scanned using a high-frequency transducer. The results demonstrated that OA images could differentiate between inflamed cells, which were targeted with OA contrast agents, and control cells.

Gold nanorods are commonly synthesized using a toxic surfactant CTAB. An important approach to ensure biocompatibility of nanorods is to replace CTAB with PEG [30, 134, 135]. With PEGylated gold nanorods, the group of Oraevsky [15, 134, 136, 137] successfully demonstrated OA imaging as a noninvasive technique in small animal models. Figure 5.13 shows 3D OAT of a mouse treated with an intravenous injection of PEGylated gold nanorods solution.

5.4.4 Gold Nanocages

With a more complicated shape, gold nanocages represent a class of optically tunable nanoparticles with cuboidal symmetry and "hollow" shape [102, 138, 139]. Hollow gold nanoparticles are made via galvanic replacement chemistry from the template of silver sols [140–142]. Silver redox potential (Ag^+/Ag 0.8V vs. standard hydrogen electrode (SHE)) has a lower redox potential than gold ($AuCl_4^-/Au$ 0.99V vs. SHE) [139, 143, 144], leading to the following replacement reaction between Ag nanocubes and $HAuCl_4$:

$$3Ag^0_{(s)} + AuCl^-_{4(aq)} = Au^0_{(s)} + 3Ag^+_{(aq)} + 4Cl^-_{(aq)}$$

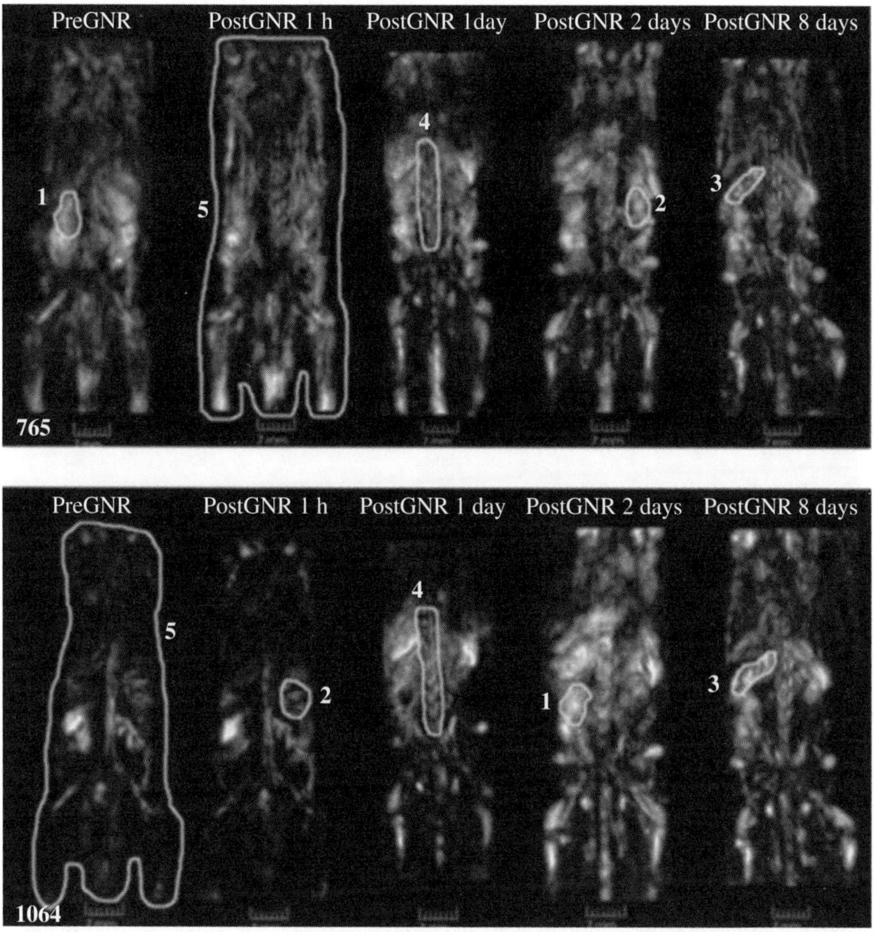

FIGURE 5.13 Three-dimensional OAT dorsoventral reconstruction from a mouse before and after intravenous injection of gold nanorod–PEG solution for 765 and 1064 nm laser illumination. The locations of the image areas used for analysis of organ-specific changes on each view are outlined: (1) left kidney, (2) right kidney, (3) spleen, (4) spine, and (5) whole body. (Adapted with permission from Ref. [15]. © SPIE-International Society for Optical Engineers.)

Formation of metal nanostructures with hollow interiors can be attained from various types of metals [143]. The major steps in manufacturing hollow nanoparticles are shown in Figure 5.14 with TEM images of silver nanoparticles and corresponding hollow gold nanostructures obtained by aqueous $HAuCl_4$ solution. Gold nanocages feature single-crystal structure, hollow interiors, ultrathin and porous walls, and high molar absorptivity [145]. Due to their strong optical absorption, nanocages can be melted even by a camera flash discharge [146].

Hollow gold nanocages have been explored in a number of biomedical applications, often in cancer detection and treatment [147]. Gold nanocages are particularly applicable for *in vivo* OA imaging because of the compact size, bioinertness, and

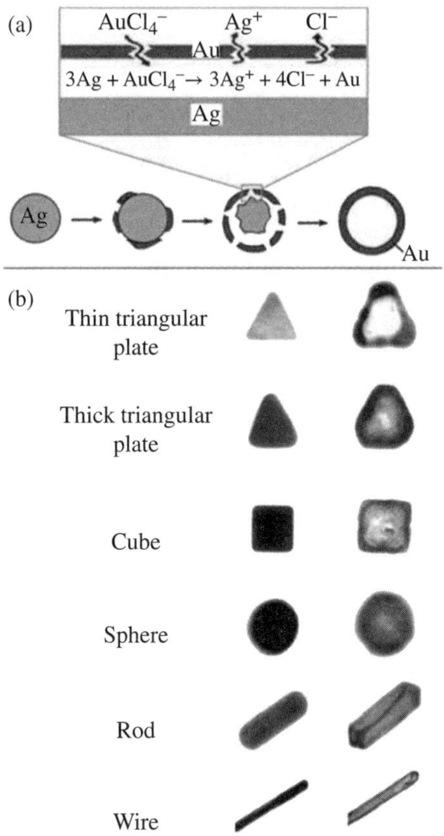

FIGURE 5.14 Schematic illustration of the experimental procedure that generates hollow gold nanoparticles by templating against silver templates (a); TEM images of silver nanoparticles with different typical morphologies and their corresponding gold hollow nanostructures (b). (Adapted with permission from Ref. [143]. © Wiley-VCH Verlag GmbH & Co. KGaA.)

well-defined surface chemistry, as well as their strong and highly wavelength-tunable optical absorption in the NIR optical window. The review [148] discussed synthesis, optical properties, and applications of gold nanocages in the context of cancer diagnostics. A recent *in vivo* study showed that gold nanocages can effectively enhance OA images when injected into the blood stream during the imaging of a rat's cerebral cortex [149]. By adjusting the amount of $HAuCl_4$ added, the SPR peaks of the resultant nanocages can be precisely tuned throughout the visible and NIR regions [145]. In addition, the SPR peaks of nanocages can be tuned to any wavelength in the NIR region by controlling the size and/or wall thickness [148].

A significant feature of gold nanocages as contrast agents resides in their much larger absorption cross sections (almost five orders of magnitude) than those of conventional organic dyes, relatively low cytotoxicity of gold, and the ease of bioconjugation with tumor-specific ligands [147, 150]. Figure 5.15 shows the OA images of a

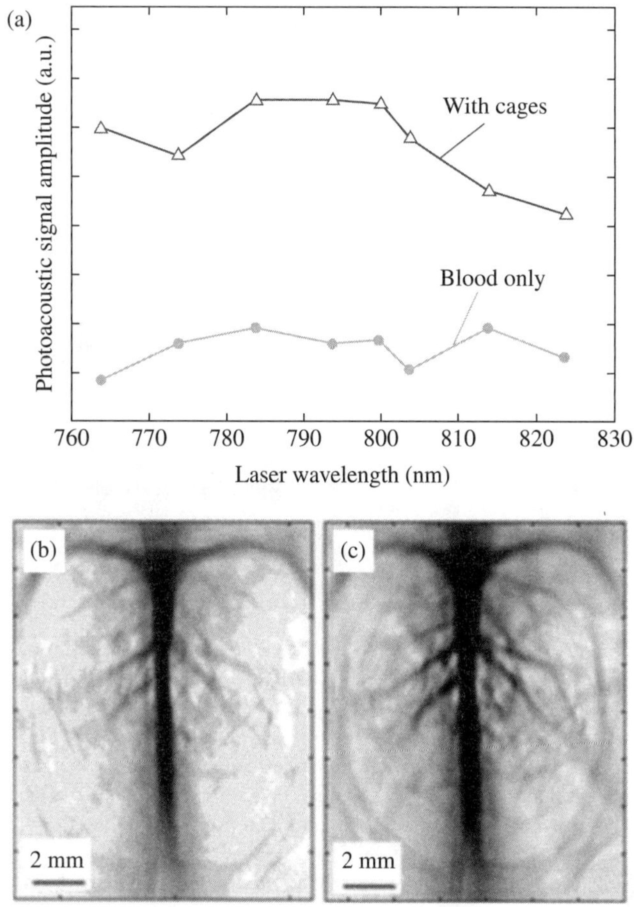

FIGURE 5.15 (a) The measured OA signal amplitude generated with and without gold nanocages in rat blood at several wavelengths. Noninvasive OA imaging of a rat's cerebral cortex (b) prior to the injection of nanocages and (c) 2 h after the final injection, which is the peak enhancement point. (Reproduced with permission from Ref. [151]. © American Chemical Society.)

rat's cerebral cortex after the injection of gold nanocages. The images clearly demonstrate that the rat brain vasculature could be visualized with great clarity and enhanced contrast.

As shown in Figure 5.16, the SLN, located quite deep below the skin's surface, could be imaged with OA technology using gold nanocages. After the injection, PA signals increased with time, which indicated gradual accumulations of the nanocages in the tissue of interest (SLN).

One of the advantages of using gold nanocages for applications in biomedical imaging lies in the well-established surface chemistry of Au, including functionalization with a variety of targeting moieties to further enhance the contrast of the targeted region.

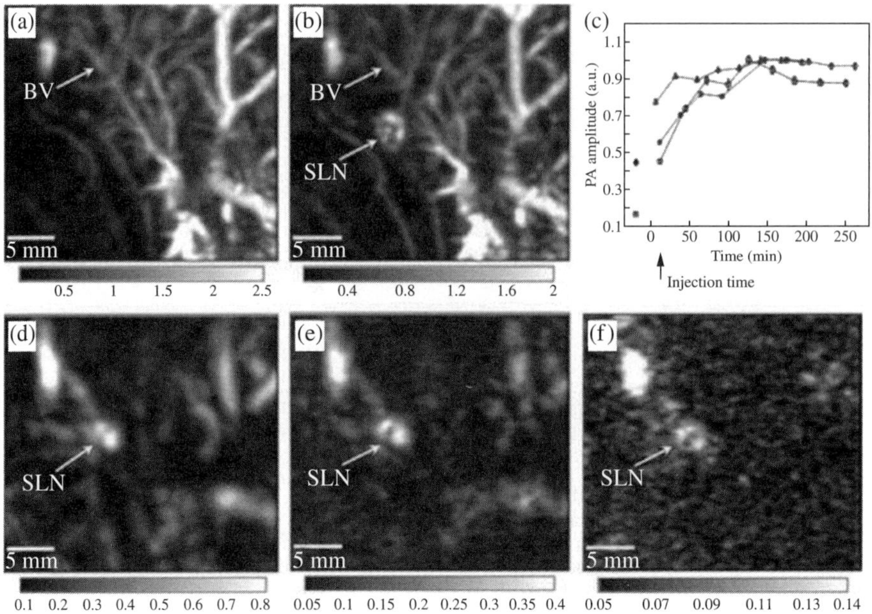

FIGURE 5.16 OA images of the axillary region of a rat taken (a) before and (b) approximately 28 min after the injection of gold nanocages. (c) The changes of OA signal amplitude as a function of the postinjection time. After the injection, OA signals increased with time, which means gradual accumulations of the nanocages. (d–f) Depth capability of SLN mapping with gold nanocages. The OA images were acquired after the injection of nanocages for (d) 126 min with a total imaging depth of 10 mm by placing a layer of chicken breast tissue on the axillary region, (e) 165 min with a total imaging depth of 21 mm by adding another layer of chicken breast tissue, and (f) 226 min with a total imaging depth of 33 mm by using three layers of chicken breast tissue. The bars represent the optical absorption. BV, blood vessel. SLN, sentinel lymph node. (Reproduced with permission from Ref. [152]. © American Chemical Society.)

A study [153] successfully demonstrated the use of gold nanocages as a contrast agent for quantitative PA imaging of melanomas *in vivo*. An additional advantage (which is common to other gold nanoparticles) is the gold nanocages' ability to effectively transfer light energy into the heat destructive to many tumors. Thus, the absorption of NIR light by the immunotargeted nanocages resulted in the selective PT destruction of cancer cells *in vitro* [154].

5.4.5 Hollow Gold Nanospheres and Nanoshells

Another type of plasmonic nanoparticle often used as a contrast agent for OAT and imaging is hollow gold nanospheres/nanoshells (HAuNS). OA imaging of living mouse brain vasculature using HAuNS has been demonstrated by Lu *et al.* [155]. Figure 5.17 shows noninvasive OA imaging of a mouse brain *in vivo* employing PEG-HAuNS and NIR light at a wavelength of 800 nm. At higher magnification on

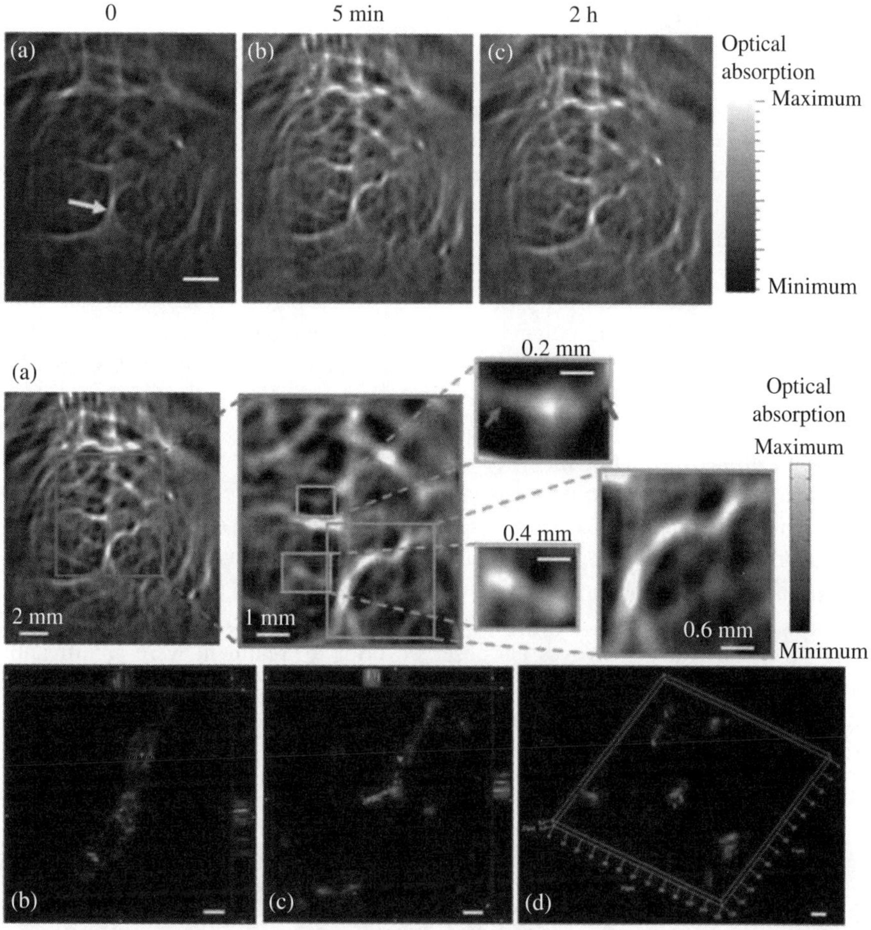

FIGURE 5.17 Noninvasive OAT imaging of a mouse brain *in vivo* employing PEG-HAuNS and NIR light at a wavelength of 800 nm. Upper (A–C) is an optoacoustic image acquired (A) before, (B) 5 min after, and (C) 2 h after the intravenous injection of PEG-HAuNS. *Arrow*, middle cerebral artery. Bottom images (A–D) demonstrate enhanced OA signals revealed a clear and detailed structure of large (yellow-framed picture) and small (green-framed picture) blood vessels in the mouse brain at higher magnification 2 h after intravenous injection of PEG-HAuNS (A). *Arrows* represent the small blood vessels with a diameter of about 100 μm, which can be seen in the contrast-enhanced images. (B–D) Distribution of PEG-HAuNS in brain vessels 2 h after injection (Bar = 10 μm). Brain vessels were stained with anti-CD31 antibody (light gray), while the scattering signals of gold particles were detected under a dark field (pseudogreen). Z-stack images showed the particles located on the luminal side of brain blood vessels (B and C). In brain capillaries, 3D reconstruction images show that the particles colocalized or stayed adjacent to the brain capillary endothelial cells (D). (Adapted with permission from Ref. [155]. © Elsevier.)

the bottom images, OAT revealed not only the structure of large vessels but also small blood vessels just few hours after PEG-HAuNS administration. Brain blood vessels in the superficial cortex as small as approximately 100 μm in diameter could be clearly seen. In order to confirm the distribution of PEG-HAuNS in the brain, the blood vessels were stained with anti-CD31 antibody, while the scattering signal of the gold nanoparticles was detected under a dark field. As shown in Figure 5.17, the HAuNS were located on the luminal side of the blood vessels 2 h after intravenous injection. In brain capillaries, most of the nanoparticles colocalized or stayed adjacent to the brain capillary endothelial cells.An interesting and promising example of novel contrast agents for OAT is a nanoconstruct called gold nanobeacons. Gold nanobeacons are lipid- or polymer-encapsulated colloidal particles incorporating tiny gold nanoparticles (2–4 nm) with an overall size in three tunable sizes (90, 150, and 390 nm). Several publications from the Lanza group [29, 156–158] summarized current data for the development of a family of colloidal GNBs for nanomedicine and OA applications. The latter was successfully used for the detection of lymph nodes and microvessels in an angiogenesis model [158]. As stated earlier, angiogenesis is an essential microanatomical component of many tumors. Angiogenesis is also an indicator of cardiovascular progression that may be effective as a quantitative marker of cardiovascular plaque progression or response to therapy [158–160]. Figure 5.18 illustrates the marked enhancement of angiogenic vessel imaging with PA contrast over time. Individual microvessels observed prior to treatment were significantly enhanced with αvβ3 nanobeacon. These developing neovessels were probably too immature and incomplete to support the blood flow. Similar images in the saline control animals showed no change in vascular anatomy over the same time course.

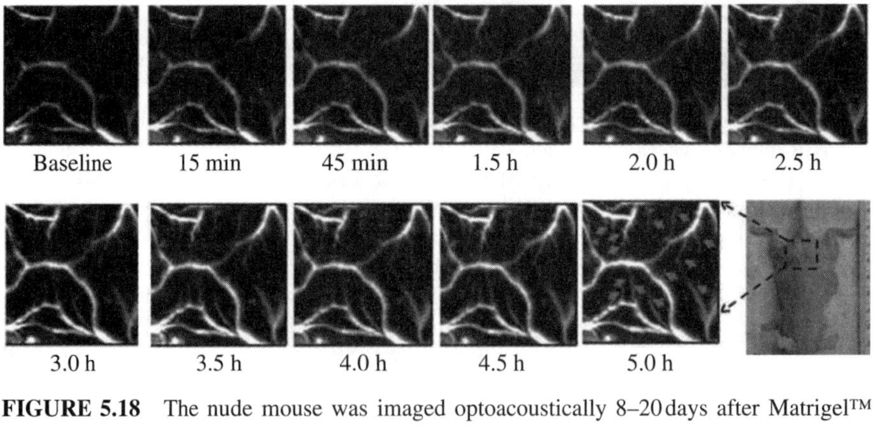

FIGURE 5.18 The nude mouse was imaged optoacoustically 8–20 days after Matrigel™ 0.75 ml aliquot subcutaneous implantation. Control image is baseline. After the control image was taken, targeted gold nanobeacons (αvβ3-GNB-M) were injected into the tail vein. OA images were acquired with an interval of approximately 0.5 up to 5 h. Gray arrows point to the angiogenic sprout (not visible in baseline). For all OA images, $\lambda = 767$ nm, and scale bar = 5 mm. Digital photograph of a mouse implanted with Matrigel™ plug. (Reproduced with permission from Ref. [158]. © John Wiley and Sons.)

5.4.6 Gold Nanoprisms

Recently, a novel class of plasmonic nanoparticle gold nanoprisms (AuNPr) has emerged as signal amplifiers in MSOT, and their applicability to visualize gastrointestinal cancer has been demonstrated [161]. PEGylated versions of AuNPrs were successfully internalized by HT-29 gastrointestinal cancer cells *in vitro*. The particles have been reported to feature good biocompatibility and a surface plasmon band centered at 830 nm, a suitable wavelength for OA imaging. Both *in vitro* and *in vivo* results showed that AuNPrs have the capacity to penetrate tumors and provide a high-resolution signal amplifier for OA imaging.

5.4.7 Silica-Coated Plasmonic Nanoparticles

Several reviews have analyzed different aspects of synthesis of multifunctional plasmonic nanostructures such as dispersibility in aqueous solutions, photoluminescence, and resistance to degradation [96, 103, 104, 121, 143, 145, 162, 163]. Good dispersibility of these nanoparticles is a key for biological applications and can be improved by surface modification. Typically, organic coatings are present on the nanoparticle surface from its wet synthesis, which include surfactants, citrate [164, 165], polymers [166], or starch [167]. However, nanoparticles coated with organics may not be stable over time, and therefore, often, an extra inorganic surface coating is sought [168]. Silica (SiO_2) coating can be applied by wet chemistry on both gold and silver nanoparticles [168–170] (Fig. 5.19). The thickness and the homogeneity of the surface coating can be well controlled. Importantly, silica coating is rather porous allowing the transport of ions [169, 170]. Particularly, gold core silica nanoparticles show promise for OA imaging [171, 172].

Several groups have successfully demonstrated silica covering of gold nanorods for prevention of destruction during high-fluence laser illumination and for enhanced OA response [172–176]. Thermal equilibration between electrons and the lattice in electron–phonon coupling was thought to cause temperatures in excess of melting, causing structural rearrangements [177]. The groups of Li [178] and Emelianov [172] have shown that the threshold for photon-induced reshaping can be raised by encapsulating gold nanorods in an amorphous silica matrix. Both groups use a single-step method where an alkoxide-based silica precursor, tetraethyl orthosilicate

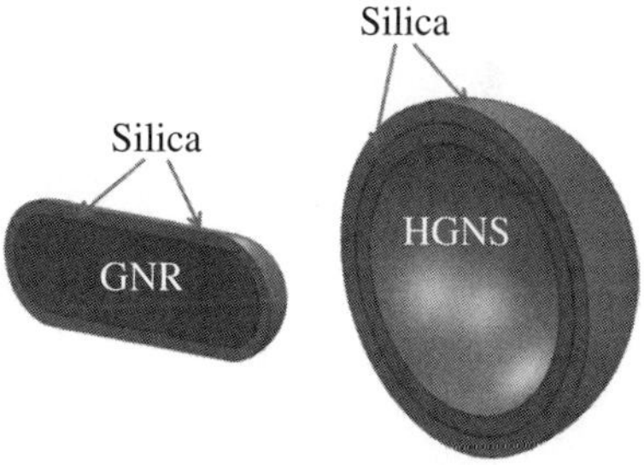

FIGURE 5.19 Schematic of gold nanorods and hollow gold nanoshells with silica cover.

(TEOS). The effect of the silica coating is to improve the stability of the nanorods due to its mechanical rigidity while not influencing the optical properties due to its transparency at NIR wavelengths. Indeed, studies have shown that the silica-coated gold nanorods retained their morphologies up to three times higher fluences than bare nanorods [172].

Silica deposition on gold nanorods and HAuNS is usually conducted with a procedure developed by Stöber [179] and later reproduced and further optimized by several research groups [178, 180, 181]. Silica covering involves hydrolysis and successive condensation of TEOS [$Si(C_2H_5O)_4$] in an alcoholic medium in the presence of ammonium hydroxide (NH_4OH) as catalyst. The reaction mechanism is shown in Equations 5.1 and 5.2. The first step is hydrolysis in which ethoxy groups are replaced by OH groups. In the second step, silicon hydroxides undergo polycondensation to form SiO_2.

$$\mathrm{Si}\left(\mathrm{OC_2H_5}\right)_4 + 4\mathrm{H_2O} \rightarrow \mathrm{Si}\left(\mathrm{OH}\right)_4 + 4\mathrm{C_2H_5OH} \tag{5.1}$$

$$\mathrm{Si}\left(\mathrm{OH}\right)_4 \rightarrow \mathrm{SiO_2} + 2\mathrm{H_2O} \tag{5.2}$$

The thickness of the layer is dependent upon the concentrations of TEOS and NH_4OH, the reaction time, and the number of TEOS additions [172, 174, 180].

Applications with a gold-based contrast agent, such as nanorods or HAuNS, require two main features: high optical absorbance and effective heat transfer to surrounding media. Since silica coating does not change absorbance properties of nanorods, Chen *et al.* explored silica-covered gold nanorods for 3D OA imaging system [172]. Silica has a significantly higher thermal diffusivity compared to water; thus, the thickness-dependent thermal stability enhancement can be attributed to a faster heat distribution over a larger area [172]. Changes in the thermal conductivity could lead to an improved cooling due to a more efficient heat transfer [172]. Conversely, low thermal conductivity of the underlying porous layer reduces the amount of heat escaping from the substrate and contributes to the efficient PA emission from Au nanoparticle arrays [182].

The amplitude of the PA signal is determined by the Grüneisen parameter of the solvent, which directs heat transfer from the nanoparticles to the liquid [174]. The effect is also dependent upon the optical absorption of the nanoparticles, laser fluence, and interfacial heat conductivity between the two materials [174]. In References [172] and [174], the authors suggest that the phenomenon can be explained by a modified interfacial thermal resistance due to the silica layer. The authors of this chapter believe that the silica layer prevents overheating of the surrounding water to increase the efficiency of OA response. Our alternative hypothesis is based on slowing the heat transfer through gold/silica/water due to low thermal conductivity and mismatch of the main energy carriers. It prevents overheating of water surrounding the gold nanoparticle and improves heat and optical energy transfer during the laser pulse as it is shown in Reference [183]. Regardless of the mechanism, these data suggest that silica-covered gold nanoparticles are efficient OA amplifiers.

5.5 CONCLUSIONS

From this short review where we briefly analyzed the design, characterizations, and application of different types of OA imaging contrast agents, we can summarize that nanoparticle-based contrast agents greatly extend OA as an imaging technique. Nanoparticles are best designed with peak absorption in the NIR region, where optical attenuation of tissue is relatively low, affording deep light penetration. Targeting peptides and antibodies can be conjugated to the nanoparticle surface for cell-specific contrast and molecular imaging. Some types of contrast agents are non-photobleaching and provide a long-circulating lifetime with good optical absorption. OA imaging could be combined with PT therapy, since both techniques share similar mechanisms of interaction with light.

Certain aspects of OA imaging with nanoparticles need further investigation. A comparison between contrast agents of different materials, shapes, and sizes on OA imaging is lacking. Presently, we are unable to specify which kind of nanoparticle is best for each OA modality (imaging, tomography, or sensing). Another challenge is an analysis of nanoparticles *in vivo*, for example, in the blood stream after *in vivo* administration, in tumors, or in other tissues. Future study should be directed toward optimization of the size and shape of nanoparticle-based OA contrast agents to ensure the most efficient transformation of light energy into high pressure.

Overall, this review presents OA imaging contrast agents and amplifiers, which show a significant promise for preclinical research in cancer and cardiovascular diagnostics, as the OA research community continues to discover new approaches and invent new technologies.

ACKNOWLEDGMENTS

This work was supported by the National Institutes of Health, grants: 1R43ES021629, R44CA110137, and R44CA110137-S1.

REFERENCES

[1] Oraevsky AA. Gold and silver nanoparticles as contrast agents for optoacoustic imaging. In: Wang LV, editor. *Photoacoustic Imaging and Spectroscopy*. New York: Taylor and Francis Group; 2009. p 373–386.

[2] Li C, Wang LV. Photoacoustic tomography and sensing in biomedicine. Phys Med Biol 2009;**54**:R59–R97.

[3] Oraevsky, AA: "Optoacoustic Tomography: From Fundamentals to Diagnostic Imaging of Breast Cancer", *in Biomedical Photonics Handbook*, Second Edition: Fundamentals, Devices, and Techniques, ed. by T. Vo-Dinh, CRC Press, Boca Raton, Florida, 2014; **222**: pp. 715–757

[4] Ntziachristos V. Going deeper than microscopy: the optical imaging frontier in biology. Nat Methods 2010;**7**:603–614.

[5] Wang LV, Hu S. Photoacoustic tomography: *in vivo* imaging from organelles to organs. Science 2012;**335**:1458–1462.

[6] Welch AJ, van Gemert MLC. Overview of optical and thermal laser-tissue interaction and nomenclature. In: Welch AJ, van Gemert MLC, editors. *Optical-Thermal Response of Laser-Irradiated Tissue*. New York: Plenum Press; 1995. p 1–12.

[7] Weissleder R. A clearer vision for *in vivo* imaging. Nat Biotechnol 2001;**19**:316–317.

[8] Niederhauser JJ, Jaeger M, Lemor R, Weber P, Frenz M. Combined ultrasound and optoacoustic system for real-time high-contrast vascular imaging *in vivo*. IEEE Trans Med Imaging 2005;**24**:436–440.

[9] Wang LV. Multiscale photoacoustic microscopy and computed tomography. Nat Photon 2009;**3**:503–509.

[10] Su R, Ermilov SA, Liopo AV, Oraevsky AA. Laser optoacoustic tomography: towards new technology for biomedical diagnostics. Nucl Instrum Methods Phys Res A 2013; **720**:58–61.

[11] Zhang HF, Maslov K, Stoica G, Wang LV. Functional photoacoustic microscopy for high-resolution and noninvasive *in vivo* imaging. Nat Biotechnol 2006;**24**:848–851.

[12] Beard P. Biomedical photoacoustic imaging. Interf Focus 2011;**1**:602–631.

[13] Brecht HP, Su R, Fronheiser M, Ermilov SA, Conjusteau A, Oraevsky AA. Whole-body three-dimensional optoacoustic tomography system for small animals. J Biomed Opt 2009;**14**:064007.

[14] Tsyboulski DA, Liopo AV, Su R, Ermilov SA, Bachilo SM, Weisman RB, Oraevsky AA. Enabling *in vivo* measurements of nanoparticle concentrations with three-dimensional optoacoustic tomography. *J Biophotonics* 2013. doi:10.1002/jbio.201200233.

[15] Su R, Ermilov SA, Liopo AV, Oraevsky AA. Three-dimensional optoacoustic imaging as a new noninvasive technique to study long-term biodistribution of optical contrast agents in small animal models. J Biomed Opt 2012;**17**:101506.

[16] Emelianov SY, Li PC, O'Donnell M. Photoacoustics for molecular imaging and therapy. Phys Today 2009;**62**:34–39.

[17] Mallidi S, Luke GP, Emelianov S. Photoacoustic imaging in cancer detection, diagnosis, and treatment guidance. Trends Biotechnol 2011;**29**:213–221.

[18] Lin AW, Lewinski NA, West JL, Halas NJ, Drezek RA. Optically tunable nanoparticle contrast agents for early cancer detection: model-based analysis of gold nanoshells. J Biomed Opt 2005;**10**:064035.

[19] Yang X, Maurudis A, Gamelin J, Aguirre A, Zhu Q, Wang LV. Photoacoustic tomography of small animal brain with a curved array transducer. J Biomed Opt 2009;**14**:054007.

[20] Razansky D, Distel M, Vinegoni C, Ma R, Perrimon N, Köster RW, Ntziachristos V. Multispectral opto-acoustic tomography of deep-seated fluorescent proteins *in vivo*. Nat Photon 2009;**3**:412–417.

[21] Ntziachristos V, Razansky D. Molecular imaging by means of multispectral optoacoustic tomography (MSOT). Chem Rev 2010;**110**:2783–2794.

[22] Torchilin VP. PEG-based micelles as carriers of contrast agents for different imaging modalities. Adv Drug Deliv Rev 2002;**54**:235–252.

[23] Shim SY, Lim DK, Nam JM. Ultrasensitive optical biodiagnostic methods using metallic nanoparticles. Nanomedicine (Lond) 2008;**3**:215–232.

[24] Fass L. Imaging and cancer: a review. Mol Oncol 2008;**2**:115–152.

[25] Luke GP, Yeager D, Emelianov SY. Biomedical applications of photoacoustic imaging with exogenous contrast agents. Ann Biomed Eng 2012;**40**:422–437.

[26] Pramanik M, Swierczewska M, Green D, Sitharaman B, Wang LV. Single-walled carbon nanotubes as a multimodal-thermoacoustic and photoacoustic-contrast agent. J Biomed Opt 2009;**14**:034018.

[27] Yang X, Stein EW, Ashkenazi S, Wang LV. Nanoparticles for photoacoustic imaging. Wiley Interdiscip Rev Nanomed Nanobiotechnol 2009;**1**:360–368.

[28] Choi YE, Kwak JW, Park JW. Nanotechnology for early cancer detection. Sensors (Basel) 2010;**10**:428–455.

[29] Pan D, Pramanik M, Senpan A, Ghosh S, Wickline SA, Wang LV, Lanza GM. Near infrared photoacoustic detection of sentinel lymph nodes with gold nanobeacons. Biomaterials 2010;**31**:4088–4093.

[30] Manohar S, Ungureanu C, Van Leeuwen TG. Gold nanorods as molecular contrast agents in photoacoustic imaging: the promises and the caveats. Contrast Media Mol Imaging 2011;**6**:389–400.

[31] Rajian JR, Carson PL, Wang X. Quantitative photoacoustic measurement of tissue optical absorption spectrum aided by an optical contrast agent. Opt Express 2009;**17**:4879–4889.

[32] Wang X, Pang Y, Ku G, Xie X, Stoica G, Wang LV. Noninvasive laser-induced photoacoustic tomography for structural and functional *in vivo* imaging of the brain. Nat Biotechnol 2003;**21**:803–806.

[33] Wang X, Pang Y, Ku G, Stoica G, Wang LV. Three-dimensional laser-induced photoacoustic tomography of mouse brain with the skin and skull intact. Opt Lett 2003;**28**: 1739–1741.

[34] Kolkman RG, Klaessens JH, Hondebrink E, Hopman JC, de Mul FF, Steenbergen W, Thijssen JM, van Leeuwen TG. Photoacoustic determination of blood vessel diameter. Phys Med Biol 2004;**49**:4745–4756.

[35] Siphanto RI, Thumma KK, Kolkman RG, van Leeuwen TG, de Mul FF, van Neck JW, van Adrichem LN, Steenbergen W. Serial noninvasive photoacoustic imaging of neovascularization in tumor angiogenesis. Opt Express 2005;**13**:89–95.

[36] Fang H, Maslov K, Wang LV. Photoacoustic Doppler effect from flowing small light-absorbing particles. Phys Rev Lett 2007;**99**:184501.

[37] Kruger RA, Kiser WL, Reinecke DR, Kruger GA, Miller KD. Thermoacoustic molecular imaging of small animals. Mol Imaging 2003;**2**:113–123.

[38] Oraevsky AA, Su R, Ermiliov SA, Liopo AV, Oraevsky AA, Wang LV. Optoacoustic 3D visualization of changes in physiological properties of mouse tissues from live to postmortem. Proc SPIE 2012;**8223**:82230K–82237K.

[39] Zhang Y, Hong H, Cai W. Photoacoustic imaging. Cold Spring Harb Protoc 2011;**2011**:pdb. top065508.

[40] Stein EW, Maslov K, Wang LV. Noninvasive, *in vivo* imaging of blood-oxygenation dynamics within the mouse brain using photoacoustic microscopy. J Biomed Opt 2009; **14**:020502.

[41] Li C, Aguirre A, Gamelin J, Maurudis A, Zhu Q, Wang LV. Real-time photoacoustic tomography of cortical hemodynamics in small animals. J Biomed Opt 2010;**15**:010509.

[42] Talbert RJ, Holan SH, Viator JA. Photoacoustic discrimination of viable and thermally coagulated blood using a two-wavelength method for burn injury monitoring. Phys Med Biol 2007;**52**:1815–1829.

[43] Aizawa K, Sato S, Saitoh D, Ashida H, Obara M. Photoacoustic monitoring of burn healing process in rats. J Biomed Opt 2008;**13**:064020.

[44] Ku G, Fornage BD, Jin X, Xu M, Hunt KK, Wang LV. Thermoacoustic and photoacoustic tomography of thick biological tissues toward breast imaging. Technol Cancer Res Treat 2005;**4**:559–566.

[45] Cai W, Chen X. Multimodality molecular imaging of tumor angiogenesis. J Nucl Med 2008;**49** (Suppl 2):113S–128S.

[46] Cai W, Niu G, Chen X. Imaging of integrins as biomarkers for tumor angiogenesis. Curr Pharm Des 2008;**14**:2943–2973.

[47] Hoelen CG, de Mul FF, Pongers R, Dekker A. Three-dimensional photoacoustic imaging of blood vessels in tissue. Opt Lett 1998;**23**:648–650.

[48] Riley PA. Melanin. Int J Biochem Cell Biol 1997;**29**:1235–1239.

[49] Dawson JB, Barker DJ, Ellis DJ, Grassam E, Cotterill JA, Fisher GW, Feather JW. A theoretical and experimental study of light absorption and scattering by *in vivo* skin, 1980. Phys Med Biol 1980;**25**:695–709.

[50] Oh JT, Li ML, Zhang HF, Maslov K, Stoica G, Wang LV. Three-dimensional imaging of skin melanoma *in vivo* by dual-wavelength photoacoustic microscopy. J Biomed Opt 2006;**11**:34032.

[51] Weight RM, Viator JA, Dale PS, Caldwell CW, Lisle AE. Photoacoustic detection of metastatic melanoma cells in the human circulatory system. Opt Lett 2006;**31**:2998–3000.

[52] Wang LV. Prospects of photoacoustic tomography. Med Phys 2008;**35**:5758–5767.

[53] Bouchard B, Fuller BB, Vijayasaradhi S, Houghton AN. Induction of pigmentation in mouse fibroblasts by expression of human tyrosinase cDNA. J Exp Med 1989;**169**:2029–2042.

[54] Stritzker J, Kirscher L, Scadeng M, Deliolanis NC, Morscher S, Symvoulidis P, Schaefer K, Zhang Q, Buckel L, Hess M, Donat U, Bradley WG, Ntziachristos V, Szalay AA. Vaccinia virus-mediated melanin production allows MR and optoacoustic deep tissue imaging and laser-induced thermotherapy of cancer. Proc Natl Acad Sci U S A 2013;**110**: 3316–3320.

[55] Krumholz A, Vanvickle-Chavez SJ, Yao J, Fleming TP, Gillanders WE, Wang LV. Photoacoustic microscopy of tyrosinase reporter gene *in vivo*. J Biomed Opt 2011;**16**: 080503.

[56] Bourke MG, Salwa S, Harrington KJ, Kucharczyk MJ, Forde PF, de Kruijf M, Soden D, Tangney M, Collins JK, O'Sullivan GC. The emerging role of viruses in the treatment of solid tumours. Cancer Treat Rev 2011;**37**:618–632.

[57] Donat U, Weibel S, Hess M, Stritzker J, Hartl B, Sturm JB, Chen NG, Gentschev I, Szalay AA. Preferential colonization of metastases by oncolytic vaccinia virus strain GLV-1h68 in a human PC-3 prostate cancer model in nude mice. PLoS One 2012;**7**: e45942.

[58] Glass K, Ito S, Wilby PR, Sota T, Nakamura A, Bowers CR, Vinther J, Dutta S, Summons R, Briggs DE, Wakamatsu K, Simon JD. Direct chemical evidence for eumelanin pigment from the Jurassic period. Proc Natl Acad Sci U S A 2012;**109**:10218–10223.

[59] Borisjuk L, Rolletschek H, Neuberger T. Surveying the plant's world by magnetic resonance imaging. Plant J 2012;**70**:129–146.

[60] Urabe K, Aroca P, Tsukamoto K, Mascagna D, Palumbo A, Prota G, Hearing VJ. The inherent cytotoxicity of melanin precursors: a revision. Biochim Biophys Acta 1994;**1221**:272–278.

[61] Giepmans BN, Adams SR, Ellisman MH, Tsien RY. The fluorescent toolbox for assessing protein location and function. Science 2006;**312**:217–224.

[62] Burgholzer P, Grün H, Sonnleitner A. Photoacoustic tomography: sounding out fluorescent proteins. Nat Photon 2009;**3**:378–379.

[63] Li L, Zemp RJ, Lungu G, Stoica G, Wang LV. Photoacoustic imaging of lacZ gene expression *in vivo*. J Biomed Opt 2007;**12**:020504.

[64] Manchester M, Singh P. Virus-based nanoparticles (VNPs): platform technologies for diagnostic imaging. Adv Drug Deliv Rev 2006;**58**:1505–1522.

[65] Pavlik C, Biswal NC, Gaenzler FC, Morton MD, Kuhn LT, Claffey KP, Zhu Q, Smith MB. Synthesis and fluorescent characteristics of imidazole-indocyanine green conjugates. Dyes Pigments 2011;**89**:9–15.

[66] Gupta S, Chatni MR, Rao AL, Vullev VI, Wang LV, Anvari B. Virus-mimicking nanoconstructs as a contrast agent for near infrared photoacoustic imaging. Nanoscale 2013;**5**:1772–1776.

[67] Wang X, Ku G, Wegiel MA, Bornhop DJ, Stoica G, Wang LV. Noninvasive photoacoustic angiography of animal brains *in vivo* with near-infrared light and an optical contrast agent. Opt Lett 2004;**29**:730–732.

[68] Ku G, Wang LV. Deeply penetrating photoacoustic tomography in biological tissues enhanced with an optical contrast agent. Opt Lett 2005;**30**:507–509.

[69] Kim G, Huang SW, Day KC, O'Donnell M, Agayan RR, Day MA, Kopelman R, Ashkenazi S. Indocyanine-green-embedded PEBBLEs as a contrast agent for photoacoustic imaging. J Biomed Opt 2007;**12**:044020.

[70] Lavelle L, Michel JP, Gingery M. The disassembly, reassembly and stability of CCMV protein capsids. J Virol Methods 2007;**146**:311–316.

[71] Koudelka KJ, Destito G, Plummer EM, Trauger SA, Siuzdak G, Manchester M. Endothelial targeting of cowpea mosaic virus (CPMV) via surface vimentin. PLoS Pathog 2009;**5**:e1000417.

[72] Young M, Willits D, Uchida M, Douglas T. Plant viruses as biotemplates for materials and their use in nanotechnology. Annu Rev Phytopathol 2008;**46**:361–384.

[73] Destito G, Yeh R, Rae CS, Finn MG, Manchester M. Folic acid-mediated targeting of cowpea mosaic virus particles to tumor cells. Chem Biol 2007;**14**:1152–1162.

[74] Shashkov EV, Everts M, Galanzha EI, Zharov VP. Quantum dots as multimodal photoacoustic and photothermal contrast agents. Nano Lett 2008;**8**:3953–3958.

[75] Gao Y, Cui Y, Chan JK, Xu C. Stem cell tracking with optically active nanoparticles. Am J Nucl Med Mol Imaging 2013;**3**:232–246.

[76] Cai W, Hsu AR, Li ZB, Chen X. Are quantum dots ready for *in vivo* imaging in human subjects? Nanoscale Res Lett 2007;**2**:265–281.

[77] Liang X, Deng Z, Jing L, Li X, Dai Z, Li C, Huang M. Prussian blue nanoparticles operate as a contrast agent for enhanced photoacoustic imaging. Chem Commun (Camb) 2013;**49**:11029–11031.

[78] Cai W, Rao J, Gambhir SS, Chen X. How molecular imaging is speeding up antiangiogenic drug development. Mol Cancer Ther 2006;**5**:2624–2633.

[79] Zharov VP, Galanzha EI, Shashkov EV, Kim JW, Khlebtsov NG, Tuchin VV. Photoacoustic flow cytometry: principle and application for real-time detection of circulating single nanoparticles, pathogens, and contrast dyes *in vivo*. J Biomed Opt 2007;**12**:051503.

[80] Mallidi S, Larson T, Tam J, Joshi PP, Karpiouk A, Sokolov K, Emelianov S. Multiwavelength photoacoustic imaging and plasmon resonance coupling of gold nanoparticles for selective detection of cancer. Nano Lett 2009;**9**:2825–2831.

[81] de la Zerda A, Kim JW, Galanzha EI, Gambhir SS, Zharov VP. Advanced contrast nanoagents for photoacoustic molecular imaging, cytometry, blood test and photothermal theranostics. Contrast Media Mol Imaging 2011;**6**:346–369.

[82] Liu Z, Cai W, He L, Nakayama N, Chen K, Sun X, Chen X, Dai H. *In vivo* biodistribution and highly efficient tumour targeting of carbon nanotubes in mice. Nat Nanotechnol 2007;**2**:47–52.

[83] Kim JW, Galanzha EI, Shashkov EV, Moon HM, Zharov VP. Golden carbon nanotubes as multimodal photoacoustic and photothermal high-contrast molecular agents. Nat Nanotechnol 2009;**4**:688–694.

[84] Pramanik M, Song KH, Swierczewska M, Green D, Sitharaman B, Wang LV. *In vivo* carbon nanotube-enhanced non-invasive photoacoustic mapping of the sentinel lymph node. Phys Med Biol 2009;**54**:3291–3301.

[85] Cai W, Chen X. Anti-angiogenic cancer therapy based on integrin alphavbeta3 antagonism. Anticancer Agents Med Chem 2006;**6**:407–428.

[86] Tuchin VV, Tarnok A, Zharov VP. *In vivo* flow cytometry: a horizon of opportunities. Cytometry A 2011;**79**:737–745.

[87] de La Zerda A, Zavaleta C, Keren S, Vaithilingam S, Bodapati S, Liu Z, Levi J, Smith BR, Ma TJ, Oralkan O, Cheng Z, Chen XY, Dai HJ, Khuri-Yakub BT, Gambhir SS. Carbon nanotubes as photoacoustic molecular imaging agents in living mice. Nat Nanotechnol 2008;**3**:557–562.

[88] de la Zerda A, Liu Z, Bodapati S, Teed R, Vaithilingam S, Khuri-Yakub BT, Chen X, Dai H, Gambhir SS. Ultrahigh sensitivity carbon nanotube agents for photoacoustic molecular imaging in living mice. Nano Lett 2010;**10**:2168–2172.

[89] Xiang L, Yuan Y, Xing D, Ou Z, Yang S, Zhou F. Photoacoustic molecular imaging with antibody-functionalized single-walled carbon nanotubes for early diagnosis of tumor. J Biomed Opt 2009;**14**:021008.

[90] Galanzha EI, Shashkov EV, Kelly T, Kim JW, Yang L, Zharov VP. *In vivo* magnetic enrichment and multiplex photoacoustic detection of circulating tumour cells. Nat Nanotechnol 2009;**4**:855–860.

[91] Galanzha EI, Shashkov EV, Kokoshka MS, Myhill JA, Zharov VP. *In vivo* noninvasive detection of metastatic cells in vasculature and sentinel lymph nodes by photoacoustic cytomery. Laser Surg Med 2008;**S20**:81–82.

[92] Kang B, Yu D, Dai Y, Chang S, Chen D, Ding Y. Cancer-cell targeting and photoacoustic therapy using carbon nanotubes as "bomb" agents. Small 2009;**5**:1292–1301.

[93] Madani SE, Naderi N, Dissanayake O, Tan A, Seifalian AM. A new era of cancer treatment: carbon nanotubes as drug delivery tools. Int J Nanomed 2011;**6**:2963–2979.

[94] Moore TL, Pitzer JE, Podila R, Wang X, Lewis RL, Grimes SW, Wilson JA, Skjervold E, Brown JM, Rao A, Alexis F. Multifunctional polymer-coated carbon nanotubes for safe drug delivery. Part Part Syst Charact 2013;**30**:365–373.

[95] Tsyboulski DA, Liopo AV, Su R, Ermilov SA, Bachilo S, Weisman RB, Oraevsky AA. Enabling *in vivo* measurements of nanoparticle concentrations with three-dimensional optoacoustic tomography. J. BioPhotonics 2013; 2014: 7(8): 581–588.

[96] Dykman L, Khlebtsov N. Gold nanoparticles in biomedical applications: recent advances and perspectives. Chem Soc Rev 2012;**41**:2256–2282.

[97] Nchl CL, Hafner JH. Shape-dependent plasmon resonances of gold nanoparticles. J Mater Chem 2008;**18**:2415.

[98] Sau TK, Rogach AL. Nonspherical noble metal nanoparticles: colloid-chemical synthesis and morphology control. Adv Mater 2010;**22**:1781–1804.

[99] Chen LC, Jiang L, Wang Y, Qian J, He S. Multi-layers polyelectrolyte-coated gold nanorods as multifunction optical contrast agents for cancer cell imaging. Biomed Biotechnol 2010;**11**:417–422.

[100] Anderson RR, Parrish JA. The optics of human skin. J Invest Dermatol 1981;**77**:13–19.

[101] Cai W, Chen X. Nanoplatforms for targeted molecular imaging in living subjects. Small 2007;**3**:1840–1854.

[102] El-Sayed MA. Some interesting properties of metals confined in time and nanometer space of different shapes. Acc Chem Res 2001;**34**:257–264.

[103] Hu M, Chen J, Li ZY, Au L, Hartland GV, Li X, Marquez M, Xia Y. Gold nanostructures: engineering their plasmonic properties for biomedical applications. Chem Soc Rev 2006;**35**:1084–1094.

[104] Khlebtsov NG, Dykman LA. Optical properties and biomedical applications of plasmonic nanoparticles. J Quant Spectrosc Radiat Transfer 2010;**111**:1–35.

[105] Cai W, Gao T, Hong H, Sun J. Applications of gold nanoparticles in cancer nanotechnology. Nanotechnol Sci Appl 2008;**1**:17–32.

[106] Eghtedari M, Liopo AV, Copland JA, Oraevsky AA, Motamedi M. Engineering of hetero-functional gold nanorods for the *in vivo* molecular targeting of breast cancer cells. Nano Lett 2009;**9**:287–291.

[107] Liopo A, Conjusteau A, Tsyboulski D, Ermolinsky B, Kazansky A, Oraevsky A. Biocompatible gold nanorod conjugates for preclinical biomedical research. J Nanomed Nanotechnol 2012;**S2**:1–41.

[108] Dreaden EC, Mackey MA, Huang X, Kang B, El-Sayed MA. Beating cancer in multiple ways using nanogold. Chem Soc Rev 2011;**40**:3391–3404.

[109] Chamberland DL, Agarwal A, Kotov N, Brian Fowlkes J, Carson PL, Wang X. Photoacoustic tomography of joints aided by an Etanercept-conjugated gold nanoparticle contrast agent-an ex vivo preliminary rat study. Nanotechnology 2008;**19**:095101.

[110] Yang Q, Cui H, Cai S, Yang X, Forrest ML. *In vivo* photoacoustic imaging of chemotherapy-induced apoptosis in squamous cell carcinoma using a near-infrared caspase-9 probe. J Biomed Opt 2011;**16**:116026.

[111] Jain PK, Eustis S, El-Sayed MA. Plasmon coupling in nanorod assemblies: optical absorption, discrete dipole approximation simulation, and exciton-coupling model. J Phys Chem B 2006;**110**:18243–18253.

[112] Averitt RD, Sarkar D, Halas NJ. Plasmon resonance shifts of Au-coated Au2 S nanoshells: insight into multicomponent nanoparticle growth. Phys Rev Lett 1997;**78**:4217.

[113] Oldenburg SJ, Averitt RD, Wescott SL, Halas NJ. Nanoengineering of optical resonances. Chem Phys Lett 1998;**288**:5.

[114] Brongersma ML. Nanoscale photonics: nanoshells: gifts in a gold wrapper. Nat Mater 2003;**2**:296–297.

[115] Hirsch LR, Stafford RJ, Bankson JA, Sershen SR, Rivera B, Price RE, Hazle JD, Halas NJ, West JL. Nanoshell-mediated near-infrared thermal therapy of tumors under magnetic resonance guidance. Proc Natl Acad Sci U S A 2003;**100**:13549–13554.

[116] Loo C, Lowery A, Halas N, West J, Drezek R. Immunotargeted nanoshells for integrated cancer imaging and therapy. Nano Lett 2005;**5**:709–711.

[117] Wang YW, Xie XY, Wang XD, Ku G, Gill KL, O'Neal DP, Stoica G, Wang LV. Photoacoustic tomography of a nanoshell contrast agent in the *in vivo* rat brain. Nano Lett 2004;**4**:1689–1692.

[118] Xu M, Wang LV. Photoacoustic imaging in biomedicine. Rev Sci Instrum 2006;**77**:041101.

[119] Oraevsky AA, Oraevsky AN. Plasmon resonance in ellipsoid nanoparticles. Quantum Electron 2002;**32**:79–82.

[120] Guzatov DV, Oraevsky AA, Oraevsky AN. Plasmon resonance in ellipsoidal nanoparticles with shells. Quantum Electron 2003;**33**:817–822.

[121] Vigderman L, Khanal BP, Zubarev ER. Functional gold nanorods: synthesis, self-assembly, and sensing applications. Adv Mater 2012;**24**:4811–4841. 5014.

[122] Oraevsky A, Karabutov A, Savateeva EA. Enhancement of optoacoustic tissue contrast with absorbing nanoparticles. Proc SPIE 2001;**4434**:60–69.

[123] Oraevsky A, Henrichs PM. High contrast optoacoustic imaging and sensing using nanoparticles. US Patent 7,500,953; 2003.

[124] Li PC, Huang SW, Wei CW, Chiou YC, Chen CD, Wang CRC. Photoacoustic flow measurements by use of laser-induced shape transitions of gold nanorods. Opt Lett 2005;**30**: 3341–3343.

[125] Wei CW, Huang SW, Wang CR, Li PC. Photoacoustic flow measurements based on wash-in analysis of gold nanorods. IEEE Trans Ultrason Ferroelectr Freq Control 2007;**54**:1131–1141.

[126] Agarwal A, Huang SW, O'Donnell M, Day KC, Day M, Kotov N, Ashkenazi S. Targeted gold nanorod contrast agent for prostate cancer detection by photoacoustic imaging. J Appl Phys 2007;**102**:701–706.

[127] Li PC, Wang CR, Shieh DB, Wei CW, Liao CK, Poe C, Jhan S, Ding AA, Wu YN. *In vivo* photoacoustic molecular imaging with simultaneous multiple selective targeting using antibody-conjugated gold nanorods. Opt Express 2008;**16**:18605–18615.

[128] Song KH, Kim C, Maslov K, Wang LV. Noninvasive *in vivo* spectroscopic nanorod-contrast photoacoustic mapping of sentinel lymph nodes. Eur J Radiol 2009;**70**: 227–231.

[129] Cui H, Staley J, Yang X. Integration of photoacoustic imaging and high-intensity focused ultrasound. J Biomed Opt 2010;**15**:021312.

[130] Taruttis A, Herzog E, Razansky D, Ntziachristos V. Real-time imaging of cardiovascular dynamics and circulating gold nanorods with multispectral optoacoustic tomography. Opt Express 2010;**18**:19592–19602.

[131] Wei CW, Liao CK, Chen YY, Wang CRC, Ding AA, Shiehd DB, Li PC. *In vivo* photoacoustic imaging with multiple selective targeting using bioconjugated gold nanorods. Proc SPIE 2008;**6856**:1–111.

[132] Li PC, Wei CW, Liao CK, Chen CD, Pao KC, Wang CRC, Wu YN, Shieh DB. Photoacoustic imaging of multiple targets using gold nanorods. IEEE Trans Ultrason Ferroelectr Freq Control 2007;**54**:1642–1647.

[133] Kim K, Huang S-W, Ashkenazi S, O'Donnell M, Agarwal A, Kotov NA, Denny MF, Kaplan MJ. Photoacoustic imaging of early inflammatory response using gold nanorods. Appl Phys Lett 2007;**90**:223901–223903.

[134] Liopo AV, Conjusteau A, Chumakova OV, Ermilov SA, Su R, Oraevsky AA. Highly purified biocompatible gold nanorods for contrasted optoacoustic imaging of small animal models. Nanosci Nanotechnol Lett 2012;**4**:681–686.

[135] Su R, Liopo AV, Brecht HP, Ermilov SA, Oraevsky AA. Gold nanorod distribution in mouse tissues after intravenous injection monitored with optoacoustic tomography. Proc SPIE 2011;**7899**:78994B.

[136] Liopo AV, Conjusteau A, Oraevsky AA. PEG-coated gold nanorod monoclonal antibody conjugates in preclinical research with optoacoustic tomography, photothermal therapy, and sensing. Proc SPIE 2012;**8223**:822344.

[137] Su R, Liopo AV, Brecht HP, Ermilov SA, Larin K, Oraevsky AA. Optoacoustic tomography in preclinical research: *in vivo* distribution of highly purified PEG-coated gold nanorods. Proc SPIE 2011;**8089**:808902.

[138] Rao CNR, Kulkarni GU, Thomas PJ, Edwards PP. Metal nanoparticles and their assemblies. Chem Soc Rev 2000;**29**:27–35.

[139] Sun Y, Xia Y. Shape-controlled synthesis of gold and silver nanoparticles. Science 2002;**298**:2176–2179.

[140] Wu G, Mikhailovsky A, Khant HA, Zasadzinski JA. Synthesis, characterization, and optical response of gold nanoshells used to trigger release from liposomes. Methods Enzymol 2009;**464**:279–307.

[141] Prevo BG, Esakoff SA, Mikhailovsky A, Zasadzinski JA. Scalable routes to gold nanoshells with tunable sizes and response to near-infrared pulsed-laser irradiation. Small 2008;**4**:1183–1195.

[142] Hao E, Schatz GC. Electromagnetic fields around silver nanoparticles and dimers. J Chem Phys 2004;**120**:357–366.

[143] Sun Y, Mayers B, Xia Y. Metal nanostructures with hollow interiors. Adv Mater 2003;**15**:641–646.

[144] Lu X, Au L, McLellan J, Li ZY, Marquez M, Xia Y. Fabrication of cubic nanocages and nanoframes by dealloying Au/Ag alloy nanoboxes with an aqueous etchant based on Fe(NO3)3 or NH4OH. Nano Lett 2007;**7**:1764–1769.

[145] Chen J, Saeki F, Wiley BJ, Cang H, Cobb MJ, Li ZY, Au L, Zhang H, Kimmey MB, Li X, Xia Y. Gold nanocages: bioconjugation and their potential use as optical imaging contrast agents. Nano Lett 2005;**5**:473–477.

[146] Chen JY, Wiley B, Li ZY, Campbell D, Saeki F, Cang H, Au L, Lee J, Li XD, Xia YN. Gold nanocages: engineering their structure for biomedical applications. Adv Mater 2005;**17**:2255–2261.

[147] Skrabalak SE, Au L, Lu X, Li X, Xia Y. Gold nanocages for cancer detection and treatment. Nanomedicine (Lond) 2007;**2**:657–668.

[148] Li W, Brown PK, Wang LV, Xia Y. Gold nanocages as contrast agents for photoacoustic imaging. Contrast Media Mol Imaging 2011;**6**:370–377.

[149] Yang X, Wang LV. Photoacoustic tomography of a rat cerebral cortex with a ring-based ultrasonic virtual point detector. J Biomed Opt 2007;**12**:060507.

[150] Chen J, Yang M, Zhang Q, Cho EC, Cobley CM, Kim C, Glaus C, Wang LV, Welch MJ, Xia Y. Gold nanocages: a novel class of multifunctional nanomaterials for theranostic applications. Adv Funct Mater 2010;**20**:3684–3694.

[151] Yang X, Skrabalak SE, Li ZY, Xia Y, Wang LV. Photoacoustic tomography of a rat cerebral cortex *in vivo* with au nanocages as an optical contrast agent. Nano Lett 2007;**7**:3798–3802.

[152] Song KH, Kim CH, Cobley CM, Xia YN, Wang LV. Near-infrared gold nanocages as a new class of tracers for photoacoustic sentinel lymph node mapping on a rat model. Nano Lett 2009;**9**:183–188.

[153] Kim C, Song KH, Gao F, Wang LV. Sentinel lymph nodes and lymphatic vessels: noninvasive dual-modality *in vivo* mapping by using indocyanine green in rats—volumetric spectroscopic photoacoustic imaging and planar fluorescence imaging. Radiology 2010;**255**:442–450.

[154] Chen J, Wang D, Xi J, Au L, Siekkinen A, Warsen A, Li ZY, Zhang H, Xia Y, Li X. Immuno gold nanocages with tailored optical properties for targeted photothermal destruction of cancer cells. Nano Lett 2007;**7**:1318–1322.

[155] Lu W, Huang Q, Ku G, Wen X, Zhou M, Guzatov D, Brecht P, Su R, Oraevsky A, Wang LV, Li C. Photoacoustic imaging of living mouse brain vasculature using hollow gold nanospheres. Biomaterials 2010;**31**:2617–2626.

[156] Pan D, Pramanik M, Senpan A, Allen JS, Zhang H, Wickline SA, Wang LV, Lanza GM. Molecular photoacoustic imaging of angiogenesis with integrin-targeted gold nanobeacons. FASEB J 2011;**25**:875–882.

[157] Pan D, Pramanik M, Senpan A, Yang X, Song KH, Scott MJ, Zhang H, Gaffney PJ, Wickline SA, Wang LV, Lanza GM. Molecular photoacoustic tomography with colloidal nanobeacons. Angew Chem Int Ed Engl 2009;**48**:4170–4173.

[158] Pan D, Pramanik M, Wickline SA, Wang LV, Lanza GM. Recent advances in colloidal gold nanobeacons for molecular photoacoustic imaging. Contrast Media Mol Imaging 2011;**6**:378–388.

[159] Winter P, Neubauer A, Caruthers S, Harris T, Robertson J, Williams T, Schmieder A, Hu G, Allen J, Lacy E, Wickline S, Lanza G. Endothelial $\alpha\nu\beta3(3)$-integrin targeted fumagillin nanoparticles inhibit angiogenesis in atherosclerosis. Arterioscler Thromb Vasc Biol 2006;**26**:2103–2109.

[160] Winter PM, Schmieder AH, Caruthers SD, Keene JL, Zhang H, Wickline SA, Lanza GM. Minute dosages of alpha(nu)beta3-targeted fumagillin nanoparticles impair Vx-2 tumor angiogenesis and development in rabbits. FASEB J 2008;**22**:2758–2767.

[161] Bao C, Beziere N, del Pino P, Pelaz B, Estrada G, Tian F, Ntziachristos V, de la Fuente JM, Cui D. Gold nanoprisms as optoacoustic signal nanoamplifiers for *in vivo* bioimaging of gastrointestinal cancers. Small 2013;**9**:68–74.

[162] Murphy CJ, Sau TK, Gole AM, Orendorff CJ, Gao J, Gou L, Hunyadi SE, Li T. Anisotropic metal nanoparticles: synthesis, assembly, and optical applications. J Phys Chem B 2005;**109**:13857–13870.

[163] Sotiriou GA. Biomedical applications of multifunctional plasmonic nanoparticles. Wiley Interdiscip Rev Nanomed Nanobiotechnol 2013;**5**:19–30.

[164] Henglein A, Giersig M. Formation of colloidal silver nanoparticles: capping action of citrate. J Phys Chem B 1999;**103**:9533–9539.

[165] Brown KR, Walter DG, Natan MJ. Seeding of colloidal Au nanoparticle solutions. 2. Improved control of particle size and shape. Chem Mater 2000;**12**:306–313.

[166] Pastoriza-Santos I, Liz-Marzán LM. Formation of PVP-Protected metal nanoparticles in DMF. Langmuir 2002;**18**:2888–2894.

[167] Raveendran P, Fu J, Wallen SL. Completely "green" synthesis and stabilization of metal nanoparticles. J Am Chem Soc 2003;**125**:13940–13941.

[168] Liz-Marzán LM, Giersig M, Mulvaney P. Synthesis of nanosized gold–silica core–shell particles. Langmuir 1996;**12**:4329–4335.

[169] Kobayashi Y, Katakami H, Mine E, Nagao D, Konno M, Liz-Marzan LM. Silica coating of silver nanoparticles using a modified stober method. J Colloid Interf Sci 2005;**283**: 392–396.

[170] Han Y, Jiang J, Lee SS, Ying JY. Reverse microemulsion-mediated synthesis of silica-coated gold and silver nanoparticles. Langmuir 2008;**24**:5842–5848.

[171] Wilson K, Homan K, Emelianov S. Biomedical photoacoustics beyond thermal expansion using triggered nanodroplet vaporization for contrast-enhanced imaging. Nat Commun 2012;**3**:618.

[172] Chen YS, Frey W, Kim S, Homan K, Kruizinga P, Sokolov K, Emelianov S. Enhanced thermal stability of silica-coated gold nanorods for photoacoustic imaging and image-guided therapy. Opt Express 2010;**18**:8867–8878.

[173] Fernandez-Lopez C, Mateo-Mateo C, Alvarez-Puebla RA, Perez-Juste J, Pastoriza-Santos I, Liz-Marzan LM. Highly controlled silica coating of PEG-capped metal nanoparticles and preparation of SERS-encoded particles. Langmuir 2009;**25**:13894–13899.

[174] Chen YS, Frey W, Aglyamov S, Emelianov S. Environment-dependent generation of photoacoustic waves from plasmonic nanoparticles. Small 2012;**8**:47–52.

[175] Bell CS, Yu SS, Giorgio TD. The multistrata nanoparticle: an FeOx/Au core/shell enveloped in a silica-Au shell. Small 2011;**7**:1157–1162.

[176] Bayer CL, Chen YS, Kim S, Mallidi S, Sokolov K, Emelianov S. Multiplex photoacoustic molecular imaging using targeted silica-coated gold nanorods. Biomed Opt Express 2011;**2**:1828–1835.

[177] Link S, El-Sayed MA. Shape and size dependence of radiative, nonradiative, and photothermal properties of gold nanocrystals. Int Rev Phys Chem 2000;**19**:409–453.

[178] Chen LC, Wei CW, Souris JS, Cheng SH, Chen CT, Yang CS, Li PC, Lo LW. Enhanced photoacoustic stability of gold nanorods by silica matrix confinement. J Biomed Opt 2010;**15**:016010.

[179] Stöber W, Fink A, Bohn E. Controlled growth of monodisperse silica spheres in the micron size range. J Colloid Interf Sci 1968;**26**:62–69.

[180] Kalele S, Gosavi SW, Urban J, Kulkarnil SK. Nanoshell particles: synthesis, properties and applications. Curr Sci 2006;**91**:1039–1053.

[181] Alwi R, Telenkov S, Mandelis A, Leshuk T, Gu F, Oladepo S, Michaelian K. Silica-coated super paramagnetic iron oxide nanoparticles (SPION) as biocompatible contrast agent in biomedical photoacoustics. Biomed Opt Express 2012;**3**:2500–2509.

[182] Namura K, Suzuki M, Nakajima K, Kimura K. Photoacoustic emission from Au nanoparticles arrayed on thermal insulation layer. Opt Express 2013;**21**:8689–8700.

[183] Egerev S, Ermilov S, Ovchinnikov O, Fokin A, Guzatov D, Klimov V, Kanavin A, Oraevsky A. Acoustic signals generated by laser-irradiated metal nanoparticles. Appl Opt 2009;**48**:C38–C45.

6

NANOPARTICLES FOR BIOIMAGING: ANALYTICAL CHARACTERIZATION AND MEASUREMENTS

Kate Nelson

Nano Research Facility, Washington University School of Medicine, St. Louis, MO, USA

Patrick Winter

Cincinnati Children's Hospital, Imaging Research Center, Cincinnati, OH, USA

Monica Shokeen, Steven Wang, and Mikhail Y. Berezin

Department of Radiology, Washington University School of Medicine, St. Louis, MO, USA

6.1 INTRODUCTION

Rapid development of nanoparticles for biomedical applications has, to a large extent, been based on a solid analytical foundation (both theoretical and instrumental) built in the previous decades. Such widespread methods such as transmission electron microscopy (TEM), scanning electron microscopy (SEM), and inductively coupled plasma mass spectrometry (ICP-MS) appeared long before the rise of nanoparticles and have been quickly adopted by the bioimaging communities that utilize nanoparticles in their research. Strict limitations to toxicity, hydrodynamic size, and surface charge that govern the absorption, biodistribution, and route elimination led to the refinement of the existing techniques to ensure the biocompatibility of the imaging nanoparticles to humans. In parallel, the advances in instrumentation have led to a number of methods that evaluate the mechanisms of nanoparticle cellular uptake and transport to predict *in vivo* behavior of the nanoparticles.

Nanotechnology for Biomedical Imaging and Diagnostics: From Nanoparticle Design to Clinical Applications, First Edition. Edited by Mikhail Y. Berezin.

The characterization of nanoparticles for imaging usually starts with size measurements and elemental analysis to warrant that the nanoparticles are formed as planned and stay intact upon storage. Hence, Sections "Elemental analysis" and "Size analysis" will provide an account of the general characterization methods that are applied to almost all classes of nanoparticles. We will then describe several more specialized methods for measuring the strength of the reporting signals for the most common classes of imaging nanoparticles: nuclear (Section "Radioactivity measurement of nanoparticles"), magnetic resonance imaging (MRI) (Section "Magnetic properties of nanoparticles"), and optical (Section "Optical techniques"). The signal strength is used to compare the nanoparticles and to evaluate the dosage for preclinical and clinical imaging studies. Sufficient details will be provided regarding the corresponding hardware to help the reader identify the right equipment for the specific project (Table 6.1). Finally, we highlight several emerging techniques that hold great potential in nanoparticle characterization. Evaluation of specificity and activity of the nanoparticles toward their biological targets will not be considered in this chapter; the readers are referred to other chapters of this book that provide a compelling account of these topics.

6.2 ELEMENTAL ANALYSIS

The determination of composition and concentration of a nanoparticle sample is often necessary for biomedical applications where dosage and toxicity effects may be of concern. A variety of elemental analysis methods are used to both determine the presence of an element in a sample and quantify how much of said element is present. While these methods are most commonly employed for inorganic particle types, they may also be used for some organic particles. In addition to characterizing a nanoparticle sample itself, these elemental analysis methods can be used to characterize the retention of nanoparticles within tissues and biological fluids, making them useful for bioretention, toxicity, and localization studies.

Elemental analysis by atomic spectroscopy techniques is commonly used for biomedical particles containing metals such as gold, silver, copper, manganese, gadolinium, iron, etc. The most frequently employed atomic spectroscopy techniques are inductively coupled plasma mass-spectrometry (ICP-MS) and inductively coupled plasma optical emission spectrometry (ICP-OES). Both techniques are capable of simultaneous detection and high-precision quantification of most metal and nonmetal elements in a liquid sample. Typically, nanoparticle samples are dissolved in an acid solution before analysis. The intensity of the signal produced by analytes of interest in the dissolved sample is compared to the signals of a set of external calibration standards of known concentration in order to determine the concentration in mass per volume units of the sample. When concentration data on particles is combined with size and morphology information, it is often possible to estimate the concentration in terms of particles per volume.

TABLE 6.1 Imaging Nanoparticles Characteristics and Most Common Analytical Methods

Method	Abbreviation	Concentration	Particle and hydrodynamic size	Elemental analysis and composition	Surface charge	Shape	Reporting activity	Quality control
Inductively coupled plasma mass spectrometry	ICP-MS	+++		+++				+
Inductively coupled plasma optical emission spectrometry	ICP-OES			+++				
Energy-dispersive X-ray spectroscopy	EDS			+++				
Transmission electron microscopy	TEM		+++					
Scanning electron microscopy	SEM		+++					
Dynamic light scattering	DLS		+++					
Zeta potential					+++			+
Steady-state absorption		+++					+++	+++
Steady-state emission		+++					+++	+++
Fluorescence anisotropy			++			+		+
Fluorescence lifetime	FLT						++	
Fluorescence correlation spectroscopy	FCS	+++	++					
Flow cytometry			+					+
Chromatography			+++					
Raman spectroscopy				++				
Fourier transform near-infrared spectroscopy	FT-NIR			++				
Nuclear magnetic resonance	NMR		+	+++			++	+
Quartz crystal microbalances		+						
Electrophoresis, gel, and capillary		+	++		+			+++
Radiography		+++						+++

+++Highly applicable; ++Applicable; +Occasionally used.

6.2.1 ICP-MS

This method is widely regarded as the gold standard for elemental composition and concentration analysis due to its low (ng/l) detection limits, ability to quickly analyze nearly all metal and nonmetal elements and isotopes, and tolerance for both simple and complex sample matrices. Generally, a liquid sample is aspirated and introduced into an argon plasma (6,000–10,000 K), which breaks the sample down into constituent atoms, some of which are ionized and then flow into a vacuum chamber where a quadrupole mass spectrometer separates the ions based on their mass-to-charge ratio before they hit an electron multiplier and are detected as electrical signal. An overall view of the instrument is shown in Figure 6.1. Since ICP-MS relies on the production of elemental ions, it is not useful for the analysis of elements that prefer to form negative ions, such as chlorine and bromine, and cannot be used to analyze the noble gases or carbon, nitrogen, oxygen, and hydrogen. As all samples are broken down into constituent elements, ICP-MS is not useful for characterization of organic molecule-based nanoparticles, except in those instances when the molecule has an associated characteristic, or tracer, element. (For instance, sulfur can be detected in order to estimate the amount of thiol in a sample.) As it has low detection limits, trace quantities of nanoparticles retained within the body can be detected, making it an ideal candidate for bioretention studies.

6.2.2 ICP-OES

This method is a very simple and robust atomic spectroscopy technique that serves as a complementary tool to ICP-MS. Like ICP-MS, a liquid sample is aspirated and introduced into argon plasma, which breaks the sample down into constituent atoms;

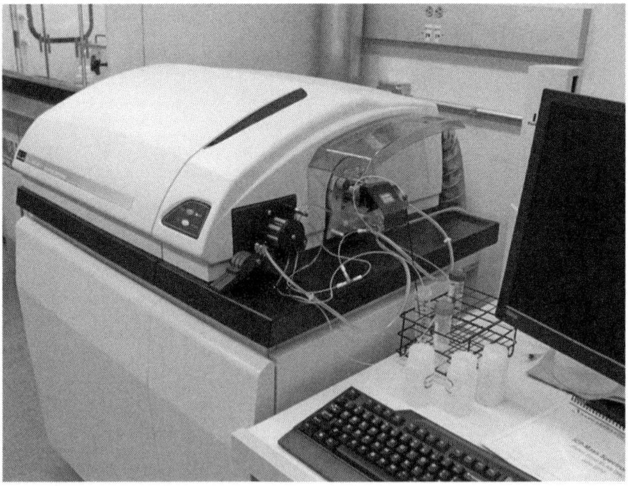

FIGURE 6.1 ELAN DRC II inductively coupled plasma mass spectrometer (ICP-MS) from Perkin Elmer.

however, in ICP-OES, elements are detected by the optical emissions emitted from the excited atoms in the sample and not by mass-to-charge ratio. Each element releases emissions at certain characteristic wavelengths; in an ICP-OES, the full spectrum of emissions is collected through a viewing window, and the wavelengths are subsequently separated using a series of gratings and prisms and then detected on a CCD as electrical signal. The detection limits on the ICP-OES are typically at least one order of magnitude higher than those on ICP-MS, and it also cannot be used to measure isotopic ratios; however, as ICP-OES can tolerate sample matrices with higher total dissolved solids, it can be useful for samples in concentrated saline and buffer solutions. The ICP-OES is also a good option for analysis of elements like sulfur and iron that experience significant interferences on ICP-MS. An overall view of the instrument is shown in Figure 6.2.

One drawback associated with atomic spectroscopy analysis of nanoparticles in liquid suspensions and within biological samples is that when the sample is dissolved for analysis, it becomes impossible to distinguish between the mass of an element that was bound in particle form and the mass of the same element that may have been dissolved in the sample. Typically, the vast majority of the mass of an element in a particle sample is coming from the particles and not ions dissolved in the solution; however, the contribution of ions produced by a dissolving particle is now getting some attention in the toxicity field. Recent advances in single particle inductively coupled plasma mass spectrometry (SP–ICP–MS) are beginning to address this issue of differentiation between dissolved and particle-bound mass. With a sufficiently fast detector speed, it is possible to detect single nanoparticles in a dilute solution as pulses of signal and differentiate them from ions of the same element, which produce a constant signal.

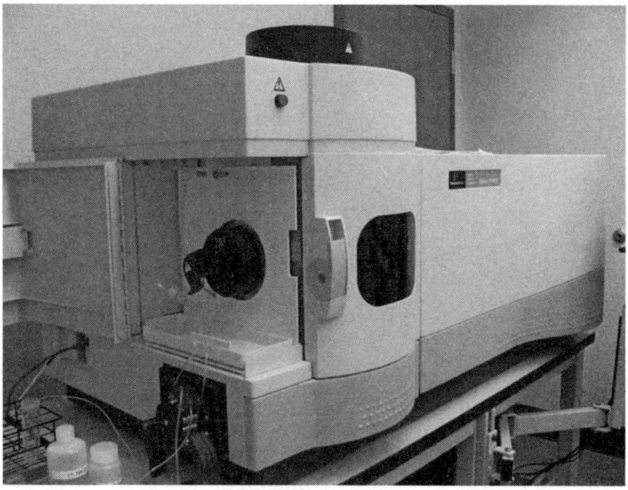

FIGURE 6.2 Optima 7300 DV inductively coupled plasma optical emission spectrometer (ICP-OES) from Perkin Elmer.

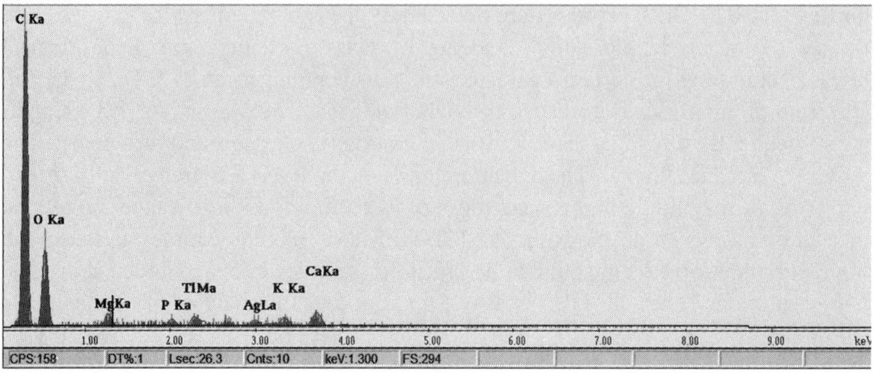

FIGURE 6.3 An example of an energy-dispersive X-ray spectrograph collected using an EDAX Genesis system.

A very different elemental analysis method is *energy-dispersive X-ray spectroscopy (EDS)*. This technique is commonly used in concert with electron microscopy where the electron beam from a scanning electron microscope or transmission electron microscope is used to excite a solid sample, which then emits X-rays that are detected as an electrical signal. The X-rays are emitted from the sample with energy characteristic to each element found within the sample. An example of an energy-dispersive X-ray spectrograph collected using an EDS system is shown in Figure 6.3. This technique is able to detect most elements heavier than beryllium, including carbon, nitrogen, and oxygen. The detection limits for EDS are significantly higher than those for atomic spectroscopy techniques as elements that compose less than 10% of a sample by weight may be indistinguishable from background noise. Quantitative analysis by EDS allows for the relative percent composition of a sample to be determined. EDS signals are sensitive to sample height, and the best results are achieved on smooth polished surfaces. Particle samples should be dispersed evenly on a flat surface for analysis and may exhibit a 5% or greater measurement variance. EDS is often used for elemental mapping of solid samples; however, this feature is rarely useful for nanoparticle studies as the spatial resolution of EDS is measured in microns and not nanometers.

6.3 SIZE ANALYSIS

As material properties at the nanoscale often differ from properties at micro- and macroscales, great attention is often paid to particle sizing analysis. Imaging techniques can provide information both about the size and morphology of particles, while statistical techniques such as dynamic light scattering (DLS) provide more information about a particle's stability in a fluid and the distribution of particle sizes in the sample.

6.3.1 TEM

TEM is one of the most popular tools for characterizing nanoparticles as it can provide high-resolution images of particles ranging in size from less than one nanometer to hundreds of nanometers. In a TEM, electrons are fired from an electron beam at voltages ranging from 20kV to about 120kV toward an ultrathin sample (hundreds of nanometers or less) within a vacuum chamber, the electrons interact with the sample, and an image is formed from the electrons that are transmitted through the sample and are focused onto an imaging plane located below the sample holder. An example of a TEM instrument is shown in Figure 6.4. In most modern TEMs, the focused image is collected using a CCD camera. As this technique utilizes electrons that have been transmitted through a sample, the image produced is a two-dimensional (2D) profile, which for many particle samples allows for easy measurement of particle diameter or edge length. However, as TEM is a high-resolution technique, the field of view is very limited, making collection of statistically relevant numbers of particle sizes a time-consuming task. For this reason, TEM is most often used to confirm morphology and sizing results obtained via different methods.

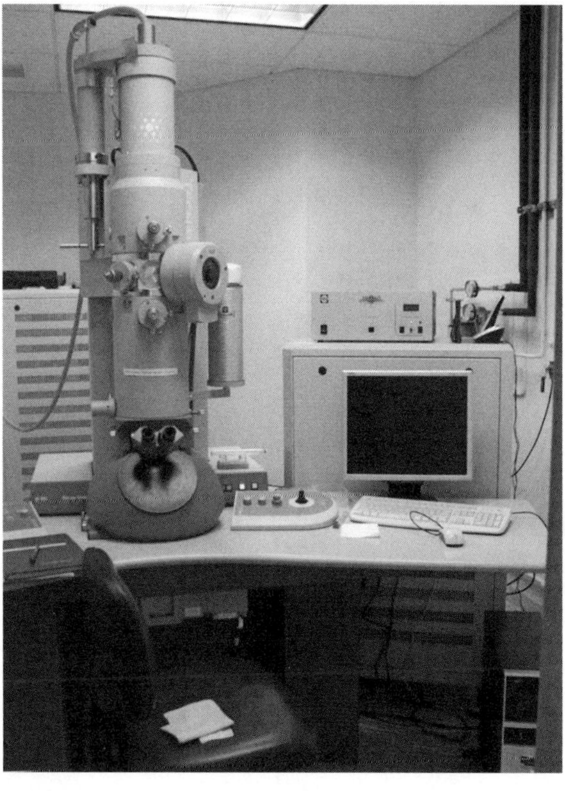

FIGURE 6.4 Tecnai Spirit transmission electron microscope from FEI Company.

Most of the contrast produced by samples on TEM is due to varying thickness and density of the sample. Higher atomic number elements and thicker particles allow for less transmission of electrons and therefore create a darker region in the image. In the case of organic particles that are composed primarily of low atomic number elements, producing enough contrast to image the particles can be an issue. One way to address low contrast in organic samples is to use a lower accelerating voltage. Another, more commonly used technique, is negative staining in which a sample is treated with a solution containing a high atomic number element that will absorb electrons (uranium, gold, lead, osmium). Successful negative staining depends on the chemistry of the stain and sample interaction. A successfully stained sample may appear either light with a dark background or dark with a light background (an example is shown in Figs. 6.5 and 6.6).

Particle sample preparation for TEM is usually a fairly simple matter involving drop casting or dip coating of particles in a solution onto a standard carbon-coated copper grid and allowing them to dry. For organic particles, drop casting or dip coating of particles is often followed by a negative staining procedure. For particles that cannot tolerate drying, a cryogenic sample holder can be used to image particles frozen in liquid nitrogen.

TEM is also used for imaging of intracellular features of cell and tissue samples. It is sometimes possible to image particles present in cell and tissue samples; however, such imaging requires extensive sample preparation including fixation, polymerization, ultrathin sectioning, and negative staining of the sample. As most tissue and cell samples are measured in microns or millimeters, locating nanoparticles can be a time-consuming affair. Likewise, this technique is not useful for particles that may be difficult to distinguish from naturally occurring organelles and cellular features.

FIGURE 6.5 Transmission electron micrograph of gold "nanocages" drop cast on a TEM grid.

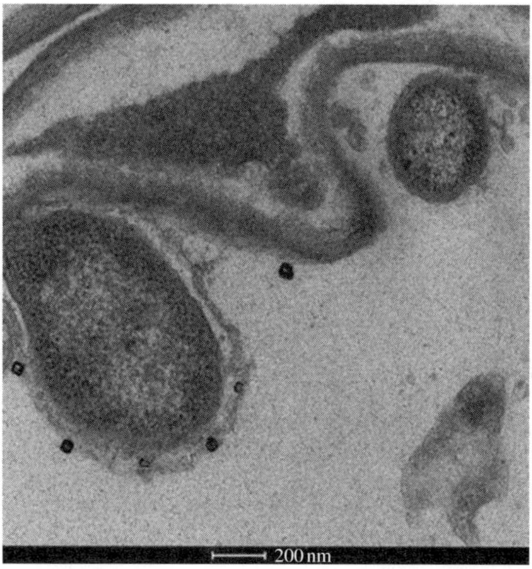

FIGURE 6.6 Transmission electron micrograph of gold "nanocages" within the root system of a royal bean plant. The plant sample underwent fixation, polymerization, ultrathin sectioning, and negative staining prior to imaging.

6.3.2 SEM

SEM is a surface imaging technique that is also widely used for characterizing nanoparticle size and shape. In a SEM, electrons are fired from an electron beam at voltages ranging from less than 1 kV to about 30 kV toward a sample in a vacuum chamber; the electrons collide with the sample and are reflected back toward detectors located above and to the sides of the sample where a CCD camera collects an image. An example of the instruments is shown in Figure 6.7. Most SEM imaging is accomplished by collecting the signal from low-energy secondary electrons. These images have a three-dimensional (3D) appearance as contrast is produced primarily by contours and surface features of the sample that alter the trajectory of the reflected secondary electrons and hence alter the intensity of detected electrons at any one point on the sample. While SEM can provide subnanometer spatial resolution, it is most useful for imaging particles tens of nanometers to several microns in size (an example is shown in Fig. 6.8). The magnification range on an SEM extends from about 10 times to greater than 100,000 times and can provide a much larger field of view than TEM and so is more useful for determining statistically relevant particle sizes.

Particle sample preparation for SEM involves securing particles to a flat, conductive substrate. For particles in a dry powder form, carbon tape is often used to secure particles to a standard aluminum stub. Particles in liquid suspensions are usually drop cast or spin coated on a piece of silicon wafer (Fig. 6.9) and allowed to dry before imaging. SEM is very sensitive to charging effects; therefore, nonconductive samples that would accumulate charge are usually sputter coated with a thin layer of

FIGURE 6.7 Nova NanoSEM 230 scanning electron microscope from FEI Company.

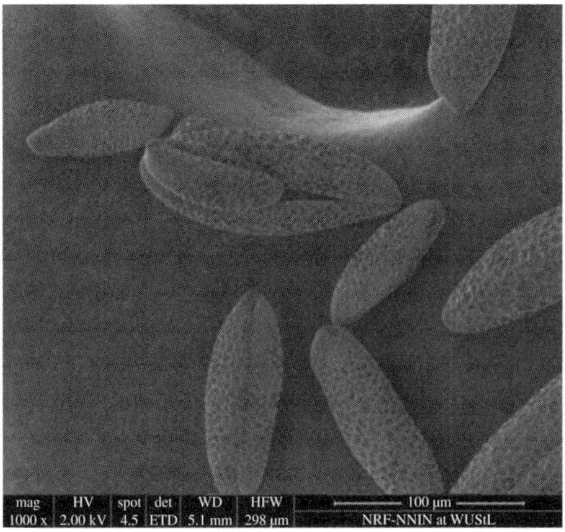

FIGURE 6.8 Scanning electron micrograph of plant pollen on carbon tape sputter coated with gold before imaging.

a conductive material such as gold. In cases where sputter coating is not desirable, charging effects on nonconductive sample may be reduced by using lower accelerating voltages or by using the low-vacuum or environmental mode available on some SEMs (Fig. 6.10). In low-vacuum mode, charging effects on nonconductive samples are eliminated by introducing water or solvent vapor into the sample chamber at the expense of some resolution and contrast.

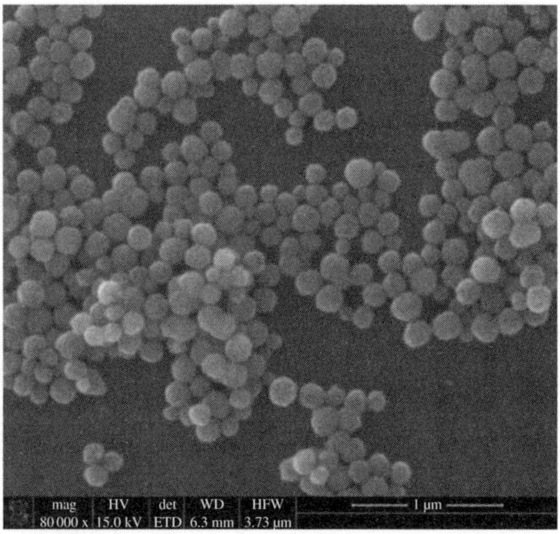

FIGURE 6.9 Scanning electron micrograph of magnetic nanoparticles drop cast on a piece of silicon wafer.

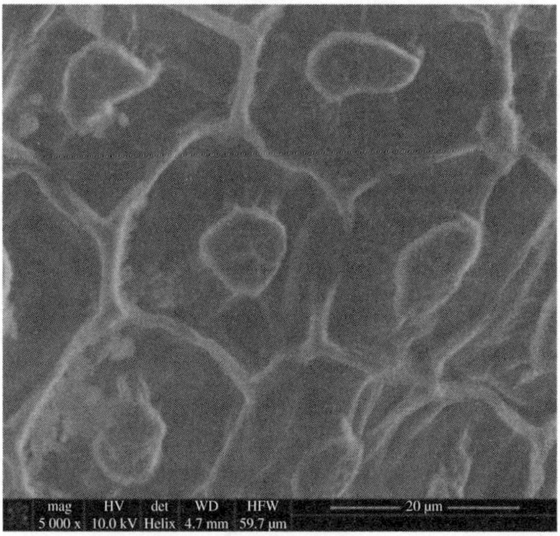

FIGURE 6.10 Scanning electron micrograph of the surface of an *Arabidopsis thaliana* seed imaging without sputter coating in low-vacuum mode.

6.3.3 DLS

DLS is a statistical particle sizing technique that can provide size distributions of the hydrodynamic radius of particles in the subnanometer to micron size range. In DLS, particles suspended in solution are illuminated with a laser; as the particles move due to Brownian motion, they scatter the laser light, creating fluctuations in the intensity of light that reaches the detector. As the speed of the Brownian motion of particles is dependent upon their size, by monitoring the fluctuations in light intensity on microsecond timescales, size information can be obtained. The intensity data is fit to a model that calculates the translational diffusion coefficient, which using the Stokes–Einstein equation can be used to estimate the hydrodynamic radius of a particle. As Stokes–Einstein assumes a spherical particle, DLS is not a useful technique for obtaining sizing information of particles with large aspect ratios. Monitoring the light scattering of a particle suspension provides information about large numbers of particles, giving both a particle size distribution and a statistically valid average size.

Sample preparation for DLS measurements involves creating a stable particle suspension. Water, organic solvents, and biologically relevant fluids such as saline and buffer solutions can all be used to suspend particles. Sonication, vortex mixing, and dilution can all be used to help create a stable particle suspension. DLS works equally well for inorganic and organic particles; however, samples that flocculate and settle out of solution, fluoresce, or are extremely polydisperse create artifacts that can lead to poor data quality. Polydisperse samples can be filtered or centrifuged to remove large aggregates or contaminants. For analysis of fluorescent particles, optical filters can be installed on a DLS system to eliminate the detection of emissions from the particles, or an alternate wavelength laser can be used for illuminating the sample.

While DLS is very easy to use, due to the large amount of mathematical processing that is conducted by the system, the results are often not well understood by users and can be easily misinterpreted. Many modern DLS systems (such as shown in Fig. 6.11) may provide at least four different interpretations of the same sizing data including intensity-based particle size distribution (Fig. 6.12), volume-based particle size distribution, number-based particle size distribution, and a "z-average." It is entirely possible for all four interpretations to be valid yet provide different results for a single sample; therefore, those utilizing the technique should be familiar with the assumptions used and applications of the different forms of data interpretation.

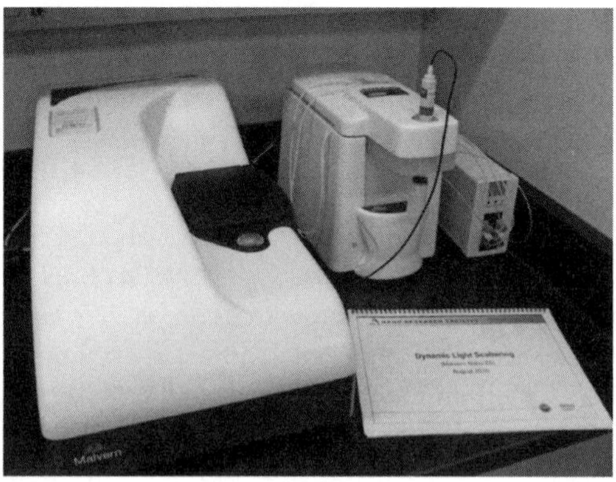

FIGURE 6.11 Zetasizer Nano ZS dynamic light scattering system from Malvern Instruments.

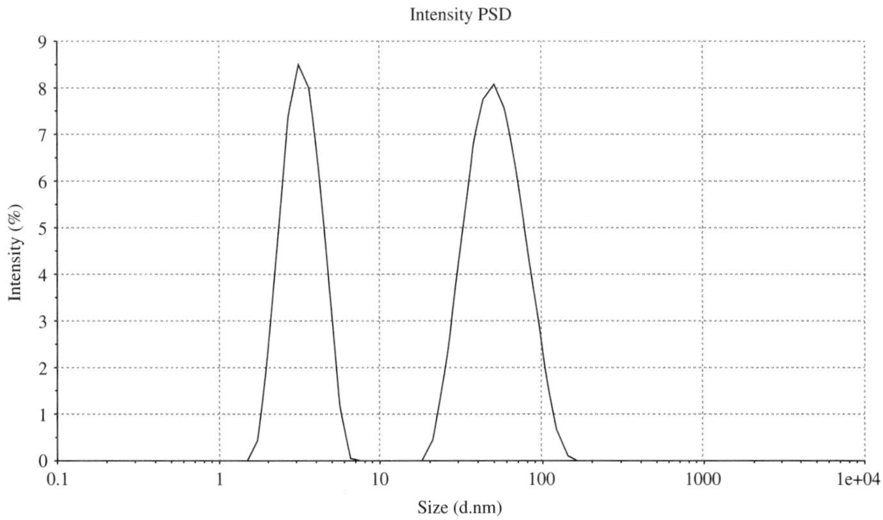

FIGURE 6.12 An example of an intensity-based particle size distribution obtained using dynamic light scattering.

6.4 SURFACE ANALYSIS TECHNIQUES

One of the most commonly used surface analysis tools is scanning probe microscopy (SPM). SPM is a tool set that allows for the characterization of surface topology as well as physical properties such as surface roughness, sample hardness, electrostatic forces, and magnetic forces. Atomic force microscopy (AFM) refers to SPM techniques in which the force of interaction between a very sharp tip/probe and the sample provides information about the physical structure and/or properties of a sample. To obtain information about a sample, an AFM system monitors laser light that is reflected off a cantilever that has a sharp probe at the end and onto a photodetector. As the probe on the cantilever interacts with the sample, the cantilever may bend and/or twist, causing the position of the laser light on the photodetector to change. This deflection in the laser light is interpreted by the system to provide information on topography and various interactive forces. When measuring topography, AFM systems are usually capable of angstrom-level vertical resolution; however, horizontal resolution is limited by the shape of the probe employed. Most general-purpose AFM probes have a tip diameter of about 10 nm; however, recent advances have made new probes with ultrathin tips (such as those with a carbon nanotube tip) much more common. An example of an AFM system is shown in Figure 6.13.

One of the strengths of AFM is that it requires very little sample preparation. Samples may be wet or dry and need only be relatively flat and secured to a substrate. Mica sheets are typically used as a substrate for particle sample preparation as mica has a very smooth surface. Two commonly used methods of preparing suspended

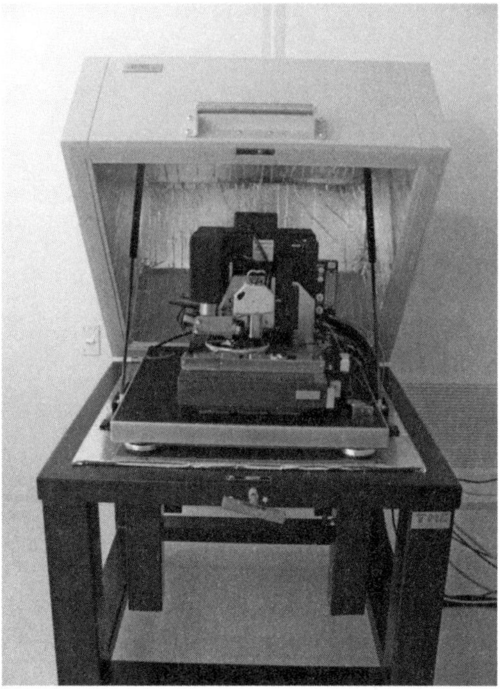

FIGURE 6.13 Veeco NanoMan scanning probe microscope.

particle samples for AFM analysis are drop casting and spin coating. In cases where particles do not adhere well to the mica surface, the mica can be treated with an ionic solution such as magnesium chloride to enhance particle attachment. In comparison to other tools used for visualization of particles, AFM is usually more time-consuming, requiring minutes instead of seconds for a single scan, and can be more difficult for amateurs to utilize effectively.

An important property of materials in solutions, zeta potential is a measurement of the charge exhibited by a particle within a specific solution environment. Zeta potential can be defined as the potential on a particle at the edge of the electric double layer and is directly related to the electrophoretic mobility of a particle. Measurements of electrophoretic mobility and zeta potential are commonly conducted using DLS systems where the speed of particles moving within an electric field is measured by monitoring fluctuations in light scattered by the moving particles. It is worth noting that zeta potential is different from surface charge and is heavily dependent on the solution environment, including pH and ionic strength. It does, however, provide critical information on the stability of a particle within a solution environment and the favorability of electrostatic interactions with other materials. For example, a particle that has a highly negative zeta potential and hence remains as a stable dispersion in water may have a neutral zeta potential in phosphate buffer solution and therefore tend to aggregate.

6.5 RADIOACTIVITY MEASUREMENT OF NANOPARTICLES

Radioactivity measurement involves detection of "radioactive decay" of the radionuclide. Common radioactive decay process involves alpha (α, ^{4}He) decay, beta (β, e^-) decay, positron (β^+) decay, or gamma (γ, photon) decay. The radioactive decay can be detected by a variety of radiation detectors such as survey meters, dose calibrators (gas-filled detectors), inorganic and liquid scintillation detectors, and semiconductor (Ge, Si) detectors. Nuclear medicine imaging modalities such as planar imaging, single-photon emission computed tomography (SPECT) imaging, and positron emission tomography (PET) imaging involve the use of radioactive decay property of radionuclides for detection and therapy of disease. A gamma camera obtains an image of the distribution of a radiopharmaceutical in the body (or organ) by detection of emitted γ-rays. Gamma cameras are used for planar and SPECT imaging. Common gamma-emitting radionuclides are Ga-67 ($t_{1/2}$=78.26 h), Tc-99m ($t_{1/2}$=6.02 h), In-111 (2.83 d), I-123 (13.2 h), I-131 (8.04 d), and Tl-201 (3.04 d). PET relies on the detection of coincident 511 keV photon pairs generated when a positron annihilates with an electron in tissue. Positron-emitting radionuclides commonly used clinically and preclinically are C-11 ($t_{1/2}$=20.38 m), N-13 ($t_{1/2}$=9.965 m), O-15 ($t_{1/2}$=122.24 s), F-18 ($t_{1/2}$=109.77 m), Cu-64 ($t_{1/2}$=12.7 h), Ga-68 ($t_{1/2}$=68 m), Zr-89 (78.5 h), Br-76 (16.2 h), and Rb-82 ($t_{1/2}$=1.3 m). Both PET and SPECT are considered highly sensitive, quantitative, and translational to clinic.

Radioactivity measurement of nanoparticles is a critical step before using the radiolabeled nanoparticle for *in vitro* or *in vivo* applications. The radioactivity measurement quantifies the radiolabeling efficiency (percent labeling yield), distinguishes the "free" radioactive label from the conjugated radioactive label, and determines the specific activity (amount of radioactivity per unit of mass of a radionuclide or labeled compound) of the radiolabeled nanoparticle. Common units of specific activity are Ci/mmol or Bq/mmol, and specific activity is inversely proportional to the half-life ($t_{1/2}$) of the radionuclide. Accurate measurement of radioactivity is considered key to developing effective radiolabeling conditions. In principle, higher specific activity is desirable as high specific activity translates to high-detection sensitivity toward targeted proteins *in vivo* under the confines of "tracer" principle. However, one of the challenges with labeling nanoparticles or small molecules with radionuclides is the presence of significant amounts of cold ("nonradiometal") impurities that interfere with effective radiolabeling at high specific activities.

The high specific activity of the radionuclide would translate into high specific activity of the radioconjugated nanoparticle. A common method used to determine the effective specific activity (ESA) for the commonly used PET radionuclide Cu-64 is by radiolabeling titrations with 1,4,7,10-tetraazacyclododecane-1,4,7,10-tetraacetic acid (DOTA) and/or 1,4,8,11-tetraazacyclotetradecane-1,4,8,11-tetraacetic acid (TETA). Briefly, the titration involves mixing macroscopic quantities of copper, using ^{64}Cu as a tracer, with DOTA or TETA on an equimolar basis. The minimum DOTA/TETA concentration where 100% labeling occurs is assumed to be equal to the concentration of Cu(II) present [1]. Recently, an elegant LC-MS approach was proposed to quantify nonradioactive transition metal impurities in metal radionuclides [2].

Since there is a huge variability in nanoparticle composition (organic vs. inorganic/metal based), standard radiolabeling and purification steps are not applicable across the board. However, the end-term characterization goals are the same. One of the critical challenges with determining the specific activity of radiolabeled nanoparticles is that occasionally determining the "exact mass" of the column eluted/purified radiolabeled nanoparticle solution is not exactly known, primarily because the molar absorptivity of the nanoparticles at low concentrations is not enough for ultraviolet (UV) detection. Researchers typically approach this challenge by making the "theoretical mass calculation" based on experience and calibration curves constructed by using known standards.

Standard steps in the characterization of organic nanoparticle radiolabeled with a radionuclide involve the use of chromatography techniques [3]. A radioactive/UV chromatogram can be utilized in the following fashion:

1. A fast protein liquid chromatography (FPLC) or radio thin-layer chromatography (R-TLC) chromatogram of the mixture of radionuclide and nanoparticle before purification (Fig. 6.14). Ideally, a radioactive peak and the corresponding UV peak (FPLC) will appear for both nanoparticle radioactive conjugate and "free" unconjugated radioactive peak. This scan will determine the radiolabeling efficiency.
2. The chromatogram should be recollected after purification (such as passage of the reaction mixture over a Zeba desalting column). This scan will determine the purity of the radiolabeled nanoparticle solution.

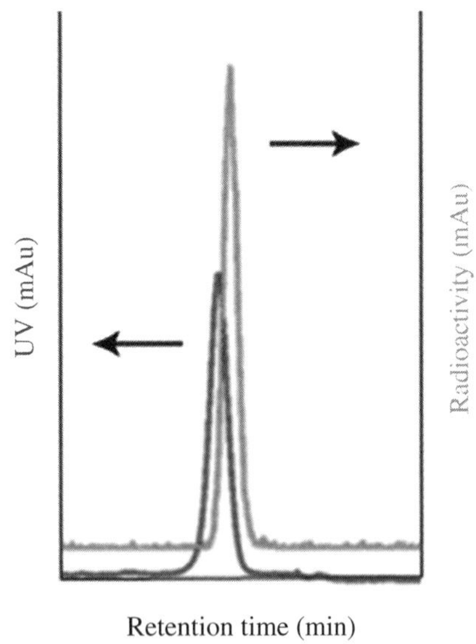

FIGURE 6.14 A schematic of radio and UV spectrum of a radiolabeled nanoparticle conjugate.

6.6 MAGNETIC PROPERTIES OF NANOPARTICLES

An MRI system consists of a wide variety of components working together to form the desired images [4]. These components include the magnet, shim coils, gradient coils, radio-frequency (RF) coils, amplifier systems, RF transmission and receiving electronics, and computer systems that synchronize the components, control the electronics, and process the imaging data. The MRI magnet produces a large uniform magnetic field, denoted as B_0. The maximum field strength currently used in clinical imaging is 3.0T, whereas experimental scanners may exceed 7T. Inside the bore of the magnet, gradient coils produce spatially variable magnetic fields that perturb B_0 along the x-, y-, and z-directions. The RF coil is also positioned within the magnet bore to generate an oscillating magnetic field perpendicular to B_0, denoted as B_1. As a basic analogy, the RF coil acts like an antenna that generates and receives the signals that will be used to construct the image. When exposed to the large B_0 magnetic field, protons in the body act like little bar magnets and line up parallel to B_0. The B_1 field rotates the proton magnetization, typically by 90° or 180°, and then detects the proton signals as they relax back to equilibrium. By convention, the B_0 field is aligned along the z-axis, and the B_1 field rotates the proton signal into the x–y plane. The z-axis is often called the longitudinal axis, while the x–y plane is called the transverse plane. Relaxation of the protons is governed by two mechanisms, longitudinal and transverse relaxations, which are commonly utilized to generate MRI contrast both in clinical applications and research studies.

6.6.1 Relaxation

Following rotation of the proton magnetization into the x–y plane, the mechanism by which protons relax back to equilibrium along the z-axis is called longitudinal relaxation or spin–lattice relaxation. Longitudinal relaxation is an exponential process given by

$$M_z = M_z^0\left(1 - \mathrm{e}^{-t/T_1}\right)$$

where M_z^0 is the equilibrium proton magnetization, M_z is the magnetization along the z-axis at time t, and T_1 is the time constant for longitudinal relaxation. Another relaxation process is called transverse relaxation or spin–spin relaxation. This process occurs when protons exchange energy but do not loose energy to their surroundings. This results in loss of proton magnetization in the transverse plane, but does not necessary generate magnetization along the z-axis. Transverse relaxation is an exponential process given by

$$M_{xy} = M_{xy}^0 * e^{-t/T_2}$$

where M_{xy}^0 is the initial transverse magnetization, M_{xy} is the transverse magnetization at time t, and T_2 is the time constant for transverse relaxation. In addition to T_2

relaxation, inhomogeneities in the main magnetic field also contribute to the decay of transverse magnetization. The time constant for the combination of these two factors is called T_2* and is given by

$$\frac{1}{T_2^*} = \frac{1}{T_2} + \frac{1}{T_2^{\text{inhomo}}}$$

To generate an image, the MRI scanner plays out an imaging pulse sequence. This is a train of RF pulses, magnetic field gradients, and signal acquisitions required to generate the proton signal, record the spatially encoded signals, and reconstruct a set of 2D or 3D images. To generate the proton signal, the RF coil produces a B_1 field that rotates the magnetization in the body from the *z*-axis into the *x*–*y* plane. To spatially localize the proton signal, the magnetic field gradients are turned on during the B_1 field, after the B_1 field, or during signal acquisition. Turning the gradients on and off requires a certain period of time, during which the proton signal decays due to T_2* relaxation. To generate a measurable proton signal for the MRI scanner to record, an echo must be formed using either a spin echo pulse sequence or a gradient echo sequence.

A spin echo is formed by applying a 180° pulse with the RF coil after the 90° pulse. The 180° pulse reverses the proton signals within the transverse plane, refocusing the signal loss due to the inhomogeneities in the B_0 field. As a result, an echo signal is formed at the echo time (TE), which is two times longer than the time between the 90° and 180° pulses. Under certain circumstances, applying a 180° pulse is not feasible, and a gradient echo pulse sequence is required. The gradient echo is formed by applying a negative gradient after the 90° pulse to dephase the proton signal in the *x*–*y* plane, followed by a positive gradient to rephase the signal and form the echo. Typically, a gradient echo can be formed with a shorter TE than is possible with a spin echo sequence, and they are often employed with fast imaging pulse sequences. The B_0 field inhomogeneities are not refocused due to the lack of a 180° pulse, causing gradient echo images to be T_2* weighted.

In order to collect all the data to reconstruct an image, a single echo is not sufficient. All pixels in the image need to be spatially encoded in three dimensions, which requires the pulse sequence to be repeated many times. However, if the pulse sequence is repeated too fast, the signal will be very low due to inadequate T_1 relaxation back along the *z*-axis. The time between successive applications of 90° pulses is referred to as the repetition time (TR).

As stated earlier, the relaxation mechanisms are often employed to generate MRI contrast. T_1-weighted images are designed to accentuate the differences in T_1 relaxation times between various tissues. T_1-weighted images can be produced with a pulse sequence with a short TR and short TE. The short TE minimizes the effects of variations in T_2 values from different tissues. However, the short TR alters the signal intensity of tissues with different T_1 values. Tissues with short T_1 times can fully relax back to equilibrium within the short TR and appear bright on the final image. Tissues with long T_1 values cannot fully relax before the next 90° pulse and appear dark.

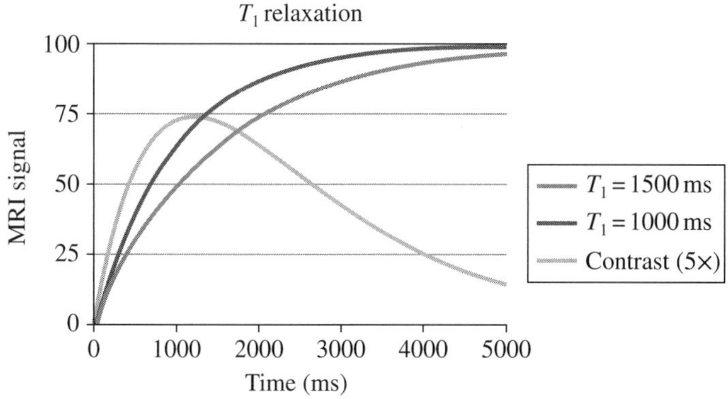

FIGURE 6.15 MRI signal intensity (arbitrary units) for tissues with T_1 values of 1500 ms or 1000 ms over a range of TR values. The tissue with the shorter T_1 relaxes back to equilibrium faster than the tissue with the longer T_1. Subtracting the MRI signals provides the image contrast between the two tissues (expanded fivefold to aid visualization). The maximum image contrast occurs with a TR of 1200 ms.

To measure the T_1 relaxation time of a tissue, images are collected with different TR values. The MRI signal is plotted versus the TR, and the data is fit to the longitudinal relaxation equation. As shown in Figure 6.15, a tissue with a T_1 of 1000 ms relaxes back to equilibrium faster than a tissue with a T_1 of 1500 ms. Subtracting the signal with a T_1 of 1000 ms from the signal with a T_1 of 1500 ms yields the image contrast (Fig. 6.15, expanded by fivefold to aid visualization). The maximum contrast between these tissues would be obtained with an image using a TR of 1200 ms.

T_2-weighted images have been developed to highlight differences in the T_2 relaxation of different tissues. T_2-weighted images are created when both the TR and TE are long. The long TR allows all the spins to relax back to equilibrium before the next imaging cycle, minimizing T_1 relaxation effects. However, the long TE allows distinction to be made between tissues with different T_2 times. Tissues with a short T_2 will lose a great deal of signal during TE and will appear dark. However, tissues with a long T_2 will not relax as much during TE and will appear bright.

To measure T_2 relaxation times, images are collected with different TE values. The MRI signal is plotted against the TE, and the data is fit to the transverse relaxation equation. Figure 6.16 shows that the tissue signal will decay more rapidly with a T_2 of 100 ms compared to 150 ms. Subtracting these two signals demonstrates that the image contrast (Fig. 6.16, expanded by fivefold to aid visualization) is maximized in an image with a TE of 120 ms.

Images collected with a long TR and a short TE will produce a proton density-weighted image. The long TR allows all the spins to return to their relaxed state before the start of the next sequence. In addition, the short TE minimizes the effects of T_2 relaxation. Proton density images tend to display poor tissue contrast because most parts of the body have essentially the same concentration of protons.

As with many medical imaging modalities, intravenous contrast agents are often used in MRI to enhance the signal of certain tissues in the body. MRI contrast agents

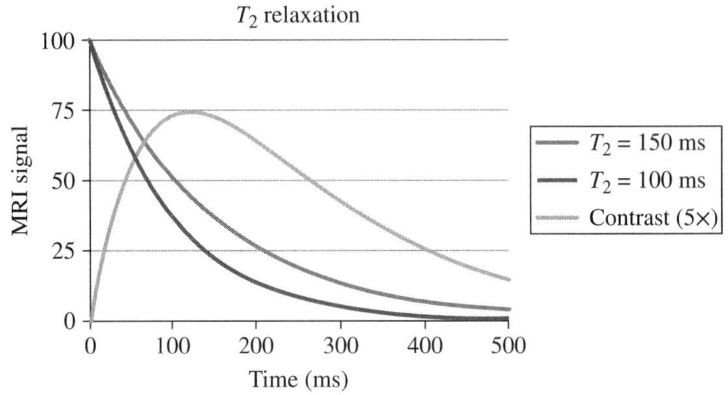

FIGURE 6.16 MRI signal intensity (arbitrary units) for tissues with T_2 values of 150 ms or 100 ms over a range of TE values. The tissue with the shorter T_2 decays quicker than the tissue with the longer T_2. Subtracting the MRI signals provides the image contrast between the two tissues (expanded fivefold to aid visualization). The maximum image contrast occurs with a TE of 120 ms.

are broadly separated into two separate categories, T_1 or T_2 agents. T_1 agents usually consist of a paramagnetic metal, while T_2 agents consist of superparamagnetic iron oxide nanoparticles. The most common paramagnetic contrast agent used in MRI is gadolinium–diethylenetriamine pentaacetic acid (Gd-DTPA) [5–7]. This agent decreases the relaxation times of nearby hydrogen nuclei. In proton density images, the contrast agent does not significantly alter the image. In T_1-weighted images, the T_1 is reduced, and tissues exposed to the contrast agent appear brighter. In T_2-weighted images, the contrast agent decreases the T_2 of tissues containing contrast agent, and so they appear darker. The additional contrast may be used to detect hemorrhage, disruption of the blood–brain barrier, or the increased vascularity associated with tumors.

6.6.2 Relaxivity

The ability of a contrast agent to alter the MRI relaxation times is denoted by its relaxivity [8]. Contrast agents alter the T_1 relaxation time by

$$\frac{1}{T_1^c} = \frac{1}{T_1} + r_1[P]$$

where T_1^c is the relaxation time with the contrast agent, T_1 is the relaxation time without the contrast agent, r_1 is the longitudinal relaxivity, and $[P]$ is the contrast agent concentration. The T_2 relaxivity is similarly given by

$$\frac{1}{T_2^c} = \frac{1}{T_2} + r_2[P]$$

The relaxivity of an MRI contrast agent is measured by recording the relaxation times (T_1 and T_2) at different concentrations of the agent. The relaxation times are converted into relaxation rates ($R_{1,2} = 1/T_{1,2}$) and plotted against the concentration. A linear fit of relaxation rate versus concentration yields a line with a slope corresponding to the relaxivity in terms of $mM^{-1}\,s^{-1}$. Relaxivity can be measured relative to the metal concentration (gadolinium or iron) or the concentration of the carrier molecule (peptide, particle, liposome, etc.). The relaxivity value is dependent on several experimental factors, including the B_0 field strength, temperature, and medium. These confounding factors can make it difficult to compare relaxivity values reported by different laboratories using various experimental conditions. Oftentimes, the relaxivity of a novel contrast agent is compared with a clinically available standard, such as Gd-DTPA (for paramagnetic agents) or feridex (for iron oxide agents), in order to provide a normalizing value that can be reproduced at other labs.

Iron oxide nanoparticles have very high T_2 relaxivities, about 5–10 times higher than their T_1 relaxivity [9–13]. Therefore, they are typically detected as a loss of signal on T_2-weighted MRI. A number of physical and chemical properties can influence the relaxivity of iron oxide nanoparticles, such as the particle size or surface coating [14]. Particles with increasing diameters (3, 4, 5, and 6 nm) display a progressive increase in the T_2 relaxivity (29, 42, 48, and 61 $mM^{-1}\,s^{-1}$). Modification of the particle surface with a polymer coating increases the T_2 relaxivity of a 6 nm iron oxide nanoparticle from 61 to 119 $mM^{-1}\,s^{-1}$.

Gadolinium is a paramagnetic metal that has been used extensively in MRI contrast agents. Paramagnetic metals generally display similar T_1 and T_2 relaxivities. Since T_1 relaxation times in the body tend to be much longer than T_2 relaxation times, these metals typically produce a more apparent shortening of T_1. Therefore, they are often used to generate increased signal intensity on T_1-weighted MRI. Gadolinium agents have been grafted onto a wide range of nanoparticle constructs, including liposomes, micelles, dendrimers, polymers, viral particles, and liquid perfluorocarbon (PFC) particles. A wide variety of factors can influence the relaxivity of paramagnetic contrast agents, including the size of the particle carrier, the number of gadolinium ions on each particle, binding to a biological target, or the accessibility of water to the gadolinium.

Several reports have demonstrated that PFC nanoparticles can be formulated with gadolinium chelates for MRI detection [15, 16]. Anchoring the gadolinium onto a nanoparticle carrier can improve the relaxivity of the contrast agent via two mechanisms. First, the relatively large size of the particle slows the tumbling rate of the metal, improving the chemical interaction between the gadolinium ion and the surrounding water molecules. As a result, the relaxivity of Gd-DTPA at 1.5 T increases from 4.5 $mM^{-1}\,s^{-1}$ when diluted in water [17] to 17.9 $mM^{-1}\,s^{-1}$ when incorporated onto PFC nanoparticles [18]. Second, utilizing a nanoparticle allows multiple copies of the gadolinium chelate to be loaded onto a single carrier, effectively multiplying the paramagnetic effect induced in the target tissue each time a particle binds to the biomarker of interest. PFC nanoparticles can be loaded with nearly 100,000 Gd-DTPA molecules per particle, leading to an overall relaxivity of 1,690,000 $mM^{-1}\,s^{-1}$ at 1.5 T [18]. With such a high relaxivity, the minimum concentration of PFC nanoparticles required for MR image detection, defined as the concentration required to produce a contrast-to-noise ratio of 5, is around 100 pM at a typical clinical field strength (1.5 T) and 25 pM at a higher field strength used for high-resolution imaging in animal research (4.7 T) [18]. The advantages of coupling numerous paramagnetic chelates onto a nanoparticle carrier have also been described for micelles [19], dendrimers [20], liposomes [21, 22], and other particle systems.

The relaxivity of a contrast agent can change when the molecule binds to a biological target. For example, ABLAVAR is an MRI contrast agent approved for use in angiography that binds to albumin in the blood pool. When ABLAVAR binds to albumin, the relaxivity increases by a factor of five compared to other clinical contrast agents [23]. In a similar manner, EP-2104R is a contrast agent that binds to fibrin in clots [24]. The relaxivity of EP-2104R is 11.1 $mM^{-1} s^{-1}$ in solution but increases to 24.9 $mM^{-1} s^{-1}$ when bound to a fibrin clot [25, 26].

Some MRI contrast agents are specifically designed to alter their relaxation behavior based on environmental conditions, such as temperature [27], pH [28], oxygenation [29], ion concentration [30, 31], or enzyme activity [32, 33]. These effects are often achieved by developing an agent that alters its chemical structure based on the local conditions. These conditions can alter the molecular bonds or relative position of chemical groups, changing the accessibility of water to the metal. Alternatively, agents have been designed to bind to plasma proteins or cluster together only in the presence of certain ions or enzymes, yielding a dramatic increase in their relaxivity under these conditions.

Liposomes have been designed to selectively release an encapsulated payload upon thermal activation. The payload could be a therapeutic drug and/or an MRI contrast agent for visualizing liposome permeability. A thermosensitive liposome that encapsulated a clinical gadolinium contrast agent displayed a very small T_1 relaxivity of 0.5 $mM^{-1} s^{-1}$ at low temperatures [34]. The relaxivity quickly rose to 3 $mM^{-1} s^{-1}$ as the temperature increased between 40 and 43°C, but did not change further up to 50°C. These results demonstrate that the liposome membrane rapidly becomes very porous at 40–43°C, allowing controlled release of the encapsulated payload.

Many factors play into the *in vivo* compatibility and utility of MRI contrast agents based on nanoparticle carriers. The chemical composition of the particle system can drastically affect the toxicity, biodistribution, elimination rates, and stability of MRI contrast agents. Furthermore, the metal content, particle size, metal loading, and water interaction can impact the relaxivity of the contrast agent. Both the biocompatibility and MRI visibility need to be optimized for *in vivo* detection of the contrast agent. If the nanoparticle is designed to specifically target a biomarker to achieve molecular imaging of disease, the targeting agent must also display high affinity for the selected target and minimal binding to other cellular components. Nanoparticle-based MRI contrast agents have been developed to take advantage of the high payload ability and long circulation time to home to pathological tissues, such as tumors, atherosclerotic plaques, myocardial infarctions, etc. These agents offer great promise for noninvasive tracking of disease progress and/or therapeutic effects. Ultimately, the clinical application of selectively targeted MRI contrast agents could be used to improve patient outcomes in a wide variety of the most pressing healthcare challenges facing society as a whole.

6.7 OPTICAL TECHNIQUES

6.7.1 Instrumentation Requirements for Nanoparticle Characterization

Recording optical spectra of clear solutions in the visible range is a routine operation, and many types of spectrophotometers are available for this task. Due to the transparency of biological tissue to NIR photons, optical nanoparticles are designed to absorb and emit in this spectral range (750–900 nm). Two major impediments to characterize optical nanoparticles include (i) scattering from the nanoparticles that disrupt the measuring beam in conventional spectrophotometers and (ii) low sensitivity of commonly used silicon (Si) detectors at wavelengths above 800 nm. In the last two decades, several instrumentation solutions to these two problems have been proposed and realized commercially.

6.7.1.1 Absorption Spectrophotometers Separating scattering from absorption measurements is one of the major challenges in nanoparticle characterization [35]. Chromophore-specific features of a spectrum are often buried in the high absorbance background originating from the scattering component, such as shown in the spectrum of indocyanine green (ICG) in milk (see Fig. 6.18). A classic (but still in use) strategy developed by B. Chance in 1951 is a dual-wavelength approach where two detectors are set up at 90° to each other [36]. One detector is coaligned with the excitation light to provide "chromophore + scatter" data, and the other provides "scatter" data (Fig. 6.17). Subtracting the latter from the former yields the "chromophore" spectrum only. However, this method has some restrictions: (i) absorbance changes should be less than 0.01, (ii) the reference wavelength should be chosen so as to minimize interference from other absorbing components, and (iii) the wavelengths should be closely spaced [37]. Overall, the dual-wavelength approach does not eliminate scattering completely.

FIGURE 6.17 An integrating cavity absorption meter consists of a two-layer cavity with walls made from a highly reflective material. With this accessory, the absorption is decoupled from scattering, allowing accurate measurements of molar absorptivity. (The photos of Clarity are kindly provided by Olis, Inc.)

A more reliable strategy is to utilize an integrating sphere that virtually eradicates scattering and reveals pure absorption. An integrating sphere is a globe covered on the inside with a highly reflective material such as magnesium oxide, Teflon, or other fluoropolymer such as Spectralon® (introduced in 1986 by Labsphere, Inc., United States). The latter with enough thickness offers greater than 99% diffuse reflectance over the range of 400–1500 nm [38] and is therefore well suitable for NIR measurements. In this technique, the spectral response of the sample placed in the integrating sphere is subtracted from the spectral response of the integrating sphere across the whole spectrum (Fig. 6.18). This method has been used to measure absorbance spectra of silver and NIR nanoparticles [39, 40]. The

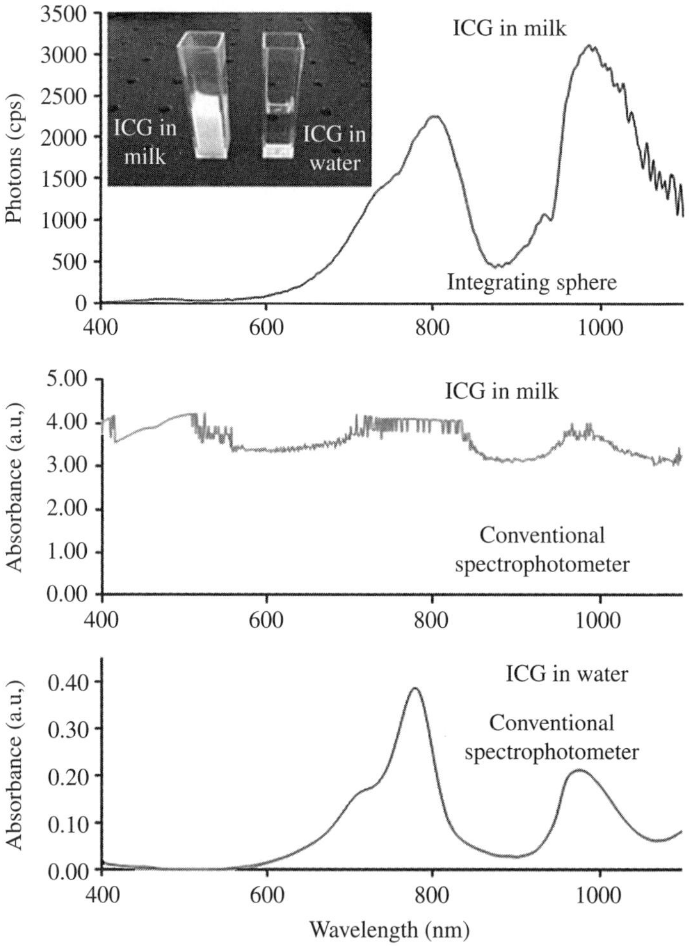

FIGURE 6.18 Spectrum of indocyanine green (ICG) in milk using an integrating sphere (top) or conventional spectrophotometer (middle). ICG abs. maximum is approximately 790 nm, and water peak is seen at approximately 960 nm. The spectral features of ICG are masked by strong scattering from milk. ICG in water is shown at bottom.

advantage of the integrating sphere is especially apparent when there is a need to measure absorption spectra of nanoparticles in very turbid media such as a cell suspension or in blood. A variation of the method uses an integrating cavity absorption meter [41] shown in Figure 6.17. Since the light path in the cavity is significantly longer due to multiple reflections and reabsorption events, molar absorptivities are corrected as described by Hodgkinson *et al.* [42] and Fry *et al.* [43] with the latter implemented in some modular spectrophotometers (such as Clarity from Olis, Inc., United States).

6.7.1.2 Fluorescence Spectrophotometers Low sensitivity in NIR is not critical for absorption spectrophotometers, allowing absorption measurement in NIR with acceptable accuracy (many conventional spectrophotometers measure up to 1100 nm). However, several considerations have to be addressed before a conventional fluorescence spectrophotometer can be modified to detect NIR emission. Basically, all of the wavelength-dependent components of the instrument have to be evaluated for performance in the NIR spectral range. These components consist of the excitation source, gratings in the excitation and emission spectrometers, detector, polarizers, and fiber optics. Due to various graded NIR parts and modules, such systems are often customized and generally associated with higher cost. An example of an instrument for nanoparticle research (NanoLog, Horiba, Inc.) is given in Figure 6.19 with some of the modules described in the following text.

6.7.1.3 Excitation The excitation source in a typical fluorescence spectrophotometer is a Xe arc lamp. This lamp has a substantial number of sharp peaks between 700 and 1000 nm (Fig. 6.20) that could introduce abnormalities into the spectra. Hence, it requires parallel recording of the light input using a photodiode and subsequent correction of the emission spectra. Alternatively, a tungsten lamp can be utilized as an excitation source because of its broad and smooth output in the NIR range (up to 2000 nm).

6.7.1.4 Gratings Conventional excitation gratings on fluorescence systems are usually diffraction-type gratings that are optimized (blazed) in the UV range (typically 300 nm), while the emission gratings are blazed in the visible range (typically 500 nm). For NIR performance, both gratings should be replaced with blazed gratings at longer wavelengths. A NIR configuration includes the excitation grating blazed at greater than equal to 500 nm and the emission grating blazed in red or NIR (500–1000 nm). Some advanced models are equipped with kinematically interchangeable gratings for higher flexibility for allowing the conversion of a standard system into a NIR-dedicated unit manually or via software interfaces.

6.7.1.5 Detectors Most research emission spectrophotometers are currently equipped with a R928 or R928P (for photon counting) Si-type photomultiplier tube (PMT). Introduced in 1974 by Hamamatsu, Inc. (Japan), this PMT with relatively high sensitivity to red and NIR ranges made an "epoch-making event" [44] in optical spectroscopy. However, the performance of R928(P) rapidly drops above

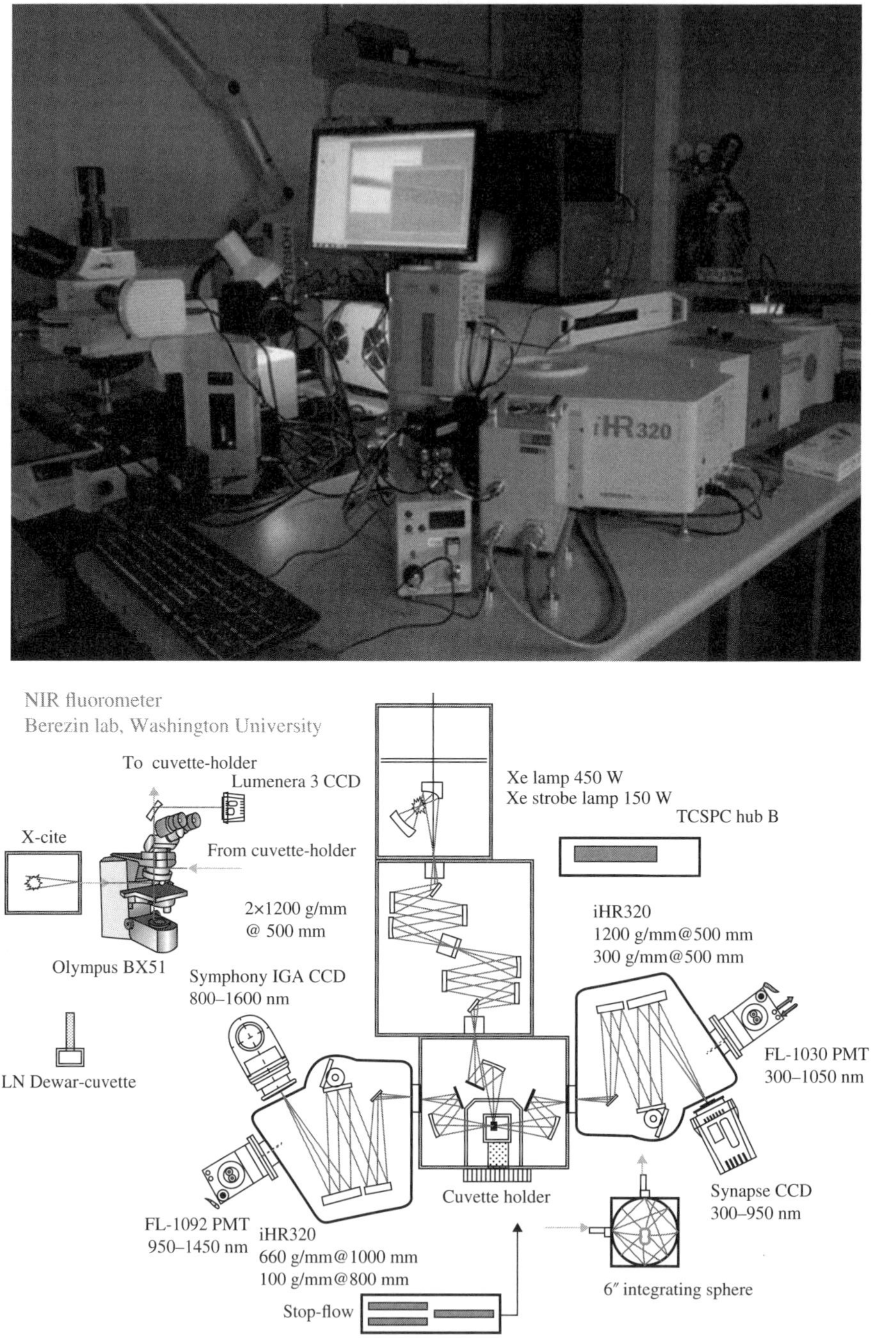

FIGURE 6.19 NIR spectrophotometer "NanoLog" for nanoparticle research located in the Optical Spectroscopy Core, Washington University, St. Louis. The system features a T-shaped geometry; Xe lamps; a double grating excitation monochromator for highest stray-light rejection; cuvette holder with 90° angle and front face detection for measuring fluorescence in solution, solid, and opaque systems; and imaging spectrographs with a kinematic turret for multiple gratings and multiple NIR detectors. (The schematics were kindly provided by Horiba, Inc.) *(See insert for color representation of the figure.)*

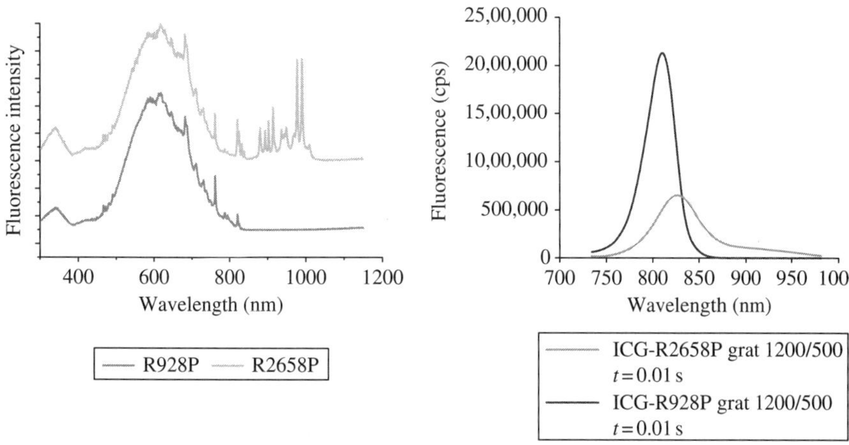

FIGURE 6.20 Performance of two commonly used NIR PMTs: R928P and R2658P. Left: no signal is seen with R928P beyond 850 nm, while R2658P remains sensitive. Spectra are normalized and shifted vertically for clarity. Right: NIR emission of ICG recorded with R928P and R2658P under the same conditions.

800 nm, and beyond 900 nm, the quantum efficiency approaches zero. To further improve the sensitivity in NIR, Hamamatsu introduced an indium gallium arsenide (InGaAs) R2658(P) PMT in the late 1980s [45]. Although it has lower quantum efficiency in the visible range, this detector is far more sensitive in NIR (up to 1010 nm), which allows one to measure fluorescence in a previously inaccessible region. Alternatively, germanium-based (Ge) NIR detectors can be used [46, 47]. Since Ge detectors exhibit a higher level of thermally induced noise than an InGaAs detector, this detector needs to be enclosed in a dewar with liquid nitrogen.

6.7.1.6 Fiber Optics Selection of fiber optics is important for a growing segment of fiber optically based spectrophotometers and peripheral devices such as integrating spheres and microscopes. Fiber-optic material (plastic, glass, or plastic–glass silica), dopings, the presence of silanol (≡Si–OH) hydroxyl groups, and cladding and coating material can significantly attenuate the amount of light and induce autofluorescence and Raman peaks [48]. High-purity glass fibers with low concentrations of silanols normally exhibit good transmission in NIR [49]. Liquid lightguides filled with special NIR transparent liquids is another option.

6.7.2 Brightness of Nanoparticles

Optical characterization of nanoparticles starts from measurements of the brightness of a probe. The brightness of a probe is defined as a product of two major parameters: molar absorptivity ε and fluorescence quantum yield Φ. For a nanoparticle labeled with multiple fluorophores, the number also known as a degree of labeling, N_f, is incorporated into the equation

$$B = \varepsilon \times \Phi \times N_f$$

The degree of labeling could be determined either from material balance during the synthesis or directly from the optical spectra of the nanoparticles [50, 51]. Quenching of the fluorophore on the nanoparticles is universal and has been described in a great number of publications [52–54]. Such quenching is a result of the aforementioned H-type aggregation [55] as well as other effects [56] and decreases the quantum yield to virtually zero. This leads to a very low brightness of the imaging nanoparticles.

6.7.2.1 Molar Absorptivity of Nanoparticles Molar absorptivity of organic fluorophores is generally determined from linear standard absorption plots using the Beer–Lambert law. For nanoparticles with several attached fluorophores (fluorescently labeled polymers, dendrimers, dye-loaded liposomes, and nanocapsules), the absorption properties are nonlinear, and the spectra differ for each individual fluorophore type. This is especially apparent in NIR cyanine dyes that tend to form H-aggregates (H for hypsochromic shift), which have distorted (broadened) absorption spectra [55] (an example is shown in Fig. 6.21). Such spectral behavior is similar to the known self-assembly of dyes in aqueous solutions where at high concentrations the individual monomers are aligned parallel to each other in a face-to-face arrangement [57, 58]. Due to the similarity between nanoparticles and concentrated solutions, unknown molar absorptivity of optical nanoparticles can be deduced from a set of concentration-dependent molar absorptivity spectra. This can be done by matching the shape of the nanoparticle spectrum to the spectrum of a sample with a known concentration. Alternatively, a ratio of the H-band intensity to the intensity of the absorption maximum can be used.

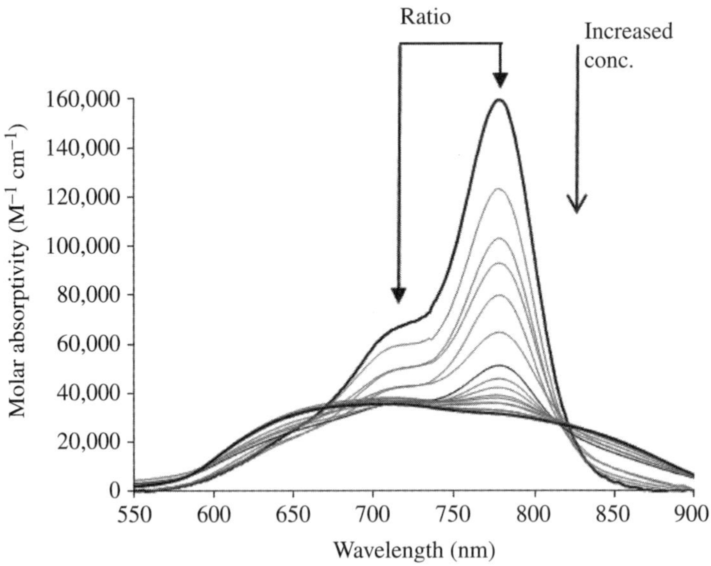

FIGURE 6.21 Molar absorptivities of a NIR dye in water with increasing concentration of the dye from 0.1 to 20 μM concentration.

6.7.2.2 Fluorescence Quantum Yield—Relative Method Fluorescence quantum yields of the nanoparticles carrying organic fluorophores are measured the same way as individual fluorophores using either relative or absolute measurements. The relative method relies on a reference dye that already has a defined quantum yield and evaluates a sample dye using the following equation [59]:

$$\Phi_{\text{sample}} = \Phi_{\text{ref}} \times \left[\frac{\text{Grad}_{\text{sample}}}{\text{Grad}_{\text{ref}}}\right]\left[\frac{\eta^2_{\text{sample}}}{\eta^2_{\text{ref}}}\right]$$

where Φ is quantum yield, Grad is a slope of the curve total number of emitted photons to absorption at the wavelength of excitation, η is the refracting index, and ref corresponds to reference dye and sample to the dye (nanoparticle) of interest.

In the relative method, both absorption and emission of the reference dye and the sample are measured under identical conditions: the type of a cuvette (material, equal optical path), excitation wavelength, slit size, integration time, detector, emission range, and polarization have to be used for both the reference and the sample. For emission, correction factors should be used if the reference and the sample emit in significantly different spectral ranges (±50 nm). Caution has to be taken when NIR dyes are measured using this method. Poor performance of standard detectors above 800 nm might lead to an erroneous conclusion that a dye with emission above 850 nm has low quantum yield if the reference dye emits at shorter wavelengths. Another source of error lies in the difference of anisotropy between the reference and the sample. To eliminate the potential discrepancy caused by the difference in anisotropy, the polarizer at emission path should be set at the "magic angle" (54.7° from the vertical). The detailed step-by-step procedure of measuring quantum yields is given in several publications [60, 61].

6.7.2.3 Fluorescence Quantum Yield—Absolute Method Relative quantum yield relies on the knowledge of Φ_{ref} for a reference dye with similar absorption/emission properties. If such a reference is not available, the QY must be measured by absolute methods [62]. The most common current technique for measuring absolute quantum yield utilizes an integrating sphere [63–66]. The main purpose of the sphere is to minimize any special anisotropy of the emission and to eliminate refractive index and polarization errors. The integrating spheres are often designed to fit in the cuvette compartment of a fluorometer and are available for most of the conventional fluorometers as an accessory [67]. In many high-end instruments, the integrating sphere is external and connected to the instrument through a fiber-optic bundle placed in the cuvette compartment (Fig. 6.22). External design allows for larger integrating spheres (typically 150 mm) with very homogeneous light flux and minimal hot spots to achieve higher accuracy. Modern integrating spheres have several ports for measuring quantum yields of samples in both liquid and solid form [64, 68].

With the commercial availability of the integrating spheres, the absolute method has increasingly been used to measure quantum yield of a variety of

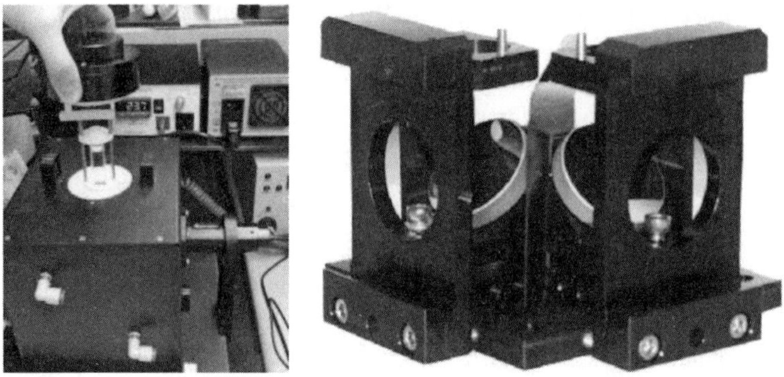

FIGURE 6.22 External integrating sphere for quantum yield measurements with fiber-optic bundle that fits a spectrophotometer compartment of a NanoLog, shown in Figure 6.19.

nanoparticles including silica-covered nanocrystals [69], unconverted nanoparticles [70], core–shell nanoparticles [56], polystyrene beads [71], and fluorescent nanopigments [72].

6.7.2.4 Photostability Photostability of nanoparticles can be, in principle, measured by many of the methods, but are rarely measured, often because it requires custom equipment with uniform illumination [61] where the wavelength is selected and the knowledge of energy is known. Most of the fluorophores are reasonably stable *in vivo*.

6.7.3 Special Fluorescence Techniques

6.7.3.1 Fluorescence Lifetime In the last decade, fluorescence lifetime (FLT) became an important parameter in nanoparticle research because of its sensitivity to the environment and because of rapid expansion in FLT technique from spectroscopy to lifetime imaging [73]. Recent availability of commercial FLT imaging systems from microscopy (Becker & Hickl GmbH (Germany), ISS, Inc. (United States), PicoQuant GmbH (Germany), Lambert Instruments (the Netherlands)) to small animal imagers (Advanced Research Technologies, Inc. (Canada)) has made FLT not only a spectroscopic technique but also a valuable optical modality.

FLT is defined as the time required by a population of excited fluorophores to decrease exponentially to N/2.718 (ca. 36.8% of its original number) via the loss of energy through fluorescence and nonradiative processes. The lifetime of commonly used fluorophores varies significantly from hundreds of picoseconds to nanoseconds for organic compounds, tens of nanoseconds for quantum dots, and microseconds or seconds for phosphorescence [74]. As an intrinsic property of a fluorophore, the lifetime does not depend on the method of measurement (time domain, frequency domain, time-gated measurements) and, in general, is independent on initial perturbation conditions such as wavelength of excitation, duration of light exposure, and one- or multiphoton excitation. In addition, FLT is a parameter largely independent of the fluorescence intensity and fluorophore concentration [75] but is sensitive to a great variety of biological factors that include polarity, pH, viscosity, and temperature [76, 77].

FLT characterization of the nanoparticles typically involves measuring the lifetimes of a free dye, the dye associated with the nanoparticles, and the dye in the biological environment. The alteration of the lifetime suggests an alteration of the environment. For example, the increase of the lifetime of ICG from 0.2 to 0.55 ns indicated the encapsulation of the dye; a subsequent increase to 0.65 ns evidenced the release of the dye into from the nanoparticles into the bloodstream [40]. Other examples include validation of dye integration into PFC nanoparticles [78], evaluation of liposome stability and leakage [79], and *in vivo* nanoparticle biodegradability [80].

6.7.3.2 Fluorescence Anisotropy Fluorescence anisotropy is used to measure hydrodynamic size of the nanoparticle [81] in order to verify the conjugation efficiency of the fluorophores to nanoparticles and is used as a quality control tool for fluorescently labeled nanoparticles [82, 83]. This optical technique has been widely utilized in biochemistry and drug discovery [84] and has recently became a tool in biomedical nanoparticle research. In fluorescence anisotropy, the fluorophore is irradiated with linearly polarized light. The resultant fluorescence intensity is measured through a polarization filter placed in front of the detector and is oriented either parallel or perpendicular to the incident polarized light. Fluorescent anisotropy r can then be determined from the following equation [85]:

$$r = \frac{I_{\text{II}} - I_{\perp}}{I_{\text{II}} + 2I_{\perp}}$$

where I_{II} is the intensity the fluorescence emission parallel to the vertically polarized light and $I_{\perp}$ is the intensity the fluorescence emission perpendicular to the vertically polarized light.

Anisotropy directly relates to the rotational correlation time (θ) of the molecules, which is proportional to the hydrodynamic volume V of the nanoparticle in solution according to a well-known Perrin equation [86], thus connecting the measurable value r with the size of the nanoparticle:

$$r = \frac{r_o}{1 + \tau / \theta}; \theta = \frac{\eta V}{RT}$$

where r is anisotropy, r^0 is limiting anisotropy (maximum possible, usually 0.4), τ is FLT, θ is rotational correlation time of a macromolecule, η is viscosity in poise, V is volume of macromolecule, R is gas constant = 8.31×10^7 erg/mol K, and T is temperature, K.

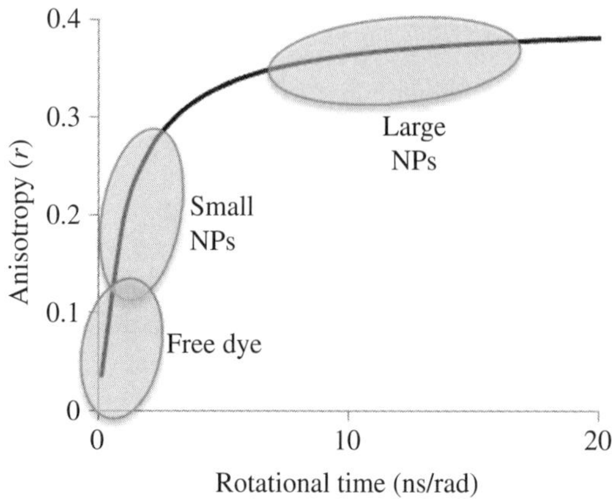

FIGURE 6.23 Fluorescence anisotropy as a function of rotational correlation time. Larger and slower nanoparticles show higher anisotropy than free dye and small nanoparticles.

In the ideal case where a nanoparticle has a round shape, the relationship between r and θ is shown in Figure 6.23. Free organic fluorophores have a low level of anisotropy due to their small hydrodynamic volume and, therefore, short rotational correlation time. Fluorophores attached to nanoparticles with a much larger size are expected to show a higher level of anisotropy. Direct measurements of the rotational time and the nanoparticles size can be achieved with time-resolved fluorescence anisotropy [81]. Rotational times as low as 1.4 μs and particles with just few nanometers in diameter can be reliably measured. The drawback with nanoparticle measurements, especially with larger particles greater than 50 nm, is that the anisotropy value is sensitive to the excitation light that scatters off the nanoparticle sample. The scattering can significantly increase the measured anisotropy value, sometimes reaching a unity ($r \sim 1$). To avoid errors, anisotropy is measured with the wavelength of emission set far from the wavelength of excitation (up to 100 nm).

Sahoo *et al.* [87] utilized fluorescence anisotropy to confirm the labeling of magnetic nanoparticles with rhodamine 110. Free dye in water exhibited almost no fluorescence anisotropy ($r \sim 0$), but it increased up to 0.4 when the dye was bound to the nanoparticles. The mobility of fluorophores in the liposomal bilayers provided information regarding the fluidity of the lipid bilayers [88]. Fluorescence anisotropy has been also utilized to establish successful incorporation of an NIR fluorescent dye into a nanoparticle [40] and has validated the attachment of nanoparticles to fluorophore-labeled DNA [89]. Anisotropy values of quantum dots are rarely measured because quantum dots usually have a spherical symmetry that results in anisotropy values close to zero. Indeed, fluorescence anisotropy values for a variety of quantum dots from 500 to 1100 nm made from different materials and in variety of solvents are slightly above zero (M.Y. Berezin, personal communication). A deviation from the zero value of anisotropy from quantum dots has been attributed to a change in quantum dots' shape [90] and can be utilized to assess the breaking of the electronic band symmetry.

6.7.3.3 Fluorescence Correlation Spectroscopy Fluorescence correlation spectroscopy (FCS) is a sensitive technique that utilizes statistical analysis of the fluctuations in fluorescence to assess diffusion or fluctuations of the molecules [91]. Initially introduced to measure the binding of dyes to DNA, the technique has been expanded to characterize the photochemistry of dyes, to monitor conformational perturbations of fluorescent proteins and DNA, and, more recently, to characterize the size and translational mobility of fluorescent nanoparticles [92–94].

The primary parameter measured in FCS is fluorescence intensity. The changes in fluorescence intensity over a relatively short period of time reflect the concentration fluctuations of a molecular system. The changes are recorded in the form of an autocorrelation function, $G(t)$ (Fig. 6.24), using a confocal microscope equipped with a correlation card that calculates the autocorrelation function from the intensity signal. In FCS, the autocorrelation function is the cross-correlation of a fluorescence intensity signal with itself and is obtained by comparing a measured value at a time with florescence intensity at a later time (delayed by τ). The autocorrelation function reveals the time required for a particle to diffuse through an open confocal volume size of less than 1 fl (femtoliter). Two signals that are recorded at nearly the same time (τ is small, i.e., microseconds) are expected to have a high correlation value. Those signals recorded far apart (τ is large, i.e., milliseconds) result in a lower correlation value. With everything equal, larger particles are expected to have higher correlation than smaller particles leading to the corresponding shift (to the right) in the autocorrelation function such as shown in Figure 6.24.

FCS curve fitting provides the correlation time and nanoparticle number. Data analysis is straightforward but often is complicated by the polydispersity of nanoparticles,

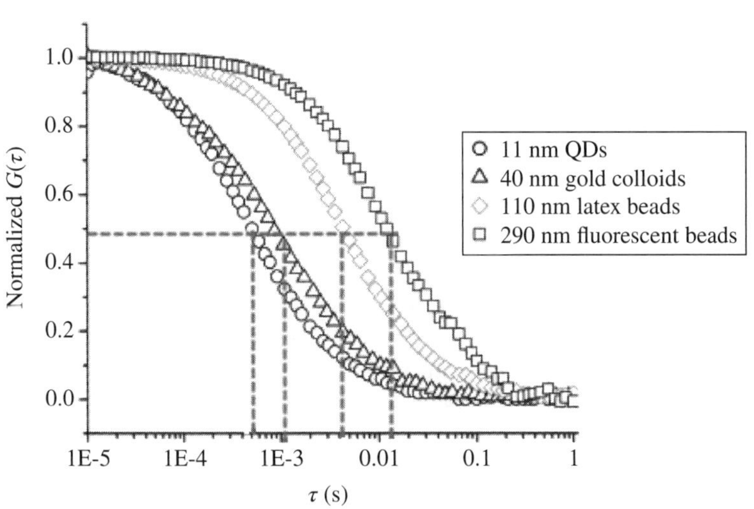

FIGURE 6.24 Autocorrelation functions for particles ranging in diameter from 11 nm quantum dots, 40 nm gold colloids, 110 nm nonfluorescent latex beads, and 290 nm fluorescent beads. (Reprinted with permission from Ref. [95]. © American Chemical Society.)

photobleaching, and triplet states. In practice, the prepared nanoparticles are polydispersed not only in size but also in brightness. Thus, a multicomponent fitting procedure is commonly performed. Such fitting algorithms are provided by most commercial FCS systems.

Conceptually, the FCS approach is similar to a more common DLS technique for measuring the size of the nanoparticles; however, FCS is much more sensitive. When analyzing the particles of sizes less than 10 nm using DLS, the uncertainty of a signal becomes large, and the interference from other particles in solution makes it difficult to obtain accurate results [95]. Besides, DLS requires relatively high sample concentrations (μM–mM) and large volumes (mL) compared to FSC (concentration pM–nM, volume < fl (femtoliter)). The drawback of FCS is that the method requires highly photostable fluorophores; otherwise, the system considers bleached fluorophores as having "left the confocal volume" resulting in a lower size measurement.

Using FCS, Balaji *et al.* measured the hydrodynamic radius of gold nanorods that have weak fluorescence but are resistant to photobleaching [96]. FCS has been utilized to reveal the mechanism of nanorod growth [97] and the size of photoluminescent carbon dots [98]. FCS has also been applied to measure the size and shape of reverse micelles loaded with Cy3, which have revealed prolate ellipsoid-shaped particles [99].

6.8 MISCELLANEOUS METHODS

6.8.1 Electrophoresis

Electrophoretic techniques separate charged molecules and nanoparticles under an applied uniform electric field where they migrate toward the electrode with opposite polarity. The rate of this migration (mobility ν) is linearly proportional to the applied electric field E and electrophoretic mobility μ_e:

$$\nu = \mu_e E$$

where E is an applied electric field and μ_e is the electrophoretic mobility proportional to the particle's charge and inversely proportional to particle's size and shape.

Quite commonly found in biological and biochemical research, electrophoretic techniques have only recently (in the last decade) been adopted by nanoparticle researchers. In gel electrophoresis, for example, particles with different sizes, charges, and even shapes can be distinguished due to distinct bands resulting from different migration mobilities. An illustration of this method for several optical nanoparticles is shown in Figure 6.25. Hence, a great number of publications selected gel electrophoresis for routine characterization of gold nanoparticles [100], quantum dots [101], polymeric nanocapsules [40], and viral nanoparticles [102].

One common application of gel electrophoresis is to control nanoparticle synthesis in order to evaluate the conjugation efficiency of a targeting moiety. Method

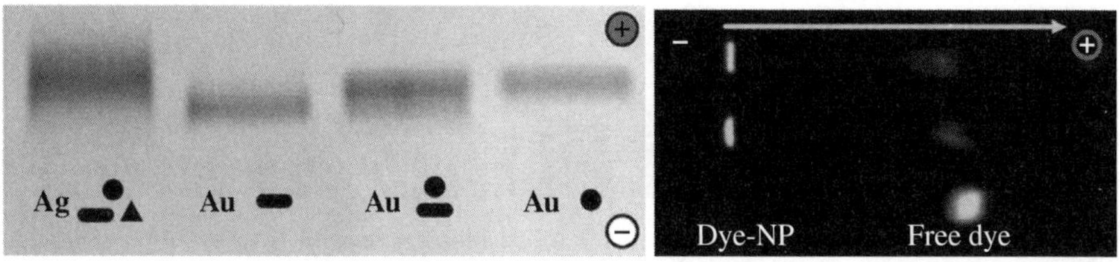

FIGURE 6.25 Left: separation of gold and silver nanoparticles according to their size and shape by agarose gel electrophoresis after coating them with a charged polymer layer. (Reprinted with permission from Ref. [103]. © American Chemical Society.) Right: gel electrophoresis of NIR-labeled nanoparticles after synthesis before purifications, negatively charged free dye migrates to the cathode, while labeled nanoparticles remain at the site of loading.

development includes the optimization of electrophoretic conditions through adjustments of the potential between the electrodes, agarose/polyacrylamide density, and buffers to achieve sufficient separation of an unreacted targeting molecule from the nanoparticle. Relatively large nanoparticles lack mobility and retain at the loading point of the gel, while smaller size targeting moieties such as peptides, proteins, and antibodies migrate through the gel matrix toward one of the electrodes (Fig. 6.25). The gel bands can be then quantified by any optical imager if a fluorescent label is used or by other means. Such methods have been used to monitor the synthesis and the stability of encapsulated ICG and used to assess the purity of the nanoparticle [40].

6.8.2 Chromatography

Chromatographic methods, where the separation is based on the differences in the partition coefficients between mobile and stationary phases for all components of the mixture, have long been used for nanoparticle purification. The method is occasionally used to evaluate the size of the nanoparticles [104].

6.8.3 Quartz Crystal Microbalances

This method provides a simple yet cost-effective and high-resolution mass sensing technique based upon the piezoelectric effect [105] and is typically utilized for measuring nanoparticle concentration [106]. The past two decades have witnessed an explosive growth in the application of the quartz crystal microbalance (QCM) technique to the study of a wide range of molecular systems at the solution–surface interface, in particular biopolymer and biochemical systems, but its full potential in nanoparticle research has yet to be demonstrated.

6.8.4 Flow Cytometry

Flow cytometry is essentially a single particle analysis, which provides rapid information regarding the size, shape, complexity, as well as optical properties of thousands of particles within several seconds. The principle of the method is based on the interaction of particles with a beam of light. Three major components of these interactions—forward scattering, side scattering, and fluorescence intensity (in the

FIGURE 6.26 Principle of measuring size of a nanoparticle with flow cytometry. Smaller particles produce less front scattering.

most common configuration)—are recorded. The magnitude of the forward scatter is to the first approximation proportional to the size of the particle. Small particles will produce small forward scattering, while large particles produce larger scattering (Fig. 6.26). Particles as small as 58 nm can be detected, allowing to distinguish the differences in the size distributions between nanoparticles even if the difference in diameter was only 25 nm [107]. Flow cytometry has thus been utilized for the characterization of gold nanorods, viral nanoparticles [108], and quantum dots.

REFERENCES

[1] McCarthy DW, Shefer RE, Klinkowstein RE, Bass LA, Margeneau WH, Cutler CS, Anderson CJ, Welch MJ. Efficient production of high specific activity 64Cu using a biomedical cyclotron. Nucl Med Biol 1997;**24**:35–43.

[2] Zeng D, Anderson CJ. Rapid and sensitive LC-MS approach to quantify non-radioactive transition metal impurities in metal radionuclides. Chem Commun (Camb) 2013;**49**: 2697–2699.

[3] Shokeen M, Fettig NM, Rossin R. Synthesis, in vitro and *in vivo* evaluation of radiolabeled nanoparticles. Q J Nucl Med Mol Imaging 2008;**52**:267–277.

[4] Hendee WR, Morgan CJ. Magnetic resonance imaging. Part I—physical principles. West J Med 1984;**141**:491–500.

[5] Chandra T, Pukenas B, Mohan S, Melhem E. Contrast-enhanced magnetic resonance angiography. Magn Reson Imaging Clin N Am 2012;**20**:687–698.

[6] Ferre JC, Shiroishi MS, Law M. Advanced techniques using contrast media in neuroimaging. Magn Reson Imaging Clin N Am 2012;**20**:699–713.

[7] Jackson A, O'Connor JP, Parker GJ, Jayson GC. Imaging tumor vascular heterogeneity and angiogenesis using dynamic contrast-enhanced magnetic resonance imaging. Clin Cancer Res 2007;**13**:3449–3459.

[8] Hao D, Ai T, Goerner F, Hu X, Runge VM, Tweedle M. MRI contrast agents: basic chemistry and safety. J Magn Reson Imaging 2012;**36**:1060–1071.

[9] Cromer Berman SM, Walczak P, Bulte JW. Tracking stem cells using magnetic nanoparticles. Wiley Interdiscip Rev Nanomed Nanobiotechnol 2011;**3**:343–355.

[10] Kanwar RK, Chaudhary R, Tsuzuki T, Kanwar JR. Emerging engineered magnetic nanoparticulate probes for targeted MRI of atherosclerotic plaque macrophages. Nanomedicine (Lond) 2012;**7**:735–749.

[11] Lee N, Hyeon T. Designed synthesis of uniformly sized iron oxide nanoparticles for efficient magnetic resonance imaging contrast agents. Chem Soc Rev 2012;**41**:2575–2589.

[12] Tassa C, Shaw SY, Weissleder R. Dextran-coated iron oxide nanoparticles: a versatile platform for targeted molecular imaging, molecular diagnostics, and therapy. Acc Chem Res 2011;**44**:842–852.

[13] Tu C, Louie AY. Nanoformulations for molecular MRI. Wiley Interdiscip Rev Nanomed Nanobiotechnol 2012;**4**:448–457.

[14] Hu F, Joshi HM, Dravid VP, Meade TJ. High-performance nanostructured MR contrast probes. Nanoscale 2010;**2**:1884–1891.

[15] Lanza GM, Winter PM, Caruthers SD, Hughes MS, Hu G, Schmieder AH, Wickline SA. Theragnostics for tumor and plaque angiogenesis with perfluorocarbon nanoemulsions. Angiogenesis 2010;**13**:189–202.

[16] Caruthers SD, Cyrus T, Winter PM, Wickline SA, Lanza GM. Anti-angiogenic perfluorocarbon nanoparticles for diagnosis and treatment of atherosclerosis. Wiley Interdiscip Rev Nanomed Nanobiotechnol 2009;**1**:311–323.

[17] Stanisz GJ, Henkelman RM. Gd-DTPA relaxivity depends on macromolecular content. Magn Reson Med 2000;**44**:665–667.

[18] Morawski AM, Winter PM, Crowder KC, Caruthers SD, Fuhrhop RW, Scott MJ, Robertson JD, Abendschein DR, Lanza GM, Wickline SA. Targeted nanoparticles for quantitative imaging of sparse molecular epitopes with MRI. Magn Reson Med 2004; **51**:480–486.

[19] Mulder WJ, Castermans K, van Beijnum JR, Oude Egbrink MG, Chin PT, Fayad ZA, Lowik CW, Kaijzel EL, Que I, Storm G, Strijkers GJ, Griffioen AW, Nicolay K. Molecular imaging of tumor angiogenesis using alphavbeta3-integrin targeted multimodal quantum dots. Angiogenesis 2009;**12**:17–24.

[20] Zhu W, Okollie B, Bhujwalla ZM, Artemov D. PAMAM dendrimer-based contrast agents for MR imaging of Her-2/neu receptors by a three-step pretargeting approach. Magn Reson Med 2008;**59**:679–685.

[21] Maiseyeu A, Mihai G, Kampfrath T, Simonetti OP, Sen CK, Roy S, Rajagopalan S, Parthasarathy S. Gadolinium-containing phosphatidylserine liposomes for molecular imaging of atherosclerosis. J Lipid Res 2009;**50**:2157–2163.

[22] Paulis LE, Jacobs I, van den Akker NM, Geelen T, Molin DG, Starmans LW, Nicolay K, Strijkers GJ. Targeting of ICAM-1 on vascular endothelium under static and shear stress conditions using a liposomal Gd-based MRI contrast agent. J Nanobiotechnol 2012;**10**:25.

[23] Hartmann M, Wiethoff AJ, Hentrich HR, Rohrer M. Initial imaging recommendations for Vasovist angiography. Eur Radiol 2006;**16** (Suppl 2):B15–B23.

[24] Overoye-Chan K, Koerner S, Looby RJ, Kolodziej AF, Zech SG, Deng Q, Chasse JM, McMurry TJ, Caravan P. EP-2104R: a fibrin-specific gadolinium-Based MRI contrast agent for detection of thrombus. J Am Chem Soc 2008;**130**:6025–6039.

[25] Spuentrup E, Botnar RM, Wiethoff AJ, Ibrahim T, Kelle S, Katoh M, Ozgun M, Nagel E, Vymazal J, Graham PB, Gunther RW, Maintz D. MR imaging of thrombi using EP-2104R, a fibrin-specific contrast agent: initial results in patients. Eur Radiol 2008;**18**: 1995–2005.

[26] Vymazal J, Spuentrup E, Cardenas-Molina G, Wiethoff AJ, Hartmann MG, Caravan P, Parsons EC Jr. Thrombus imaging with fibrin-specific gadolinium-based MR contrast agent EP-2104R: results of a phase II clinical study of feasibility. Invest Radiol 2009;**44**: 697–704.

[27] Rieke V, Butts Pauly K. MR thermometry. J Magn Reson Imaging 2008;**27**:376–390.

[28] Martinez GV, Zhang X, Garcia-Martin ML, Morse DL, Woods M, Sherry AD, Gillies RJ. Imaging the extracellular pH of tumors by MRI after injection of a single cocktail of T1 and T2 contrast agents. NMR Biomed 2011;**24**:1380–1391.

[29] Yoo B, Pagel MD. An overview of responsive MRI contrast agents for molecular imaging. Front Biosci 2008;**13**:1733–1752.

[30] Lubag AJ, De Leon-Rodriguez LM, Burgess SC, Sherry AD. Noninvasive MRI of beta-cell function using a Zn2+-responsive contrast agent. Proc Natl Acad Sci U S A 2011;**108**: 18400–18405.

[31] Major JL, Boiteau RM, Meade TJ. Mechanisms of ZnII-activated magnetic resonance imaging agents. Inorg Chem 2008;**47**:10788–10795.

[32] Gulaka PK, Yu JX, Liu L, Mason RP, Kodibagkar VD. Novel S-Gal analogs as H MRI reporters for *in vivo* detection of beta-galactosidase. Magn Reson Imaging 2013;**31**: 1006–1011.

[33] Napolitano R, Pariani G, Fedeli F, Baranyai Z, Aswendt M, Aime S, Gianolio E. Synthesis and relaxometric characterization of a MRI gd-based probe responsive to glutamic Acid decarboxylase enzymatic activity. J Med Chem 2013;**56**:2466–2477.

[34] de Smet M, Heijman E, Langereis S, Hijnen NM, Grull H. Magnetic resonance imaging of high intensity focused ultrasound mediated drug delivery from temperature-sensitive liposomes: an *in vivo* proof-of-concept study. J Control Release 2011;**150**:102–110.

[35] He GS, Zhu J, Yong K-T, Baev A, Cai H-X, Hu R, Cui Y, Zhang X-H, Prasad PN. Scattering and absorption cross-section spectral measurements of gold nanorods in water. J Phys Chem B 2010;**114**:2853–2860.

[36] Chance B. Rapid and sensitive spectrophotometry. III. A double beam apparatus. Rev Sci Instrum 1951;**22**:634–638.

[37] Cowles JC. Theory of dual-wavelength spectrophotometry for turbid samples. J Opt Soc Am 1965;**55**:690–692.

[38] Georgiev GT, Butler JJ. Long-term calibration monitoring of Spectralon diffusers BRDF in the air-ultraviolet. Appl Opt 2007;**46**:7892–7899.

[39] Evanoff DD, Chumanov G. Size-controlled synthesis of nanoparticles. 2. Measurement of extinction, scattering, and absorption cross sections. J Phys Chem B 2004;**108**: 13957–13962.

[40] Gustafson TP, Dergunov SA, Akers WJ, Cao Q, Magalotti S, Achilefu S, Pinkhassik E, Berezin MY. Blood triggered rapid release porous nanocapsules. RSC Adv 2013;**3**: 5547–5555.

[41] Fry ES, Kattawar GW, Pope RM. Integrating cavity absorption meter. Appl Opt 1992;**31**:2055–2065.

[42] Hodgkinson J, Masiyano D, Tatam RP. Using integrating spheres as absorption cells: path-length distribution and application of Beer's law. Appl Opt 2009;**48**:5748–5758.

[43] Fry ES, Kattawar GW, Strycker BD, Zhai P-W. Equivalent path lengths in an integrating cavity: comment. Appl Opt 2010;**49**:575–577.

[44] Hakamata T, editor. *Photomultiplier Tubes-Basics and Applications*. 3rd ed. Hamamatsu City: Hamamatsu Photonics K.K.; 2006.

[45] Rubelowsky K, Kaminski R. Practical applications of infrared fluorescence spectroscopy. In: SPIE, editor. *Fluorescence Detection III*. 1989. p 2–7.

[46] Fischer GM, Jungst C, Isomaki-Krondahl M, Gauss D, Moller HM, Daltrozzo E, Zumbusch A. Asymmetric PPCys: strongly fluorescing NIR labels. Chem Commun (Camb) 2010;**46**:5289–5291.

[47] Berezin MY, Akers WJ, Guo K, Fischer GM, Daltrozzo E, Zumbusch A, Achilefu S. Long fluorescence lifetime molecular probes based on near infrared pyrrolopyrrole cyanine fluorophores for *in vivo* imaging. Biophys J 2009;**97**:L22–L24.

[48] Udovich JA, Kirkpatrick ND, Kano A, Tanbakuchi A, Utzinger U, Gmitro AF. Spectral background and transmission characteristics of fiber optic imaging bundles. Appl Opt 2008;**47**:4560–4568.

[49] Plotnichenko VG, Sokolov VO, Dianov EM. Hydroxyl groups in high-purity silica glass. J Non-Cryst Solids 2000;**261**:186–194.

[50] Laux E-M, Behnke T, Hoffmann K, Resch-Genger U. Keeping particles brilliant – simple methods for the determination of the dye content of fluorophore-loaded polymeric particles. Anal Meth 2012;**4**:1759–1768.

[51] Sun G, Berezin MY, Fan J, Lee H, Ma J, Zhang K, Wooley KL, Achilefu S. Bright fluorescent nanoparticles for developing potential optical imaging contrast agents. Nanoscale 2010;**2**:548–558.

[52] Chou LYT, Chan WCW. Fluorescence-tagged gold nanoparticles for rapidly characterizing the size-dependent biodistribution in tumor models. Adv Healthcare Mater 2012;**1**: 714–721.

[53] Gustafson TP, Cao Q, Wang ST, Berezin MY. Design of irreversible optical nanothermometers for thermal ablations. Chem Commun (Camb) 2013;**49**:680–682.

[54] Cohen S, Margel S. Engineering of near IR fluorescent albumin nanoparticles for *in vivo* detection of colon cancer. J Nanobiotechnol 2012;**10**:36.

[55] Galande AK, Hilderbrand SA, Weissleder R, Tung C-H. Enzyme-targeted fluorescent imaging probes on a multiple antigenic peptide core. J Med Chem 2006;**49**:4715–4720.

[56] Fleige E, Ziem B, Grabolle M, Haag R, Resch-Genger U. Aggregation phenomena of host and guest upon the loading of dendritic core-multishell nanoparticles with solvatochromic dyes. Macromolecules 2012;**45**:9452–9459.

[57] Zhou JF, Chin MP, Schafer SA. *Aggregation and Degradation of Indocyanine Green.* Bellingham, CA: SPIE; 1994. p 495–505.

[58] Emerson ES, Conlin MA, Rosenoff AE, Norland KS, Rodriguez H, Chin D, Bird GR. The geometrical structure and absorption spectrum of a cyanine dye aggregate. J Phys Chem 1967;**71**:2396–2403.

[59] Williams ATR, Winfield SA, Miller JN. Relative fluorescence quantum yields using a computer-controlled luminescence spectrometer. Analyst 1983;**108**:1067–1071.

[60] Fery-Forgues S, Lavabre D. Are fluorescence quantum yields so tricky to measure? A demonstration using familiar stationery products. J Chem Educ 1999;**76**:1260–1264.

[61] Rurack K, Spieles M. Fluorescence quantum yields of a series of red and near-infrared dyes emitting at 600–1000 nm. Anal Meth 2011;**83**:1232–1242.

[62] Crosby GA, Demas JN. Measurement of photoluminescence quantum yields. Review. J Phys Chem 1971;**75**:991–1024.

[63] Semonin OE, Johnson JC, Luther JM, Midgett AG, Nozik AJ, Beard MC. Absolute photoluminescence quantum yields of IR-26 dye, PbS, and PbSe quantum dots. J Phys Chem Lett 2010;**1**:2445–2450.

[64] Würth C, Pauli J, Lochmann C, Spieles M, Resch-Genger U. Integrating sphere setup for the traceable measurement of absolute photoluminescence quantum yields in the near infrared. Anal Meth 2011;**84**:1345–1352.

[65] Suzuki K, Kobayashi A, Kaneko S, Takehira K, Yoshihara T, Ishida H, Shiina Y, Oishi S, Tobita S. Reevaluation of absolute luminescence quantum yields of standard solutions

using a spectrometer with an integrating sphere and a back-thinned CCD detector. Phys Chem Chem Phys 2009;**11**:9850–9860.

[66] Faulkner DO, McDowell JJ, Price AJ, Perovic DD, Kherani NP, Ozin GA. Measurement of absolute photoluminescence quantum yields using integrating spheres—which way to go? Laser Photonics Rev 2012;**6**:802–806.

[67] Porrès L, Holland A, Pålsson L-O, Monkman A, Kemp C, Beeby A. Absolute measurements of photoluminescence quantum yields of solutions using an integrating sphere. J Fluoresc 2006;**16**:267–273.

[68] Ishida H, Tobita S, Hasegawa Y, Katoh R, Nozaki K. Recent advances in instrumentation for absolute emission quantum yield measurements. Coordin Chem Rev 2010;**254**: 2449–2458.

[69] Timmerman D, Valenta J, Dohnalova K, de Boer WD, Gregorkiewicz T. Step-like enhancement of luminescence quantum yield of silicon nanocrystals. Nat Nanotechnol 2011;**6**:710–713.

[70] Boyer JC, van Veggel FC. Absolute quantum yield measurements of colloidal NaYF4: Er3+, Yb3+ upconverting nanoparticles. Nanoscale 2010;**2**:1417–1419.

[71] Huber A, Behnke T, Würth C, Jaeger C, Resch-Genger U. Spectroscopic characterization of coumarin-stained beads: quantification of the number of fluorophores per particle with solid-state 19f-nmr and measurement of absolute fluorescence quantum yields. Anal Meth 2012;**84**:3654–3661.

[72] Ferrini R, Nicolet O, Huber S, Zuppiroli L, Chappellet S, Lovey C, Perrenoud J, Pauchard M. Fluorescent nanopigments: quantitative assessment of their quantum yield. J Appl Phys 2010;**107**:114323–114329.

[73] Berezin MY, Achilefu S. Fluorescence lifetime measurements and biological imaging. Chem Rev 2010;**110**:2641–2684.

[74] McGown LB, Nithipatikom K. Molecular fluorescence and phosphorescence (Reprinted from Analytical Instrumentation Handbook. 2nd ed. Revised and Expanded; 1997. p 351–391). Appl Spectrosc Rev 2000;**35**:353–393.

[75] Chen Y, Periasamy A. Characterization of two-photon excitation fluorescence lifetime imaging microscopy for protein localization. Microsc Res Technol 2004;**63**:72–80.

[76] Berezin MY, Guo K, Akers W, Northdurft RE, Culver JP, Teng B, Vasalatiy O, Barbacow K, Gandjbakhche A, Griffiths GL, Achilefu S. Near-infrared fluorescence lifetime pH-sensitive probes. Biophys J 2011;**100**:2063–2072.

[77] Berezin MY, Lee H, Akers W, Achilefu S. Near infrared dyes as lifetime solvatochromic probes for micropolarity measurements of biological systems. Biophys J 2007;**93**: 2892–2899.

[78] Akers WJ, Kim C, Berezin M, Guo K, Fuhrhop R, Lanza GM, Fischer GM, Daltrozzo E, Zumbusch A, Cai X, Wang LV, Achilefu S. Noninvasive photoacoustic and fluorescence sentinel lymph node identification using dye-loaded perfluorocarbon nanoparticles. ACS Nano 2011;**5**:173–182.

[79] Patel H, Tscheka C, Heerklotz H. Characterizing vesicle leakage by fluorescence lifetime measurements. Soft Matter 2009;**5**:2849–2851.

[80] Almutairi A, Akers WJ, Berezin MY, Achilefu S, Frechet JM. Monitoring the biodegradation of dendritic near-infrared nanoprobes by *in vivo* fluorescence imaging. Mol Pharmaceutics 2008;**5**:1103–1110.

[81] Apperson K, Karolin J, Martin RW, Birch DJS. Nanoparticle metrology standards based on the time-resolved fluorescence anisotropy of silica colloids. Meas Sci Technol 2009;**20**:025310.

[82] Kumar R, Roy I, Ohulchanskyy TY, Goswami LN, Bonoiu AC, Bergey EJ, Tramposch KM, Maitra A, Prasad PN. Covalently dye-linked, surface-controlled, and bioconjugated organically modified silica nanoparticles as targeted probes for optical imaging. ACS Nano 2008;**2**:449–456.

[83] Gustafson TP, Cao Q, Achilefu S, Berezin MY. Defining a polymethine dye for fluorescence anisotropy applications in the near-infrared spectral range. Chemphyschem 2012; **13**:716–723.

[84] Jameson DM, Ross JA. Fluorescence polarization/anisotropy in diagnostics and imaging. Chem Rev 2010;**110**:2685–2708.

[85] Lakowicz JR. *Principles of Fluorescence Spectroscopy*. 3rd ed. New York: Springer; 2006.

[86] Perrin F. The fluorescence of solutions: molecular induction, polarization and duration of emission and photochemistry. Ann Phys 1929;**12**:169–275.

[87] Sahoo Y, Goodarzi A, Swihart MT, Ohulchanskyy TY, Kaur N, Furlani EP, Prasad PN. Aqueous ferrofluid of magnetite nanoparticles: fluorescence labeling and magnetophoretic control. J Phys Chem B 2005;**109**:3879–3885.

[88] Bothun GD. Hydrophobic silver nanoparticles trapped in lipid bilayers: size distribution, bilayer phase behavior, and optical properties. J Nanobiotechnol 2008;**6**:13.

[89] Yang R, Jin J, Chen Y, Shao N, Kang H, Xiao Z, Tang Z, Wu Y, Zhu Z, Tan W. Carbon nanotube-quenched fluorescent oligonucleotides: probes that fluoresce upon hybridization. J Am Chem Soc 2008;**130**:8351–8358.

[90] Grecco HE, Lidke KA, Heintzmann R, Lidke DS, Spagnuolo C, Martinez OE, Jares-Erijman EA, Jovin TM. Ensemble and single particle photophysical properties (two-photon excitation, anisotropy, FRET, lifetime, spectral conversion) of commercial quantum dots in solution and in live cells. Microsc Res Technol 2004;**65**:169–179.

[91] Elson EL. Fluorescence correlation spectroscopy and photobleaching recovery. Annu Rev Phys Chem 1985;**36**:379–406.

[92] Rocker C, Potzl M, Zhang F, Parak WJ, Nienhaus GU. A quantitative fluorescence study of protein monolayer formation on colloidal nanoparticles. Nat Nanotechnol 2009;**4**:577–580.

[93] Dong C, Huang X, Ren J. Characterization of water-soluble luminescent quantum dots by fluorescence correlation spectroscopy. Ann N Y Acad Sci 2008;**1130**:253–261.

[94] Benezra M, Penate-Medina O, Zanzonico PB, Schaer D, Ow H, Burns A, DeStanchina E, Longo V, Herz E, Iyer S, Wolchok J, Larson SM, Wiesner U, Bradbury MS. Multimodal silica nanoparticles are effective cancer-targeted probes in a model of human melanoma. J Clin Invest 2011;**121**:2768–2780.

[95] Zhang P, Li L, Dong C, Qian H, Ren J. Sizes of water-soluble luminescent quantum dots measured by fluorescence correlation spectroscopy. Anal Chim Acta 2005;**546**: 46–51.

[96] Balaji PS, Murthy AVR, Tiwari N, Kulkarni S. Fluorescence correlation spectroscopy of gold nanoparticles. Spectrosc Lett 2012;**45**:22–28.

[97] Sharma S, Pal N, Chowdhury PK, Sen S, Ganguli AK. Understanding growth kinetics of nanorods in microemulsion: a combined fluorescence correlation spectroscopy,

dynamic light scattering, and electron microscopy study. J Am Chem Soc 2012;**134**: 19677–19684.

[98] Li Q, Ohulchanskyy TY, Liu R, Koynov K, Wu D, Best A, Kumar R, Bonoiu A, Prasad PN. Photoluminescent carbon dots as biocompatible nanoprobes for targeting cancer cells in vitro. J Phys Chem B 2010;**114**:12062–12068.

[99] McPhee JT, Scott E, Levinger NE, Van Orden A. Cy3 in AOT reverse micelles II. Probing intermicellar interactions using fluorescence correlation spectroscopy. J Phys Chem B 2011;**115**:9585–9592.

[100] Hasenoehrl C, Alexander CM, Azzarelli NN, Dabrowiak JC. Enhanced detection of gold nanoparticles in agarose gel electrophoresis. Electrophoresis 2012;**33**:1251–1254.

[101] Xiong L, Shuhendler AJ, Rao J. Self-luminescing BRET-FRET near-infrared dots for *in vivo* lymph-node mapping and tumour imaging. Nat Commun 2012;**3**:1193.

[102] Steinmetz NF, Hong V, Spoerke ED, Lu P, Breitenkamp K, Finn MG, Manchester M. Buckyballs meet viral nanoparticles: candidates for biomedicine. J Am Chem Soc 2009;**131**:17093–17095.

[103] Hanauer M, Pierrat S, Zins I, Lotz A, Sönnichsen C. Separation of nanoparticles by gel electrophoresis according to size and shape. Nano Lett 2007;**7**:2881–2885.

[104] Gray EP, Bruton TA, Higgins CP, Halden RU, Westerhoff P, Ranville JF. Analysis of gold nanoparticle mixtures: a comparison of hydrodynamic chromatography (HDC) and asymmetrical flow field-flow fractionation (AF4) coupled to ICP-MS. J Anal Atom Spectrom 2012;**27**:1532–1539.

[105] Marx KA. Quartz crystal microbalance: a useful tool for studying thin polymer films and complex biomolecular systems at the solution–surface interface. Biomacromolecules 2003;**4**:1099–1120.

[106] Reipa V, Purdum G, Choi J. Measurement of nanoparticle concentration using quartz crystal microgravimetry. J Phys Chem B 2010;**114**:16112–16117.

[107] Nakamura M, Ishimura K. Rapid size evaluation of nanoparticles using flow cytometry. Adv Sci Lett 2010;**3**:130–137.

[108] Robertson KL, Liu JL. Engineered viral nanoparticles for flow cytometry and fluorescence microscopy applications. Wiley Interdiscip Rev Nanomed Nanobiotechnol 2012;**4**:511–524.

PART II

IMAGING MODALITIES: FROM CONCEPTS TO APPLICATIONS

7

RADIO-LABELED NANOPARTICLES FOR BIOMEDICAL IMAGING

Tolulope Aweda, Deborah Sultan, and Yongjian Liu

Department of Radiology, Washington University School of Medicine, St. Louis, MO, USA

7.1 INTRODUCTION

In the field of biomedical imaging, developing delivery systems specific to the target site is an ongoing objective for both therapeutic and imaging applications. Cancer targeting systems have utilized monoclonal antibodies (mAbs) [1–3] or peptides that bind antigens or receptors, which are expressed on the targeted cell surface [4–6]. Other strategies have included the use of antibody–fused streptavidin with a biotin-probe [7–9] engineered proteins that bind specifically to synthetic tags [10–12]. These imaging systems are often limited by suboptimal pharmacokinetics, low sensitivity, and poor targeting efficiency, especially for disease models with low biomarker expression [11]. An alternative is the fast-growing field of nanomedicine, which offers advantages including tuned *in vivo* biodistribution, enhanced targeting efficiency, greater imaging sensitivity, and potential for identification of diseases at an early stage [13, 14]. It is becoming increasingly clear that nanomedicine holds tremendous potential for personalized diagnosis and treatment [15, 16].

Typically, nanoparticles range in size from 1 to 100 nm, exhibiting unique size-dependent physical and chemical properties including optical, magnetic, catalytic, thermodynamic, and electrochemical traits [17]. Nanoparticles can generally be grouped into one of three constructs based on their structure: (1) inorganic nanoparticles including quantum dots, iron oxide nanoparticles, gold nanostructures, and upconversion nanophosphors [18–22]; (2) polymer nanoparticles such as core–shell

Nanotechnology for Biomedical Imaging and Diagnostics: From Nanoparticle Design to Clinical Applications, First Edition. Edited by Mikhail Y. Berezin.

dendrimers and amphiphilic nanoparticles [23, 24]; and (3) lipid nanoparticles that include liposomes and solid lipid nanoparticles [25–29]. These nanoconstructs, owing to their high surface area-to-volume ratio, can be further modified to have multivalency and multifunctionality, thus altering the drug delivery capacity, cellular response, and sensitivity for molecular imaging. Furthermore, one can manipulate the particles for multimodality imaging, providing both functional and anatomic information for synergistic applications [30–34]. Among various molecular imaging modalities, radionuclide-based positron emission tomography (PET) and single-photon emission computed tomography (SPECT) are commonly used for nanoparticle-based diagnostic applications. Straightforward radiochemistry, high imaging sensitivity, and translational potential make radiolabeled nanoparticles an invaluable tool [35]. To date, a great number of nanoconstructs have been radiolabeled with a variety of radioisotopes, resulting in significant contributions to drug discovery, oncology, cardiovascular imaging, neurology, lung-related diseases, and clinical diagnostics [30, 31, 33, 36, 37].

7.2 RADIOLABELED NANOPARTICLES FOR BIOMEDICAL IMAGING

Nanoparticles are currently one of the most widely investigated constructs for delivering radioactive doses for imaging and therapeutic studies in preclinical and clinical settings. The capacity of radiolabeled nanoparticles to improve the detection limits and resolution of imaging applications is constantly being improved upon. Radiolabeling of nanoparticles can be accomplished by labeling the surface of the nanoparticle, labeling the core of the nanoparticle, or radiolabeling a payload encapsulated within the nanoparticle [38]. Multivalency of nanoparticles is important in achieving high specific activity labeling, thus allowing administration of trace amounts of compound for diagnosis [35]. Additionally, the binding kinetics must fall into ranges suitable for the desired biological application, especially when a stable bond is required between the probe and target [39]. For clinical applications, care must be taken in choosing radioisotopes with a translational capability [40].

7.3 RADIOLABELED NANOPARTICLE FOR SPECT IMAGING

7.3.1 γ-Ray-Emitting Radionuclides

SPECT is a powerful nuclear imaging technique that utilizes gamma-emitting radioisotopes to visualize both physiological and functional information at a molecular level. Radioisotopes with energies ranging between 30 and 370 keV, and optimally between 100 and 200 keV [28], are commonly employed. Among them, technetium-99m (^{99m}Tc), produced from a ^{99}Mo/^{99m}Tc generator, is the most widely used isotope for nuclear imaging. Currently, about 85% of diagnostic imaging is carried out using ^{99m}Tc. Other commonly used SPECT radionuclides are summarized in Table 7.1.

TABLE 7.1 Commonly Used SPECT Radionuclides for Nanoparticles[a]

Nuclide	$T_{1/2}$	Decay	Main photon, keV (%)	Production
^{67}Ga	3.26 d	EC	93 (39.2), 185 (21.2), 300 (16.8)	^{68}Zn (p, 2n) ^{67}Ga
^{99m}Tc	6.01 h	IT	140 (89.1)	^{99}Mo/^{99m}Tc generator
^{111}In	2.8 d	EC	171 (90.7), 245 (94.1)	^{111}Cd (p, n) ^{111}In ^{112}Cd(p, 2n) ^{111}In
^{123}I	13.27 h	EC	159 (83.3)	^{124}Te (p, 2n) ^{123}I ^{124}Xe (p, 2n)^{123}I
^{125}I	59.41 d	EC	35.5 (6.68%)	^{124}Xe (n, γ) $^{125m/125g}$Xe → ^{125}I
^{201}Tl	72.91 h	EC	167 (10.0)	^{203}Tl (p, 3n)^{201}Pb:^{201}Tl
^{133}Xe	5.24 d	β^-	81 (38.0)	^{235}U fission

EC, electron capture (capture of an atomic electron converting a proton to a neutron and emission of a neutrino); IT, isomeric transition (gamma decay from a long-lived metastable state); β^-, electron emission (conversion of a neutron to a proton and emission of an electron and an antineutrino).
[a]Based on Ref. [41].

7.3.2 Radiolabeled Liposomes for SPECT Imaging

Liposomes, a member of the lipid nanoparticle construct, are frequently used for imaging with SPECT radioisotopes. Liposomes are vesicles made of spherical lipid bilayers. The lipid bilayer contains natural phospholipids and/or mixed lipid chains. Encapsulated within the bilayers is often an aqueous center that can deliver hydrophilic molecules. Hydrophobic molecules are transported within the bilayer itself. Liposomal anatomy allows fusing with similarly hydrophobic structures, such as cell membranes, thus allowing both hydrophobic and hydrophilic molecules to be delivered simultaneously. This unique property resulted in liposomes being among the first generation of nanoparticles to be utilized clinically.

Liposomes can be used to deliver radiolabeled drugs or imaging probes to regions of interest [42]. The surfaces of liposomes can be modified with targeting ligands (active targeting) or ligands can be encapsulated within the liposome and released at the point of interest by changes in physiological environment [43]. Different surface or encapsulation modifications of liposomes have been developed and tuned, resulting in improved pharmacokinetics, reduced toxicity, and increased compatibility for drug delivery with selective targeting to tumor site [44]. Non-PEGylated, or "naked," liposomes have shown short blood circulation half-lives, rapidly clearing out through the reticuloendothelial system (RES) (liver and spleen). In order for diagnostics and therapy to be effective, longer-circulating liposomes were developed by attaching polyethylene glycol (PEG) chains onto the lipid bilayer, which act to shield the liposome from recognition by plasma proteins and cell surface receptors [45]. These long-circulating liposomes (LCLs) are sterically stabilized by the incorporated PEGs and thus are able to circumvent recognition by the RES. The longer blood circulation time provides enhanced entrapment of LCLs in tumors with improved pharmacokinetics [46–48].

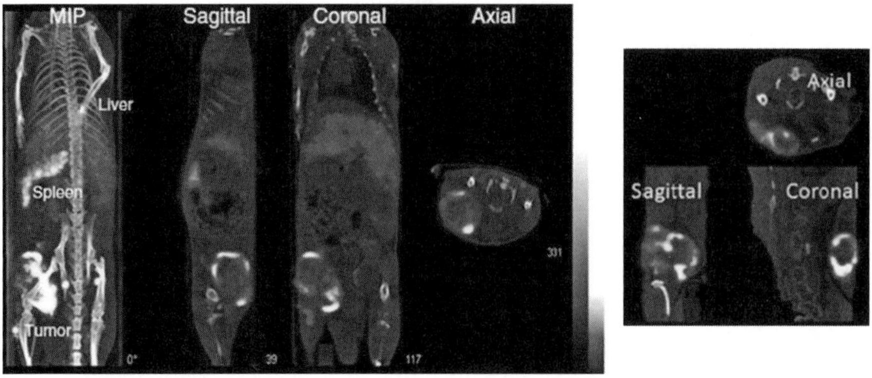

FIGURE 7.1 *In vivo* SPECT/CT images of sarcoma 180-bearing mice 24 h after administration of ^{111}In-labeled liposomes. Case #1 (left panel): whole-body images containing maximum intensity projection (MIP) and representative sagittal, coronal, and axial slice images containing the tumor. Case #2 (right panel): representative sagittal, coronal, and axial slice images of the tumor site in another mouse. The SPECT projection data were acquired under the optimized conditions for *in vivo* imaging. The acquisition time was 150–300 s per projection with 24 (6×4) directions, total = 1–2 h. (With permission from Ref. [50]. © Springer.) (*See insert for color representation of the figure.*)

A number of liposome probes have been radiolabeled with ^{99m}Tc, ^{111}In, and ^{67}Ga. PEGylated liposomes encapsulated with ^{111}In–diethylenetriaminepentaacetic acid (DTPA) have been used to target solid tumors [49, 50]. Radiolabeling was performed by a transchelation reaction between a lipophilic oxine and hydrophilic DTPA ligands. The sensitive spatial resolution of the SPECT camera, coupled with effective targeting of the ^{111}In-DTPA-labeled liposomes, showed intratumoral heterogeneity of the radiolabeled liposome (Fig. 7.1) [50, 51]. Although there was elevated accumulation in the liver and spleen, the tumor accumulation 24–48 h post-injection (p.i.) was between 0.26 and 1.2 MBq/g of tumor (5–10% ID/g). By using ^{111}In-DTPA-labeled PEGylated liposomes, Harrington *et al.* also observed high uptake in head and neck cancers (33.0 ± 15.8% ID/kg), moderate uptake in the lung tumors (18.3 ± 5.7% ID/kg), and low uptake in breast cancers (5.3 ± 2.6% ID/kg). Their data demonstrated that PEGylated liposomes can be utilized as therapeutic agents for solid tumors, especially head and neck tumors. The toxicity level was minimal with only one out of 17 patients exhibiting some acute reactions that subsided after a few minutes. The major clearance route was through the liver, which reached a maximum retention at 4 h p.i., and the spleen [49]. For active targeting of PEGylated liposomes, Elbayoumi and Torchilin prepared ^{111}In-labeled liposome systems modified with mAb2C5 for gamma-imaging of tumors. The mAb specifically interacts with the surface-bound nucleosomes of tumor cells, but not normal cells. Biodistribution data showed that the accumulation in the tumor of the ^{111}In-labeled liposome systems was almost double that of the surrounding tissue. Additionally, gamma-imaging showed enhanced tumor visualization with the ^{111}In-labeled liposomes as compared to the control radio formulations [52]. In a related mice study, more ^{111}In loading of PEGylated liposomes was achieved via the liposome-incorporated polychelating amphiphilic polymer (PAP), which resulted in better and faster gamma-imaging compared to nontargeted liposomal controls [53].

Liposomes labeled with ^{99m}Tc have also shown promise in monitoring therapeutic drug delivery. In one such study, ^{99m}Tc was introduced into the aqueous layer of liposomes of similar size and composition to the drug Doxil, a PEGylated liposomal-encapsulated form of doxorubicin. Mice were injected with either the ^{99m}Tc-labeled liposome, Doxil, or both with subsequent quantification of tumor delivery post-hyperthermia treatment. This allowed visualization of the drug's effectiveness in reaching its target site [54]. In another study, the pharmacokinetics of intratumoral injections of liposomal-based drugs was quantified with ^{99m}Tc-labeled liposomes. Rat head and neck tumors had intratumoral injections of either ^{99m}Tc-labeled liposomes or a ^{99m}Tc-labeled small-molecule N,N-bis(2-mercaptoethyl)-N′, N′-diethylethylenediamine (BMEDA). At 20h p.i., the liposomal-encapsulated ^{99m}Tc showed a greater tumor retention ($39.2 \pm 10.6\%$, $n=4$) compared to the unencapsulated form ($18.7 \pm 3.3\%$, $n=3$) [55]. Additionally, better intratumoral ^{99m}Tc activity diffusion was observed for ^{99m}Tc-liposomes when compared with ^{99m}Tc-BMEDA as shown in the pinhole microSPECT images.

Radiolabeled liposomes can also be used to monitor therapeutic drug delivery to infection sites. ^{99m}Tc was radiolabeled on liposomes using the chelator *N*-hydroxysuccinimidyl hydrazine nicotinate hydrochloride (HYNIC), which was first conjugated to the liposome and then incubated with the ^{99m}Tc. This robust labeling did not require subsequent purification. High retention at the disease site was observed ($1.74 \pm 0.38\%$ ID/g, 24h post injection) with minimal uptake in the kidney. This new labeling technique showed better *in vitro* and *in vivo* stability as compared to an older chelation approach using hexamethylpropyleneamine oxime (HMPAO)-liposome [56].

Researchers have taken advantage of the multifunctional capabilities of liposomes. Zielhuis *et al.* successfully labeled liposomes with both a SPECT radiotracer, either ^{99m}Tc or ^{166}Ho, and paramagnetic gadolinium. In addition to SPECT imaging, these multilabeled liposomes allowed for measurements of MRI relaxivity. Potential therapeutic advantages could be seen from the beta emissions of ^{166}Ho. Thus, a single liposome could effectively offer two different imaging modalities as well as therapeutic gains [57].

7.3.3 Radiolabeled Iron Oxides for SPECT Imaging

Iron oxide, a member of the inorganic nanoparticle construct, has garnered attention due to its magnetic properties as well as it low toxicity in humans. Under the influence of an external alternating magnetic field (AMF), iron oxide exhibits thermoablative properties and thus, when targeted to solid tumors, can be used in a therapeutic manner. De Nardo *et al.* looked at ^{111}In-DOTA-labeled chimeric L6 (ChL6) mAb conjugated to PEGylated, dextran-coated, 20nm superparamagnetic iron oxide (SPIO) particles with one to two mAbs per nanoparticle. The study included considerations of radioimmunotargeting, therapeutic potential, and toxicity of the immunotargeted iron oxide nanoprobe in mouse breast cancer xenografts. This immunotargeted nanocarrier showed significantly increased blood circulation half-lives and tumor uptake compared to previous nanoprobes, though the liver and spleen uptakes were twice as much as the ^{111}In-ChL6 with no iron oxide. SPECT imaging of excised tumors 48 h p.i. showed that the radiolabeled iron oxide probe was delivered to the tumor in sufficient amounts (14% ID/g) without degradation. Toxicity levels were minimal and only observed at significantly higher AMF amplitudes of 1300 Oe [58, 59]. Interestingly, with the increase of size (30–100 nm), although higher thermoablative effect could be observed, the tumor accumulation of radiolabeled SPIO was significantly decreased [60]. Also, SPIO was also explored for SPECT/MR application to study the *in vivo* mapping of sentinel lymph node (SLN) as a means of staging metastasis in breast cancer and malignant melanoma via ^{99m}Tc radiolabeling. The biodistribution showed 100% ID/g accumulation of ^{99m}Tc-labeled superparamagnetic iron oxide nanoparticles (SPIONs) in the SLN and less than 2% ID/g in the liver and spleen, which was also confirmed by SPECT/MR images. Autoradiography showed the heterogeneous distribution of ^{99m}Tc-SPIONs in the lymph nodes [61].

SPIONs with bare or weakly bound coating on their surface can be surface modified directly with bisphosphonates (BPs). BPs are drugs that bind strongly to metabolically active bone as used in oncology and osteoporosis [62]. BP can be further functionalized with fluorophores, radionuclides, and biological tags, thus serving as anchor agents on SPIONs [63–65]. A clinically approved, SPIO-based liver MRI agent, Endorem/Feridex, was conjugated to a ^{99m}Tc-BP tracer via the dipicolylamine-alendronate (DPA-ale) group. The radiolabeled Endorem/Feridex showed excellent *in vitro* and *in vivo* stability according to nanoSPECT while maintaining its MRI contrast capabilities. This multimodal SPECT/MRI agent, ^{99m}Tc-DPA-ale-Endorem, showed rapid uptake by the RES with 96.9 ± 0.9% ID found in the liver, 1.3 ± 0.4% ID in the spleen, and 1.8 ± 0.7% ID in other organs. The uptake in bone as measured in the femur was about 2% ID. These values are in good agreement with the nanoparticle-bound ^{99m}Tc-DPA-ale showing that the radiolabeled BP and the nanoparticle colocalized in the RES organs with little or no degradation *in vivo*. Due to the versatility of the anchoring agent, BP, radiolabeled SPIO-based contrast particles can be extended to PET/MRI applications for accurate quantification of the biodistribution of the contrast agent [62].

7.3.4 Radiolabeled Polymeric Nanoparticles for SPECT Imaging

Dendrimers, a member of the polymer nanoparticle construct, have tunable properties and an elevated capacity for multiple targeting, diagnostic imaging, and therapeutic agents per molecule. They consist of highly branched monomer units with sizes ranging from 10 to 100 nm. These monomers are synthetically layered upon each other in a stepwise controlled fashion, resulting in a monodispersed molecular assembly. Dendrimers contain a core–shell surrounded by layers of covalently linked monomers called generations (G) with terminal surface groups (Z) [66]. Besides their excellent size, shape, and surface chemistry control, which allow for improved manipulation of their *in vivo* biodistribution and pharmacokinetics, dendrimers are often soluble, biocompatible, and nonimmunogenic [66–68]. Similar to other targeting nanoparticles, dendrimers can be used for both passive targeting, where their size is coupled with the leaky vasculature of the tumor, direct accumulation, and active targeting, where the surface is functionalized with a moiety that recognizes a cell surface receptor.

Poly(amido)amine (PAMAM), one of the most commonly utilized dendrimers, has been functionalized with targeting groups for *in vivo* and *in vitro* applications. Zhang *et al.* have conjugated folic acid to PAMAM generation 5 (G5) dendrimers and radiolabeled them with ^{99m}Tc via DTPA chelation. The PAMAM dendrimers were partially acetylated to improve their solubility and increase cellular uptake specificity. The dendrimer conjugate, ^{99m}Tc-PAMAM-FA, was studied in KB tumor-bearing mice. *In vitro* stability studies in phosphate-buffered saline and newborn calf serum at 37°C showed that 90% and 78%, respectively, of the dendrimer was retained after 6 h. *In vivo* studies showed that the ^{99m}Tc-PAMAM-FA conjugate had more than 70% stability in the blood. *In vivo* biodistribution of the ^{99m}Tc-PAMAM-FA conjugate in KB-bearing mice at 2, 4, and 6 h p.i. showed an expected increase in accumulation of the radiolabeled conjugate at the tumor (13.34 ± 1.36% ID/g, 6 h p.i.). Kidney accumulation was influenced by the presence of folate-binding proteins that regulate uptake. Some accumulation was also observed in the liver (9.48 ± 0.51% ID/g) and heart (6.88 ± 0.47% ID/g). MicroSPECT images taken between 2 and 6 h p.i. confirm the increased accumulation of ^{99m}Tc-PAMAM-FA at the tumor and kidney. In a subsequent study, tumor uptake was increased by spacing the PAMAM dendrimer with a PEG linker, which allowed for greater structurally flexibility and selective binding of folic acid to the metastatic tumor while mostly evading the mononuclear phagocyte system (MPS) [67, 69]. These studies show that folic acid-conjugated dendrimers have great promise for diagnostic imaging in cancer. In a similar study, acetylated PAMAM was labeled with ^{99m}Tc and conjugated to either biotin or avidin. The nanoconjugates were evaluated *in vitro* in HeLa cells, and their biodistribution quantified in normal healthy mice. A relatively fast blood clearance to tissues and excretory organs was observed for the radiolabeled dendrimer conjugates. Unlike the ^{99m}Tc-PAMAM-FA, high uptake in the RES organs (liver and spleen) was observed in the microSPECT mice images. This could be due to nonspecific interactions between the hepatic cells and the net positive charge on the dendrimers, the relatively larger size of the resultant PAMAM-avidin molecule, or the interaction between the receptors on the hepatocytes and mannose residues on the avidin molecule [68].

7.3.5 Radiolabeled Carbon-Based Nanostructures for SPECT Imaging

Carbon-based nanostructures are made of cylindrical graphite sheets rolled at specific and discrete angles, often less than 100 nm in size. The graphite sheet, or graphene, can be rolled as either single- or multiwalled nanotubes (SWNT or MWNT). These carbon nanotubes (CNTs) are either metallic or semiconductive, possessing thermal and electrical properties. They have been utilized in nanotechnology, electronics, and optics as biosensors. They are also being investigated as drug delivery vehicles, diagnostic biosensors, and thermal ablative agents, as modifications of the tubes allow for these applications [70]. CNTs can be functionalized to improve their solubility, lower their *in vivo* toxicity, and improve their pharmacokinetic properties [71, 72]. CNTs can be radiolabeled by surface modification and subsequent labeling or by encapsulation of the radionuclide within their inner walls [73–76]. CNTs have also been reported to localize on cells via different routes based on their layering and size. Small CNTs with lengths less than 400 nm can be taken up into cells by diffusion, while larger CNTs enter cells by endocytosis [77]. Based on the number of layers, MWNTs are often localized on the cell surface, while the SWNTs are internalized into the nucleus (short SWNTs) or cytoplasm (long SWNTs) [78, 79].

One of the first studies that used CNTs for *in vivo* imaging was done by surface modification of SWNT and MWNT to create a free $–NH_3^+$ group. This amine group was then covalently conjugated to DTPA and subsequently chelated to ^{111}In with greater than 80% radiolabeling yield [76]. The ^{111}In-DTPA-MWNT showed good stability in serum at 24 h p.i. at 37°C. The distribution study using microSPECT/CT showed that within 1 min of injection, the ^{111}In-DTPA-MWNT began accumulating in the kidney and bladder with maximal accumulation seen in these organs after 30 min. At 24 h p.i., there was minimal radioactivity left in the kidney, indicating high renal clearance of ^{111}In-DTPA-MWNT. Interestingly, the nonfunctionalized MWNT accumulated mostly in the liver and lungs [75]. A recent approach filled the interior of SWNTs with radiohalogens such as ^{125}I, while modifying the surface with the sugar *N*-acetylglucosamine (GlcNAc^D). This formed a high radionuclide payload conjugate of GlcNAc^D–Na^{125}I-SWNTs. In contrast to the injected free Na^{125}I control, which accumulated mainly in the high-affinity organs such as thyroid and stomach, the Na^{125}I-SWNTs were localized predominantly in the lungs with no uptake in the thyroid, stomach, or bladder. This preferential and persistent uptake in the lungs at 4 and 24 h and 7 days p.i. indicated interactions of the sugar group with lung-localized proteins. An *in vitro* cytotoxicity assay of Na^{125}I-SWNTs on human lung epithelial cells revealed no toxicity at concentrations up to 125 μg/ml. This study shows that encapsulated high-density radionuclides in SWNT can be safely directed to the site of action without leakage for organ-specific diagnostic imaging [74].

7.3.6 Radiolabeled Gold Nanoparticles for SPECT Imaging

Gold nanoparticles (AuNPs), owing to their optical, conductive, and electronic properties, have been widely used in biomedical applications. Because they can absorb light in the visible/near-infrared (NIR) region to produce surface plasmons or waves of excitation in the atoms making up a conductive material, they can be used for photothermal therapy as well as other photosensing applications. The absorption wavelength is dependent upon the size and shape of the AuNP with larger gold particles absorbing more in the red region. A typical example is observed in gold nanorods (GNRs), which have two characteristic absorption wavelengths corresponding to the transverse (480–520 nm) and longitudinal (700–900 nm) sections of the rods [80]. In general, AuNPs with surface plasmon absorption in the far visible/NIR wavelength allow for the application of light irradiation that will penetrate further into tissues with less damage as compared to UV–Vis irradiation [80–82]. The ability to design not only the nanorods but other AuNPs with different surface plasmon wavelengths has greatly expanded their practical applications in nanomedicine [83].

AuNPs have been widely used for biomedical imaging owing to their straightforward preparation and versatile surface chemistry. Lately, cyclic RGD-PEGylated AuNP probes were radiolabeled with ^{125}I (^{125}I-cRP-AuNPs) to target U87MG tumors. Transmission electron microscope (TEM) images confirmed the localization of ^{125}I-cRP-AuNPs in $\alpha_v\beta_3$-positive U87MG cells and negligible amounts in control $\alpha_v\beta_3$-negative MCF7 cells. SPECT/CT imaging clearly showed the tumor accumulation 10 min after administration with high tumor/background contrast (Fig. 7.2). Clearance studies showed fast renal elimination as confirmed by TEM of urine samples. Interestingly, the radiolabeling of ^{125}I was directly on the gold surface with good yield and stability [84–86]. There were about six ^{125}I isotopes per AuNP with serum stability tests showing more than 85% of ^{125}I remaining intact on AuNPs after a 24 h incubation at 37°C, indicating the potential of this AuNP for tumor SPECT imaging [87].

GNRs have shown promise as theranostic agents owing to their optical properties and photothermal effect. After conjugation with folic acid and a tyrosine-tagged peptide, ^{125}I-labeled folate acid-gold nanorod (FA-GNR) showed clear SPECT images at 4 and 24 h p.i in SKOV3 tumor-bearing mice. However, a strong thyroid uptake signal was observed, indicating dehalogenation of the ^{125}I from the ^{125}I-FA-GNR complex. *In vivo* photothermal treatment of the tumor was done by illumination with an 810 nm CW laser (2 W/cm^2, 600 J/cm^2) at 24 h p.i. Tumor ablation was detected in the ^{125}I-FA-GNR-treated mice after day 2, and complete obliteration was observed in 67% of the study group up until day 30 [88].

7.4 RADIOLABELED NANOPARTICLES FOR PET IMAGING

7.4.1 Positron-Emitting Radionuclides

Among molecular imaging modalities, PET, due to its high sensitivity, quantitative measurement, limitless depth of tissue penetration, and versatility of chemical probes, has emerged as a major player in nuclear imaging [89]. Positron-emitting radionuclide-labeled nanoparticles have been extensively used in both preclinical and clinical studies as a tool to explore nanoparticles' *in vivo* pharmacokinetics, imaging capability, and theranostic potential [35]. For nanoparticles with different physicochemical properties and functional groups, the PET radionuclide, as well as the radiolabeling strategy, must be carefully considered to generate an optimal imaging outcome. The characteristics of commonly used PET radionuclides for nanoparticle labeling are summarized in Table 7.2.

7.4.2 Radiolabeled Liposomes for PET Imaging

Liposomes radiolabeled with ^{64}Cu were used to probe the enhanced permeability and retention (EPR) effect in tumor-bearing mice. In this study, a remote loading approach was employed to encapsulate ^{64}Cu into liposomes to achieve improved radiolabeling stability and tumor accumulation (5.0 ± 2.0% ID/organ) [90]. Due to the long *in vivo* half-life ($t_{1/2}$ = 18 h), high tumor-to-muscle ratio (17.9 ± 8.1) was achieved with these ^{64}Cu-liposomes. In contrast to a small-molecule-based radiotracer of ^{124}I-SKI212243 targeting EGFR kinase, ^{124}I-SKI212243-conjugated liposomes showed significantly

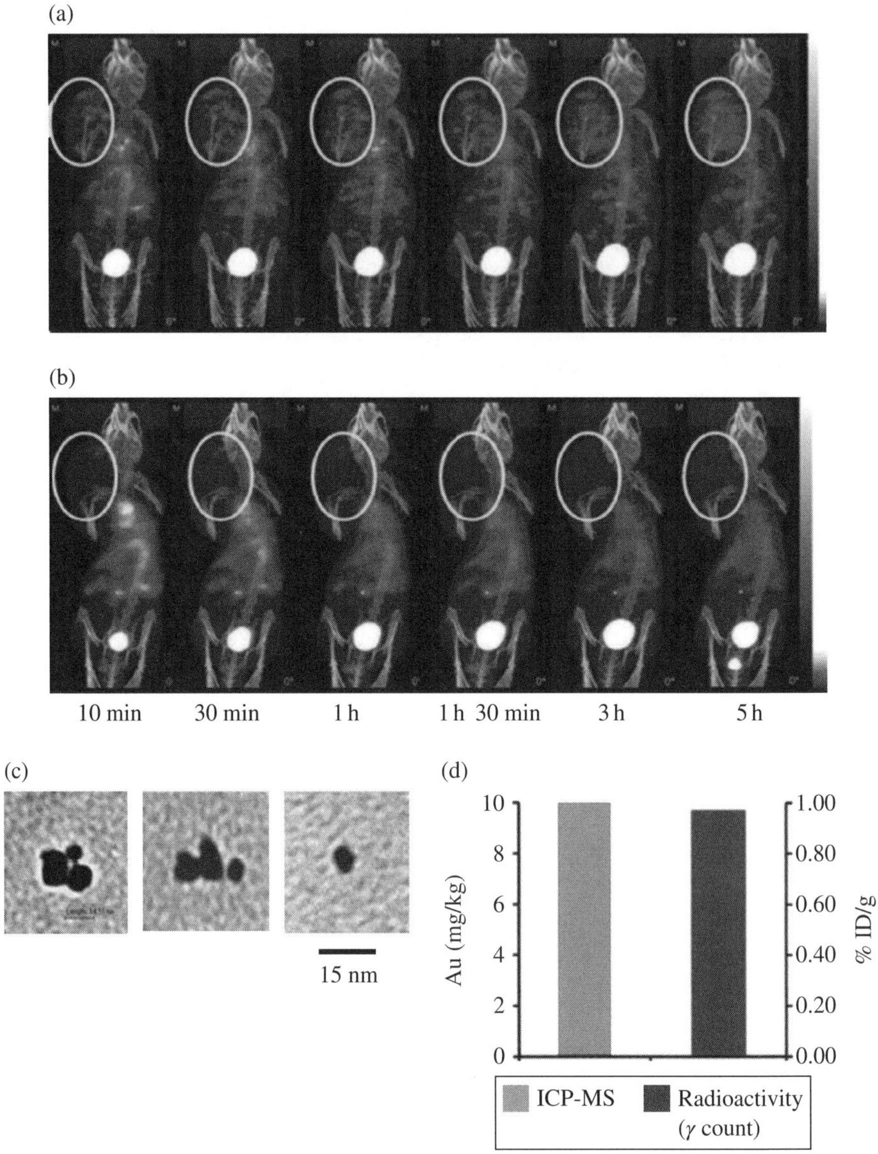

FIGURE 7.2 (a) Small animal SPECT/CT of ^{125}I-cRP-AuNPs and (b) cold-form-blocked ^{125}I-cRGD-AuNPs in nude mice bearing U87MG tumors xenografts after injection through tail vein. The SPECT/CT studies were performed serially from 0 to 5 h after injection of 11.1 MBq radiolabeled AuNPs. Yellow circles indicates tumor region. (c) TEM images of gold nanoparticles obtained from urine samples. (d) Biodistribution comparison of cRPAuNPs (inductively coupled plasma mass spectrometry [ICP-MS]) and iodine-125 (γ count). Data obtained from tumor 1 h after injection. %ID = injection dose. (With permission from Ref. [87]. © Wiley.) *(See insert for color representation of the figure.)*

TABLE 7.2 Nuclear Characteristics of Selected PET Radionuclides for Nanoparticles[a]

Radionuclide	$T_{1/2}$	Decay (%)	Main photon KeV (%)	Production
^{68}Ga	67.7 min	β^+ (89) EC (11)	511 (178.3)	$^{68}Ge/^{68}Ga$ generator
^{18}F	109.7 min	β^+ (96.7) EC (0.1)	511 (193.5)	^{18}O (p, n) ^{18}F
^{64}Cu	12.7 h	β^+ (17) EC (44)	511 (34.8)	^{64}Ni (p, n) ^{64}Cu
^{76}Br	16.2 h	β^+ (55) EC (45)	511 (109), 559 (74) 657 (15.9), 1854 (14.7)	^{76}Se (p, n) ^{76}Br ^{76}Se (d, 2n) ^{76}Br
^{86}Y	14.7 h	β^+ (33) EC (66)	511 (63.9), 1077 (82.5) 1115 (30.5), 627 (32.6)	^{86}Sr (p, n) ^{86}Y
^{89}Zr	3.3 d	β^+ (23) EC(77)	909 (100)	^{89}Y(p, n)^{89}Zr
^{124}I	4.18 d	β^+ (23) EC (77)	511 (46), 603 (62.9) 723 (10.3)	^{124}Te (p, n) ^{124}I ^{124}Te (d, 2n) ^{124}I

[a]Based in part on Ref. [41].

higher tumor accumulation (three- to six-fold increase) at 48 h p.i. Additionally, embedding the EGFR into the liposomes allowed for improved solubility of the lipophilic nanoparticle, increased blood circulation, and bioavailability for tumor targeting [91]. ^{18}F was also used for radiolabeling PEG-liposomes by the SophT method. In contrast to [^{18}F]FDG, biodistribution studies showed enhanced blood retention and high RES accumulation of [^{18}F]SteP2-labeled PEG-liposomes. Although the brain tumor uptake was lower than with [^{18}F]FDG, the PET images obtained using [^{18}F]SteP2-labeled PEG-liposomes were clearer due to lowered background as compared with [^{18}F]FDG [92].

7.4.3 Radiolabeled Iron Oxides for PET Imaging

Magnetic nanoparticles, especially iron oxide nanoparticles, have also been widely explored for PET and PET/MRI applications because of their inherently high T_2 relaxivity for enhanced contrast [93, 94]. To date, radiolabeling strategies resulting in high labeling yield and specific activity have been established for labeling iron oxide nanoparticles with positron emitters such as ^{64}Cu, ^{68}Ga, and ^{124}I. The *in vivo* biodistribution and targeting efficiency of these radiolabeled nanoparticles have been studied in a variety of animal disease models [95–98]. After coating the iron oxide surface with the carbohydrate dextran and conjugating with DTPA, ^{64}Cu-radiolabeled iron oxide nanoparticles (^{64}Cu-TNP) were used to target macrophages in an apolipoprotein E-deficient (apoE$^{-/-}$) mouse model of aneurysms. The high specific activity (3.7×10^8 Bq/mg Fe of nanoparticle) ensured lower-dose administration (1.5 mg Fe/kg body weight) than that used in oncology clinical trials (2.6 mg Fe/kg body weight) and sensitive detection of nanoparticle accumulation in various organs. The *in vivo* biodistribution of ^{64}Cu-TNP showed sufficient blood circulation ($t_{1/2} \geq 4$ h) and major accumulation in the liver and intestine. PET/CT imaging showed significant localization of ^{64}Cu-TNP in the thoracic aorta with a target-to-background ratio of 5.1 ± 0.9, indicating potential for clinical translatability. In a similar study, an ^{18}F-radiolabeled iron oxide nanoparticle (^{18}F-CLIO) was developed due to the wide availability, sensitivity, and covalent radiolabeling of this radioisotope [19]. With rapid ^{18}F click fluorination, high radiolabeling efficiency and specific activity were achieved ($6.8 \pm 0.8 \times 10^8$ Bq/mg Fe of nanoparticle). *In vivo* pharmacokinetic studies showed comparable blood retention to ^{64}Cu-TNP. In an apoE$^{-/-}$ aneurysm mouse model, PET images showed that the avid internalization by phagocytic cells led to significantly higher tracer accumulation at aneurysms relative to wild-type aorta [99].

PEG-coated iron oxide nanoparticles have also been radiolabeled with ^{64}Cu via a DOTA functionality (specific activity $= 3.7 - 7.4 \times 10^8$ Bq/mg Fe). High blood retention of $37.3 \pm 12.9\%$ ID/g at 1 h p.i. in mice was observed as confirmed by PET imaging [96]. In another study, a cyclic RGD peptide (c(RGDfC)) was conjugated to a SPIO for targeted PET/MR tumor imaging [100, 101]. This ^{64}Cu-cRGD-SPION gave low (<15% ID/g) hepatic burden up to 48 h p.i. and constant tumor uptake (~5% ID/g) across the study timeline, with the highest tumor-to-muscle ratio (11.3 ± 2.5) observed at 48 h p.i. Compared to the control ^{64}Cu-SPIO, ^{64}Cu-cRGD-SPIO had significantly ($p < 0.05$) higher tumor accumulation during the study, indicating $\alpha_v\beta_3$-specific targeting [98].

7.4.4 Radiolabeled Polymeric Nanoparticles for PET Imaging

Among various nanostructures such as iron oxide, silica, and AuNPs [102–106], amphiphilic core–shell nanostructures have received particular attention because of their tunable *in vivo* pharmacokinetics and versatile conjugation chemistry [107, 108]. Shell cross-linked knedel-like nanoparticles (SCK-NPs) are comprised of a hydrophobic polystyrene core that can be used to load hydrophobic drug molecules and a hydrophilic external shell of poly(acrylic acid-co-acrylamide) that provides additional sites for functional units such as imaging moieties. Through various synthetic strategies, especially cross-linking, SCK-NPs can be prepared with controlled size, surface charge, PEGylation density, functionality, and *in vivo* pharmacokinetics [109, 110]. The multivalency of SCK-NPs allows flexible radiolabeling (^{64}Cu, ^{76}Br, ^{124}I, and ^{18}F) for PET applications. By conducting DOTA conjugation prior to nanoparticle assembly, the amount of DOTA accessible for ^{64}Cu labeling can be accurately controlled with more than 400 copies per SCK, leading to a specific activity greater than 1.48×10^7 Bq/μg [111]. A novel strategy based on metal-free "click" chemistry has been developed for the ^{64}Cu radiolabeling of SCKs. In contrast to the Cu(I)-catalyzed click chemistry, this metal-free strategy improves ^{64}Cu labeling of the core of the nanoparticles due to the high click efficiency. When 1.85 GBq of ^{64}Cu was used for radiolabeling of the SCK-NPs containing 10% in-core azides, the specific activity could be up to 3.61×10^{13} Bq/μmol [112], indicating the great potential of this "click" labeling strategy for PET imaging.

Another type of core–shell, starlike, or comblike copolymer can be prepared with nitroxide-mediated living radical polymerization to create defined sizes and morphologies. In one example, the chelator DOTA was placed in an internal, hydrophilic environment allowing efficient ^{64}Cu radiolabeling to make a protected and high specific activity nanoscopic imaging probe. Biodistribution studies showed a distinct correlation between the length of PEG grafts and the *in vivo* circulation time; with increased PEG chain length, increased blood retention and reduced RES uptake were observed [113]. Furthermore, the cargo loading capacity of this type of nanoparticle can be adjusted while retaining comparable physicochemical properties. In a recent study, varying amounts of RGD peptide (5–50% RGD) were accurately conjugated to the shell of comblike nanoparticles for targeting $\alpha_v\beta_3$ integrin, and these RGD-combs all maintained similar sizes and radiolabeling specific activities [114, 115]. *In vitro* studies of RGD-combs showed positive correlation between RGD peptide loading and uptake in $\alpha_v\beta_3$ integrin-positive U87MG glioblastoma cells, demonstrating the importance of controlled conjugation of targeting groups [115]. Further, the comb nanoparticles were conjugated with C-type atrial natriuretic factor (CANF) to target the natriuretic peptide clearance receptor (NPRC) in a mouse angiogenesis model. By controlling the number of DOTA conjugations, high specific activity (5.4 ± 1.2 GBq/nmol) ^{64}Cu radiolabeling could be achieved, ensuring the trace administration of ^{64}Cu-DOTA-CANF-comb (7 pmol) for imaging studies. PET images showed significantly higher standardized uptake values (SUVs) at angiogenesis sites created by hind limb ischemia compared to contralateral control sites. More importantly, the SUVs of ^{64}Cu-DOTA-CANF-comb were 3.4 times higher than those obtained with DOTA-CANF peptide tracer and about triple those from the nontargeted ^{64}Cu-DOTA-comb, demonstrating the superiority of a multivalent nanoprobe over the corresponding monovalent CANF peptide for *in vivo* molecular targeting [116].

In developing nanoparticles for targeted drug delivery, key considerations include controlled-release kinetics, bioavailability, and reduced toxicity. This has led to active research in biodegradable nanoparticles [117–119]. Compared to their inorganic counterparts, polymeric nanoparticles can be prepared with biodegradable cores or cross-linkers for programmed release of therapeutic payload via enzyme or pH response degradation. This greatly enhances their biocompatibility and makes them candidates for targeted diagnosis and drug delivery. A core–shell biodegradable dendritic nanoprobe labeled with ^{76}Br has been prepared for targeting $\alpha_v\beta_3$ integrins expressed in a mouse angiogenesis model. The controlled introduction of targeting CRGDC peptide to the shell resulted in a 50-fold enhancement of *in vitro* binding affinity to $\alpha_v\beta_3$ integrins relative to the monovalent RGD peptide alone. *In vivo*, specific targeting to $\alpha_v\beta_3$ was observed with the targeted nanoprobes demonstrating a six-fold increase of receptor-mediated endocytosis at the injured site compared to the control nanoprobes [23].

Due to FDA approval for human use, the potential of poly(lactide-co-glycolide)-based biodegradable nanoparticles has been assessed for PET imaging [120]. Recently, various materials such as poly(lactide-co-glycolide), polyacrylates, and polyacrylamide have been used in the formulation of biodegradable nanoparticles [121–124]. In contrast to hydrophobic materials, polyacrylamide-based hydrogels offer excellent biocompatibility and hydrophobicity. They are also strongly endosome disrupting, which makes them a candidate for cytoplasmic delivery and imaging [125, 126]. With an acid-degradable cross-linker, the entrapped payload can be released in a pH-dependent manner inside endosomes [127]. For better cellular delivery, a cell-penetrating peptide (CPP) can be used on the nanoparticles to get various cargos into the cells without disturbing the stability of the cell membrane and with low cytotoxic effects [128]. In a recent study, it was reported that the optimal size for deposition in the deep lung for systemic delivery is approximately 1–3 μm—microparticles rather than nanoparticles [129]. Therefore, a nona-arginine-functionalized polyacrylamide-based microparticle was synthesized to study the delivery efficiency of entrapped protein into nonphagocytic lung epithelial cells (BEAS-2B). *In vitro* results showed effective delivery of encapsulated BSA–Alexa Fluor 488 into the BEAS-2B cells in both CPP- and concentration-dependent manners, as well as a time dependency [130]. This CPP-modified microparticle was then labeled with radiohalogens (^{125}I and ^{76}Br) for animal studies to assess the *in vivo* fate, lung retention, and cellular uptake after intratracheal administration. Furthermore, nanosized CPP particles were synthesized to compare size-related differences in clearance profiles. Biodistribution studies revealed that particle retention and extrapulmonary distribution were, in part, size dependent. Microparticles were rapidly cleared by mucociliary routes but, unexpectedly, also through circulation. In contrast, nanoparticles had prolonged lung retention enhanced by the CPP, which was confirmed with PET imaging analysis using ^{76}Br-radiolabeled nanoparticles (Fig. 7.3). These studies indicate the potential of microparticles for short-term applications as well as the benefits of nanoparticles for serial imaging or therapy of a persistent lung injury [131]. In contrast, a study of acute lung injury used latex nanoparticles coated with anti-intercellular adhesion molecule-1(ICAM-1) antibody and labeled with ^{64}Cu for targeting the pulmonary endothelium. Biodistribution studies showed three- to four-fold higher uptake in the lungs of mice injected with ICAM-targeted nanoparticles than those receiving control nanoparticles. PET imaging demonstrated the accumulation of radioactivity in the lungs. However, metabolic studies showed that the *in vivo* stability of this nanoprobe needs further improvement for prolonged pulmonary drug delivery [132].

7.4.5 Radiolabeled Carbon-Based Nanostructures for PET Imaging

CNTs have also been radiolabeled with PET isotopes either on the surface or by encapsulation within their inner walls. The use of radiolabeled ^{89}Zr-CNTs for PET oncological imaging has gained interest due to the unique decay properties of ^{89}Zr ($t_{1/2}$ = 78.4 h) as well as the availability of high specific activity product [133]. In an LS174T colon carcinoma model, ^{89}Zr-labeled SWCNT-([^{89}Zr]-DFO) showed rapid tumor accumulation and gradually increasing tumor-to-muscle contrast ratio over time (1.61 at 1 h to 5.08 at 96 h p.i.). In another colon carcinoma model (CT26), a ^{89}Zr-labeled cross-linked dextran nanoparticle showed primary localization in the lymph node (34 ± 16% ID/g). In tumor-bearing mice, PET imaging showed intense tumor uptake (20 ± 5% ID/g), surprisingly higher than other RES organs, indicating translational potential [134]. A new type of promising carbon biomaterial known

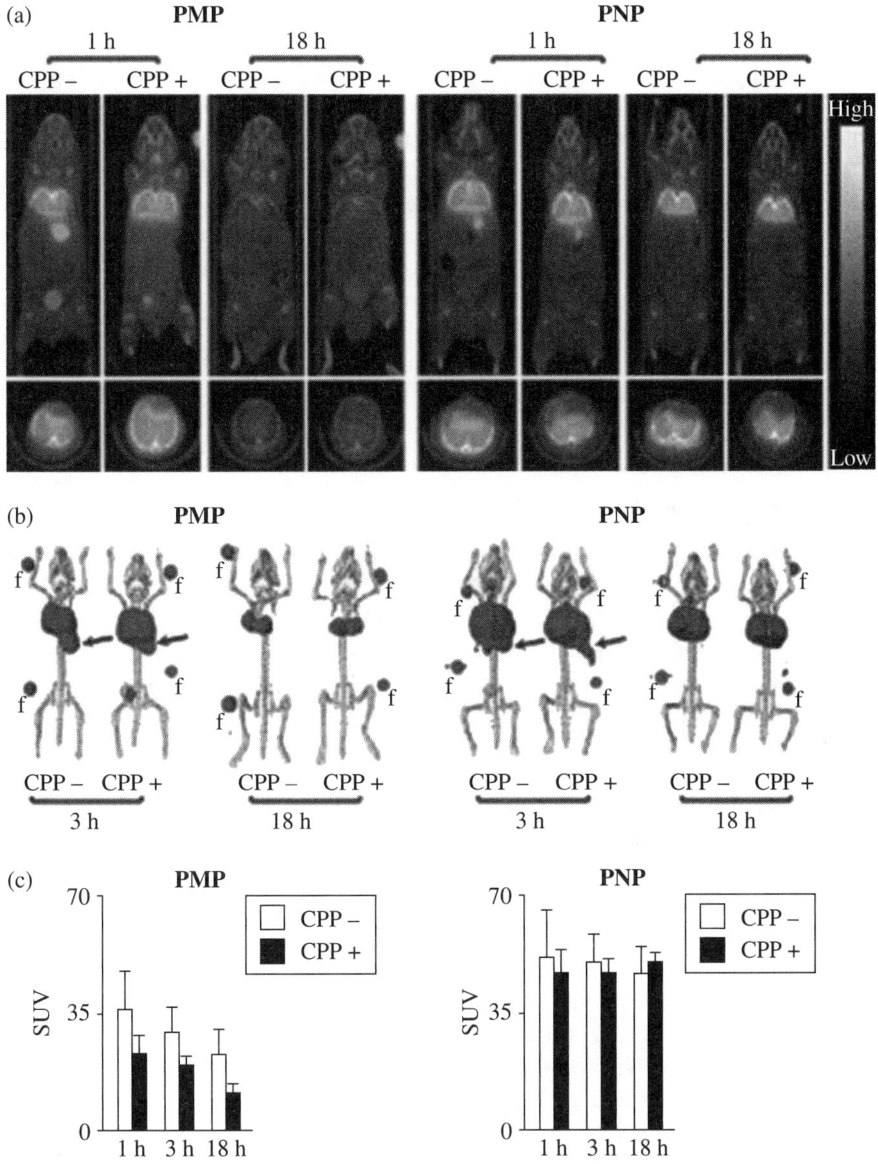

FIGURE 7.3 MicroPET/CT imaging of ^{76}Br-labeled particles following intratracheal delivery. (a) Representative midthoracic coronal (top) and transverse (bottom) slices of serial microPET and CT coregistered images obtained from mice at the indicated time following intratracheal administration of ^{76}Br-labeled particles. (b) Three-dimensional reconstructions of PET scan activity merged with X-ray CT images that were signal adjusted to reveal the skeleton. *Arrows* indicate gastrointestinal tract activity. Fiduciaries (f) used for coregistration are included. (c) The standardized uptake values (SUVs) of the PET activity within the lung region from images as in (a), shown as the mean ± SD of 3–4 mice per particle type. (Reprinted with permission from Ref. [130]. © American Chemical Society.)

as nanodiamond has been explored for biomedical applications due to its biocompatibility, ability to cross the cell membrane, and capability to be functionalized [134]. In a study by Rojas *et al.*, ^{18}F was radiolabeled on nanodiamonds via the reaction of ^{18}F-SFB ω-aminopropyl [135]. The radiolabeled nanodiamonds were further functionalized to study their *in vivo* biodistribution. Initial pharmacokinetics and PET imaging showed high retention in the lung, liver, and spleen, and significant excretion through the urinary tract. In further studies, the addition of surfactants to the nanodiamonds did not affect the biodistribution result, whereas the removal of larger nanodiamonds by filtration led to lower liver uptake with no retention in the lung or spleen. This suggests that the high pulmonary retention is most likely due to size exclusion and phagocytosis [135]. In clinical applications, ^{68}Ga-labeled carbon nanoparticles have also been used for pulmonary embolism PET/CT ventilation/perfusion (V/Q) imaging and have demonstrated superiority to conventional V/Q lung scintigraphy [136].

7.4.6 Radiolabeled AuNPs for PET Imaging

Gold nanostructures have been explored for theranostic application via PET imaging in various disease models [137, 138]. Gold nanoshells (NS) have been used for multimodality theranostics with ^{64}Cu radiolabeling and RGDfK peptide conjugation. PET/CT images of these ^{64}Cu-NS-RGDfK nanoprobes were compared with the nontargeted ^{64}Cu-NS in nude rat models bearing head and neck squamous cell carcinoma xenografts. The images showed significant tumor uptake and tumor vascular specificity, indicating active targeting and promising improved efficacy of photothermal ablation [139]. A chelator-free [^{64}Cu]CuS nanoparticle with controlled specific activity was prepared for tumor theranostic application. The PEGylated [^{64}Cu]CuS nanoparticles showed about 15% ID/g blood retention at 4 h p.i. in biodistribution studies and high tumor-to-muscle ratio (T/M = 6.55) at 24 h p.i. Interestingly, this PEG-[^{64}Cu]CuS nanoparticle also displayed photothermal properties [140].

Recently, a novel class of nanomaterials, gold nanocages (AuNCs), have been radiolabeled with ^{64}Cu with high specific activity for PET imaging applications. In contrast to 55 nm AuNCs, the smaller 30 nm AuNCs showed superior blood retention and lower hepatic and splenic uptake. In a murine EMT-6 tumor model, the smaller cages also showed a significantly higher level of tumor uptake and a greater tumor-to-muscle ratio than the large AuNCs. Quantitative PET imaging confirmed rapid accumulation and retention of AuNCs inside the tumors (Fig. 7.4). The ability to directly and quickly image the distribution of AuNCs *in vivo* provided a direction to further optimize their physicochemical properties for a range of theranostic applications [141].

7.4.7 Radiolabeled Silica Nanoparticles for PET Imaging

Silica nanoparticles, due to their biocompatibility and non-toxicity, have been radiolabeled for various biomedical applications [142, 143]. Labeled with ^{18}F, the thermally hydrocarbonized porous silicon nanoparticles passed intact through the gastrointestinal

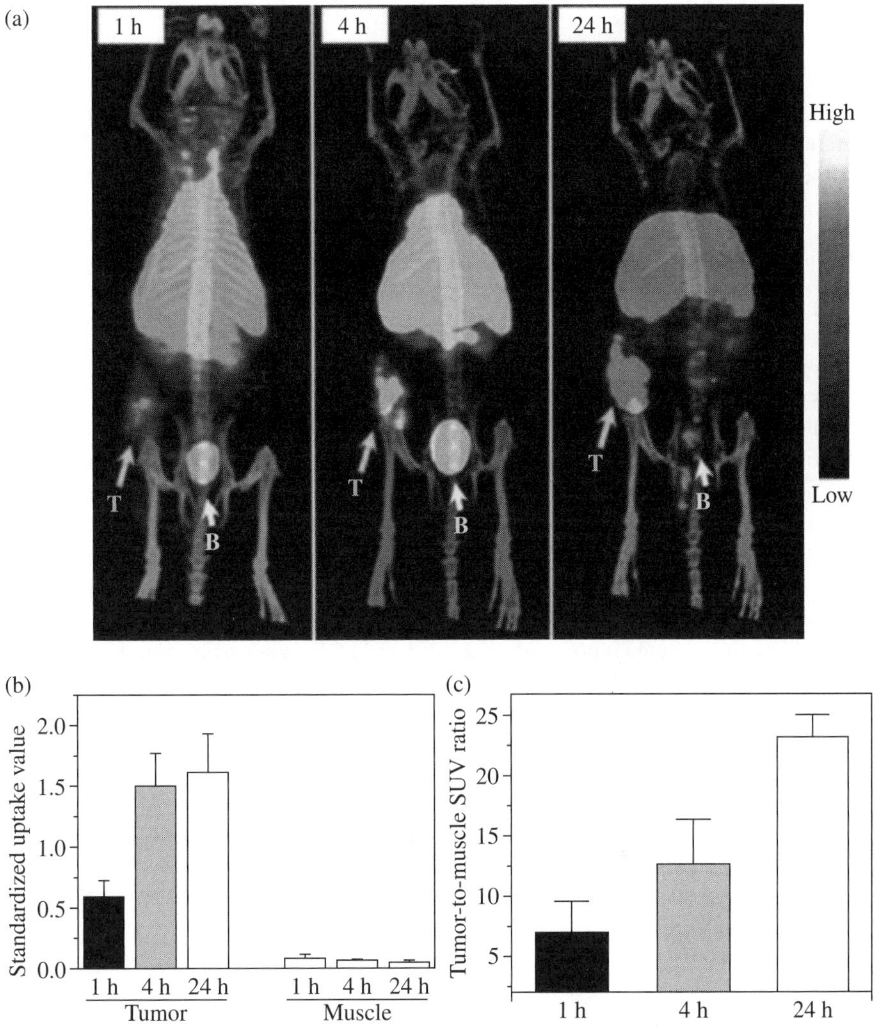

FIGURE 7.4 (a) PET/CT images of the 30 nm ^{64}Cu-DOTA-PEG-AuNCs in a mouse bearing an EMT-6 tumor at 1, 4, and 24 h postinjection (3.7 MBq injection/mouse). B, bladder; T, tumor. (b) SUVs in tumor and muscle regions obtained from PET/CT images taken at different times. (c) Tumor-to-muscle SUV ratios obtained from PET/CT images taken at different time points. (Reprinted with permission from Ref. [140]. © American Chemical Society.)

tract after oral administration and were not absorbed from a subcutaneous deposit. With intravenous injections, a fast RES clearance profile was confirmed. These silica nanoparticles exhibited excellent *in vivo* stability, low cytotoxicity, and nonimmunogenic profiles, indicating the potential for oral drug delivery [142]. In a theranostic application, cyclic arginine–glycine–aspartic acid with a tyrosine group (cRGDY) was conjugated to silica nanoparticles to achieve high *in vitro* binding affinity (IC_{50} = 1.2 nM) to melanoma cells. In an animal melanoma model, targeted ^{124}I-cRGDY-PEG-dots nanoprobes showed optimized pharmacokinetics (blood and tumor half-lives = 5.9 and 73.5 h, respectively), $\alpha_v\beta_3$-specific tumor uptake, and high tumor-to-muscle ratio (T/M = 5 at 24 h p.i.). Dosimetry studies demonstrated comparable radiation doses to other clinically used PET tracers. Toxicity studies confirmed full clearance in one week and no tissue-specific pathologic effect. A human clinical trial has been planned to investigate the potential of this targeted nanoprobe in staging metastatic disease [32].

7.5 SUMMARY

Radiolabeled nanoparticles have been widely used in biomedical research for drug delivery and diagnostic imaging. Compared to conventional small-molecule-based imaging probes, these nanoagents often provided superior pharmacokinetics, enhanced targeting efficiency, improved biocompatibility, and reduced toxicity, leading to significant progress in the development of personalized nanomedicine. However, owing to the complexity of living biological system, nanomedicine is still at the initial stage toward the future regular clinical application. More effort will need to be focused on the development of more sensitive and specific nanoagents for the detection of low-abundant biomarkers at the early stages of disease. Furthermore, the translational potential of these nanoagents must be considered and evaluated for the design of the nanostructures to be used in future clinical studies.

REFERENCES

[1] Börjesson PKE, Jauw YWS, de Bree R, Roos JC, Castelijns JA, Leemans CR, van Dongen GAMS, Boellaard R. Radiation dosimetry of 89Zr-labeled chimeric monoclonal antibody U36 as used for immuno-PET in head and neck cancer patients. J Nucl Med 2009;**50**:1828–1836.

[2] Braun S, Hepp F, Kentenich CRM, Janni W, Pantel K, Riethmüller G, Willgeroth F, Sommer HL. Monoclonal antibody therapy with edrecolomab in breast cancer patients: monitoring of elimination of disseminated cytokeratin-positive tumor cells in bone marrow. Clin Cancer Res 1999;**5**:3999–4004.

[3] Wolterink S, Moldenhauer G, Fogel M, Kiefel H, Pfeifer M, Lüttgau S, Gouveia R, Costa J, Endell J, Moebius U, Altevogt P. Therapeutic antibodies to human L1CAM: functional characterization and application in a mouse model for ovarian carcinoma. Cancer Res 2010;**70**:2504–2515.

[4] Dharap SS, Wang Y, Chandna P, Khandare JJ, Qiu B, Gunaseelan S, Sinko PJ, Stein S, Farmanfarmaian A, Minko T. Tumor-specific targeting of an anticancer drug delivery system by LHRH peptide. Proc Natl Acad Sci U S A 2005;**102**:12962–12967.

[5] Froidevaux S, Eberle AN. Somatostatin analogs and radiopeptides in cancer therapy. Pept Sci 2002;**66**:161–183.

[6] Ko SY, Guo H, Barengo N, Naora H. Inhibition of ovarian cancer growth by a tumor-targeting peptide that binds eukaryotic translation initiation factor 4E. Clin Cancer Res 2009;**15**:4336–4347.

[7] Breitz HB, Weiden PL, Beaumier PL, Axworthy DB, Seiler C, Su F-M, Graves S, Bryan K, Reno JM. Clinical optimization of pretargeted radioimmunotherapy with antibody-streptavidin conjugate and 90Y-DOTA-biotin. J Nucl Med 2000;**41**: 131–140.

[8] Knox SJ, Goris ML, Tempero M, Weiden PL, Gentner L, Breitz H, Adams GP, Axworthy D, Gaffigan S, Bryan K, Fisher DR, Colcher D, Horak ID, Weiner LM. Phase II trial of yttrium-90-DOTA-biotin pretargeted by NR-LU-10 antibody/streptavidin in patients with metastatic colon cancer. Clin Cancer Res 2000;**6**:406–414.

[9] Pagel JM, Hedin N, Subbiah K, Meyer D, Mallet R, Axworthy D, Theodore LJ, Wilbur DS, Matthews DC, Press OW. Comparison of anti-CD20 and anti-CD45 antibodies for conventional and pretargeted radioimmunotherapy of B-cell lymphomas. Blood 2003;**101**:2340–2348.

[10] Corneillie TM, Whetstone PA, Lee KC, Wong JP, Meares CF. Converting weak binders into infinite binders. Bioconjug Chem 2004;**15**:1389–1391.

[11] Goodwin DA, Meares CF. Pretargeting: general principles. Cancer 1997;**80**:2675–2680.

[12] Wei LH, Olafsen T, Radu C, Hildebrandt IJ, McCoy MR, Phelps ME, Meares C, Wu AM, Czernin J, Weber WA. Engineered antibody fragments with infinite affinity as reporter genes for PET imaging. J Nucl Med 2008;**49**:1828–1835.

[13] Loudos G, Kagadis GC, Psimadas D. Current status and future perspectives of *in vivo* small animal imaging using radiolabeled nanoparticles. Eur J Radiol 2011;**78**: 287–295.

[14] Gunasekera U, Pankhurst Q, Douek M. Imaging applications of nanotechnology in cancer. Target Oncol 2009;**4**:169–181.

[15] Heidel J, Davis M. Clinical developments in nanotechnology for cancer therapy. Pharm Res 2011;**28**:187–199.

[16] Hamoudeh M, Kamleh MA, Diab R, Fessi H. Radionuclides delivery systems for nuclear imaging and radiotherapy of cancer. Adv Drug Deliv Rev 2008;**60**:1329–1346.

[17] Sanvicens N, Marco MP. Multifunctional nanoparticles – properties and prospects for their use in human medicine. Trends Biotechnol 2008;**26**:425–433.

[18] Chen K, Li Z-B, Wang H, Cai W, Chen X. Dual-modality optical and positron emission tomography imaging of vascular endothelial growth factor receptor on tumor vasculature using quantum dots. Eur J Nucl Med Mol Imaging 2008;**35**:2235–2244.

[19] Devaraj NK, Keliher EJ, Thurber GM, Nahrendorf M, Weissleder R. 18F Labeled nanoparticles for *in vivo* PET-CT imaging. Bioconjug Chem 2009;**20**:397–401.

[20] McDevitt MR, Chattopadhyay D, Kappel BJ, Jaggi JS, Schiffman SR, Antczak C, Njardarson JT, Brentjens R, Scheinberg DA. Tumor targeting with antibody-functionalized, radiolabeled carbon nanotubes. J Nucl Med 2007;**48**:1180–1189.

[21] Choi CHJ, Alabi CA, Webster P, Davis ME. Mechanism of active targeting in solid tumors with transferrin-containing gold nanoparticles. Proc Natl Acad Sci U S A 2010;**107**:1235–1240.

[22] Huang X, Peng X, Wang Y, Wang Y, Shin DM, El-Sayed MA, Nie S. A reexamination of active and passive tumor targeting by using rod-shaped gold nanocrystals and covalently conjugated peptide ligands. ACS Nano 2010;**4**:5887–5896.

[23] Almutairi A, Rossin R, Shokeen M, Hagooly A, Ananth A, Capoccia B, Guillaudeu S, Abendschein D, Anderson CJ, Welch MJ, Fréchet JMJ. Biodegradable dendritic positron-emitting nanoprobes for the noninvasive imaging of angiogenesis. Proc Natl Acad Sci U S A 2009;**106**:685–690.

[24] Liu Z, Liu S, Wang F, Liu S, Chen X. Noninvasive imaging of tumor integrin expression using (18)F-labeled RGD dimer peptide with PEG (4) linkers. Eur J Nucl Med Mol Imaging 2009;**36**:1296–1307.

[25] Bao A, Goins B, Klipper R, Negrete G, Mahindaratne M, Phillips WT. A novel liposome radiolabeling method using 99mTc-"SNS/S" complexes: in vitro and *in vivo* evaluation. J Pharm Sci 2003;**92**:1893–1904.

[26] Boerman OC, Laverman P, Oyen WJG, Corstens FHM, Storm G. Radiolabeled liposomes for scintigraphic imaging. Prog Lipid Res 2000;**39**:461–475.

[27] Phillips WT. Delivery of gamma-imaging agents by liposomes. Adv Drug Deliv Rev 1999;**37**:13–32.

[28] Phillips WT, Goins BA, Bao A. Radioactive liposomes. Wiley Interdiscip Rev Nanomed Nanobiotechnol 2009;**1**:69–83.

[29] Sofou S. Surface-active liposomes for targeted cancer therapy. Nanomedicine 2007;**2**: 711–724.

[30] Gomes CM, Abrunhosa AJ, Ramos P, Pauwels EKJ. Molecular imaging with SPECT as a tool for drug development. Adv Drug Deliv Rev 2011;**63**:547–554.

[31] Jarzyna PA, Gianella A, Skajaa T, Knudsen G, Deddens LH, Cormode DP, Fayad ZA, Mulder WJM. Multifunctional imaging nanoprobes. Wiley Interdiscip Rev Nanomed Nanobiotechnol 2010;**2**:138–150.

[32] Benezra M, Penate-Medina O, Zanzonico PB, Schaer D, Ow H, Burns A, DeStanchina E, Longo V, Herz E, Iyer S, Wolchok J, Larson SM, Wiesner U, Bradbury MS. Multimodal silica nanoparticles are effective cancer-targeted probes in a model of human melanoma. J Clin Invest 2011;**121**:2768–2780.

[33] Welch MJ, Hawker CJ, Wooley KL. The advantages of nanoparticles for PET. J Nucl Med 2009;**50**:1743–1746.

[34] Schluep T, Hwang J, Hildebrandt IJ, Czernin J, Choi CHJ, Alabi CA, Mack BC, Davis ME. Pharmacokinetics and tumor dynamics of the nanoparticle IT-101 from PET imaging and tumor histological measurements. Proc Natl Acad Sci U S A 2009;**106**: 11394–11399.

[35] Liu Y, Welch MJ. Nanoparticles labeled with positron emitting nuclides: advantages, methods, and applications. Bioconjug Chem 2012;**23**:671–682.

[36] Hong H, Zhang Y, Sun J, Cai W. Molecular imaging and therapy of cancer with radiolabeled nanoparticles. Nano Today 2009;**4**:399–413.

[37] Minchin RF, Martin DJ. Minireview: nanoparticles for molecular imaging—an overview. Endocrinology 2010;**151**:474–481.

[38] Kagadis GC, Loudos G, Katsanos K, Langer SG, Nikiforidis GC. *In vivo* small animal imaging: current status and future prospects. Med Phys 2010;**37**:6421–6442.

[39] Aweda TA, Meares CF. Combination of isothermal titration calorimetry and time-resolved luminescence for high affinity antibody–ligand interaction thermodynamics and kinetics. Methods 2012;**56**:145–153.

[40] Wadas TJ, Deng H, Sprague JE, Zheleznyak A, Weilbaecher KN, Anderson CJ. Targeting the alphavbeta3 integrin for small-animal PET/CT of osteolytic bone metastases. J Nucl Med 2009;**50**:1873–1880.

[41] Rossin R. Radiolabeled nanoplatforms: imaging hot bullets hitting their target. In: Chen Z, editor. *Nanoplatform-Based Molecular Imaging*. Hoboken, NJ: John Wiley & Sons, Inc; 2011. p 399–429.

[42] Liu Y, Miyoshi H, Nakamura M. Nanomedicine for drug delivery and imaging: a promising avenue for cancer therapy and diagnosis using targeted functional nanoparticles. Int J Cancer 2007;**120**:2527–2537.

[43] Ganta S, Devalapally H, Shahiwala A, Amiji M. A review of stimuli-responsive nanocarriers for drug and gene delivery. J Control Release 2008;**126**:187–204.

[44] Park YS. Tumor-directed targeting of liposomes. Biosci Rep 2002;**22**:267–281.

[45] Hong R-L, Huang C-J, Tseng Y-L, Pang VF, Chen S-T, Liu J-J, Chang F-H. Direct comparison of liposomal doxorubicin with or without polyethylene glycol coating in C-26 tumor-bearing mice. Clin Cancer Res 1999;**5**:3645–3652.

[46] Gabizon A, Price DC, Huberty J, Bresalier RS, Papahadjopoulos D. Effect of liposome composition and other factors on the targeting of liposomes to experimental tumors: biodistribution and imaging studies. Cancer Res 1990;**50**:6371–6378.

[47] Papahadjopoulos D, Allen TM, Gabizon A, Mayhew E, Matthay K, Huang SK, Lee KD, Woodle MC, Lasic DD, Redemann C. Sterically stabilized liposomes: improvements in pharmacokinetics and antitumor therapeutic efficacy. Proc Natl Acad Sci U S A 1991;**88**:11460–11464.

[48] Xiong X-B, Huang Y, Lu W-l, Zhang H, Zhang X, Zhang Q. Enhanced intracellular uptake of sterically stabilized liposomal doxorubicin in vitro resulting in improved antitumor activity *in vivo*. Pharm Res 2005;**22**:933–939.

[49] Harrington KJ, Mohammadtaghi S, Uster PS, Glass D, Peters AM, Vile RG, Stewart JSW. Effective targeting of solid tumors in patients with locally advanced cancers by radiolabeled pegylated liposomes. Clin Cancer Res 2001;**7**:243–254.

[50] Umeda I, Tani K, Tsuda K, Kobayashi M, Ogata M, Kimura S, Yoshimoto M, Kojima S, Moribe K, Yamamoto K, Moriyama N, Fujii H. High resolution SPECT imaging for visualization of intratumoral heterogeneity using a SPECT/CT scanner dedicated for small animal imaging. Ann Nucl Med 2012;**26**:67–76.

[51] Ogihara-Umeda I, Sasaki T, Kojima S, Nishigori H. Optimal radiolabeled liposomes for tumor imaging. J Nucl Med 1996;**37**:326–332.

[52] Elbayoumi T, Torchilin V. Tumor-targeted immuno-liposomes for delivery of chemotherapeutics and diagnostics. J Pharm Innov 2008;**3**:51–58.

[53] Erdogan S, Roby A, Torchilin VP. Enhanced tumor visualization by γ-scintigraphy with 111In-labeled polychelating-polymer-containing immunoliposomes. Mol Pharm 2006;**3**:525–530.

[54] Kleiter MM, Yu D, Mohammadian LA, Niehaus N, Spasojevic I, Sanders L, Viglianti BL, Yarmolenko PS, Hauck M, Petry NA, Wong TZ, Dewhirst MW, Thrall DE. A tracer dose of technetium-99m–labeled liposomes can estimate the effect of hyperthermia on intratumoral doxil extravasation. Clin Cancer Res 2006;**12**:6800–6807.

[55] Bao A, Phillips WT, Goins B, Zheng X, Sabour S, Natarajan M, Ross Woolley F, Zavaleta C, Otto RA. Potential use of drug carried-liposomes for cancer therapy via direct intratumoral injection. Int J Pharm 2006;**316**:162–169.

[56] Laverman P, Dams ETM, Oyen WJG, Storm G, Koenders EB, Prevost R, van der Meer JWM, Corstens FHM, Boerman OC. A novel method to label liposomes with 99mtc by the hydrazino nicotinyl derivative. J Nucl Med 1999;**40**:192–197.

[57] Zielhuis SW, Seppenwoolde JH, Mateus VA, Bakker CJ, Krijger GC, Storm G, Zonnenberg BA, van het Schip AD, Koning GA, Nijsen JF. Lanthanide-loaded liposomes for multimodality imaging and therapy. Cancer Biother Radiopharm 2006;**21**:520–527.

[58] DeNardo SJ, DeNardo GL, Miers LA, Natarajan A, Foreman AR, Gruettner C, Adamson GN, Ivkov R. Development of tumor targeting bioprobes (111In-chimeric L6 monoclonal antibody nanoparticles) for alternating magnetic field cancer therapy. Clin Cancer Res 2005;**11**:7087s–7092s.

[59] DeNardo SJ, DeNardo GL, Natarajan A, Miers LA, Foreman AR, Gruettner C, Adamson GN, Ivkov R. Thermal dosimetry predictive of efficacy of 111In-ChL6 nanoparticle AMF–induced thermoablative therapy for human breast cancer in mice. J Nucl Med 2007;**48**:437–444.

[60] Natarajan A, Gruettner C, Ivkov R, DeNardo GL, Mirick G, Yuan A, Foreman A, DeNardo SJ. Nanoferrite particle based radioimmunonanoparticles: binding affinity and *in vivo* pharmacokinetics. Bioconjug Chem 2008;**19**:1211–1218.

[61] Madru R, Kjellman P, Olsson F, Wingårdh K, Ingvar C, Ståhlberg F, Olsrud J, Lätt J, Fredriksson S, Knutsson L, Strand S-E. 99mTc-labeled superparamagnetic iron oxide nanoparticles for multimodality SPECT/MRI of sentinel lymph nodes. J Nucl Med 2012;**53**:459–463.

[62] Torres Martin de Rosales R, Tavaré R, Glaria A, Varma G, Protti A, Blower PJ. 99mTc-bisphosphonate-iron oxide nanoparticle conjugates for dual-modality biomedical imaging. Bioconjug Chem 2011;**22**:455–465.

[63] Lalatonne Y, Paris C, Serfaty JM, Weinmann P, Lecouvey M, Motte L. Bis-phosphonates-ultra small superparamagnetic iron oxide nanoparticles: a platform towards diagnosis and therapy. Chem Commun (Camb) 2008:2553–2555.

[64] Lalatonne Y, Monteil M, Jouni H, Serfaty JM, Sainte-Catherine O, Lièvre N, Kusmia S, Weinmann P, Lecouvey M, Motte L. Superparamagnetic bifunctional bisphosphonates nanoparticles: a potential MRI contrast agent for osteoporosis therapy and diagnostic. J Osteoporosis 2010;**2010**:747852.

[65] Benyettou F, Lalatonne Y, Sainte-Catherine O, Monteil M, Motte L. Superparamagnetic nanovector with anti-cancer properties: γFe2O3@Zoledronate. Int J Pharm 2009;**379**: 324–327.

[66] Tomalia DA, Reyna LA, Svenson S. Dendrimers as multi-purpose nanodevices for oncology drug delivery and diagnostic imaging. Biochem Soc Trans 2007;**35**:61–67.

[67] Zhang Y, Sun Y, Xu X, Zhu H, Huang L, Zhang X, Qi Y, Shen Y-M. Radiosynthesis and micro-SPECT imaging of 99mTc-dendrimer poly(amido)-amine folic acid conjugate. Bioorg Med Chem Lett 2010;**20**:927–931.

[68] Xu X, Zhang Y, Wang X, Guo X, Zhang X, Qi Y, Shen Y-M. Radiosynthesis, biodistribution and micro-SPECT imaging study of dendrimer–avidin conjugate. Bioorg Med Chem 2011;**19**:1643–1648.

[69] Zhang Y, Sun Y, Xu X, Zhang X, Zhu H, Huang L, Qi Y, Shen Y-M. Synthesis, biodistribution, and microsingle photon emission computed tomography (SPECT) imaging study of technetium-99m labeled pegylated dendrimer poly(amidoamine) (PAMAM)–folic acid conjugates. J Med Chem 2010;**53**:3262–3272.

[70] Utreja P, Jain S, Tiwary AK. Novel drug delivery systems for sustained and targeted delivery of anti-cancer drugs: current status and future prospects. Curr Drug Deliv 2010;**7**:152–161.

[71] Sahoo NG, Bao H, Pan Y, Pal M, Kakran M, Cheng HKF, Li L, Tan LP. Functionalized carbon nanomaterials as nanocarriers for loading and delivery of a poorly water-soluble anticancer drug: a comparative study. Chem Commun (Camb) 2011;**47**:5235–5237.

[72] Yang ST, Luo J, Zhou Q, Wang H. Pharmacokinetics, metabolism and toxicity of carbon nanotubes for biomedical purposes. Theranostics 2012;**2**:271–282.

[73] Wu W, Li R, Bian X, Zhu Z, Ding D, Li X, Jia Z, Jiang X, Hu Y. Covalently combining carbon nanotubes with anticancer agent: preparation and antitumor activity. ACS Nano 2009;**3**:2740–2750.

[74] Hong SY, Tobias G, Al-Jamal KT, Ballesteros B, Ali-Boucetta H, Lozano-Perez S, Nellist PD, Sim RB, Finucane C, Mather SJ, Green MLH, Kostarelos K, Davis BG. Filled and glycosylated carbon nanotubes for *in vivo* radioemitter localization and imaging. Nat Mater 2010;**9**:485–490.

[75] Lacerda L, Soundararajan A, Singh R, Pastorin G, Al-Jamal KT, Turton J, Frederik P, Herrero MA, Li S, Bao A, Emfietzoglou D, Mather S, Phillips WT, Prato M, Bianco A, Goins B, Kostarelos K. Dynamic imaging of functionalized multi-walled carbon nanotube systemic circulation and urinary excretion. Adv Mater 2008;**20**:225–230.

[76] Singh R, Pantarotto D, Lacerda L, Pastorin G, Klumpp C, Prato M, Bianco A, Kostarelos K. Tissue biodistribution and blood clearance rates of intravenously administered carbon nanotube radiotracers. Proc Natl Acad Sci U S A 2006;**103**:3357–3362.

[77] Antonelli A, Serafini S, Menotta M, Sfara C, Pierige F, Giorgi L, Ambrosi G, Rossi L, Magnani M. Improved cellular uptake of functionalized single-walled carbon nanotubes. Nanotechnology 2010;**21**:22.

[78] Kang B, Chang S, Dai Y, Yu D, Chen D. Cell response to carbon nanotubes: size-dependent intracellular uptake mechanism and subcellular fate. Small 2010;**6**:2362–2366.

[79] Kang B, Li J, Chang S, Dai M, Ren C, Dai Y, Chen D. Subcellular tracking of drug release from carbon nanotube vehicles in living cells. Small 2012;**8**:777–782.

[80] Kennedy LC, Bickford LR, Lewinski NA, Coughlin AJ, Hu Y, Day ES, West JL, Drezek RA. A new era for cancer treatment: gold-nanoparticle-mediated thermal therapies. Small 2011;**7**:169–183.

[81] Weissleder R. A clearer vision for *in vivo* imaging. Nat Biotech 2001;**19**:316–317.

[82] Liu L, Ding H, Yong K-T, Roy I, Law W-C, Kopwitthaya A, Kumar R, Erogbogbo F, Zhang X, Prasad P. Application of gold nanorods for plasmonic and magnetic imaging of cancer cells. Plasmonics 2011;**6**:105–112.

[83] Xia Y, Li W, Cobley CM, Chen J, Xia X, Zhang Q, Yang M, Cho EC, Brown PK. Gold nanocages: from synthesis to theranostic applications. Acc Chem Res 2011;**44**:914–924.

[84] Cheng W, Dong S, Wang E. Iodine-induced gold-nanoparticle fusion/fragmentation/aggregation and iodine-linked nanostructured assemblies on a glass substrate. Angew Chem Int Ed Engl 2003;**42**:449–452.

[85] Shao X, Zhang H, Rajian JR, Chamberland DL, Sherman PS, Quesada CA, Koch AE, Kotov NA, Wang X. 125I-Labeled gold nanorods for targeted imaging of inflammation. ACS Nano 2011;**5**:8967–8973.

[86] Shao X, Agarwal A, Rajian JR, Kotov NA, Wang X. Synthesis and bioevaluation of 125I-labeled gold nanorods. Nanotechnology 2011;**22**:135102.

[87] Kim Y-H, Jeon J, Hong SH, Rhim W-K, Lee Y-S, Youn H, Chung J-K, Lee MC, Lee DS, Kang KW, Nam J-M. Tumor targeting and imaging using cyclic RGD-PEGylated gold nanoparticle probes with directly conjugated iodine-125. Small 2011;**7**:2052–2060.

[88] Jang B, Park S, Kang SH, Kim JK, Kim S-K, Kim I-H, Choi Y. Gold nanorods for target selective SPECT/CT imaging and photothermal therapy *in vivo*. Quant Imaging Med Surg 2012;**2**:1–11.

[89] Hagooly A, Rossin R, Welch MJ. Small Molecule Receptors as Imaging Targets. In: Semmler W, Schwaiger M, editors. *Molecular Imaging II*. Berlin, Heidelberg: Springer; 2008. p 93–129.

[90] Petersen AL, Binderup T, Rasmussen P, Henriksen JR, Elema DR, Kjaer A, Andresen TL. 64Cu loaded liposomes as positron emission tomography imaging agents. Biomaterials 2011;**32**:2334–2341.

[91] Medina OP, Pillarsetty N, Glekas A, Punzalan B, Longo V, Gonen M, Zanzonico P, Smith-Jones P, Larson SM. Optimizing tumor targeting of the lipophilic EGFR-binding radiotracer SKI 243 using a liposomal nanoparticle delivery system. J Control Release 2011;**149**:292–298.

[92] Oku N, Yamashita M, Katayama Y, Urakami T, Hatanaka K, Shimizu K, Asai T, Tsukada H, Akai S, Kanazawa H. PET imaging of brain cancer with positron emitter-labeled liposomes. Int J Pharm 2011;**403**:170–177.

[93] McCarthy JR, Weissleder R. Multifunctional magnetic nanoparticles for targeted imaging and therapy. Adv Drug Deliv Rev 2008;**60**:1241–1251.

[94] Heesakkers RA, Hovels AM, Jager GJ, van den Bosch HC, Witjes JA, Raat HP, Severens JL, Adang EM, van der Kaa CH, Futterer JJ, Barentsz J. MRI with a lymph-node-specific contrast agent as an alternative to CT scan and lymph-node dissection in patients with prostate cancer: a prospective multicohort study. Lancet Oncol 2008;**9**:850–856.

[95] Choi JS, Park JC, Nah H, Woo S, Oh J, Kim KM, Cheon GJ, Chang Y, Yoo J, Cheon J. A hybrid nanoparticle probe for dual-modality positron emission tomography and magnetic resonance imaging. Angew Chem Int Ed Engl 2008;**47**:6259–6262.

[96] Glaus C, Rossin R, Welch MJ, Bao G. *In vivo* evaluation of (64)Cu-labeled magnetic nanoparticles as a dual-modality PET/MR imaging agent. Bioconjug Chem 2010;**21**: 715–722.

[97] Stelter L, Pinkernelle JG, Michel R, Schwartlander R, Raschzok N, Morgul MH, Koch M, Denecke T, Ruf J, Baumler H, Jordan A, Hamm B, Sauer IM, Teichgraber U. Modification of aminosilanized superparamagnetic nanoparticles: feasibility of multimodal detection using 3T MRI, small animal PET, and fluorescence imaging. Mol Imaging Biol 2010;**12**:25–34.

[98] Yang X, Hong H, Grailer JJ, Rowland IJ, Javadi A, Hurley SA, Xiao Y, Yang Y, Zhang Y, Nickles RJ, Cai W, Steeber DA, Gong S. cRGD-functionalized, DOX-conjugated, and (64)Cu-labeled superparamagnetic iron oxide nanoparticles for targeted anticancer drug delivery and PET/MR imaging. Biomaterials 2011;**32**:4151–4160.

[99] Nahrendorf M, Keliher E, Marinelli B, Leuschner F, Robins CS, Gerszten RE, Pittet MJ, Swirski FK, Weissleder R. Detection of macrophages in aortic aneurysms by nanoparticle positron emission tomography-computed tomography. Arterioscler Thromb Vasc Biol 2011;**31**:750–757.

[100] Beer AJ, Kessler H, Wester HJ, Schwaiger M. PET imaging of integrin alphavbeta3 expression. Theranostics 2011;**1**:48–57.

[101] Tabatabai G, Tonn JC, Stupp R, Weller M. The role of integrins in glioma biology and anti-glioma therapies. Curr Pharm Des 2011;**17**:2402–2410.

[102] Anderson CJ, Bulte JW, Chen K, Chen X, Khaw BA, Shokeen M, Wooley KL, VanBrocklin HF. Design of targeted cardiovascular molecular imaging probes. J Nucl Med 2010;**51** (Suppl 1):3S–17S.

[103] Buxton DB. Current status of nanotechnology approaches for cardiovascular disease: a personal perspective. Wiley Interdiscip Rev Nanomed Nanobiotechnol 2009;**1**:149–155.

[104] Nahrendorf M, Sosnovik DE, French BA, Swirski FK, Bengel F, Sadeghi MM, Lindner JR, Wu JC, Kraitchman DL, Fayad ZA, Sinusas AJ. Multimodality cardiovascular molecular imaging, Part II. Circ Cardiovasc Imaging 2009;**2**:56–70.

[105] Nahrendorf M, Zhang H, Hembrador S, Panizzi P, Sosnovik DE, Aikawa E, Libby P, Swirski FK, Weissleder R. Nanoparticle PET-CT imaging of macrophages in inflammatory atherosclerosis. Circulation 2008;**117**:379–387.

[106] Sinusas AJ, Bengel F, Nahrendorf M, Epstein FH, Wu JC, Villanueva FS, Fayad ZA, Gropler RJ. Multimodality cardiovascular molecular imaging, part I. Circ Cardiovasc Imaging 2008;**1**:244–256.

[107] Gaucher G, Dufresne MH, Sant VP, Kang N, Maysinger D, Leroux JC. Block copolymer micelles: preparation, characterization and application in drug delivery. J Control Release 2005;**109**:169–188.

[108] Nishiyama N, Kataoka K. Current state, achievements, and future prospects of polymeric micelles as nanocarriers for drug and gene delivery. Pharmacol Ther 2006;**112**:630–648.

[109] Sun G, Hagooly A, Xu J, Nystrom AM, Li Z, Rossin R, Moore DA, Wooley KL, Welch MJ. Facile, efficient approach to accomplish tunable chemistries and variable biodistributions for shell cross-linked nanoparticles. Biomacromolecules 2008;**9**:1997–2006.

[110] Xu J, Sun G, Rossin R, Hagooly A, Li Z, Fukukawa KI, Messmore BW, Moore DA, Welch MJ, Hawker CJ, Wooley KL. Labeling of polymer nanostructures for medical imaging: importance of crosslinking extent, spacer length, and charge density. Macromolecules 2007;**40**:2971–2973.

[111] Sun G, Xu J, Hagooly A, Rossin R, Li Z, Moore DA, Hawker CJ, Welch MJ, Wooley KL. Strategies for optimized radiolabeling of nanoparticles for *in vivo* PET imaging. Adv Mater 2007;**19**:3157–3162.

[112] Zeng D, Lee NS, Liu Y, Zhou D, Dence CS, Wooley KL, Katzenellenbogen JA, Welch MJ. (64)Cu core-labeled nanoparticles with high specific activity via metal-free click chemistry. ACS Nano 2012;**6**:5209–5219.

[113] Fukukawa K, Rossin R, Hagooly A, Pressly ED, Hunt JN, Messmore BW, Wooley KL, Welch MJ, Hawker CJ. Synthesis and characterization of core-shell star copolymers for *in vivo* PET imaging applications. Biomacromolecules 2008;**9**:1329–1339.

[114] Pressly ED, Rossin R, Hagooly A, Fukukawa K, Messmore BW, Welch MJ, Wooley KL, Lamm MS, Hule RA, Pochan DJ, Hawker CJ. Structural effects on the biodistribution and positron emission tomography (PET) imaging of well-defined (64)Cu-labeled nanoparticles comprised of amphiphilic block graft copolymers. Biomacromolecules 2007;**8**:3126–3134.

[115] Shokeen M, Pressly ED, Hagooly A, Zheleznyak A, Ramos N, Fiamengo AL, Welch MJ, Hawker CJ, Anderson CJ. Evaluation of multivalent, functional polymeric nanoparticles for imaging applications. ACS Nano 2011;**5**:738–747.

[116] Liu Y, Pressly ED, Abendschein D, Hawker CJ, Woodard GE, Woodard PK, Welch MJ. Targeting angiogenesis using a C-type atrial natriuretic factor conjugated nanoprobe and positron emission tomography. J Nucl Med 2011;**52**:1956–1963.

[117] Gan CW, Chien S, Feng SS. Nanomedicine: enhancement of chemotherapeutical efficacy of docetaxel by using a biodegradable nanoparticle formulation. Curr Pharm Des 2010;**16**:2308–2320.

[118] Kumari A, Yadav SK, Yadav SC. Biodegradable polymeric nanoparticles based drug delivery systems. Colloid Surf B Biointerf 2010;**75**:1–18.

[119] Soppimath KS, Aminabhavi TM, Kulkarni AR, Rudzinski WE. Biodegradable polymeric nanoparticles as drug delivery devices. J Control Release 2001;**70**:1–20.

[120] Courant T, Roullin VG, Cadiou C, Delavoie F, Molinari M, Andry MC, Chuburu F. Development and physicochemical characterization of copper complexes-loaded PLGA nanoparticles. Int J Pharm 2009;**379**:226–234.

[121] Roy I, Vij N. Nanodelivery in airway diseases: challenges and therapeutic applications. Nanomedicine 2010;**6**:237–244.

[122] Rytting E, Nguyen J, Wang X, Kissel T. Biodegradable polymeric nanocarriers for pulmonary drug delivery. Expert Opin Drug Deliv 2008;**5**:629–639.

[123] Semete B, Booysen L, Lemmer Y, Kalombo L, Katata L, Verschoor J, Swai HS. *In vivo* evaluation of the biodistribution and safety of PLGA nanoparticles as drug delivery systems. Nanomedicine 2010;**6**:662–671.

[124] Semete B, Booysen LI, Kalombo L, Venter JD, Katata L, Ramalapa B, Verschoor JA, Swai H. *In vivo* uptake and acute immune response to orally administered chitosan and PEG coated PLGA nanoparticles. Toxicol Appl Pharmacol 2010;**249**:158–165.

[125] Murthy N, Thng YX, Schuck S, Xu MC, Frechet JM. A novel strategy for encapsulation and release of proteins: hydrogels and microgels with acid-labile acetal cross-linkers. J Am Chem Soc 2002;**124**:12398–12399.

[126] Murthy N, Xu M, Schuck S, Kunisawa J, Shastri N, Frechet JM. A macromolecular delivery vehicle for protein-based vaccines: acid-degradable protein-loaded microgels. Proc Natl Acad Sci U S A 2003;**100**:4995–5000.

[127] Standley SM, Kwon YJ, Murthy N, Kunisawa J, Shastri N, Guillaudeu SJ, Lau L, Frechet JM. Acid-degradable particles for protein-based vaccines: enhanced survival rate for tumor-challenged mice using ovalbumin model. Bioconjug Chem 2004;**15**:1281–1288.

[128] Lundberg P, Langel U. A brief introduction to cell-penetrating peptides. J Mol Recognit 2003;**16**:227–233.

[129] Shoyele SA, Cawthorne S. Particle engineering techniques for inhaled biopharmaceuticals. Adv Drug Deliv Rev 2006;**58**:1009–1029.

[130] Cohen JL, Almutairi A, Cohen JA, Bernstein M, Brody SL, Schuster DP, Frechet JM. Enhanced cell penetration of acid-degradable particles functionalized with cell-penetrating peptides. Bioconjug Chem 2008;**19**:876–881.

[131] Liu Y, Ibricevic A, Cohen JA, Cohen JL, Gunsten SP, Frechet JM, Walter MJ, Welch MJ, Brody SL. Impact of hydrogel nanoparticle size and functionalization on *in vivo* behavior for lung imaging and therapeutics. Mol Pharm 2009;**6**:1891–1902.

[132] Rossin R, Muro S, Welch MJ, Muzykantov VR, Schuster DP. *In vivo* imaging of 64Cu-labeled polymer nanoparticles targeted to the lung endothelium. J Nucl Med 2008;**49**:103–111.

[133] Holland JP, Sheh Y, Lewis JS. Standardized methods for the production of high specific-activity zirconium-89. Nucl Med Biol 2009;**36**:729–739.

[134] Keliher EJ, Yoo J, Nahrendorf M, Lewis JS, Marinelli B, Newton A, Pittet MJ, Weissleder R. (89)Zr-labeled dextran nanoparticles allow *in vivo* macrophage imaging. Bioconjug Chem 2011;**22**:2383–2389.

[135] Rojas S, Gispert JD, Martin R, Abad S, Menchon C, Pareto D, Victor VM, Alvaro M, Garcia H, Herance JR. Biodistribution of amino-functionalized diamond nanoparticles. *In vivo* studies based on 18F radionuclide emission. ACS Nano 2011;**5**:5552–5559.

[136] Hofman MS, Beauregard JM, Barber TW, Neels OC, Eu P, Hicks RJ. 68Ga PET/CT ventilation-perfusion imaging for pulmonary embolism: a pilot study with comparison to conventional scintigraphy. J Nucl Med 2011;**52**:1513–1519.

[137] Chen J, Glaus C, Laforest R, Zhang Q, Yang M, Gidding M, Welch MJ, Xia Y. Gold nanocages as photothermal transducers for cancer treatment. Small 2010;**6**:811–817.

[138] Khlebtsov N, Dykman L. Biodistribution and toxicity of engineered gold nanoparticles: a review of in vitro and *in vivo* studies. Chem Soc Rev 2011;**40**:1647–1671.

[139] Xie H, Diagaradjane P, Deorukhkar AA, Goins B, Bao A, Phillips WT, Wang Z, Schwartz J, Krishnan S. Integrin alphavbeta3-targeted gold nanoshells augment tumor vasculature-specific imaging and therapy. Int J Nanomed 2011;**6**:259–269.

[140] Zhou M, Zhang R, Huang M, Lu W, Song S, Melancon MP, Tian M, Liang D, Li C. A chelator-free multifunctional [64Cu]CuS nanoparticle platform for simultaneous micro-PET/CT imaging and photothermal ablation therapy. J Am Chem Soc 2010;**132**: 15351–15358.

[141] Wang Y, Liu Y, Luehmann H, Xia X, Brown P, Jarreau C, Welch M, Xia Y. Evaluating the pharmacokinetics and *in vivo* cancer targeting capability of au nanocages by positron emission tomography imaging. ACS Nano 2012;**6**:5880–5888.

[142] Bimbo LM, Sarparanta M, Santos HA, Airaksinen AJ, Makila E, Laaksonen T, Peltonen L, Lehto VP, Hirvonen J, Salonen J. Biocompatibility of thermally hydrocarbonized porous silicon nanoparticles and their biodistribution in rats. ACS Nano 2010;**4**:3023–3032.

[143] Kumar R, Roy I, Ohulchanskky TY, Vathy LA, Bergey EJ, Sajjad M, Prasad PN. *In vivo* biodistribution and clearance studies using multimodal organically modified silica nanoparticles. ACS Nano 2010;**4**:699–708.

8

MRI WITH GADOLINIUM-BASED NANOPARTICLES

FRANÇOIS GUÉRARD, GEOFFREY L. RAY, AND MARTIN W. BRECHBIEL

Radioimmune & Inorganic Chemistry Section, Radiation Oncology Branch, NCI, National Institutes of Health, Bethesda, MD, USA

8.1 INTRODUCTION

Magnetic resonance imaging (MRI) is a powerful tool that has provided scientists and clinicians with the ability to acquire *in vivo* images of anatomy and physiology in animals and humans. This imaging tool has become increasingly popular in experimental molecular imaging and clinical radiology because it is noninvasive and capable of producing three-dimensional representations of opaque organisms with high spatial and high temporal resolution. Over the past two decades, MRI has become a valuable technique for medical diagnostics providing excellent soft tissue contrast with submillimeter resolution on clinical scanners while not having to employ of ionizing radiation.

MRI, with the help of a contrast agent (CA), can eliminate painful, invasive procedures and provide diagnostic information earlier in the clinical evaluation. Currently, 40–50% of all clinical MR exams utilize contrast media. While instrumentation and techniques continue to improve image quality, CAs can dramatically highlight anatomic and pathologic features of interest. Over the last 25 years, gadolinium (Gd^{3+})-based MR CAs have been developed to enhance signal of water in tissues by altering proton relaxation properties, and the ideal characteristics of Gd^{3+} made it the most used CA in MRI examinations. However, improvements of the sensitivity of Gd^{3+}-based CA are still required. A number of Gd^{3+}-based nanoparticles under investigation could fulfill the need of more sensitive and safer agents and could be part of the next generation of CA for MRI.

Nanotechnology for Biomedical Imaging and Diagnostics: From Nanoparticle Design to Clinical Applications, First Edition. Edited by Mikhail Y. Berezin.

In this chapter, we first describe the mechanisms involved in the production of contrasted images with Gd^{3+}-based CAs and detail the parameters to consider when designing Gd^{3+} complexes. The advantages and limitations of the currently approved CAs are discussed. These considerations should provide the reader the important points that have guided the design of the nanoparticles reviewed in this chapter.

8.2 GADOLINIUM AS A CONTRAST AGENT IN MRI

CAs are species that alter the de-excitation signal of the surrounding water molecule that has been submitted to a magnetic field. The de-excitation, or relaxation of protons (present in the body mostly in water and fat), after absorption of a specific pulsed radio frequency (RF), can be divided into two time components: the longitudinal relaxation time (T_1) and the transverse relaxation time (T_2). T_1 and T_2 values of water differ within the kind of tissues (for instance, blood, fat, or gray matter produce signals that can be differentiated). Also, depending on the nature of the CA, T_1 and/or T_2 can be altered. The measurement of the differences in relaxation times allows reconstitution of images with contrasted area depending on the nature of tissues or the presence or absence of the CA in the area of interest. It is beyond the scope of this chapter to detail the mechanisms involved in the production of images in an MRI experiment. The theory of MRI can be found in recent books [1, 2]. The progresses in MRI diagnosis are also strongly dependent on the advances in instrumentation. The primary parameter that impacts the sensitivity of MRI is the magnetic field strength of the scanner, which is expressed in tesla (T). While most of the scanners currently used in clinics are of 1.5–3 T, higher magnetic fields can now be reached. 15–20 T scanners have become available, mostly for animal studies but also for human, especially in neurological [3], musculoskeletal [4], or cardiac [5] imaging where high sensitivity and resolution are required. These hardware improvements come along with software evolutions and with the design of new RF pulse sequences, allowing visualization of tissues of interest with increasing accuracy and sensitivity by, for example, suppression of unwanted tissue signal during image acquisition or by making possible the production of three-dimensional images. Recent advances and perspectives in the technology of MRI have been reviewed by Blamire [6].

8.2.1 Gadolinium and Relaxivity

The relaxation of a nucleus is mostly caused by the inhomogeneities in the local magnetic field and the dipole–dipole interactions with the surrounding nuclei. Introduction of a paramagnetic species such as Gd^{3+} in the environment of a proton nucleus can accelerate the relaxation process. The efficiency of a paramagnetic substance to increase the proton relaxation rate is termed relaxivity. The relaxation rate is proportional to the relaxivity of the paramagnetic species r_i and to the concentration of the paramagnetic species $[M]$:

$$\frac{1}{T_{i,\mathrm{obs}}} = \frac{1}{T_{i,\mathrm{d}}} + \frac{1}{T_{i,\mathrm{p}}} = \frac{1}{T_{i,\mathrm{d}}} + r_i\left[M\right]$$

with i = 1 or 2 for longitudinal or transverse relaxation, respectively, $1/T_{i,\text{obs}}$ the observed relaxation rate, $1/T_{i,\text{d}}$ the diamagnetic relaxation rate, and $1/T_{i,\text{p}}$ the paramagnetic relaxation rate. r_i is generally expressed in $\text{mM}^{-1}\,\text{s}^{-1}$. Relaxivity is governed by a number of parameters described by the Solomon–Bloembergen–Morgan (SBM) theory [7].

The influence of a paramagnetic species on the proton nucleus of water decreases with distance, and its effects on the relaxation can be distinguished according to three relaxation spheres. The inner sphere corresponds to the water directly bound to the paramagnetic metal, the second sphere corresponds to the water molecules bound to the chelating agent–metal complex (by hydrogen bonds), and the outer sphere describes the water from the bulk (Fig. 8.1). Because of water molecule exchanges between the three spheres, the modification on the relaxation rate will not affect only one water molecule, but it will be observed on a larger number of water molecules. Relaxivity is governed by the residence time of the coordinated water molecule, τ_m; the rotational tumbling time of the metal–water complex, τ_r; and the diffusional correlation time, τ_D (Fig. 8.1).The first use of a CA for MRI was reported in 1978 by Lauterbur *et al.* and involved a Mn^{2+} salt [8]. With its seven unpaired 4f electrons, the lanthanide ion Gd^{3+} exhibits excellent characteristics and has rapidly become the most widely used CA in MRI. Furthermore, the symmetric electronic state of Gd^{3+} ($^8S_{7/2}$ ground state) plays an important part in its efficiency, increasing the longitudinal and transverse relaxation rate without shifting the proton resonance frequency.

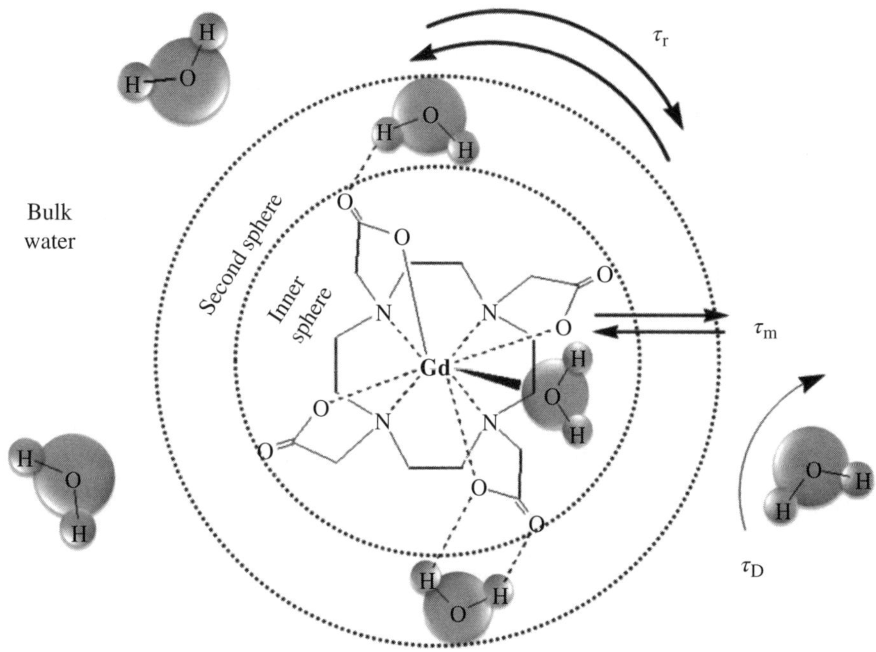

FIGURE 8.1 Representation of the inner sphere, the second sphere, and the outer relaxation sphere.

Dysprosium (Dy^{3+}) and holmium (Ho^{3+}) have larger magnetic moments than Gd^{3+} leading to a higher relaxivity, but this comes with an undesirable shift of the proton resonance frequency induced by their asymmetric electronic ground state [9].

Paramagnetic ions such as Gd^{3+} belong to the T_1 class of agents as they affect longitudinal relaxation rate more than transverse relaxation rate. Superparamagnetic agents such as large iron oxide particles belong to the T_2 class of agents.

With an ionic radius of 0.99Å, equivalent to Ca^{2+} (1.00Å), Gd^{3+} competes with Ca^{2+} in biological systems, which makes it highly toxic if used in the free form [10–12]. This toxicity is why the free metal ion cannot be injected directly to the patient; complexation with a chelating agent is necessary. Several studies have focused on the development of stable sequestering agents, which are the subject of the next section.

8.2.2 Complexation Chemistry of Gadolinium: A Compromise between Stability and Relaxivity

With its high charge (+3) and its relatively small ionic radius (0.99Å), Gd^{3+} belongs to the hard acids according to the HSAB principle of Pearson [13]. Consequently, it binds strongly to hard bases such as nitrogen or negatively charged oxygen atoms. In aqueous media, Gd^{3+} tends to form eight- or nine-coordinated complexes. For these reasons, the molecules developed and clinically used today are generally derivatives of two main structures: the linear triamine DTPA and the 12-membered tetraazamacrocycle (cyclen) with five or four carboxylate pending arms, respectively, forming eight-coordinated complexes with the ninth coordination site available to be occupied by a water molecule (Chart 8.1). The principles of designing chelating agents for various kinds of metals, including the lanthanides, have been extensively reviewed [14–16]. They are generally used for purification purposes (e.g., extraction of heavy metals from water) or medical applications (for extraction of contaminant metals in blood, in nuclear medicine, or as CAs in MRI). For medical applications, the chelating agent must strongly bind the metals of interest to avoid release of the free ions that are often highly toxic to the biological system (by their radiations or chemical properties). The challenge is all the more difficult since the injected substances are highly diluted in the human body, exposing the complex to faster transchelation by

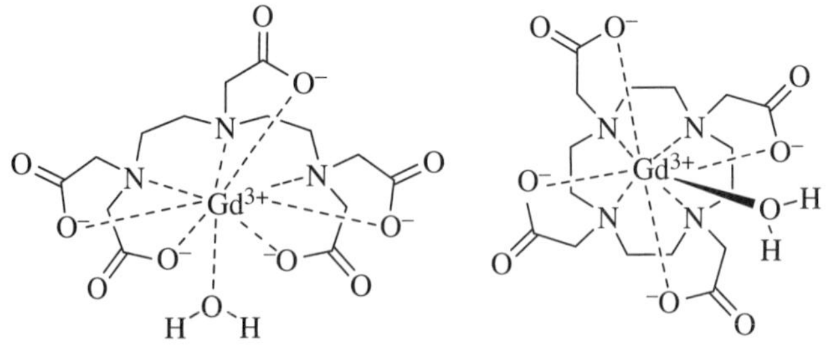

Gd-DTPA (Magnevist®) Gd-DOTA (Dotarem®)

CHART 8.1 Examples of approved contrast agents for clinical use.

replacement of the metal by another one in much higher concentration such as Fe^{3+}, Cu^{2+}, or Zn^{2+} or by release of the cation of interest into another endogenous chelator. More than the thermodynamic stability, above all the kinetic inertness must be high to achieve adequate *in vivo* stability to obviate concerns regarding toxicity. Kinetic inertness can be determined by competition reaction of the complex in the presence of other cations such as Zn^{2+}, the main competitor in the case of Gd^{3+} as suggested by Laurent *et al.* [17]. The chelating agent must include enough donor atoms (generally seven or eight nitrogen and negative oxygen atoms are involved), and the size of the chelate ring should be five for Gd^{3+} to form stable complexes [14]. Furthermore, higher preorganized ligands such as macrocycles generally lead to higher stability than the more flexible open-chain analogs. This has been demonstrated for Gd^{3+} by several comparative studies of cyclic and acyclic chelates in animals and humans [18].

However, another difficulty that has to be overcome when designing Gd complexes for MRI is that the relaxivity must not suffer from the gain in stability. Theoretically, the best relaxivity would be obtained by the octa- or nonahydrated Gd^{3+}. The hydration number, q, is a major parameter governing the relaxivity. The q value is equal to only one in each of the approved CAs. As expected, studies with heptacoordinating ligands such as DTTA or AAZTA (Chart 8.2) leading to complexes with $q=2$ have shown that the relaxivity was approximately two to four times higher than CAs with only one coordinated water molecule [19–21]. Alternatives to polyaminocarboxylic acid-based ligands have been proposed. The ligands developed in Raymond's group contain bidentate hydroxypyridinone (HOPO) units leading to hexacoordinated Gd^{3+} with two or three bound water molecules (Chart 8.2) [22, 23]. High relaxivities (up to $13\,mM^{-1}\,s^{-1}$ at 20 MHz) and high stabilities were observed.

However, hydration number is not the only parameter governing relaxivity. The residence time of the coordinated molecule, τ_m, plays a key role and is influenced by several features such as the charge of the complex or accessibility to the water binding site [24]. Efforts are made to obtain short τ_m close to the optimal value of

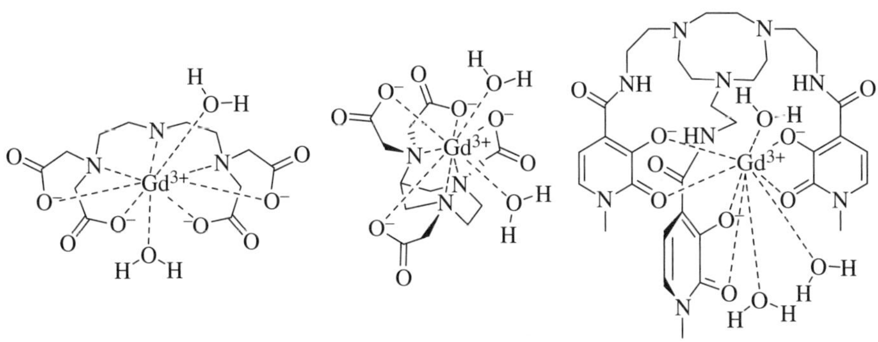

Gd-DTTA Gd-AAZTA Gd-TACN-1-Me-3,2-HOPO

CHART 8.2 Examples of complexes with two or three water molecules bound to the gadolinium ion.

$[Gd(H_2O)_8]^{3+}$ ($\tau_m = 1.2$ ns). τ_m can be measured by ^{17}O NMR and short values of approximately 2 ns have been obtained with HOPO-like Gd^{3+} complexes [23], suggesting possible applications in high magnetic fields. Higher relaxivities can also be obtained by optimizing the hydration number and the residence time in the second sphere (q_{ss} and τ_{mss}). For instance, the presence of phosphonates or phenolates which promotes hydrogen bond formation can increase relaxivity in DOTA and DTPA derivatives [25–28]. Studies have shown the high contribution that can be achieved by the second sphere as molecules with $q=0$ with optimized second sphere characteristics could even exhibit relaxivities close to $q=1$ complexes [29, 30]. Finally, it has been observed that increasing the rotation correlation time, τ_r, enhances relaxivity. This can be achieved by increasing the molecular volume/size (and consequently molecular weight) of the molecule, as demonstrated by the enhancement of the relaxivity from 4 $mM^{-1}s^{-1}$ for $[Gd(dota)(H_2O)]^-$ to 23.5 $mM^{-1}s^{-1}$ for $[Gd(dota\text{-}glu_{12})(H_2O)]^-$ (25°C, 20 MHz) [31]. Increasing the rigidity of the complex is also a method that can be used to increase τ_r [32, 33]. Specific examples of τ_r enhancement in the case of nanoparticles are discussed in the second part of this chapter.

8.2.3 Approved Gadolinium-Based Contrast Agents: Clinical Applications and Limitations

Magnevist™ was the first CA approved for clinical use in 1988. There are currently nine approved Gd^{3+}-based CAs (Table 8.1). They are all derivatives of DTPA or DOTA with a relatively low molecular weight (~600–800 Da). Their relaxivities range from 4.1 to 6.9 $mM^{-1}s^{-1}$ except for the most recently introduced MS-325 whose relaxivity is 27.7 $mM^{-1}s^{-1}$ (1.5 T, 37°C), because it binds serum albumin *in vivo* leading to a macromolecular species with a low tumbling rate of the complex. They differ in charge, shape, and chemical functions, conferring specific clearance rates and routes of excretion and thus different applications. They can be classified according to three main classes: extracellular fluid (ECF) agents or extracellular space (ECS) agents, liver agents, and intravascular or blood pool agents. ECF agents are hydrophilic, distribute nonspecifically in the whole ECF space, and are excreted exclusively by glomerular filtration through the kidneys with a half-life of approximately 1.5 h. They allow visualization of abnormalities in blood vessels and inflamed or diseased tissues. Because these molecules do not normally cross the blood–brain barrier, they can be used to visualize abnormalities causing permeability to these molecules in this area (Fig. 8.2a). MultiHance™ and Eovist™ are DTPA-based CAs with a benzyl group in their structure conferring higher lipophilicity. They are cleared by the hepatobiliary system and by the kidneys, allowing for their use as ECF and liver CAs (Fig. 8.2b). Ablavar is a DTPA-based CA with a biphenylcyclohexyl group that binds reversibly to serum albumin resulting in a macromolecule *in vivo* with a long rotational correlation time and, consequently, a high relaxivity [34]. A phosphodiester in its structure limits hepatic uptake that would be induced by the presence of the phenyl groups. This affinity for albumin makes this CA a blood pool agent with long blood half-life and applications in MR angiography [35] (Fig. 8.2c).

TABLE 8.1 Description of Approved Gadolinium-based Contrast Agents

Chemical name	Trade name	Plasma relaxivity ($mM^{-1}s^{-1}$) at 1.5T, 37°C[a]	Formal charge	Use
Gd-DTPA	Magnevist	4.1	2^-	ECF
Gd-DTPA-BMA	Omniscan	4.3	0	ECF
Gd-DTPA-BMEA	OptiMARK	4.7	0	ECF
Gd-DOTA	Dotarem	3.6	1^-	ECF
Gd-HPDO3A	ProHance	4.1	0	ECF
Gd-DO3A-butrol	Gadovist	5.2	0	ECF
Gd-EOB-DTPA	Eovist (United States), Primovist (EU)	6.9	2^-	ECF + liver
Gd-BOPTA	MultiHance	6.3	2^-	ECF + liver
MS-325 (gadofosveset)	Ablavar (United States), Vasovist (EU)	27.7	3^-	Blood pool

[a]Values from Aime and Caravan [18].

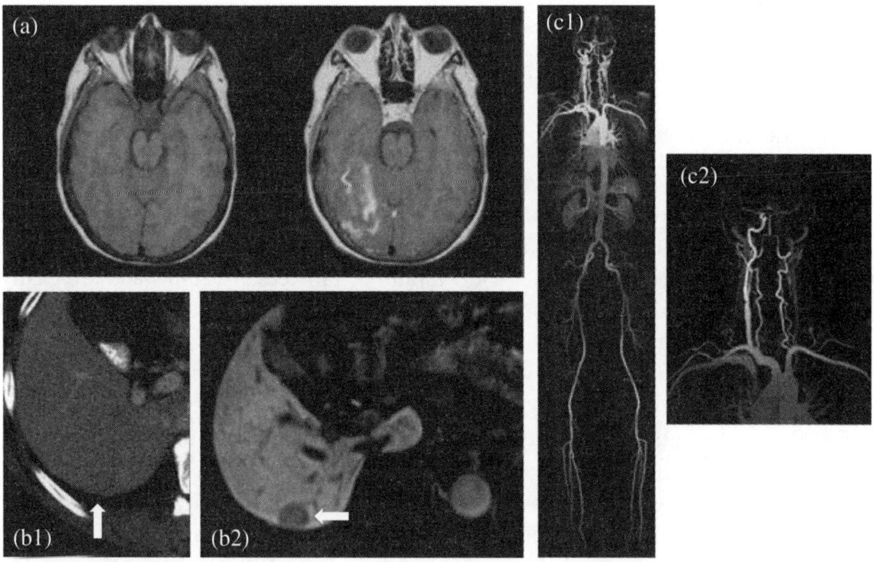

FIGURE 8.2 Examples of contrasted pictures obtained with the three classes of approved Gd-based contrast agents (CAs). (a) MRI of the brain of a patient after an acute stroke event. The left picture is a T_1-weighted acquisition without CA. The right picture shows a cerebral hemorrhage after administration of Gd-DTPA (Reprinted with permission from Ref. [36]. © Wolters Kluwer Health). (b1) Visualization of a grade I hepatocellular carcinoma in the liver without CA. (b2) 20 min after administration of Gd-EOB-DTPA, the high contrast allows a better diagnosis of the tumor (Reprinted with permission from Ref. [37]. © Hindawi Publishing Corporation). (c1) Whole-body angiography after administration of MS-325 produces strong contrast, particularly in the supra-aortic/thoracic region. (c2) A zoom from the same picture into the neck region shows the occlusion of the left carotid artery (Reprinted with permission from Ref. [38]. © Wolters Kluwer Health).

Although more than 30% of MRI exams use one of the CAs cited earlier, their characteristics continue to confine their use to limited applications. The main issue is their rapid clearance time that does not allow visualization of tissues out of the vasculature. Larger sizes of CAs have been proposed to address this issue. With molecular weight higher than 20 kDa, permeability and retention in leaky vasculature is enhanced, allowing better visualization of additional pathologies such as cancer or vascular diseases. Furthermore, as described in Section "Complexation chemistry of gadolinium: a compromise between stability and relaxivity" of this chapter, higher molecular weights induce generally higher relaxivity. Combined with the high payload of Gd^{3+} they can carry, they reduce the necessary dose of CA for image acquisition. With diameter greater than 1 nm, these agents have reduced renal excretion, allowing prolonged retention time as compared to the small approved chelate, which then increases the need to address stability and retention of chelated Gd^{3+}. Polymers, dendrimers, liposomes, and other nanoparticles have the advantage to be modifiable to provide them specific targeting properties or distribution. The second part of this chapter describes their preparation and characteristics as well as their potential applications in MRI.

8.3 GADOLINIUM-BASED NANOPARTICLES

8.3.1 Linear Polymers

8.3.1.1 Enhanced Relaxivity and Circulation Time The use of polymers has rapidly appeared as a popular strategy to produce new CAs with high relaxivities and prolonged circulation. Commercially available poly-L-lysine (P-Lys) or polyethylene glycol (PEG) can be obtained in a wide variety of molecular weights and are easily chemically modified to modulate their biodistribution properties. Furthermore, a high number of Gd^{3+} chelates can be introduced to get high molecular relaxivity.

P-Lys have been the most widely explored linear polymers. They are easily conjugated to DTPA or DOTA on the ε-amino group of the lateral chain of the lysines, resulting in the introduction of a high number of Gd^{3+} per molecule [39, 40] (Chart 8.3). It was shown that relaxivity was improved as compared to the mono Gd^{3+} chelate; however, relaxivity was not influenced significantly by the size of the polymer [39]. As shown in animal studies [41, 42], P-Lys-based agents are predominantly excreted through the kidneys and exhibit prolonged circulation time as the molecular weight increases, leading to constant enhancement in tissue signal over a 1 h period [43]. These characteristics make them interesting as blood pool agents. Gd^{3+}-DTPA-P-Lys was used in MR angiography in rabbits to monitor blood flow in the extremities, allowing visualization of vessels smaller than 1 mm [44], or to visualize lung lesions exhibiting abnormal blood flow [45–47]. Furthermore, Gd^{3+}-DTPA-P-Lys was shown to accumulate in tumors and produced a contrast enhancement of approximately 300% lasting for several days in rat [48].

PEGs have been studied for their low immunogenicity and long circulation time. Dresser *et al.* compared DTPA-PEG of various sizes for visualization of

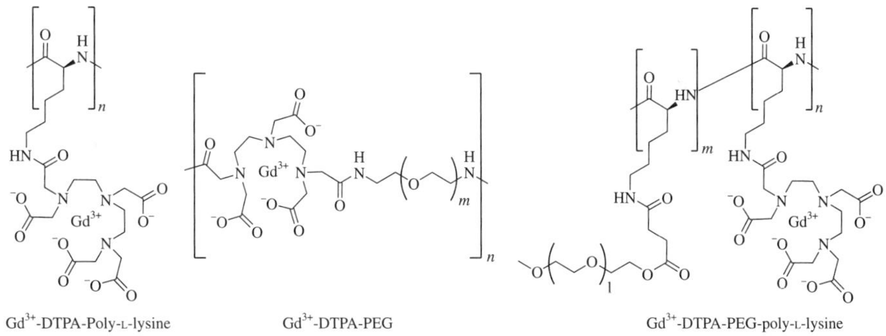

CHART 8.3 Polymers discussed in the text.

tumors in rabbits. They were synthesized by reacting DTPA dianhydride with α,ω-diaminopolyethylene glycols to yield polymers ranging from approximately 11 to 83 kDa incorporating ranges of 6–7 up to 35–36 Gd^{3+}-DTPA chelates (Chart 8.3). For molecular weights lower than 20 kDa, no significant differences were observed as compared with the monomer Gd^{3+}-DTPA; however, longer circulation times with better and faster tumor contrast enhancements were observed above 20 kDa with the best results being obtained for the 21.3 kDa polymer [49]. PEGs have especially been shown to be useful in combination with other polymers to improve their solubility and immunogenicity. PEG covalently bound to P-Lys (Chart 8.3) was proposed as blood pool agent [50]. This copolymer exhibited a 14 h blood half-life and a constant vascular enhancement for 2 h in rats. The relaxivity was also improved due to the higher rigidity of the polymer and possible involvement of the PEG units in the complexation of Gd^{3+}. With necessary dose to obtain contrasted images being 10 fold lower than the clinically used Gd-DTPA and with longer-lasting contrast, this agent was expected to be useful for whole-body, central nervous system, or peripheral MR angiography. However, the very long residence time of this copolymer in the body (half-life = 88 h) and a lack of knowledge on the resulting toxicity and total clearance are limitations of this system.

A wide variety of other linear polymers have been studied such as polyaspartate with up to 40 Gd^{3+} per molecule [51], DTPA dianhydride copolymerized with various monomers such as tartaric acid leading to macromolecules with high hydrophilicity [52], or alkyldiamine-based polymers exhibiting a high rigidity resulting from hydrophobic interaction [53]. Alternatively to DTPA, a HOPO-based polymer with remarkable molecular relaxivity (111 $mM^{-1}\,s^{-1}$ at 60 MHz and 22°C) was produced via ring-opening metathesis [54]. In an interesting study by Lad and coworkers, the influence of several properties in the design of a good blood pool agent such as the molecular weight, relaxivity, metal content, viscosity, and chelate stability was evaluated. The best results were obtained with a DOTA dendrimer that was the result of the incorporation of optimized features, that is, high rigidity, low viscosity, high number of Gd^{3+} chelate, and high stability provided by the DOTA. These combined characteristics contributed to a long blood half-life and a high relaxivity [55].

8.3.1.2 Targeted and Functionalized Polymers Biodistribution properties of Gd^{3+} polymers for visualization of specific area can be modulated by functionalization or by conjugation to targeting moieties. Albumin received considerable attention in applications to improve blood retention and relaxivity of many compounds, including polymers. Indeed, the first studies involving the monomeric Gd^{3+}-DTPA conjugated to albumin highlighted a longer blood retention time and a higher relaxivity than the single Gd^{3+}-DTPA due to the slow tumbling rate of the resulting macromolecule [56]. In the same way, various polymers such as polylysine have been conjugated to albumin to enhance their blood pool behavior [40]. Although no details were given on the *in vivo* behavior, Pashkunova-Martic *et al.* studied the relaxivity of several copolymers of DTPA bound to albumin and to lectin, a protein that binds specifically to endothelial and epithelial cells [57]. Antibodies directed against specific cancer cells have also been studied as polymer carriers but with limited success: P-Lys has been particularly studied, allowing the introduction of 30–65 Gd^{3+} per antibody [58, 59], but high loss of immunoreactivity was observed along with poor clearance and accumulation in nonspecific tissues limiting their use.

Alternatively, it was shown in several studies that selection of the copolymer and of the nature of the polymer bond could influence the biodistribution. Yan *et al.* studied several polyaspartamides with Gd^{3+}-DTPA complexes containing sulfadiazine groups and selected the best of the series as a CA specific to hepatoma in mouse [60]. For long-term cell tracking, Tachibana *et al.* proposed the use of poly(vinyl alcohol) copolymerized with Gd^{3+}-DOTA [61]. The resulting compound exhibited high hydrosolubility with long retention inside living cells in contrast to rapid clearance from blood [61]. *N*-(2-hydroxypropyl)methacrylamide copolymerized with mannosamine and conjugated to Gd^{3+}-DOTA was used for the targeting and imaging of activated macrophages [62] or breast cancer cells when conjugated to RGDfK, a peptide with specificity to $\alpha_v\beta_3$ receptors [63].

Novel strategies currently under investigation are the development of "activatable" Gd polymers exhibiting high relaxivity but only under specific conditions of biological interest. For instance, pH-sensitive agents have been studied to map changes in pH that is generally associated to various pathologies such as tumors, strokes, and infections. For instance, a Gd^{3+}-DO3A-polyornitine was shown to exhibit a linearly increasing relaxivity from $23\,mM^{-1}\,s^{-1}$ at pH 4 to $32\,mM^{-1}\,s^{-1}$ at pH 8 [64]. Alternatively, a CA sensitive to an enzyme was proposed by Bogdanov *et al.* In this strategy called MR signal amplification (MRamp), a Gd-DTPA chelate linked to a catechol derivative is administered in the monomeric form. Then, it polymerizes in the presence of exogenous horseradish peroxidase (HRP) to generate the polymer *in situ* exhibiting enhanced relaxivity. This method was shown to be sensitive enough to detect nanomolar amounts of peroxidase in cell culture and allowed visualization of carcinoma in mouse when HRP conjugated to a tumor-specific antibody was injected prior to administration of the CA [65].

8.3.1.3 Biodegradable Polymers Many encouraging results were obtained with polymers that exhibit higher relaxivities and visualization of new targets as compared to the currently approved CAs. However, their introduction into the clinic has been hampered by the slow excretion of these high-molecular-weight compounds

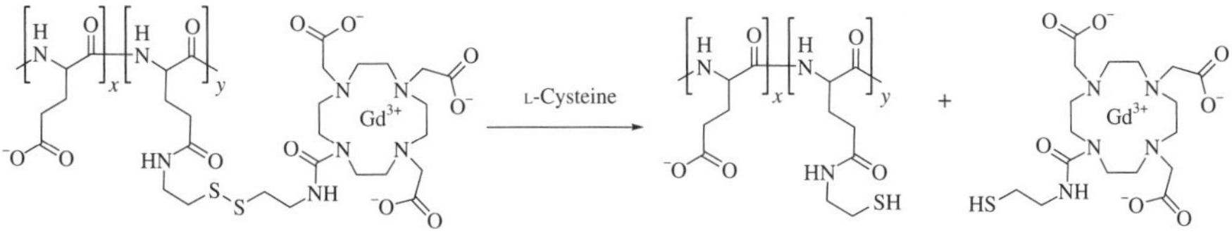

Poly(L-glutamic acid)-cystamine-[Gd(III)-DO3A]

CHART 8.4 Example of a biodegradable polymer and its fragments after incubation with L-cysteine.

and the potential toxicity due to the released Gd^{3+}. This slow excretion rate issue is why efforts are being made to develop polymers that are degraded *in vivo* into smaller parts for faster elimination.

Most of the biodegradable polymers for MRI were developed in Lu's group [66–68]. They are derivatives of the polymers described previously with disulfide bonds in their structure allowing cleavage into smaller residues by endogenous thiols such as cysteine and glutathione (Chart 8.4). For instance, the lateral chain of poly-(L-glutamic acid) was conjugated to DO3A with a cystamine spacer and labeled with Gd^{3+}. The resulting polymer exhibited enhanced blood pool contrast and fast excretion of the Gd^{3+}-DO3A chelate through renal filtration [66]. Similarly, PEGylated Gd^{3+}-DTPA copolymerized with L-cystine exhibited good blood pool agent characteristics with fast elimination [67]. However, a degradation that would occur too fast is not desirable, and it was demonstrated that the rate of disulfide bond cleavage could be controlled by introducing steric hindrance around it to limit accessibility by endogenous thiols [68]. Alternatively to disulfide bond investigations, Wen *et al.* studied the influence of the Gd^{3+}-DTPA linkage on the degradability of poly-(L-glutamic acid). A higher degradation rate was observed with a benzyl linker as compared to a hexyl in the presence of cathepsin B, a lysosomal enzyme [69]. Another interesting approach is the use of protein polymers proposed by Karfeld-Sulzer *et al.* that consist of repeated amino acid units produced by genetic engineering conjugated to Gd^{3+}-DTPA on the amino groups of the lysines [70]. Size of the polymer and lysine spacing could be controlled to modulate the relaxivity (up to 461 $mM^{-1} s^{-1}$ at 60 MHz, 37°C). *In vitro* studies indicated degradation by plasmin, a naturally occurring enzyme that cleaves on the carboxyl side of lysine residues.

Although not strictly belonging to the linear polymers, carbohydrate derivatives do deserve mention in this section. With molecular weights up to 150 kDa, carbohydrates such as dextran, starch, and inulin have been investigated, and, as for the linear polymers described previously, high relaxivities and long circulation times were obtained, making them good blood pool agents for visualization of various lesions and diseases. Carbohydrates have the advantage of being naturally present in the body in various shapes. That is why their Gd^{3+} derivatives exhibit low immunogenicity and low toxicity. Furthermore, they are readily metabolized by various processes allowing their excretion. A good example is the use of hyaluronan proposed recently by Zhu and Artemov. Hyaluronan is a linear glycosaminoglycan approved by the FDA for various medical applications. It is degraded by hyaluronidase in lymph nodes and liver. When conjugated to Gd^{3+}-DTPA, it exhibited enhanced vasculature visualization and was excreted primarily through the kidneys [71]. Many results have been published on carbohydrate-based CAs, and the reader is referred to the recent review by Villaraza [72] for more information about their preparation and applications.

8.3.2 Dendrimers

8.3.2.1 Preparation and Characteristics Dendrimers are polymers built from a small polyfunctional core by repeated iterative branching reactions, each reaction accounting for a generation (denoted G1 for the first generation, G2 for the second one, and so on through the series—see Villaraza [72] for a detailed description on nomenclature of dendrimers in this application). The first example of such an iterative synthetic procedure, named cascade synthesis by the authors, was described in 1978 by Buhleier *et al.* [73]. Synthesis strategies permit high modulation in dendrimer size, shape, and functionalities, and their potential medical applications have increased exponentially over the last three decades [74]. Their use as a platform for Gd^{3+} hosting for MRI CA development was first suggested in 1994 by Wiener *et al.* [75], allowing the introduction of up to 180 gadolinium ions into a single molecule with production of high relaxivity per particle (5800 $mM^{-1} s^{-1}$ as compared to 5.4 $mM^{-1} s^{-1}$ for the monomeric Gd^{3+}-DTPA at 20 MHz and 20°C).

The processes used in dendrimer syntheses involve simple reagents, efficient reactions, and relatively easy purification allowing the production of accurately size-controlled particles with very narrow polydispersities (polydispersity index (PI) generally below 1.01) and a high number of functional group sites allowing, for instance, introduction of a high number of Gd^{3+} chelates in the case of MRI applications. Two main strategies based on two different expansion methods are used for their preparation. The first strategy to be described was the divergent approach consisting in expanding the branches from a central core in a cascade-like manner (Fig. 8.3). Starting from a mono- or a diamine core, each generation was produced by Michael addition of acrylonitrile, subsequent reduction of the nitrile producing reactive amines for a new iteration that in turn gave birth to a new generation [73]. Similar expansion of polyalcohol to produce molecules named arborols was described by Newkome [76]. At the same time, the first three-dimensional dendrimer was described by Tomalia *et al.* [77]. Exhibiting a radial symmetry, these macromolecules called starburst dendrimers were produced from an ammonia or an ethylenediamine core, each generation being produced via Michael addition of an acrylate ester followed by alkylation with alkylene diamine. Consequently to this three-dimensional growth, the number of terminal amines increased exponentially with each generation, 256 terminal amines being obtained at the sixth generation with an ethylenediamine core. These dendrimers are named poly(amidodiamine), or PAMAMs, are commercially available, and are also the most studied dendrimers as platform for development of CAs in MRI. Alternatively, poly(propylene imine) (PPI) dendrimers with a diaminobutane core have been synthesized with a PI of 1.002 for the first five generations [78–80]. In the second strategy called convergent approach, the synthesis is started from the periphery to end at the core (Fig. 8.3). In this way, the number of reaction

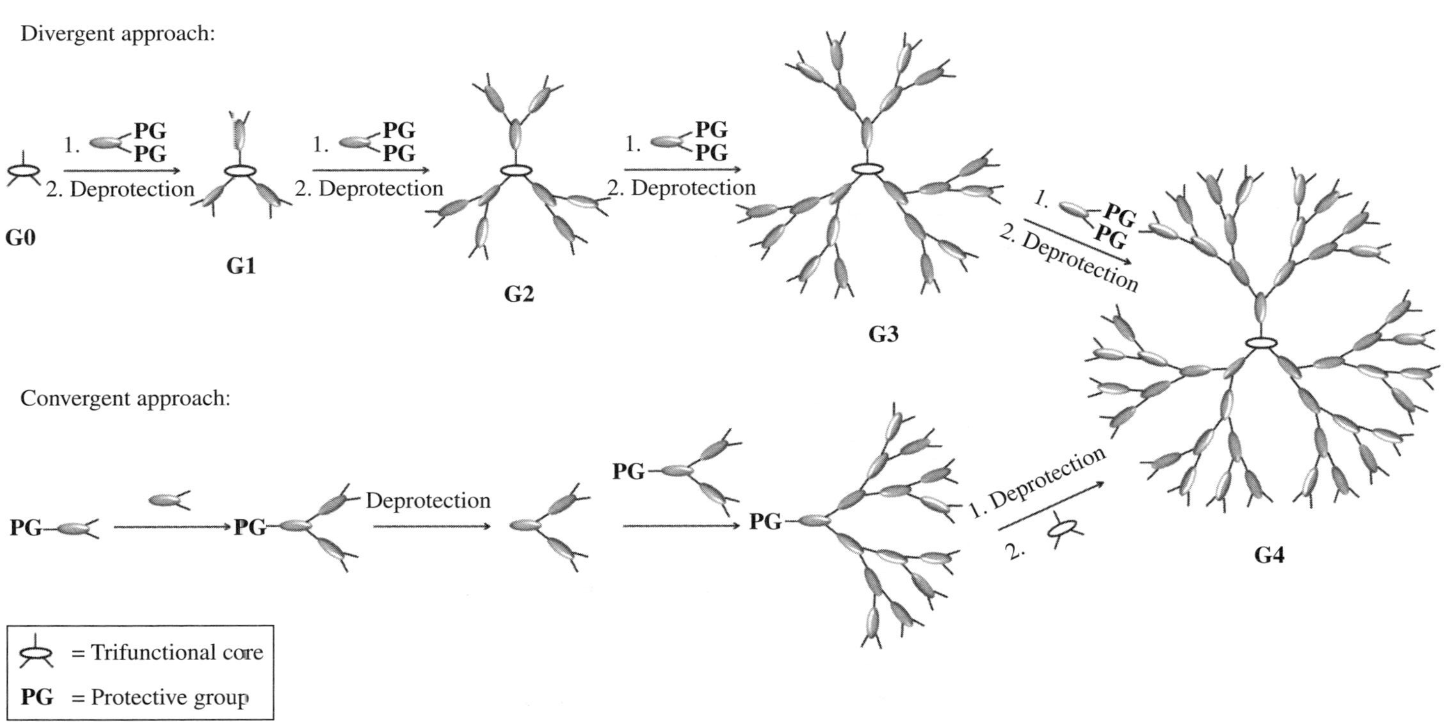

FIGURE 8.3 Strategies for the synthesis of a G4 dendrimer via the divergent and the convergent approaches.

sites decreases at each step of the synthesis, reducing the risk of incomplete conversion or side reactions. Furthermore, purifications can be performed more easily at every step (even though difficulty increases for high number of generation dendrimers). The first examples of dendrimers produced by the convergent approach were polyethers obtained from 3,5-dihydroxybenzyl alcohol grafted on a multifunctional core [81]. Since then, many other dendrimers have been produced by this approach, as reviewed by Grayson and Fréchet [82].

In solution, the volume occupied by a dendrimer increases cubically with generations, while the mass increases exponentially. This typical expansion confers to dendrimer properties that differ from linear polymers, especially as the size increases. For instance, while the intrinsic viscosity increases linearly with the size of linear polymers, it reaches a maximum at a certain generation for dendrimers [83]. Furthermore, from the MRI point of view, the higher rigidity of these structures leads to higher relaxivity as the number of generation increases, while the size of linear polymers has a limited influence on relaxivity because of their high flexibility.

Several studies show that the relaxivity per Gd^{3+} increases with generations. For instance, Bryant and coworkers measured the relaxivity of G5 to G10 PAMAM dendrimers with Gd^{3+}-DOTA conjugated to the free amino groups at the surface (i.e., from 118 kDa with 96 Gd^{3+} to 3000 kDa with 1860 Gd^{3+}). The ion relaxivity increased from 30 $mM^{-1}s^{-1}$ for G=5 to 36 $mM^{-1}s^{-1}$ for G=7 (20 MHz, 23°C). However, a plateau was reached beyond G=7. It was observed that increasing the temperature in turn increased relaxivity, indicating that the relaxivity was limited by a slow water exchange [84]. The use of a complex with a higher water exchange, ethylenepropylenetriamine-N,N,N′,N″,N″-pentaacetic acid (EPTPA; see Chart 8.5), was proposed to circumvent this issue [85]. The results obtained with PAMAM dendrimers from G5 to G9 demonstrated a beneficial effect of this ligand, even if a decrease in relaxivity was observed for G9. Nonetheless, the fact that decreasing the temperature leads to an increase in the relaxivity indicated that the water exchange rate was no longer the limiting parameter. Furthermore, the same study highlighted the fact that the internal motion of dendrimers also influences relaxivity, as demonstrated by the improved relaxivity obtained at lower pH resulting from protonation of the amines and subsequent higher rigidity of the structure. The latter results were confirmed by a study published by Rudovsky *et al.* who demonstrated that the relaxivity of a negatively

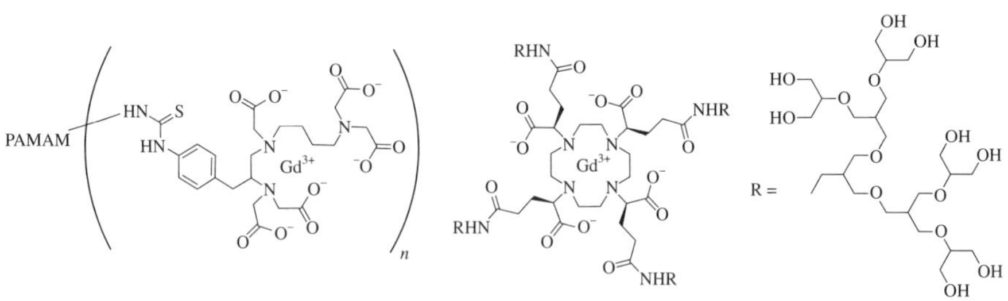

EPTPA

Dendrimer with Gd-chelate at the barycenter

CHART 8.5 Dendrimers with improved characteristics discussed in the text.

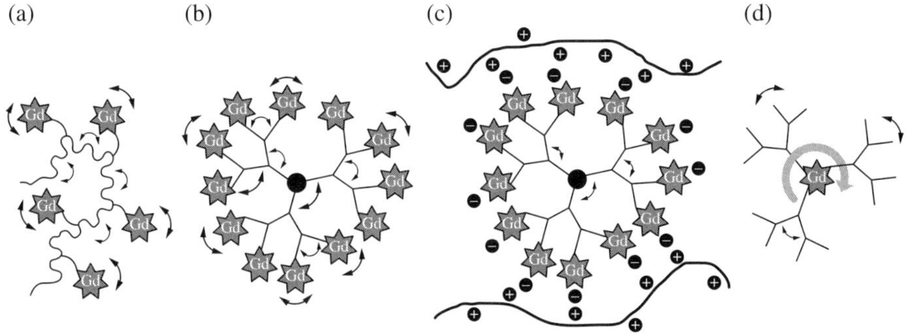

FIGURE 8.4 Polymers classified by increasing relaxivity. (a) The high flexibility of linear polymers limits the relaxivity per gadolinium. (b) Dendrimers are less flexible and exhibit higher relaxivity but still undergo internal motion. (c) Relaxivity can be improved by introducing rigidification (with a cationic polymer and negatively charged Gd complexes in the example earlier). (d) When the Gd complex is at the barycenter, it rotates at the rate of the whole molecule, which leads to higher relaxivity.

charged PAMAM dendrimer was improved in the presence of a cationic poly(arginine), introduced in order to induce internal rigidification by anionic/cationic interactions (Fig. 8.4) [86]. The initial r_1 value of $20.4\,mM^{-1}\,s^{-1}$ of the free dendrimer reached $34\,mM^{-1}\,s^{-1}$ in the presence of poly(arginine) (20MHz, 25°C). Ultimately, higher relaxivity could be obtained by placing the Gd^{3+} complex at the barycenter of the dendrimer as proposed by Fulton *et al.* [87]. In this system, the Gd^{3+} chelate is at the core and is connected to polyalcohol branches (Chart 8.5). This system results in a Gd-OH_2 vector coupled more effectively to the rotational motion of the whole molecule leading to improved relaxivity.

Interestingly, a new approach to introduce the Gd^{3+} into the dendrimer developed by Nwe *et al.* was shown to influence highly the relaxivity. While the Gd^{3+} is classically introduced in the last step of the synthesis into the chelating agent already conjugated to the dendrimer, the novelty was to conjugate the preformed Gd^{3+}–ligand complex to the dendrimer [88]. This method leads to a better control of the free DTPA/complexed DTPA ratio with virtually 0% of free ligand on the final dendrimer with the preligation method, while approximately 20% of ligands remain unoccupied with the classical method. The relaxivity per Gd increased from $13.9\,mM^{-1}\,s^{-1}$ for the postligation method to $26.9\,mM^{-1}\,s^{-1}$ (at 3T and 22°C) with this new method. It was hypothesized by the authors that carboxylates from the free unoccupied DTPA moieties in the vicinity of the Gd^{3+} complex were responsible for disturbing the water exchange when dendrimers are formed by the classical method, which is not the case with the preligation method.

8.3.2.2 Dendrimers Biodistribution and Applications in MRI

Passive Distribution The biodistribution of dendrimers and their potential applications in MRI are primarily governed by their size as observed in many studies with PAMAM dendrimers (Fig. 8.5). They exhibit prolonged circulation time as

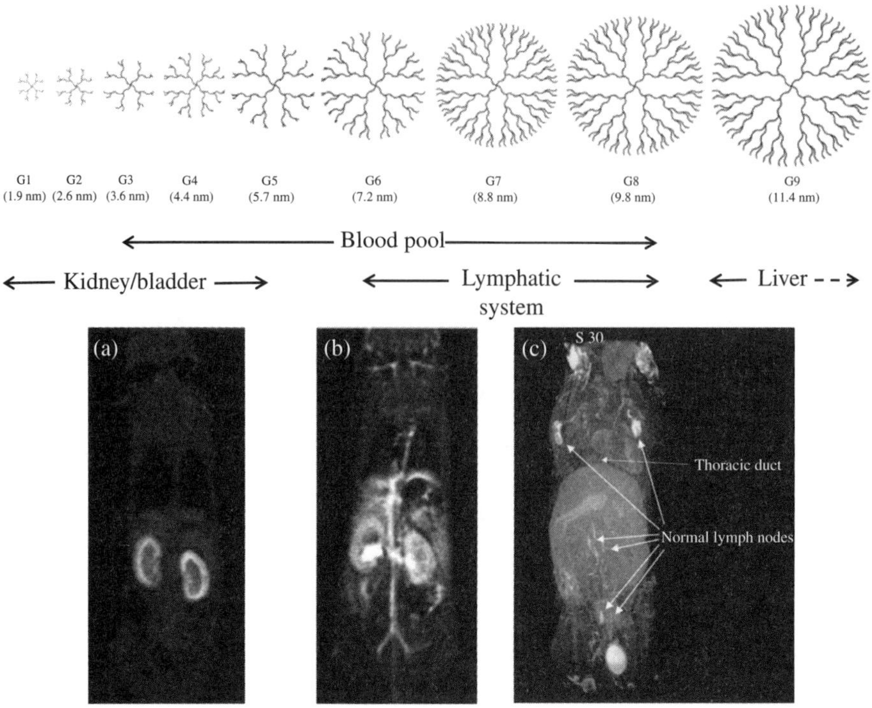

FIGURE 8.5 Top: Size-dependent passive biodistribution of a series of PAMAM-Gd-DTPA dendrimers. Bottom: (a) G4 with visualization of the kidneys and (b) G6 with visualization of vasculature, 14 min postinjection (Reprinted with permission from Ref. [89]. © Wiley-Liss, Inc.) (c) Visualization of the lymphatic system with G8 dendrimer 45 min postinjection (Reprinted with permission from Ref. [91]. © American Association for Cancer Research.)

the generation number increases. The smaller generation PAMAM-DTPA dendrimers (G1–G4) exhibit rapid clearance from the body via elimination through the kidneys, whereas G5 and G6 are retained longer, improving significantly visualization of fine vasculature [89]. Higher generations exhibit less renal uptake with an increased elimination via the liver. The G8 dendrimer was shown to produce better enhancement over time in tumor vasculature than G6 [90]. Interestingly, the circulation times of G6 and G8 were both shown to be long enough for visualization of deep lymph nodes and vessels [91, 92]. Similar size-related behaviors were also observed with PPI-DTPA dendrimers as shown in a study by Langereis *et al.* [93].

Several studies have reported the importance of the nature of the dendrimers on their biodistribution. For instance, G4 PPI dendrimers with their higher hydrophobicity than PAMAM have been shown to accumulate significantly more in mice liver, allowing visualization of micrometastatic tumors of 0.3 mm diameter [94]. The nature of the core can also be critical in the development of a better blood pool agent as demonstrated by a study by Kobayashi *et al.* showing that the replacement of the ammonia core by an ethylenediamine leads to longer blood retention and slower renal uptake [95]. Ultimately, circulation and excretion properties can be modulated by modification in the surface of dendrimers. Examples are PEGylation of the surface of PAMAM-DTPA dendrimers, which leads to longer blood clearance and faster excretion with lower liver and kidney accumulation [96, 97]. Also, faster excretion was obtained by coinjection of lysine [98], by the use of a lysine-based dendrimer [99], or by administration of an avidin chase after injection of the dendrimer previously modified with biotin on its surface [100]. Interestingly, it was demonstrated that the nature of the ligand could be chosen to optimize elimination rate of the dendrimers as shown by Nwe *et al.* who compared DTPA with DOTA on two identical G4 PAMAM dendrimers and concluded that DOTA was more suitable ($T_{1/2}$ = 16 min for DOTA vs. 29 min for DTPA) [101]. Alternatively, cleavable cystamine linkers were proposed to conjugate the Gd^{3+} complex to the dendrimer in order to accelerate the elimination of the metallic complex [102, 103]. These studies are particularly important to lower the potential toxicity of these long-lasting particles in the body in the perspective of clinical uses.

Selective and Targeted Dendrimers The first attempts to improve selectivity of dendrimers were proposed with the use of an antibody with tumor targeting properties. The first study reported a G2 PAMAM dendrimer coupled to monoclonal antibody (mAb) 2E4 that was efficiently loaded without loss of immunoreactivity [104]. Similarly, OST7, a murine monoclonal IgG1, conjugated to a G4 PAMAM-DTPA dendrimer exhibited no loss of immunoreactivity. Biodistribution studies indicated accumulation in the tumor site and better blood clearance than the antibody labeled directly with DTPA without the dendrimer. Recently, Nwe *et al.* reported the preparation of G4 and G5 dendrimers with a cystamine core that, after cleavage of the disulfide bond, were coupled as half dendrimers to $F(ab')_2$ fragments of panitumumab premodified with a maleimide linker [105]. The relaxivity was higher than the monocomplex alone but lower than the dendrimer without antibody. Even though the studies cited previously reported promising results on the relaxivity and on the *in vitro* and *in vivo* behavior of these dendrimer–antibody systems, no data are available yet on their ability to effectively enhance the visualization of the targeted tumors. Alternatively, the (strept)avidin–biotin system has been employed in pretargeting strategies similarly to that used for the site-specific delivery of radionuclide developed by Paganelli [106]. A three-step strategy was proposed by Zhu *et al.* consisting in injecting a biotinylated antibody (trastuzumab), followed by avidin that targets biotin on the antibody. In the last step, the dendrimer with chelated Gd^{3+} and also conjugated to biotin is injected and accumulates on the antibody–avidin adduct [107]. Although accumulation was observed in tumors, no significant improvement of the MR contrast was observed, probably because of a saturation of the binding sites on the avidin (four binding sites available per avidin) by the biotinylated antibody. This system was also used in a two-step approach by Sano *et al.* who injected the antibody conjugated to streptavidin first , followed by the biotinylated Gd^{3+} dendrimer 3 days later. Good relaxivity was measured, and signal enhancement of FM3A tumors in mice was clearly observed [108].

Dendrimers have also been conjugated to several small targeting molecules such as folate that is directed against folate receptor expressing ovarian cancers [109–111], CLT1 peptide targeting the fibrin–fibronectin complexes in tumor [112], or an RGD-cyclopeptide targeting integrin $\alpha_V\beta_3$ [113]. Also, PEGylation of the surface of a dendrimer combined with the conjugation to peptide T7 targeting transferin highly expressed in brain and liver tumor led to enhancement of the imaging of tumors in the liver, but not in the brain [114].

8.3.3 Lipid-Based Nanoparticles

8.3.3.1 Structure and Relaxivity Liposomes and their derivatives can be obtained with various sizes and degrees of functionalization, allowing delivery of drugs or imaging agents to many targets of interest (see Chapter 3). The first type of liposome-based MR CA described contained the paramagnetic species in the aqueous inner core (Fig. 8.6a). However, Gd^{3+}-DTPA entrapped in such systems produced limited ionic relaxivity, lower than the free Gd^{3+}-DTPA. These disappointing results were attributed to the reduced water exchange at the Gd^{3+} due to the slow diffusion of water molecules through the lipidic layers. It was also shown that relaxivity was highly influenced by the liposome size with smaller liposomes exhibiting better results due to a larger surface area to volume ratio or by the membrane composition and its permeability [115]. Nonetheless, this low relaxivity could be advantageously used for pH-dependent imaging systems as proposed by Løkling *et al.* who developed liposomes that are stable at physiological pH but disaggregate at lower pH, thus releasing the Gd^{3+}-DTPA that once free would exhibit a higher relaxivity to the surrounding tissues [116].

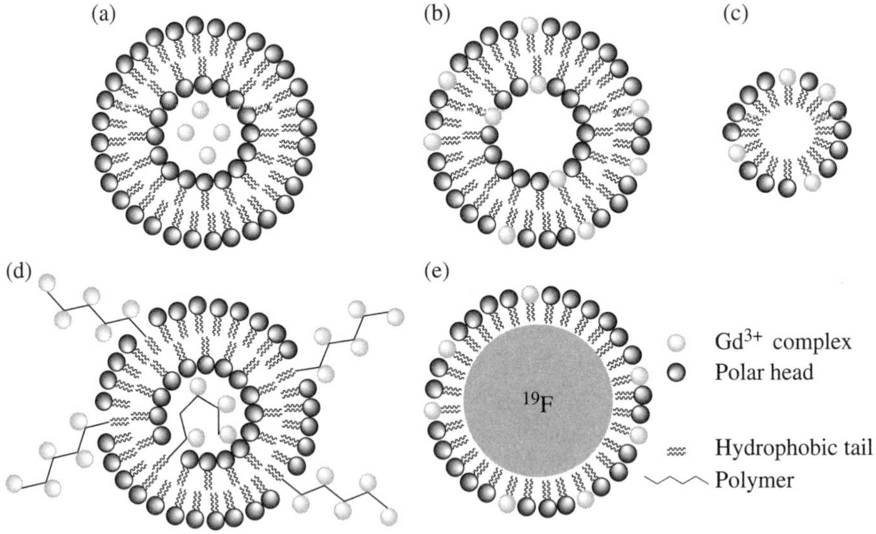

FIGURE 8.6 Schematic representation of (a) liposome with Gd^{3+} complex in the aqueous core, (b) liposome with Gd^{3+} complex incorporated into the lipid bilayer, (c) micelles, (d) liposome with Gd^{3+} complex at the exterior, and (e) perfluorocarbon nanoparticles.

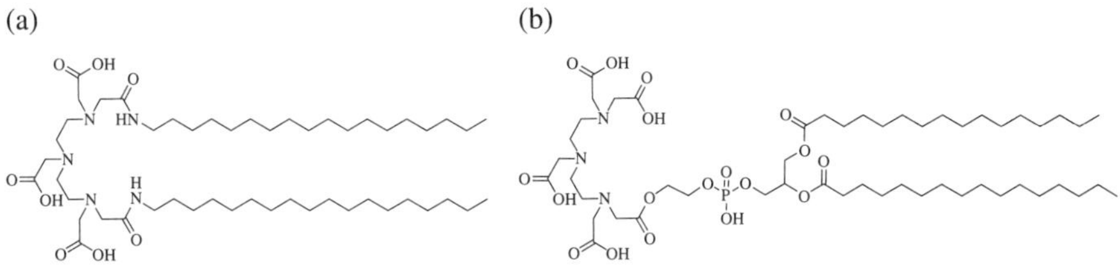

CHART 8.6 Examples of amphiphilic DTPA derivatives: (a) stearylamine and (b) phosphatidylethanolamine.

Efforts to improve the relaxivity led to the development of structures incorporating the Gd^{3+} chelate within the membrane, providing a facilitated access of water to the Gd^{3+} coordination site at the surface of the liposome (Fig. 8.6b). These structures were made possible by the synthesis of Gd^{3+} chelates bound to fatty acid chains such as stearyl derivatives (Chart 8.6). The resulting relaxivity was clearly improved with values—two to five times that of the free Gd^{3+} complex [117]. However, the size of the particle did not influence the relaxivity [118], probably because of the possible diffusion of the Gd^{3+} chelate across the membrane. The use of more rigid liposomes was proposed by Storrs *et al.* by cross-linking the components of the lipidic layers via photochemical reaction after assembly. This resulted in a higher rigidity and a relaxivity independent on the size of the liposome but dependent on the linker length [119]. Also, restricting the local motion of the Gd^{3+} chelate was proposed by optimization of the paramagnetic amphiphilic species to produce high relaxivities [120]. However, as noticed by Schühle *et al.*, a loss of relaxivity still remains for the portion of Gd^{3+} chelates oriented toward the core of the liposome, which are not exposed to the water at the surface. This issue could be circumvented by the incorporation of a calixarene in the amphiphilic Gd^{3+}-chelate structure. This resulted in a highly sterically hindered paramagnetic head which orients preferentially toward the outside layer during the assembly. By using this strategy, more than 80% of the paramagnetic head was estimated to be available in the outside layer as compared to approximately 50% with the usual procedures, and improvement in the relaxivity properties was observed [121]. Alternatively, micelles can be formed from amphiphilic Gd^{3+} chelates with alkyl chains longer than 10 carbons [122]. In these structures, all the complexes face outward, without limitation of access to water (Fig. 8.6c), resulting in a relaxivity comparable to other macromolecular CAs [123]. Gadolinium can also be loaded at the exterior of the liposome as proposed by Torchilin and Trubetskoy [124]. These nanoparticles were prepared with the use of polychelating amphiphilic polymers consisting of a hydrophilic chain (polylysine) bearing multiple DTPA chelates and a hydrophilic terminal group for incorporation in the liposome membrane (Fig. 8.6d). This system allowed the introduction of high numbers of Gd^{3+} as a way to improve the relaxivity per particle. Also, the use of perfluorocarbon nanoparticles has been proposed with potential applications in molecular imaging. With a lipidic fluorinated liquid core (composed, for instance, of perfluorooctyl bromide or perfluoro-15-crown-5-ether), the amphiphilic species in the membrane faces all with the polar head directed toward the exterior (Fig. 8.6e). Their size is about 250 nm, and very high relaxivities (~2.5×10^6 mM^{-1} s^{-1} at 1.5T) have been obtained [125]. The presence of a high concentration of ^{19}F in the core makes them interesting for quantitative imaging and spectroscopy in addition to the ^1H MRI [126].

8.3.3.2 Passive Distribution The pharmacokinetics of lipidic nanoparticles depends on several factors such as their size, or the nature of the components of the membranes that govern surface charge, permeability, and the degree of stabilization. Gd^{3+} liposomes have been reported to accumulate in the liver and spleen regardless of the size and allowed detection of tumors in the liver with better results for the smaller particles [127]. Because of their large size, they exhibit a longer blood circulation time than single Gd^{3+} chelates, especially when derivatized with PEG, allowing visualization of fine vasculature [128, 129]. Also, paramagnetic lipid-based agents are the object of a growing interest for the visualization of atherosclerotic plaque by various mechanisms [130–133].

8.3.3.3 Tissue-/Cell-Specific Contrast Enhancement with Lipid Nanoparticles Micelles and liposomes incorporating peptides or antibodies on their surface have been the object of several investigations. Morelli's group developed a series of peptide-conjugated liposomes and micelles with potential targeting ability for somatostatin receptors and interesting relaxivity properties [134, 135], although no investigation on the *in vivo* behavior has been published yet. Similarly, RGD peptide-conjugated liposomes were developed by Kok *et al. In vitro* studies highlighted the ability of these CAs to internalize in endothelial cells overexpressing the $\alpha_v\beta_3$ integrin cell surface receptor [136]. Antibody-conjugated liposomes confirmed also their potential utility through *in vitro* studies by Mulder *et al.* who demonstrated the binding properties of an antiselectin CA in cell culture [137] or *in vivo* by conjugation to the mAb 2C5 with significant tumor contrast enhancement 4h postinjection [138]. Another recent site-specific liposome under investigation is a folate-targeted construct exhibiting faster and stronger contrast enhancement in the targeted tumors than the nontargeted one [139].

Alternatively, systems have been developed with ability to exhibit their contrast enhancement under specific conditions and allow particular physiological phenomenon detection. For instance, a particularly ingenious system was proposed recently by Figueiredo *et al.* consisting of the use of negatively charged liposomes aggregated together with the positively charged protamine. In the presence of the enzyme of interest, for example, serine protease in this case, digestion of the protamine and subsequent release of the liposomes lead to an enhanced relaxivity [140]. Also, thermally responsive paramagnetic lipidic structures have been developed allowing three-dimensional tissue thermometry for monitoring in thermotherapy [141].

8.3.4 Carbon-Based Nanomaterials

8.3.4.1 Gadofullerenes Fullerenes are molecules made entirely of carbon in a closed spherical shape. They have rapidly become interesting for sequestration of metal ions due to the high inertness of the resulting metallofullerene structure. Indeed, unlike metals complexed with chelating agents that are always in an equilibrium state susceptible to reverse mechanisms with potential release of the free ion in the organism, fullerenes entrap the metal and isolate it from physiological conditions, avoiding virtually any risk of release of the free metal. The idea to use these structures to entrap highly toxic radiometals for nuclear medicine as well as Gd^{3+} for MRI rapidly grew to give birth to a new field of investigations. The first report of such structure described the incorporation of Gd^{3+} into the endohedral fullerene C_{82}, and the resulting compound was designated as $Gd@C_{82}$ [142]. The synthesis involved an electrochemical reaction, named the arc process, with Gd_2O_3-impregnated graphite rods. However, with the resulting species being highly insoluble in water, efforts were made to increase the hydrophilicity by modification of the surface of the fullerenes with functional groups such as hydroxyl (yielding $Gd@C_{82}(OH)_n$) [143]. Similarly, water solubility was obtained by the introduction of β-alanine that could also be used for bioconjugation purposes [144]. Also, a fullerene incorporating three Gd^{3+} as the nitride, modified with PEG on the surface and denoted $Gd_3N@C_{80}$-[DiPEG5000(OH)$_n$], was reported [145].

The first reports on the relaxivity of such water-soluble gadofullerenes described surprisingly high relaxivities, up to 20-fold higher than the Gd^{3+}-DTPA complex [144]. These results were unexpected because the entrapped Gd^{3+} is supposed to be completely isolated from the bulk water as the fullerene surface does not allow passage of water or ions under physiological conditions. Several investigations reported r_1 values highly dependent to pH, concentration, temperature, magnetic field strength, and the nature of the groups attached at the surface [146]. The mechanisms involved in the production of such high r_1 are not yet fully understood, but several explanations have been exposed. Electron energy loss spectroscopy (EELS) experiments have highlighted the fact that the metallofullerene itself was a paramagnetic structure, due to the three-electron transfer from the Gd^{3+} to the fullerene cage. Consequently, water relaxation can occur at the large surface of the gadofullerene [147]. Furthermore, gadofullerenes are highly subjected to pH-dependent aggregation resulting in a higher rotational correlation time as demonstrated by studies at various pH [148, 149].

The discovery of the interesting properties of gadofullerenes being still recent, only limited data on the *in vitro* and *in vivo* behavior of these particles are available. $Gd@C_{60}[C(COOH)_2]_{10}$ exhibited favorable characteristics *in vitro* for the development of targeted intracellular MRI CAs as suggested by a study by Sitharaman *et al.* [143]. This gadofullerene was shown to internalize in MSC and NIH3T3 cells with production of high contrast and could be used for tracking mammalian cells. *In vivo* studies were performed with $Gd@C_{82}(OH)_n$, which, injected at a 20-fold lower dose than the Gd^{3+}-DTPA complex, produced strong signals in the lung, liver, spleen, and kidney. This unfavorable high uptake in the reticular–endothelial system could be reduced with the use of $Gd@C_{60}[C(COOH)_2]_{10}$, which does not form aggregates and is excreted via the kidney 1 h after injection, similarly to the low-molecular-weight Gd^{3+} chelates [150]. Also, fullerenes incorporating three Gd^{3+} ions have been tested for the visualization of brain tumor in rats after direct infusion into the brain. The very high relaxivity (in the range of 150–200 $mM^{-1}s^{-1}$ at 2.4 T) produced good contrast with only 1/30 and 1/50 of the necessary dose for Gd^{3+}-DTPA with

$Gd_3N@C_{80}$-[DiPEG5000(OH)$_n$] and $Gd_3N@C_{80}(OH)_{26}(CH_2CH_2COOM)_{16}$, respectively (M = H or Na) [145, 151].

8.3.4.2 Gadonanotubes The use of carbon nanotubes as carriers for Gd^{3+} was first suggested by Sitharaman *et al.* [152]. Nanotubes in the 20–100 nm size range seem suitable for cellular uptake and exhibit favorable biocompatibility. Furthermore, they can be derivatized at the exterior while sequestering the substance of interest in the interior. The so-called gadonanotubes were produced by sonication of ultrasmall (US) tubes in aqueous $GdCl_3$. The resulting compounds exhibited highly improved relaxivities as compared to the approved Gd-based CA, with r_1 values 40-fold higher under clinically used magnetic field conditions and up to 90-fold higher at lower fields. The mechanisms by which the relaxivity is strengthened appear to be different from the fullerene ones because in the case of gadonanotubes, no aggregation of the particles occurs. Furthermore, the Gd^{3+} is directly exposed to water in this case, and this is the very efficient access to water in the nanotubes, which are known to be good transporters of water, that would contribute to enhancing relaxivity. Interestingly, pH dependency of the relaxivity was observed with a dramatic decrease from pH 7 to 7.4, suggesting possible applications in pH mapping [153]. *In vitro* studies have highlighted the internalization properties in various cell lines such as stem cells [154] or macrophage cells, [155] and functionalization of the surface of the tubes with serine were efficient for labeling MCF-7 human breast cancer cells [156]. *In vivo* studies were initiated with gadonanotubes incorporating biodegradable polymers in order to assess the biocompatibility of these new materials [157, 158]. These preliminary results suggested limited toxicity of these compounds, but more experiments will be necessary to make sure that the use of this new class of CAs is safe for human use.

8.3.5 Other New Emerging Gd Nanoparticles

8.3.5.1 Gold Nanoparticles Gold nanocrystals can be easily functionalized on their surface by thiol derivatives. When loaded with Gd^{3+}, they can be used as multimodal agents with the properties of Gd^{3+} for MRI and the properties of the gold core for X-ray imaging and radiotherapy [159]. The process used in the preparation of paramagnetic gold nanoparticles (GNPs) involves the reduction of a gold salt ($HAuCl_4$) by the thiol-containing Gd^{3+}-DTPA derivative [160] (Fig. 8.7). This method leads to the production of 2–2.5 nm GNP coated with approximately 150 Gd^{3+} complexes and $r_1 = 585\,mM^{-1}\,s^{-1}$. The use of a Gd^{3+} complex bearing a thiol at both extremities, as proposed by Park *et al.*, allowed the introduction of the ligand in polymerized form with a very large gadolinium payload (~10,000 Gd/particle, 5–7 nm in size; see Fig. 8.7) [161, 162]. Investigations to improve the relaxivity of GNPs showed that bigger particles tend to decrease the tumbling rate of the Gd^{3+} complex, thus increasing r_1. Also, it was shown that overcoating the GNPs with multilayers of oppositely charged polyelectrolytes restricted the motion of the Gd^{3+} chelates and improved the relaxivity [163]. *In vivo* studies highlighted the potential of these particles as blood pool agents without undesirable accumulation in the

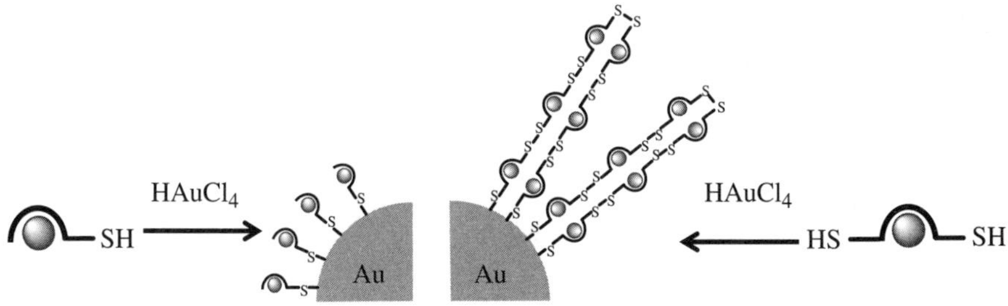

FIGURE 8.7 Preparation of Gd-loaded gold nanoparticles with monochelate (left) or polymerizable chelate (right).

lungs, spleen, or liver [162]. They have also been derivatized for targeting by means of antibodies [164], biotin [165], or DNA for cellular uptake [166].

8.3.5.2 Viral Capsids When the genome of viruses is removed, the resulting virus capsids become noninfectious and unable to replicate. But, because they can still access cellular materials, they have been suggested by Douglas and Young as carriers for drug delivery or bioimaging applications [167]. Virus capsids are made of proteic subunits, with a vacant interior that can be used to entrap substances of interest and the exterior that can be modified chemically. Furthermore, genetic engineering allows modifications of the subunit for modulation of the architecture or metal binding [168]. The first example of Gd^{3+}-loaded virus was described by Allen *et al.* who used cowpea chlorotic mottle virus (CCMV). The presence of 180 binding sites for Ca^{2+} in the interior was used to introduce approximately 140 Gd^{3+} per particle with a resulting relaxivity of 202 $mM^{-1}\,s^{-1}$ at 61 MHz, which is approximately 25-fold higher than monomeric Gd^{3+} chelates [169]. However, the low binding affinity of Gd^{3+} ($K_D = 31\,\mu M$) precludes any *in vivo* application. In order to improve the Gd^{3+} sequestration in such structures, the surface of the capsids has been conjugated to the well-known ligands DTPA [170] and DOTA [171], with 100–500 Gd^{3+} incorporated per viral particle. The measured relaxivity was two- to threefold that of Gd^{3+}-DTPA. However, these values are too low if we consider the overall weight of the viral capsid (~2000 kDa in this case) and the resulting dose that would be necessary to obtain a good contrast enhancement. Improvements under investigation are the incorporation of the Gd^{3+} complex on the interior surface of the capsid, which was shown to produce 30–50% increase in the relaxivity with HOPO-like ligands [172]. Also, when the rigidity of the linker between the capsid surface and the HOPO ligand was increased, better relaxivity was obtained [173]. The behavior of these particles in cell culture and *in vivo* should provide in the near future indications on their immunogenicity and on their potential applications.

8.3.5.3 Gadolinium Oxide Nanoparticles While most of the gadolinium-based CAs incorporate the paramagnetic species in the ionic form (Gd^{3+}), the idea to use the oxide form Gd_2O_3 has emerged recently. The first species were prepared as a Gd_2O_3 core capped with diethylene glycol (DEG). The relaxivity was two to three times higher than Gd^{3+}-DTPA, with better results when the concentration was decreased [174]. However, it was shown in aqueous media that these nanoparticles tend to agglomerate with subsequent decrease in the relaxivity [175]. Nonetheless, improved contrast was obtained in the visualization of brain tumors [176] or in the detection of glioblastoma cancer cells *in vivo* [177]. These particles can furthermore be modified on their surface by incorporation of various functional groups such as organic acids [178] or PEG [179], suggesting potential control over stability and biodistribution. Alternatively, the Gd_2O_3 core has been incorporated in polysiloxane shells allowing also derivatization on the surface for the design of targeted or multimodal probes [180, 181]. Apoferritin was also described by Sanchez *et al.* as a potential sequestering agent for Gd_2O_3 with relaxivities 10 to 25 folds higher than Gd^{3+}-DTPA, depending on the field strength [182].

8.4 ALTERNATIVES TO GADOLINIUM FOR MRI

The second most used metal for MRI is iron in superparamagnetic iron oxides to produce T_2 or T_2*-weighted pictures. Their preparation and use have already been described in Chapter 2. The last part of this chapter focuses on alternative metal ions to Gd^{3+} as T_1 CAs and also on the recently discovered chemical exchange saturation transfer (CEST) effect.

8.4.1 Manganese and Other Alternative Metals

Gd^{3+} is by far the most studied metal ion as T_1 CA because of its ideal characteristics as described in Section 8.2. However, the use of Gd^{3+}-based CAs has recently been linked to the apparition of serious side effects for patients with impaired renal function named nephrogenic systemic fibrosis (NSF). This pathology is characterized by thickening and hardening of the skin of the extremities and affects also numerous organ systems such as the liver, lungs, muscles, and heart [183]. More than 90% of the NSF cases reported have appeared after administration of a Gd^{3+}-based CA. Also, NSF is much more frequent when acyclic complex is used as compared to the more stable macrocyclic ligands, suggesting an effect of the free Gd^{3+} released from the complex. Another concern is the increasing amount of gadolinium in wastewater effluents as a result of the increasing clinical use of Gd^{3+}-based CAs [184]. For these reasons, metal ions such as Mn^{2+}, Mn^{3+}, Fe^{3+}, and Cu^{2+} have received an increasing interest in recent years, since they are naturally present in the organism and their behavior in biological systems is well documented.

Although not as effective as Gd^{3+}, Mn^{2+} exhibits interesting properties as T_1 CA, thanks to its five unpaired electrons. Besides, Mn^{2+} in the $MnCl_2$ form was among the very first species to be studied as a CA for MRI [8]. However, although Mn^{2+} is naturally present in the organism at low concentration, it is toxic at relatively low doses when injected intravenously (LD_{50}=0.3 mmol kg^{-1} in mice). But it is much less toxic when administered orally (13.6 mmol kg^{-1}), and LumenHance®, an oral preparation of $MnCl_2$, is used routinely in clinical exams for visualization of the gastrointestinal system. For intravenous (IV) administration of Mn^{2+}, a sequestering agent is required, and dipyridoxal diphosphate (DPDP) is the ligand used in Teslascan®, the only Mn^{2+} complex approved for clinical use. The complex exhibits a relatively good relaxivity in water (2.8 $mM^{-1}\,s^{-1}$ at 20 MHz and 40°C), and provides selective uptake by normal hepatocytes, allowing determination of liver function and identification of cancerous tumors [185]. Similarly to Gd^{3+}, Mn^{2+} has been incorporated to nanosized carriers. The resulting highly Mn^{2+}-loaded compounds combined with the slow tumbling rate minimize the drawback of the lower relaxivity of Mn^{2+} species. Liposomes and other lipid-based compounds have been used to entrap $MnCl_2$ or Mn^{2+} complexes to produce very high relaxivities, comparable to the ones obtained with Gd^{3+}-loaded structures. Also, dendritic structures and manganese oxide nanoparticles are the object of increasing attention. Their preparation and characteristics have been reviewed recently by Pan *et al.* [186]. These studies show that Mn^{2+} is, in some cases, a plausible alternative to Gd^{3+}.

Another paramagnetic species under investigation is Mn^{3+}. Complexed in porphyrins, it exhibits high relaxivities and can be used to target various cancer cells. In this case, Mn^{3+} is used advantageously over Gd^{3+} because of the higher stabilities of the complexes formed, Gd^{3+} being released from porphyrins in solution [187]. As a nanoparticle component, Mn^{3+} porphyrins have been coupled to dextran to obtain macromolecular structures exhibiting tumor cell membrane affinity with relaxivities 50–75% higher than Gd^{3+}-DTPA [188]. Another possibility could be the use of the high-spin d^5 Fe^{3+}. It is too toxic to be injected in the free form ($LD_{50i.v.}$ = 0.88 mmol kg^{-1} in mice), but the lower oral toxicity (LD_{50} = 7.9 mmol kg^{-1}) allows its use for gastrointestinal organ visualization with Geritol®, a ferric ammonium citrate-based preparation. For IV administration of Fe^{3+}, various ligands deriving from natural iron chelators have been studied. The limited relaxivities obtained have also been improved by the use of polymeric structures such as dextran [189]. Finally, because of its only one unpaired electron resulting in a low relaxivity (r_1 = 0.9 $mM^{-1}\,s^{-1}$ for $CuCl_2$ in water at 20 MHz and 37°C), Cu^{2+} has received much less attention than Gd^{3+} and the other species cited previously. However, its relatively low toxicity allows injection of higher doses to offset this low relaxivity. A few chelates have been investigated for this application, but the relaxivities obtained seem too low to be able to replace Gd^{3+} chelates (r_1 values <0.1 $mM^{-1}\,s^{-1}$). However, when incorporated in polymers, Cu^{2+} exhibits more reasonable relaxivities (values between 0.5 and 2 $mM^{-1}\,s^{-1}$) [189]. Nonetheless, success with Cu^{2+} would be attainable only with high loads of the paramagnetic ion. Pan *et al.* prepared a nanoparticle made of a Cu^{2+} core encapsulated in a lipid layer [190]; approximately 14,000 copper atoms were measured per particle with a total relaxivity of 66,000 ± 2,200 $mM^{-1}\,s^{-1}$. Furthermore, the incorporation of biotin at the surface of the nanoparticle provided targeting properties. The size of the particles (~210 nm) could make them useful as a blood pool agent.

8.4.2 CEST/PARACEST Agents

Recently, a new MRI technique named chemical exchange saturation transfer (CEST) has emerged. Image contrast with this method is obtained in a fundamentally different way than discussed in this chapter. With CEST, the CA does not influence proton relaxation, but instead reduces the magnetization of the water signal with limited effect on the longitudinal relaxation rate. This new technique is expected to address some of the drawbacks and limitations of the classical MRI technique, namely, (i) the necessity of a precontrast image before administration of the CA; (ii) the impossibility to turn off the contrast after administration of the CA and the fact that its elimination from the body must be complete before acquisition of a new image or administration of another CA; (iii) large CA doses are still required; (iv) environment-responsive CAs (such as enzymes, pH, or temperature) have been developed, but because the concentration of the CA *in vivo* is usually unknown, it makes difficult to quantify the environmental parameter of interest; and (v) only one target per exam can be detected with targeted CA.

The CEST effect was first applied in 2000 by Ward *et al.* [191]. In a typical CEST contrasted MRI experiment, the exchangeable protons of the CA are saturated by a continuous frequency-selective RF saturation pulse. The exchange of protons between the CA and water from the bulk results in a decrease in the intensity of the bulk signal along with the rise of a signal that differs in frequency from the bulk (this difference, named chemical shift, is denoted $\Delta\omega$ and expressed in Hz or in ppm). The physical principles and use of the CEST effect in MRI are described in detail in a recent review by Hancu *et al.* [192].

Interestingly, this technique is sensitive enough to make use of endogenous molecules bearing exchangeable protons (such as in –OH or $-NH_2$ functions in glucose derivatives or proteins) to image their own presence or environmental factors related to them. For instance, the CEST effect with endogenous CA has been used to detect pH changes during ischemia in rat brain through the exchangeable amide proton of cellular proteins and peptides [193] or to measure the concentration of glycosaminoglycans (a polysaccharide bearing a high number of –OH) in cartilages and other tissues [194]. The advantage of this technique is that it can be completely noninvasive. However, CA of particular interest such as sugars, amino acids, or nucleotides can also be administered for specific applications. Thus, macromolecular agents such as polymers bearing a high number of exchangeable protons are of great interest [192]. The CAs cited previously are diamagnetic and are denoted DIACEST agents. They offer many advantages such as being noninvasive for the endogenous ones or biocompatible for the exogenous ones. However, the small chemical shift they induce can sometimes render difficult detect ion of the CEST signal. That is why PARACEST agents incorporating paramagnetic ions and able to induce higher $\Delta\omega$ are also under investigation.

While Gd^{3+} exhibits the best characteristics as a T_1-shortening agent, other lanthanides are preferred as PARACEST agents because of their better hyperfine shifting constant (C_j). C_j is one of the parameters that influence the $\Delta\omega$ obtained, and higher C_j are required for a higher $\Delta\omega$. Thus, complexes of Eu^{3+}, Yb^{3+}, or Tm^{3+} have been preferred. Another parameter that differs from T_1-shortening agents is that slow water exchange rates are beneficial in the case of PARACEST agents. Several PARACEST agents have been investigated, and they could be used to monitor physiological phenomenon. For instance, Eu-D2BAM-2BiPyAm is a Eu^{3+} complex that exhibits slow water exchange. But *in vitro*, in the presence of Zn^{2+}, water exchange is accelerated, which makes this CA "on" in the absence of Zn^{2+} and "off" in the presence of Zn^{2+} [195].

Also, in order to improve sensitivity, macromolecular PARACEST agents were developed, such as the pH-sensitive Yb^{3+}-DOTAM complex grafted on dendrimers [196]. Lanthanide-loaded liposomes are also under investigation for this kind of applications. Named LipoCEST agents, these nanoparticles are the object of an increasing interest because of their high versatility with potential to map physiological phenomenon at low concentrations [72, 192].

While the use of CEST is promising, this relatively young technique still requires technical development. Especially, in the case of PARACEST, *in vivo* experimentations are limited to a few studies that show that better sensitivity is required before human uses become applicable.

8.5 CONCLUSION

From this review of the recent literature, it appears clearly that Gd^{3+}-based nanoparticles have a high potential to improve MRI diagnoses. Their high relaxivity can improve the sensitivity of the technique and reduce the necessary dose per exam. Furthermore, their high versatility gives access to new targets and organs that are not easily imaged with the currently approved CAs. However, concerns about the potential toxicity of these longer-lasting particles in the body have also risen together with their development. One of the current biggest challenges for the scientists is the design of nanoparticles that are highly efficient but also safe for human use. And this last criterion will have to be addressed with certainty before clinical use can be approved. This makes the challenge all the more difficult for the scientist as compromises between relaxivity, stability, biodistribution, and safety of use have to be found. The recent development of biocompatible and biodegradable nanoparticles shows that this goal could be reached shortly. The last advances in the design of efficient CAs with other paramagnetic species than Gd^{3+} such as Mn^{2+}, Mn^{3+}, Fe^{3+}, or even Cu^{2+} show that alternatives could be available in the future, especially for patients at risk regarding NSF related to the use of Gd^{3+}-based CAs. Also, the recent discovery of the CEST/PARACEST technique shows that new targets could be attainable at the molecular level. The combination of all these advances, together with the development of more and more powerful scanners, shows that the MRI technique is only at its beginning and that the diagnosis possibilities we can expect from it in the future are virtually unlimited.

REFERENCES

[1] Prasad PV. *Magnetic Resonance Imaging: Methods and biologic Applications.* Totowa: Humana Press; 2006.

[2] Vlaardingerbroek MT, Boer JA. *Magnetic Resonance Imaging: Theory and Practice.* Berlin: Springer Verlag; 2003.

[3] Moenninghoff C, Maderwald S, Theysohn J, Kraff O, Ladd M, Hindy N, Nes J, Forsting M, Wanke I. Imaging of adult astrocytic brain tumours with 7 T MRI: preliminary results. Eur Radiol 2010;**20**:704–713.

[4] Chang G, Wang L, Schweitzer M, Regatte R. 3D 23Na MRI of human skeletal muscle at 7 Tesla: initial experience. Eur Radiol 2010;**20**:2039–2046.

[5] von Knobelsdorff-Brenkenhoff F, Frauenrath T, Prothmann M, Dieringer M, Hezel F, Renz W, Kretschel K, Niendorf T, Schulz-Menger J. Cardiac chamber quantification using magnetic resonance imaging at 7 Tesla—a pilot study. Eur Radiol 2010;**20**: 2844–2852.

[6] Blamire A. The technology of MRI—the next 10 years? Br J Radiol 2008;**81**:601–617.

[7] Caravan P, Ellison JJ, McMurry TJ, Lauffer RB. Gadolinium(III) chelates as MRI contrast agents: structure, dynamics, and applications. Chem Rev 1999;**99**:2293–2352.

[8] Lauterbur PC, Mendonca-Dias M, Rudin A. Augmentation of tissue water proton spin-lattice relaxation rates by *in vivo* addition of paramagnetic ions. Front Biol Energ 1978;**1**:752–759.

[9] Kellar KE, Fossheim SL, Koenig SH. Magnetic field dependence of solvent proton relaxation by solute dysprosium (III) complexes. Invest Radiol 1998;**33**:835–840.

[10] Ersoy H, Rybicki FJ. Biochemical safety profiles of gadolinium-based extracellular contrast agents and nephrogenic systemic fibrosis. J Magn Reson Imaging 2007;**26**: 1190–1197.

[11] Bellin M-F. MR contrast agents, the old and the new. Eur J Radiol 2006;**60**:314–323.

[12] Cacheris WP, Quay SC, Rocklage SM. The relationship between thermodynamics and the toxicity of gadolinium complexes. Magn Reson Imaging 1990;**8**:467–481.

[13] Pearson RG. Hard and soft acids and bases. J Am Chem Soc 1963;**85**:3533–3539.

[14] Hancock RD. Chelate ring size and metal ion selection. The basis of selectivity for metal ions in open-chain ligands and macrocycles. J Chem Educ 1992;**69**:615.

[15] Liu S, Edwards DS. Bifunctional chelators for therapeutic lanthanide radiopharmaceuticals. Bioconjug Chem 2001;**12**:7–34.

[16] Merbach AE, Tóth É. *The Chemistry of Contrast Agents in Medical Magnetic Resonance Imaging*. Chichester: Wiley; 2001.

[17] Laurent S, Elst LV, Muller RN. Comparative study of the physicochemical properties of six clinical low molecular weight gadolinium contrast agents. Contrast Media Mol Imaging 2006;**1**:128–137.

[18] Aime S, Caravan P. Biodistribution of gadolinium-based contrast agents, including gadolinium deposition. J Magn Reson Imaging 2009;**30**:1259–1267.

[19] Gianolio E, Giovenzana GB, Longo D, Longo I, Menegotto I, Aime S. Relaxometric and modelling studies of the binding of a lipophilic Gd-AAZTA complex to fatted and defatted human serum albumin. Chem Eur J 2007;**13**:5785–5797.

[20] Briley-Saebo KC, Geninatti-Crich S, Cormode DP, Barazza A, Mulder WJM, Chen W, Giovenzana GB, Fisher EA, Aime S, Fayad ZA. High-relaxivity gadolinium-modified high-density lipoproteins as magnetic resonance imaging contrast agents. J Phys Chem B 2009;**113**:6283–6289.

[21] Silverio S, Torres S, Martins AF, Martins JA, Andre JP, Helm L, Prata MIM, Santos AC, Geraldes CFGC. Lanthanide chelates of (bis)-hydroxymethyl-substituted DTTA with potential application as contrast agents in magnetic resonance imaging. Dalton Trans 2009;**28**:4656–4670.

[22] Datta A, Raymond KN. Gd–hydroxypyridinone (HOPO)-based high-relaxivity magnetic resonance imaging (MRI) contrast agents. Acc Chem Res 2009;**42**:938–947.

[23] Werner EJ, Botta M, Aime S, Raymond KN. Effect of a mesitylene-based ligand cap on the relaxometric properties of Gd(III) hydroxypyridonate MRI contrast agents. Contrast Media Mol Imaging 2009;**4**:220–229.

[24] Hermann P, Kotek J, Kubicek V, Lukes I. Gadolinium(iii) complexes as MRI contrast agents: ligand design and properties of the complexes. Dalton Trans 2008;**23**:3027–3047.

[25] Jacques V, Dumas S, Sun W-C, Troughton JS, Greenfield MT, Caravan P. High-relaxivity magnetic resonance imaging contrast agents part 2: optimization of inner- and second-sphere relaxivity. Invest Radiol 2010;**45**:613–624.

[26] Vitha T, Kubicek V, Kotek J, Hermann P, Vander Elst L, Muller RN, Lukes I, Peters JA. Gd(iii) complex of a monophosphinate-bis(phosphonate) DOTA analogue with a high relaxivity; Lanthanide(iii) complexes for imaging and radiotherapy of calcified tissues. Dalton Trans 2009;**17**:3204–3214.

[27] Lebduskova P, Sour A, Helm L, Toth E, Kotek J, Lukes I, Merbach AE. Phosphinic derivative of DTPA conjugated to a G5 PAMAM dendrimer: an 17O and 1H relaxation study of its Gd(iii) complex. Dalton Trans 2006;**28**:3399–3406.

[28] Dumas S, Jacques V, Sun W-C, Troughton JS, Welch JT, Chasse JM, Schmitt-Willich H, Caravan P. High relaxivity magnetic resonance imaging contrast agents part 1: impact of single donor atom substitution on relaxivity of serum albumin-bound gadolinium complexes. Invest Radiol 2010;**45**:600–612.

[29] Tei L, Botta M, Lovazzano C, Barge A, Milone L, Aime S. 1H and 17O NMR relaxometric study in aqueous solution of Gd(III) complexes of EGTA-like derivatives bearing methylenephosphonic groups. Magn Reson Chem 2008;**46**:S86–S93.

[30] Caravan P, Greenfield MT, Li X, Sherry AD. The Gd3+ complex of a fatty acid analogue of DOTP binds to multiple albumin sites with variable water relaxivities. Inorg Chem 2001;**40**:6580–6587.

[31] Fulton DA, Elemento EM, Aime S, Chaabane L, Botta M, Parker D. Glycoconjugates of gadolinium complexes for MRI applications. Chem Commun (Camb) 2006;**10**:1064–1066.

[32] Rudovsky J, Kotek J, Hermann P, Lukes I, Mainero V, Aime S. Synthesis of a bifunctional monophosphinic acid DOTA analogue ligand and its lanthanide(iii) complexes. A gadolinium(iii) complex endowed with an optimal water exchange rate for MRI applications. Org Biomol Chem 2005;**3**:112–117.

[33] Tallec G, Fries PH, Imbert D, Mazzanti M. High relaxivity and stability of a hydroxyquinolinate-based tripodal monoaquagadolinium complex for use as a bimodal MRI/optical imaging agent. Inorg Chem 2011;**50**:7943–7945.

[34] Caravan P, Cloutier NJ, Greenfield MT, McDermid SA, Dunham SU, Bulte JWM, Amedio JC, Looby RJ, Supkowski RM, Horrocks WD, McMurry TJ, Lauffer RB. The interaction of MS-325 with human serum albumin and its effect on proton relaxation rates. J Am Chem Soc 2002;**124**:3152–3162.

[35] Goyen M. Gadofosveset-enhanced magnetic resonance angiography. Vasc Health Risk Manag 2008;**4**:1.

[36] Grossman RI, Rubin DL, Hunter G, Haughton VM, Lee D, Sze G, Kuhn MJ, Maravilla K, Tu R, Heindel W, Wippold JI, Leeds N, Zelch J, Jinkins JR, Grodd W, Truwit C, Kanal E, Provenzale JM, Ramsey R, Simon J, Brunberg JA, Stevens GR, Kristy RM. Magnetic resonance imaging in patients with central nervous system pathology: a comparison of OptiMARK (Gd-DTPA-BMEA) and magnevist (Gd-DTPA). Invest Radiol 2000;**35**:412–419.

[37] Chanyaputhipong J, Low S-CA, Chow PK. Gadoxetate acid-enhanced MR imaging for HCC: a review for clinicians. Int J Hepatol 2011;**2011**:489342.

[38] Klessen C, Hein PA, Huppertz A, Voth M, Wagner M, Elgeti T, Kroll H, Hamm B, Taupitz M, Asbach P. First-pass whole-body magnetic resonance angiography (MRA) using the blood-pool contrast medium gadofosveset trisodium: comparison to gadopentetate dimeglumine. Invest Radiol 2007;**42**:659–664.

[39] Spanoghe M, Lanens D, Dommisse R, Van der Linden A, Alderweireldt F. Proton relaxation enhancement by means of serum albumin and poly-l-lysine labeled with DTPA-Gd3+: Relaxivities as a function of molecular weight and conjugation efficiency. Magn Reson Imaging 1992;**10**:913–917.

[40] Sieving PF, Watson AD, Rocklage SM. Preparation and characterization of paramagnetic polychelates and their protein conjugates. Bioconjug Chem 1990;**1**:65–71.

[41] Schuhmann-Giampieri G, Schmitt-Willich H, Frenzel T, Press W-R, Weinmann H-J. *In vivo* and in vitro evaluation of Gd-DTPA-polylysine as a macromolecular contrast agent for magnetic resonance imaging. Invest Radiol 1991;**26**:969–974.

[42] Van Hecke P, Marchal G, Bosmans H, Johannik K, Jiang Y, Vogler H, Van Ongeval C, Baert A, Speck U. NMR imaging study of the pharmacodynamics of polylysine-gadolinium-DTPA in the rabbit and the rat. Magn Reson Imaging 1991;**9**:313–321.

[43] Vexler VS, Clément O, Schmitt-Willich H, Brasch RC. Effect of varying the molecular weight of the MR contrast agent Gd-DTPA-polylysine on blood pharmacokinetics and enhancement patterns. J Magn Reson Imaging 1994;**4**:381–388.

[44] Marchal G, Bosmans H, Van Hecke P, Speck U, Aerts P, Vanhoenacker P, Baert AL. MR angiography with gadopentetate dimeglumine-polylysine: evaluation in rabbits. Am J Roentgenol 1990;**155**:407–411.

[45] Berthezene Y, Vexler V, Jerome H, Sievers R, Moseley ME, Brasch RC. Differentiation of capillary leak and hydrostatic pulmonary edema with a macromolecular MR imaging contrast agent. Radiology 1991;**181**:773–777.

[46] Berthezène Y, Vexler V, Price DC, Wisner-Dupon J, Moseley ME, Aicher KP, Brasch RC. Magnetic resonance imaging detection of an experimental pulmonary perfusion deficit using a macromolecular contrast agent: polylysine-gadolinium-DTPA40. Invest Radiol 1992;**27**:346–351.

[47] Berthezène Y, Vexler V, Kuwatsuru R, Rosenau W, Mühler A, Clément O, Price DC, Brasch RC. Differentiation of alveolitis and pulmonary fibrosis with a macromolecular MR imaging contrast agent. Radiology 1992;**185**:97–103.

[48] Opsahl LR, Uzgiris EE, Vera DR. Tumor imaging with a macromolecular paramagnetic contrast agent: gadopentetate dimeglumine-polylysine. Acad Radiol 1995;**2**:762–766.

[49] Desser TS, Rubin DL, Muller HH, Qing F, Khodor S, Zanazzi G, Young SW, Ladd DL, Wellons JA, Kellar KE. Dynamics of tumor imaging with Gd-DTPA—polyethylene glycol polymers: dependence on molecular weight. J Magn Reson Imaging 1994;**4**:467–472.

[50] Bogdanov AA, Weissleder R, Frank HW, Bogdanova AV, Nossif N, Schaffer BK, Tsai E, Papisov MI, Brady TJ. A new macromolecule as a contrast agent for MR angiography: preparation, properties, and animal studies. Radiology 1993;**187**:701–706.

[51] Cavagna F, Luchinat C, Scozzafava A, Xia Z. Polymetallic macromolecules are potential contrast agents of improved efficiency. Magn Reson Med 1994;**31**:58–60.

[52] Lucas RL, Benjamin M, Reineke TM. Comparison of a tartaric acid derived polymeric MRI contrast agent to a small molecule model chelate. Bioconjug Chem 2007;**19**:24–27.

[53] Kellar KE, Henrichs PM, Hollister R, Koenig SH, Eck J, Wei D. High relaxivity linear Gd (DTPA)-polymer conjugates: the role of hydrophobic interactions. Magn Reson Med 1997;**38**:712–716.

[54] Allen MJ, Raines RT, Kiessling LL. Contrast agents for magnetic resonance imaging synthesized with ring-opening metathesis polymerization. J Am Chem Soc 2006;**128**: 6534–6535.

[55] Ladd DL, Hollister R, Peng X, Wei D, Wu G, Delecki D, Snow RA, Toner JL, Kellar K, Eck J, Desai VC, Raymond G, Kinter LB, Desser TS, Rubin DL. Polymeric gadolinium chelate magnetic resonance imaging contrast agents: design, synthesis, and properties. Bioconjug Chem 1999;**10**:361–370.

[56] Ogan MD, Schmiedl U, Moseley ME, Grodd W, Paajanen H, Brasch RC. Albumin labeled with Gd-DTPA: an intravascular contrast-enhancing agent for magnetic resonance blood pool imaging: preparation and characterization. Invest Radiol 1987;**22**:665–671.

[57] Pashkunova-Martic I, Kremser C, Galanski M, Schluga P, Arion V, Debbage P, Jaschke W, Keppler B. Lectin conjugates as biospecific contrast agents for MRI. Coupling of lycopersicon esculentum agglutinin to linear water-soluble DTPA-loaded oligomers. Mol Imaging Biol 2011;**13**:432–442.

[58] Shreve P, Aisen AM. Monoclonal antibodies labeled with polymeric paramagnetic ion chelates. Magn Reson Med 1986;**3**:336–340.

[59] GÖhr-Rosenthal S, Schmitt-Willich H, Ebert W, Conrad J. The demonstration of human tumors on nude mice using gadolinium-labelled monoclonal antibodies for magnetic resonance imaging. Invest Radiol 1993;**28**:789–795.

[60] Yan G-P, Liu M-L, Li LY. Polyaspartamide gadolinium complexes containing sulfadiazine groups as potential macromolecular MRI contrast agents. Bioconjug Chem 2005;**16**:967–971.

[61] Tachibana Y, Enmi J-i, Mahara A, Iida H, Yamaoka T. Design and characterization of a polymeric MRI contrast agent based on PVA for *in vivo* living-cell tracking. Contrast Media Mol Imaging 2010;**5**:309–317.

[62] Zarabi B, Nan A, Zhuo J, Gullapalli R, Ghandehari H. Macrophage targeted N-(2-Hydroxypropyl)methacrylamide conjugates for magnetic resonance imaging. Mol Pharm 2006;**3**:550–557.

[63] Zarabi B, Borgman M, Zhuo J, Gullapalli R, Ghandehari H. Noninvasive monitoring of HPMA copolymer–RGDfK conjugates by magnetic resonance imaging. Pharm Res 2009;**26**:1121–1129.

[64] Aime S, Botta M, Frullano L, Crich SG, Giovenzana GB, Pagliarin R, Palmisano G, Sisti M. Contrast agents for magnetic resonance imaging: a novel route to enhanced relaxivities based on the interaction of a GdIII chelate with Poly-β-cyclodextrins. Chem Eur J 1999;**5**:1253–1260.

[65] Bogdanov A, Mazzanti ML. Molecular magnetic resonance contrast agents for the detection of cancer: past and present. Semin Oncol 2011;**38**:42–54.

[66] Lu Z-R, Wang X, Parker DL, Goodrich KC, Buswell HR. Poly(l-glutamic acid) Gd(III)-DOTA conjugate with a degradable spacer for magnetic resonance imaging. Bioconjug Chem 2003;**14**:715–719.

[67] Mohs AM, Wang X, Goodrich KC, Zong Y, Parker DL, Lu Z-R. PEG-g-poly(GdDTPA-co-l-cystine): a biodegradable macromolecular blood pool contrast agent for mr imaging. Bioconjug Chem 2004;**15**:1424–1430.

[68] Zong Y, Wang X, Goodrich KC, Mohs AM, Parker DL, Lu Z-R. Contrast-enhanced MRI with new biodegradable macromolecular Gd(III) complexes in tumor-bearing mice. Magn Reson Med 2005;**53**:835–842.

[69] Wen X, Jackson EF, Price RE, Kim EE, Wu Q, Wallace S, Charnsangavej C, Gelovani JG, Li C. Synthesis and characterization of poly(l-glutamic acid) gadolinium chelate: a new biodegradable mri contrast agent. Bioconjug Chem 2004;**15**:1408–1415.

[70] Karfeld-Sulzer LS, Waters EA, Davis NE, Meade TJ, Barron AE. Multivalent protein polymer mri contrast agents: controlling relaxivity via modulation of amino acid sequence. Biomacromolecules 2010;**11**:1429–1436.

[71] Zhu W, Artemov D. Biocompatible blood pool MRI contrast agents based on hyaluronan. Contrast Media Mol Imaging 2011;**6**:61–68.

[72] Villaraza AJ, Bumb A, Brechbiel MW. Macromolecules, dendrimers, and nanomaterials in magnetic resonance imaging: the interplay between size, function, and pharmacokinetics. Chem Rev 2010;**110**:2921–2959.

[73] Buhleier E, Wehner W, Vögtle F. Cascade and nonskid-chain-like synthesis of molecular cavity topologies. Synthesis 1978;**2**:155–158.

[74] Astruc D, Boisselier E, Ornelas C. Dendrimers designed for functions: from physical, photophysical, and supramolecular properties to applications in sensing, catalysis, molecular electronics, photonics, and nanomedicine. Chem Rev 2010;**110**:1857–1959.

[75] Wiener E, Brechbiel MW, Brothers H, Magin R, Gansow O, Tomalia D, Lauterbur P. Dendrimer-based metal chelates: a new class of magnetic resonance imaging contrast agents. Magn Reson Med 1994;**31**:1–8.

[76] Newkome GR, Yao Z, Baker GR, Gupta VK. Micelles. Part 1. Cascade molecules: a new approach to micelles. A [27]-arborol. J Org Chem 1985;**50**:2003–2004.

[77] Tomalia D, Baker H, Dewald J, Hall M, Kallos G, Martin S, Roeck J, Ryder J, Smith P. A new class of polymers: starburst-dendritic macromolecules. Polym J 1985;**17**:117–132.

[78] Brabander-van den Berg dE, Meijer E. Poly (propylenimine) dendrimers: large-scale synthesis via heterogeneously catalyzed hydrogenation. Angew Chem Int Ed Engl 1993;**32**:1308.

[79] Wörner C, Mülhaupt R. Polynitrile- and polyamine-functional poly(trimethylene imine) dendrimers. Angew Chem Int Ed Engl 1993;**32**:1306–1308.

[80] Hummelen JC, Van Dongen JL, Meijer E. Electrospray mass spectrometry of poly (propylene imine) dendrimers—the issue of dendritic purity or polydispersity. Chem Eur J 1997;**3**:1489–1493.

[81] Hawker C, Fréchet JM. A new convergent approach to monodisperse dendritic macromolecules. *J Chem Soc Chem Commun* 1990:1010–1013.

[82] Grayson SM, Fréchet JMJ. Convergent dendrons and dendrimers: from synthesis to applications. Chem Rev 2001;**101**:3819–3868.

[83] Bosman AW, Janssen HM, Meijer EW. About dendrimers: structure, physical properties, and applications. Chem Rev 1999;**99**:1665–1688.

[84] Bryant LH, Brechbiel MW, Wu C, Bulte JWM, Herynek V, Frank JA. Synthesis and relaxometry of high-generation (G = 5, 7, 9, and 10) PAMAM dendrimer-DOTA-gadolinium chelates. J Magn Reson Imaging 1999;**9**:348–352.

[85] Laus S, Sour A, Ruloff R, Tóth É, Merbach AE. Rotational dynamics account for pH-dependent relaxivities of PAMAM dendrimeric, Gd-based potential MRI contrast agents. Chem Eur J 2005;**11**:3064–3076.

[86] Rudovsky J, Hermann P, Botta M, Aime S, Lukes I. Dendrimeric Gd(iii) complex of a monophosphinated DOTA analogue: optimizing relaxivity by reducing internal motion. Chem Commun (Camb) 2005;**18**:2390–2392.

[87] Fulton DA, O'Halloran M, Parker D, Senanayake K, Botta M, Aime S. Efficient relaxivity enhancement in dendritic gadolinium complexes: effective motional coupling in medium molecular weight conjugates. Chem Commun (Camb) 2005;**4**:474–476.

[88] Nwe K, Xu H, Regino CAS, Bernardo M, Ileva L, Riffle L, Wong KJ, Brechbiel MW. A new approach in the preparation of dendrimer-based bifunctional diethylenetriaminepentaacetic acid mr contrast agent derivatives. Bioconjug Chem 2009;**20**: 1412–1418.

[89] Sato N, Kobayashi H, Hiraga A, Saga T, Togashi K, Konishi J, Brechbiel MW. Pharmacokinetics and enhancement patterns of macromolecular MR contrast agents with various sizes of polyamidoamine dendrimer cores. Magn Reson Med 2001;**46**: 1169–1173.

[90] Yordanov AT, Kobayashi H, English SJ, Reijnders K, Milenic D, Krishna MC, Mitchell JB, Brechbiel MW. Gadolinium-labeled dendrimers as biometric nanoprobes to detect vascular permeability. J Mater Chem 2003;**13**:1523–1525.

[91] Kobayashi H, Kawamoto S, Star RA, Waldmann TA, Tagaya Y, Brechbiel MW. Micro-magnetic resonance lymphangiography in mice using a novel dendrimer-based magnetic resonance imaging contrast agent. Cancer Res 2003;**63**:271–276.

[92] Sena LM, Fishman SJ, Jenkins KJ, Xu H, Brechbiel MW, Regino CAS, Kosaka N, Bernardo M, Choyke PL, Kobayashi H. Magnetic resonance lymphangiography with a nano-sized gadolinium-labeled dendrimer in small and large animal models. Nanomedicine 2010;**5**:1183–1191.

[93] Langereis S, de Lussanet QG, van Genderen MHP, Meijer EW, Beets-Tan RGH, Griffioen AW, van Engelshoven JMA, Backes WH. Evaluation of Gd(III)DTPA-terminated poly(propylene imine) dendrimers as contrast agents for MR imaging. NMR Biomed 2006;**19**:133–141.

[94] Kobayashi H, Saga T, Kawamoto S, Sato N, Hiraga A, Ishimori T, Konishi J, Togashi K, Brechbiel MW. Dynamic micro-magnetic resonance imaging of liver micrometastasis in mice with a novel liver macromolecular magnetic resonance contrast agent DAB-Am64-(1B4M-Gd)64. Cancer Res 2001;**61**:4966–4970.

[95] Kobayashi H, Sato N, Kawamoto S, Saga T, Hiraga A, Haque TL, Ishimori T, Konishi J, Togashi K, Brechbiel MW. Comparison of the macromolecular MR contrast agents with ethylenediamine-core versus ammonia-core generation-6 polyamidoamine dendrimer. Bioconjug Chem 2000;**12**:100–107.

[96] Kobayashi H, Kawamoto S, Saga T, Sato N, Hiraga A, Ishimori T, Konishi J, Togashi K, Brechbiel MW. Positive effects of polyethylene glycol conjugation to generation-4 polyamidoamine dendrimers as macromolecular MR contrast agents. Magn Reson Med 2001;**46**:781–788.

[97] Kojima C, Turkbey B, Ogawa M, Bernardo M, Regino CAS, Bryant LH, Choyke PL, Kono K, Kobayashi H. Dendrimer-based MRI contrast agents: the effects of PEGylation on relaxivity and pharmacokinetics. Nanomedicine 2011;**7**:1001–1008.

[98] Kobayashi H, Sato N, Kawamoto S, Saga T, Hiraga A, Ishimori T, Konishi J, Togashi K, Brechbiel MW. Novel intravascular macromolecular MRI contrast agent with generation-4 polyamidoamine dendrimer core: accelerated renal excretion with coinjection of lysine. Magn Reson Med 2001;**46**:457–464.

[99] Ogawa M, Regino CAS, Marcelino B, Williams M, Kosaka N, Bryant LH, Choyke PL, Kobayashi H. New nanosized biocompatible mr contrast agents based on lysine-dendri-graft macromolecules. Bioconjug Chem 2010;**21**:955–960.

[100] Kobayashi H, Kawamoto S, Star RA, Waldmann TA, Brechbiel MW, Choyke PL. Activated clearance of a biotinylated macromolecular mri contrast agent from the blood pool using an avidin chase. Bioconjug Chem 2003;**14**:1044–1047.

[101] Nwe K, Bernardo M, Regino CAS, Williams M, Brechbiel MW. Comparison of MRI properties between derivatized DTPA and DOTA gadolinium–dendrimer conjugates. Bioorg Med Chem 2010;**18**:5925–5931.

[102] Xu R, Wang Y, Wang X, Jeong E-K, Parker DL, Lu Z-R. *In vivo* evaluation of a pamam-cystamine-(Gd-DO3A) conjugate as a biodegradable macromolecular MRI contrast agent. Exp Biol Med 2007;**232**:1081–1089.

[103] Xu R, Kaneshiro TL, Jeong E-K, Parker DL, Lu Z-R. Synthesis and evaluation of nanoglobule-cystamine-(Gd-DO3A), a biodegradable nanosized magnetic resonance contrast agent for dynamic contrast-enhanced magnetic resonance urography. Int J Nanomed 2010;**5**:707.

[104] Wu C, Brechbiel MW, Kozak RW, Gansow OA. Metal-chelate-dendrimer-antibody constructs for use in radioimmunotherapy and imaging. Bioorg Med Chem Lett 1994;**4**:449–454.

[105] Nwe K, Milenic DE, Ray GL, Kim Y-S, Brechbiel MW. Preparation of cystamine core dendrimer and antibody–dendrimer conjugates for MRI angiography. Mol Pharm 2011;**9**:374–381.

[106] Lesch HP, Kaikkonen MU, Pikkarainen JT, Yla-Herttuala S. Avidin-biotin technology in targeted therapy. Expert Opin Drug Deliv 2010;**7**:551–564.

[107] Zhu W, Okollie B, Bhujwalla ZM, Artemov D. PAMAM dendrimer-based contrast agents for MR imaging of Her-2/neu receptors by a three-step pretargeting approach. Magn Reson Med 2008;**59**:679–685.

[108] Sano K, Temma T, Azuma T, Nakai R, Narazaki M, Kuge Y, Saji H. A pre-targeting strategy for mr imaging of functional molecules using dendritic Gd-based contrast agents. Mol Imaging Biol 2011;**13**:1196–1203.

[109] Konda SD, Wang S, Brechbiel MW, Wiener EC. Biodistribution of a 153Gd-folate dendrimer, generation = 4, in mice with folate-receptor positive and negative ovarian tumor xenografts. Invest Radiol 2002;**37**:199–204.

[110] Chen W-T, Thirumalai D, Shih T-F, Chen R-C, Tu S-Y, Lin C-I, Yang P-C. Dynamic contrast-enhanced folate-receptor-targeted MR imaging using a Gd-loaded PEG-dendrimer–folate conjugate in a mouse xenograft tumor model. Mol Imaging Biol 2010;**12**:145–154.

[111] Swanson SD, Kukowska-Latallo JF, Patri AK, Chen C, Ge S, Cao Z, Kotlyar A, East AT, Baker JR. Targeted gadolinium-loaded dendrimer nanoparticles for tumor-specific magnetic resonance contrast enhancement. Int J Nanomed 2008;**3**:201.

[112] Tan M, Wu X, Jeong E-K, Chen Q, Lu Z-R. Peptide-targeted nanoglobular Gd-DOTA monoamide conjugates for magnetic resonance cancer molecular imaging. Biomacromolecules 2010;**11**:754–761.

[113] Boswell CA, Eck PK, Regino CAS, Bernardo M, Wong KJ, Milenic DE, Choyke PL, Brechbiel MW. Synthesis, characterization, and biological evaluation of integrin $\alpha v\beta 3$-targeted PAMAM dendrimers. Mol Pharm 2008;**5**:527–539.

[114] Han L, Li J, Huang S, Huang R, Liu S, Hu X, Yi P, Shan D, Wang X, Lei H. Peptide-conjugated polyamidoamine dendrimer as a nanoscale tumor-targeted T1 magnetic resonance imaging contrast agent. Biomaterials 2011;**32**:2989–2998.

[115] Fossheim SL, Fahlvik AK, Klaveness J, Muller RN. Paramagnetic liposomes as MRI contrast agents: influence of liposomal physicochemical properties on the in vitro relaxivity. Magn Reson Imaging 1999;**17**:83–89.

[116] Økling KE, Skurtveit R, Bjørnerud A, Fossheim SL. Novel pH-sensitive paramagnetic liposomes with improved MR properties. Magn Reson Med 2004;**51**:688–696.

[117] Kabalka G, Davis M, Moss T, Buonocore E, Hubner K, Holmberg E, Maruyama K, Huang L. Gadolinium-labeled liposomes containing various amphiphilic Gd-DTPA derivatives: targeted MRI contrast enhancement agents for the liver. Magn Reson Med 1991;**19**:406–415.

[118] Tilcock C, Ahkong QF, Koenig SH, Brown RD, Davis M, Kabalka G. The design of liposomal paramagnetic mr agents: effect of vesicle size upon the relaxivity of surface-incorporated lipophilic chelates. Magn Reson Med 1992;**27**:44–51.

[119] Storrs RW, Tropper FD, Li HY, Song CK, Kuniyoshi JK, Sipkins DA, Li KC, Bednarski MD. Paramagnetic polymerized liposomes: synthesis, characterization, and applications for magnetic resonance imaging. J Am Chem Soc 1995;**117**:7301–7306.

[120] Kielar F, Tei L, Terreno E, Botta M. Large relaxivity enhancement of paramagnetic lipid nanoparticles by restricting the local motions of the gdiii chelates. J Am Chem Soc 2010;**132**:7836–7837.

[121] Schühle DT, van Rijn P, Laurent S, Vander Elst L, Muller RN, Stuart MCA, Schatz J, Peters JA. Liposomes with conjugates of a calix[4]arene and a Gd-DOTA derivative on the outside surface; an efficient potential contrast agent for MRI. Chem Commun (Camb) 2010;**46**:4399–4401.

[122] Gløgård C, Hovland R, Fossheim SL, Aasen AJ, Klaveness J. Synthesis and physicochemical characterisation of new amphiphilic gadolinium DO3A complexes as contrast agents for MRI. J Chem Soc Perkin Trans 2000;**2**:1047–1052.

[123] André JP, Tóth É, Fischer H, Seelig A, Mäcke HR, Merbach AE. High relaxivity for monomeric Gd (DOTA)-based MRI contrast agents, thanks to micellar self-organization. Chem A Eur J 1999;**5**:2977–2983.

[124] Trubetskoy VS, Torchilin VP. New approaches in the chemical design of Gd-containing liposomes for use in magnetic resonance imaging of lymph nodes. J Liposome Res 1994;**4**:961–980.

[125] Winter PM, Caruthers SD, Yu X, Song S-K, Chen J, Miller B, Bulte JWM, Robertson JD, Gaffney PJ, Wickline SA, Lanza GM. Improved molecular imaging contrast agent for detection of human thrombus. Magn Reson Med 2003;**50**:411–416.

[126] Kok MB, de Vries A, Abdurrachim D, Prompers JJ, Grüll H, Nicolay K, Strijkers GJ. Quantitative 1H MRI, 19F MRI, and 19F MRS of cell-internalized perfluorocarbon paramagnetic nanoparticles. Contrast Media Mol Imaging 2011;**6**:19–27.

[127] Unger EC, Winokur T, MacDougall P, Rosenblum J, Clair M, Gatenby R, Tilcock C. Hepatic metastases: liposomal Gd-DTPA-enhanced MR imaging. Radiology 1989;**171**: 81–85.

[128] Ayyagari AL, Zhang X, Ghaghada KB, Annapragada A, Hu X, Bellamkonda RV. Long-circulating liposomal contrast agents for magnetic resonance imaging. Magn Reson Med 2006;**55**:1023–1029.

[129] Bui T, Stevenson J, Hoekman J, Zhang S, Maravilla K, Ho RJ. Novel Gd nanoparticles enhance vascular contrast for high-resolution magnetic resonance imaging. PLoS One 2010;**5**:e13082.

[130] Mulder WJM, Douma K, Koning GA, van Zandvoort MA, Lutgens E, Daemen MJ, Nicolay K, Strijkers GJ. Liposome-enhanced MRI of neointimal lesions in the ApoE-KO mouse. Magn Reson Med 2006;**55**:1170–1174.

[131] Briley-Saebo KC, Mulder WJM, Mani V, Hyafil F, Amirbekian V, Aguinaldo JGS, Fisher EA, Fayad ZA. Magnetic resonance imaging of vulnerable atherosclerotic plaques: current imaging strategies and molecular imaging probes. J Magn Reson Imaging 2007;**26**:460–479.

[132] Maiseyeu A, Mihai G, Kampfrath T, Simonetti OP, Sen CK, Roy S, Rajagopalan S, Parthasarathy S. Gadolinium-containing phosphatidylserine liposomes for molecular imaging of atherosclerosis. J Lipid Res 2009;**50**:2157–2163.

[133] Beilvert A, Cormode DP, Chaubet F, Briley-Saebo KC, Mani V, Mulder WJM, Vucic E, Toussaint J-F, Letourneur D, Fayad ZA. Tyrosine polyethylene glycol (PEG)-micelle magnetic resonance contrast agent for the detection of lipid rich areas in atherosclerotic plaque. Magn Reson Med 2009;**62**:1195–1201.

[134] Morisco A, Accardo A, Gianolio E, Tesauro D, Benedetti E, Morelli G. Micelles derivatized with octreotide as potential target-selective contrast agents in MRI. J Pept Sci 2009;**15**:242–250.

[135] Accardo A, Morisco A, Gianolio E, Tesauro D, Mangiapia G, Radulescu A, Brandt A, Morelli G. Nanoparticles containing octreotide peptides and gadolinium complexes for MRI applications. J Pept Sci 2011;**17**:154–162.

[136] Kok MB, Strijkers GJ, Nicolay K. Dynamic changes in 1H-MR relaxometric properties of cell-internalized paramagnetic liposomes, as studied over a five-day period. Contrast Media Mol Imaging 2011;**6**:69–76.

[137] Mulder WJM, Strijkers GJ, Griffioen AW, van Bloois L, Molema G, Storm G, Koning GA, Nicolay K. A liposomal system for contrast-enhanced magnetic resonance imaging of molecular targets. Bioconjug Chem 2004;**15**:799–806.

[138] Erdogan S, Medarova ZO, Roby A, Moore A, Torchilin VP. Enhanced tumor MR imaging with gadolinium-loaded polychelating polymer-containing tumor-targeted liposomes. J Magn Reson Imaging 2008;**27**:574–580.

[139] Kamaly N, Kalber T, Thanou M, Bell JD, Miller AD. Folate receptor targeted bimodal liposomes for tumor magnetic resonance imaging. Bioconjug Chem 2009;**20**:648–655.

[140] Figueiredo S, Moreira JN, Geraldes CF, Aime S, Terreno E. Supramolecular protamine/Gd-loaded liposomes adducts as relaxometric protease responsive probes. Bioorg Med Chem 2011;**19**:1131–1135.

[141] Shuhendler AJ, Staruch R, Oakden W, Gordijo CR, Rauth AM, Stanisz GJ, Chopra R, Wu XY. Thermally-triggered 'off–on–off' response of gadolinium-hydrogel–lipid hybrid nanoparticles defines a customizable temperature window for non-invasive magnetic resonance imaging thermometry. J Control Release 2012;**157**:478–484.

[142] Wilson LJ. Medical applications of fullerenes and metallofullerenes. Electrochem Soc Interf 1999;**8**:24–28.

[143] Mikawa M, Kato H, Okumura M, Narazaki M, Kanazawa Y, Miwa N, Shinohara H. Paramagnetic water-soluble metallofullerenes having the highest relaxivity for MRI contrast agents. Bioconjug Chem 2001;**12**:510–514.

[144] Shu C-Y, Gan L-H, Wang C-R, X-l P, H-b H. Synthesis and characterization of a new water-soluble endohedral metallofullerene for MRI contrast agents. Carbon 2006;**44**:496–500.

[145] Fatouros PP, Corwin FD, Chen Z-J, Broaddus WC, Tatum JL, Kettenmann B, Ge Z, Gibson HW, Russ JL, Leonard AP, Duchamp JC, Dorn HC. In vitro and *in vivo* imaging studies of a new endohedral metallofullerene nanoparticle 1. Radiology 2006;**240**:756–764.

[146] Bolskar RD. Gadofullerene MRI contrast agents. Nanomedicine 2008;**3**:201–213.

[147] Kato H, Suenaga K, Mikawa M, Okumura M, Miwa N, Yashiro A, Fujimura H, Mizuno A, Nishida Y, Kobayashi K. Syntheses and EELS characterization of water-soluble multi-hydroxyl Gd@ C_{82} fullerenols. Chem Phys Lett 2000;**324**:255–259.

[148] Tóth É, Bolskar RD, Borel A, González G, Helm L, Merbach AE, Sitharaman B, Wilson LJ. Water-soluble gadofullerenes: toward high-relaxivity, pH-responsive MRI contrast agents. J Am Chem Soc 2004;**127**:799–805.

[149] Shu C-Y, Zhang E-Y, Xiang J-F, Zhu C-F, Wang C-R, Pei X-L, Han H-B. Aggregation studies of the water-soluble gadofullerene magnetic resonance imaging contrast agent: [Gd@C82O6(OH)16(NHCH2CH2COOH)8]x. J Phys Chem B 2006;**110**:15597–15601.

[150] Bolskar RD, Benedetto AF, Husebo LO, Price RE, Jackson EF, Wallace S, Wilson LJ, Alford JM. First soluble M@C60 derivatives provide enhanced access to metallofullerenes and permit *in vivo* evaluation of Gd@C60[C(COOH)2]10 as a MRI contrast agent. J Am Chem Soc 2003;**125**:5471–5478.

[151] Shu C, Corwin FD, Zhang J, Chen Z, Reid JE, Sun M, Xu W, Sim JH, Wang C, Fatouros PP, Esker AR, Gibson HW, Dorn HC. Facile preparation of a new gadofullerene-based magnetic resonance imaging contrast agent with high 1H relaxivity. Bioconjug Chem 2009;**20**:1186–1193.

[152] Sitharaman B, Kissell KR, Hartman KB, Tran LA, Baikalov A, Rusakova I, Sun Y, Khant HA, Ludtke SJ, Chiu W, Laus S, Toth E, Helm L, Merbach AE, Wilson LJ. Superparamagnetic gadonanotubes are high-performance MRI contrast agents. Chem Commun (Camb) 2005;**5**:3915–3917.

[153] Hartman KB, Laus S, Bolskar RD, Muthupillai R, Helm L, Toth E, Merbach AE, Wilson LJ. Gadonanotubes as ultrasensitive ph-smart probes for magnetic resonance imaging. Nano Lett 2008;**8**:415–419.

[154] Tran LA, Krishnamurthy R, Muthupillai R, da Graça Cabreira-Hansen M, Willerson JT, Perin EC, Wilson LJ. Gadonanotubes as magnetic nanolabels for stem cell detection. Biomaterials 2010;**31**:9482–9491.

[155] Tang AM, Ananta JS, Zhao H, Cisneros BT, Lam EY, Wong ST, Wilson LJ, Wong KK. Cellular uptake and imaging studies of gadolinium-loaded single-walled carbon nanotubes as MRI contrast agents. Contrast Media Mol Imaging 2011;**6**:93–99.

[156] Hassan AA, Chan BT-Y, Tran LA, Hartman KB, Ananta JS, Mackeyev Y, Hu L, Pautler RG, Wilson LJ, Lee AV. Serinc-derivatized gadonanotubes as magnetic nanoprobes for intracellular labeling. Contrast Media Mol Imaging 2010;**5**:34–38.

[157] van der Zande M, Sitharaman B, Walboomers XF, Tran L, Ananta JS, Veltien A, Wilson LJ, Álava JI, Heerschap A, Mikos AG. *In vivo* magnetic resonance imaging of the distribution pattern of gadonanotubes released from a degrading poly (lactic-co-glycolic acid) scaffold. Tissue Eng Part C Methods 2010;**17**:19–26.

[158] Sitharaman B, Van Der Zande M, Ananta JS, Shi X, Veltien A, Walboomers XF, Wilson LJ, Mikos AG, Heerschap A, Jansen JA. Magnetic resonance imaging studies

on gadonanotube-reinforced biodegradable polymer nanocomposites. J Biomed Mater Res A 2010;**93A**:1454–1462.

[159] Hainfeld JF, Dilmanian FA, Slatkin DN, Smilowitz HM. Radiotherapy enhancement with gold nanoparticles. J Pharm Pharmacol 2008;**60**:977–985.

[160] Debouttière PJ, Roux S, Vocanson F, Billotey C, Beuf O, Favre-Réguillon A, Lin Y, Pellet-Rostaing S, Lamartine R, Perriat P. Design of gold nanoparticles for magnetic resonance imaging. Adv Funct Mater 2006;**16**:2330–2339.

[161] Park J-A, Reddy PAN, Kim H-K, Kim I-S, Kim G-C, Chang Y, Kim T-J. Gold nanoparticles functionalised by Gd-complex of DTPA-bis(amide) conjugate of glutathione as an MRI contrast agent. Bioorg Med Chem Lett 2008;**18**:6135–6137.

[162] Alric C, Taleb J, Duc GL, Mandon C, Billotey C, Meur-Herland AL, Brochard T, Vocanson F, Janier M, Perriat P, Roux S, Tillement O. Gadolinium chelate coated gold nanoparticles as contrast agents for both x-ray computed tomography and magnetic resonance imaging. J Am Chem Soc 2008;**130**:5908–5915.

[163] Warsi MF, Chechik V. Strategies for increasing relaxivity of gold nanoparticle based MRI contrast agents. Phys Chem Chem Phys 2011;**13**:9812–9817.

[164] Lim YT, Cho MY, Choi BS, Lee JM, Chung BH. Paramagnetic gold nanostructures for dual modal bioimaging and phototherapy of cancer cells. Chem Commun (Camb) 2008;**40**:4930–4932.

[165] Warsi MF, Adams RW, Duckett SB, Chechik V. Gd-functionalised Au nanoparticles as targeted contrast agents in MRI: relaxivity enhancement by polyelectrolyte coating. Chem Commun (Camb) 2010;**46**:451–453.

[166] Song Y, Xu X, MacRenaris KW, Zhang X-Q, Mirkin CA, Meade TJ. Multimodal gadolinium-enriched DNA–gold nanoparticle conjugates for cellular imaging. Angew Chem Int Ed Engl 2009;**48**:9143–9147.

[167] Douglas T, Young M. Host–guest encapsulation of materials by assembled virus protein cages. Nature 1998;**393**:152–155.

[168] Pokorski JK, Steinmetz NF. The art of engineering viral nanoparticles. Mol Pharm 2010;**8**:29–43.

[169] Allen M, Bulte JWM, Liepold L, Basu G, Zywicke HA, Frank JA, Young M, Douglas T. Paramagnetic viral nanoparticles as potential high-relaxivity magnetic resonance contrast agents. Magn Reson Med 2005;**54**:807–812.

[170] Anderson EA, Isaacman S, Peabody DS, Wang EY, Canary JW, Kirshenbaum K. Viral nanoparticles donning a paramagnetic coat: conjugation of MRI contrast agents to the MS2 capsid. Nano Lett 2006;**6**:1160–1164.

[171] Prasuhn JDE, Yeh RM, Obenaus A, Manchester M, Finn MG. Viral MRI contrast agents: coordination of Gd by native virions and attachment of Gd complexes by azide-alkyne cycloaddition. Chem Commun (Camb) 2007;**12**:1269–1271.

[172] Hooker JM, Datta A, Botta M, Raymond KN, Francis MB. Magnetic resonance contrast agents from viral capsid shells: a comparison of exterior and interior cargo strategies. Nano Lett 2007;**7**:2207–2210.

[173] Garimella PD, Datta A, Romanini DW, Raymond KN, Francis MB. Multivalent, high-relaxivity MRI contrast agents using rigid cysteine-reactive gadolinium complexes. J Am Chem Soc 2011;**133**:14704–14709.

[174] Engström M, Klasson A, Pedersen H, Vahlberg C, Käll P-O, Uvdal K. High proton relaxivity for gadolinium oxide nanoparticles. Magn Reson Mater Phy 2006;**19**:180–186.

[175] Faucher L, Gossuin Y, Hocq A, Fortin M-A. Impact of agglomeration on the relaxometric properties of paramagnetic ultra-small gadolinium oxide nanoparticles. Nanotechnology 2011;**22**:295103.

[176] Park JY, Baek MJ, Choi ES, Woo S, Kim JH, Kim TJ, Jung JC, Chae KS, Chang Y, Lee GH. Paramagnetic ultrasmall gadolinium oxide nanoparticles as advanced T1 MRI contrast agent: account for large longitudinal relaxivity, optimal particle diameter, and *in vivo* T1 MR images. ACS Nano 2009;**3**:3663–3669.

[177] Faucher L, Guay-Bégin A-A, Lagueux J, Côté M-F, Petitclerc É, Fortin M-A. Ultra-small gadolinium oxide nanoparticles to image brain cancer cells *in vivo* with MRI. Contrast Media Mol Imaging 2011;**6**:209–218.

[178] Söderlind F, Pedersen H, Petoral RM Jr, Käll P-O, Uvdal K. Synthesis and characterisation of Gd2O3 nanocrystals functionalised by organic acids. J Colloid Interf Sci 2005;**288**:140–148.

[179] Ahrén M, Selegard L, Klasson A, Soderlind F, Abrikossova N, Skoglund C, Bengtsson T, Engstrom M, Kall P-O, Uvdal K. Synthesis and characterization of PEGylated Gd2O3 nanoparticles for MRI contrast enhancement. Langmuir 2010;**26**:5753–5762.

[180] Bridot J-L, Faure A-C, Laurent S, Rivière C, Billotey C, Hiba B, Janier M, Josserand V, Coll J-L, Vander Elst L, Muller R, Roux S, Perriat P, Tillement O. Hybrid gadolinium oxide nanoparticles: multimodal contrast agents for *in vivo* imaging. J Am Chem Soc 2007;**129**:5076–5084.

[181] Kryza D, Taleb J, Janier M, Marmuse L, Miladi I, Bonazza P, Louis C, Perriat P, Roux S, Tillement O, Billotey C. Biodistribution study of nanometric hybrid gadolinium oxide particles as a multimodal SPECT/MR/optical imaging and theragnostic agent. Bioconjug Chem 2011;**22**:1145–1152.

[182] Sanchez P, Valero E, Galvez N, Dominguez-Vera JM, Marinone M, Poletti G, Corti M, Lascialfari A. MRI relaxation properties of water-soluble apoferritin-encapsulated gadolinium oxide-hydroxide nanoparticles. Dalton Trans 2009;**5**:800–804.

[183] Zou Z, Zhang HL, Roditi GH, Leiner T, Kucharczyk W, Prince MR. Nephrogenic systemic fibrosis: review of 370 biopsy-confirmed cases. JACC Cardiovasc Imaging 2011;**4**:1206–1216.

[184] Kümmerer K, Helmers E. Hospital effluents as a source of gadolinium in the aquatic environment. Environ Sci Technol 2000;**34**:573–577.

[185] Rocklage SM, Cacheris WP, Quay SC, Hahn FE, Raymond KN. Manganese (II) N, N'-dipyridoxylethylenediamine-N, N'-diacetate 5, 5'-bis (phosphate). Synthesis and characterization of a paramagnetic chelate for magnetic resonance imaging enhancement. Inorg Chem 1989;**28**:477–485.

[186] Pan D, Schmieder AH, Wickline SA, Lanza GM. Manganese-based MRI contrast agents: past, present, and future. Tetrahedron 2011;**67**:8431–8444.

[187] Lyon R, Faustino P, Cohen J, Katz A, Mornex F, Colcher D, Baglin C, Koenig S, Hambright P. Tissue distribution and stability of metalloporphyrin MRI contrast agents. Magn Reson Med 1987;**4**:24–33.

[188] Zhang Z, He R, Yan K, Guo Q-N, Lu Y-G, Wang X-X, Lei H, Li Z-Y. Synthesis and in vitro and *in vivo* evaluation of manganese(III) porphyrin–dextran as a novel MRI contrast agent. Bioorg Med Chem Lett 2009;**19**:6675–6678.

[189] Schwert DD, Davies JA, Richardson N. *Non-Gadolinium-Based MRI Contrast Agents*. Berlin: Springer; 2002. p 165–199.

[190] Pan D, Caruthers SD, Senpan A, Yalaz C, Stacy AJ, Hu G, Marsh JN, Gaffney PJ, Wickline SA, Lanza GM. Synthesis of NanoQ, a copper-based contrast agent for high-resolution magnetic resonance imaging characterization of human thrombus. J Am Chem Soc 2011;**133**:9168–9171.

[191] Ward KM, Aletras AH, Balaban RS. A new class of contrast agents for MRI based on proton chemical exchange dependent saturation transfer (CEST). J Magn Reson 2000;**143**:79–87.

[192] Hancu I, Dixon WT, Woods M, Vinogradov E, Sherry AD, Lenkinski RE. CEST and PARACEST MR contrast agents. Acta Radiol 2010;**51**:910–923.

[193] Zhou J, Payen J-F, Wilson DA, Traystman RJ, van Zijl PC. Using the amide proton signals of intracellular proteins and peptides to detect pH effects in MRI. Nat Med 2003;**9**:1085–1090.

[194] Ling W, Regatte RR, Navon G, Jerschow A. Assessment of glycosaminoglycan concentration *in vivo* by chemical exchange-dependent saturation transfer (gagCEST). Proc Natl Acad Sci U S A 2008;**105**:2266–2270.

[195] Trokowski R, Ren J, Kálmán FK, Sherry AD. Selective sensing of zinc ions with a PARACEST contrast agent. Angew Chem Int Ed Engl 2005;**44**:6920–6923.

[196] Pikkemaat JA, Wegh RT, Lamerichs R, van de Molengraaf RA, Langereis S, Burdinski D, Raymond AYF, Janssen HM, de Waal BFM, Willard NP, Meijer EW, Grüll H. Dendritic PARACEST contrast agents for magnetic resonance imaging. Contrast Media Mol Imaging 2007;**2**:229–239.

9

IN VIVO MOLECULAR FLUORESCENCE IMAGING

YASAMAN ARDESHIRPOUR, VICTOR CHERNOMORDIK, MOINUDDIN HASSAN, AND AMIR H. GANDJBAKHCHE

Section on Functional and Analytical Biophotonics, Program of Pediatrics Imaging and Tissue Sciences, Eunice Kennedy Shriver National Institute of Child Health and Human Development, National Institutes of Health, Bethesda, MD, USA

DAN SACKETT

Section of Cell Biophysics, Program in Physical Biology, Eunice Kennedy Shriver National Institute of Child Health and Human Development, National Institutes of Health, Bethesda, MD, USA

9.1 INTRODUCTION

In the last few decades, many academic and industrial groups have focused their research on creating new approaches and concepts in nanomaterials. The healthcare expects to benefit most from these approaches by using nanoparticles both for drug delivery and as imaging agents [1, 2]. Currently the traditional therapy of many diseases, including cancer, suffers from late diagnostics and inadequate drug delivery. Increasing the dose of drugs to overcome their poor delivery often results in systemic toxicity, generating adverse side effects. The other weakness of current traditional therapy is the lack of timely monitoring of the therapeutic response especially at the early stages of treatment. Nanoparticles can overcome these problems.

By definition, nanoparticles are a group of materials with a unit size in the 1–100 nm range. Due to the laws of quantum mechanics that apply to the materials at nano-scale range and the increased in their surface to volume ratio; optical, magnetic,

Nanotechnology for Biomedical Imaging and Diagnostics: From Nanoparticle Design to Clinical Applications, First Edition. Edited by Mikhail Y. Berezin.

electrical, and chemical behaviors of nano-scale range materials are different from the same materials at larger scale range. Nanotechnology is a good candidate to solve fundamental problems of the current drug industry such as poor solubility and lack of target specificity. Since the size of the smallest capillaries is around 5–6 μm, nanoparticles are a promising agent for intravenous drug and imaging agent delivery. In cancer, in particular, due to leakiness in the tumor vascularization, these particles can easily penetrate the tumor and be accumulated inside [3]. Nanoparticles can be used to internalize drugs through epithelial and endothelial barriers. For example, due to the blood–brain barrier, many therapeutics and contrast agents cannot reach the brain; therefore, smaller molecules and nanoparticles are needed to pass through these barriers [4]. Another advantage of nanoparticles is their capability to deliver drugs specifically to particular cells, locations, or organs. They are also capable of codelivering two or more drugs at the same time.

Recent discoveries in cancer and other disease-related biomarkers and the successful treatments by drugs that selectively target those biomarkers (such as monoclonal antibodies (mAbs) [5, 6]) have created a new demand for developing better molecular imaging probes and improving the *in vivo* imaging systems. These imaging techniques can be used to trace the drug delivery path inside the body and to monitor the efficacy of therapeutic agents, especially at the early stages of treatment. In clinical studies, the current gold standards for the detection of disease-specific biomarkers are primarily based on *ex vivo* methods, such as immunohistochemistry (IHC), fluorescence *in situ* hybridization (FISH), and enzyme-linked immunosorbent assay (ELISA) [7–9]. All these methods require several biopsies from the lesion. However, usually due to heterogeneity of the lesions, a few biopsies cannot cover the whole lesion, and some parts may be left uncharacterized. Also, during the therapeutic cycles, the number of times that a biopsy can be taken is limited. The current goal is to replace these invasive methods with noninvasive techniques to image the tissue while it is still intact and in its natural environment.

In many cases, there is a need to use contrast agents to improve the signal-to-noise ratio and specificity to the targets, since it is difficult to distinguish between the tumor and benign tissues only from their structural differences. Targeted molecular probes can be used to differentiate the tumor from normal tissues, based on their molecular specifications, *in vivo* [10, 11] or in clinical surgery [12–14].

The design of each contrast agent depends strongly on the imaging technique. In optical imaging, each imaging technique requires different characteristics for its contrast agent. For example, optical coherence tomography (OCT) images the back reflection photons from different layers of the tissue; therefore, it needs contrast agents that can improve the scattering properties of the target. On the other hand, high absorption and high quantum yield are desired properties for the contrast agents in photoacoustic tomography [15–17] and fluorescence imaging, respectively.

This chapter focuses mainly on *in vivo* optical imaging techniques for fluorescent nanoparticles as exogenous contrast agents. These noninvasive *in vivo* methods allow the detection of the emitted photons from fluorescent contrast agents inside the tissues. Exogenous fluorescent probes can be designed as nontargeted contrast agents, which circulate intravenously inside the body and accumulate in malignant lesions due to the leakiness of their vascularization or as targeted probes, which can attach to specific disease biomarkers [18, 19]. The targeted probes can bind to the disease-related biomarkers specifically and stay longer, thus improving the signal-to-background ratio significantly. Fluorescent probes, based on their structure, can also provide new information about tissue microenvironment, such as pH, temperature, or hypoxia [20–24].

The characteristics of a suitable imaging agent in general and of a fluorescent imaging agent in particular can be summarized as follows [25]:

- High accumulation in the target region with good specificity: Have a stable binding to a specific biomarker of the disease
- Fast washout time from the blood and normal tissues and other organs (however, the clearance of the probes from the body has to be slow enough to allow accumulation in the target site for sufficient binding of the probes to the receptors)
- High sensitivity: Generates sufficient contrast and signal-to-background ratio at low concentration levels (in nanomolar to micromolar range) in the region of interest (ROI) in the minimum amount of time
- Has minimum effect on the normal tissue and organ, low level of toxicity, and side effects
- Does not interfere with drugs and therapies [19]

Fluorescence imaging has the potential to monitor multiple biomarkers simultaneously by using different fluorescent probes with different emission wavelengths [26, 27]. Multicolor imaging can be used to characterize simultaneously several disease biomarkers and/or to evaluate the effect of multiple drugs *in vivo*. It can also be used to study the pharmacokinetics of several drugs or probes with different clearance rates at the same time, which is unique among imaging modalities.

This chapter is organized as follows: Section "Basics of fluorescence" introduces the physics behind fluorescence imaging. In Section "General behavior of light in biological tissue", the optical tissue properties and tissue photon interaction are discussed. Techniques being employed in diffuse optical fluorescence imaging instrumentations are reviewed in Section "Diffuse fluorescence optical imaging instruments." Forward models, in Section "Modeling of light propagation in tissue," and planar and tomography reconstruction algorithms, in Section "Imaging algorithms" are briefly introduced. They are followed by a summary in Section "Summary."

9.2 BASICS OF FLUORESCENCE

Some materials can emit photons at a different wavelength when they absorb light or other electromagnetic radiation. Fluorescence can be described as absorption of a photon at one wavelength and re-emission at a different, usually longer wavelength. Molecules that are able to absorb and emit photons are known as fluorophores. Figure 9.1

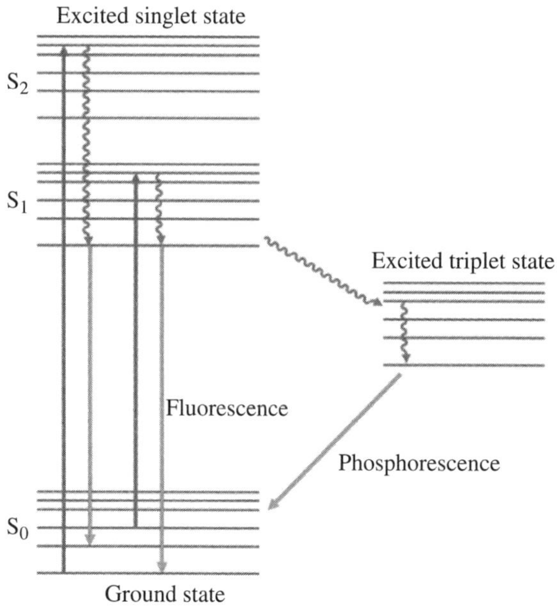

FIGURE 9.1 Jablonski diagram.

shows the Jablonski diagram, which demonstrates the different transitions of an electron after absorption of a photon [28].

A photon can be absorbed by an electron in the ground state and transfer the electron to a higher energy state in the excitation band. This transition is very fast and usually takes about 10^{-15} s. If the electron in the excited state spins in the opposite direction of the electron in the ground state, it can return back to the ground state rapidly by emission of a photon. This phenomenon is called fluorescence, and its emission time is between 10^{-9} and 10^{-7} s. If the excited electron is in triplet state and has the same spin direction of the electron in the ground state, then its return to the ground state is forbidden and usually takes about 10^{-3}–1 s until it returns to ground state and releases its energy as a photon. This phenomenon is called phosphorescence. The energy of electron transitions between different vibrational levels in a single energy state is released as thermal energy.

In most cases, emitted photons have longer wavelength and lower energy than the absorbed photons. However, when the excitation light is very intense, it is possible for one electron to absorb two photons. This two-photon absorption can lead to emission of a photon with a shorter wavelength than the originally absorbed photons.

Excitation wavelength is usually independent of the emission wavelength. Fluorescence spectrum, when plotted as a function of wavelengths, is generally a mirror image of its absorption (excitation) spectrum (Fig. 9.2). This is due to the similar spacing in the energy structures of the ground and excited energy levels and the intermediate vibrational relaxation of electron in the subbands of the ground and excited states.

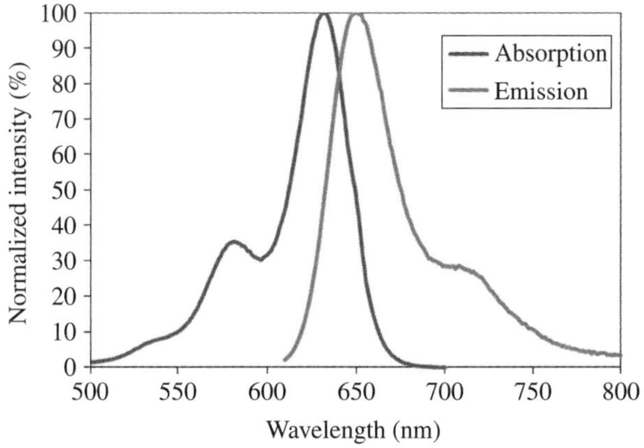

FIGURE 9.2 Excitation and emission spectrum of Alexa Fluor 633 fluorescent dye.

The characteristics of a fluorophore are usually defined by molar extinction coefficient (molar absorptivity), maximum excitation (absorption) wavelength, maximum emission wavelength, quantum yield, and lifetime. The maximum excitation wavelength determines the penetration depth of photons inside the tissue. The *molar extinction coefficient* demonstrates the light absorption capability of a fluorophore. Fluorophores with a higher molar extinction coefficient have stronger capability to absorb photons. The *fluorescence quantum yield* of a fluorophore is the ratio of the number of emitted to the number of absorbed photons. Fluorescent probes need to have high quantum yield to generate a brighter fluorescence signal. *Fluorescence intensity* is proportional to the product of its extinction coefficient and quantum yield. *Fluorescence lifetime* is the average time that an excited fluorophore stays in the excitation state before its transition to the ground state followed by the emission of a photon. Excited state lifetimes can change with fluorophore microenvironment.

Fluorescence quenching refers to any events that can decrease the fluorescence intensity of a fluorophore. It can be caused by a wide variety of processes and is often categorized as dynamic and static quenching. Dynamic quenching happens when a fluorophore is in contact with another molecule that acts as a quencher and when quenching happens by transient collisional interaction between excited state of fluorophore and ground state of quencher. In this case, the excited state of fluorophore overlaps with the quencher's ground state, and its electron at the excited state transfers its energy nonradiatively to the quencher's ground state and moves the electron to quencher's excited state. In this case, the energy can transfer from the quencher's excited state to its ground state as nonradiative (dark) or as a photon at the quencher's emission wavelength. The final product acts as a fluorescence agent with excitation emission spectra of an original fluorophore and emission spectra of a quencher (Fig. 9.3). Dynamic quenching is extremely dependent on the molecular distance between fluorophore and quencher. It typically occurs in distances up to 100Å. Oxygen, halogens, amines, iodide ions, and electron-deficient molecules like acrylamide are common chemical quenchers.

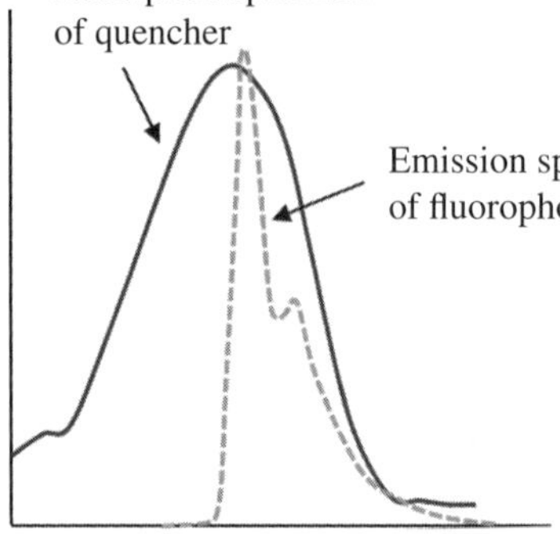

FIGURE 9.3 Dynamic quenching occurs when the absorption spectrum of a quencher overlaps the emission spectrum of a fluorophore. To generate photons at the emission wavelength of a quencher, the distance between the fluorophore's and the quencher's molecules should be smaller than 100Å.

Static quenching happens when a complex formation in ground state of a fluorophore and a quencher occurs. In this case, the complex is a completely new material with its own unique properties and excitation and emission spectra that are different from the original fluorophore's and quencher's spectra.

Dynamic quenching is the basis for fluorescence resonance energy transfer (FRET) assays and fluorescent probes that can turn on in specific conditions and turn off in others. The fluorophore and quencher are usually referred to as donor and acceptor molecules. Dynamic quenching is the basis for activatable optical contrast agents for molecular imaging [25, 29]. Using activatable fluorescent probes that can turn on and turn off can improve the image quality and signal-to-background ratio.

The parameters that have to be considered in designing a fluorescent probe for biomedical applications, in addition to its optical properties, are stability, toxicity, and molecular size. The molecular size of a fluorescent probe is important since it determines the pharmacokinetics and clearance time of the probe from the blood circulation and normal tissues.

Biological tissues contain many endogenous fluorophores such as amino acids, lipofuscins, melanin, collagen, elastin, nicotinamide adenine dinucleotide (NADH), and flavins [30, 31]. These fluorophores mainly fluoresce in ultraviolet or visible wavelengths. In near-infrared light, the autofluorescence of the tissue is negligible, and in many *in vivo* systems, it is below the noise level. Therefore, using exogenous fluorophores that emit in near-infrared spectrum has a benefit of no interference with autofluorescence of tissue.

An alternative approach to fluorescence probes is phosphorescence probes [32, 33]. Phosphorescence probes have a broader temporal response (microsecond range) compared to fluorescent probes (nanosecond range). Due to their longer temporal response, the separation of the phosphorescence signal from the excitation and autofluorescence signals (background noise) can be obtained by simple time gating. In addition, implementation of a time-resolved phosphorescence imaging system is simpler and less expensive than a time-resolved fluorescence system.

9.3 GENERAL BEHAVIOR OF LIGHT IN BIOLOGICAL TISSUE

This section describes the mechanisms that govern light transport through biological tissues. The main optical characteristics of biological tissue defined by absorption coefficient, scattering coefficient, and refractive index are discussed in the following text.

9.3.1 Absorption

Absorption is a process of absorbing the energy of a photon by a molecule to transfer its electron from a lower energy level to a higher energy level. Mean absorption path length is the mean free path that a photon travels in a medium before absorption. Absorption coefficient μ_a is defined as reciprocal of mean absorption path length and is the probability of a photon being absorbed by a medium per unit path length.

If the media contains several chromophores with different extinction coefficient, the mean absorption coefficient can be defined as [34]

$$\mu_a = \sum_n \varepsilon_n(\lambda) C_n \tag{9.1}$$

where n is the number of chromophores, $\varepsilon_n(\lambda)$ and C_n are the extinction coefficient and concentration of each chromophore, respectively. According to the Bear–Lambert law, light attenuates as it propagates in an absorbing media (nonscattering) based on the following equations:

$$\frac{dI}{I} = -\mu_a dx, \qquad I(x) = I_0 \exp(-\mu_a x) \tag{9.2}$$

where $I(x)$ is the light intensity at the distance x from the source, along the light propagation direction. I_0 is the light intensity at the source location ($x=0$).

Optical absorption in biological tissue (in NIR region) originates primarily from oxy- and deoxyhemoglobin, melanin, lipid, and water [35, 36]. Figure 9.4 shows the absorption spectra of some of these components in the NIR window.

Since the absorption coefficient of tissue chromophores is low in the NIR region (650–900 nm) compared to other wavelengths, light can penetrate deeply (a few centimeters) into tissue. At wavelengths lower than 650 nm, absorption by hemoglobin limits the penetration of photons into tissue, and at wavelengths higher than 900 nm, absorption by water dominates.

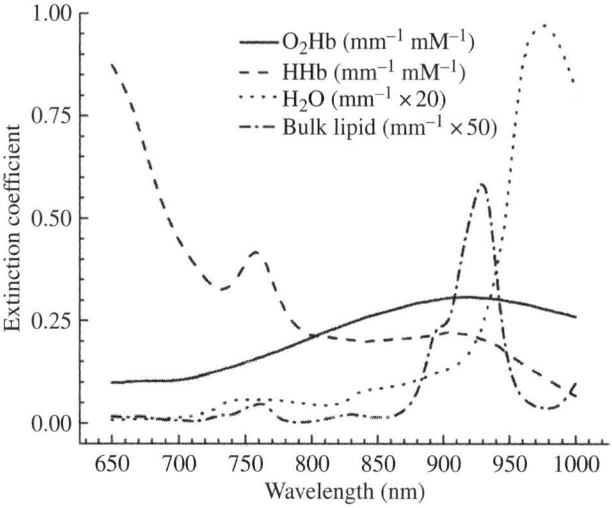

FIGURE 9.4 Absorption spectra of main chromophores in biological tissue in the NIR window. (Reprinted with permission from Ref. [37]. © IOS Press.)

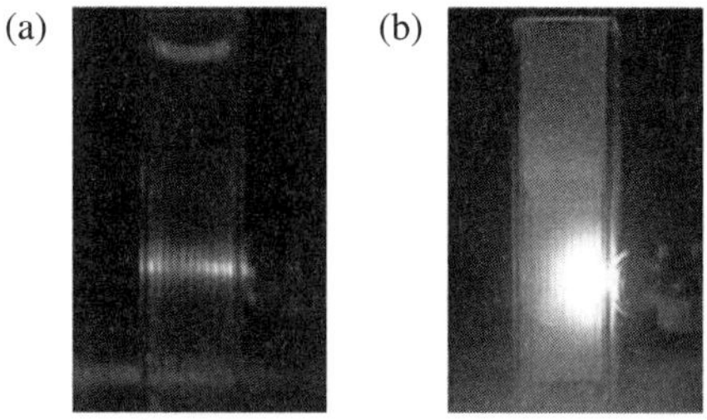

FIGURE 9.5 Light propagation in (a) nonscattering and (b) high scattering media.

9.3.2 Scattering

In nonscattering media, light propagates along straight lines. Therefore, any occlusion in the media can be imaged or reconstructed with a simple projection or back projection algorithm similar to X-ray or computed tomography (CT). However, due to the highly scattering nature of biological tissues, photons change their propagation directions randomly throughout the media, and finding the occlusions inside the media is not straightforward as in the case of nonscattering media. Figure 9.5 shows the light propagation path in (i) nonscattering and (ii) high scattering media. Random propagation of light in high scattering media significantly decreases the resolution in the deep tissue images.

The light source can illuminate the media in different modes, that is, a continuous-wave (CW), short pulse, or frequency-modulated signal. The effect of scattering media on a CW is the same as absorption and attenuates the received photons to the detector. The photons propagate in all directions inside the media, and not all of them can reach the detector.

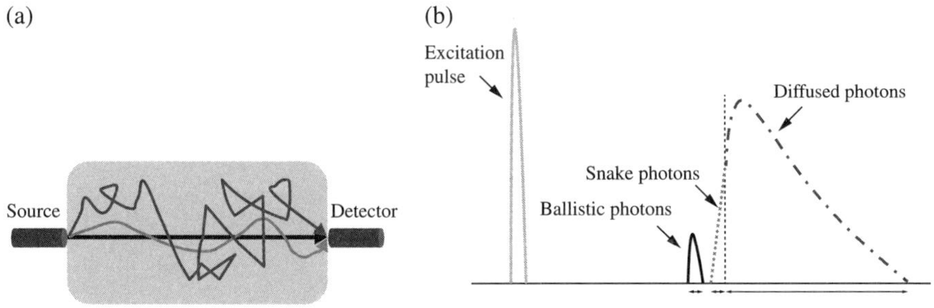

FIGURE 9.6 (a) Ultrashort laser pulse propagation through scattering medium. (b) The photons received by a detector can be categorized as ballistic photons (photons travel along a straight line), snake photons (photons travel slightly off a straight line path and go through very small scattering), and diffused photons (photons undergo random walks and arrive last).

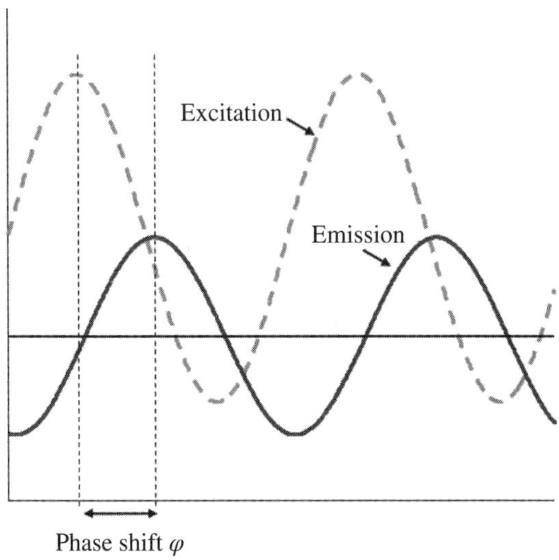

FIGURE 9.7 Effect of scattering media on a frequency-modulated optical signal.

Figure 9.6 shows the effect of the scattering property of media on a short incident light pulse. The scattering broadens the pulse in time domain (TD) and the photons received by a detector with different delay times. The received photons can be categorized as ballistic photons (photons travel along a straight line), snake photons (photons travel slightly off a straight line and go through few scattering events), and diffused photons (photons go through random walks and multiple scattering and arrive last).

In frequency domain (FD), the scattering property of media can shift the phase of the frequency-modulated signal. It also attenuates the detected signal, since the emitted photons scatter in all directions and not all of them can reach the detector (Fig. 9.7).

The mean scattering free path is the mean distance a photon travels in each scattering event in a medium. Scattering coefficient μ_s is defined as the reciprocal of mean scattering path length and is the probability of a photon being scattered in a medium per unit path length.

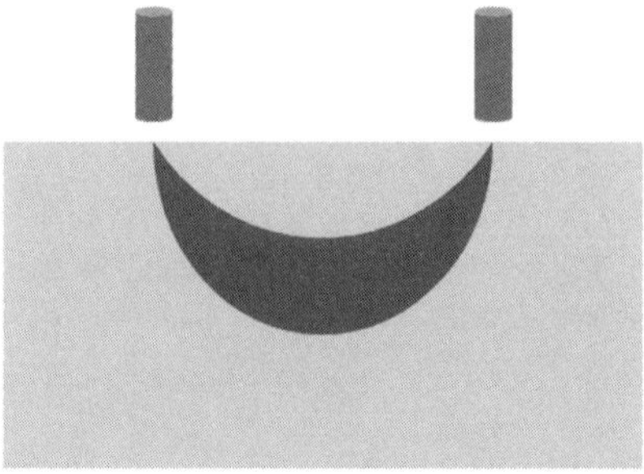

FIGURE 9.8 Reflection light path in high scattering media.

The scattering of light by a spherical particle can be modeled by the Mie theory. This model can be simplified by Rayleigh approximation for the particles whose size is much smaller than the wavelength of light (Rayleigh region). Scattering of light by particles much larger than the light wavelength is more directional (anisotropic) and is usually classified as the Mie region.

In addition to scattering coefficient, the other parameter affecting the scattering of the photons in a medium is the scattering anisotropy (g). The scattering anisotropy is the mean cosine of the scattering angle in a medium. It has a value between −1 and 1. If the optical anisotropy is 0, the photon scattering is isotropic. The photons propagate in the forward/backward direction, if the optical anisotropy has a value equal to 1/−1. Figure 9.8 shows the light propagation path from a source to a detector through a medium with optical properties similar to biological tissues. Often, the Henyey–Greenstein phase function is used to model the direction of the photon packets in scattering media. In this model, the scattering angle, θ, is determined using the following formula [38]:

$$\cos\theta = \begin{cases} \dfrac{1}{2g}\left[1+g^2-\left(\dfrac{1-g^2}{1-g+2g\xi}\right)^2\right] & \text{if } g \neq 0 \\ 2\xi - 1 & \text{if } g = 0 \end{cases} \tag{9.3}$$

where $\xi = \varphi/2\pi$ and the azimuthal angle φ is generally assumed to be uniformly distributed between 0 and 2π.

In biological tissue, the scattering is often defined by another parameter as reduced scattering coefficient

$$\mu_s' = \mu_s(1-g)\left(\text{cm}^{-1}\right) \tag{9.4}$$

The purpose of using μ_s' is to describe the diffusion of photons as a random walk of step size of $1/\mu_s'$ (cm) where each step goes through an isotropic scattering. Such a description is equivalent to photon migration, involving many small steps $1/\mu_s$ with a partial deflection angle θ.

On the other hand, the reduced scattering coefficient can be modeled based on the Mie scattering theory as [34, 39]

$$\mu_s' = A\lambda^{-b} \tag{9.5}$$

where A is the scattering prefactor, b is the scattering power, and λ is the wavelength of the light source. A and b are functions of the size, refractive index, and concentration of scatters in biological tissues as well as the refractive index of the surrounding media. The value of b varies from 4 to 0.37 for scatters with the radius of 0.05–1 μm.

9.3.3 Refractive Index

The refractive index determines the speed and wavelength of light in the medium. Changes in refractive index in the tissue can cause scattering, refraction, or reflection of light passing through the tissue. In homogenous media, refractive index $n(\lambda)$ can be modeled as

$$n(\lambda) = n_R(\lambda) - in_I(\lambda) \tag{9.6}$$

where $n_R(\lambda)$ and $n_I(\lambda)$ determine the speed and attenuation of light in the tissue, respectively. The speed and wavelength of light change with refractive index as follows:

$$\lambda = \frac{\lambda_0}{n_R} \text{ and } C = \frac{C_0}{n_R} \tag{9.7}$$

where λ_0 and c_0 are the wavelength and speed of light in a vacuum, respectively. The refractive indices of different biological tissues are in the range of 1.33–1.62.

9.4 DIFFUSE FLUORESCENCE OPTICAL IMAGING INSTRUMENTS

Four essential components exist in any fluorescence imaging system: a light source, a detector, required optics (including fluorescence emission filter), and electronic parts. The commonly used options for the light source are light-emitting diodes (LEDs) and lasers. An emission filter is used to eliminate the excitation signal from the emitted fluorescence signal. The commonly used detectors are avalanche photodiodes (APD), photomultiplier tubes (PMT), and charge-coupled device (CCD) cameras. The choice and specifications of source, detector, and required electronic parts depend on many factors including measurement schemes, size, and cost of the system.

In terms of measurement schemes, diffuse fluorescence imaging instruments can be classified into three different categories: CW systems, FD systems, and TD systems. CW imaging can only capture the fluorescence intensity and does not provide any information on fluorescence lifetime. Therefore, to reconstruct the fluorescence lifetime, either TD or FD techniques have to be used. Fluorescence lifetime can provide useful clinical information because it is potentially sensitive to local biochemical environment, for

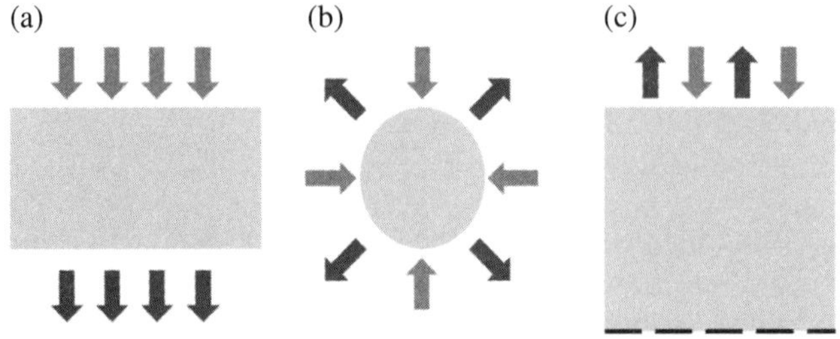

FIGURE 9.9 Different imaging probe configurations, (a) transmission mode, (b) transmission and reflection mode, and (c) reflection mode.

example, temperature and pH, or molecular interactions [40, 41]. On the other hand, its value does not depend on the concentration of the fluorophores or the intensity of the excitation light (before saturation) [28]. The sensitivity of fluorescence lifetime to biochemical environment is based on the structure of the dye. To evaluate specific properties of the probe environment in tissue, it is important to choose a proper dye with lifetime sensitivity that is dominated by only one specific factor.

In terms of imaging probe geometry, light can transmit through the tissue in reflection mode, in transmission mode, or in both reflection and transmission mode (Fig. 9.9). In transmission mode (transillumination), the source and detector are placed on opposite sides of the tissue; and the detector captures the transmitted photons through the tissue. In reflection mode (epi-illumination), the source and detectors are placed on the same side of the tissue, and the reflected fluorescence signal from the tissue is captured. In many medical applications, using transillumination is not possible, and the source and detectors have to be placed on the same side of the object. It is also possible to position the source and detectors (e.g., in a ring geometry) so that the fluorescence signal can be captured in both reflection and transmission mode.

To detect the fluorescence signals coming from different depths in epi-illumination mode, multiple sources and detectors are required to be positioned at different distances. This is due to the banana-shaped pattern of light propagation in high scattering media (Fig. 9.8). As the distance between source and detector increases, the center of the path covers a deeper area inside the tissue. Owing to the diffusion of excitation and emission light in tissue, epi-illumination can image fluorescence activities at depths ranging from millimeters to a few centimeters, based on the distance between sources and detectors and the excitation and emission wavelengths of the fluorophore.

Data can be collected either by scanning of the ROI by the source–detector pairs or by the area detection, using a camera. A mechanical scanner or a mirror with galvanometer is usually used to scan the ROI.

The most common fluorescence imaging scheme is planar imaging. This method projects all the fluorescence light on the tissue surface without considering the depth of fluorescence target. It is fast and easy to implement and works very well on superficial fluorescence targets. However, in deep tissue imaging, even though the fluorescence intensity is a linear function of fluorophore concentration, it depends nonlinearly on optical properties of tissue and the fluorophore depth inside the media.

Therefore, three-dimensional (3D) localization of the fluorophore and quantitative information about fluorophore concentrations are hard to estimate. For example, two fluorescence targets located at different depths in a high scattering media project with different intensity in epi-illumination geometry. On the other hand, the fluorescence target can be shadowed or attenuated by an absorbing occlusion on a shallower depth. Different planar imaging methods are introduced in Section "Modeling of light propagation in tissue."

A more accurate method for reconstruction of fluorophore concentrations inside a medium is fluorescence tomography. Tomography reconstructs the intensity and lifetime of fluorophores inside the tissue by using a light propagation model and information about positions of the source and detector as well as optical properties of the medium. More information on tomography methods will be provided in Section "Imaging algorithms."

9.4.1 CW Domain

The most common fluorescence imaging technique is CW fluorescence imaging. This method uses a CW or very-low-frequency-modulated light source to provide the excitation light. The intensity of the reflected and\or transmitted fluorescence signal is detected by a CCD camera, APD, or PMT. Implementation of this method is the least expensive and easiest to arrange compared to other fluorescence imaging techniques. The disadvantage of this method is that it only captures the intensity information of the fluorescence signals. Captured fluorescence intensity data are sensitive to the fluctuations of the excitation light, distance of the probe from the fluorophore, and optical parameters of the system.

9.4.2 TD

TD instruments use femtosecond or picosecond laser pulse sources to measure the temporal distribution of emitted fluorescence light, known as temporal point spread function (TPSF). The detector should provide a very fast time response and can be implemented by a time-gated intensified charge-coupled device (ICCD) or a fast PMT with a time-correlated single-photon counter (TCSPC). TD systems can be used to reconstruct both fluorescence intensity and lifetime.

Figure 9.10 shows the schematic of a small animal fluorescence imager consisting of a CW and a TD fluorescence system [42]. A cooled CCD camera with a field of view of $12 \times 12\,cm^2$ captures the fluorescence images in CW mode with a band-pass filter ($800 \pm 20\,nm$) to find the location of the tumor (the ROI) and locate the probe of the TD system on the ROI. The same camera can capture the white-light image of the small animal after removing the band-pass filter.

The excitation source is a tunable Ti–sapphire pulse laser with a pulse width of 100 fs and repetition rate of 80 MHz (Tsunami, Spectra Physics, Mountain View, CA). The laser peak can be tuned based on the excitation peak of fluorescent dye. The femtosecond laser pulse scans the target in a raster pattern through a scanning head with the one source and four detector fibers at 2–8 mm distances. The integration

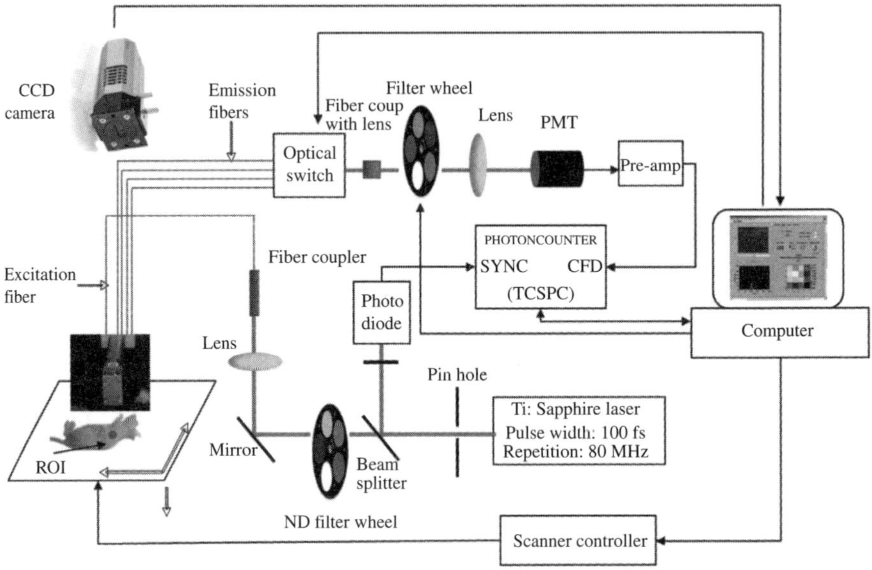

FIGURE 9.10 Schematic of a time-resolved small animal imaging system. (With permission from Ref. [42]. © Society of Molecular Imaging.)

time for each pixel can be selected based on the maximum intensity of the detected fluorescence signal and the saturation point of the PMT. The reflected fluorescence signal is filtered by a high-pass emission filter at maximum emission wavelength of fluorophore and is collected by a PMT. Detected photons are counted by a TCSPC. System initialization, data acquisition, and scanning are controlled by the Labview software.

9.4.3 FD

The performance of FD systems can be linked to that of the TD systems via the Fourier transform [43]. In the FD technique, the excitation light intensity is sinusoidally modulated at hundreds of megahertz (MHz). Passing modulated light through tissue changes the amplitude and phase of the fluorescence signal. FD systems can be used to reconstruct both fluorescence intensity and lifetime.

Figure 9.11 shows the schematic of a FD fluorescence imaging system. The excitation source is a frequency-modulated laser diode at 100.02 MHz frequency. After being captured by optical fibers, the emitted fluorescence signal, filtered by an emission filter, is collected by a PMT, and its frequency is downconverted by a mixer and a local oscillator at 100 MHz frequency. The output signal of mixer is filtered by a band-pass filter with 20 kHz center frequency to extract the amplitude and phase of the fluorescence signal and remove the DC and high-frequency components. A second mixer extracts the phase difference between source and detector oscillators and generates a 20 kHz SYNC signal as a reference. System initialization and data acquisition are controlled by the Labview software.

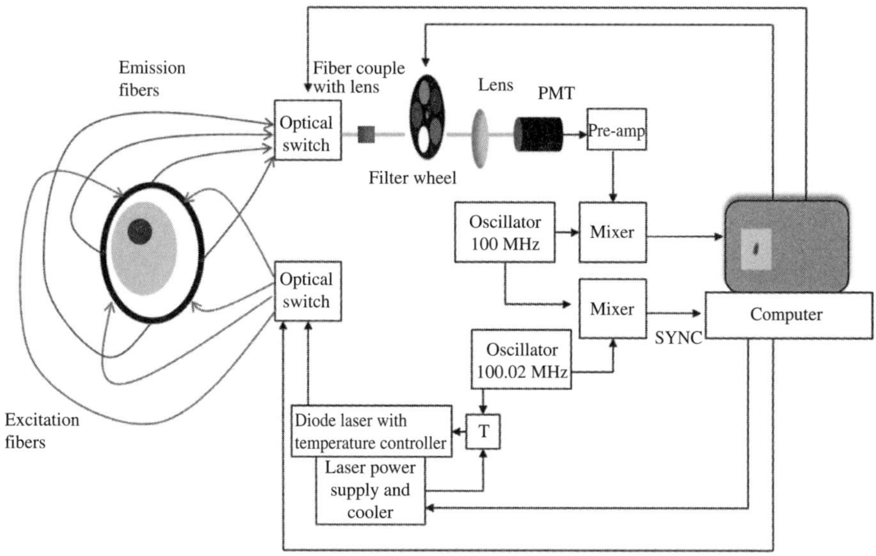

FIGURE 9.11 Schematic of a frequency-modulated system.

In practice, implementation of the FD is usually limited to one or a few modulation frequencies. Therefore, the TD data can provide more information about the media, since a short laser pulse contains a broader range of the modulation frequencies, including the zero-frequency component. On the other hand, instrumentation for time-resolved imaging is considerably more expensive, and data collection time is longer relative to a FD system. Therefore, in applications that need very fast imaging, the FD technique is more suitable.

9.5 MODELING OF LIGHT PROPAGATION IN TISSUE

This section briefly discusses forward models of light propagation in tissue followed by light propagation of fluorescence signal in high scattering media and different simulation algorithms to implement these models.

9.5.1 Forward Model for Light Propagation in Tissue

In highly scattering media, the light propagates in random directions. However, the path of photon propagation from a source to a detector can be statistically modeled with mathematical formulas. Photon transport in scattering media, like tissues, can be modeled with the radiative transport equation (RTE)

$$\frac{1}{c}\frac{\partial L(r,\hat{s},t)}{\partial t} + \nabla . L(r,\hat{s},t)\hat{s} = -(\mu_s + \mu_a)L(r,\hat{s},t) + \mu_s \iint_{4\pi} L(r,\hat{s}',t) f(\hat{s},\hat{s}') d\Omega' + Q(r,\hat{s},t) \tag{9.8}$$

where $L(r,\hat{s},t)$ is the radiance, $\hat{s}$ is the unit vector pointing in the direction of interest, ω is a solid angle, and μ_s and μ_a are the scattering and absorption coefficients, respectively. $f(\hat{s},\hat{s}')$ is the normalized probability for a photon traveling in the direction of $\hat{s}$ and being scattered in the direction $\hat{s}'$, which satisfies $\iint_{4\pi}(\hat{s},\hat{s}')\,d\Omega' = 1$. $Q(r,\hat{s},t)$ is the source and defines as the power injected from the direction $\hat{s}$ into a unit solid angle centered at r [44].

By integrating the transport equation over all solid angles, Equation 9.8 can be simplified as

$$\frac{1}{c}\frac{\partial\varphi(r,t)}{\partial t} + \nabla.j(r,t) = -\mu_a\varphi(r,t) + S(r,t) \tag{9.9}$$

where the source S, photon density (fluence rate) $\varphi(r,t)$, and the flux j are defined as

$$S(r,t) = \iint_{4\pi} Q(r,\hat{s},t)\,d\Omega \tag{9.10}$$

$$\phi(r,t) = \iint_{4\pi} L(r,\hat{s},t)\,d\Omega \tag{9.11}$$

$$j(r,t) = \iint_{4\pi} L(r,\hat{s},t)\hat{s}\,d\Omega \tag{9.12}$$

For some geometries, the general closed-form solutions of the RTE are not available, and the closed forms, based on diffusion approximation, are usually applied. If the reduced scattering coefficient is much larger than the absorption coefficient (in other words, if many scattering events happen before an absorption event, i.e., $\mu_a << \mu_s'$), the transport equation can be simplified to the diffusion equation

$$\nabla.D\nabla\varphi(r,t) - \mu_a\varphi(r,t) + \frac{1}{c}\frac{\partial\varphi(r,t)}{\partial t} + S(r,t) = 0 \tag{9.13}$$

where $D = \left(1/3[(1-g)\mu_s + \mu_a]\right) = \left(1/3(\mu_s' + \mu_a)\right)$ is the photon diffusion coefficient, g is the mean cosine of the scattering angle, and μ_s' is the reduced scattering coefficient. In the FD, the diffusion formula can be written as

$$-\nabla.D\nabla\phi(r,\omega) + \left(\mu_a - \frac{j\omega}{c}\right)\phi(r,\omega) = S(r,\omega) \tag{9.14}$$

where ω is the angular modulation frequency. For CW mode, $\omega = 0$ and the diffusion equation simplifies to

$$-\nabla.D\nabla\varphi(r) + \mu_a.\varphi(r) = S(r) \tag{9.15}$$

The diffusion approximation is not accurately very close to the source and/or boundaries and very close source–detector pairs [45].

9.5.2 Boundary Conditions

Based on the Fresnel reflection theory, if the reflection indices of the strongly scattering medium and the boundary medium are different, a fraction of the radiation energy will be reflected back into the scattering medium. If we set the irradiance at the boundary equal to the integral of the reflected radiance, the extrapolated boundary conditions [46] in terms of the fluence rate and its normal derivative will be equal to

$$\hat{n}.D\nabla\phi(\bar{r},t)+q\phi(\bar{r},t)=0 \tag{9.16}$$

where $\hat{n}$ is the outward normal unit vector at the boundary and q is a positive number related to the internal reflection at the boundary and can be defined by

$$q=0.5\frac{1-R_{\text{eff}}}{1+R_{\text{eff}}} \tag{9.17}$$

R_{eff} is the effective reflection coefficient and depends on the refractive index of two media. In a semi-infinite absorption boundary condition, all the photons are absorbed by the boundary and nothing is reflected back to the media, $R_{\text{eff}}=0$ and $q=0.5$.

9.5.3 Forward Model for Fluorescence Light Propagation in Tissue

The forward model of a fluorescence light propagation in tissue consists of two stages: (i) excitation of fluorophores inside tissue (by absorbing the excitation photons), followed by (ii) emission of fluorescence light. Therefore, two coupled Boltzmann or diffusion equations are needed to describe the propagation of (i) excitation photons and (ii) emission photons in the tissue [47]:

$$\left[\nabla.D_{\text{x}}(r)\nabla-(\mu_{\text{ax}}(r)+\mu_{\text{af}}(r))+\frac{1}{c}\frac{\partial}{\partial t}\right]\varphi_{\text{x}}(r,r_{\text{s}},t)=-S(r_{\text{s}},t) \tag{9.18}$$

$$\left[\nabla.D_{\text{m}}(r)\nabla-\mu_{\text{am}}(r)-\frac{1}{c}\frac{\partial}{\partial t}\right]\varphi_{\text{m}}(r_{\text{d}},r,t)=-\frac{\eta(r)\mu_{\text{af}}(r)}{\tau(r)}\left[\varphi_{\text{x}}(r,r_{\text{s}},t)\otimes\exp\left(\frac{-t}{\tau}\right)U(t)\right] \tag{9.19}$$

where the subscripts x and m indicate the variables measured at excitation λ_{x} and emission λ_{f} wavelengths, respectively. $\varphi_{\text{x}}(r,r_{\text{s}},t)$ is photon density function at time t and position r in response to the excitation source $S(r_{\text{s}},t)$ with wavelength λ_{x} at location r_{s}. $\varphi_{\text{m}}(r_{\text{d}},r,t)$ is photon density detected at location r_{d} and time t in response to fluorescence signal with wavelength λ_{f} emitted from location r. μ_{ax}, μ_{am}, and μ_{af} are the absorption coefficients of media at excitation and emission wavelengths and of fluorophore, respectively. $\eta(r)$ is the quantum efficiency and $\tau(r)$ is the fluorescence lifetime at location r. $D_x = \dfrac{1}{3(\mu_{ax}(r) + \mu'_{sx}(r))}$ and $D_m = \dfrac{1}{3(\mu_{am}(r) + \mu'_{sm}(r))}$ are the diffusion coefficients of the media at excitation and emission wavelengths. In Equation 9.19, symbol ⊗ is the temporal convolution operator, $U(t)$ is the unit-step function, and c is the speed of light in the media.

In the FD, by taking the Fourier transform of Equations 9.18 and 9.19, the diffusion formula can be written as

$$\left[\nabla.D_x(r)\nabla - \left(\mu_{ax}(r) + \mu_{af}(r) - \frac{1}{c}j\omega\right)\right]\varphi_x(r, r_s, \omega) = -S(r_s, \omega) \tag{9.20}$$

$$\left[\nabla.D_m(r)\nabla - \left(\mu_{am}(r) - \frac{1}{c}j\omega\right)\right]\varphi_m(r_d, r, \omega) = -\frac{\eta(r)\mu_{af}(r)}{1 + j\omega\tau}\left[\varphi_x(r, r_s, \omega)\right] \tag{9.21}$$

The above formula can be simplified for CW imaging by applying $\omega = 0$:

$$\left[\nabla.D_x(r)\nabla - \left(\mu_{ax}(r) + \mu_{af}(r)\right)\right]\varphi_x(r, r_s) = -S(r_s) \tag{9.22}$$

$$\left[\nabla.D_m(r)\nabla - \mu_{am}(r)\right]\varphi_m(r_d, r) = -\eta(r)\mu_{af}(r)\left[\varphi_x(r, r_s)\right] \tag{9.23}$$

In all cases, Equation 9.16 in section "Boundary Conditions" can be used to model the boundary condition.

9.5.4 Simulation Algorithms of Forward Models

In the last three decades, many computational algorithms based on analytical, numerical, and statistical models have been developed. These algorithms can be divided into three different categories. The first category includes the fastest algorithms and is based on analytical approximation of diffusion equations, such as Born or Rytov approximations [43, 48, 49]. The algorithms in the second category are based on the numerical models of diffusion approximation or radiative transport equation, using finite element method [50–54], boundary element method [55], or finite difference method [56]. These solutions are more computational intensive but are capable of modeling any arbitrary geometry and boundary conditions.

The third category includes algorithms based on statistical modeling, for example, Monte Carlo simulations and random walk theory [57, 58]. These algorithms track and record the path of individual photons in turbid media. Their results are very close to the radiative transfer equation; however, since they are based on statistical models, they require tracking a significant number of photons, a process that can be computationally intensive and time-consuming. They can be used to model any arbitrary geometry with different spatial and temporal resolutions. In these methods, a specific weight is attributed to each photon (initialized as one). As the photon moves in the turbid medium, its weight changes due to absorption. The remaining weight scatters with the photon to other parts of the medium. Local rules for photon transport in scattering media are expressed as probability distributions that describe the step size of photon movement and the angles of its deflection in each scattering event. Each photon will undergo many steps in the turbid media until it is absorbed by the media (its weight becomes zero) or exits the boundary.

9.6 IMAGING ALGORITHMS

Fluorescence imaging is usually performed based on fluorescence projection (planar imaging) or fluorescence tomography. This section briefly introduces each of these imaging algorithms. In general, the quality of any image reconstruction depends upon parameters such as signal-to-noise ratio, background fluorescence signal, optical properties of the media, and depth of the target from the optical probe.

9.6.1 Planar Imaging

Planar imaging or topography is based on projection of the detected fluorescence signal on the tissue boundary. It does not consider the effect of depth and optical properties of the tissue in a reconstructed image. This method is good for fluorescence targets that are located superficially under the skin whose depth and tissue scattering and absorption properties do not have a significant effect on fluorescence signal.

9.6.1.1 Intensity In this mode, a camera usually captures two images, one with and one without an emission filter. These two images are superimposed to obtain the fluorescence image on an anatomical (white-field) image of the tissue (Fig. 9.12). Even though this method can show the location corresponding to the fluorophores' accumulation, the fluorescence intensity cannot be used alone to quantify the

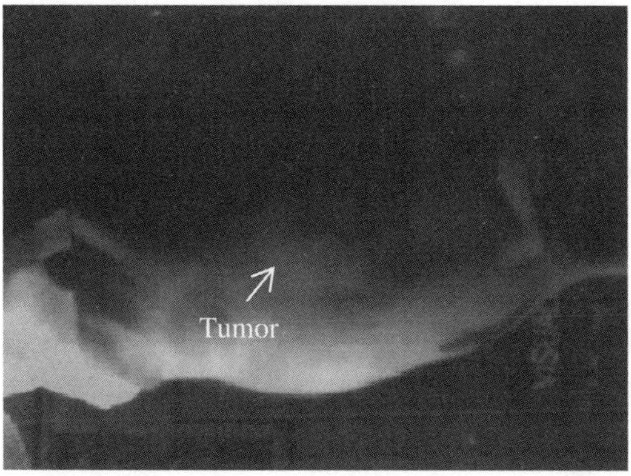

FIGURE 9.12 Superimpose of planar white-field and fluorescence images after injection of HER2-targeted fluorescence probe in a mouse with high HER2-expressed tumor. (*See insert for color representation of the figure.*)

fluorescence concentration. Fluorescence intensity is very sensitive to the intensity variation of excitation source, optical properties of the media, depth, changes in system parameters, and concentration of the injected dye.

9.6.1.2 Pharmacokinetic Information Uptake and clearance rate information of fluorescent probes in biological tissues can help to overcome the limitations that were mentioned in section "Intensity," by using fluorescence intensity alone. In 2007, Hillman [59] showed that temporal techniques can be used to decouple the fluorescence signals of each organ in a mouse body based on the differences in the accumulation and clearance rate of fluorescence probes in different organs.

In nontargeted fluorescence dye, the temporal response (pharmacokinetics) is usually based on monitoring the washout time of the dye from disease tissues (such as tumors) compared to normal tissues.

In targeted fluorescent probes that can bind to specific disease biomarkers and stay there for longer times, the fluorophore concentration can be calculated based on the compartmental ligand–receptor model [60]. Assuming that the dissociation rate of bound fluorophores is very low, the intensity of the fluorescence signal inside the tumor can be written as

$$\begin{aligned} I_{\mathrm{f}} &= I_{\text{free ligands in blood}} + I_{\text{free ligands in the tumor}} + I_{\text{bound ligand–receptor}} \\ &= \alpha F_{\mathrm{bl}} \exp\left(-\frac{t}{\tau}\right) + \beta F_{\mathrm{T}} + \gamma B_{\max}[1-\exp(-k_{\mathrm{on}} t F_{\mathrm{T}})] \end{aligned} \tag{9.24}$$

where α, β, and γ are constants; F_{bl}, F_{T}, and B are the concentrations of free ligands in the blood and free and bound ligands in the tissue; and $B_{\max}$ is the concentration of the specific receptors in the tumor region, respectively. The clearance time of the fluorophores from the blood circulation is τ, and ligand–receptor binding rate is k_{on}. In this approximation, it is assumed that after the initial time t_1, the concentration of local free ligands in the tumor stays constant and the intensity of the free ligands in blood is considered to be the same as the intensity measured at the contralateral site. Therefore, by subtracting the measurements at the contralateral site from the tumor site, the two remaining components are the free and bound ligands to the receptors in the tumor region, and the model time dependence of the bound ligands can be simplified to the form

$$y(t) = y_0 + a\left[1-\exp(-bt)\right] \tag{9.25}$$

To eliminate the system variations between different experiments, the measurement data at different time points should be normalized to the first measurement data, when binding is almost negligible and the accumulation of the free ligands in the tumor is stabilized. The derivative $\left.{}^{dy}\!/_{dt}\right|_{t=0} = ab$ presents the normalized rate of accumulation (NRA). By using a fitting algorithm, NRA can be estimated from a time series of fluorescence intensity measurements.

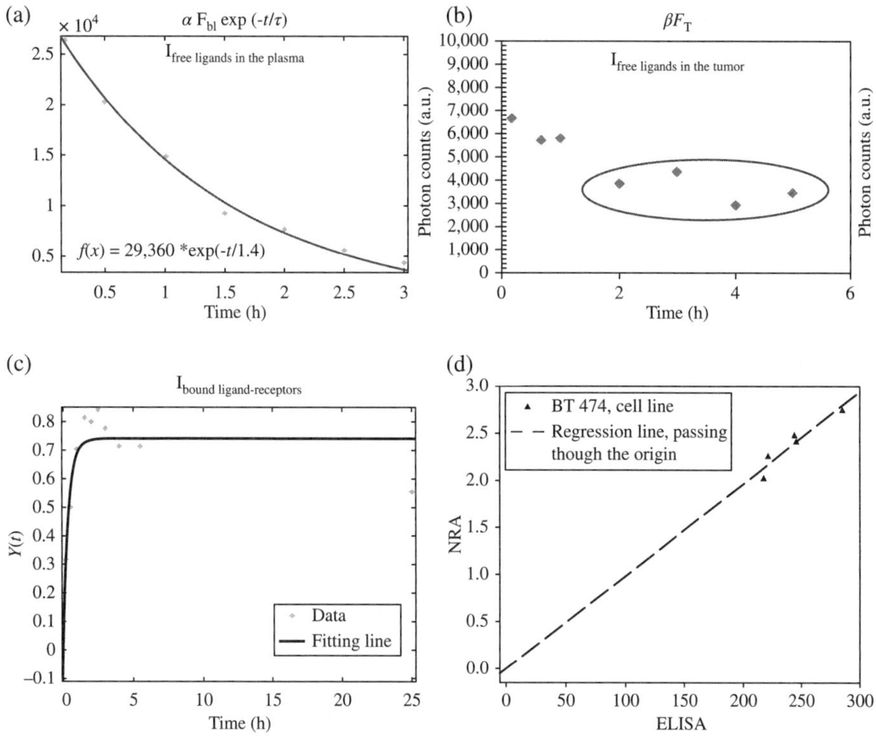

FIGURE 9.13 Pharmacokinetic information obtained from imaging a mouse with high HER2 expression tumor (BT474) after injection of HER2-specific affibody conjugated to DyLight 750. (a) Intensity measured at contralateral site as a function of time and its exponential fitting curve. (b) Concentration of free affibody probe inside the tumor. (c) NRA fitting. (d) Comparison of NRA of five tumors with a high level of HER2 receptors (BT474 tumor) with the values obtained by ELISA after excision of the tumor. (With permission from Ref. [60]. © Society of Molecular Imaging.)

Figure 9.13 shows the results obtained from imaging a mouse with high HER2 expression tumor after injection of HER2-specific affibody conjugated with DyLight 750 fluorophores. The affibody probe binds to HER2 receptors and stays inside the tumor but clears out from the blood circulation and normal tissues. Figure 9.13a shows the fluorescence intensity, measured at the contralateral site as a function of time and its single exponential fitting curve. Figure 9.13b presents the concentration of free affibody probes in the tumor. The results shown in Figure 9.13b were obtained from another mouse with the same type of tumor but after the injection of a nonspecific HER affibody probe to monitor the pharmacokinetics of free dyes inside the tumor. This figure shows the difference between the data obtained from the tumor and contralateral sites and demonstrates the accumulation of free affibody probes inside the tumor due to the leakiness of its vascularization. These results show that the concentration of free affibody probes is almost constant inside the tumor and can be modeled with the earliest measurement data when the binding of fluorescence probe to receptors is still negligible. The final NRA fitting on the fluorescence signal is shown in Figure 9.13c. Finally, Figure 9.13d shows the comparison of the NRA of several tumors with a high level of HER2 receptors (BT474 tumor) with the readings of the *ex vivo* ELISA after the tumor excision. The data show a good linear correlation between the NRA and HER2 expressions, obtained by ELISA. The ELISA was performed according to the protocol provided by the manufacturer (Calbiochem, Gibbstown, NJ), and HER2 concentration is expressed in nanograms of HER2 per milligram of total protein.

9.6.1.3 Lifetime Imaging Fluorescence lifetime images can be obtained either by TD or FD techniques. In the FD method, two fluorescence lifetimes are defined as, phase lifetime (τ_p) and modulation lifetime (τ_M). The definitions of these two lifetimes are as follows:

$$\tau_p = \frac{\tan(\varphi)}{\omega}; \tau_M = \left(\frac{1}{\omega}\right)\sqrt{\left(\frac{1}{M^2} - 1\right)} \tag{9.26}$$

where φ is the phase shift, ω is the angular modulation frequency of the excitation light, and M is the attenuation in modulation depth of the received signal [28, 61].

If the medium contains only one type of fluorophore, the phase and modulation lifetimes are equal. However, if the medium contains different fluorophores, then the value of these two parameters can be different and can be used to decouple the images of different fluorophores. In the TD method, the measured signal can be written as

$$I(t) = IRF \otimes I_0 e^{-t/\tau} \tag{9.27}$$

where IRF is the impulse response function of the system, $\otimes$ is the convolution, I_0 is the intensity of the excitation light, and τ is the fluorescence lifetime. The fluorescence lifetime can be estimated by curve-fitting procedures, such as a least-squares algorithm. If the medium contains more than one kind of fluorophore, a multiexponential curve-fitting algorithm should be used to reconstruct the images of different fluorophores.

The effect of measurement noise is less important for the TD method, since its lifetime calculation is based on the slope of the detected signal, and the slope of a signal is less susceptible to noise than its amplitude [62, 63].

Equations 9.26 and 9.27 are valid if the effects of optical properties of the media on measured fluorescence lifetime τ' are negligible. If the fluorophore is embedded deeper than several scattering lengths $l_s \sim 1/\mu_s'$ in the turbid medium, measured values of τ are larger than intrinsic fluorescence lifetime τ. In this case, the time-resolved fluorescence intensity distribution presents the convolution of the exponential decay curve, similar to Equation 9.27, and time-resolved and Green functions, describing photon migration from the source to the fluorophore inside the turbid medium and from the fluorophore to detector (see, e.g., Refs. [64–66]). In shallow tumors such as skin or head and neck tumors, the exponential approximation of Equation 9.27 can still be applied. However, for applications dealing with the deeply embedded

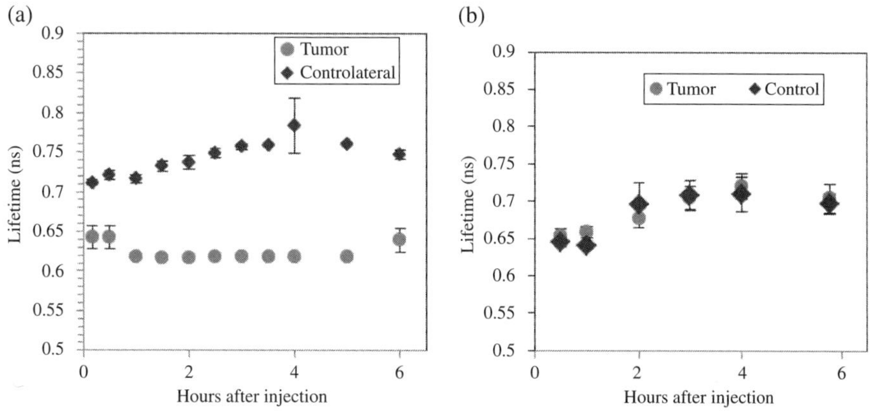

FIGURE 9.14 Difference between fluorescence lifetime of tumor and contralateral sites for (a) targeted fluorescence probe and (b) nontargeted fluorescence probe to HER2 receptors in tumors with high HER2 expression (BT474) (From Ref. [67].)

tumors, the effect of the photon migration on the observed time-resolved fluorescence intensities should be taken into account, using other reconstruction algorithms such as fluorescence diffuse tomography.

Fluorescence lifetime can also be used as an alternative method to diagnose tumors with specific biomarkers. Figure 9.14 shows the difference between fluorescence lifetime of tumor and contralateral site for (i) bound fluorescence probe and (ii) unbound fluorescence probe to HER2 receptors in tumors with high HER2 expression (BT474).

9.6.2 Diffuse Fluorescence Optical Tomography

Fluorescence diffuse optical tomography (FDOT) uses an approach similar to X-ray CT to localize and reconstruct 3D images of concentration and lifetime of the fluorophores inside the tissue based on the measurements collected on the tissue boundary. The principal difference between FDOT and X-ray tomography algorithms is the character of light propagation through the tissue. While in X-ray tomography photons propagate through the tissue along the straight line, in the diffuse optical regime, the photon undergoes multiple scattering events between the source and detector, resulting in random changes of the propagation direction. As a result, the latter case should be modeled by the radiative transfer equation or diffusion approximation [68, 69].

The common optical tomography algorithm is based on perturbation approximation. To generate a recursive formula for image reconstruction, photon density (Eqs. 9.16–9.23) can be written as a Taylor series of the unknown optical properties of the media, for example, the fluorescence yield and lifetime:

$$\varphi(\tilde{\alpha}) = \varphi(\alpha) + \frac{\partial \varphi}{\partial \alpha}\Delta\alpha + \cdots \tag{9.28}$$

where $\Delta\alpha$ is the update of optical parameters and $\varphi(\tilde{\alpha})$ and $\varphi(\alpha)$ are the measured and calculated values for photon density function, respectively. If the $\Delta\alpha$ s are small, the higher order differential components in Equation 9.28 can be disregarded. In discrete form, it can be written (as averaged values in each imaging voxel) as

$$\Delta\varphi_i = \varphi_i(\tilde{\alpha}) - \varphi_i(\alpha) = \sum_{j=1}^{N}\left(\frac{\partial\varphi_i}{\partial\alpha_j}\Delta\alpha_j\right), \quad i = 1,\ldots,M, j = 1,\ldots,N \qquad (9.29)$$

where M is the total number of measurements and N is the total number of imaging voxels.

Equation 9.29 can be converted to matrix equation as

$$\Delta\phi = \mathrm{W}\Delta\alpha \qquad (9.30)$$

where W is the weight (Jacobian) matrix and represents the sensitivity of photon density function, corresponding to each source–detector pair, to optical parameters of different voxels ($W_{ij} = \partial\varphi_i/\partial\alpha_j$).

For a point source and detector, photon density function can be written as

$$\varphi(r_\mathrm{d}, r_\mathrm{s}, t) = S_0 \int_\Omega G_\mathrm{em}(r_\mathrm{d}, r', t) \otimes O(t) \otimes G_\mathrm{ex}(r', r_\mathrm{s}, t)dr' \qquad (9.31)$$

where $G(r', r'', t)$ is the Green function of photon propagation to position r' in response to a delta source located at initial point r''. Subscripts "*ex*" and "*em*" represent the calculated or measured variables at the excitation and emission wavelengths, respectively. $O(t) = {}^{\eta}/_{\tau} \exp\left({}^{-t}/_{\tau}\right)$ is the pulse response of fluorescence target with quantum yield of η and lifetime of τ. Ω represents the target region where fluorescent probes are located.

For FD imaging systems, Equation 9.31 can be written as

$$\varphi(r_\mathrm{d}, r_\mathrm{s}, \omega) = \frac{S_0}{4\pi D_\mathrm{ex} D_\mathrm{em}} \frac{\eta\varepsilon}{1 - i\omega\tau} \int_\Omega G_\mathrm{em}(r_\mathrm{d}, r', \omega) G_\mathrm{ex}(r', r_\mathrm{s}, \omega) N(r') d\Omega \qquad (9.32)$$

where ω, η, and ε are the modulation frequency, quantum yield, and extinction coefficient of fluorophore, respectively. $N(r)$ is the concentration of fluorophores. FD equation can be used in CW systems by inserting $\omega = 0$.

A comprehensive discussion on Green functions for different boundary conditions and imaging geometries is provided in Refs. [69–71]. Green functions for diffusion equation in the FD and an infinite medium can be written as

$$G(r, r') = A\frac{\exp\left(ik|r - r'|\right)}{4\pi cD} \qquad (9.33)$$

where k is

$$k = \sqrt{\frac{-c\mu_a + i\omega}{cD}} \tag{9.34}$$

To reduce the effect of the mismatch between measured and calculated data as well as of unknown system parameters, such as different gains and time/phase delays for each source–detector pair and heterogeneity of background media on the reconstruction image, the fluorescence intensity has to be normalized to the background signal. Better results have been obtained by using the normalized ratio of fluorescence emission signal to the unfiltered diffused signal, coming from the excitation light (Eq. 9.35). In this case, the unknown system parameters that are common in both fluorescence and unfiltered diffused signal will be canceled out. Unfiltered diffused signal has better signal-to-noise ratio and provides more information about the heterogeneity and optical properties of the background tissues than just the background fluorescence signal:

$$\left[\frac{\varphi_{\text{em\_measured}}(i)}{\varphi_{\text{ex\_measured}}(i)} - \frac{\varphi_{\text{em\_calculated}}(i)}{\varphi_{\text{ex\_calculated}}(i)}\right] = [W(i,j)].[\Delta\alpha(j)] \quad i = 1,\ldots,M \text{ and } j = 1,\ldots,N \tag{9.35}$$

where $\varphi_{\text{em\_measured}}(i)$ and $\varphi_{\text{em\_calculated}}(i)$ are the measured and calculated fluorescence field and $\varphi_{\text{ex\_measured}}(i)$ and $\varphi_{\text{ex\_calculated}}(i)$ are the measured and the calculated excitation field, respectively, all associated with the source–detector pair i [72, 73].

Different optimization algorithms can be used to evaluate the unknown fluorescence parameters $\Delta\alpha$ in Equation 9.30. The most common are Newton/Marquardt–Levenberg algorithms and the total least-squares method [71, 74].

The primary difficulty of diffuse optical tomography (DOT) is related to the fact that multiple scattering dominates NIR light propagation in tissue, making 3D localization of targets and accurate quantification of their optical parameters difficult. Scattering properties of the media and a limited number of measurements compared to the number of voxels make Equation 9.35 ill posed and underdetermined. Coregistration of fluorescence imaging with high-resolution imaging modalities such as CT, MRI, and US can provide anatomical structures to reduce the uncertainties of unknown variables in the fluorescence reconstruction algorithms and to improve the two-dimensional (2D) and 3D images of the fluorescence probes [75–80].

Since absorption and scattering properties of the tissue can affect the detected fluorescence yield and lifetime in the deep tissue imaging, they should be taken into account in the imaging reconstruction algorithms. These properties of the tissue can be estimated by DOT methods [81]. In general, the quality of image reconstruction depends upon many parameters, including signal-to-noise ratio, measurement geometry, and depth of the target from the source and detector. A more thorough review of fluorescence optical tomography is provided in [47, 82].

9.7 SUMMARY

In this chapter, we have presented a general overview of existing *in vivo* fluorescence imaging techniques. The current demands for better imaging tools in medical diagnostics and monitoring of the efficacy of treatments require synthesis of better fluorescent contrast agents and development of more complicated *in vivo* optical and endoscopic equipment and faster and more efficient computational algorithms for fluorescence optical imaging. In cancer imaging, fluorescence optical tomography has shown a great promise in diagnostics of the optical signature of cancer lesions. Targeted fluorescent probes are expected to potentially create new imaging and treatment paradigms in cancer diagnostics and therapy in the near future.

ACKNOWLEDGMENTS

The authors acknowledge the intramural program of the Eunice Kennedy Shriver, National Institute of Child Health and Human Development, National Institutes of Health.

REFERENCES

[1] Cuenca AG, Jiang H, Hochwald SN, Delano M, Cance WG, Grobmyer SR. Emerging implications of nanotechnology on cancer diagnostics and therapeutics. *Cancer* 2006;**107**:459–466.

[2] Schroeder A, Heller D, Winslow MM, Dahlman JE, Pratt GW, Langer R, Jacks T, Anderson DG. Treating metastatic cancer with nanotechnology. *Nat Rev Cancer* 2012;**12**:39–50.

[3] Ferrari M. Cancer nanotechnology: opportunities and challenges. *Nat Rev Cancer* 2005;**5**:161–171.

[4] Bhaskar S, Tian F, Stoeger T, Kreyling W, de la Fuente JM, Grazu V, Borm P, Estrada G, Ntziachristos V, Razansky D. Multifunctional nanocarriers for diagnostics, drug delivery and targeted treatment across blood-brain barrier: perspectives on tracking and neuroimaging. *Part Fibre Toxicol* 2010;**7**:3.

[5] Adams GP, Weiner LM. Monoclonal antibody therapy of cancer. *Nat Biotechnol* 2005;**23**:1147–1157.

[6] Nelson AL, Dhimolea E, Reichert JM. Development trends for human monoclonal antibody therapeutics. *Nat Rev Drug Discov* 2010;**9**:767–774.

[7] Laudadio J, Quigley DI, Tubbs R, Wolff DJ. HER2 testing: a review of detection methodologies and their clinical performance. *Expert Rev Mol Diagn* 2007;**7**:53–64.

[8] Menard S, Pupa SM, Campiglio M, Tagliabue E. Biologic and therapeutic role of HER2 in cancer. *Oncogene* 2003;**22**:6570–6578.

[9] Moelans CB, de Weger RA, Van der Wall E, van Diest PJ. Current technologies for HER2 testing in breast cancer. *Crit Rev Oncol Hematol* 2011;**80**:380–392.

[10] Au KM, Lu Z, Matcher SJ, Armes SP. Polypyrrole nanoparticles: a potential optical coherence tomography contrast agent for cancer imaging. *Adv Mater* 2011;**23**:5792–5795.

[11] Barton JK, Hoying JB, Sullivan CJ. Use of microbubbles as an optical coherence tomography contrast agent. *Acad Radiol* 2002;**9** (Suppl 1):S52–S55.

[12] van den Berg NS, van Leeuwen FW, van der Poel HG. Fluorescence guidance in urologic surgery. *Curr Opin Urol* 2012;**22**:109–120.

[13] Liu DZ, Mathes DW, Zenn MR, Neligan PC. The application of indocyanine green fluorescence angiography in plastic surgery. *J Reconstr Microsurg* 2011;**27**:355–364.

[14] Wagnieres GA, Star WM, Wilson BC. *In vivo* fluorescence spectroscopy and imaging for oncological applications. *Photochem Photobiol* 1998;**68**:603–632.

[15] Zhang Q, Iwakuma N, Sharma P, Moudgil BM, Wu C, McNeill J, Jiang H, Grobmyer SR. Gold nanoparticles as a contrast agent for *in vivo* tumor imaging with photoacoustic tomography. *Nanotechnology* 2009;**20**:395102.

[16] Ku G, Wang LV. Deeply penetrating photoacoustic tomography in biological tissues enhanced with an optical contrast agent. *Opt Lett* 2005;**30**:507–509.

[17] Pilatou MC, Marani E, de Mul FF, Steenbergen W. Photoacoustic imaging of brain perfusion on albino rats by using evans blue as contrast agent. *Arch Physiol Biochem* 2003;**111**:389–397.

[18] Bloch S, Lesage F, McIntosh L, Gandjbakhche A, Liang K, Achilefu S. Whole-body fluorescence lifetime imaging of a tumor-targeted near-infrared molecular probe in mice. *J Biomed Opt* 2005;**10**:054003.

[19] Lee SB, Hassan M, Fisher R, Chertov O, Chernomordik V, Kramer-Marek G, Gandjbakhche A, Capala J. Affibody molecules for *in vivo* characterization of HER2-positive tumors by near-infrared imaging. *Clin Cancer Res* 2008;**14**:3840–3849.

[20] Berezin MY, Guo K, Akers W, Northdurft RE, Culver JP, Teng B, Vasalatiy O, Barbacow K, Gandjbakhche A, Griffiths GL, Achilefu S. Near-infrared fluorescence lifetime ph-sensitive probes. *Biophys J* 2011;**100**:2063–2072.

[21] Cui L, Zhong Y, Zhu W, Xu Y, Du Q, Wang X, Qian X, Xiao Y. A new prodrug-derived ratiometric fluorescent probe for hypoxia: high selectivity of nitroreductase and imaging in tumor cell. *Org Lett* 2011;**13**:928–931.

[22] Nakata E, Yukimachi Y, Kariyazono H, Im S, Abe C, Uto Y, Maezawa H, Hashimoto T, Okamoto Y, Hori H. Design of a bioreductively-activated fluorescent pH probe for tumor hypoxia imaging. *Bioorg Med Chem* 2009;**17**:6952–6958.

[23] Okuda K, Okabe Y, Kadonosono T, Ueno T, Youssif BG, Kizaka-Kondoh S, Nagasawa H. 2-nitroimidazole-tricarbocyanine conjugate as a near-infrared fluorescent probe for *in vivo* imaging of tumor hypoxia. *Bioconjug Chem* 2012;**23**:324–329.

[24] Youssif BG, Okuda K, Kadonosono T, Salem OI, Hayallah AA, Hussein MA, Kizaka-Kondoh S, Nagasawa H. Development of a hypoxia-selective near-infrared fluorescent probe for non-invasive tumor imaging. *Chem Pharm Bull (Tokyo)* 2012;**60**:402–407.

[25] Kobayashi H, Ogawa M, Alford R, Choyke PL, Urano Y. New strategies for fluorescent probe design in medical diagnostic imaging. *Chem Rev* 2010;**110**:2620–2640.

[26] Kosaka N, Ogawa M, Sato N, Choyke PL, Kobayashi H. *In vivo* real-time, multicolor, quantum dot lymphatic imaging. *J Invest Dermatol* 2009;**129**:2818–2822.

[27] Andersson-Engels S, Johansson J, Svanberg S. Medical diagnostic system based on simultaneous multispectral fluorescence imaging. *Appl Opt* 1994;**33**:8022–8029.

[28] Lakowicz JR. *Principles of Fluorescence Spectroscopy*. 2nd ed. New York: Kluwer Academic/Plenum; 1999.

[29] Lee H, Akers W, Bhushan K, Bloch S, Sudlow G, Tang R, Achilefu S. Near-infrared ph-activatable fluorescent probes for imaging primary and metastatic breast tumors. *Bioconjug Chem* 2011;**22**:777–784.

[30] Richards-Kortum R, Sevick-Muraca E. Quantitative optical spectroscopy for tissue diagnosis. *Annu Rev Phys Chem* 1996;**47**:555–606.

[31] Monici M. Cell and tissue autofluorescence research and diagnostic applications. *Biotechnol Annu Rev* 2005;**11**:227–256.

[32] Apreleva SV, Wilson DF, Vinogradov SA. Tomographic imaging of oxygen by phosphorescence lifetime. *Appl Opt* 2006;**45**:8547–8559.

[33] Lebedev AY, Cheprakov AV, Sakadzic S, Boas DA, Wilson DF, Vinogradov SA. Dendritic phosphorescent probes for oxygen imaging in biological systems. *ACS Appl Mater Interf* 2009;**1**:1292–1304.

[34] Corlu A, Durduran T, Choe R, Schweiger M, Hillman EMC, Arridge SR, Yodh AG. Uniqueness and wavelength optimization in continuous-wave multispectral diffuse optical tomography. *Opt Lett* 2003;**28**:2339–2341.

[35] Cerussi AE, Berger AJ, Bevilacqua F, Shah N, Jakubowski D, Butler J, Holcombe RF, Tromberg BJ. Sources of absorption and scattering contrast for near-infrared optical mammography. *Acad Radiol* 2001;**8**:211–218.

[36] Durduran T, Choe R, Culver JP, Zubkov L, Holboke MJ, Giammarco J, Chance B, Yodh AG. Bulk optical properties of healthy female breast tissue. *Phys Med Biol* 2002;**47**:2847–2861.

[37] Kukretia S, Cerussi A, Tromberg B, Gratton E. Intrinsic near-infrared spectroscopic markers of breast tumors. *Dis Markers* 2008;**25**:281–290.

[38] Binzoni T, Leung TS, Gandjbakhche AH, Rufenacht D, Delpy DT. The use of the Henyey-Greenstein phase function in Monte Carlo simulations in biomedical optics. *Phys Med Biol* 2006;**51**:N313–N322.

[39] Weber JR, Cuccia DJ, Johnson WR, Bearman GH, Durkin AJ, Hsu MK, Lin A, Binder DK, Wilson D, Tromberg BJ. Multispectral imaging of tissue absorption and scattering using spatial frequency domain imaging and a computed-tomography imaging spectrometer. *J Biomed Opt* 2011;**16**:011015.

[40] Berezin MY, Achilefu S. Fluorescence lifetime measurements and biological imaging. *Chem Rev* 2010;**110**:2641–2684.

[41] Szmacinski H, Lakiwicz JR. *Topics in Fluorescence Spectroscopy Lifetime-based Sensing*. New York: Plenum; 1994.

[42] Hassan M, Riley J, Chernomordik V, Smith P, Pursley R, Lee SB, Capala J, Gandjbakhche AH. Fluorescence lifetime imaging system for *in vivo* studies. *Mol Imaging* 2007;**6**:229–236.

[43] Arridge SR, Cope M, Delpy DT. The theoretical basis for the determination of optical pathlengths in tissue—temporal and frequency-analysis. *Phys Med Biol* 1992;**37**:1531–1560.

[44] Ishimaru A. *Wave Propagation and Scattering in Random Media*. Volume **12**, New York: Wiley-IEEE Press; 1999.

[45] Kim AD. Correcting the diffusion approximation at the boundary. *J Opt Soc Am A* 2011;**28**:1007–1015.

[46] Haskell RC, Svaasand LO, Tsay TT, Feng TC, McAdams MS, Tromberg BJ. Boundary conditions for the diffusion equation in radiative transfer. *JOSA A* 1994;**11**:2727–2741.

[47] Zhu Q, Dehghani H, Tichauer KM, Holt RW, Vishwanath K, Leblond F, Pogue BW. A three-dimensional finite element model and image reconstruction algorithm for time-domain fluorescence imaging in highly scattering media. *Phys Med Biol* 2011;**56**:7419–7434.

[48] Boas DA, Oleary MA, Chance B, Yodh AG. Scattering of diffuse photon density waves by spherical inhomogeneities within turbid media—analytic solution and applications. *Proc Natl Acad Sci U S A* 1994;**91**:4887–4891.

[49] Farrell TJ, Patterson MS, Wilson B. A diffusion theory model of spatially resolved, steady-state diffuse reflectance for the noninvasive determination of tissue optical properties *in vivo*. *Med Phys* 1992;**19**:879–888.

[50] Arridge SR, Schweiger M, Hiraoka M, Delpy DT. A finite-element approach for modeling photon transport in tissue. *Med Phys* 1993;**20**:299–309.

[51] Jiang HB. Frequency-domain fluorescent diffusion tomography: a finite-element-based algorithm and simulations. *Appl Opt* 1998;**37**:5337–5343.

[52] Dehghani H, Srinivasan S, Pogue BW, Gibson A. Numerical modelling and image reconstruction in diffuse optical tomography. *Philos Trans R Soc A-Math Phys Eng Sci* 2009;**367**:3073–3093.

[53] Paulsen KD, Jiang HB. Spatially varying optical property reconstruction using a finite-element diffusion equation approximation. *Med Phys* 1995;**22**:691–701.

[54] Jacques SL, Pogue BW. Tutorial on diffuse light transport. *J Biomed Opt* 2008;**13**:041302.

[55] Srinivasan S, Pogue BW, Carpenter C, Yalavarthy PK, Paulsen K. A boundary element approach for image-guided near-infrared absorption and scatter estimation. *Med Phys* 2007;**34**:4545–4557.

[56] Arridge SR, Schweiger M. Image reconstruction in optical tomography. *Philos Trans R Soc B-Biol Sci* 1997;**352**:717–726.

[57] Wang LH, Jacques SL, Zheng LQ. MCML—Monte-Carlo modeling of light transport in multilayered tissues. *Comput Meth Programs Biomed* 1995;**47**:131–146.

[58] Flock ST, Patterson MS, Wilson BC, Wyman DR. Monte Carlo modeling of light propagation in highly scattering tissue I: model predictions and comparison with diffusion theory. *IEEE Trans Biomed Eng* 1989;**36**:1162–1168.

[59] Hillman EM, Moore A. All-optical anatomical co-registration for molecular imaging of small animals using dynamic contrast. *Nat Photon* 2007;**1**:526–530.

[60] Chernomordik V, Hassan M, Lee SB, Zielinski R, Gandjbakhche A, Capala J. Quantitative analysis of Her2 receptor expression *in vivo* by near-infrared optical imaging. *Mol Imaging* 2010;**9**:192–200.

[61] Chen YC, Clegg RM. Fluorescence lifetime-resolved imaging. *Photosynth Res* 2009;**102**:143–155.

[62] Riley J, Hassan M, Chernomordik V, Gandjbakhche A. Choice of data types in time resolved fluorescence enhanced diffuse optical tomography. *Med Phys* 2007;**34**:4890–4900.

[63] Kumar AT, Raymond SB, Bacskai BJ, Boas DA. Comparison of frequency-domain and time-domain fluorescence lifetime tomography. *Opt Lett* 2008;**33**:470–472.

[64] Kumar AT, Raymond SB, Boverman G, Boas DA, Bacskai BJ. Time resolved fluorescence tomography of turbid media based on lifetime contrast. *Opt Express* 2006;**14**:12255–12270.

[65] Han SH, Farshchi-Heydari S, Hall DJ. Analytical method for the fast time-domain reconstruction of fluorescent inclusions in vitro and *in vivo*. *Biophys J* 2010;**98**:350–357.

[66] Chernomordik V, Gandjbakhche AH, Hassan M, Pajevic S, Weiss GH. A CTRW-based model of time-resolved fluorescence lifetime imaging in a turbid medium. *Opt Commun* 2010;**283**:4832–4839.

[67] Ardeshirpour Y, Chernomordik V, Zielinski R, Capala J, Griffiths G, Vasalatiy O, Smirnov AV, Knutson JR, Lyakhov I, Achilefu S, Gandjbakhche A, Hassan M. *In vivo* fluorescence lifetime imaging monitors binding of specific probes to cancer biomarkers. *PLoS One* 2012;**7**:e31881.

[68] Arridge SR. Optical tomography in medical imaging. *Inverse Probl* 1999;**15**:R41–R93.

[69] Arridge SR, Schotland JC. Optical tomography: forward and inverse problems. *Inverse Probl* 2009;**25**:123010.

[70] Arridge SR. Photon-measurement density functions. Part I: analytical forms. *Appl Opt* 1995;**34**:7395–7409.

[71] Arridge SR, Hebden JC. Optical imaging in medicine. Part 2: modelling and reconstruction. *Phys Med Biol* 1997;**42**:841–853.

[72] Vinegoni C, Fumene Feruglio P, Cortez-Retamozo V, Razansky D, Medoff BD, Ntziachristos V, Sbarbati A, Pittet M, Weissleder R. Imaging of molecular probe activity with Born-normalized fluorescence optical projection tomography. *Opt Letters* 2010;**35**:1088–1090.

[73] Vinegoni C, Razansky D, Figueiredo JL, Fexon L, Pivovarov M, Nahrendorf M, Ntziachristos V, Weissleder R. Born normalization for fluorescence optical projection tomography for whole heart imaging. *J Vis Exp* 2009;**28**:1389.

[74] Arridge SR, Lionheart WR. Nonuniqueness in diffusion-based optical tomography. *Opt Lett* 1998;**23**:882–884.

[75] Davis SC, Pogue BW, Springett R, Leussler C, Mazurkewitz P, Tuttle SB, Gibbs-Strauss SL, Jiang SS, Dehghani H, Paulsen KD. Magnetic resonance-coupled fluorescence tomography scanner for molecular imaging of tissue. *Rev Sci Instrum* 2008;**79**:064302.

[76] Gruber JD, Paliwal A, Krishnaswamy V, Ghadyani H, Jermyn M, O'Hara JA, Davis SC, Kerley-Hamilton JS, Shworak NW, Maytin EV, Hasan T, Pogue BW. System development for high frequency ultrasound-guided fluorescence quantification of skin layers. *J Biomed Opt* 2010;**15**:026028.

[77] Razansky D, Ntziachristos V. Hybrid photoacoustic fluorescence molecular tomography using finite-element-based inversion. *Med Phys* 2007;**34**:4293–4301.

[78] Sampath L, Kwon S, Hall MA, Price RE, Sevick-Muraca EM. Detection of cancer metastases with a dual-labeled near-infrared/positron emission tomography imaging agent. *Transl Oncol* 2010;**3**:307–317.

[79] Schulz RB, Ale A, Sarantopoulos A, Freyer M, Soehngen E, Zientkowska M, Ntziachristos V. Hybrid system for simultaneous fluorescence and x-ray computed tomography. *IEEE Trans Med Imaging* 2010;**29**:465–473.

[80] Lin Y, Gao H, Nalcioglu O, Gulsen G. Fluorescence diffuse optical tomography with functional and anatomical a priori information: feasibility study. *Phys Med Biol* 2007;**52**:5569–5585.

[81] Tan Y, Jiang H. DOT guided fluorescence molecular tomography of arbitrarily shaped objects. *Med Phys* 2008;**35**:5703–5707.

[82] Gao F, Li J, Zhang L, Poulet P, Zhao H, Yamada Y. Simultaneous fluorescence yield and lifetime tomography from time-resolved transmittances of small-animal-sized phantom. *Appl Opt* 2010;**49**:3163–3172.

10

PHOTOACOUSTIC AND ULTRASOUND IMAGING WITH NANOSIZED CONTRAST AGENTS

MANSIK JEON AND CHULHONG KIM

Departments of Electrical Engineering and Creative IT Engineering, Pohang University of Science and Technology, Pohang, Republic of Korea

10.1 INTRODUCTION

Molecular imaging has attracted great attention in biomedicine in the last few decades as a valuable tool for personalized medicine. Existing pure optical molecular imaging modalities, such as confocal microscopy [1–5], two-photon microscopy [6–9], optical coherence tomography [10–16], and fluorescence (FL) tomography [17–21], have had a great impact on preclinical and clinical research. However, the intrinsic optical opaqueness of biological tissues prevents these modalities from maintaining high spatial resolution beyond the one optical transport mean free path (~1 mm) [22]. Photoacoustic tomography (PAT) has been in the spotlight during the last decade because PAT overcomes limitations of commercially available optical and ultrasound (US) imaging modalities and shows great promises in molecular imaging. PAT, a hybrid imaging modality by taking advantages of strong optical contrast and high ultrasonic resolution, was developed to overcome the fundamental impediments of existing techniques [23–27]. At the same time, the contrast agents for PAT have been heavily investigated to improve the sensitivity and specificity of molecular PAT. In this chapter, we will mainly discuss principles of photoacoustic (PA) imaging with a short overview of nanosized PA contrast agents.

Nanotechnology for Biomedical Imaging and Diagnostics: From Nanoparticle Design to Clinical Applications, First Edition. Edited by Mikhail Y. Berezin.

10.2 PRINCIPLES OF PAT

The basic principles of PAT are illustrated in Figure 10.1 [22, 28–30]. When a pulsed light (typically on the order of a nanosecond) irradiates biological tissues and light is absorbed by objects, these objects thermoelastically expand. Under thermal and stress confinements, sudden wideband pressure waves are generated, referred to as PA waves. For instance, the thermal and stress relaxations times of an object with a size of 15 μm are 1.7 ms and 0.01 μs, respectively. In addition, the time-varying excitation only produces PA pressure waves. The amplitude of an initial PA pressure (p_0) wave is as follows:

$$p_0 = \Gamma \eta_{th} A_e \tag{10.1}$$

All three parameters in the initial PA pressure amplitude are highly related to specific PA image contrasts. First, the temperature dependence of the Grüneisen parameter (Γ, dimensionless), the thermomechanical conversion efficiency, provides a means of noninvasively obtaining PA maps of temperature distributions in tissues with an accuracy of 0.16°C [31, 32]. Second, the conversion efficiency (η_{th}) between specific optical absorption (A_e) and heat is another critical parameter. A majority of fluorescence (FL) contrast agents emit FL light due to radiative decay of absorbed energy depending on their intrinsic optical properties (i.e., FL quantum yield) [33–35]. By contrast, the energy released from the nonradiative decay (i.e., 1-FL quantum yield) can be converted into heat and consequently produce the PA pressure waves. Last, most importantly, PAT is extremely sensitive to specific optical absorption (A_e), which is equivalent to the product of the optical absorption coefficient of the object (μ_a) and the local light fluence (F). Thus, small fractional change in optical absorption coefficient directly results in equal change in PA signal.

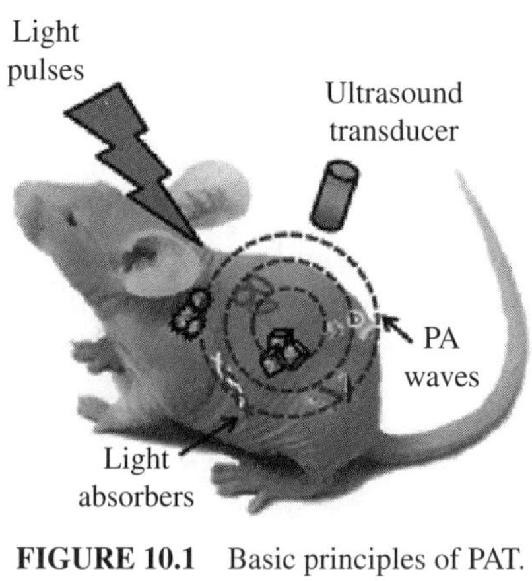

FIGURE 10.1 Basic principles of PAT.

The produced PA waves easily travel in biological tissues, which are much more transparent for US than light (by ~10^3 times). Furthermore, because the sound velocity in tissues (~1480 m/s) is much slower than that of light (3×10^8 m/s), the time-dependent PA signals can be easily detected by conventional US detectors. The temporal information based on the arrival times of the PA waves is directly associated with the spatial distribution of the objects in the tissue. There are two ways to localize the PA sources: hardware and software focusing. In hardware focusing, a focusing acoustic lens is typically used, whereas an inverse algorithm is used in software focusing.

The combination of these intrinsic properties brings several advantages: (i) The maximum penetration depth can be extended beyond the one optical transport mean free path and reach up to approximately 8 cm in tissues [36]. (ii) The spatial resolution and penetration depth are scalable according to US parameters and optical wavelengths. More importantly, the depth-to-resolution ratio of PAT remains constant (~200) in all imaging depths [23, 27]. (iii) By tuning optical wavelengths, PAT can probe different chemicals [26, 37]. (iv) Unlike optical coherence tomography and US imaging, no speckle artifacts are visible in PA images [38]. (v) Existing clinical US imaging systems can be adapted to provide dual-modality US and PA imaging capabilities [36, 39–41]. (vi) PAT is completely free of harmful ionizing radiation. Thus, the clinical translation potential of PAT is very high. Based on these features, PAT can noninvasively supply structural (i.e., vascular structures [42], solid tumors and angiogenesis [43–49], and internal organs [50, 51]), functional (i.e., total hemoglobin concentration, oxygen saturation of hemoglobin [24, 25], blood flow in microvasculature [52–54], oxygen partial pressure [55], pH [56–58], and metabolic rate of oxygen consumption [59, 60]), and molecular (i.e., probing disease-specific molecules at a molecular level using biomarkers and exogenous contrast agents) information of tissues [26].

10.3 PAT MODALITIES

PA imaging systems can be categorized into two types [23, 30]: (1) inverse algorithm-based photoacoustic computed tomography (PACT) [25] and (2) point-by-point detection-based photoacoustic microscopy (PAM) [50, 61, 62]. Both types can be implemented in an endoscopic probe [63–68]. The first type, PACT, utilizes array-based US detection systems for parallel detection and inverse algorithms to reconstruct images in real time. As the first attempt, a circular scanning-based PACT system using a single-element unfocused US transducer was developed for *in vivo* PA imaging (Fig. 10.2a) [25, 69]. Although the system was successfully used for providing anatomical [43], functional [25, 69], and molecular [37] information of brains in small animals *in vivo*, it took 15 min to acquire one two-dimensional (2D) cross-sectional image because of the mechanical scanning of the transducer. Later, US array systems were adapted to enhance the image acquisition speed. To mimic 360° full circular scanning with a single-element transducer, a full-ring ultrasonic transducer (512 elements) array was developed, and one 2D image acquisition speed

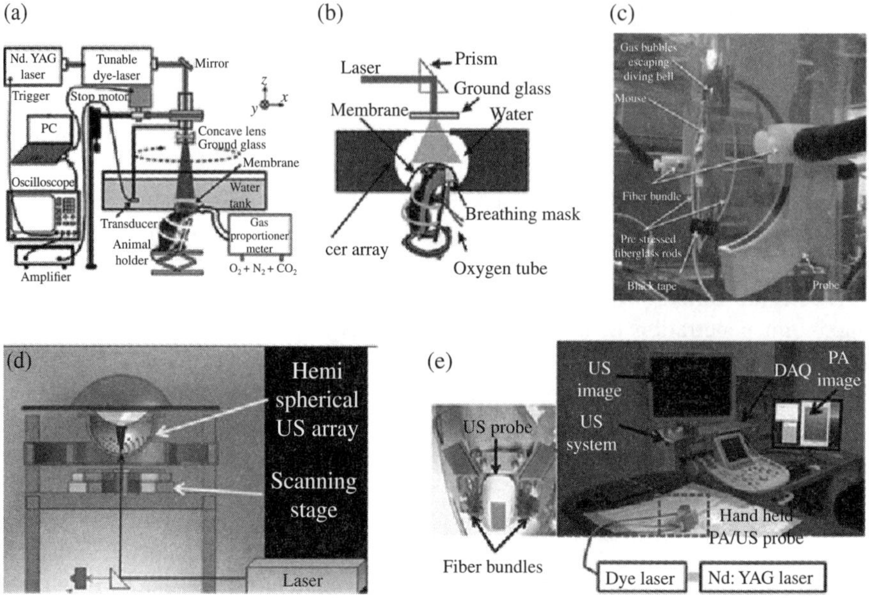

FIGURE 10.2 Various types of PACT systems implemented by (a) a single-element ultrasound transducer with 360° full circular mechanical scanning (Reprinted with permission from Ref. [69]. © SPIE, (b) a full-ring ultrasound array (Reprinted with permission from Ref. [70]. © SPIE, (c) an arc-shaped ultrasound array (Reprinted with permission from Ref. [71]. © SPIE (Reprinted with permission from Ref. [72]. © Optical Society of America, (d) a hemisphere ultrasound array, and (e) a handheld probe modified from a clinical ultrasound array system. (Reprinted with permission from Ref. [40]. © The American Association of Physicists Medicine.)

was greatly improved to a frame rate of 0.9 Hz (Fig. 10.2b) [70, 73]. Using this system, functional brain [70] and whole-body imaging [74] of small animals have been successfully performed. As another example, an arc array (64 elements and a central frequency of 5 MHz) was implemented to provide whole-body imaging of rodents (Fig. 10.2c) [71]. The acquisition time of one three-dimensional (3D) image was 8 min. To further enhance clinical translation, clinically friendly PACT systems have been recently developed such as tomographic PA breast scanners and linear-array-based PA probes. Figure 10.2d shows the schematic of a tomographic PA and US breast scanner consisting of 128 US elements with a center frequency of 5 MHz [72]. This system can be potentially used to screen early stage of breast cancer patients and map breast angiography. In addition, a handheld PA and US probe (Fig. 10.2e), modified from a clinical US system (Philips iU22), is currently used in phase I clinical studies to monitor early response of chemotherapy and identify sentinel lymph nodes in breast cancer patients [40, 41]. The second type, PAM, uses a single-element focused US transducer to directly collect one-dimensional (1D) depth-sensitive images (referred to as A-lines) and 2D/3D images with mechanical or optical raster scanning [50, 61, 62, 75]. Hardware-based acoustic or optical focusing provides the transverse resolution, whereas the acoustic bandwidth

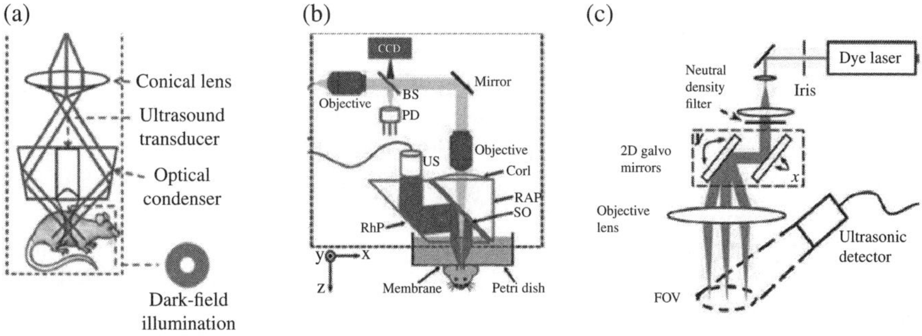

FIGURE 10.3 Various reflection-mode PAM systems implemented by (a) a dark-field confocal configuration (Reprinted with permission from Ref. [76]. © American Chemical Society.), (b) optical focusing with mechanical raster scanning (From Ref. [77]. © Optical Society of America.), and (c) optical focusing with optical raster scanning. (From Ref. [75]. © Optical Society of America.)

of the transducer controls the axial resolution. Although no mathematical image reconstruction algorithm is required and its associated artifacts are eliminated, the imaging speed and field of view (FOV) are quite limited in this type. Figure 10.3 shows various types of PAM systems. For instance, the schematic of dark-field confocal acoustic-resolution PAM is shown in Figure 10.3a [76]. A donut-shaped light illumination pattern (i.e., dark-field confocal configuration) formed by a conical lens and optical condenser greatly enhances the penetration depth and signal-to-noise ratio (SNR). When a 5 MHz ultrasonic transducer was used, the axial and transverse resolutions were 150 and 590 μm, respectively. The penetration depth was beyond 30 mm. When a 50 MHz ultrasonic transducer was used, the axial and transverse resolutions were 15 and 45 μm, respectively. The penetration depth was approximately 3 mm. To improve the transverse resolution to the order of a μm, hardware-based optical focusing was applied. This mode is called as optical-resolution PAM. The transverse resolution can be enhanced up to the diffraction-limited optical resolution. Since the transverse resolution is based on optical focusing, the penetration depth is limited to one optical transport mean free path (~1 mm), similar to conventional optical microscopy. Figure 10.3b shows the diagram of a reflection-mode optical-resolution PAM [77]. To form 2D or 3D images, mechanical or optical raster scanning is required. The mechanical scanning-based PAM system has a high SNR, while it suffers from a low imaging speed. Although the imaging speed can be improved by scanning optically, the system had a low SNR because the system used an unfocused US transducer (Fig. 10.3c) [75].

Typically, the generated PA waves are detected by using piezoelectric transducers. However, by employing a transparent Fabry–Perot polymer film etalon, the PA waves can be probed optically [78, 79]. The thickness changes of the etalon due to the PA waves disturbed the reflectivity of the etalon, and this led to change in the reflected light intensity of a probe laser beam. The advantages of this system are broader PA detection bandwidth and detection aperture.

10.4 INTRINSIC CONTRASTS FOR PAT

PAT has been successfully provided anatomical and functional information of biological tissues using intrinsic contrasts. Major endogenous absorbents include two types of hemoglobin (i.e., oxyhemoglobin and deoxyhemoglobin), melanin, lipid, and water [80]. In addition, by using different optical wavelengths, PAT can spectrally distinguish each chromophore. For example, two types of hemoglobin and melanin dominantly absorb visible light, while water and lipid absorb near-infrared (NIR) light. Figure 10.4 shows the agent-free morphological PA images. A single melanoma (Fig. 10.4a) and red blood cell (Fig. 10.4b) are clearly imaged by using a subwavelength optical-resolution PAM system *in vitro*, which confirms the capability of agent-free PA imaging [81]. As shown in Figure 10.4c, PAT clearly visualizes the morphologies of a B16 melanoma skin cancer and its surrounding vasculatures in small animals *in vivo* based on these intrinsic contrasts [76]. Further, PAT can provide the whole-body small animal images acquired with endogenous contrasts and an array-based PACT system. Figure 10.4d shows volumetric PA rendering of

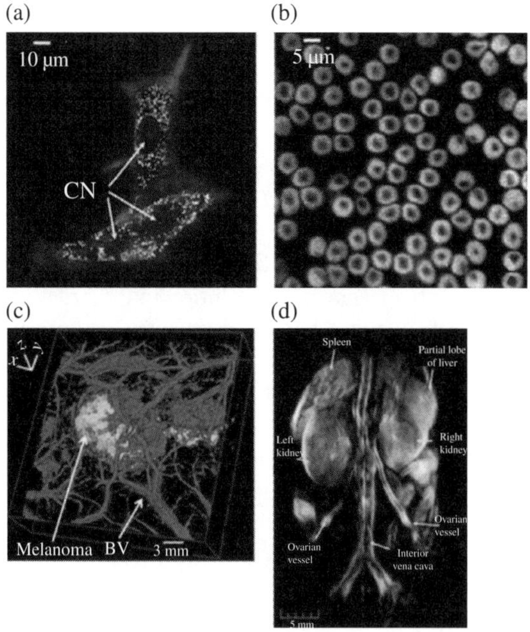

FIGURE 10.4 PA images of intrinsic chromophores. *In vitro* PA images of (a) B16 melanoma cells and (b) red blood cells. CN, cell nuclei (Reprinted with permission from Ref. [81]. © Optical Society of America.), (c) Volumetric anatomical PA image of B16 melanoma (778 nm) and its feeding blood vessels (570 nm) (Reprinted with permission from Ref. [76]. © American Chemical Society.). BV, blood vessels, (d) *In vivo* noninvasive volumetric PA image of the whole nude mouse body obtained at 755 nm. (Permission from Ref. [71]. © SPIE.)

intact internal organs (e.g., spleen, liver, kidney, and major blood vessels) in a nude mouse *in vivo* [71].

Various functional parameters, such as the total hemoglobin concentration (HbT), oxygen saturation of hemoglobin (SO_2), blood flow, and metabolic rate of oxygen consumption (MRO_2), can be directly measured by PAT using only intrinsic contrast [60]. Label-free PAT is the only available imaging modality that can solely provide all these information *in vivo*. These variables can be used to study brain hemodynamic and activation, tumor hypermetabolism, wound healing process, and gene expression.

Both vessel-by-vessel mapping of HbT and SO_2 can be photoacoustically acquired by considering that the two types of hemoglobin are dominant absorbers in tissues and have different molar extinction spectra. Because the initial PA amplitudes are proportional to the local energy deposition, the absorption coefficient can be replaced with the measured PA amplitude. At the isosbestic light wavelengths, where oxyhemoglobin and deoxyhemoglobin have the same molar extinction coefficients, the PA contrast is only sensitive to HbT. Further, to measure SO_2, the concentration of each type of hemoglobin needs to be estimated by using at least two optical wavelengths. As another parameter, the transverse blood flow velocity can be photoacoustically measured by using PA Doppler bandwidth broadening. If one uses all aforementioned photoacoustically measured parameters, MRO_2 can be calculated, a measurement of the rate of direct oxygen consumption within a region of interest (ROI). With the well-defined input and output blood vessels within the ROI, MRO_2 can be estimated as follows:

$$\mathrm{MRO}_2 = \varepsilon \times \mathrm{HbT} \times \left(\mathrm{SO}_{2\text{-in}} \times A_{\mathrm{in}} \times V_{\mathrm{in}} - \mathrm{SO}_{2\text{-out}} \times A_{\mathrm{out}} \times V_{\mathrm{out}}\right) / W \qquad (10.2)$$

where ε is the constant oxygen binding capacity of hemoglobin (1.36 ml O_2/g hemoglobin), HbT is the total hemoglobin concentration (in grams of hemoglobin/ml blood), SO_2 is the oxygen saturation of hemoglobin (in %), A is the cross-sectional area (in mm^2) of the blood vessels, V is the blood flow speed (in mm/s), and W is the weight of the ROI (in grams). Further, oxygen extraction fraction (OEF), defined as

$$\frac{\left(\mathrm{SO}_{2\text{-in}} - \mathrm{SO}_{2\text{-out}}\right)}{\mathrm{SO}_{2\text{-in}}}$$

can be calculated by PAT. OEF represents the fraction of oxygen molecules across the capillary wall. Figure 10.5 shows metabolic PA measurements of U87 glioblastoma xenograft in a nude mouse *in vivo*. Early-stage tumor hyperoxia and angiogenesis were observed (Fig. 10.5b and c). PA quantification of blood flow, OEF, and MRO_2 was compared between 0 and 7 days after tumor inoculation (Fig. 10.5d). Furthermore, the PA SO_2 values were quantified within the intra- and extratumoral vasculatures (Fig. 10.5e). These results prove that PAT is the perfect imaging modality to quantify the functional parameters of biological tissues in various diseases.

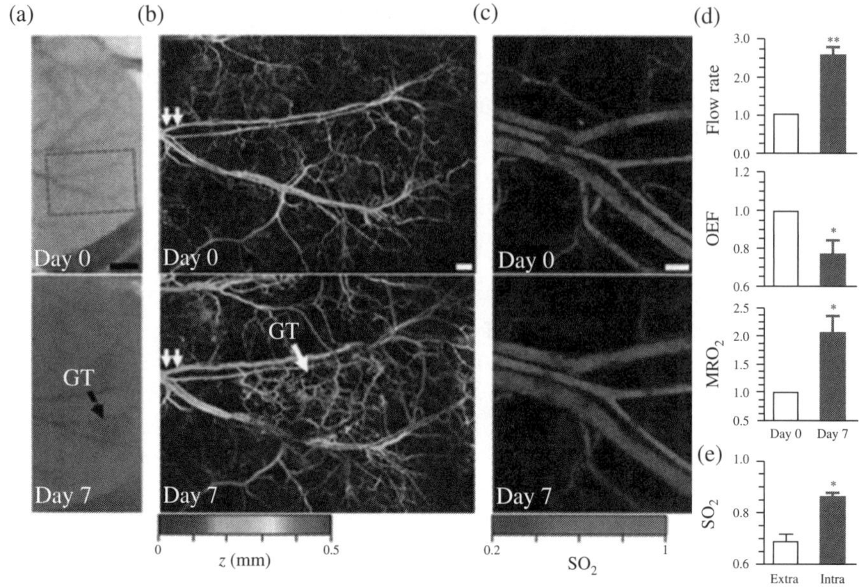

FIGURE 10.5 Label-free metabolic PA quantification of MRO_2 *in vivo*. (a) Photographs of a mouse ear before and 7 days after inoculation of U87 glioblastoma cells. (b) *In vivo* PA images of microvasculatures in the tumor region. (c) *In vivo* PA images of SO_2 in the tumor region. (d) Metabolic PA quantification of volumetric blood flow rate, OEF, and MRO_2 at 7 days postinoculation compared to 0 day. (e) Comparison of the averaged SO_2 values in the intra- and extratumoral vasculatures. (Reprinted with permission from Ref. [60]. © SPIE.)
$^*p<0.05$
$^{**}p<0.01$

10.5 EXOGENOUS CONTRASTS FOR PAT

Although the aforementioned intrinsic parameters provide considerable PA contrasts, these contrasts deliver strong PA signals in the visible spectral range, where the optical scattering and absorption of biological tissues are significant. This ultimately prevents light penetration. Therefore, NIR light (i.e., ~650 to ~900 nm), which can maximally penetrate in tissues due to significantly reduced optical absorption and scattering, is widely used for deep tissue PA imaging [82]. However, not many intrinsic contrasts are available in this spectral range. Thus, NIR absorbents have been extensively explored to enhance the capability of PAT. The sensitivity, specificity, and contrast of PAT are dramatically improved by exogenous contrast agents. Additionally, some optically transparent biological tissues in both visible and NIR regions such as lymphatics and urinary bladders can be visualized with injection of the contrast agents [83, 84]. Optical absorbents including small molecules, micro- and nanoparticles, reporter genes, and FL proteins have been successfully adapted in PAT. Organic dyes such as methylene blue (MB) [39, 40, 83] and indocyanine green (ICG) [41, 85] have been widely used in PAT as contrast agents, but the small molecules suffer from fast circulation times, relatively small optical absorption cross-sectional areas, and/or chemical stains. To surmount the limitations of small molecules, more recently, gold nanostructures have been heavily investigated as contrast agents for PAT [86–89]. Nanotechnology has garnered much attention due to its potential application in biomedicine [90–93]. Nanostructure-based theranostic systems bring many advantages: (i) the nanostructures can be functionalized using standard procedures in nanotechnology. The type, the size, and the number of linkers on the surface of nanostructures can be tuned according to their specific applications. (ii) They can selectively deliver diagnostic or therapeutic agents to the disease sites at effective concentration and consequently boost the efficacy of existing imaging and treatment outcomes. (iii) The release kinetics of the agents or drugs can be controlled by using internal or external manipulation. (iv) Most importantly, the optical properties of the nanoplatforms, in the case of plasmonic gold nanostructures, can generate excessive heating due to localized surface plasmon resonance (LSPR) under laser irradiation [94]. This effect is known as the photothermal effect [95–98], which considerably assists PAT and killing of tumor cells.

10.5.1 Plasmonic Gold Nanocages

Among many different types of gold nanostructures (e.g., gold nanoshells [99, 100], gold nanorods [101–111], gold nanospheres [112, 113], gold nanostars [114], gold nanobeacons [115–118], etc.), gold nanocages (GNCs) have been actively used as a PA contrast agent due to their strong and tunable optical absorption in the NIR spectral range, bioinertness, low heavy metal toxicity, and easiness to be functionalized the surface [76, 86, 119, 120]. Generally, the GNCs can be synthesized using the galvanic replacement reaction between the Ag nanocubes and chloroauric acid in aqueous solution [121–123]. Figure 10.6 shows the optical and physical properties of GNCs. As shown in Figure 10.6a, the LSPR peak of GNCs can be adjustable by controlling the ratio between wall thickness and edge length via the reaction of Au nanocube solution with different volumes of $HAuCl_4$ solution [122]. A TEM image of GNCs with an edge length of 55 nm is shown in Figure 10.6b [76]. The theoretical calculation based on a discrete dipole approximation method confirmed that GNCs were a highly absorbing material instead of a scattering one (Fig. 10.6c) [124]. The experimentally measured absorption cross section was approximately $6 \times 10^{-15}\,m^2$. Further, the ratio between the absorption and extinction cross sections was 82% in both experimental and mathematical simulation results. These results indicate that the absorption cross section of the GNCs is 10^5 larger than that of the small dye molecules, making the GNCs ideal as a contrast agent for PAT. From a phantom study, PA signals could be detectable from the GNCs at a particle concentration of 4.5 pM and a SNR of 9. This detection sensitivity corresponds to approximately 9×10^{-21} moles of GNCs per imaging voxel.

As *in vivo* preclinical applications, GNCs conjugated with [Nle^4, D-Phe^7]-α-melanocyte-stimulating hormone (MSH) was utilized to photoacoustically delineate the boundaries of B16 melanomas in small animals *in vivo* (Fig. 10.7a) [76]. The enhancement in PA signal within the melanomas reached approximately 38% with an intravenous injection of [Nle^4, D-Phe^7]-α-MSH-GNCs at the dose of 1 pmol. In

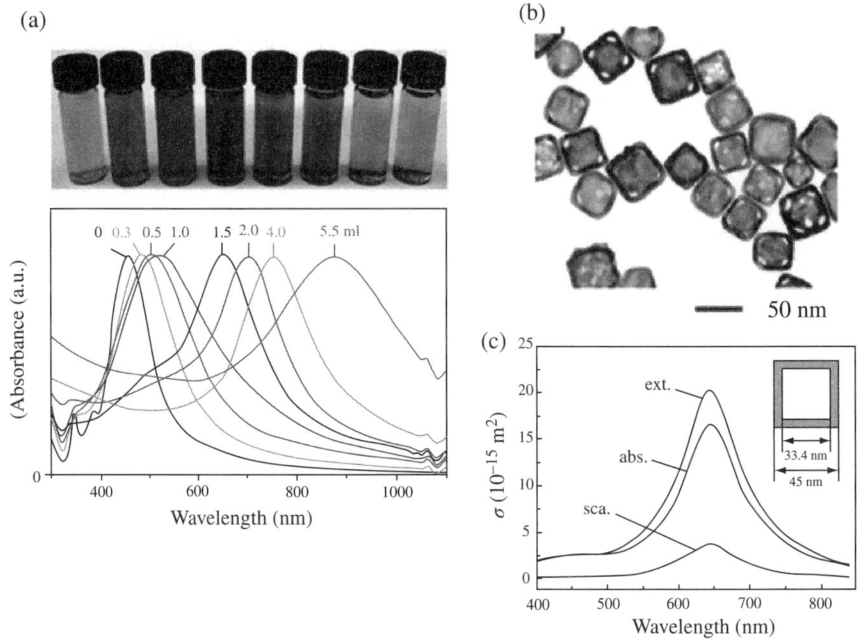

FIGURE 10.6 Optical and physical properties of gold nanocages (GNCs). (a) Various types of GNCs prepared by reacting Au nanocube solution with different volumes of $HAuCl_4$ solution (Reprinted with permission from Ref. [122]. © American Chemical Society.). (b) TEM image of GNCs with an edge length of 55 nm (Reprinted with permission from Ref. [76]. © American Chemical Society.). (c) Calculated optical spectrum of GNCs with an outer edge length of 45.0 nm and wall thickness of 5.8 nm. (Reprinted with permission from Ref. [124]. © American Chemical Society.)

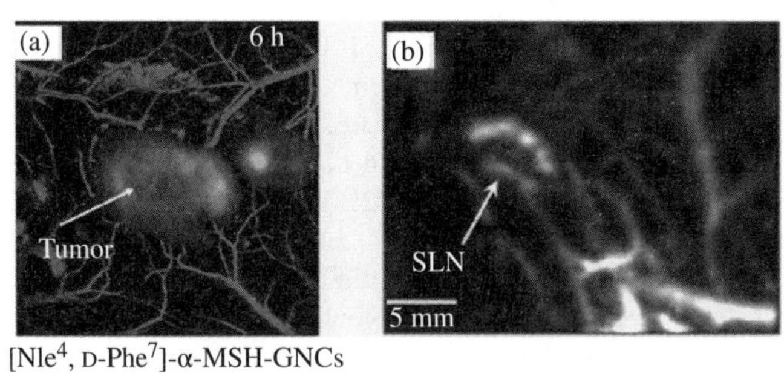

FIGURE 10.7 *In vivo* photoacoustic (PA) imaging of GNCs in small animals. (a) *In vivo* noninvasive PA images of B16 melanomas in a nude mouse acquired at 6 h post-intravenous injection of [Nle4,D-Phe7]-α-MSH-GNCs (Reprinted with permission from Ref. [76]. © American Chemical Society.). (b) *In vivo* PA images of sentinel lymph nodes of a rat acquired after the intradermal injection of GNCs. (Reprinted with permission from Ref. [119]. © American Chemical Society.)

addition, GNCs were successfully utilized as a nonionizing optical lymph node tracer for noninvasive PA mapping sentinel lymph nodes in small animals *in vivo* (Fig. 10.7b) [119]. After intradermal injection of 0.2 pmol GNCs, the PA signal was enhanced by 630%.

10.5.2 Biodegradable Plasmonic Gold Nanobeacons and Organic Porphysomes [115–118, 125, 126]

In spite of these promising results and biocompatibility, GNCs are not biodegradable. The previous literature indicates that greater than 40% of intravenously injected PEGylated GNCs accumulated in internal organs such as the livers and spleens, and its long-term toxicity has not been studied [127, 128]. Thus, a concern regarding long-term heavy metal poisoning is still an ongoing debate. Therefore, this may ultimately hinder inorganic nanostructures from being used widely in clinical practice. In contrast, organic nanoparticles (including liposomes, micelles, and polymersomes) have been widely used in many clinical therapeutic practices due to their biocompatibility, biodegradability, and drug delivery capacity [129]. Unfortunately, they are not useful for PAT because the light absorption of these organic nanoparticles is significantly low and thus negligible in the NIR spectral region. To enhance the biodegradability of nanoplatforms while maintaining strong optical absorption, two types of nanoparticles have been tested: (1) colloidal gold nanobeacons formed by a cluster of small gold nanoparticles (~2–4 nm) coated with large phospholipid layer (~150 nm) [115–118] and (2) porphysomes self-assembled from phospholipid–porphyrin conjugates, providing liposome-like structure [125, 126]. In the first type, smaller gold nanoparticles (2–4 nm) were encapsulated by soft and hollow colloidal phospholipid particles as shown in Figure 10.8 [117]. Although gold nanobeacons still used metallic gold nanoparticles, the size of them was small enough to be excreted from a body. Further, gold nanobeacons strongly absorb light in both visible and NIR regimes, and thus it can serve as a PA contrast agent. To test the *in vivo* PA imaging capability of gold nanobeacons, axillary vasculatures were monitored before and after intravenous injection of a solution of gold nanobeacons. The PA signals in the blood vessels quantified from the PA image (Fig. 10.8b, right) acquired at 156 min postinjection were enhanced by approximately 60% compared to that measured from the control PA image (Fig. 10.8b, left). Thus, gold nanobeacons can be potentially used for molecular PAT as exogenous contrast agents. In the second type, porphysomes are self-assembled from monomers, efficient drug delivery vehicles, highly biodegradable, and highly biocompatible. In addition, porphysomes have strong and tunable optical absorption coefficients and are effective agents for PAT. Figure 10.8c shows a schematic and electron micrograph of a pyropheophorbide–lipid porphysome. From a phantom study, PA signals could be detectable from the porphysomes at a particle concentration of 25 pM. This detection sensitivity corresponds to approximately 50×10^{-21} moles of porphysomes per imaging voxel. The axillary region of a rat was photoacoustically imaged before and after intradermal injection of porphysomes (2.3 pmol) in small animals *in vivo*. Secondary lymph vessels, sentinel lymph node, and inflowing lymph vessel are clearly visible in the PA image (Fig. 10.8d).

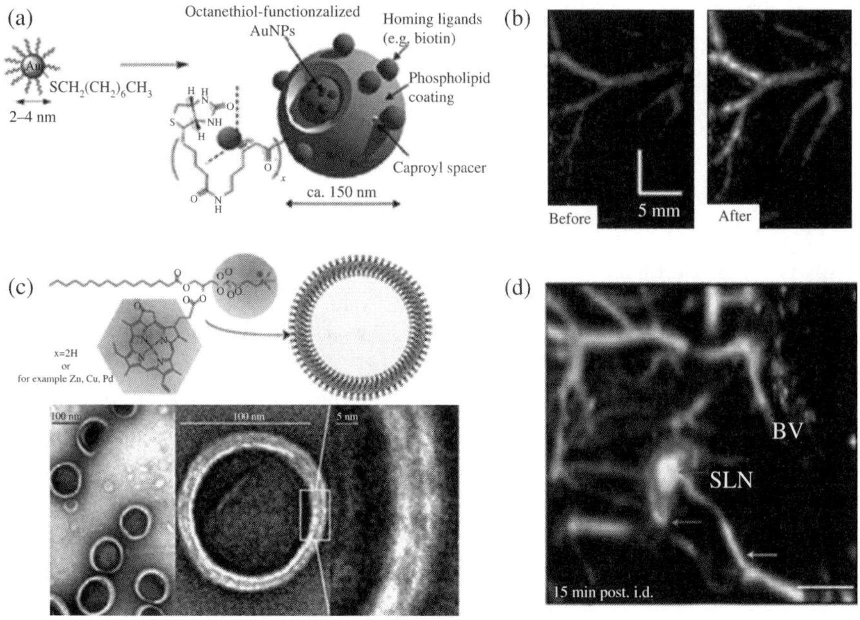

FIGURE 10.8 *In vivo* photoacoustic (PA) imaging of potentially biodegradable contrast agents in small animals *in vivo*. (a) Gold nanobeacons (GNBs) synthesized from octanethiol-functionalized gold nanoparticles (AuNPs) with phospholipid coating. (b) *In vivo* PA images acquired before and after intravenous injection of GNBs (Reprinted with permission from Ref. [117]. © John Wiley & Sons. (c) Schematic and TEM images of a pyropheophorbide–lipid porphysome. (d) *In vivo* PA images acquired after intradermal injection of porphysomes. (Reprinted with permission from Ref. [126]. © Macmillan Publishers Ltd.)

10.5.3 Dye-Enhanced Single-Walled Carbon Nanotubes

As one of the explored nanosized contrast agents, single-walled carbon nanotubes (SWCNTs) have been considerably examined as a theranostic agent in preclinical studies [130–137]. The advantages of SWCNTs include the large surface area, easy fabrication and bioconjugation, strong mechanical properties, and strong optical absorption. In addition, unlike other plasmonic gold nanoparticles, SWCNTs have the broader optical absorption spectrum. Thus, any optical wavelength can be possibly chosen according to specific applications. Based on these desirable optical characteristics, SWCNT has been successfully used as a PA contrast agent. However, there is still an ongoing debate on the toxicity of carbon nanotubes for biomedical applications. Thus, increasing PA sensitivity is the key issue for developing new carbon nanotube-based agents, and then the injection dose can be greatly reduced. Consequently, its ongoing concern on toxicity can be relatively relieved. Recently, de la Zerda *et al.* demonstrated SWCNTs conjugated to ICG (SWCNT-ICG) for molecular PA imaging of brain tumors in living mice [137]. Figure 10.9a shows the schematic of SWCNT-ICG conjugate, and the corresponding optical spectra are

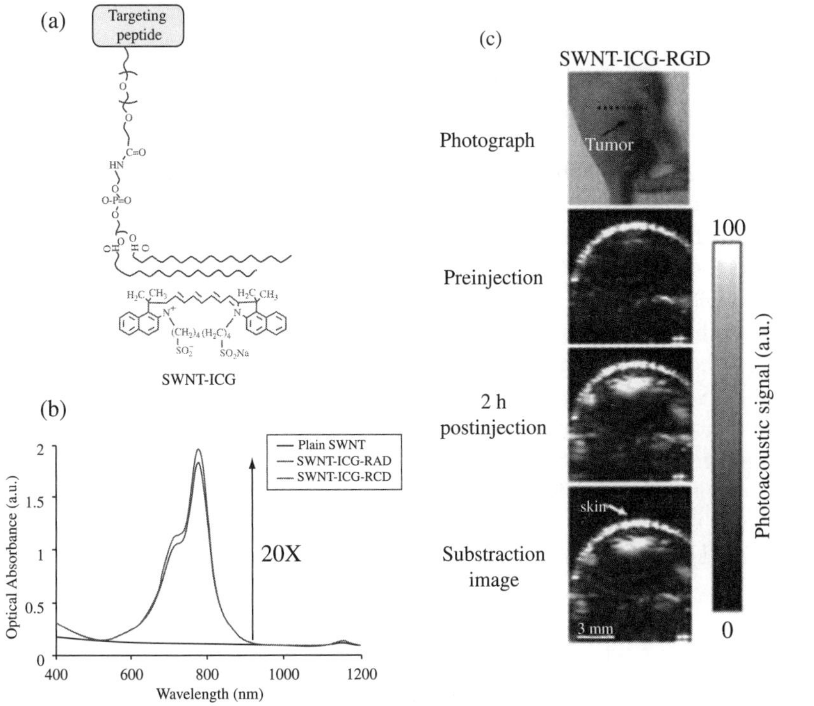

FIGURE 10.9 *In vivo* photoacoustic (PA) imaging of U87 tumors in living mice using single-walled carbon nanotube-indocyanine green conjugated with RGD peptide (SWCNT-ICG-RGD). (a) Schematic of a SWCNT-ICG particle. (b) Optical spectra of plain SWCNT and SWCNT-ICG-RGD. (c) *In vivo* PA images of U87 tumors in living mice acquired with SWCNT-ICG-RGD injection. (Reprinted with permission from Ref. [137]. © American Chemical Society.)

shown in Figure 10.9b. When SWCNTs were conjugated with ICG, the optical absorption was enhanced by 20-fold. The *in vivo* PA sensitivity of SWCNT-ICG (~0.8 nM) was approximately 100 times higher than that of plain SWCNTs (~100 nM). A subcutaneous U87 brain tumor xenograft was photoacoustically imaged after injection of SWCNT-ICG-RGD, which targeted $\alpha_v\beta_3$ integrins overexpressed in the tumor (Fig. 10.9c). The PA signal was enhanced by greater than 300% with SWCNT-ICG-RGD injection, whereas the improvement was limited to only 80% with SWCNT-RGD injection under the same injection dose.

10.5.4 Dual-Modality Photoacoustic and Ultrasound Contrast Agents

Due to the hybrid nature of PAT (i.e., combination of light and US), PAT has been easily integrated with existing US imaging systems [138, 139]. The combined system shared the same US detectors. At the same time, the dual-modality contrast agents for PA and US imaging have been investigated to improve the sensitivity and specificity. Microbubbles filled with gas are successfully adapted in clinical practice as US contrast agents [140, 141]. Due to the frequencies (on the order of MHz) of clinical US imaging systems, the sizes of these bubbles are on the order of micrometers. Since typical microbubbles do not intrinsically absorb light because of its optical transparency, additional PA contrast agents were added into microbubbles. Figure 10.10a shows the schematic of encapsulated-ink poly(lactic-co-glycolic acid) (PLGA) microbubbles and nanobubbles [138, 142]. The average sizes of microbubbles and nanobubbles were approximately 1 and 0.3 μm, respectively. Figure 10.10b and c are US and PA images obtained from a tissue-simulating phantom, respectively. Objects 1 through 4 were composed of encapsulated-ink microbubbles at concentrations of 2.5, 5.0, 10, and 15 mg/ml, respectively, while objects 5 through 8 were made of encapsulated-ink nanobubbles at concentrations of 2.5, 5.0, 10, and 15 mg/ml, respectively. All eight objects are clearly shown in both the US and PA images. Linear increases of the US and PA signals were observed as a function of concentration. As the second example, Figure 10.10d shows the TEM image of encapsulated gold nanorod human serum albumin (HSA)-shelled microbubbles (AuNR-HSA) [139]. The US and PA images acquired from a sample containing AuNR-HSA are shown in Figure 10.10e and f, respectively. As another example, liquid perfluorocarbon (PFC) nanodroplets with encapsulated plasmonic

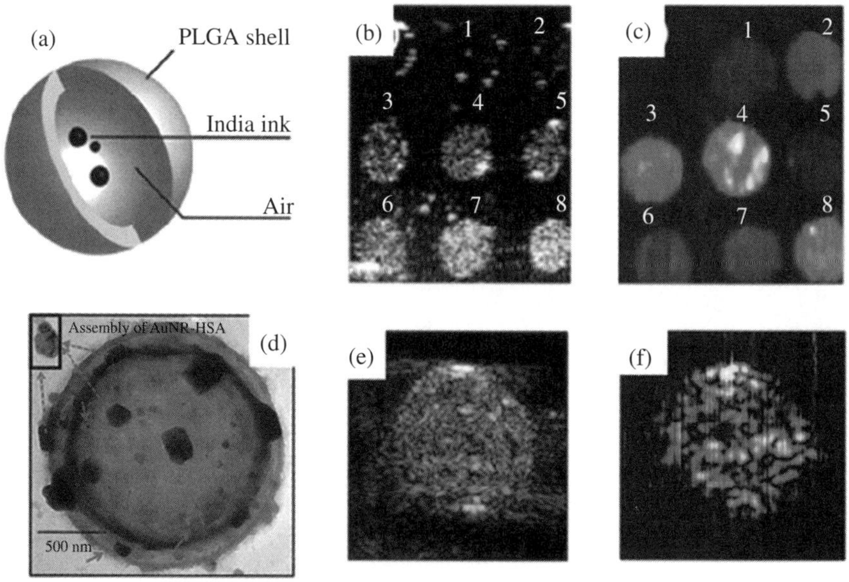

FIGURE 10.10 Dual-modality photoacoustic (PA) and ultrasound (US) contrast agents. (a) Schematic of PLGA microbubbles (MBs) and nanobubbles (NBs) encapsulating black india ink. (b) US and (c) PA images of a phantom containing tumor simulators made of encapsulated-ink MBs and NBs with various concentrations. 1 through 4: MBs at concentrations of 2.5, 5.0, 10, and 15 mg/ml, respectively. 5 through 8: NBs at concentrations of 2.5, 5.0, 10, and 15 mg/ml, respectively. (d) TEM images of albumin-shelled MBs with encapsulated gold nanorods (AuMBs). (e) US and (f) PA images of a phantom containing AuMBs. (Reprinted with permission from Refs. [138, 139]. © SPIE.)

nanoparticles were successfully developed as a dual-modality PA and US contrast agent [143]. Upon laser irradiation and its consequent liquid-to-gas phase transition of liquid PFC, significant PA signals were generated. The generated gaseous phase enhances the US contrast.

10.5.5 Dual-Modality Photoacoustic and Fluorescence Contrast Agents and Activatable Probe

As another dual-modality imaging capability of hybrid PAT, FL imaging has been combined with PAT [126, 144–146]. As mentioned previously, the conversion efficiency between specific optical absorption and heat of a given molecule determines the amplitudes of initially generated PA pressure waves (see section "Principles of PAT"). Thus, it is expected that a material with high nonradiative quantum yield is a good candidate as a PA contrast agent. In other words, if the quantum yield between radiative and nonradiative energy release is properly controlled, the dual-modality PA and FL imaging is highly feasible using a single contrast agent. For example, owing to its moderate FL quantum yield (e.g., ~12% in DMSO and <2% in water), ICG (an FDA-approved organic dye) was used as a multimodal contrast agent for both FL and PA imaging. Recently, Walter *et al.* demonstrated that FL quenching of NIR fluorescent dyes in nanoparticles significantly enhanced the PA contrast, a way to convert a highly

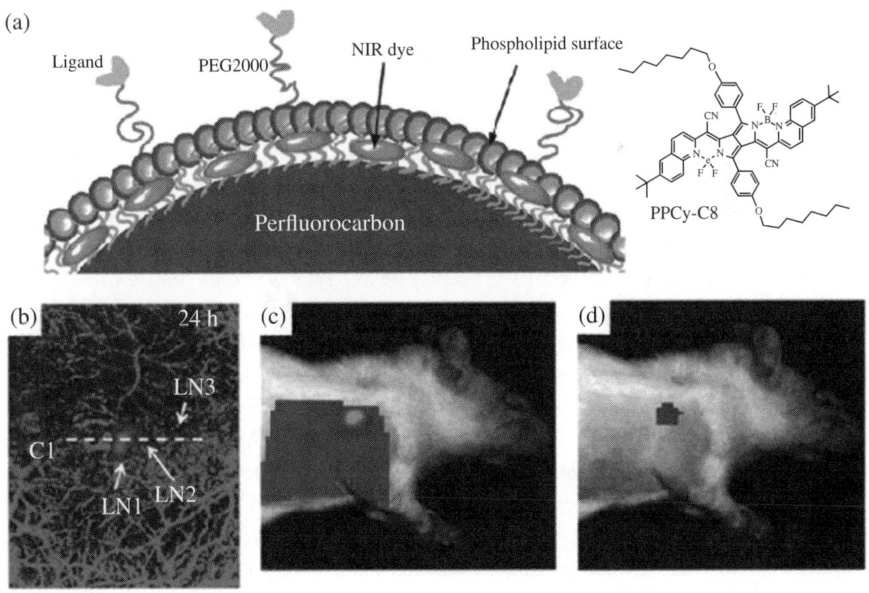

FIGURE 10.11 Dual-modality photoacoustic (PA) and fluorescence (FL) contrast agents. (a) Schematic of dye-loaded perfluorocarbon nanoparticles and its chemical structure. (b) *In vivo* noninvasive PA mapping of sentinel lymph node in small animals. *In vivo* (c) intensity- and (d) lifetime-based FL mapping of the sentinel lymph node in a rat. (Reprinted with permission from Ref. [144]. © American Chemical Society.) (*See insert for color representation of the figure.*)

fluorescent dye into a strong PA contrast material [144]. Figure 10.11a shows the schematic of PFC nanoparticles loaded with an NIR dye, pyrrolopyrrole cyanine (PPCy-C8), and the chemical structure of the dye. Specially, PFC is biocompatible and has a half-life of 3 days in the body. Thus, PFC can completely be removed from the body via expired air. From *in vitro* studies, PPCy-C8-loaded PFC nanoparticles produced a significantly stronger PA signal (~9 times) compared to whole blood, while no and significantly lower PA signals were generated from pure PFC nanoparticles and free dye, respectively. As an *in vivo* preclinical application, PPCy-C8-loaded PFC nanoparticles were successfully used as a PA lymph node tracer. Lymph nodes were detected by PAT at 3 h postinjection with a PA contrast of 8.7, and the contrast was enhanced to 12.3 at 24 h postinjection. FL imaging was also performed. *In vivo* FL intensity and FL lifetime imaging of lymph nodes with injection of the PPCy-C8-loaded PFC nanoparticles are shown in Figure 10.11b and c. The FL intensity with dye-loaded PFC nanoparticles was approximately 10 times lower than that with the free dye. FL lifetime of free PPCy-C8 in the lymph node was greater than 3 ns, while that of the dye-loaded PFC nanoparticles was 0.22 ns. This result implies that the FL property of the dye was quenched in the PFC nanoparticles.

Further, this type of dual-modality PA and FL imaging contrast agent can be physiologically activated, referred to as an activatable PA probe. Figure 10.12a shows the diagram of GNCs and organic dye molecules (FITC) connected together through an enzyme-cleavable peptide [146]. To fabricate the matrix metalloprotease (MMP)-sensitive agent, a FITC-labeled C-terminus peptide was conjugated to the surface of GNCs. By controlling the LSPR peak of the GNCs (770 nm) away from the FL emission wavelength of the dye molecule, the FL light emitted from the released dyes was detected with high sensitivity. The FL quenching efficiency for the dye was high. Prior to enzyme cleavage, the FL intensity of the solution was 18. After incubation with MMP-2 enzyme for 3 h, the FL dramatically increased to 69. As another approach, Levi *et al.* developed an activatable PA probe combining two separate organic molecules having two distinct optical absorption peaks (Fig. 10.12b) [145]. When the linker between two dyes was cleaved by the MMP-2 enzyme, the dye conjugated with cell-penetrating peptide (CPP) stayed in cells, whereas the other one diffused away. The first dye was BHQ3 with an absorption wavelength of 672 nm and an extinction coefficient of 42,700 $M^{-1} cm^{-1}$, while the second one was Alex 750 with an excitation wavelength of 749 nm and an extinction coefficient of 290,000 $M^{-1} cm^{-1}$. Before cleavage, strong PA signals were generated from the conjugate, BHQ3–activatable photoacoustic probe–Alex 750 (B-APP-A), at both 650 and 750 nm. After the linker was cleaved by MMP-2 enzyme, a strong PA signal was observed at 650 nm. The subtracted PA image (i.e., PA image acquired at 650 nm—PA image obtained at 750 nm) clearly visualized the enzymatic process (Fig. 10.12c). As another successful activatable contrast agent, aforementioned porphysomes have significant photoswitching capability between PA and FL properties. Intact porphysomes are quenched but can become fluorescent upon dissociation. The nanoscale photoswitching is significant, with 1000-fold FL increase per particle [126].

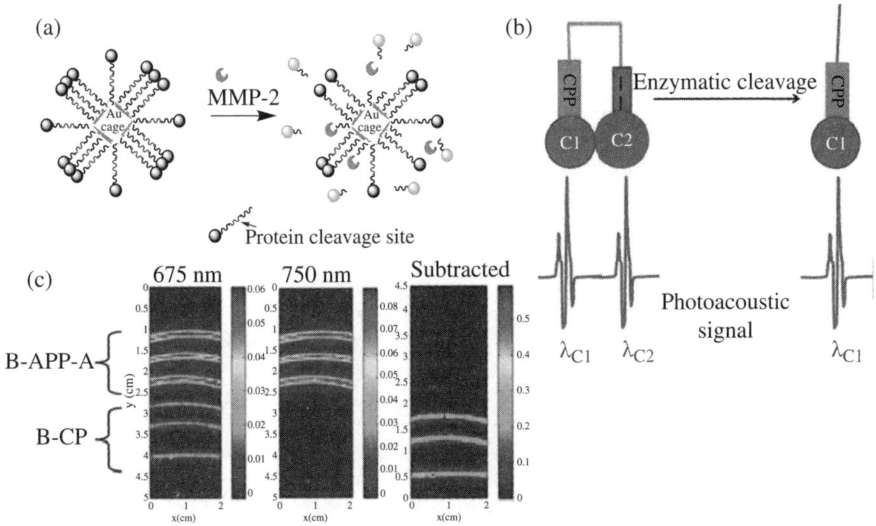

FIGURE 10.12 Enzyme-sensitive activatable photoacoustic (PA) contrast agents. (a) Schematic of a GNC-based activatable probe. GNCs and fluorescent dyes are linked through enzyme-cleavable peptides (based on Ref. [146]). (b) Schematic of an activatable PA probe consisting of two chromophores linked through enzyme-cleavable peptides. Initially, two strong PA signals were generated at two optical wavelengths. When cleaved, the PA signal was induced at only one of the two wavelengths. (c) Dual-wavelength PA imaging of the activatable probes. B-APP-A (BHQ3-APP-Alex 750, uncleaved) and B-CP (cleaved). APP, activatable photoacoustic probe. (Adapted with permission from Ref. [145]. © American Chemical Society.)

10.5.6 Multifunctional Magnetomotive Photoacoustic Contrast Agent

As another approach to boost PA sensitivity, multifunctional magnetomotive PA (mmPA) contrast agents were successfully developed [147–149]. Strong background PA signals from intrinsic contrasts can be significantly suppressed by either magnetic modulation or magnetic enrichment. For the first method, magnetic nanoparticle-gold core–shell nanostructures (MNP-Au) were manipulated as shown in Figure 10.13a [148]. The principle of mmPA imaging was that MNP-Au nanoparticles were displaced under a magnetic field and then they returned back to the original positions when the field was off. By processing a series of PA images based on the magnetic-induced motion, PA signals stemming from MNP-Au nanoparticles were rectified, while all background signals were rejected. This process greatly improved the PA sensitivity and specificity of the nanoparticles. Figure 10.13b shows that mmPA imaging significantly improved contrast sensitivity and specificity. Although an inclusion containing gold nanorods generated strong PA signals in a conventional PA imaging mode, the PA signals were completely suppressed in the mmPA imaging mode. However, the strong PA signals were produced from an inclusion with MNP-Au nanoparticles in both conventional and mmPA imaging modes and also stronger (20 dB) than that with MNPs alone. In addition, these MNPs can be used as an MRI contrast agent as well.

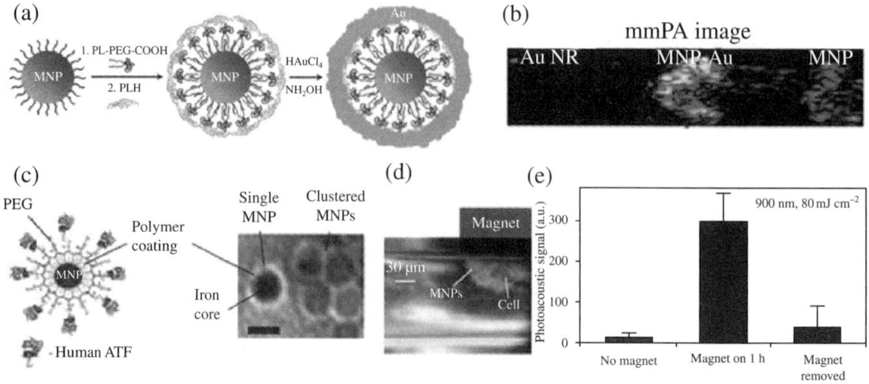

FIGURE 10.13 Magnetomotive photoacoustic (PA) contrast agents. (a) Schematic of magnetic nanoparticle-gold core–shell nanostructures (MNP-Au) (Reprinted with permission from Ref. [147]. © Macmillan Publishers Ltd.). (b) Magnetomotive PA imaging of gold nanorods (AuNR), MNP-Au, and magnetic nanoparticles (MNPs). (c) Schematic (left) and TEM image (right) of MNPs. (d) *In vitro* magnetic trapping of MNP-labeled cancer cells at a flow velocity of 5 cm/s. (e) *In vivo* PA detection of magnetic enrichment of circulating tumor cells in tumor-bearing mice. (Reprinted with permission from Ref. [148]. © Macmillan Publishers Ltd.) (*See insert for color representation of the figure.*)

For the second method, MNPs were aggregated under a magnetic field, the local concentration of MNPs within an ROI was increased, and finally, the PA signals were significantly enhanced, referred to as magnetic enrichment [147]. This approach was applied to magnetically capture rare circulating tumor cells (CTCs) in the systemic bloodstream in living mice and photoacoustically detect the CTCs with high contrast. The illustration and TEM image of MNPs are shown in Figure 10.13c. Magnetic capturing of CTCs conjugated with MNPs was clearly identified as shown in an optical microscopy image *in vitro* (Fig. 10.13d). PA signals measured from CTCs in abdominal skin vessels of tumor-bearing mice were compared before, during, and after magnet placement. The PA amplitude increased 88-fold (Fig. 10.13e). After the magnet was removed, the PA signals were dramatically reduced due to the release of captured CTCs attached to MNPs. As another example, plasmonic gold nanostars cored with superparamagnetic iron oxide were effectively used as a sentinel lymph node tracer in PAT [114].

As another example, Kircher *et al.* have recently developed a novel triple-modality MRI–PA imaging–Raman imaging nanoparticle (MPR nanoparticle) to clearly visualize the brain tumor margins [149]. The MPR nanoparticle comprised a 60 nm gold core covered with the Raman molecular tag, and a 30 nm silica coating protected the Raman-active layer. Further, the particles were coated with Gd^{3+} ions resulting in the MPR nanoparticle (Fig. 10.14a). The clinical scenario includes (i) preoperative MRI detection of brain tumors after injection of MPR nanoparticles and its consequent surgical planning, (ii) intraoperative PA image-guided surgical resection of bulky tumors, and (iii) Raman image-guided microsurgery of residual tumors (Fig. 10.14a).

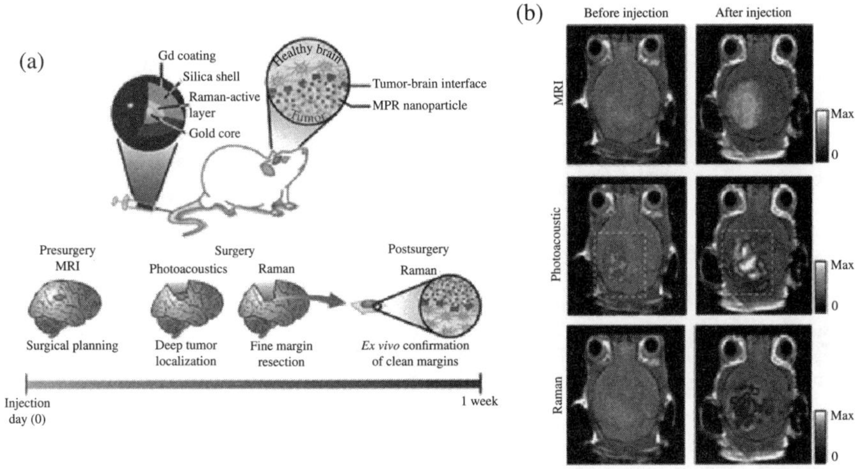

FIGURE 10.14 (a) Schematic and clinical scenario of brain tumor molecular imaging using magnetic resonance imaging–photoacoustic imaging–Raman imaging nanoparticle (MPR nanoparticle). (b) Triple-modality identification of brain tumors in living mice with MPRs. (Reprinted with permission from Ref. [149]. © Macmillan Publishers Ltd.)

All three images, acquired after injection of MPR nanoparticles, clearly visualize the brain tumor through the intact skin and skulls of a living mouse (Fig. 10.14b). The MRI contrast-to-noise ratio increased by approximately 536%, while the PA signal increased by 75%. Further, the Raman signal was 1.0 ± 0.09 AU after injection, while the signal was not detectable before injection.

10.5.7 Radiolabeled Photoacoustic Contrast Agent [150, 151]

Although PAT overcomes the shallow penetration depth of pure optical imaging by probing several centimeters in tissues, the PA sensitivity decreases as the imaging depth increases and the strong PA background signals from intrinsic contrasts reduce the PA image contrast and sensitivity. This issue can be overcome by combining PAT with another imaging tool that can provide high SNR in deep tissues and validate the findings of each modality. When PAT combines with nuclear imaging, the strong SNR and accurate quantification can be derived from nuclear imaging, whereas PAT can provide improved local information with high spatial resolution. To study the feasibility of the dual-modality imaging capability, radiolabeled optical absorbents were developed for PAT and single-photon emission computed tomography (SPECT). As the first approach, gold nanorods were radiolabeled with ^{125}I and conjugated with antitumor necrosis factor (TNF-α) [151]. PA and nuclear imaging were performed to identify the distribution of gold nanorods in articular tissues in small animals' tail joints *in situ*. A microCT image clearly shows the sagittal section of the rat tail joints in Figure 10.15a. The distribution of ^{125}I–gold nanorod–TNF-α conjugate and soft tissue structures were clearly visualized by PAT, while the hard tissue structures

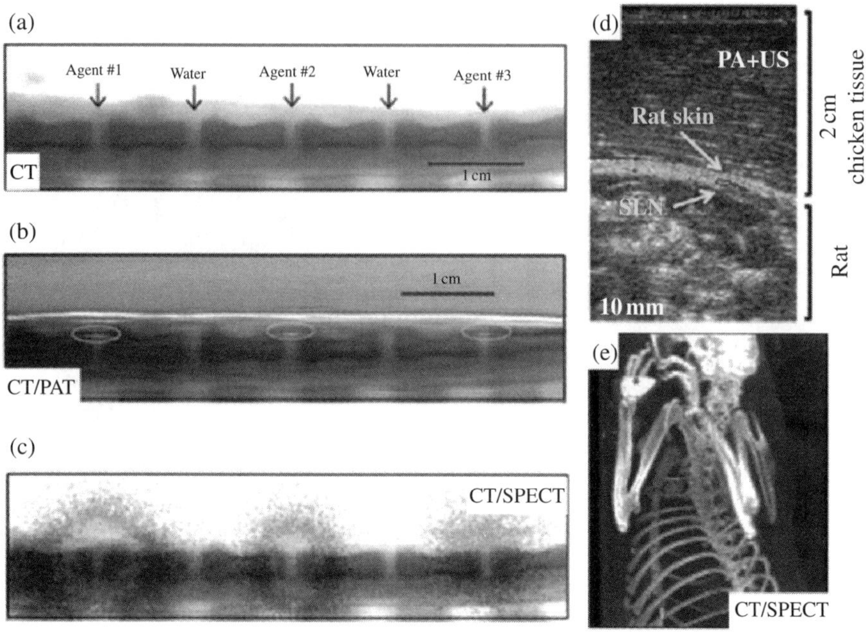

FIGURE 10.15 Multimodal imaging of radiolabeled optical absorbents in small animals. (a) MicroCT image of a sagittal section of joints in a rat tail after intra-articular injection of gold nanorods labeled with ^{125}I. (b) Overlaid PA and CT image of the joint section. (c) Overlaid CT and SPECT image of the joint. (d) Overlaid PA and US image of a sentinel lymph node after intradermal injection of methylene blue. (e) SPECT/CT projection image of the rat axillary region acquired at 1 h postinjection of ^{125}I methylene blue. (Reprinted with permission from Refs. [150, 151]. © SPIE and © Elsevier.)

(i.e., bone) were clearly shown in the microCT image (Fig. 10.15b). In Figure 10.15c, SPECT image was overlaid with the microCT image visualizing both the distribution of contrast agent and the hard tissue morphologies. As one can see, the contrast of nuclear imaging was stronger, whereas the resolution of PAT was better to show microstructures. As the second approach, MB was radiolabeled with ^{125}I as a PA and nuclear lymph node tracer [150]. PA and nuclear imaging were performed to map axillary lymph nodes in small animals *in vivo*. An overlaid PA and US image (Fig. 10.15d) provides both functional (MB uptake in the sentinel lymph node) and structural information of surrounding tissues, respectively. The SPECT image clearly visualizes the localized sentinel lymph node uptaken by ^{125}I MB in the axillary region at 1 h postinjection, while the CT image shows the surrounding structures (Fig. 10.15e).

10.5.8 Organic Dyes

Highly optically absorbing organic dyes (i.e., having high nonradiative quantum yield) have been widely applied as a PA contrast agent to various preclinical and clinical applications. ICG [85], MB [83], and Evans blue (EB) [152, 153] have

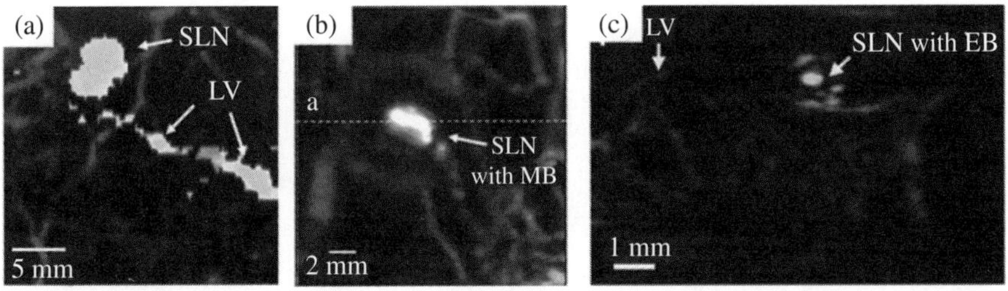

FIGURE 10.16 *In vivo* PA images of sentinel lymph nodes (SLNs) in small animals after intradermal injection of (a) indocyanine green (ICG), (b) methylene blue (MB), and (c) Evans blue (EB). (Reprinted with permission from Refs. [83, 85, 153]. © SPIE, © The American Association of Physicists Medicine, © Radiological Society of North America.)

already been used in clinics, which is the strong advantage of using these organic dyes. The molar extinction coefficients of ICG, MB, and EB are 1.5×10^5, 7.4×10^4, and $7.8 \times 10^4\,cm^{-1}\,M^{-1}$, respectively. The peak absorption wavelengths of ICG and MB are 790 and 667 nm, respectively, within the NIR window (650–900 nm), while that of EB is 626 nm. However, the drawbacks of these dyes include the short circulation time in a bloodstream, smaller optical absorption cross sections, and difficult surface modification. ICG, MB, and EB have been successfully used for noninvasive mapping of sentinel lymph nodes in small animals *in vivo* (Fig. 10.16).

10.5.9 Fluorescence Proteins [154]

Imaging FL proteins using FL microscopy plays an important role in biological researches. As aforementioned, however, the applications of FL proteins are limited because of the shallow imaging depth of optical imaging modalities. Some FL proteins having small fluorescent quantum yields (i.e., high nonradiative quantum yield) can be great contenders as PA contrast agents. Since PAT can achieve deep tissue imaging, molecular PAT combined with FL proteins has great potential in biological studies. For the first time, the expression of mCherry in an adult transgenic zebra fish was photoacoustically mapped *in vivo* [155]. However, visible optical wavelengths, where the peak optical absorption of mCherry is, were used, and the imaging depth was consequently limited in this approach. Recently, Filonov *et al.* have developed an NIR FL protein, bacteriophytochrome-based NIR FL proteins (named iRFP), as a PA probe [154]. iRFP possesses a very high molar extinction coefficient of $1{,}05{,}000\,cm^{-1}\,M^{-1}$ and a low fluorescent quantum yield of 6%. To prove the strong PA contrast of iRFP, the PA signals measured from purified iRFP, E2-Crimson, mNeptune, mKate2, eqFP670, and TagRFP657 proteins were compared with lysed oxygenated blood *in vitro* (Fig. 10.17a). The PA amplitude of iRFP at 680 nm was higher than that at 600 nm and was the highest among all proteins. Figure 10.17 shows the spectrally decoded PA images of a mouse mammary gland tumor *in vivo* without (Fig. 10.17b) and with (Fig. 10.17c) the tissue overlay. The boundary of the tumor positioned at a depth of 4 mm in tissue was clearly identified.

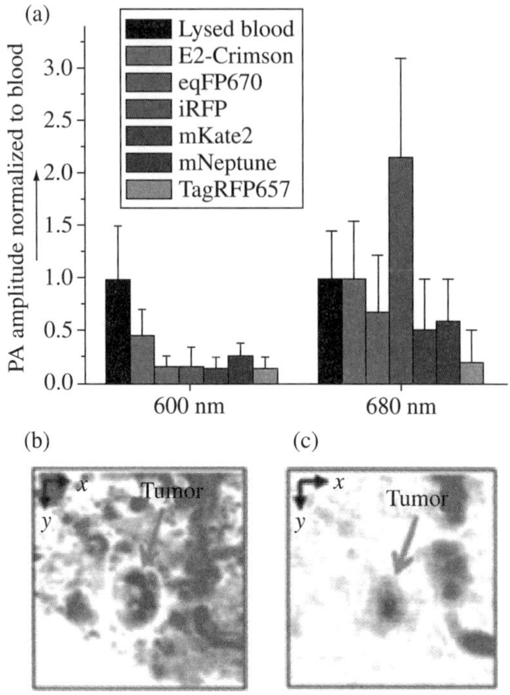

FIGURE 10.17 *In vivo* PA images of mouse mammary gland tumor. (a) Comparison of PA signals of various proteins such as purified iRFP, purified E2-Crimson, eqFP670, mKate2, mNeptune, and TagRFP657 at 600 and 680 nm. *In vivo* spectrally resolved PA images of the tumor and surrounding blood vessels acquired without (b) and with (c) overlaid 4 mm-thick tissue. Blue represents the tumor, whereas red represents the blood vessels. (Reprinted with permission from Ref. [154]. © John Wiley & Sons.)

10.6 CONCLUSIONS

PAT is a novel premier hybrid biomedical imaging technique, combining strong optical absorption contrast and high ultrasonic resolution. In this chapter, the following topics of PAT have been discussed: (i) basic principles of PAT, (ii) two types of PAT modalities (i.e., inverse algorithm-based PACT and raster scanning-based PAM), (iii) intrinsic contrasts for morphological and functional PAT, and (iv) exogenous contrasts for PAT. Although many contrast agents manipulated by nanotechnology have been investigated and the results were promising in preclinical PA applications, so far, any nanostructures have not been approved by the FDA, which is a key hurdle for clinical translation. The key criterion for the clinical agent must be biodegradability and biocompatibility. For example, biocompatible metallic nanoplatforms should be removed from the body so that the long-term heavy metal toxicity must be excluded. Otherwise, the absence of any potential hazards should be proved with significant long-term metal poisoning studies. Once these problems are resolved, the potential for clinical translation of these agents with PAT is very high. These promising clinical applications of PAT include imaging primary tumor margins, early diagnosis, or staging metastasis.

REFERENCES

[1] Marchese-Ragona SP, Haydon PG. Near-field scanning optical microscopy and near-field confocal optical spectroscopy: emerging techniques in biology. Ann N Y Acad Sci 1997;**820**:196–206; discussion 197–206.

[2] Patel DV, McGhee CN. *In vivo* confocal microscopy of human corneal nerves in health, in ocular and systemic disease, and following corneal surgery: a review. Br J Ophthalmol 2009;**93**:853–860.

[3] Reynaud K, Nogueira D, Cortvrindt R, Kurzawa R, Smitz J. Confocal microscopy: principles and applications to the field of reproductive biology. Folia Histochem Cytobiol 2001;**39**:75–85.

[4] Rollan A, Ward T, McHale AP. Real time confocal laser scanning microscopy: potential applications in space medicine and cell biology. Acta Astronaut 1998;**42**:37–50.

[5] Stelzer EH, Wacker I, De Mey JR. Confocal fluorescence microscopy in modern cell biology. Semin Cell Biol 1991;**2**:145–152.

[6] Farkas ER, Smith RL, Baumgart T, Webb WW. Probing chain-order in model membranes via fluorescence polarization using two-photon microscopy. *Biophys J* 2007:420a.

[7] Guild JB, Xu C, Webb WW. Pulse dispersion by high numerical aperture objectives for nonlinear microscopy using two-photon fluorescence excitation. Biophys J 1996;**70**: Wp284.

[8] Huang SH, Heikal AA, Webb WW. Two-photon fluorescence spectroscopy and microscopy of NAD(P)H and flavoprotein. Biophys J 2002;**82**:2811–2825.

[9] Williams RM, Webb WW. Plasma membrane heterogeneities during cytokinesis of RBL-2H3 cells characterized using two photon excited fluorescence microscopy. Biophys J 1996;**70**:Tu419.

[10] Huang D, Swanson EA, Lin CP, Schuman JS, Stinson WG, Chang W, Hee MR, Flotte T, Gregory K, Puliafito CA, *et al.* Optical coherence tomography. Science 1991;**254**: 1178–1181.

[11] Tearney GJ, Brezinski ME, Bouma BE, Boppart SA, Pitris C, Southern JF, Fujimoto JG. *In vivo* endoscopic optical biopsy with optical coherence tomography. Science 1997;**276**: 2037–2039.

[12] Boppart SA. Advances in contrast enhancement for optical coherence tomography. Conf Proc IEEE Eng Med Biol Soc 2006;**1**:121–124.

[13] Jeon M, Kim J, Jung U, Lee C, Jung W, Boppart SA. Full-range k-domain linearization in spectral-domain optical coherence tomography. Appl Opt 2011;**50**:1158–1163.

[14] Jung W, Boppart SA. Optical coherence tomography for rapid tissue screening and directed histological sectioning. Anal Cell Pathol (Amst) 2012;**35**:129–143.

[15] Jung W, Kim J, Jeon M, Chaney EJ, Stewart CN, Boppart SA. Handheld optical coherence tomography scanner for primary care diagnostics. IEEE Trans Biomed Eng 2011;**58**:741–744.

[16] Oldenburg A, Toublan F, Suslick K, Wei A, Boppart S. Magnetomotive contrast for *in vivo* optical coherence tomography. Opt Express 2005;**13**:6597–6614.

[17] Graves EE, Culver JP, Ripoll J, Weissleder R, Ntziachristos V. Singular-value analysis and optimization of experimental parameters in fluorescence molecular tomography. J Opt Soc Am A Opt Image Sci Vis 2004;**21**:231–241.

[18] Nothdurft RE, Patwardhan SV, Akers W, Ye Y, Achilefu S, Culver JP. *In vivo* fluorescence lifetime tomography. J Biomed Opt 2009;**14**:024004.

[19] Patwardhan S, Bloch S, Achilefu S, Culver J. Time-dependent whole-body fluorescence tomography of probe bio-distributions in mice. Opt Express 2005;**13**:2564–2577.

[20] Patwardhan SV, Bloch S, Achilefu S, Culver JP. Quantitative small animal fluorescence tomography using an ultrafast gated image intensifier. Conf Proc IEEE Eng Med Biol Soc 2006;**1**:2675–2678.

[21] Solomon M, White BR, Nothdruft RE, Akers W, Sudlow G, Eggebrecht AT, Achilefu S, Culver JP. Video-rate fluorescence diffuse optical tomography for *in vivo* sentinel lymph node imaging. Biomed Opt Express 2011;**2**:3267–3277.

[22] Wang LV, Wu H. *Biomedical Optics: Principles and Imaging*. Hoboken: John Wiley & Sons, Inc.; 2007.

[23] Wang LV. Multiscale photoacoustic microscopy and computed tomography. Nat Photonics 2009;**3**:503–509.

[24] Zhang HF, Maslov K, Stoica G, Wang LV. Functional photoacoustic microscopy for high-resolution and noninvasive *in vivo* imaging. Nat Biotechnol 2006;**24**:848–851.

[25] Wang X, Pang Y, Ku G, Xie X, Stoica G, Wang LV. Noninvasive laser-induced photoacoustic tomography for structural and functional *in vivo* imaging of the brain. Nat Biotechnol 2003;**21**:803–806.

[26] Kim C, Favazza C, Wang LV. *In vivo* photoacoustic tomography of chemicals: high-resolution functional and molecular optical imaging at new depths. Chem Rev 2010; **110**:2756–2782.

[27] Wang LV, Hu S. Photoacoustic tomography: *in vivo* imaging from organelles to organs. Science 2012;**335**:1458–1462.

[28] Wang LV. Prospects of photoacoustic tomography. Med Phys 2008;**35**:5758–5767.

[29] Yao J, Wang LV. Photoacoustic tomography: fundamentals, advances and prospects. Contrast Media Mol Imaging 2011;**6**:332–345.

[30] Wang LV. Tutorial on photoacoustic microscopy and computed tomography. IEEE J Sel Top Quant 2008;**14**:171–179.

[31] Pramanik M, Wang LV. Thermoacoustic and photoacoustic sensing of temperature. J Biomed Opt 2009;**14**:054024.

[32] Shah J, Park S, Aglyamov S, Larson T, Ma L, Sokolov K, Johnston K, Milner T, Emelianov SY. Photoacoustic imaging and temperature measurement for photothermal cancer therapy. J Biomed Opt 2008;**13**:034024.

[33] Andreussi O, Corni S, Mennucci B, Tomasi J. Radiative and nonradiative decay rates of a molecule close to a metal particle of complex shape. J Chem Phys 2004;**121**:10190–10202.

[34] Estrada LA, Cai X, Neckers DC. Nonradiative decay mechanism of fluoren-9-ylidene malononitrile ambipolar derivatives. J Phys Chem A 2011;**115**:2184–2195.

[35] Galloway CM, Etchegoin PG, Le Ru EC. Ultrafast nonradiative decay rates on metallic surfaces by comparing surface-enhanced Raman and fluorescence signals of single molecules. Phys Rev Lett 2009;**103**:063003.

[36] Kim C, Erpelding TN, Jankovic L, Wang LV. Performance benchmarks of an array-based hand-held photoacoustic probe adapted from a clinical ultrasound system for non-invasive sentinel lymph node imaging. Philos Trans R Soc Lond A 2011;**369**: 4644–4650.

[37] Li ML, Oh JT, Xie XY, Ku G, Wang W, Li C, Lungu G, Stoica G, Wang LV. Simultaneous molecular and hypoxia imaging of brain tumors *in vivo* using spectroscopic photoacoustic tomography. Proc IEEE 2008;**96**:481–489.

[38] Guo Z, Li L, Wang LV. On the speckle-free nature of photoacoustic tomography. Med Phys 2009;**36**:4084–4088.

[39] Erpelding TN, Kim C, Pramanik M, Jankovic L, Maslov K, Guo Z, Margenthaler JA, Pashley MD, Wang LV. Sentinel lymph nodes in the rat: noninvasive photoacoustic and US imaging with a clinical US system. Radiology 2010;**256**:102–110.

[40] Kim C, Erpelding TN, Jankovic L, Pashley MD, Wang LV. Deeply penetrating *in vivo* photoacoustic imaging using a clinical ultrasound array system. Biomed Opt Express 2010;**1**:278–284.

[41] Kim D, Kim E, Lee J, Hong S, Sung W, Lim N, Park CG, Kim K. Direct synthesis of polymer nanocapsules: self-assembly of polymer hollow spheres through irreversible covalent bond formation. J Am Chem Soc 2010;**132**:9908–9919.

[42] Zhang HF, Maslov K, Wang LV. *In vivo* imaging of subcutaneous structures using functional photoacoustic microscopy. Nat Protoc 2007;**2**:797–804.

[43] Ku G, Wang XD, Xie XY, Stoica G, Wang LHV. Imaging of tumor angiogenesis in rat brains *in vivo* by photoacoustic tomography. Appl Opt 2005;**44**:770–775.

[44] Siphanto RI, Thumma KK, Kolkman RGM, van Leeuwen TG, de Mul FFM, van Neck JW, van Adrichem LNA, Steenbergen W. Serial noninvasive photoacoustic imaging of neovascularization in tumor angiogenesis. Opt Express 2005;**13**:89–95.

[45] Xiang LZ, Xing D, Gu HM, Zhou FF, Yang DW, Zeng LM, Yang SH. *In vivo* monitoring of neovascularization in tumour angiogenesis by photoacoustic tomography. Chinese Phys Lett 2007;**24**:751–754.

[46] Bauer DR, Olafsson R, Montilla LG, Witte RS. 3-D photoacoustic and pulse echo imaging of prostate tumor progression in the mouse window chamber. J Biomed Opt 2011;**16**:026012.

[47] Jose J, Manohar S, Kolkman RG, Steenbergen W, van Leeuwen TG. Imaging of tumor vasculature using Twente photoacoustic systems. J Biophotonics 2009;**2**:701–717.

[48] Lao Y, Xing D, Yang S, Xiang L. Noninvasive photoacoustic imaging of the developing vasculature during early tumor growth. Phys Med Biol 2008;**53**:4203–4212.

[49] Xi L, Grobmyer SR, Wu L, Chen R, Zhou G, Gutwein LG, Sun J, Liao W, Zhou Q, Xie H, Jiang H. Evaluation of breast tumor margins *in vivo* with intraoperative photoacoustic imaging. Opt Express 2012;**20**:8726–8731.

[50] Song KH, Wang LV. Deep reflection-mode photoacoustic imaging of biological tissue. J Biomed Opt 2007;**12**:060503.

[51] Song KH, Wang LV. Noninvasive photoacoustic imaging of the thoracic cavity and the kidney in small and large animals. Med Phys 2008;**35**:4524–4529.

[52] Proskurnin MA, Zhidkova TV, Volkov DS, Sarimollaoglu M, Galanzha EI, Mock D, Nedosekin DA, Zharov VP. *In vivo* multispectral photoacoustic and photothermal flow cytometry with multicolor dyes: a potential for real-time assessment of circulation, dye-cell interaction, and blood volume. Cytometry A 2011;**79**:834–847.

[53] Yao J, Maslov KI, Shi Y, Taber LA, Wang LV. *In vivo* photoacoustic imaging of transverse blood flow by using Doppler broadening of bandwidth. Opt Lett 2010;**35**:1419–1421.

[54] Yao J, Maslov KI, Wang LV. *In vivo* photoacoustic tomography of total blood flow and potential imaging of cancer angiogenesis and hypermetabolism. Technol Cancer Res Treat 2012;**11**:301–307.

[55] Wang Y, Hu S, Maslov K, Zhang Y, Xia Y, Wang LV. *In vivo* integrated photoacoustic and confocal microscopy of hemoglobin oxygen saturation and oxygen partial pressure. Opt Lett 2011;**36**:1029–1031.

[56] Chatni MR, Yao J, Danielli A, Favazza CP, Maslov KI, Wang LV. Functional photoacoustic microscopy of pH. J Biomed Opt 2011;**16**:100503.

[57] Horvath TD, Kim G, Kopelman R, Ashkenazi S. Ratiometric photoacoustic sensing of pH using a "sonophore". Analyst 2008;**133**:747–749.

[58] Hanyz I, Wrobel D. The influence of pH on charged porphyrins studied by fluorescence and photoacoustic spectroscopy. Photochem Photobiol Sci 2002;**1**:126–132.

[59] Liu T, Wei Q, Wang J, Jiao S, Zhang HF. Combined photoacoustic microscopy and optical coherence tomography can measure metabolic rate of oxygen. Biomed Opt Express 2011;**2**:1359–1365.

[60] Yao J, Maslov KI, Zhang Y, Xia Y, Wang LV. Label-free oxygen-metabolic photoacoustic microscopy *in vivo*. J Biomed Opt 2011;**16**:076003.

[61] Maslov K, Zhang HF, Hu S, Wang LV. Optical-resolution photoacoustic microscopy for *in vivo* imaging of single capillaries. Opt Lett 2008;**33**:929.

[62] Maslov K, Stoica G, Wang LV. *In vivo* dark-field reflection-mode photoacoustic microscopy. Opt Lett 2005;**30**:625–627.

[63] Yang JM, Maslov K, Yang HC, Zhou Q, Shung KK, Wang LV. Photoacoustic endoscopy. Opt Lett 2009;**34**:1591–1593.

[64] Emelianov S, Wang B, Su J, Karpiouk A, Yantsen E, Sokolov K, Amirian J, Smalling R, Sethuraman S. Intravascular ultrasound and photoacoustic imaging. Conf Proc IEEE Eng Med Biol Soc 2008;**2008**:2–5.

[65] Hsieh BY, Chen SL, Ling T, Guo LJ, Li PC. Integrated intravascular ultrasound and photoacoustic imaging scan head. Opt Lett 2010;**35**:2892–2894.

[66] Karpiouk AB, Wang B, Emelianov SY. Development of a catheter for combined intravascular ultrasound and photoacoustic imaging. Rev Sci Instrum 2010;**81**:014901.

[67] Wang B, Su JL, Amirian J, Litovsky SH, Smalling R, Emelianov S. Detection of lipid in atherosclerotic vessels using ultrasound-guided spectroscopic intravascular photoacoustic imaging. Opt Express 2010;**18**:4889–4897.

[68] Wei W, Li X, Zhou Q, Shung KK, Chen Z. Integrated ultrasound and photoacoustic probe for co-registered intravascular imaging. J Biomed Opt 2011;**16**:106001.

[69] Wang X, Xie X, Ku G, Wang LV, Stoica G. Noninvasive imaging of hemoglobin concentration and oxygenation in the rat brain using high-resolution photoacoustic tomography. J Biomed Opt 2006;**11**:024015.

[70] Li C, Aguirre A, Gamelin J, Maurudis A, Zhu Q, Wang LV. Real-time photoacoustic tomography of cortical hemodynamics in small animals. J Biomed Opt 2010;**15**:010509.

[71] Brecht HP, Su R, Fronheiser M, Ermilov SA, Conjusteau A, Oraevsky AA. Whole-body three-dimensional optoacoustic tomography system for small animals. J Biomed Opt 2009;**14**:064007.

[72] Kruger RA, Lam RB, Reinecke DR, Del Rio SP, Doyle RP. Photoacoustic angiography of the breast. Med Phys 2010;**37**:6096–6100.

[73] Gamelin J, Maurudis A, Aguirre A, Huang F, Guo PY, Wang LV, Zhu Q. A real-time photoacoustic tomography system for small animals. Opt Express 2009;**17**:10489–10498.

[74] Xia J, Guo Z, Maslov K, Aguirre A, Zhu Q, Percival C, Wang LV. Three-dimensional photoacoustic tomography based on the focal-line concept. J Biomed Opt 2011;**16**:090505.

[75] Xie Z, Jiao S, Zhang HF, Puliafito CA. Laser-scanning optical-resolution photoacoustic microscopy. Opt Lett 2009;**34**:1771–1773.

[76] Kim C, Cho EC, Chen J, Song KH, Au L, Favazza C, Zhang Q, Cobley CM, Gao F, Xia Y, Wang LV. *In vivo* molecular photoacoustic tomography of melanomas targeted by bioconjugated gold nanocages. ACS Nano 2010;**4**:4559–4564.

[77] Hu S, Maslov K, Wang LV. Second-generation optical-resolution photoacoustic microscopy with improved sensitivity and speed. Opt Lett 2011;**36**:1134–1136.

[78] Nuster R, Gruen H, Reitinger B, Burgholzer P, Gratt S, Passler K, Paltauf G. Downstream Fabry-Perot interferometer for acoustic wave monitoring in photoacoustic tomography. Opt Lett 2011;**36**:981–983.

[79] Zhang E, Laufer J, Beard P. Backward-mode multiwavelength photoacoustic scanner using a planar Fabry-Perot polymer film ultrasound sensor for high-resolution three-dimensional imaging of biological tissues. Appl Opt 2008;**47**:561–577.

[80] OMLC. 2001. Available at http://omlc.ogi.edu/spectra/. Accessed June 3, 2014.

[81] Zhang C, Maslov K, Wang LV. Subwavelength-resolution label-free photoacoustic microscopy of optical absorption *in vivo*. Opt Lett 2010;**35**:3195–3197.

[82] Wikipedia.org. 2012. Near-infrared window in biological tissue. Available at http://en.wikipedia.org. Accessed June 3, 2014.

[83] Song KH, Stein EW, Margenthaler JA, Wang LV. Noninvasive photoacoustic identification of sentinel lymph nodes containing methylene blue *in vivo* in a rat model. J Biomed Opt 2008;**13**:054033.

[84] Kim C, Jeon M, Wang LV. Nonionizing photoacoustic cystography *in vivo*. Opt Lett 2011;**36**:3599–3601.

[85] Kim C, Song KH, Gao F, Wang LHV. Sentinel lymph nodes and lymphatic vessels: noninvasive dual-modality *in vivo* mapping by using indocyanine green in rats-volumetric spectroscopic photoacoustic imaging and planar fluorescence imaging. Radiology 2010;**255**:442–450.

[86] Li W, Brown PK, Wang LV, Xia Y. Gold nanocages as contrast agents for photoacoustic imaging. Contrast Media Mol Imaging 2011;**6**:370–377.

[87] Yang X, Stein EW, Ashkenazi S, Wang LV. Nanoparticles for photoacoustic imaging. Wiley Interdiscip Rev Nanomed Nanobiotechnol 2009;**1**:360–368.

[88] Yuan Z, Jiang H. Photoacoustic tomography for imaging nanoparticles. Methods Mol Biol 2010;**624**:309–324.

[89] Zhang Q, Iwakuma N, Sharma P, Moudgil BM, Wu C, McNeill J, Jiang H, Grobmyer SR. Gold nanoparticles as a contrast agent for *in vivo* tumor imaging with photoacoustic tomography. Nanotechnology 2009;**20**:395102.

[90] Koo YE, Reddy GR, Bhojani M, Schneider R, Philbert MA, Rehemtulla A, Ross BD, Kopelman R. Brain cancer diagnosis and therapy with nanoplatforms. Adv Drug Deliv Rev 2006;**58**:1556–1577.

[91] Moghimi SM, Hunter AC, Murray JC. Nanomedicine: current status and future prospects. FASEB J 2005;**19**:311–330.

[92] Kukowska-Latallo JF, Candido KA, Cao Z, Nigavekar SS, Majoros IJ, Thomas TP, Balogh LP, Khan MK, Baker JR Jr. Nanoparticle targeting of anticancer drug improves therapeutic response in animal model of human epithelial cancer. Cancer Res 2005;**65**: 5317–5324.

[93] McNeil SE. Nanotechnology for the biologist. J Leukoc Biol 2005;**78**:585–594.

[94] Haes AJ, Van Duyne RP. A nanoscale optical biosensor: sensitivity and selectivity of an approach based on the localized surface plasmon resonance spectroscopy of triangular silver nanoparticles. J Am Chem Soc 2002;**124**:10596–10604.

[95] Gu L, Vardarajan V, Koymen AR, Mohanty SK. Magnetic-field-assisted photothermal therapy of cancer cells using Fe-doped carbon nanoparticles. J Biomed Opt 2012;**17**:018003.

[96] Wang J, Chen Y, Chen X, Hao J, Yan M, Qiu M. Photothermal reshaping of gold nanoparticles in a plasmonic absorber. Opt Express 2011;**19**:14726–14734.

[97] Yang K, Hu L, Ma X, Ye S, Cheng L, Shi X, Li C, Li Y, Liu Z. Multimodal imaging guided photothermal therapy using functionalized graphene nanosheets anchored with magnetic nanoparticles. Adv Mater 2012;**24**:1868–1872.

[98] Zandberg WF, Bakhtiari AB, Erno Z, Hsiao D, Gates BD, Claydon T, Branda NR. Photothermal release of small molecules from gold nanoparticles in live cells. Nanomedicine 2011;**8**:908–915.

[99] Li ML, Wang JC, Schwartz JA, Gill-Sharp KL, Stoica G, Wang LHV. In-vivo photoacoustic microscopy of nanoshell extravasation from solid tumor vasculature. J Biomed Opt 2009;**14**:010507.

[100] Wang YW, Xie XY, Wang XD, Ku G, Gill KL, O'Neal DP, Stoica G, Wang LV. Photoacoustic tomography of a nanoshell contrast agent in the *in vivo* rat brain. Nano Lett 2004;**4**:1689–1692.

[101] Manohar S, Ungureanu C, Van Leeuwen TG. Gold nanorods as molecular contrast agents in photoacoustic imaging: the promises and the caveats. Contrast Media Mol Imaging 2011;**6**:389–400.

[102] Bayer CL, Chen YS, Kim S, Mallidi S, Sokolov K, Emelianov S. Multiplex photoacoustic molecular imaging using targeted silica-coated gold nanorods. Biomed Opt Express 2011;**2**:1828–1835.

[103] Chen LC, Wei CW, Souris JS, Cheng SH, Chen CT, Yang CS, Li PC, Lo LW. Enhanced photoacoustic stability of gold nanorods by silica matrix confinement. J Biomed Opt 2010;**15**:016010.

[104] Chen YS, Frey W, Kim S, Homan K, Kruizinga P, Sokolov K, Emelianov S. Enhanced thermal stability of silica-coated gold nanorods for photoacoustic imaging and image-guided therapy. Opt Express 2010;**18**:8867–8877.

[105] Chen YS, Frey W, Kim S, Kruizinga P, Homan K, Emelianov S. Silica-coated gold nanorods as photoacoustic signal nanoamplifiers. Nano Lett 2011;**11**:348–354.

[106] Ha SH, Carson A, Agarwal A, Kotov NA, Kim K. Detection and monitoring of the multiple inflammatory responses by photoacoustic molecular imaging using selectively targeted gold nanorods. Biomed Opt Express 2011;**2**:645–657.

[107] Kim K, Huang SW, Ashkenazi S, O'Donnell M, Agarwal A, Kotov NA, Denny MF, Kaplan MJ. Photoacoustic imaging of early inflammatory response using gold nanorods. Appl Phys Lett 2007;**90**:223901.

[108] Li PC, Huang SW, Wei CW, Chiou YC, Chen CD, Wang CRC. Photoacoustic flow measurements by use of laser-induced shape transitions of gold nanorods. Opt Letters 2005;**30**:3341–3343.

[109] Li PC, Wang CRC, Shieh DB, Wei CW, Liao CK, Poe C, Jhan S, Ding AA, Wu YN. *In vivo* photoacoustic molecular imaging with simultaneous multiple selective targeting using antibody-conjugated gold nanorods. Opt Express 2008;**16**:18605–18615.

[110] Li PC, Wei CW, Liao CK, Chen CD, Pao KC, Wang CRC, Wu YN, Shieh DB. Photoacoustic imaging of multiple targets using gold nanorods. IEEE Trans Ultrason Ferroelectr Freq Control 2007;**54**:1642–1647.

[111] Wei C, Huang SW, Wang CRC, Li PC. Photoacoustic flow measurements based on wash-in analysis of gold nanorods. IEEE Trans Ultrason Ferroelectr Freq Control 2007;**54**:1131–1141.

[112] Lu W, Huang Q, Ku G, Wen X, Zhou M, Guzatov D, Brecht P, Su R, Oraevsky A, Wang LV, Li C. Photoacoustic imaging of living mouse brain vasculature using hollow gold nanospheres. Biomaterials 2010;**31**:2617–2626.

[113] Lu W, Melancon MP, Xiong C, Huang Q, Elliott A, Song S, Zhang R, Flores LG 2nd, Gelovani JG, Wang LV, Ku G, Stafford RJ, Li C. Effects of photoacoustic imaging and photothermal ablation therapy mediated by targeted hollow gold nanospheres in an orthotopic mouse xenograft model of glioma. Cancer Res 2011;**71**:6116–6121.

[114] Kim C, Song HM, Cai X, Yao J, Wei A, Wang LV. *In vivo* photoacoustic mapping of lymphatic systems with plasmon-resonant nanostars. J Mater Chem 2011;**21**:2841–2844.

[115] Pan D, Pramanik M, Senpan A, Allen JS, Zhang H, Wickline SA, Wang LV, Lanza GM. Molecular photoacoustic imaging of angiogenesis with integrin-targeted gold nanobeacons. FASEB J 2011;**25**:875–882.

[116] Pan D, Pramanik M, Senpan A, Ghosh S, Wickline SA, Wang LV, Lanza GM. Near infrared photoacoustic detection of sentinel lymph nodes with gold nanobeacons. Biomaterials 2010;**31**:4088–4093.

[117] Pan D, Pramanik M, Senpan A, Yang X, Song KH, Scott MJ, Zhang H, Gaffney PJ, Wickline SA, Wang LV, Lanza GM. Molecular photoacoustic tomography with colloidal nanobeacons. Angew Chem Int Ed Engl 2009;**48**:4170–4173.

[118] Pan D, Pramanik M, Wickline SA, Wang LV, Lanza GM. Recent advances in colloidal gold nanobeacons for molecular photoacoustic imaging. Contrast Media Mol Imaging 2011;**6**:378–388.

[119] Song KH, Kim CH, Cobley CM, Xia YN, Wang LV. Near-infrared gold nanocages as a new class of tracers for photoacoustic sentinel lymph node mapping on a rat model. Nano Lett 2009;**9**:183–188.

[120] Cai X, Li WY, Kim CH, Yuan YC, Wang LHV, Xia YN. *In vivo* quantitative evaluation of the transport kinetics of gold nanocages in a lymphatic system by noninvasive photoacoustic tomography. ACS Nano 2011;**5**:9658–9667.

[121] Chen J, McLellan JM, Siekkinen A, Xiong Y, Li ZY, Xia Y. Facile synthesis of gold-silver nanocages with controllable pores on the surface. J Am Chem Soc 2006;**128**: 14776–14777.

[122] Skrabalak SE, Chen J, Sun Y, Lu X, Au L, Cobley CM, Xia Y. Gold nanocages: synthesis, properties, and applications. Acc Chem Res 2008;**41**:1587–1595.

[123] Xia Y, Li W, Cobley CM, Chen J, Xia X, Zhang Q, Yang M, Cho EC, Brown PK. Gold nanocages: from synthesis to theranostic applications. Acc Chem Res 2011;**44**:914–924.

[124] Cho EC, Kim C, Zhou F, Cobley CM, Song KH, Chen JY, Li ZY, Wang LHV, Xia YN. Measuring the optical absorption cross sections of Au-Ag nanocages and Au nanorods by photoacoustic imaging. J Phys Chem C 2009;**113**:9023–9028.

[125] Lovell JF, Jin CS, Huynh E, MacDonald TD, Cao W, Zheng G. Enzymatic regioselection for the synthesis and biodegradation of porphysome nanovesicles. Angew Chem Int Ed Engl 2012;**51**:2429–2433.

[126] Lovell JF, Jin CS, Huynh E, Jin H, Kim C, Rubinstein JL, Chan WC, Cao W, Wang LV, Zheng G. Porphysome nanovesicles generated by porphyrin bilayers for use as multimodal biophotonic contrast agents. Nat Mater 2011;**10**:324–332.

[127] Chen J, Glaus C, Laforest R, Zhang Q, Yang M, Gidding M, Welch MJ, Xia Y. Gold nanocages as photothermal transducers for cancer treatment. Small 2010;**6**:811–817.

[128] Cobley CM, Au L, Chen J, Xia Y. Targeting gold nanocages to cancer cells for photothermal destruction and drug delivery. Expert Opin Drug Deliv 2010;**7**:577–587.

[129] Peer D, Karp JM, Hong S, FaroKHzad OC, Margalit R, Langer R. Nanocarriers as an emerging platform for cancer therapy. Nat Nanotechnol 2007;**2**:751–760.

[130] Avti PK, Hu S, Favazza C, Mikos AG, Jansen JA, Shroyer KR, Wang LV, Sitharaman B. Detection, mapping, and quantification of single walled carbon nanotubes in histological specimens with photoacoustic microscopy. PLoS One 2012;**7**:e35064.

[131] De la Zerda A, Zavaleta C, Keren S, Vaithilingam S, Bodapati S, Liu Z, Levi J, Smith BR, Ma TJ, Oralkan O, Cheng Z, Chen X, Dai H, Khuri-Yakub BT, Gambhir SS. Carbon nanotubes as photoacoustic molecular imaging agents in living mice. Nat Nanotechnol 2008;**3**:557–562.

[132] Kang B, Yu D, Dai Y, Chang S, Chen D, Ding Y. Cancer-cell targeting and photoacoustic therapy using carbon nanotubes as "bomb" agents. Small 2009;**5**:1292–1301.

[133] Kim JW, Galanzha EI, Shashkov EV, Moon HM, Zharov VP. Golden carbon nanotubes as multimodal photoacoustic and photothermal high-contrast molecular agents. Nat Nanotechnol 2009;**4**:688–694.

[134] Pramanik M, Swierczewska M, Green D, Sitharaman B, Wang LV. Single-walled carbon nanotubes as a multimodal-thermoacoustic and photoacoustic-contrast agent. J Biomed Opt 2009;**14**:034018.

[135] Xiang L, Yuan Y, Xing D, Ou Z, Yang S, Zhou F. Photoacoustic molecular imaging with antibody-functionalized single-walled carbon nanotubes for early diagnosis of tumor. J Biomed Opt 2009;**14**:021008.

[136] Zhou F, Wu S, Yuan Y, Chen WR, Xing D. Mitochondria-targeting photoacoustic therapy using single-walled carbon nanotubes. Small 2012;**8**:1543–1550.

[137] de la Zerda A, Liu Z, Bodapati S, Teed R, Vaithilingam S, Khuri-Yakub BT, Chen X, Dai H, Gambhir SS. Ultrahigh sensitivity carbon nanotube agents for photoacoustic molecular imaging in living mice. Nano Lett 2010;**10**:2168–2172.

[138] Kim C, Qin R, Xu JS, Wang LV, Xu R. Multifunctional microbubbles and nanobubbles for photoacoustic and ultrasound imaging. J Biomed Opt 2010;**15**:010510.

[139] Wang YH, Liao AH, Chen JH, Chris Wang CR, Li PC. Photoacoustic/ultrasound dual-modality contrast agent and its application to thermotherapy. J Biomed Opt 2012;**17**: 045001.

[140] Miller DL. Frequency relationships for ultrasonic activation of free microbubbles, encapsulated microbubbles, and gas-filled micropores. J Acoust Soc Am 1998;**104**: 2498–2505.

[141] Zhao YZ, Liang HD, Mei XG, Halliwell M. Preparation, characterization and *in vivo* observation of phospholipid-based gas-filled microbubbles containing hirudin. Ultrasound Med Biol 2005;**31**:1237–1243.

[142] Xu RX. Multifunctional microbubbles and nanobubbles for photoacoustic imaging. Contrast Media Mol Imaging 2011;**6**:401–411.

[143] Wilson K, Homan K, Emelianov S. Biomedical photoacoustics beyond thermal expansion using triggered nanodroplet vaporization for contrast-enhanced imaging. Nat Commun 2012;**3**:618.

[144] Akers WJ, Kim C, Berezin M, Guo K, Fuhrhop R, Lanza GM, Fischer GM, Daltrozzo E, Zumbusch A, Cai X, Wang LV, Achilefu S. Noninvasive photoacoustic and fluorescence sentinel lymph node identification using dye-loaded perfluorocarbon nanoparticles. ACS Nano 2011;**5**:173–182.

[145] Levi J, Kothapalli SR, Ma TJ, Hartman K, Khuri-Yakub BT, Gambhir SS. Design, synthesis, and imaging of an activatable photoacoustic probe. J Am Chem Soc 2010;**132**: 11264–11269.

[146] Xia X, Yang M, Oetjen LK, Zhang Y, Li Q, Chen J, Xia Y. An enzyme-sensitive probe for photoacoustic imaging and fluorescence detection of protease activity. Nanoscale 2011;**3**:950–953.

[147] Galanzha EI, Shashkov EV, Kelly T, Kim JW, Yang L, Zharov VP. *In vivo* magnetic enrichment and multiplex photoacoustic detection of circulating tumour cells. Nat Nanotechnol 2009;**4**:855–860.

[148] Jin YD, Jia CX, Huang SW, O'Donnell M, Gao XH. Multifunctional nanoparticles as coupled contrast agents. Nat Commun 2010;**1**:41.

[149] Kircher MF, de la Zerda A, Jokerst JV, Zavaleta CL, Kempen PJ, Mittra E, Pitter K, Huang R, Campos C, Habte F, Sinclair R, Brennan CW, Mellinghoff IK, Holland EC, Gambhir SS. A brain tumor molecular imaging strategy using a new triple-modality MRI-photoacoustic-Raman nanoparticle. Nat Med 2012;**18**:829–834.

[150] Akers WJ, Edwards WB, Kim C, Xu B, Erpelding TN, Wang LV, Achilefu S. Multimodal sentinel lymph node mapping with single-photon emission computed tomography (SPECT)/computed tomography (CT) and photoacoustic tomography. Transl Res 2012;**159**:175–181.

[151] Agarwal A, Shao X, Rajian JR, Zhang H, Chamberland DL, Kotov NA, Wang X. Dual-mode imaging with radiolabeled gold nanorods. J Biomed Opt 2011;**16**:051307.

[152] Yao JJ, Maslov K, Hu S, Wang LHV. Evans blue dye-enhanced capillary-resolution photoacoustic microscopy *in vivo*. J Biomed Opt 2009;**14**:054049.

[153] Song L, Kim C, Maslov K, Shung KK, Wang LHV. High-speed dynamic 3D photoacoustic imaging of sentinel lymph node in a murine model using an ultrasound array. Med Phys 2009;**36**:3724–3729.

[154] Filonov GS, Krumholz A, Xia J, Yao J, Wang LV, Verkhusha VV. Deep-tissue photoacoustic tomography of a genetically encoded near-infrared fluorescent probe. Angew Chem Int Ed Engl 2012;**51**:1448–1451.

[155] Razansky D, Distel M, Vinegoni C, Ma R, Perrimon N, Koster RW, Ntziachristos V. Multispectral opto-acoustic tomography of deep-seated fluorescent proteins *in vivo*. Nat Photonics 2009;**3**:412–417.

11

SURFACE-ENHANCED RAMAN SCATTERING-BASED BIOIMAGING

LIMEI TIAN AND SRIKANTH SINGAMANENI

Department of Mechanical Engineering and Materials Science, Washington University, St. Louis, MO, USA

11.1 OVERVIEW

Raman spectroscopy is a unique nondestructive tool for probing the structure and properties of a wide variety of organic and inorganic materials. Raman spectroscopy is based on the inelastic scattering of light that provides molecular fingerprint related to the vibrational, rotational, and other low-frequency transitions of the molecules [1, 2]. Despite the rich molecular information provided by Raman spectroscopy, the technique was not considered to be a handy analytical tool (let alone imaging tool) due to the extremely weak signal intensity of normal Raman scattering, a result of the extremely small scattering cross section for most molecules except some special cases (e.g., resonance Raman scattering). Hence, for the longest period after its discovery, Raman scattering has remained limited to "bulk" investigations.

Surface-enhanced Raman scattering (SERS), which involves dramatic enhancement of the Raman scattering of an analyte that is adsorbed on or in close proximity to a metal surface, overcomes the fundamental problem associated with normal Raman scattering (i.e., weak signal). SERS is emerging as a powerful technique for the trace-level detection of various biological and chemical species and believed to make a huge impact on life sciences, environmental monitoring, and homeland security [3–9]. Micro-Raman microscopy involves acquiring spatially resolved Raman spectra by combining the conventional Raman spectrometer with a microscopic tool,

Nanotechnology for Biomedical Imaging and Diagnostics: From Nanoparticle Design to Clinical Applications, First Edition. Edited by Mikhail Y. Berezin.

typically an optical microscope [10]. With the introduction of commercial Raman systems, this technique has emerged as a powerful and handy analytical tool for the characterization of a wide variety of nanostructures over the last decade.

In this chapter, we present recent advances in the application of SERS in bioimaging. We do not intend to provide a comprehensive literature review but instead highlight several recent results as key examples that demonstrate the breadth of the applications of SERS. Following a brief introduction to Raman imaging using micro-Raman spectroscopy (μRS), we present an overview of the instrumentation underlining the similarities and differences compared to conventional Raman spectroscopy and recent development in deep noninvasive Raman spectroscopy for probing diffusely scattering media, such as living tissue. In the following section, we mainly highlight important considerations in the design and fabrication of bright SERS probes and experimental approaches of SERS-based bioimaging with selected examples. We conclude with a brief summary and outlook of the SERS-based bioimaging.

11.2 INTRODUCTION

Spectral information obtained using conventional Raman spectroscopy is spatially averaged over a large area (few millimeter square area), whereas micro-Raman spectroscopy (μRS) enables the collection of spectral information with submicron lateral and vertical resolution. With the use of a high magnification objective, the laser beam can be focused into a spot with an effective diameter (considering a Gaussian footprint of the beam) as small as approximately 200 nm, providing excellent resolution limited only by the far-field diffraction limit. The unique combination of high chemical specificity, excellent lateral and vertical resolution, and nondestructive nature of μRS makes it a powerful analytical tool for the characterization of nanostructured materials.

As mentioned earlier, surface-enhanced Raman scattering (SERS) involves the dramatic enhancement (under ideal conditions more than 10^{10} times) of Raman scattering from molecules adsorbed on or in close proximity to a nanostructured metal surface [11]. Accidentally discovered nearly 35 years ago, SERS continues to attract broad scientific and technological interest owing to its ever-expanding application landscape. A nonexhaustive list of applications includes trace chemical and biological detection, environmental monitoring, surface analysis, pharmaceutical tracking, and molecular bioimaging [12–17].

Optical bioimaging, which involves detection, localization, and monitoring of molecular processes of cells and target tissues using inherent contrast offered by the biological species or exogenous contrast agents, is a rapidly emerging field. SERS offers numerous advantages for *in vivo* and *ex vivo* bioimaging:

1. On a per particle basis, SERS probes are significantly brighter compared to near-infrared (NIR)-emitting semiconducting quantum dots (QDs), which have been extensively investigated for bioimaging applications. Even simple designs involving individual gold nanoparticles tagged with resonant Raman reporters are nearly 200 times brighter compared to QDs [18].

2. The full width at half maximum (FWHM) of Raman bands (1–2 nm) are nearly 20–30 times narrower compared to the emission bands of QDs (40–60 nm), making spectral multiplexing an order of magnitude higher for SERS probes compared to that attainable with QDs [19].
3. The excitation and emission of SERS probes can be easily tuned to the NIR therapeutic window (650–900 nm), where the endogenous absorption coefficient of tissue is nearly two orders magnitude lower compared to that in the visible parts of electromagnetic (EM) spectrum [19]. Furthermore, absence of interference from water and autofluorescence of the tissues in NIR is yet another significant advantage of SERS compared to photoluminescence-based optical imaging techniques.
4. SERS probes offer remarkable photostability compared to organic fluorophores and QDs, which suffer from either photobleaching or blinking. SERS probes, typically made of gold, are significantly less cytotoxic compared to NIR emitting QDs comprised of CdSe and CdTe.

11.3 RAMAN INSTRUMENTATION FOR BIOIMAGING

11.3.1 Overview of the Instrumentation

Since the first experimental observation of Raman scattering by Sir Chandrasekhara Venkata Raman using sunlight as the excitation source and the eye as detector, there have been remarkable advancements in Raman instrumentation to current handheld spectrometers and advanced confocal Raman microscopes combined with complementary microscopic techniques such as atomic force microscope [19–21]. A highly simplified schematic of micro-Raman spectrometer is shown in Figure 11.1 [10]. The design is based on a confocal optical microscope equipped with an objective with a large numerical aperture (NA) (as high as 1.4) and a pinhole in the intermediate focal plane. An intense light source such as a laser with a narrow line width (~0.2 nm)

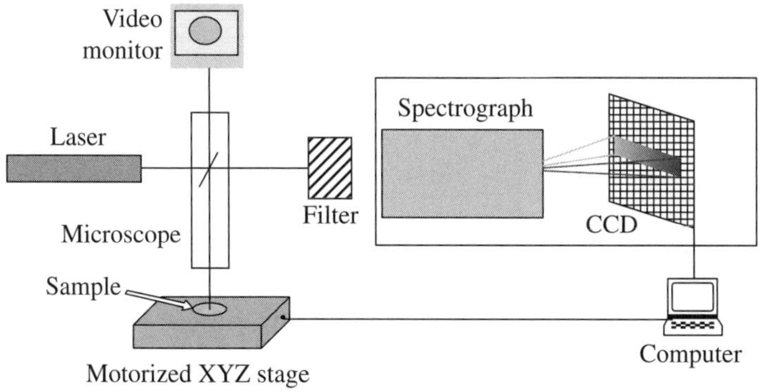

FIGURE 11.1 Simplified schematic illustration of a setup of micro-Raman spectrometer. (Reprinted with permission from Ref. [10]. © Elsevier.)

is used to illuminate the surface under investigation. The light source can be built into the spectrometer or an external source can be used for illumination. Modern Raman spectrometers generally include holographic gratings for improved excitation light rejection, notch filter for reflected and Rayleigh scattered light rejection, and a liquid nitrogen or Peltier-cooled charge-coupled device (CCD) to reduce variations in dark current at the detector. A small local area of the sample is probed, where the CCD detects only photons coming from a narrow spatial domain of the material [22]. The filtered light is analyzed as a function of the probed position where the signal is read by the detector and the intensity of each frequency is measured by an individual pixel on the array. The intensity of the Raman signal is a function of four factors: material properties (absorptivity, reflectivity, and the intrinsic strength of the Raman modes), source laser power, the width of the spectrometer admission slit, and the width of the resolution slit. Based on the intensity and frequency of the Raman signal received, spatially resolved structure and properties of the material are identified.

Owing to the submicron spatial resolution achieved with μRS, precise areas of interest can be chosen to identify and analyze physical and chemical properties of the materials under investigation or location of predesigned bright SERS probes, which form exogenous contrast agents for SERS-based bioimaging. A motorized *X*, *Y*, and *Z* stage with a nominal resolution as small as approximately 10 nm is used for scanning the surface for lateral and depth information. Raman spectral mapping complements the visual image obtained with the microscope by providing information about the variation in chemical (or physical) properties of a heterogeneous surface.

Modern μRS systems are accompanied with powerful spectral acquisition and analysis software, which enables the creation of 1D (cross section), 2D, and 3D maps of various features from the 1D, 2D, or 3D array of spatially resolved Raman spectra. Various features that can be routinely mapped include intensity variations of specific peaks (by plotting the user-defined peak intensity or integrated area under the peak), intensity ratio of two different bands, peak position (by user-defined peak fitting routines such as Gaussian and Lorentzian), and peak widths. The obtained images can be further processed to highlight the spatial variations of the acquired spectra. For example, Boolean maps, which present a binary representation of the desired attribute (such as intensity, peak width range), are obtained by specifying the threshold values of the parameter. The Boolean maps are extremely helpful to visualize the distribution of exogenous contrast agents in a living animal or tissue.

11.3.2 Resolution Criteria

Dictated by the diffraction limit of light, the lateral resolution of a far-field optical microscope is defined as the smallest distance between two features that can still be resolved in the image and is given by Rayleigh criterion as shown in Equation 11.1 [23]:

$$R = 0.61\left(\frac{\lambda}{\mathrm{NA}}\right) \tag{11.1}$$

where λ is the wavelength of incident light and NA (=$n\sin\theta$) is the numerical aperture of an optical system, n is the refractive index of the medium, and θ is the angle subtended by the optics. The depth of the field or vertical resolution of μRS is approximated by Conrady expression as given in Equation 11.2 [24]:

$$\Delta Z = \left(\frac{\lambda}{n\sin^2\theta}\right) = \lambda/\mathrm{NA}^2 \tag{11.2}$$

Under normal conditions ($n=1$ for air, $\lambda=632.8$ nm, NA = 0.95), the typical lateral and depth resolutions would be about 400 and 700 nm, respectively. Using the smallest visible wavelength (~400 nm) and a high NA ($n=1.515$, NA ~ 1.4), one can estimate the highest lateral resolution as 200 nm (based on Abbe criterion) and the smallest field depth as 400 nm. However, if a series of spectra are recorded at very close equidistant locations, a reduced spot size (considering the Gaussian profile of the beam) is obtained through a convolution of the spot profile.

11.3.3 Raman Instrumentation Tailored for Bioimaging

To date, very few of instrumentation developments are solely dedicated to Raman-based *in vivo* bioimaging applications. Most of *in vivo* demonstrations so far have relied on either handheld Raman spectrometers or confocal Raman microscopes (discussed earlier), which are suboptimal for bioimaging applications. Often, commercial Raman microspectrometers that are designed for probing relatively flat substrates arc customized for bioimaging applications [20]. For example, for performing Raman mapping on small animals using a confocal Raman microscope, a low magnification objective (12×) with long working distance, intentional defocusing of the laser beam, and high sensitivity settings by sacrificing spectra resolution are necessary to attain satisfactory results. A high NA, commonly used in Raman microscope setup, results in a small laser spot illumination, which limits the permissible laser power that can be used before tissue damage. Moreover, only a very small amount of scattered photons emanate from the small laser spot due to a highly diffusive propagation of light photons through tissue so that high NA objectives can only efficiently collect light from this small spot size. A 12× open-field lens with a defocused beam provides a spot size of 20 × 200 μm, which is much larger than that using higher NA. While this approach of customizing suboptimal instrumentation suffices proof-of-concept demonstrations, advanced Raman instrumentation tailored for bioimaging applications is critical for enabling rapid progress of this technique in preclinical and clinical settings.

11.3.3.1 Portable and Ergonomic Devices for Live Subjects and Surgery Intraoperative handheld device is an excellent tool for a surgeon to identify tumor margins and satellite nodules during surgical procedures. We would like to quickly point out a recent report, which made significant progress along this direction using portable Raman. Nie and coworkers have developed a handheld spectroscopic device

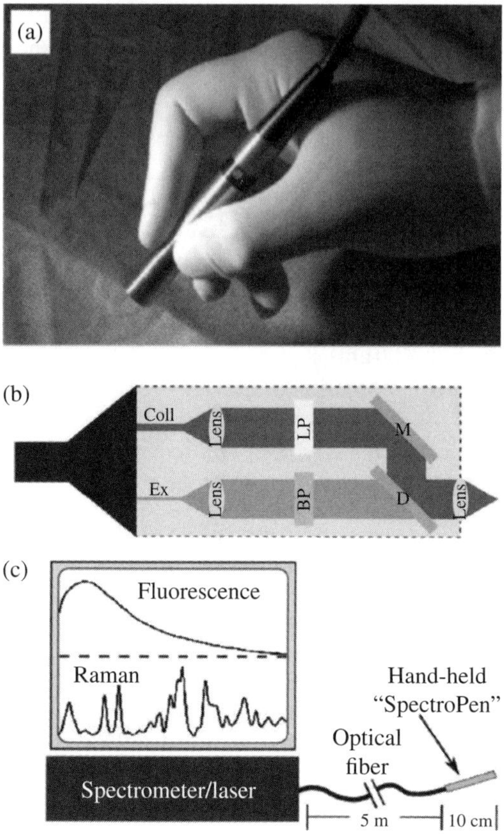

FIGURE 11.2 (a) Photograph showing the SpectroPen held by an operator in a surgical setting. (b) Optical beam paths of the SpectroPen. A 785 nm laser diode (200 mW output) serves as an excitation source. The illustration also shows excitation fiber (Ex), band-pass filter (BP), dichroic filter (D), collection fiber (Coll), long-pass filter (LP), and reflective mirror (M). (c) Schematic illustration of the hybrid spectrometer designed for wavelength-resolved fluorescence and Raman scattering. (Reprinted with permission from Ref. [21]. © American Chemical Society.)

called SpectroPen that enables wavelength-resolved fluorescence and Raman measurements for *in vivo* and intraoperative SERS-based bioimaging applications, as showed in Figure 11.2 [21]. This SpectroPen device effectively removes silica Raman peaks from the fiber optics and reflected excitation light, allowing resolvable fluorescence and Raman signals when contrast agents were buried 5–10 mm deep in fresh animal tissues. After surgery, the SpectroPen device provides further evaluation of both positive and negative tumor margins around the surgical cavity, raising new possibilities for real-time tumor detection and image-guided surgery.

11.3.3.2 Seeing through Diffuse Human Tissue A different line of efforts is focused on enhancing the depth of imaging. It is known that commonly employed backscattering geometry suffers from limited depth of penetration (only several hundred micrometers in living tissue) even in NIR range due to highly diffusive nature of living tissue. A significant improvement on the penetration depth achievable in Raman spectroscopy of turbid media was recently accomplished by utilizing the diffuse component of light. The diffuse component is of highest importance to the spectroscopic investigations of deep layers of turbid media due to its deepest penetration among all components of light (see Fig. 11.3a). In tissue, the diffuse component penetrates to the depths of up to several centimeters as opposed to the hundreds micrometers or few millimeters of the ballistic and snake components in the NIR region of the spectrum. Two principal noninvasive concepts, temporal and spatial approach, were developed according to the photon migration concept [25].

The investigation of stratified layers within a turbid sample using temporal approach is based on impulsive Raman excitation and fast temporal gating of the Raman signal. This concept relies on the simple fact that photons emanating from deeper layers of the measured turbid medium have to diffusely migrate larger distances through than the photons generated at shallower depths. Temporal resolved Raman light was first used to image an object in a turbid medium by Wu *et al.* [27]. Later, Matousek *et al.* demonstrated that the Raman spectrum of a deeply buried layer in a turbid sample could be fully recovered [28].

An alternate method known as spatially offset Raman spectroscopy (SORS) involves spatially offsetting the collection of Raman scattered light from the excitation regions (i.e., laser incident regions), which enables the collection of diffuse component of scattered light as opposed to conventional ballistic and snake photons in backscattering geometry, thus significantly enhancing the depth of probing. Recently, this method has been integrated with SERS resulting in spatially offset SERS (SESORS), which enables significantly deeper probing into mammalian tissue [26, 29]. Multiplexed Raman signals emanating from SERS probes buried 25 mm deep inside a porcine tissue were successfully collected and analyzed by offsetting the excitation and collection by 180°, that is, transmission Raman spectroscopy, a special case of SORS.

11.3.3.3 Rapid Line Raman Mapping The conventional Raman point mapping provides information of complete spectra of each pixel in the scanning area as the focused beam is rastered across the sample area. The point scanning technique is time-consuming, especially if longer exposure time is required for low concentration samples. Raman line mapping dramatically decreases the time to record an image by line focusing the laser and recording spectra across the full area of the CCD. The high efficiency of line mapping was obtained by compromising with reduced resolution along the laser line, which can be traded against Y binning and scanning speed. Moreover, the power density on any one area is reduced by a factor of 40–60 times due to the defocus of laser and rapid scanning, compared to that in point mapping. Surface-enhanced resonance Raman scattering (SERRS) mapping has been achieved on bone marrow-derived immune cells treated with plasmonic nanoparticles by multiple-wavelength line scanning [30]. Thanks to the high efficiency of the SERRS method, line mapping was reported to collect up to 1000

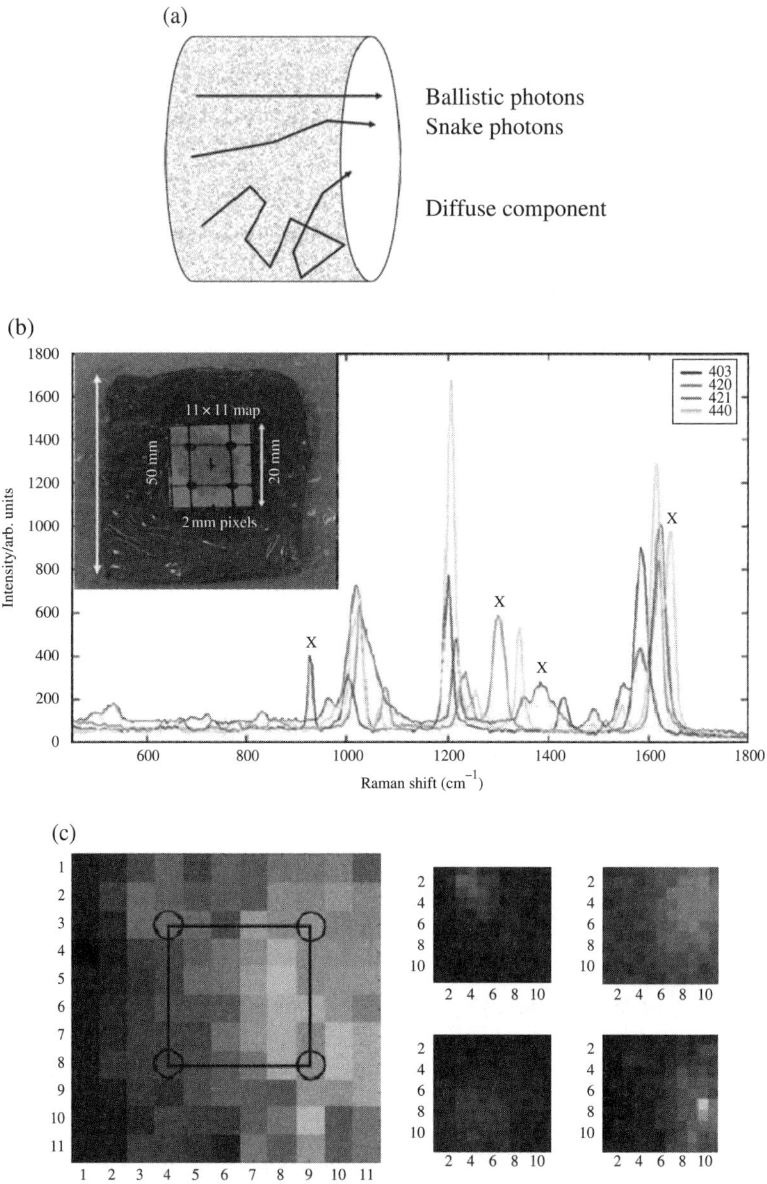

FIGURE 11.3 (a) Three main light components within turbid media. (b) SERS spectra of Cabot nanoparticles (NP) with different labels. The Xs signify the identity peaks used to measure the NP flavor signal in the SESORS spectrum. Inset image is a 20×50×50 mm porcine block with grid used to align injection points of NP. (Reproduced with permission from Ref. [25]. © Royal Society of Chemistry.) (c) SERS mapping measured in a 11×11 grid, pixel size 2 mm. Left image shows all different signals plotted together with the injection points marked, and right image shows the SERS signal of each labels. (Reproduced with permission from Ref. [26]. © Royal Society of Chemistry.) *(See insert for color representation of the figure.)*

spectra per minute, 30 times faster than point mapping approach. Moreover, the power density on any one area of the sample is reduced by a factor of 40–60 times due to the defocusing of laser and rapid scanning compared to an equivalent response in point mapping.

11.4 *IN VITRO* AND *IN VIVO* RAMAN-BASED BIOIMAGING

SERS-based bioimaging can be broadly classified into two types: labeled and label-free. In the label-free approach, plasmonic nanostructures are employed to enhance the intrinsic Raman signal of the biological species under investigation. On the other hand, labeled approach involves the use of exogenous contrast, which is directed to the target biological species either by active or passive targeting followed by their mapping (or detection).

11.4.1 Label-Free SERS-Based Bioimaging

For label-free SERS-based assays, biomolecules (analytes) directly locate on or in close proximity to a nanostructured metal surface and thus are recognized by their enhanced Raman signals without external labels needed. Detection can be achieved with the use of individual or assembled plasmonic nanoparticles, which exhibit intense local EM fields. We would like to highlight a recent report that the molecular changes of cell components, such as the nucleus, could be monitored during the different phased of its full cell cycle by its Rayleigh images and Raman vibration spectra in real time [31]. Figure 11.4 showed that the different phases of the cell cycle could be identified based on the Raman peaks attributed to DNA and protein. The label-free approach is simple and can provide molecular information about target cells, but signals can often be complicated due to the interference from the surrounding cellular matrix or impurities whose Raman signals could also be enhanced.

11.4.2 Labeled SERS-Based Bioimaging

The other type of SERS-based bioimaging involves plasmonic nanoparticles adsorbed or covalently conjugated with Raman reporters (molecules with large Raman cross sections), which serve as exogenous contrast agents. The labeled approach is highly specific because it exhibits little to no signal interference from other Raman-active species as the SERS enhancement decays exponentially from the surface of the nanoparticles, making any other contributions virtually undetectable compared to the Raman-active molecules adsorbed on the nanoparticle surface. Gold nanostructures with Raman reporters (called SERS probes henceforth) have been extensively investigated for cancer [18, 32–35] and other disease [36, 37] biomarker detection, as well as bacterial [38–41] and viral [42, 43] microorganism [42–44] detection.

Common with any other optical bioimaging modality, which employs exogenous contrast agents, ultrabright probes are paramount for deep and high-resolution

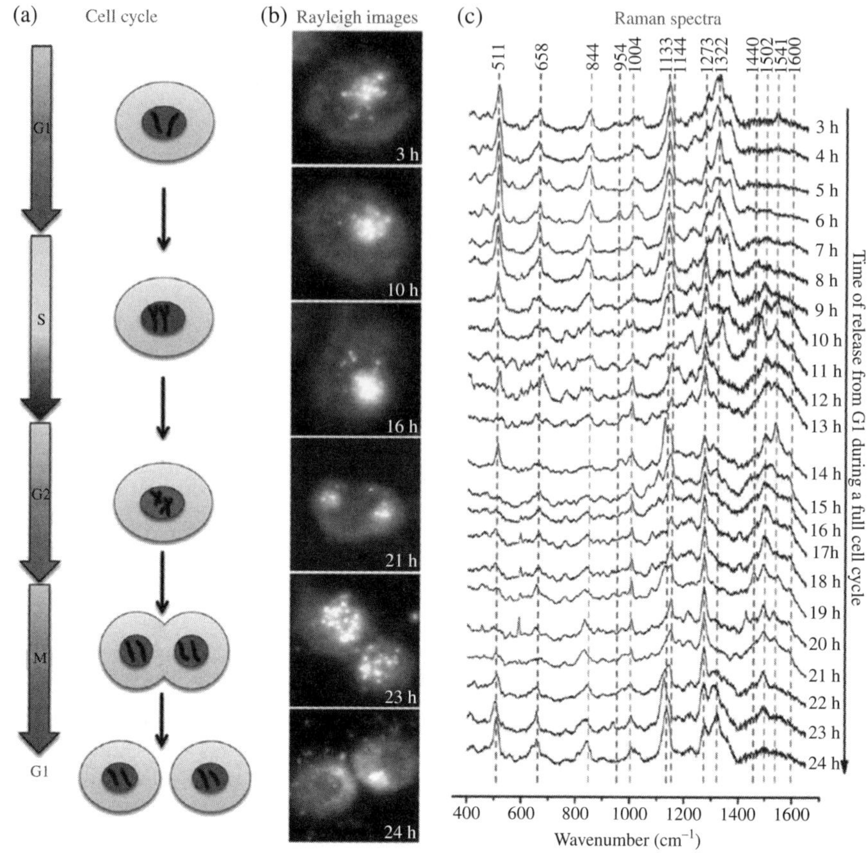

FIGURE 11.4 Real-time dark-field images and spectra of the targeted nucleus during the complete cycle of a single HSC-3 cell: (a) cell cycle, (b) plasmonically enhanced Rayleigh images, (c) corresponding SERS spectra showing that different phases of the cell cycle could be identified based on the Raman peaks attributed to DNA and protein. (Reprinted with permission from Ref. [31]. © American Chemical Society.)

SERS-based bioimaging. Ideally, SERS probes intended for *in vivo* bioimaging should (i) provide large, uniform, and stable SERS signal with NIR excitation (650–900 nm), (ii) be biocompatible and stable in complex physiological fluids and exhibit long blood circulation time, and (iii) be amenable for biofunctionalization for targeted delivery. The key considerations in the design of such efficient SERS probes include the following.

11.4.2.1 Composition of Plasmonic Nanostructures Along with other factors such as size and shape of nanostructures, composition (of both bulk and surface) of plasmonic nanostructures plays a determining role in the biocompatibility of SERS probes for *in vivo* applications. Silver nanoparticles exhibit stronger plasmon fields than gold nanoparticles of same shape and size based on the fact that the plasmon band of silver nanoparticles does not partially overlap with interband electronic transitions observed with gold nanoparticles [45]. This overlap decreases the degree of coherence motion of free electrons, which produce surface plasmon field. Although silver exhibits significantly stronger localized surface plasmon resonance (LSPR) and higher SERS enhancement factor (EF) compared to gold, the superior biocompatibility of gold makes it a better choice for *in vivo* and *in vitro* SERS-based bioimaging [18, 46].

11.4.2.2 Plasmonic Engineering Plasmonic engineering involves the rational design of size, shape, and assembly state of the nanostructures to achieve large SERS enhancement and in turn bright SERS probes. Most of the efforts related to the development of SERS probes are centered on plasmonic engineering. Various sizes of shape-controlled gold nanostructures such as spheres, rods, cubes, and stars have been employed as SERS probes for bioimaging [18, 47, 48]. Shape offers a convenient handle to tune the optical properties of metal nanostructures. For example, the longitudinal LSPR of gold nanorods can be tuned from visible to parts of NIR by varying the aspect ratio of nanorods [47]. Nanostructures with sharp apexes such as nanostars or nanopopcorns provide additional enhancement due to concentration of EM field at sharp apexes (nanoantenna effect) [48–50]. These shape-controlled nanoparticles have been successfully applied in various *in vitro* and *in vivo* imaging applications.

Lightly aggregated or assembled nanoparticles, which host intense EM hot spots at their interstices, are yet another-attractive class of SERS probes [51–53]. Recently, we reported a novel class of shape-controlled plasmonic nanostructures, namely, core–satellite structures, which was achieved through self-assembly using simple molecular cross-linkers (see Fig. 11.5) [16].

Several computational and experimental studies have clearly demonstrated that the SERS EF at these EM hot spots is upward of 10^8, which is nearly four to six orders of magnitude higher than the corresponding individual nanostructures [16, 54]. Despite significant efforts toward controlled assembly of plasmonic nanostructures, realizing scalable nanoparticle assemblies that offer highly uniform, isotropic, and stable SERS enhancement remains challenging. Controlled assembly using colloidal chemistry is a promising field for high-yield production of metal nanoparticle clusters with small interparticle gaps. To date, most of the reported methods have used nucleic acid or other organic molecules as linkers, which lead to long separation distances and thus weak plasmon coupling [55, 56]. Moreover, only simple clusters, such as dimers and trimers, have been efficiently synthesized [57, 58]. Recently, the controlled assembly of gold nanospheres into well-defined nanoparticle clusters has been reported to indicate the outstanding optical performance and therefore ultrasensitive SERS-based detection-capability [59]. In this work, block copolymers were used to produce highly symmetric gold nanoparticle clusters with large coordination numbers (up to 7). This is an area of research that has received significant interest in the last 4–5 years, and we can expect some important developments in the years to come.

Recently, core–shell plasmonic nanostructures with 1 nm interior gaps, where the Raman reporter is precisely placed between the two concentric gold layers, have been demonstrated [60]. The obvious advantage of such SERS probes is the intense

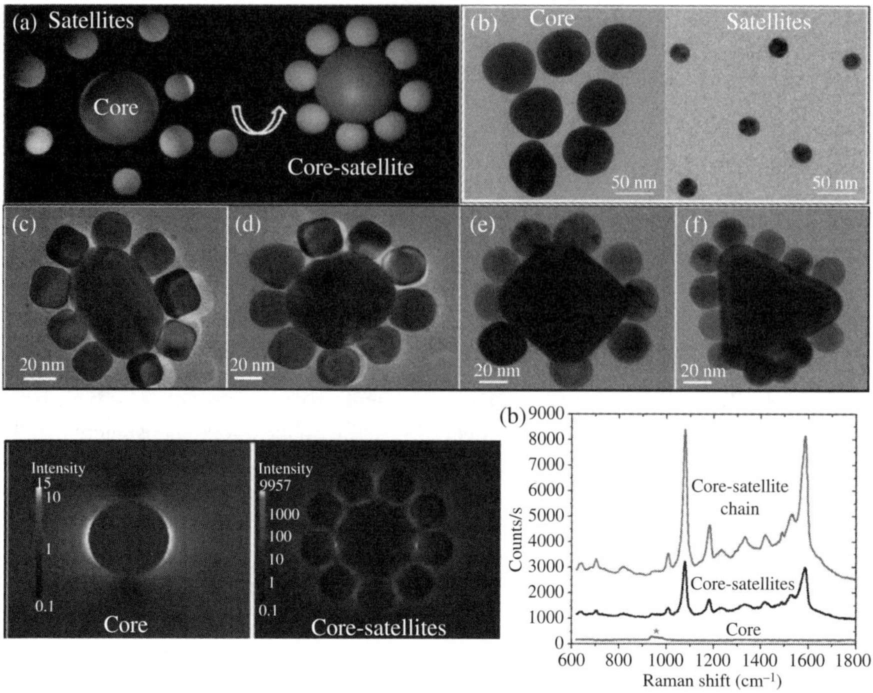

FIGURE 11.5 Structure of plasmonic planet–satellite analogs and chain structure using shape-controlled nanostructures as building blocks. Upper panel: (a) illustration of the core and satellites assembling in solution phase to form organized core–satellite clusters. (b) TEM images showing as synthesized gold nanoparticles with (left) 65 nm diameter, employed as cores, and 35 nm diameter, employed as satellites. Middle panel: representative TEM images showing the core–satellite structures observed upon successful assembly of shape-controlled nanostructures. (c) Truncated nanocubes around nanosphere. (d) Truncated nanocubes around truncated nanocube. (e) Nanospheres around sharp nanocube. (f) Nanospheres around nanoprism. TEM images showing further assembly of core–satellite clusters into chains. Lower panel: (a) electric field distribution around the cores, core–satellites, and core–satellite chain obtained using FDTD simulations. The core–satellite cluster and the chain exhibit large enhancement (~700 times) in electric field intensity compared to the cores. (b) Surface-enhanced Raman scattering spectra of the cores, core–satellite clusters, and their chains. While the Raman bands from cores were not distinguishable, clusters, and chains exhibited strong Raman bands corresponding to the molecular cross-linker used for self-assembly. (Reprinted with permission from Ref. [16]. © Wiley.)

EM hot spot within the nanostructures (formed between the core and shell) as opposed to previous approaches involving the assembly of two or more nanoparticles. The SERS EF achieved in this design is approximately 5.0×10^9. These SERS probes exhibited strong Raman signals for 633 nm excitation and virtually no signal for NIR (785 nm) excitation, which makes them nonideal for *in vivo* bioimaging applications. More recently, we reported an ultrabright SERS probe comprised of Raman reporters trapped in a subnanometer gap between gold core and shell (see Fig. 11.6) [61]. These Bilayered Raman-Intense Gold Nanostructures with Hidden

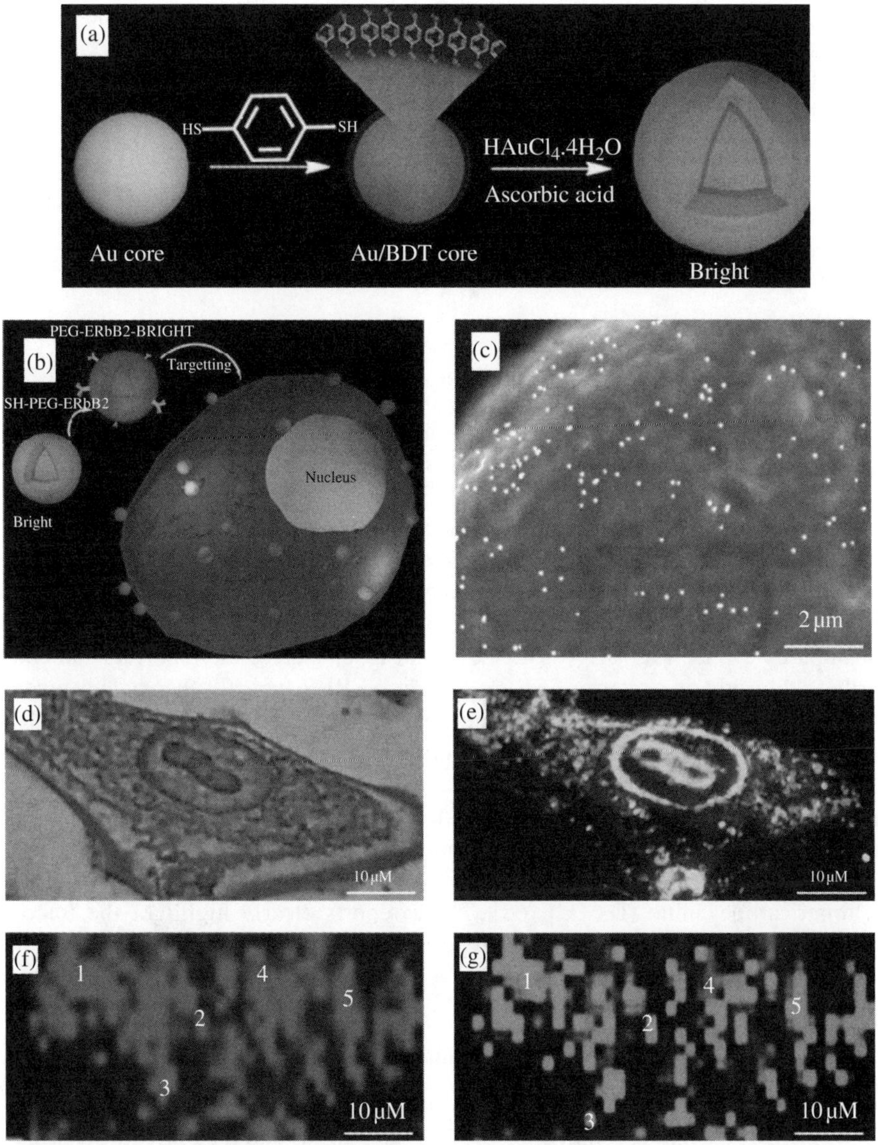

FIGURE 11.6 (a) Synthesis of BRIGHTs involves the exchange of gold nanoparticle surfactant with a Raman reporter. (b) Schematic diagram illustrating the biofunctionalization *in vitro* targeting of BRIGHTs to SKBR3 breast cancer cells. (c) SEM image of BRIGHT-targeted fixed SKBR3 cell reveals that the nanostructures largely remain as individual particles. (d) A bright-field optical micrograph of BRIGHT-targeted SKBR3 cells and (e) corresponding dark-field micrograph showing the intense Rayleigh scattering from BRIGHTs. Intensity map of 1078 cm^{-1} Raman band from BRIGHTs targeted to SKBR3 cell collected using (f) 785 nm and (g) 632.8 nm as excitation sources, respectively. (Reprinted with permission from Ref. [61]. © Wiley.) (*See insert for color representation of the figure.*)

Tags (BRIGHTs) offer a remarkably high SERS EF of 1.7×10^{11} with 785 nm excitation, which is higher than any other gold-based SERS probe reported so far. More recently, we have demonstrated core–Janus shell plasmonic nanostructures comprised of Au and Ag, which is highly Raman active, making it attractive for Raman-based bioimaging applications [62]. Those novel SERS tags are expected to enable high-resolution, non-invasive Raman-based molecular bioimaging. Particularly intense Raman signal sparks from individual nanostructures with in-built EM hot spots, in contrast to the SERS tags that rely on clusters of nanostructures to create Raman-intense EM hot spots.

11.5 RAMAN REPORTERS

Apart from plasmonic engineering, rational choice of Raman reporter molecules also plays a critical role in the brightness of the SERS probe. Recent systematic investigations suggest that carefully chosen resonant Raman reporters can provide up to two orders of magnitude higher SERS EF compared to nonresonant counterparts [63]. In this investigation, the effective differential Raman scattering cross section ($d\sigma_R/d\Omega$) of SERS nanolabels was proposed to estimate the brightness of the whole label, basing on the type of the Raman reporter and the size of nanoparticles (see Fig. 11.7). The molecular resonance of the reporters can contribute to producing intense SERRS signals. SERRS typically has enhancement several orders magnitude greater than nonresonant SERS; the maximal SERRS intensity usually coincides with the frequency of the electronic transition maximum rather than the plasmon maximum, emphasizing the difference between SERRS and SERS [64–67]. An ultrasensitive NIR Raman reporter CyNAMLA-381 was identified in an 80-member tricarbocyanine library for SERS-based *in vivo* cancer detection, showing 12-fold higher sensitivity than the standard 3,3′-diethylthiatricarbocyanine (DTTC) [68]. These reports clearly highlight the importance of rational choice of Raman reporters in improving the brightness of the SERS probes.

11.5.1 Surface Coating for Stabilization

As mentioned earlier, stability of SERS probes against desorption of Raman tags and enzymatic degradation in physiological conditions and long circulation times are critical for *in vitro* and *in vivo* bioimaging applications. Common protection/stabilization layers including glass (silica), lipid, and biocompatible polymers (e.g., poly(ethylene glycol)) render stealth character to SERS probes to escape the reticuloendothelial system and reach the target site [18, 20, 69]. Figure 11.8 shows that SERS probes stabilized by thiol-terminated polyethylene glycol, which significantly enhances the circulation time of the SERS probes [18]. These PEGylated SERS probes were used for *in vivo* tumor targeting. While there has been significant progress in proof-of-concept SERS-based bioimaging demonstrations, factors that determine the biodistribution, cellular uptake, and pharmacokinetics

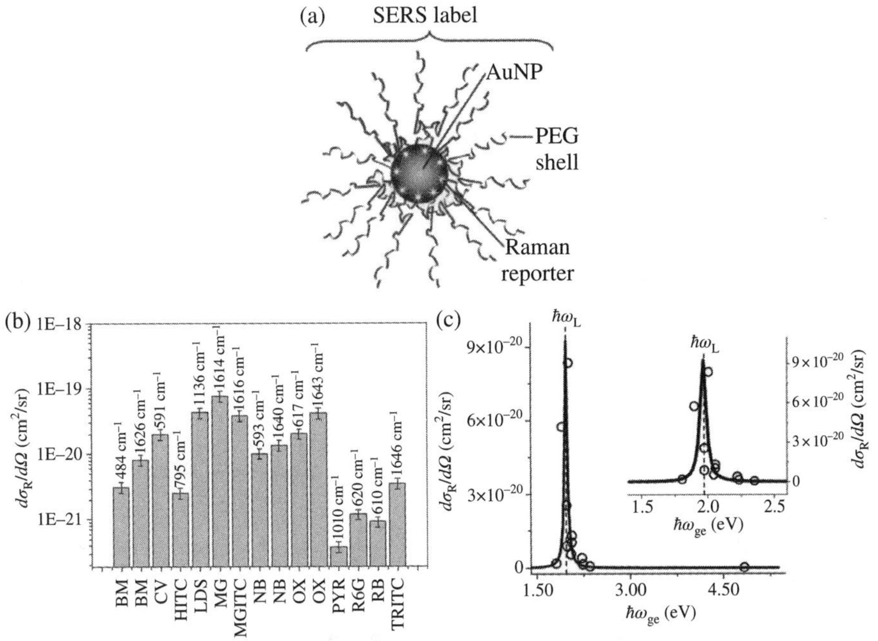

FIGURE 11.7 (a) Sketch of the structure of SERS labels. (b) $d\sigma_R/d\Omega$ for the most intense bands of 12 SERS labels at excitation wavelength of 632.8 nm. The wavenumber corresponding to the Raman band is reported above each bar. (c) Values of $d\sigma_R/d\Omega$ with the optical excitation energy $h\omega_{ge}$ of the 12 SERS labels are well fitted by a Lorentzian curve (black line) centered at the laser excitation energy $h\omega_L$ of 1.96 eV (632.8 nm). Inset shows a magnification of the region around the Lorentzian maximum. (Reprinted with permission from Ref. [63]. © Wiley.)

of SERS probes, and nanoparticles in general, remain poorly understood. This is a critical challenge that needs to be addressed for effective translation of SERS probes into clinical settings.

11.5.2 Combination with Other Techniques for Bioimaging

Raman imaging allows for highly sensitive and specific detection of SERS contrast agents, as well as the detection of multiple agents in living subjects, thanks to its unique signature of SERS spectra [19, 70–73]. It can combine Raman imaging with other imaging techniques, such as magnetic resonance imaging (MRI) and photoacoustic imaging, serving as a multiple-modality strategy. A recent work combined MRI, -photoacoustic and Raman imaging as a new triple-modality strategy to noninvasively image brain tumor margins, which enables more accurate brain tumor imaging and resection (see Fig. 11.9) [20]. The tumor visualization at a depth of 2–5 mm was achieved through the intact skin and skulls of live mice, although Raman imaging has limited penetration depth.

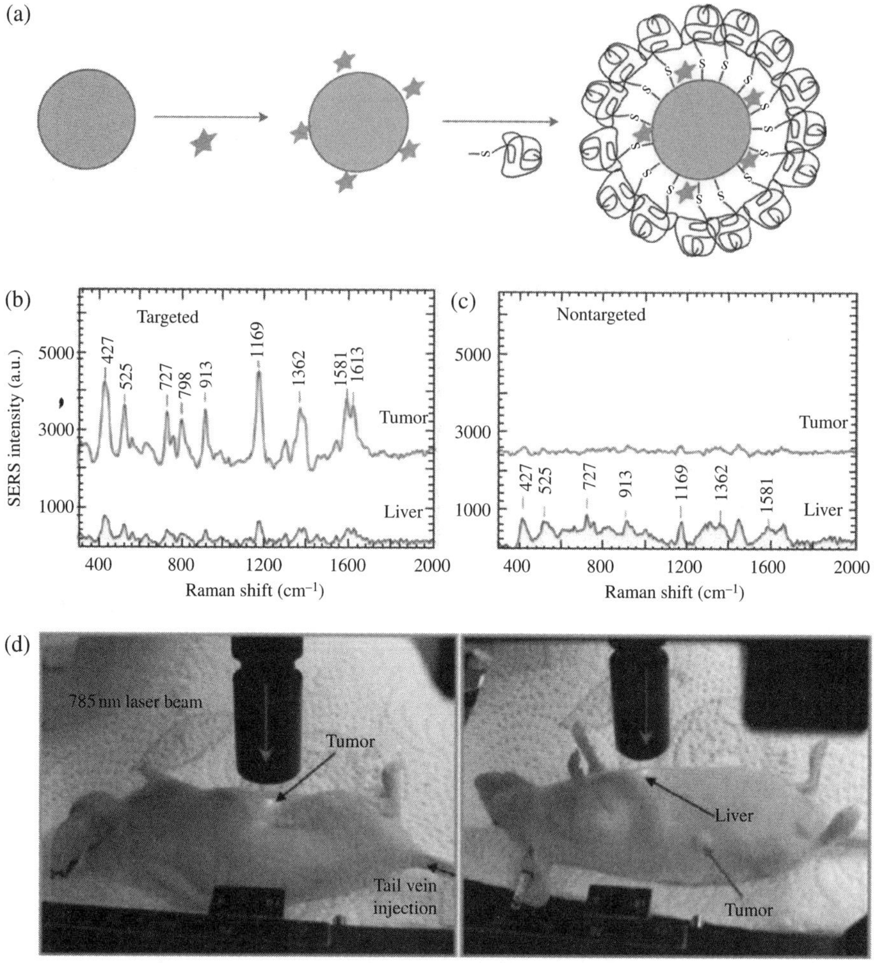

FIGURE 11.8 Design, preparation, and properties of PEGylated gold nanoparticles for *in vivo* tumor targeting and spectroscopic detection. (a) Preparation and schematic illustration of the original gold colloid, a particle encoded with a Raman reporter and stabilized with a protective layer of thiol-polyethylene glycol (thiol-PEG). *In vivo* cancer targeting and SERS detection by antibody-conjugated gold nanoparticles that recognize the tumor biomarker. SERS spectra obtained from the tumor and the liver locations by using (b) targeted and (c) nontargeted nanoparticles. (d) Photographs showing the laser beam (785 nm) focusing on the tumor site or on the anatomical location of the liver. The Raman reporter molecule used was malachite green. (Reprinted with permission from Ref. [18]. © Macmillan Publishers Ltd.)

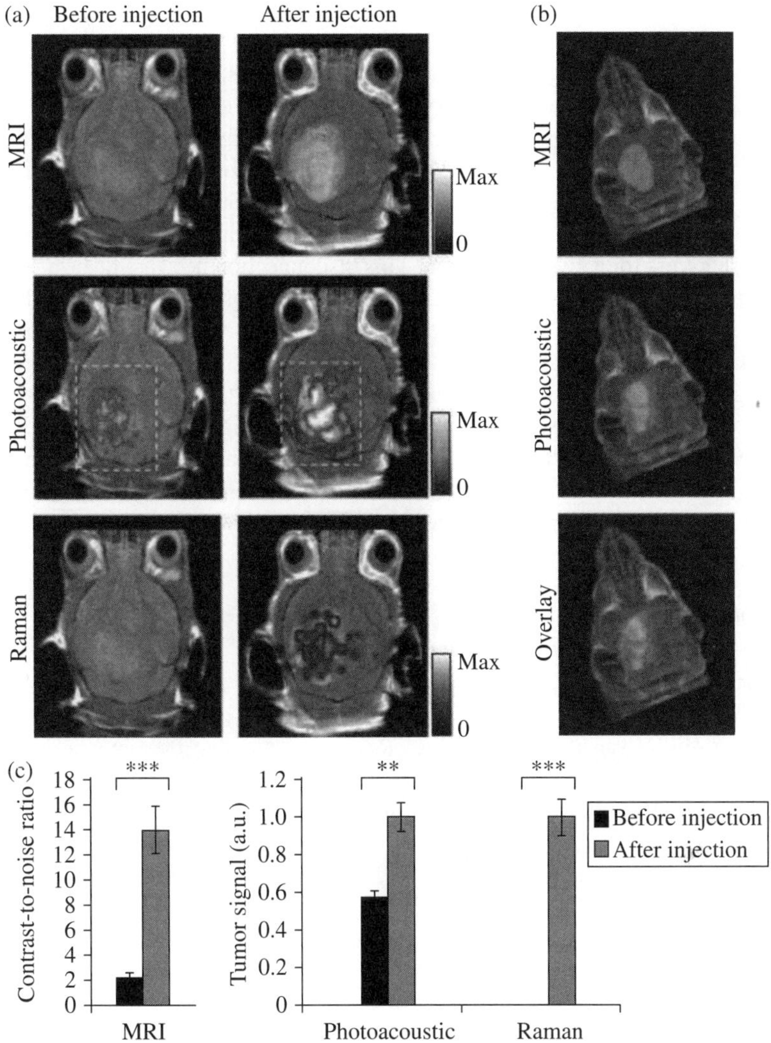

FIGURE 11.9 Triple-modality detection of brain tumors in living mice with magnetic resonance imaging–photoacoustic imaging–Raman imaging (MPR) nanoparticle. (a) Two-dimensional axial magnetic resonance imaging (MRI), photoacoustic, and Raman images. The postinjection images of all three modalities showed clear tumor visualization (dashed boxed outline highlighted the imaged area). (b) A three-dimensional (3D) rendering of magnetic resonance images with the tumor segmented (light grey; top), an overlay of the 3D photoacoustic images (light grey) over the MRI (middle), and overlay of MRI, the segmented tumor, and the photoacoustic images (bottom) showing good colocalization of the photoacoustic signal with the tumor. (c) Quantification of the signals in the tumor showing a significant increase in the MRI, photoacoustic, and Raman signals after as compared to before the injection. $n=4$ mice. Data represent mean ± s.e.m. ***P<0.001, **P<0.01 (one-sided Student's *t* test.) AU, arbitrary units. (Reprinted with permission from Ref. [20]. © Macmillan Publishers Ltd.)

11.6 SUMMARY AND FUTURE PERSPECTIVE

The future for Raman-based bioimaging certainly looks bright. Although phenomenal progress has been made in the design and synthesis of ultrabright SERS probes, there appears to be a slight disconnect between ultrabright probe development and their testing in live subjects. Most of *in vivo* applications so far relied on simple spherical nanoparticles, which are significantly dimmer compared to some of the novel probes involving advanced plasmonic engineering and rational choice of Raman reporters for a given excitation. This disconnection needs to be bridged in the future to enable rigorous testing of Raman-based bioimaging in preclinical and clinical settings. We also expect to see increased efforts in achieving multimodality and multifunctionality, which aim at synergistically integrating complementary bioimaging techniques or engineering nanostructures that can not only serve as contrast agents but also serve as therapeutic agents (e.g., drug and gene delivery, photothermal or photodynamic therapy). We believe that better understanding of bio-/nanointerface, innovative instrumentation, and engineered probes will propel Raman-based bioimaging into routine clinical applications.

REFERENCES

[1] Raman CV, Krishnan KS. A new type of secondary radiation (Reprinted from Nature, vol 121, pg 501-502, 1928). Curr Sci 1998;**74**:381.

[2] Ferraro JR, Nakamoto K. *Introductory Raman Spectroscopy*. Boston: Academic Press; 1994.

[3] Moskovits M. Surface-enhanced Raman spectroscopy: a brief retrospective. J Raman Spectrosc 2005;**36**:12.

[4] Golightly RS, Doering WE, Natan MJ. Surface-enhanced Raman spectroscopy and homeland security: a perfect match? ACS Nano 2009;**3**:2859–2869.

[5] Homola J. Surface plasmon resonance sensors for detection of chemical and biological species. Chem Rev 2008;**108**:462–493.

[6] Fang X, Ahmad SR. Detection of explosive vapour using surface-enhanced Raman spectroscopy. Appl Phys B 2009;**97**:4.

[7] Hering K, Cialla D, Ackermann K, Dorfer T, Moller R, Schneidewind H, Mattheis R, Fritzsche W, Rosch P, Popp J. SERS: a versatile tool in chemical and biochemical diagnostics. Anal Bioanal Chem 2008;**390**:113–124.

[8] Stewart ME, Anderton CR, Thompson LB, Maria J, Gray SK, Rogers JA, Nuzzo RG. Nanostructured plasmonic sensors. Chem Rev 2008;**108**:494–521.

[9] Haynes CL, Van Duyne RP. Plasmon-sampled surface enhanced Raman excitation spectroscopy. J Phys Chem B 2003;**107**:8.

[10] Gouadec G, Colomban P. Raman Spectroscopy of nanomaterials: how spectra relate to disorder, particle size and mechanical properties. Prog Cryst Growth Charact 2007;**53**:1–56.

[11] Moskovits M. Surface-enhanced spectroscopy. Rev Mod Phys 1985;**57**:783–826.

[12] Camden JP, Dieringer JA, Zhao J, Van Duyne RP. Controlled plasmonic nanostructures for surface-enhanced spectroscopy and sensing. Acc Chem Res 2008;**41**:1653–1661.

[13] Ko H, Singamaneni S, Tsukruk VV. Nanostructured surfaces and assemblies as SERS media. Small 2008;**4**:1576–1599.

[14] Campion A, Kambhampati P. Surface-enhanced Raman scattering. Chem Soc Rev 1998;**27**:241–250.

[15] Li JF, Huang YF, Ding Y, Yang ZL, Li SB, Zhou XS, Fan FR, Zhang W, Zhou ZY, Wu DY. Shell-isolated nanoparticle-enhanced Raman spectroscopy. Nature 2010;**464**:392–395.

[16] Gandra N, Abbas A, Tian L, Singamaneni S. Plasmonic planet–satellite analogues: hierarchical self-assembly of gold nanostructures. Nano Lett 2012;**12**:2645–2651.

[17] Liu T-Y, Tsai K-T, Wang H-H, Chen Y, Chen Y-H, Chao Y-C, Chang H-H, Lin C-H, Wang J-K, Wang Y-L. Functionalized arrays of Raman-enhancing nanoparticles for capture and culture-free analysis of bacteria in human blood. Nat Commun 2011;**2**:538.

[18] Qian X, Peng X-H, Ansari DO, Yin-Goen Q, Chen GZ, Shin DM, Yang L, Young AN, Wang MD, Nie S. *In vivo* tumor targeting and spectroscopic detection with surface-enhanced Raman nanoparticle tags. Nat Biotechnol 2008;**26**:83–90.

[19] Keren S, Zavaleta C, Cheng Z, de la Zerda A, Gheysens O, Gambhir SS. Noninvasive molecular imaging of small living subjects using Raman spectroscopy. Proc Natl Acad Sci U S A 2008;**105**:5844–5849.

[20] Kircher MF, de la Zerda A, Jokerst JV, Zavaleta CL, Kempen PJ, Mittra E, Pitter K, Huang R, Campos C, Habte F, Sinclair R, Brennan CW, Mellinghoff IK, Holland EC, Gambhir SS. A brain tumor molecular imaging strategy using a new triple-modality MRI-photoacoustic-Raman nanoparticle. Nat Med 2012;**18**:829–834.

[21] Mohs AM, Mancini MC, Singhal S, Provenzale JM, Leyland-Jones B, Wang MD, Nie S. Hand-held spectroscopic device for *in vivo* and intraoperative tumor detection: contrast enhancement, detection sensitivity, and tissue penetration. Anal Methods 2010;**82**:9058–9065.

[22] Sourisseau C. Polarization measurements in macro- and micro-Raman spectroscopies: molecular orientations in thin films and azo-dye containing polymer systems. Chem Rev 2004;**104**:3851–3892.

[23] Rayleigh L. Investigations in optics, with special reference to the spectroscope. Philos Mag 1879;**8**:261–274.

[24] Conrady AE. *Applied Optics and Optical Design, Part 2*. New York: Dover Publications; 1960.

[25] Matousek P. Deep non-invasive Raman spectroscopy of living tissue and powders. Chem Soc Rev 2007;**36**:1292–1304.

[26] Stone N, Kerssens M, Lloyd GR, Faulds K, Graham D, Matousek P. Surface enhanced spatially offset Raman spectroscopic (SESORS) imaging—the next dimension. Chem Sci 2011;**2**:776–780.

[27] Wu J, Wang Y, Perelman L, Itzkan I, Dasari RR, Feld MS. 3-dimensional imaging of objects embedded in turbid media with fluorescence and Raman-spectroscopy. Appl Opt 1995;**34**:3425–3430.

[28] Matousek P, Everall N, Towrie M, Parker AW. Depth profiling in diffusely scattering media using Raman spectroscopy and picosecond Kerr gating. Appl Spectrosc 2005;**59**:200–205.

[29] Stone N, Faulds K, Graham D, Matousek P. Prospects of deep Raman spectroscopy for noninvasive detection of conjugated surface enhanced resonance Raman scattering nanoparticles buried within 25 mm of mammalian tissue. Anal Methods 2010;**82**: 3969–3973.

[30] Stokes RJ, McKenzie F, McFarlane E, Ricketts A, Tetley L, Faulds K, Alexander J, Graham D. Rapid cell mapping using nanoparticles and SERRS. Analyst 2009;**134**:170–175.

[31] Kang B, Austin LA, El-Sayed MA. Real-time molecular imaging throughout the entire cell cycle by targeted plasmonic-enhanced Rayleigh/Raman spectroscopy. Nano Lett 2012;**12**:5369–5375.

[32] Qian X, Zhou X, Nie S. Surface-enhanced Raman nanoparticle beacons based on bioconjugated gold nanocrystals and long range plasmonic coupling. J Am Chem Soc 2008;**130**:14934–14935.

[33] Chon H, Lee S, Son SW, Oh CH, Choo J. Highly sensitive immunoassay of lung cancer marker carcinoembryonic antigen using surface-enhanced Raman scattering of hollow gold nanospheres. Anal Methods 2009;**81**:3029–3034.

[34] Chon H, Lim C, Ha S-M, Ahn Y, Lee EK, Chang S-I, Seong GH, Choo J. On-chip immunoassay using surface-enhanced Raman scattering of hollow gold nanospheres. Anal Methods 2010;**82**:5290–5295.

[35] Wang G, Lipert RJ, Jain M, Kaur S, Chakraboty S, Torres MP, Batra SK, Brand RE, Porter MD. Detection of the potential pancreatic cancer marker MUC4 in serum using surface-enhanced Raman scattering. Anal Methods 2011;**83**:2554–2561.

[36] El-Said WA, Kim T-H, Yea C-H, Kim H, Choi J-W. Fabrication of gold nanoparticle modified ITO substrate to detect -amyloid using surface-enhanced Raman scattering. J Nanosci Lett 2011;**11**:768–772.

[37] Dufek EJ, Ehlert B, Granger MC, Sandrock TM, Legge SL, Herrmann MG, Meikle AW, Porter MD. Competitive surface-enhanced Raman scattering assay for the 1,25-dihydroxy metabolite of vitamin D3. Analyst 2010;**135**:2811–2817.

[38] Porter MD, Lipert RJ, Siperko LM, Wang G, Narayanan R. SERS as a bioassay platform: fundamentals, design, and applications. Chem Soc Rev 2008;**37**:1001–1011.

[39] Rule KL, Vikesland PJ. Surface-enhanced resonance Raman spectroscopy for the rapid detection of Cryptosporidium parvum and Giardia lamblia. Environ Sci Technol 2009;**43**:1147–1152.

[40] Guven B, Basaran-Akgul N, Temur E, Tamer U, Boyac IH. SERS-based sandwich immunoassay using antibody coated magnetic nanoparticles for Escherichia coli enumeration. Analyst 2011;**136**:740–748.

[41] Wang Y, Ravindranath S, Irudayaraj J. Separation and detection of multiple pathogens in a food matrix by magnetic SERS nanoprobes. Anal Bioanal Chem 2011;**399**:1271–1278.

[42] Driskell JD, Kwarta KM, Lipert RJ, Porter MD, Neill JD, Ridpath JF. Low-level detection of viral pathogens by a surface-enhanced Raman scattering based immunoassay. Anal Methods 2005;**77**:6147–6154.

[43] Neng J, Harpster MH, Zhang H, Mecham JO, Wilson WC, Johnson PA. A versatile SERS-based immunoassay for immunoglobulin detection using antigen-coated gold nanoparticles and malachite green-conjugated protein A/G. Biosens Bioelectron 2010;**26**:1009–1015.

[44] Zhang H, Harpster MH, Park HJ, Johnson PA, Wilson WC. Surface-enhanced Raman scattering detection of DNA derived from the West Nile virus genome using magnetic capture of Raman-active gold nanoparticles. Anal Methods 2010;**83**:254–260.

[45] Dreaden EC, Alkilany AM, Huang XH, Murphy CJ, El-Sayed MA. The golden age: gold nanoparticles for biomedicine. Chem Soc Rev 2012;**41**:2740–2779.

[46] Nie S, Emory SR. Probing single molecules and single nanoparticles by surface-enhanced Raman scattering. Science 1997;**275**:1102–1106.

[47] von Maltzahn G, Centrone A, Park J-H, Ramanathan R, Sailor MJ, Hatton TA, Bhatia SN. SERS-coded gold nanorods as a multifunctional platform for densely multiplexed near-infrared imaging and photothermal heating. Adv Mater 2009;**21**:3175–3180.

[48] Yuan H, Liu Y, Fales AM, Li YL, Liu J, Vo-Dinh T. Quantitative surface-enhanced resonant Raman scattering multiplexing of biocompatible gold nanostars for in vitro and ex vivo detection. Anal Methods 2012;**85**:208–212.

[49] Xie J, Zhang Q, Lee JY, Wang DIC. The synthesis of SERS-active gold nanoflower tags for *in vivo* applications. ACS Nano 2008;**2**:2473–2480.

[50] Lu W, Singh AK, Khan SA, Senapati D, Yu H, Ray PC. Gold nano-popcorn-based targeted diagnosis, nanotherapy treatment, and in situ monitoring of photothermal therapy response of prostate cancer cells using surface-enhanced Raman spectroscopy. J Am Chem Soc 2010;**132**:18103–18114.

[51] Abbas A, Kattumenu R, Tian L, Nergiz SZ, Singamaneni S. Self-assembly of plasmonic nanostructures. J Nanosci Lett 2012;**2**:17.

[52] Abbas A, Kattumenu R, Tian L, Singamaneni S. Molecular linker-mediated self-assembly of gold nanoparticles: understanding and controlling the dynamics. Langmuir 2013;**29**:56–64.

[53] Abbas A, Tian L, Kattumenu R, Halim A, Singamaneni S. Freezing the self-assembly process of gold nanocrystals. Chem Commun (Camb) 2012;**48**:1677–1679.

[54] Gandra N, Singamaneni S. "Clicked" plasmonic core-satellites: covalently assembled gold nanoparticles. Chem Commun (Camb) 2012;**48**:11540–11542.

[55] Angioletti-Uberti S, Mognetti BM, Frenkel D. Re-entrant melting as a design principle for DNA-coated colloids. Nat Mater 2012;**11**:518–522.

[56] Macfarlane RJ, Lee B, Jones MR, Harris N, Schatz GC, Mirkin CA. Nanoparticle superlattice engineering with DNA. Science 2011;**334**:204–208.

[57] Costi R, Cohen G, Salant A, Rabani E, Banin U. Electrostatic force microscopy study of single Au-CdSe hybrid nanodumbbells: evidence for light-induced charge separation. Nano Lett 2009;**9**:2031–2039.

[58] Li WY, Camargo PHC, Au L, Zhang Q, Rycenga M, Xia YN. Etching and dimerization: a simple and versatile route to dimers of silver nanospheres with a range of sizes. Angew Chem Int Ed 2010;**49**:164–168.

[59] Pazos-Perez N, Wagner CS, Romo-Herrera JM, Liz-Marzán LM, García de Abajo FJ, Wittemann A, Fery A, Alvarez-Puebla RA. Organized plasmonic clusters with high coordination number and extraordinary enhancement in surface-enhanced Raman scattering (SERS). Angew Chem Int Ed Engl 2012;**51**:12688–12693.

[60] Lim D-K, Jeon K-S, Hwang J-H, Kim H, Kwon S, Suh YD, Nam J-M. Highly uniform and reproducible surface-enhanced Raman scattering from DNA-tailorable nanoparticles with 1-nm interior gap. Nat Nano 2011;**6**:452–460.

[61] Gandra N, Singamaneni S. Bilayered Raman-intense gold nanostructures with hidden tags (BRIGHTs) for high-resolution bioimaging. Adv Mater 2012. DOI: 10.1002/adma.201203415.

[62] Gandra N, Portz C, Singamaneni S. Bimetallic Janus nanostructures via programmed shell growth. Nanoscale 2013;**5**:1806–1809.

[63] Amendola V, Meneghetti M. Exploring how to increase the brightness of surface-enhanced Raman spectroscopy nanolabels: the effect of the Raman-active molecules and of the label size. Adv Funct Mater 2012;**22**:353–360.

[64] Cunningham D, Littleford RE, Smith WE, Lundahl PJ, Khan I, McComb DW, Graham D, Laforest N. Practical control of SERRS enhancement. Farad Discuss 2006;**132**:135–145.

[65] McNay G, Eustace D, Smith WE, Faulds K, Graham D. Surface-enhanced Raman scattering (SERS) and surface-enhanced resonance Raman scattering (SERRS): a review of applications. Appl Spectrosc 2011;**65**:825–837.

[66] Meyer SA, Ru ECL, Etchegoin PG. Quantifying resonant Raman cross sections with SERS. J Phys Chem A 2010;**114**:5515–5519.

[67] Gabudean AM, Focsan M, Astilean S. Gold nanorods performing as dual-modal nano-probes via metal-enhanced fluorescence (MEF) and surface-enhanced Raman scattering (SERS). J Phys Chem C 2012;**116**:12240–12249.

[68] Samanta A, Maiti KK, Soh KS, Liao X, Vendrell M, Dinish US, Yun SW, Bhuvaneswari R, Kim H, Rautela S, Chung J, Olivo M, Chang YT. Ultrasensitive near-infrared Raman reporters for SERS-based *in vivo* cancer detection. Angew Chem Int Ed Engl 2011;**50**:6089–6092.

[69] Mulvaney SP, Musick MD, Keating CD, Natan MJ. Glass-coated, analyte-tagged nanoparticles: a new tagging system based on detection with surface-enhanced Raman scattering. Langmuir 2003;**19**:4784–4790.

[70] Zavaleta CL, Kircher MF, Gambhir SS. Raman's "effect" on molecular imaging. J Nucl Med 2011;**52**:1839–1844.

[71] Zavaleta C, de la Zerda A, Liu Z, Keren S, Cheng Z, Schipper M, Chen X, Dai H, Gambhir SS. Noninvasive Raman spectroscopy in living mice for evaluation of tumor targeting with carbon nanotubes. Nano Lett 2008;**8**:2800–2805.

[72] Zavaleta CL, Smith BR, Walton I, Doering W, Davis G, Shojaei B, Natan MJ, Gambhir SS. Multiplexed imaging of surface enhanced Raman scattering nanotags in living mice using noninvasive Raman spectroscopy. Proc Natl Acad Sci U S A 2009;**106**:13511–13516.

[73] Wang YL, Seebald JL, Szeto DP, Irudayaraj J. Biocompatibility and biodistribution of surface-enhanced Raman scattering nanoprobes in zebrafish embryos: *in vivo* and multiplex imaging. ACS Nano 2010;**4**:4039–4053.

PART III

NANOTECHNOLOGY IN BIOMEDICAL IMAGING AND BEYOND

12

PANDIA®: GOLD NANORODS AND THEIR APPLICATIONS IN CANCER THERAPY AND *IN VIVO* IMAGING IN COMPANION ANIMALS AND THEIR POTENTIAL APPLICATION TO HUMANS

Christian Schoen
Nanopartz Inc., Loveland, CO, USA

Cheryl London
Department of Veterinary Biosciences, Ohio State University, Columbus, OH, USA

12.1 INTRODUCTION

Nanopartz gold nanorods (AuNRs) were intravenously injected into canines for use as a photothermal imaging tool and therapy of soft tissue sarcomas. We call this system Pandia® (Fig. 12.1). In a clinical trial, seven client-owned canines were injected with up to 420 mg of AuNRs where UV–Vis was used to measure AuNR blood concentration over a 3-day period. Once cleared, these sarcomas were treated with a 10 W 808 nm laser diode and imaged with an IR thermal camera. In many cases, complete recovery was seen.

Cancer is the major cause of death for companion animals. It is reported that 23% of all dogs and 32% of all cats die from cancer. While, on the one hand, advances in veterinary care can increase the life span of our companion animals, on the other

Nanotechnology for Biomedical Imaging and Diagnostics: From Nanoparticle Design to Clinical Applications, First Edition. Edited by Mikhail Y. Berezin.

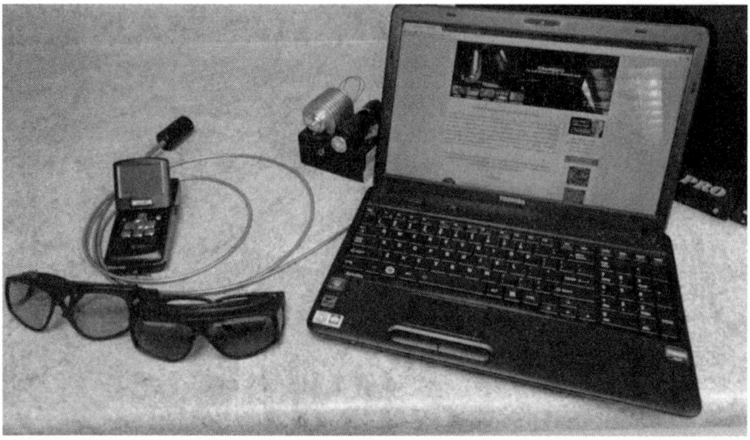

FIGURE 12.1 Pandia® system includes a 20 W 808 nm diode laser with fiber-optic delivery, injectable Ntracker™ gold nanoparticle reagent (not shown), a thermal imaging camera, and laser safety goggles.

hand, the statistics for cancer development as our pets live longer lives are even grimmer: approximately 45% of pets over ten years of age will be diagnosed with the disease [1]. Of the 74 million dogs and 90 million cats in U.S. households, it is estimated that 4 million dogs and 4 million cats will develop cancer each year [2, 3]. Current treatment methods for companion animals diagnosed with many types of cancers are similar to those for humans. Primary therapeutic regimens include one or more of the following: surgical removal, burning, freezing, radiation therapy, chemotherapy, hormone therapy, and immunotherapy. These treatments have varying degrees of success, depending on a wide array of factors, including stage of the disease at diagnosis, location of the tumor(s), age and health of the animal, and many other conditions. Additionally, conventional treatment with chemotherapy and radiation is toxic with many side effects. Pet owners are increasingly demanding treatments for their companions that will improve their outcome and extend their life. For cancers that do not respond to traditional therapies or are too difficult to treat by standard methods such as recurring soft tissue sarcomas, an approach that will significantly improve the prognosis for the animal, and thus for their owners, is desperately needed.

Thermal ablation of tumors is one such treatment that holds the promise of radically changing the outcome for many stricken companion animals while offering hope to their owners desperate for an approach that gives the very real possibility of both longer life and improved quality of life for their pets. Thermal ablation of tumors has the potential for a generalized method of cancer therapy with less chance of side effects from the treatment [4–6]. Further, there is solid clinical evidence that heat treatment potentiates both chemotherapeutics and radiation [5–8]. In particular, heat treatment inhibits DNA replication and damage repair, preferentially kills nutrient-deficient cells, increases tumor blood flow and

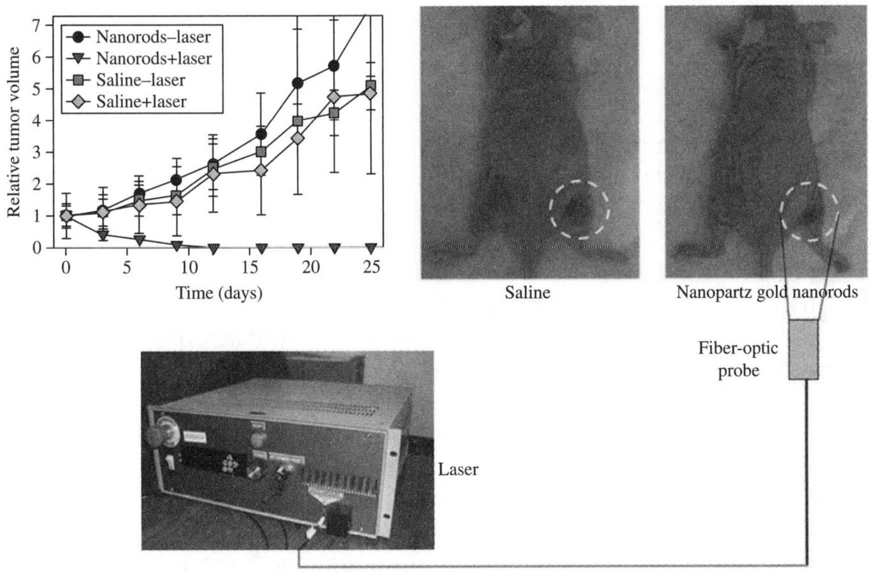

FIGURE 12.2 In published mouse studies, the shown system has demonstrated efficacy in destroying MDA-MB-435 breast cancer tumors. In this example, tumors were grown in two mice. In mouse A (left), PBS was injected, whereas in mouse B (right), Nanopartz gold nanorods were injected. After 72 h, a 2 W 808 nm diode laser was used to photothermally treat the cancer areas. For the Nanopartz-treated mouse, the tumor was completely destroyed and no mass remained after 12 days.

drug and oxygen delivery, and, most importantly, increases vessel porosity allowing enhanced delivery of nanoparticle therapeutics.

In this effort, we evaluated the use of polyethylene-coated AuNRs as a potential cancer therapy and imaging device—one that couples intravenously injected Nanopartz polyethylene-coated AuNRs with an external laser to heat and potentially ablate cancer tumors in companion animals with soft tissue sarcomas. Our initial efforts have shown success in mice in a preclinical study (Fig. 12.2) and, more recently, canines in an Ohio State University clinical trial [9, 10]. In this chapter, we demonstrate the results of imaging and treating sarcomas in companion animals.

12.2 BACKGROUND

12.2.1 History of Thermal Ablation

Throughout history, hyperthermia has been an accepted method of cancer treatment in humans, either alone or as a secondary therapy in concert with another method. One example of enhanced therapeutic benefits offered by hyperthermia is found in the treatment of head and neck cancers; using standard radiation techniques, the

5-year survival rate of those patients treated with hyperthermia was 53 versus 0% for those offered radiation alone [11].

Hyperthermia methods with clinical history include radio frequency (RF), laser, cryogenics, and ultrasound. Some basic hyperthermia techniques include endocavity irrigation and whole patient and regional perfusion techniques. However, none of these approaches enable tumor-specific energy deposition. Recently, this has been shown to be solved with plasmonic nanomaterial such as AuNRs.

The electromagnetic properties of plasmonic nanomaterials have been harnessed to develop ultrasensitive diagnostic [12, 13], spectroscopic [14, 15], and, more recently, therapeutic technologies [16–19]. In particular, tunable plasmonic nanomaterials have attracted attention for their immense optical absorption coefficients and potential as injectable nanoantennas that can target tumors and locally convert otherwise benign electromagnetic energy to thermal energy for ablation. Currently, tumor ablation approaches in clinical practice, including RF, laser, and focused ultrasound methods, lack intrinsic tumor specificity to energy absorption. The inability to selectively heat tumor tissues over surrounding compartments necessitates efforts to externally direct applied energy toward tumor tissues, making effective treatment of tumor margins and complex tumor geometries very challenging. By providing a tumor-specific heat source (808 nm laser), nanoantennas have the very real potential to considerably broaden the clinical applicability of thermal therapies by simplifying their integration with current therapeutic practices (including improving margin clearance in surgery and synergizing with regional radiation therapies) and reducing morbidity due to off-target heating. Furthermore, by pulsing the external energy source used, tumor-targeted nanoantennas can theoretically ablate with single-cell precision, thereby providing improved accuracy over standard surgical methods and opening the possibility of precisely treating complex tumor margins in sensitive tissues.

To date, Nanopartz AuNRs have shown considerable efficacy for tumor ablation using near-infrared (NIR) laser light [16, 20, 21]. These results highlight the clinical promise of Nanopartz AuNRs as photothermal therapies. From a material perspective, nanoantennas with enhanced circulation times *in vivo*, increased absorption coefficients per weight, and narrower absorption spectra would more efficiently permeate tumors after intravenous (IV) administration, amplify the photothermal contrast between antennas and normal tissue, and allow improved tumor treatment at lower laser intensities or at greater depths.

The long precedence of gold nanoparticles in clinical rheumatoid arthritis therapies, as well as in more recent trials, makes AuNRs appealing new candidates for nanoantenna-based photothermal ablation and a wide array of other biomedical applications.

12.2.2 History and Synthesis of Nanopartz Nanorods

In 2007, Nanopartz, Inc., commercialized rod-shaped gold nanoparticles. Our exclusively licensed, patented technology [22] is based on the Murphy seed-based technique [23, 24]. Nanopartz offers over 30 different highly monodisperse AuNRs

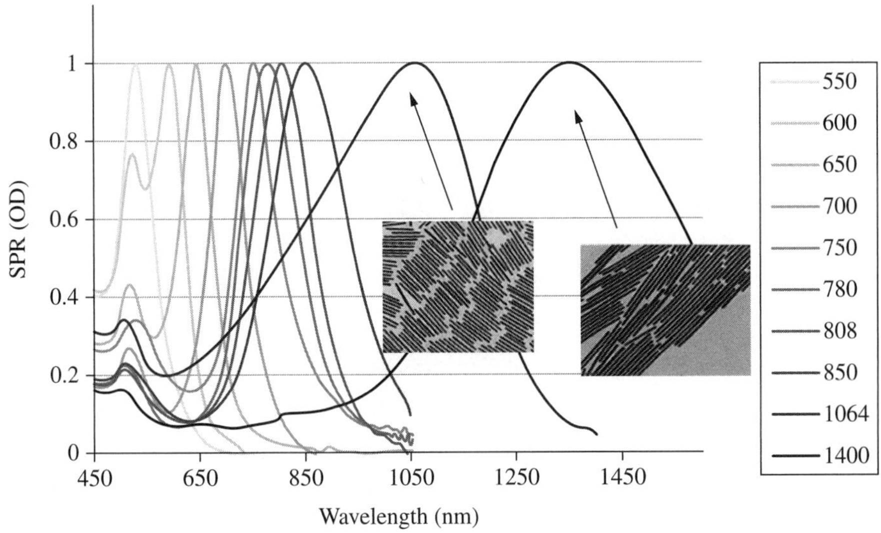

FIGURE 12.3 Nanopartz gold nanorods can be tuned across the visible spectrum into the near IR with very narrow absorptions.

with surface plasmon resonances (absorptions) tuned from 550 to 1400 nm, allowing selection of the nanorod (NR) to match that of excitation laser (Fig. 12.3).

Nanopartz materials have been cited in over 200 publications (www.google.com/scholar, search "Nanopartz"), 16 patents, and 2 journal covers. Nanopartz has also developed proprietary and trade secret methods to replace the toxic cetyltrimethylammonium bromide (CTAB) capping agent used in the production of many nanoparticle formulations with a noncytotoxic polymer that also results in long circulation times. The examples of commercial Nanopartz AuNRs are shown in Figure 12.4.

12.2.3 Nanorod Scale-Up Synthesis

During the past decade, a significant success has been achieved in the synthesis of well-defined nanocrystals [25–27], but their organization into functional macroscopic materials and devices [28–30] remains a challenge [30–32]. The majority of existing synthetic procedures are designed for a milligram scale, and very often they cannot be extrapolated to a large-scale synthesis [33–35]. A particularly important example is related to anisotropic nanostructures of gold, such as single crystalline NRs [35], penta-twinned NRs [36], and nanowires [37]. These materials have been shown to have a remarkable potential in areas ranging from nanomedicine to metamaterials and data storage [38]. Since the late 1990s, there has been an exponential growth in the number of papers reporting on various syntheses of AuNRs using electrochemical [39], photochemical [40], template-assisted [41], and wet chemistry methods [42]. The best results are generally achieved by

FIGURE 12.4 Commercial samples of gold nanorods from Nanopartz, Inc. (*See insert for color representation of the figure.*).

seed-mediated synthesis in water in the presence of a cationic surfactant CTAB and specific amounts of silver ions (~20 mol%). Under these conditions, the content of rodlike particles is almost 100%, and no tedious purification from other shapes is required.

This synthesis was introduced by Murphy *et al.* [35] and until now remains the best available technique in spite of many other reports published since then. However, even this method has a serious limitation because the yield of conversion (gold ions to metal) is only 10–15%, and up to 90% of the precious starting material (gold chloride) remains in solution after the growth of NRs stops completely. Unfortunately, attempts to improve the yield by increasing the amount of reducing agent normally lead to uncontrolled nucleation and many irregular shapes, which are very hard to separate. As a result, only milligram quantities can be produced in one batch using the best available technique [43]. Interestingly, AuNRs recently became commercially available, but their price is astronomically high. They are sold by Sigma-Aldrich, Inc. (which is a distributor of Nanopartz AuNRs), at \$306/mg, which is more than 7200 times the current price of bulk gold. The fact that many researchers from academe and industry are willing to buy AuNRs at that price clearly indicates that there is an urgent need to develop new knowledge in the synthesis of anisotropic gold structures and find an efficient way to produce their large quantities.

If reliable and inexpensive techniques to make AuNRs become available, it will significantly increase the scope and number of their real-life applications [44]. Both experimental and theoretical studies indicate that long-range periodic assemblies of anisotropic nanostructures may possess vectorial optical and electrical properties [45]. These are key elements for making macroscopically large functional structures. Many reports in the literature highlight examples of the NR assemblies, which are based on noncovalent interactions [46], polymer-induced self-corralling [47], external electric fields [48], tip-selective functionalization [49], and breath figures technique [50]. One critically important feature required for the assembly of rod-shaped particles into large colloidal superstructures is their narrow size distribution both in terms of their diameter and length. While the diameter of AuNRs is normally very uniform, the length distribution is very difficult to control. This is the main reason why superlattices made of inorganic NRs are often quite small and rarely exceed several microns in lateral dimensions.

The main goal in this effort was to develop a method to scale up nanorod batch manufacturing. The working hypothesis for this goal is that the yield of conversion in a seed-mediated synthesis can be increased from conventional approximately 10 up to 100% if the unreacted gold ions are slowly reduced on the surface of preformed NRs. This is based on preliminary data, which shows that it is possible to suppress random nucleation as long as the rate of reduction is very low. In order to keep this rate low, reducing agent is added to the growth solution in small aliquots. Each addition should be followed by several hours period in order to allow for a full consumption of the added amount of reducing agent. Under these conditions, the reduction happens only on the surface of metallic gold, which catalyzes the disproportionation of Au^{+} ions. As a result, the initial NRs would undergo a uniform amplification until all gold ions are consumed. Once the yield is improved to near-quantitative level, an increase in volume of the reaction should allow for the synthesis of gram quantities of well-defined NRs. Wet chemistry synthesis of inorganic nanostructures has a distinct advantage over the alternative techniques because it is more suitable for the large-scale preparation and chemical functionalization of nanostructures in solution. The research conducted by many groups clearly suggests that the seed-mediated synthesis of AuNRs is the most efficient method [38]. There are many variables and chemical species that may affect the outcome of the synthesis. Detailed investigations of the role of seed, CTAB, silver ions, and different reducing agents have been reported in the literature [37, 51].

While these findings are extremely valuable, nearly all of them pertain to the original small-scale synthesis (10 mL of 5×10^{-4} M solution), which cannot produce more than 0.1 mg of AuNRs even if the yield of gold ion conversion were 100%. In reality, that yield is only approximately 10%, which generates approximately 0.01 mg of NRs. Because this synthesis is a kinetically controlled process, it strongly depends on the actual scale, which makes the existing findings not amenable to a multiliter scale. This may also explain why a synthesis that could produce more than a few milligrams of isolated and pure AuNRs has not been reported to date, nearly one decade after the introduction of this method. The seed solution is normally made by reducing gold (III) ions with excess sodium borohydride ($NaBH_4$) in 0.1 M aqueous solution of CTAB. The fast reduction leads to small spherical particles of gold (seeds) that are

2–3 nm in diameter. A small amount of the seed solution is then added to the growth solution containing Au (I) ions to initiate the formation of NRs. However, the role of seed has been a subject of debate [37], and it is not clear if all resulting NRs originate from the preformed seed particles. In fact, several reports have shown that the formation of rods can be initiated by direct addition of $NaBH_4$ [52] to the growth solution (seedless method).

Knowing the amount of metallic gold and the average size of the seed particles, one is able to find the actual number of seeds that are introduced in the beginning of the synthesis. The number of rods that subsequently form in solution is determined by the same combination of TEM and ICP and compared with that of seeds. These experiments revealed the percentage of NRs that formed from seeds and those originated from *in situ* nucleation. Under the conventional conditions, the final number of rods is much lower than it could be if all ions were reduced [53]. However, attempts to increase the number of rods by introducing more seeds generally lead to poor results [54]. Instead of more rods, the reaction produces a mixture of spheres and rods. As mentioned before, the conventional seed-mediated synthesis generates a very limited number of NRs, and their growth completely stops even when there are still up to 90% of residual Au (I) ions present in solution [53]. Preliminary studies show that all these ions can be reduced on the surface of initial rods if the additional amount of reducing agent is introduced at a very low rate. This process of amplification does not change the number of rods, but it increases their mass more than 10 times. This allows one to reach a quantitative yield of conversion while preserving the purity and narrow size distribution of rods. In our procedure, the time interval between the consecutive additions of reducing agent was kept at 2–3 h in order to ensure that all reducing agent is completely consumed before the next portion is introduced. The number of initial rods is larger, and therefore, the concentration of residual Au (I) ions is lower. This eases the amplification process and decreases the probability of random nucleation, especially in the early stages. It also produces NRs with a higher aspect ratio, which is desirable for our proposed studies on colloidal crystallization. The most reliable synthesis of AuNRs is based on the seed-mediated method in the presence of silver ions [35].

However, this synthesis produces only submilligram quantities of NRs with a low yield (10–15% with respect to the amount of gold chloride used in the synthesis). In addition, the NRs made by this method do not undergo colloidal crystallization because their size distribution is not sufficiently narrow. In this work we identified the conditions for the uniform and "safe" amplification, and were able to increase the reaction mixture from 10 [43] to 7000 mL and doubled the concentration of gold ions. By doing so, we were able to amplify all NRs and isolate > 1 g of AuNRs, thereby achieving nearly 100% in the isolated yield. The resulting NRs measured on average anywhere from 10 to 25 nm in diameter and 40 to 60 nm in length. This amplification resulted in a narrower size distribution and a significant narrowing of the longitudinal peak during the amplification process. The ability to make NRs of different sizes using this technique is shown in Figures 12.5 and 12.6.

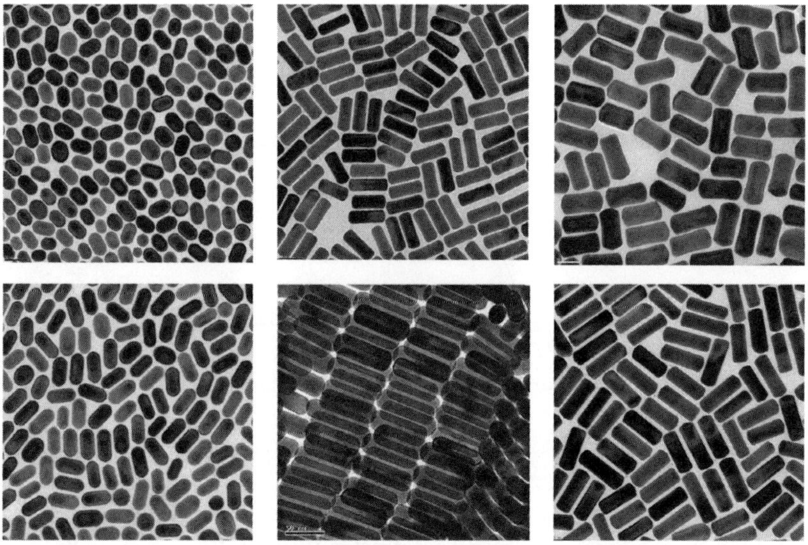

FIGURE 12.5 Assorted Nanopartz nanorods of different diameters and aspect ratios.

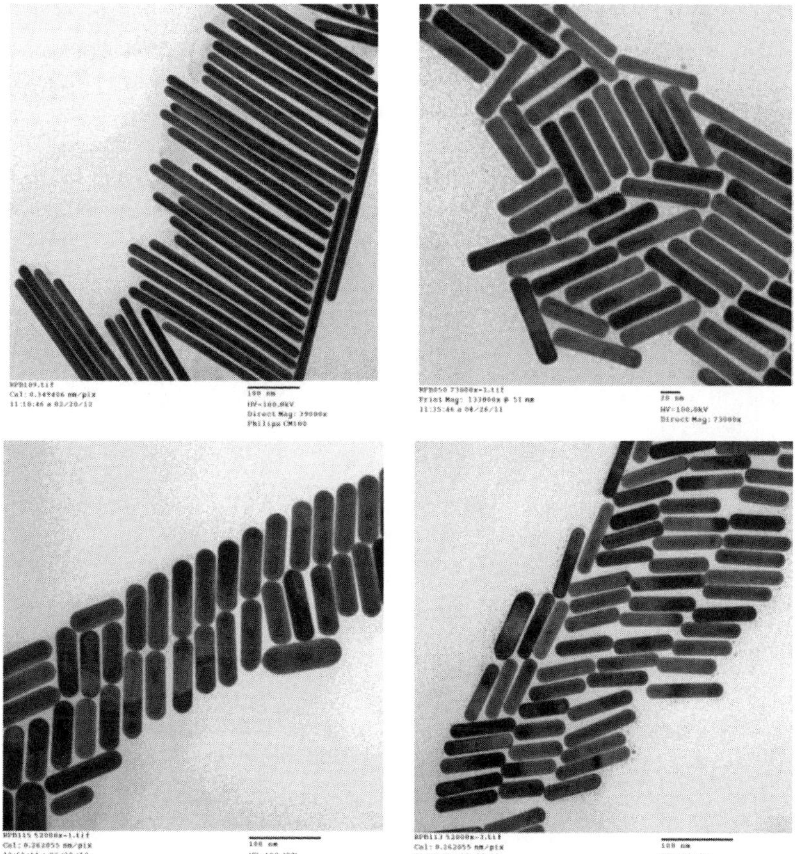

FIGURE 12.6 Assorted Nanopartz nanorods of longer aspect ratios.

12.3 PHOTOTHERMAL IMAGING AND THERAPY *IN VIVO*

12.3.1 Efficacy in Mice

There have been a number of studies conducted using Nanopartz AuNRs that have shown efficacy in cancer tumor therapy [16, 29]. Using PEGylated Nanopartz AuNRs (PEG-NRs), Maltzahn *et al.* [10] determined that:

1. PEG-NRs were highly stable *in vitro*, showing minimal spectral shifting (which would indicate particle destabilization and aggregation) even after greater than 1000 h in 0.15 mol/l NaCl or 10% human serum and highly biocompatible.
2. PEG-NRs exhibited less than 1/3 of the spectral bandwidth and approximately three times higher extinction coefficient per gram of gold than PEG-nanoshells. Additionally, under identical experimental conditions, irradiated PEG-NR solutions generated heat greater than six times more rapidly than PEG-nanoshells per gram of gold.
3. In order for PEG-NRs to passively target tumors via the enhanced permeability and retention (EPR) effect and act as nanoantennas for photothermal therapy, they must be able to traverse the systemic circulation, deter protein opsonization and reticuloendothelial system (RES) clearance, permeate through transendothelial pores in tumor blood vessels, and be retained in the tumor interstitium. After IV administration to tumor-bearing mice (20 mg Au/kg), Nanopartz PEG-NRs were found to exhibit blood half-lives of approximately 17 h and to maintain their 810 nm longitudinal plasmon resonance throughout this period, allowing spectrophotometric detection in the serum over time (Fig. 12.7).

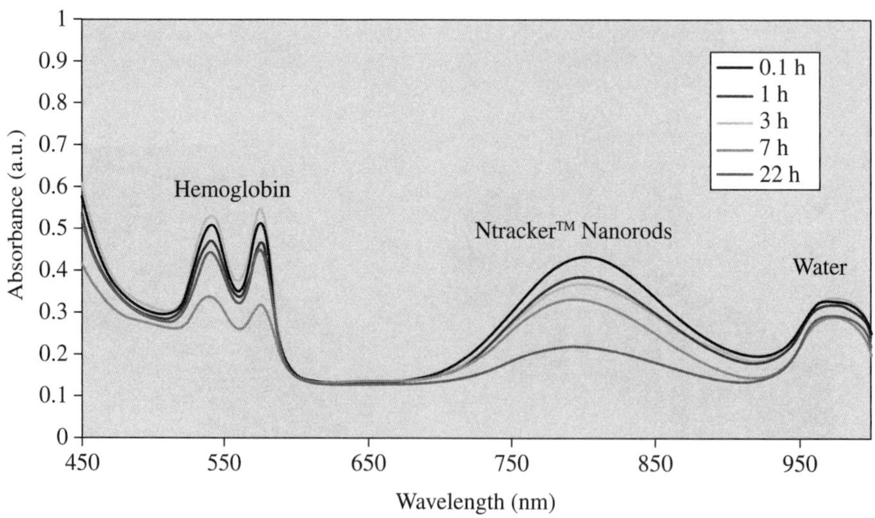

FIGURE 12.7 UV–Vis absorption of mice blood over 22 h after IV injection of Nanopartz nanorods.

4. Further, Maltzahn *et al.* tested the prediction that a single dose of PEG-NRs could destroy tumors. In this analysis, nude mice bearing bilateral human MDA-MB-435 tumors were injected with either PEG-NRs or saline. After 72 h, the right flank of each mouse was irradiated for 5 min (810 nm, 2 W/cm^2), and all tumors were measured at regular intervals over time. All irradiated, PEG-NR-targeted tumors completely disappeared by external observation within 10 days, whereas all other tumors, including those exposed to laser after saline injection, continued to grow uninhibited. To assess the survival benefit of PEG-NR-directed tumor ablation, mice bearing a single MDA-MB-435 tumor were divided between four cohorts (PEG-NRs + laser, PEG-NRs − laser, saline + laser, saline − laser), and all tumors were measured over time. By 20 days after treatment, all irradiated, PEG-NR-injected mice displayed only a minor scar with no evidence of tumor regrowth, whereas all other surviving mice harbored thriving tumors. Over the course of greater than 50 days of observation, no irradiated, PEG-NR-injected mice showed evidence of recurrence, whereas all mice in the control had to be euthanized by day 33. Body weights of PEG-NR-treated mice were monitored over time and showed no obvious losses due to tumor ablation and resorption.

12.3.2 Nanopartz Nanorod Mouse Biodistribution

By using the EPR effect, gold nanoparticles can passively and selectively concentrate in cancer tumors upward of 7% injected dose (ID)/g (refer to Table 12.1).

12.3.3 Canine Toxicology Evaluation

It is critical to the success of the platform as a potential therapy for cancer in companion animals to establish the safety of this device. Thus, in 2010, Nanopartz funded a research study by CARE Research LLC (Ft. Collins, CO), entitled "Non-GLP Single Dose (IV Slow Bolus) Dose Escalation (Up to 5 Escalations

TABLE 12.1 Biodistribution Data from Mouse Injected Gold Nanorods[a]

Tissue	% ID/g tissue	Standard deviation
Tumor	7.1	±2.9
Brain	0.5	±0.3
Bladder	0.4	±0.1
Muscle	1.7	±0.8
Heart	2.4	±0.9
Lung	4.4	±0.9
Kidney	4.0	±0.3
Liver	14.0	±4.0
Spleen	51.9	±13.0

[a] From Ref. [16].

with 7-Day Wash-Outs) with Pandia™ (gold nanorods) Study in Non-Naïve Beagle Dogs." The objective of this study was to determine the maximum tolerated dose (MTD) for Pandia® in one male and one female nonnaïve beagle dog via dose escalation, beginning at a starting dose of 2 mg/kg and ending at a maximum dose level of 32 mg/kg. No overt toxicity was observed in any of the dose levels. Based on the live-phase results of this study, it can be concluded that a single infused (16 ml/h rate) IV dose of Pandia® (maximum concentration of 32 mg/kg, 16 mg/ml, 2 ml/kg) was well tolerated by nonnaïve beagle dogs over the duration of this study (74 study days).

12.3.4 Efficacy in Client-Owned Canines

The results presented here are from an ongoing clinical study at Ohio State University using client-owned cancer dogs. The complete research effort is seven canines with similar osteosarcomas (the list of patients is given in Table 12.2). In all cases, using different AuNR concentrations and different heating temperature profiles, complete destruction of the cancer was achieved (an example is given in Fig. 12.8). Using AuNR injections from 2 to 32 mg/kg, NR optical densities in the bloodstream taken immediately after injection ranged from 1.0 to 2.0. After 3 days, in all cases, the optical density is reduced to less than 0.1. Treatments utilized a 30 W 808 nm diode laser at laser powers from 10 to 30 W creating temperature increases in the laser tumor from 15 to 100°C (Fig. 12.9).

The primary purpose of this pilot study was to assess the clinical effects of AuNRs administered to dogs with solid tumors and the antitumor activity following NIR diode laser treatment.

This was a 6-month research effort with the following timeline: day 1, baseline tumor measurement, CBC, chemistry panel, baseline tumor biopsy and serum/plasma NR levels, NR administration at 10 mg/kg and 250 mg/h (5 mg/ml), and posttreatment serum/plasma NR levels; day 3, 72 h serum/plasma NR levels and NIR diode laser therapy; day 7, recheck exam; day 14, recheck exam, CBC, chemistry panel, and tumor biopsy or tumor removal; and days 14–180, ongoing monitoring.

TABLE 12.2 Patients from Ohio State University Trial, Breed, Age, Sex, and Weight

Patient	Breed	Age (y)	Sex	Weight (kg)	Disease	Prior treatment	Nanorod dose (mg)
1	Greyhound	12	MN	28.2	STS	Surgery	282
2	Mixed breed	8	MN	37.8	STS	None	350
3	Boston terrier	6	FS	11.7	STS	None	120
4	Beagle	9	MN	17	MCT	None	170
5	Beagle	13	FS	9.1	STS	None	91
6	Labrador red	10	FS	38.3	MCT	None	383
7	Golden ret	7	FS	42.1	STS	Surgery/RT	410

MCT, mast cell tumors; STS, soft tissue sarcoma.

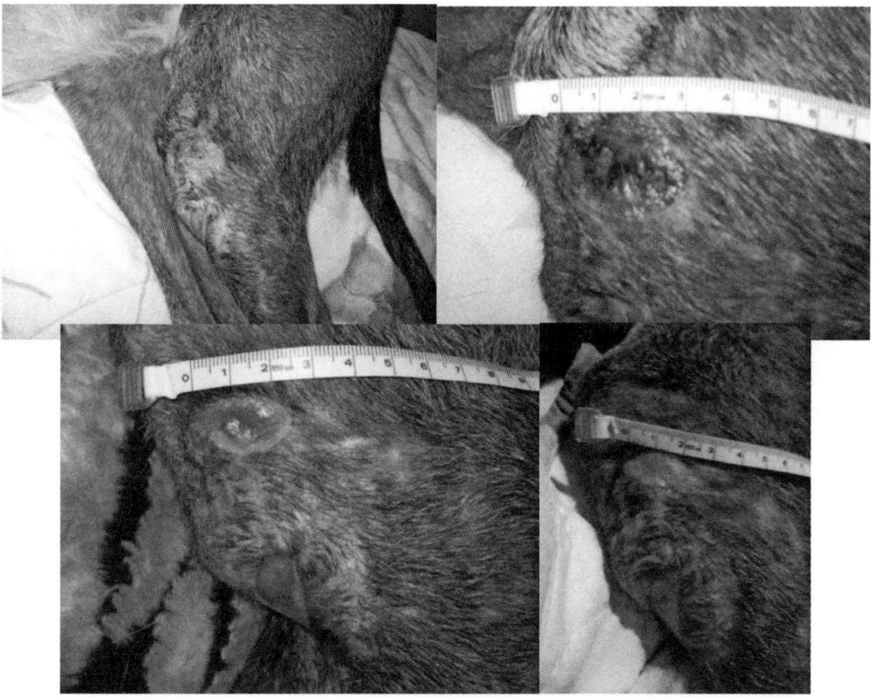

FIGURE 12.8 Nanopartz funded Clinical Case 2011-1 at Ohio State University. Greyhound, 11 years old, 40 kg with recurring soft tissue sarcoma. Two regions on hindquarter have had multiple surgeries using conventional removal techniques only to see recurrence. Nanopartz gold nanorods were injected three days prior to laser treatment. Top (left), the picture was taken 3 h after laser treatment to upper exposed using Pandia® system. Grape-sized tumor is now black in color. The tumor achieved over 100°C increase in temperature after 5 s of 15W exposure. The total exposure time was 1 min. Top (right), two days after treatment. Bottom (left), 6 days after treatment. Bottom (right), 2 weeks after treatment; biopsies were negative for cancer.

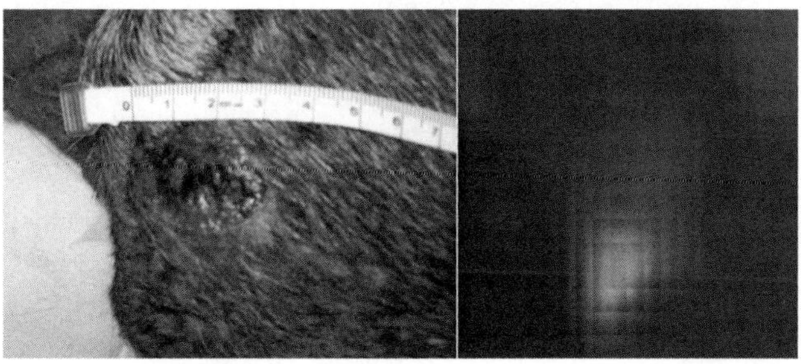

FIGURE 12.9 Thermal image of trial 2011-1 osteosarcoma at 15 W laser illumination. Temperature at center is 60°C and surrounding tissue 3 cm from center is 35°C.

The details and results of the study are given in Figures 12.10, 12.11, 12.12, and 12.13 and summarized in Table 12.3.

Our effort resulted in the following: AuNRs exhibited no clinical, hematologic, or biochemical toxicities when administered to dogs in this pilot study. NRs could be

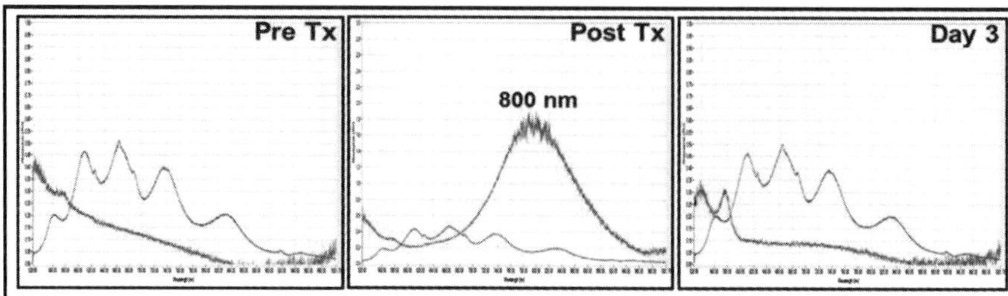

Patient	NRs post treatment (μg/ml)	NRs day 3 (μg/ml)
1	108	12
2	120	0
3	192	97
4	160	0
5	60	6
6	72	0
7	168	0

FIGURE 12.10 Top panel: (left) UV–Vis absorption of blood serum before injection, (middle) immediately after injection, and (right) 3 days after injection. Bottom panel: tabulated patient injection amount and amount still circulating in the bloodstream after 3 days.

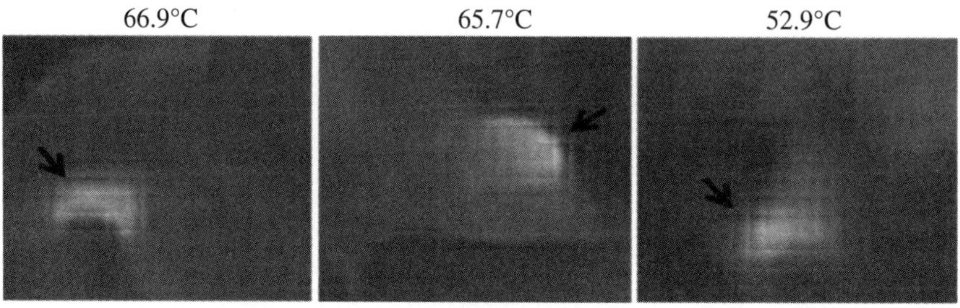

Patient	NIR diode laser treatment
1	75–100°C for 1 min
2	40–60°C for 2 min
3	53–66°C for 3 min
4	63–70°C for 1 min
5	45–55°C for 4 min
6	78–100°C for 40 s
7	45–60°C for 3 min

FIGURE 12.11 Top panel: thermal images of patients during heating phase of treatment. Bottom panel: table shows patient treatment—temperature and duration of treatment.

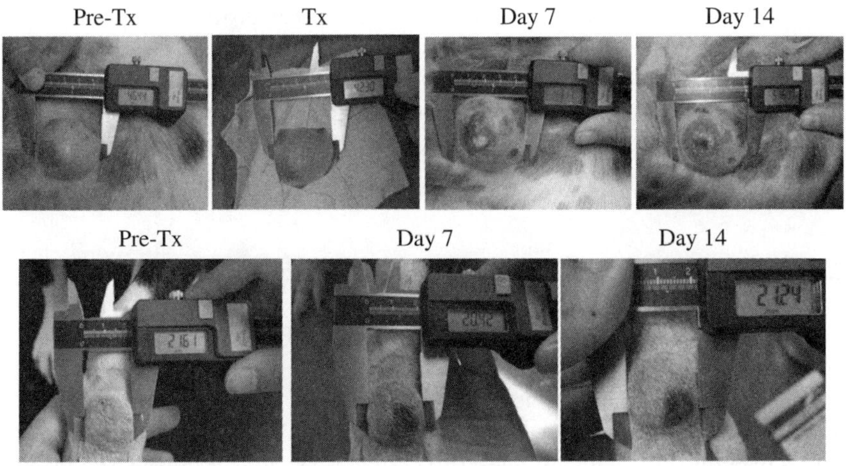

FIGURE 12.12 Patient treatment response from pretreatment to day 14.

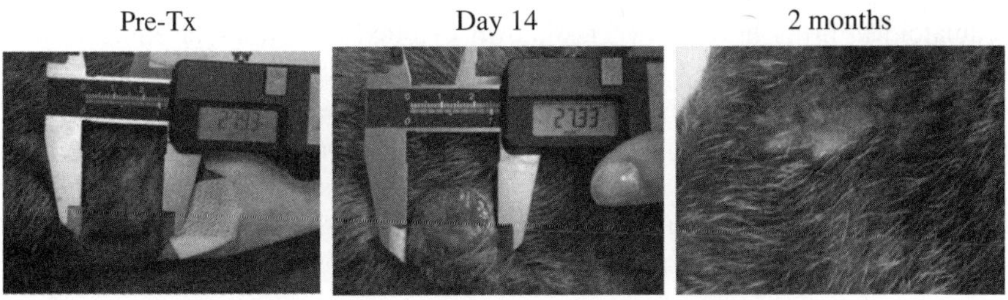

FIGURE 12.13 Dog 1 complete recovery after the treatment.

TABLE 12.3 List of Histopath Posttreatment for Seven Canines Used in the Study

Patient	Response	Histopath posttreatment
1	CR	Tumor still evident at day 14 but regressed over time
2	SD	No evidence of tumor cell death; uncharged from pretreatment
3	SD	60% necrosis with hemorrhage, tumor still present
4	SD	Fibrin, hemorrhage, skin necrosis, tumor present but completely removed
5	SD	Multifocal mild hemorrhage and necrosis but <50%; tumor present
6	SD	No evidence of tumor cell death or necrosis on histopath
7	SD	Not done yet

CR, complete response; SD, stable disease.

detected in the serum and plasma immediately after administration and were completely cleared within 72 h in most dogs. The application of the NIR diode laser resulted in rapid tumor heating. Much of tumor heating appeared to concentrate in the skin, and the use of an amplitude-modulated laser may induce the NRs undergoing sonic cavitation that will further disrupt tumor vascularity. Histopathologic evidence of necrosis was present in two per six dogs and one dog had evidence of tumor regression.

NRs can potentially be loaded with chemotherapeutic agents (doxorubicin, Taxol) using a heat-sensitive or pH-sensitive linker. In such construct, chemotherapeutic is released locally following the application of the NIR diode laser.

12.4 POTENTIAL APPLICATIONS IN HUMANS

To establish a basis of safety for the IV injection of PEG-coated AuNRs for *in vivo* photothermal cancer therapy in humans, we have analyzed (i) historical toxicity analysis of gold nanoparticles used in treating arthritis, (ii) current uses of polyethylene glycol (PEG), (iii) survey of existing published literature and credible web sources regarding toxicology studies, (iv) current toxicology results in healthy dogs, (v) future human toxicology study plans, and (vi) potential for AuNR use as improved drug delivery tools.

We found in the literature that over 40 years of historical use in the treatment of rheumatoid arthritis have shown no known cytotoxic reactions in amounts much greater and in longer duration than what is proposed for the current study. PEG is used widely in improving the efficacy of drugs for *in vivo* use and in many other applications. Mice studies have shown no known cytotoxicity and 35% clearance from major organs after 120 days. Our current toxicity studies in healthy dogs have shown neither short-term nor long-term toxic reactions. Further, published literature shows canines are currently being used as models for human cancer research [55].

12.4.1 Historical Toxicity Analysis of Old Nanoparticles Used in Treating Arthritis

According to Huerta *et al.* [56], gold compounds and penicillamine are older drugs that are used less frequently in rheumatic diseases but can be considered for patients who cannot tolerate other drugs. Gold sodium thiomalate and gold thioglucose (aurothioglucose) are injectable (e.g., intramuscularly (IM)); a test dose of 10 mg for the first week is followed by a 25 mg dose 1 week later. All subsequent IM doses of 25–50 mg are given weekly until a cumulative dose of 1 g is achieved and a therapeutic response or toxicity occurs. If a therapeutic response occurs, maintenance doses of 25–50 mg are continued weekly, then every 2 weeks, then every 3 weeks, and then monthly. Monthly therapy should be continued indefinitely to prevent a recurrence.

Our efforts in canines show that we need a one-time single injection of between 0.2 and 2 g in an adult male human. This compares to gold salts, which according to the *Merck Manual of Geriatrics* for treatment of rheumatic diseases recommends weekly injections to an accumulation of 2 g with 50 mg monthly injections, thereafter indefinitely.

According to Lawrence [57], gold salts have been used in the treatment of arthritis since the favorable report of Forestier in 1929. Oily suspensions were introduced around 1933 versus aqueous suspensions 50–200 mg weekly. When the plasma fibrinogen, estimated monthly, fell below 450 mg/100 ml on three consecutive occasions, a maintenance dose of 50 mg every second week was given. Results were assessed after 2 years' treatment. Consequently, this study showed no toxicity after 200 mg of weekly injections for 2 years.

Reference [58] presents an 18-month study rheumatology drug monitoring clinic. The normal dose regimen consisted of 50 mg weekly for 20 weeks, 50 mg fortnightly for 20 weeks, and then 50 mg monthly. Sixteen cases of either a pityriasiform or discoid eczematous rash occurring in patients with rheumatoid arthritis receiving treatment with gold (sodium aurothiomalate and auranofin) were studied. The results suggest that this is a dose-related, not allergic, reaction to gold. The development of this rash is not an absolute indication to stop treatment with gold. Control can often be affected with potent topical steroids or a reduction in the dose or frequency of treatment with gold. If the rash had been allergic in nature, its difference would be expected that the eruption would continue as long as treatment with gold is given. In seven patients, however, the rash was controlled with topical treatment without a gold dose, and in a further three patients, the rash that was present in both patients was controlled by a reduction in dose or the frequency of gold treatment.

In the study described in Reference [59], 65 patients with rheumatoid arthritis attending 2 rheumatology clinics were compared with 51 control subjects matched by age and sex. The controls consisted of 20 healthy subjects, 16 patients with osteoarthritis, and 15 with nonarticular rheumatism. Drug therapies of patients in the study varied: NSAIDs, gold, penicillamine, methotrexate, sulfasalazine, hydroxychloroquine, and azathioprine. From this research, one could not conclude from results if microalbuminuria, noted in patients with rheumatoid arthritis, is solely a drug-induced side effect.

12.4.2 Current Uses of PEG in Biomedical Applications

Polymer (PEG) passivates surface to reduce reactive surface area providing long-term circulation and organ detargeting (Fig. 12.14).

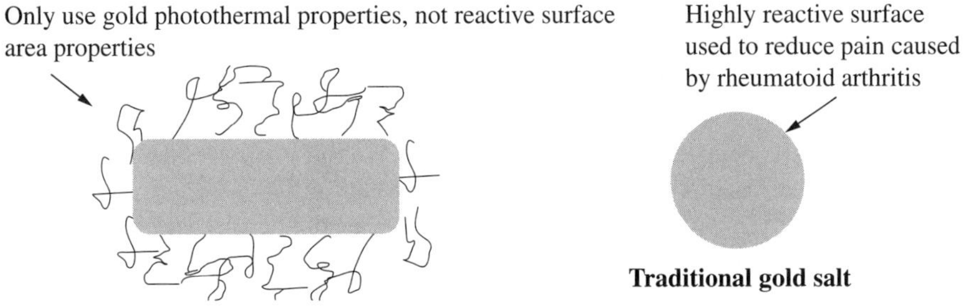

FIGURE 12.14 Comparing PEGylated nanorods to conventional citrate-stabilized spherical gold nanoparticles.

PEG is a polymer of ethylene oxide. PEG has a low toxicity and is used in a variety of products [60]. It is the basis of a number of laxatives (e.g., macrogol-containing products, such as Movicol and PEG 3350, or SoftLax, MiraLAX, or GlycoLax). It is the basis of many skin creams, frequently combined with glycerin. When attached to various protein medications, PEG allows a slowed clearance of the carried protein from the blood. This makes for a longer-acting medicinal effect and reduces toxicity, and it allows longer dosing intervals. Examples include PEG-interferon alpha, which is used to treat hepatitis C, and PEG-filgrastim (Neulasta), which is used to treat neutropenia. It has been shown that PEG can improve healing of spinal injuries in dogs [61]. One of the earlier findings that PEG can aid in nerve repair came from the University of Texas [62]. PEG is commonly used to fuse B cells with myeloma cells in monoclonal antibody production. PEG has recently been proved to give better results in constipation patients than tegaserod [63].

In research, PEG when labeled with an NIR fluorophore has been used in preclinical work as a vascular agent, lymphatic agent, and general tumor imaging agent by exploiting the EPR effect of tumors [64]. High-molecular-weight PEG, for example, PEG 8000, has been shown to be a dietary preventive agent against colorectal cancer in animal models [65]. The chemoprevention database shows it is the most effective agent to suppress chemical carcinogenesis in rats. Cancer prevention in humans has not yet been tested in clinical trials. The injection of PEG 2000 into the bloodstream of guinea pigs after spinal cord injury leads to rapid recovery through molecular repair of nerve membranes [66]. The effectiveness of this treatment to prevent paraplegia in humans after an accident is not known yet. Research is being done in the use of PEG to mask antigens on red blood cells. Various research institutes have reported that using PEG can mask antigens without damaging the functions and shape of the cell. PEG is being used in the repair of motor neurons damaged in crush or laceration incidence *in vivo* and *in vitro*. When coupled with melatonin, 75% of damaged sciatic nerves were rendered viable [67]. Most importantly, gene therapy vectors (such as viruses) can be PEG coated to shield them from inactivation by the immune system and to detarget them from organs where they may build up and have a toxic effect [68]. The size of the PEG polymer has been shown to be important, with large polymers achieving the best immune protection.

12.4.3 PEG-Coated Gold Nanoparticles in Mice

There are no reports of any examples of toxicity associated with gold nanoparticles at any dosage. Five published journal papers were found that studied the toxicology of PEG-Au nanoparticles in mice. None found any signs of cytotoxicity. Organ biodistribution depended heavily on size—35% clearance from major organs after 120 days.

In Reference [16], over 20 mice were used in the study and were injected with 100 μgAu/g of body weight. The biodistribution showed 57% spleen, 10% liver, less than 5% in kidney, lung, heart, muscle, brain approximately 7% ID/g to tumor. The mice showed no signs of neither stress nor weight loss up to and after 6 months after injection.

In Reference [69], PEG-coated gold nanoparticles (492.5 μg/g body weight) were used, and biodistribution showed an accumulation of the liver (64.4% ID) at 3 d, spleen (9.7% ID), small intestine (4.3% ID), lymph nodes (1.9% ID), large intestines (1.6% ID), muscle (1.3% ID), and kidneys (~1% ID), minimal in the brain, heart, and lung. In this study, mice showed normal hematology and blood chemistry at days 3 and 14. Histological examination of nine major organs and tissues 3 days after injection showed no evidence of inflammatory cell infiltration, cell swelling, or tissue necrosis in any mice. No signs of sickness were observed in the mice up to 6 months.

In Reference [70], the authors evaluated the bioaccumulation and toxic effects of different doses (40, 200, and 400 μg/kg/day) of 12 nm gold nanoparticles upon intraperitoneal administration in mice every day for 8 days. The gold levels in blood did not increase with the dose administered, whereas in all the organs examined, there was a proportional increase on gold, indicating efficient tissue uptake.

No evidence of toxicity was observed in any of the diverse studies performed, including survival, behavior, animal weight, organ morphology, blood biochemistry, and tissue histology. The results indicate that tissue accumulation pattern of GNPs depends on the doses administered and the accumulation of the particles does not produce subacute physiological damage.

In Reference [71], the cytotoxicity *in vitro* and biodistribution *in vivo* of Au-Au2S nanoparticles were studied by using NIH3T3 cells and KM mice, respectively. Au-Au2S nanoparticles (<300 μg/ml) showed good biocompatibility. In this study, Au-Au2S nanoparticles were preferentially taken up by the liver and spleen and ultimately eliminated mostly in the feces.

12.5 SUMMARY

Cancer is the major cause of death for companion animals. As we noted earlier, 45% of all dogs over 10 years of age and 23% of all dogs and 32% of all cats succumb to cancer [1]. Skin cancer, mastocytomas, is the number 1 cancer killer in dogs accounting for 23% of all cancers. Cats have a higher prevalence of cancer than all other companion animals. Skin cancer is the second most prevalent cancer in cats occurring in 7% of all cancer cases [72]. Mouth cancers occur in 3–4% of all cat cancer cases. It is estimated that four million dogs and four million cats will be diagnosed with cancer each year. Current methods of cancer treatment are similar to humans. These include complete surgical removal, burning or freezing, radiation therapy, chemotherapy, hormone therapy, and immunotherapy. These methods can result in incomplete removal, reduced efficacy, scarring, and other deleterious side effects.

The presented technology incorporates the use of gold nanoparticles as a method for inducing localized thermal therapy, thereby aiding destruction of cancer tumors without surgery. In use, our gold nanoparticles are injected intravenously into the animal where they target cancer tumors due to the EPR effect inherent difference between cancer tissue and healthy tissue. A laser tuned to the absorption peak of the gold nanoparticles is used to heat the area of the tumor. Only where the gold nanoparticles are localized is there a dramatic increase in temperature. In less than 5 min, the cancer tumor is ablated, leaving healthy tissue unaffected.

This method is significantly different than any other cancer procedure. Cancer procedures currently rely on surgical removal, burning or freezing, radiation therapy, chemotherapy, hormone therapy, and immunotherapy. These methods can result in incomplete removal, reduced efficacy, scarring, and other deleterious side effects. The value proposition for the customer is improved therapy and efficacy with reduced side effects. These side effects, as well as long-term scarring associated with surgical removal procedures, can potentially be avoided using the proposed technique.

Since Nanopartz is focused on being an R&D material supplier, it has become its mission to spin off technologies to new entities. In this case, Nanopartz CRx, Inc., was created to commercialize the Pandia® system. Our prior work with mice has shown significant therapeutic effects in both heating and ablation modes (refer to Fig. 12.2). Subsequent to that work, we have completed normal dog toxicity studies with gold injections of greater than 32 mg/kg with no signs of toxicity. Recent publications (*10*) have shown that doses as low 8–16 mg/kg have been highly successful in mice.

AuNRs may also potentially be used in a method that avoids the necrotic remains associated with the prior method of photodynamic therapy. Research has shown that approximately 7% of the AuNRs injected end up in the solid tumor vasculature [16]. By using a pulsed NIR laser in place of a continuous wave (CW) laser, we can rupture the NRs disrupting the blood supply to the tumor, instead of heating the tumor. This offers the potential to eliminate the necrotic mess normally associated with this technique.

The results of our scale-up success are significant because these further the current understanding of kinetically controlled syntheses of *anisotropic* nanostructures and provide a systematic study of highly unusual and yet unexplored colloidal crystallization of *rod-shaped particles*. From the broader perspective, the development of large-scale synthesis of AuNRs resolves the major issue of their accessibility, which is the main bottleneck of their numerous applications in nanotechnology and nanomedicine and advances our knowledge in self-organization phenomena that lead to anisotropic solids composed of hard inorganic and soft organic components. This combination is analogous to biomineralization products, and therefore, this research provides mechanistic and structural insights into some natural processes. The hierarchical arrangement of nanosized building blocks into microscopic colloidal crystals, which are organized in a periodic array, provides a convenient platform for various photonic and optical applications including metamaterials, SERS detection, and bioanalytical chemistry. The structures also represent an example of bottom-up assembly of materials with vectorial plasmonic, electrical, and mechanical properties. The establishment of methods for the preparation of anisotropic 3D colloidal crystals will be useful not only in the context of materials chemistry but also to a wide variety of disciplines and can provide a model of integration of nanoscale synthesis with self-assembly and microfabrication.

Consequently, we have developed a therapeutic regime that is noncytotoxic, may be injected intravenously, seeks out cancer tumors using long circulation times, can also be used for ultrasensitive imaging, and may be heated without surgery, thereby destroying the cancer tumor.

ACKNOWLEDGEMENT

This material is based upon work supported by DARPA under contract no N66001-10-C-2128. Any opinions, findings and conclusion or recommendations expressed in this material are those of the author(s) and do not necessarily reflect the views of DARPA. Special thanks to Dr. Larry Stotts for his support. Thanks to Dr. Eugene Zubarev and Dr. Shelley Coldiron for their technical input.

REFERENCES

[1] Sessa G, Weissmann G. Phospholipid spherules (liposomes) as a model for biological membranes. J Lipid Res 1968;**9**:310–318.

[2] Withrow S, Vail DM, Page RL. *Withrow and MacEwen's Small Animal Clinical Oncology*. London: Elsevier Health Sciences; 2012.

[3] Bangham A, Standish M, Watkins J. Diffusion of univalent ions across the lamellae of swollen phospholipids. J Mol Biol 1965;**13**:238–252.

[4] Meyer JL. The clinical efficacy of localized hyperthermia. Cancer Res 1984;**44**:4745s–4751s.

[5] Hall EJ, Roizin-Towle L. Biological effects of heat. Cancer Res 1984;**44**:4708s–4713s.

[6] Ahmed M, Goldberg S. Combination radiofrequency thermal ablation and adjuvant IV liposomal doxorubicin increases tissue coagulation and intratumoural drug accumulation. Int J Hyperthermia 2004;**20**:781–802.

[7] Corry PM, Dewhirst MW. Thermal medicine, heat shock proteins and cancer. Int J Hyperthermia 2005;**21**:675–677.

[8] Moyer HR, Delman KA. The role of hyperthermia in optimizing tumor response to regional therapy. Int J Hyperthermia 2008;**24**:251–261.

[9] Schoen CL, Coldiron S, London C. Clinical Evaluation of Pandia in Canine Sarcoma Patients at Ohio State University; 2011.

[10] von Maltzahn G, Park J-H, Agrawal A, Bandaru NK, Das SK, Sailor MJ, Bhatia SN. Computationally guided photothermal tumor therapy using long-circulating gold nanorod antennas. Cancer Res 2009;**69**:3892–3900.

[11] Valdagni R. Head and Neck 5 year survival, 66 patients, Clinical Trial; 1999.

[12] Elghanian R, Storhoff JJ, Mucic RC, Letsinger RL, Mirkin CA. Selective colorimetric detection of polynucleotides based on the distance-dependent optical properties of gold nanoparticles. Science 1997;**277**:1078–1081.

[13] Grubisha DS, Lipert RJ, Park H-Y, Driskell J, Porter MD. Femtomolar detection of prostate-specific antigen: an immunoassay based on surface-enhanced Raman scattering and immunogold labels. Anal. Methods 2003;**75**:5936–5943.

[14] Jackson JB, Westcott SL, Hirsch LR, West JL, Halas NJ. Controlling the surface enhanced Raman effect via the nanoshell geometry. Appl Phys Lett 2003;**82**:257–259.

[15] Qian X, Peng X-H, Ansari DO, Yin-Goen Q, Chen GZ, Shin DM, Yang L, Young AN, Wang MD, Nie S. *In vivo* tumor targeting and spectroscopic detection with surface-enhanced Raman nanoparticle tags. Nat Biotechnol 2008;**26**:83–90.

[16] Hirsch LR, Stafford RJ, Bankson JA, Sershen SR, Rivera B, Price RE, Hazle JD, Halas NJ, West JL. Nanoshell-mediated near-infrared thermal therapy of tumors under magnetic resonance guidance. Proc Natl Acad Sci U S A 2003;**100**:13549–13554.

[17] O'Neal DP, Hirsch LR, Halas NJ, Payne JD, West JL. Photo-thermal tumor ablation in mice using near infrared-absorbing nanoparticles. Cancer Lett 2004;**209**:171–176.

[18] Norman RS, Stone JW, Gole A, Murphy CJ, Sabo-Attwood TL. Targeted photothermal lysis of the pathogenic bacteria, Pseudomonas aeruginosa, with gold nanorods. Nano Lett 2007;**8**:302–306.

[19] Huang X, El-Sayed IH, Qian W, El-Sayed MA. Cancer cell imaging and photothermal therapy in the near-infrared region by using gold nanorods. J Am Chem Soc 2006;**128**: 2115–2120.

[20] Park J-H, von Maltzahn G, Xu MJ, Fogal V, Kotamraju VR, Ruoslahti E, Bhatia SN, Sailor MJ. Cooperative nanomaterial system to sensitize, target, and treat tumors. Proc Natl Acad Sci U S A 2010;**107**:981–986.

[21] Fourkal E, Velchev I, Taffo A, Ma C, Khazak V, Skobeleva N. Photo-thermal cancer therapy using gold nanorods. In: Dössel O, Schlegel W, editors. *World Congress on Medical Physics and Biomedical Engineering, September 7–12, 2009, Munich, Germany*. Berlin Heidelberg: Springer; 2009. p 761–763.

[22] Orendorff CJ, Gole A, Sau TK, Murphy CJ. Surface-enhanced Raman spectroscopy of self-assembled monolayers: sandwich architecture and nanoparticle shape dependence. Anal Methods 2005;**77**:3261–3266.

[23] Jana NR, Gearheart L, Murphy CJ. Seed-mediated growth approach for shape-controlled synthesis of spheroidal and rod-like gold nanoparticles using a surfactant template. Adv Mater 2001;**13**:1389.

[24] Gole A, Murphy CJ. Seed-mediated synthesis of gold nanorods: role of the size and nature of the seed. Chem Mater 2004;**16**:3633–3640.

[25] Mirkin CA, Letsinger RL, Mucic RC, Storhoff JJ. A DNA-based method for rationally assembling nanoparticles into macroscopic materials. Nature 1996;**382**:607–609.

[26] Alivisatos AP, Johnsson KP, Peng X, Wilson TE, Loweth CJ, Bruchez MP, Schultz PG. Organization of "nanocrystal molecules" using DNA. Nature 1996;**382**:609–611.

[27] Bruchez M, Moronne M, Gin P, Weiss S, Alivisatos AP. Semiconductor nanocrystals as fluorescent biological labels. Science 1998;**281**:2013–2016.

[28] Park H, Park J, Lim AKL, Anderson EH, Alivisatos AP, McEuen PL. Nanomechanical oscillations in a single-C60 transistor. Nature 2000;**407**:57–60.

[29] Brust M, Walker M, Bethell D, Schiffrin DJ, Whyman R. Synthesis of thiol-derivatised gold nanoparticles in a two-phase liquid-liquid system. J Chem Soc Chem Commun 1994;**7**:801–802.

[30] Collier C, Saykally R, Shiang J, Henrichs S, Heath J. Reversible tuning of silver quantum dot monolayers through the metal-insulator transition. Science 1997;**277**:1978–1981.

[31] Fendler JH. Self-assembled nanostructured materials. Chem Mater 1996;**8**:1616–1624.

[32] Mirkin CA. Programming the assembly of two- and three-dimensional architectures with DNA and nanoscale inorganic building blocks. Inorg Chem 2000;**39**:2258–2272.

[33] Li M, Schnablegger H, Mann S. Coupled synthesis and self-assembly of nanoparticles to give structures with controlled organization. Nature 1999;**402**:393–395.

[34] Ohno K, Koh K-M, Tsujii Y, Fukuda T. Synthesis of gold nanoparticles coated with well-defined, high-density polymer brushes by surface-initiated living radical polymerization. Macromolecules 2002;**35**:8989–8993.

[35] Peng X, Manna L, Yang W, Wickham J, Scher E, Kadavanich A, Alivisatos AP. Shape control of CdSe nanocrystals. Nature 2000;**404**:59–61.

[36] Jana NR, Gearheart L, Murphy CJ. Wet chemical synthesis of high aspect ratio cylindrical gold nanorods. J Phys Chem A 2001;**105**:4065–4067.

[37] Critchley K, Khanal BP, Górzny MŁ, Vigderman L, Evans SD, Zubarev ER, Kotov NA. Near-bulk conductivity of gold nanowires as nanoscale interconnects and the role of atomically smooth interface. Adv Mater 2010;**22**:2338–2342.

[38] Huang X, Neretina S, El-Sayed MA. Gold nanorods: from synthesis and properties to biological and biomedical applications. Adv Mater 2009;**21**:4880–4910.

[39] Pérez-Juste J, Liz-Marzán LM, Carnie S, Chan DYC, Mulvaney P. Electric-field-directed growth of gold nanorods in aqueous surfactant solutions. Adv Funct Mater 2004;**14**:571–579.

[40] Nishioka K, Niidome Y, Yamada S. Photochemical reactions of ketones to synthesize gold nanorods. Langmuir 2007;**23**:10353–10356.

[41] Park S, Lim J-H, Chung S-W, Mirkin CA. Self-assembly of mesoscopic metal-polymer amphiphiles. Science 2004;**303**:348–351.

[42] Perez-Juste J, Pastoriza-Santos I, Liz-Marzan LM, Mulvaney P. Gold nanorods: synthesis, characterization and applications. Coord Chem Rev 2005;**249**:1870–1901.

[43] Nikoobakht B, El-Sayed MA. Preparation and growth mechanism of gold nanorods (NRs) using seed-mediated growth method. Chem Mater 2003;**15**:1957–1962.

[44] Grzelczak M, Perez-Juste J, Mulvaney P, Liz-Marzan LM. Shape control in gold nanoparticle synthesis. Chem Soc Rev 2008;**37**:1783–1791.

[45] Titov AV, Král P. Modeling the self-assembly of colloidal nanorod superlattices. Nano Lett 2008;**8**:3605–3612.

[46] Wijaya A, Hamad-Schifferli K. Ligand customization and DNA functionalization of gold nanorods via round-trip phase transfer ligand exchange. Langmuir 2008;**24**:9966–9969.

[47] Gupta S, Zhang Q, Emrick T, Russell TP. "Self-corralling" nanorods under an applied electric field. Nano Lett 2006;**6**:2066–2069.

[48] Ryan KM, Mastroianni A, Stancil KA, Liu H, Alivisatos AP. Electric-field-assisted assembly of perpendicularly oriented nanorod superlattices. Nano Lett 2006;**6**:1479–1482.

[49] Nie Z, Fava D, Kumacheva E, Zou S, Walker GC, Rubinstein M. Self-assembly of metal-polymer analogues of amphiphilic triblock copolymers. Nat Mater 2007;**6**:609–614.

[50] Khanal BP, Zubarev ER. Rings of nanorods. Angew Chem Int Ed Engl 2007;**46**: 2195–2198.

[51] Khanal BP, Zubarev ER. Purification of high aspect ratio gold nanorods: complete removal of platelets. J Am Chem Soc 2008;**130**:12634–12635.

[52] Jana NR. Gram-scale synthesis of soluble, near-monodisperse gold nanorods and other anisotropic nanoparticles. Small 2005;**1**:875–882.

[53] Orendorff CJ, Murphy CJ. Quantitation of metal content in the silver-assisted growth of gold nanorods. J Phys Chem A 2006;**110**:3990–3994.

[54] Murphy CJ, Gole AM, Hunyadi SE, Orendorff CJ. One-dimensional colloidal gold and silver nanostructures. Inorg Chem 2006;**45**:7544–7554.

[55] Gordon I, Paoloni M, Mazcko C, Khanna C. The comparative oncology trials consortium: using spontaneously occurring cancers in dogs to inform the cancer drug development pathway. PLoS Med 2009;**6**:e1000161.

[56] Huerta MF, Farber GK, Wilder EL, Kleinman DV, Grady PA, Schwartz DA, Tabak LA. NIH roadmap interdisciplinary research initiatives. PLoS Comput Biol 2005;**1**:e59.

[57] Lawrence J. Comparative toxicity of gold preparations in treatment of rheumatoid arthritis. Ann Rheum Dis 1976;**35**:171–173.

[58] Wilkinson S, Smith A, Davis M, Mattey D, Dawes P. Pityriasis rosea and discoid eczema: dose related reactions to treatment with gold. Ann Rheum Dis 1992;**51**:881–884.

[59] Pedersen LM, Nordin H, Svensson B, Bliddal H. Microalbuminuria in patients with rheumatoid arthritis. Ann Rheum Dis 1995;**54**:189–192.

[60] Sheftel VO. *Indirect Food Additives and Polymers: Migration and Toxicology*. London: Lewis Publishers Inc.; 2000.

[61] Laverty PH, Leskovar A, Breur GJ, Coates JR, Bergman RL, Widmer WR, Toombs JP, Shapiro S, Borgens RB. A preliminary study of intravenous surfactants in paraplegic dogs: polymer therapy in canine clinical SCI. J Neurotrauma 2004;**21**:1767–1777.

[62] Krause TL, Bittner GD. Rapid morphological fusion of severed myelinated axons by polyethylene glycol. Proc Natl Acad Sci U S A 1990;**87**:1471–1475.

[63] Di Palma JA, Cleveland MV, McGowan J, Herrera JL. A randomized, multicenter comparison of polyethylene glycol laxative and tegaserod in treatment of patients with chronic constipation. Am J Gastroenter 2007;**102**:1964–1971.

[64] Kovar J, Wang Y, Simpson MA, Olive DM. Imaging lymphatics with a variety of near-infrared labeled optical agents. World Molecular Imaging; 2009. p 67–68.

[65] Corpet DE, Parnaud G, Delverdier M, Peiffer G, Taché S. Consistent and fast inhibition of colon carcinogenesis by polyethylene glycol in mice and rats given various carcinogens. Cancer Res 2000;**60**:3160–3164.

[66] Borgens RB, Bohnert D. Rapid recovery from spinal cord injury after subcutaneously administered polyethylene glycol. J Neurosci Res 2001;**66**:1179–1186.

[67] Stavisky RC, Britt JM, Zuzek A, Truong E, Bittner GD. Melatonin enhances the in vitro and *in vivo* repair of severed rat sciatic axons. Neurosci Lett 2005;**376**:98–101.

[68] Kreppel F, Kochanek S. Modification of adenovirus gene transfer vectors with synthetic polymers: a scientific review and technical guide. Mol Ther 2007;**16**:16–29.

[69] Leung, K. (2010) National Center for Biotechnology Information, N., NIH, Ed. Available at http://www.ncbi.nlm.nih.gov/bookshelf/br.fcgi?book=micad&part=Au-PEG-NP. Accessed May 31, 2014.

[70] Lasagna-Reeves C, Gonzalez-Romero D, Barria MA, Olmedo I, Clos A, Sadagopa Ramanujam VM, Urayama A, Vergara L, Kogan MJ, Soto C. Bioaccumulation and toxicity of gold nanoparticles after repeated administration in mice. Biochem Biophys Res Commun 2010;**393**:649–655.

[71] Sonavane G, Tomoda K, Makino K. Biodistribution of colloidal gold nanoparticles after intravenous administration: effect of particle size. Colloids Surf B 2008;**66**:274–280.

[72] Lazo JS. Roadmap or roadkill: a pharmacologist's analysis of the NIH molecular libraries initiative. Mol Interv 2006;**6**:240.

13

IMAGING GENETIC INFORMATION

John-Stephen Taylor

Department of Chemistry, Washington University, St. Louis, MO, USA

13.1 INTRODUCTION

Almost all disease states can be attributed to genetic or epigenetic information found within the diseased host that is different from that of the normal host, with the exception of prion diseases [1–4]. The disease-causing genetic information might reside in the genomic DNA or RNA, or in a virus, bacterium, or other pathogen, or from alterations in the methylation pattern of DNA, which in turn alters RNA expression levels. The altered or foreign genetic information can result in the synthesis of mutated or nonnative proteins or cause changes in the expression of normal proteins that in turn can affect the expression of other proteins and downstream metabolites that themselves can cause alterations in protein expression and epigenetic modifications. As a result, diseased tissue in a person can be imaged by the accumulation of a probe targeted to a disease-specific biomolecule by a process known as molecular imaging [5, 6] (Fig. 13.1). Because proteins are present in much higher concentration than the DNA or mRNA that encodes them, most imaging agents for diseases are targeted to disease-specific proteins. Of the many possible protein targets, those that are on a cell surface are the most desirable target because they are the most accessible and don't require intracellular entry of the imaging agent. The diseased cell or disease-causing organism can then be recognized by a small molecule ligand for a receptor or enzyme [7] or by a protein binding molecule, such as an antibody, minibody [8], or aptamer [9, 10].

If the diseased cell does not express a unique or overexpressed cell surface target that is specific for the disease state, then an internal target is necessary. Small molecule ligands for target proteins make excellent imaging agents because they are often

Nanotechnology for Biomedical Imaging and Diagnostics: From Nanoparticle Design to Clinical Applications, First Edition. Edited by Mikhail Y. Berezin.

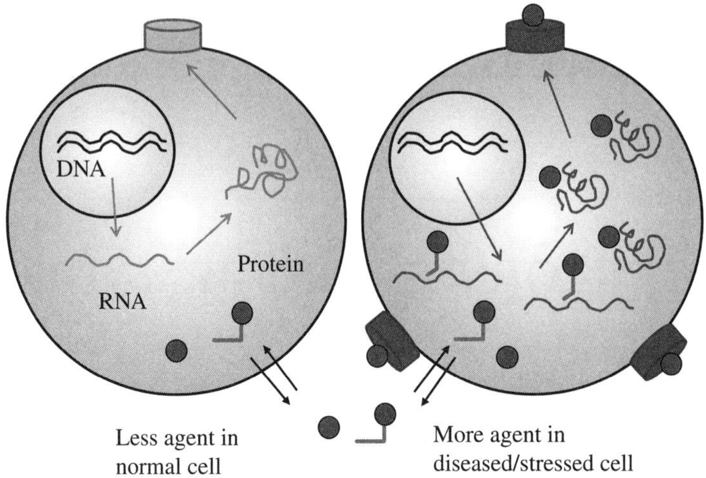

FIGURE 13.1 Idealized picture of gene expression imaging in which antisense probes accumulate in a cell expressing a unique or overexpressed mRNA sequence. If an always -on probe is used, such as a radiolabeled antisense agent, the probe must be able to enter the cytoplasm, and any unbound probe or reporter must be able to efficiently exit the cytoplasm. Turn-on probes only need to enter efficiently. The problem is that antisense agents themselves are membrane impermeable and thus must be attached to cell-penetrating carriers. Another problem is that mRNA is folded and bound to proteins, making many sites inaccessible to a probe in its folded state. *(See insert for color representation of the figure.)*

membrane permeable so that they can readily diffuse in and out of the cell and concentrate in the target cells [11], but there are still many challenges with imaging intracellular targets [12]. Attempting to detect a target protein within a cell by an antibody or aptamer is even more problematic, however, as such agents are not membrane permeable and would require some sort of auxiliary mechanism to facilitate cell entry and exit. If entering by an endocytosis mechanism, the imaging agent would still have to be able to exit the endosome to access the target biomolecule within the cytoplasm, which would require additional auxiliaries. Once the imaging agent is in the cytoplasm, unbound imaging agent or the reporting group would also have to be able to rapidly exit the cytoplasm so that a target-specific signal can be generated, which is likely to be problematic since such agents are not cell permeable. To circumvent the latter problem, researchers have been designing imaging agents that only turn-on upon binding to the target protein [13–15]. While a strategy based on targeting proteins affords many opportunities for the design of target-specific agents, a specific ligand or binding agent with the appropriate pharmacokinetics would need to be developed for each target, which would be a very time-consuming and costly process.

13.2 NUCLEIC ACIDS AS BIOMARKERS FOR DISEASE

A much more attractive and universal intracellular target from a design point of view would be to directly target genetic information associated with a disease state because information in the form of DNA or RNA can in principle be easily recognized by

Watson–Crick base pairing with an antisense oligonucleotide. This would presumably greatly simplify the development of generic delivery systems with the appropriate pharmacokinetics since the only difference between imaging agents would be the precise sequence of the antisense agent for a given nucleic acid target. Nucleic acids have already been extensively exploited for *ex vivo* diagnostic applications, and many disease-specific target sequences are known [16–19]. Potential nucleic acid targets within a cell are duplex DNA residing within the nucleus and in the mitochondria and various RNAs that are transcribed from the DNA such as messenger RNA (mRNA), ribosomal RNA (rRNA), and noncoding RNAs such as microRNAs (miRNAs) (Fig. 13.2).

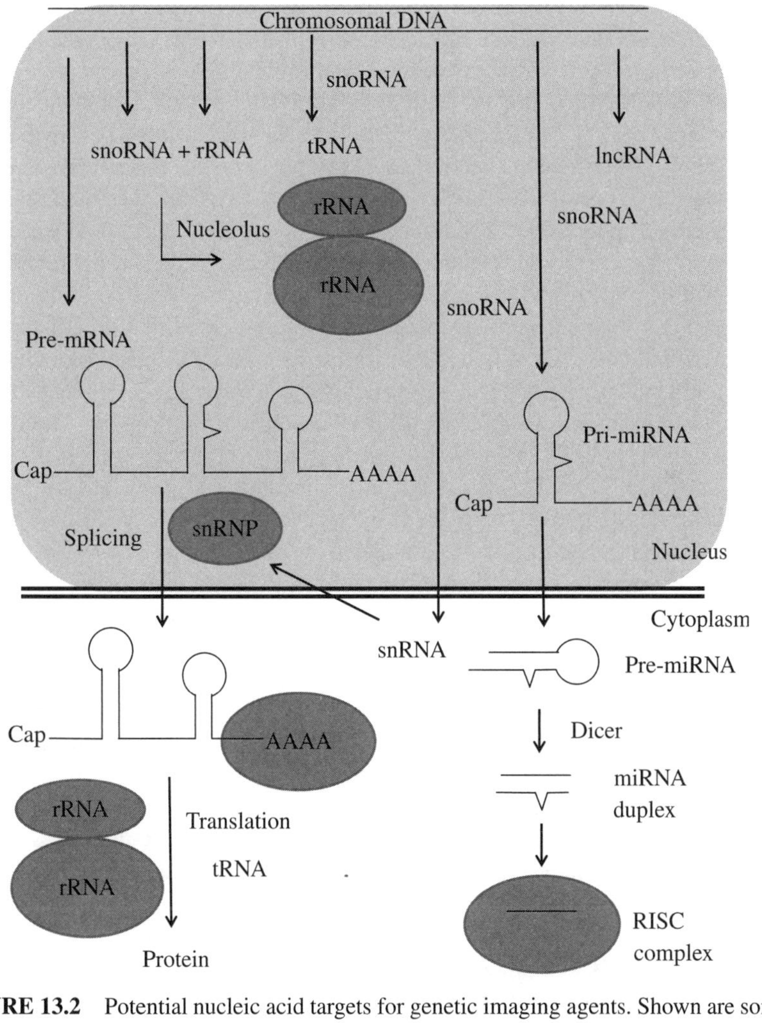

FIGURE 13.2 Potential nucleic acid targets for genetic imaging agents. Shown are some of the major nucleic acid metabolism pathways in the nucleus and cytoplasm of a cell. Not shown are targets in the mitochondrion that contains mtDNA, which encodes its own rRNA, mRNAs, tRNAs, and ncRNAs.

13.2.1 DNA as a Biomarker

While chromosomal DNA is the primary source of genetic information, it is only present in two copies, which would require an extremely sensitive imaging agent with a very low background signal. In a typical mammalian cell, a single DNA copy of a DNA sequence would correspond to about 1 pM concentration (10^{-12} M). Some sequences of DNA exist as multiple copies such as the DNA coding for rRNA, for which 300–400 copies are dispersed over five human chromosomes. There are also genetic diseases that have multiple sequence repeats that can reach thousands of bases, such as Friedreich's ataxia, fragile X syndrome, and Huntington's disease [20], which would also greatly increase the target sequence concentration.

Aside from concentration, DNA is not a particularly good target for imaging because it exists almost exclusively in a duplex form and is therefore inaccessible for base pairing by simple single-strand antisense agents. DNA can be recognized in its duplex form, however, by triplex-forming ODNs (TFOs) [21, 22], peptide nucleic acids (PNAs) [23, 24], minor groove binding polyamides [25], or engineered zinc finger proteins [26, 27]. The TFOs require polypurine tract sequences, which need to be quite long to be unique (18 or so nucleotides for humans) and are not as unique. On the other hand, polyamides and zinc finger proteins can recognize heterogeneous DNA sequences by hydrogen bonding interactions in the minor groove and can be made to recognize quite long sequences. A strategy for detecting triplet repeat DNA by hairpin PNAs has recently been described [28], and a self-assembling catalytic zinc finger system based on β-lactamase has been developed that greatly amplifies the signal from the binding of two zinc fingers to a single DNA target sequence [29]. Another problem with genomic DNA as a target is that nontranscribed DNA is tightly packaged into chromatin, and binding of a probe to the nucleosome-bound sequences (150 mer stretches of DNA) would be greatly inhibited. Further complicating matters, chromosomal DNA is sequestered in the nucleus and protected by a double nuclear membrane that must be traversed by the imaging agent (Fig. 13.3). Transport across the nuclear membrane can be facilitated by attachment of nuclear localization peptide sequences (NLS), which have been used to deliver nucleic acids into the cell nucleus through the nuclear pore [30]. The most serious concern in using chromosomal DNA as a target is that agents that bind to the DNA might also induce DNA mutations, deletions, and rearrangements that could lead to cancer, as has been observed for triplex-forming oligonucleotides [31, 32]. Overall, the low copy number, duplex nature of the DNA, and potential for mutating the DNA do not make chromosomal DNA itself an attractive target for genetic imaging agents.

Mitochondrial DNA, unlike chromosomal DNA, is present in a much higher copy number in a cell, with about 5 copies per mitochondrion, and 1000 mitochondria/cell, bringing the total to about 5000 copies/cell or 5 nM in concentration. Mitochondrial DNA, however, is a circular DNA that has only 16,532 bp compared to the 2 billion base pairs of chromosomal DNA and only encodes 13 proteins, 22 transfer RNAs (tRNAs), and 2 rRNAs [33]. As such, there is much less in the way of disease information to be imaged, though more than 300 mutations have been identified that are associated with disease symptoms. Mitochondrial DNA is not very accessible, as it is sequestered in the mitochondrion by two membrane barriers (Fig. 13.3). Signal

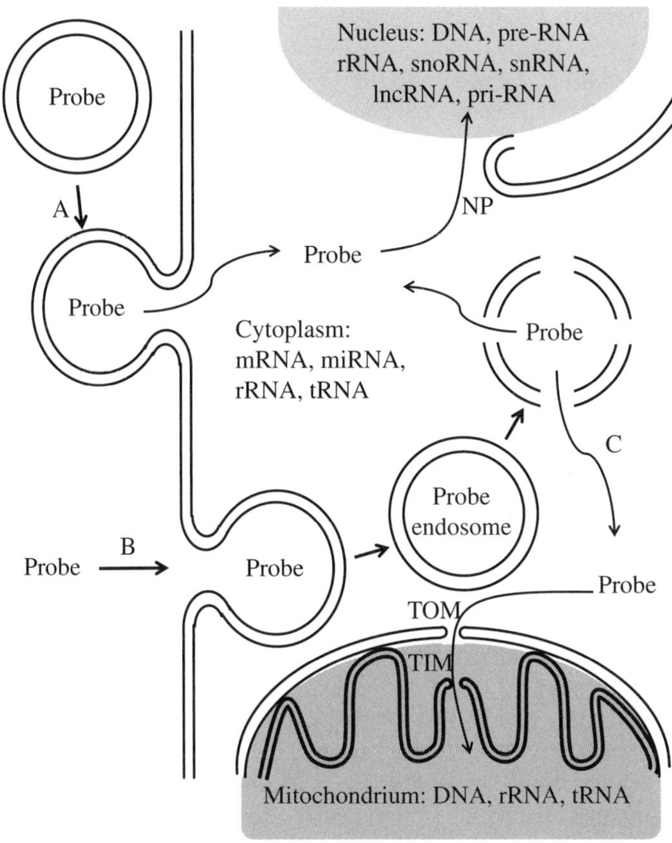

FIGURE 13.3 Biological barriers and general entry modes for nucleic acid-targeted probes. Membrane-impermeable probes can enter the cytoplasm via direct membrane fusion with a liposome or exosome (path A) or by receptor-mediated or receptor-independent endocytosis (path B) followed by endosomal escape. If the target is in the nucleus, the probe could enter through a nuclear pore (NP) via a nuclear localization peptide sequence. If the target is in the mitochondrion, then the probe could enter through outer and inner mitochondrial membrane portals (TOM and TIM).

peptides have been identified, however, that afford the opportunity for delivery of antisense imaging agents [34, 35]. As with genomic DNA, there is the possibility that DNA binding agent could cause to unwanted mutations or other alterations in the mitochondrial DNA.

13.2.2 RNA as a Biomarker

While DNA, except for special cases, would not appear to be the best target for genetic imaging agents, RNA has a number of attributes that make it a much more viable target. First, RNA is often produced in much higher copy number than the DNA from which it is transcribed, thereby affording the opportunity for a higher target signal. Second, RNA is produced in a single-strand form, as opposed to a double-strand form, making it more accessible to a simple complementary antisense probe. While it is produced in a single-strand form, self-complementary sequences within the RNA form duplexes leading to a folded structure, much of which would be inaccessible to an antisense probe [36, 37]. RNA also associates with proteins [38], which would also limit its accessibility to an antisense probe and also be initially produced and sequestered within the nucleus or mitochondria.

The most abundant RNA within a cell is rRNA, which consists of 5S, 5.8S, 18S, and 28S rRNAs and is produced in the nucleolus from 300 to 400 copies of ribosomal DNA sequence. The rRNA is rapidly packaged with proteins in the nucleolus to form ribosomes prior to export to the cytoplasm. Because of the compact nature of the ribonucleoprotein complex [39], the rRNA is not very accessible to antisense agents once assembled. The rRNA produced in the mitochondria is likewise rapidly assembled and remains sequestered within the mitochondria. While rRNA only represents an extremely small fraction of genetic information, there are a number of genetic diseases involving rRNA, and it is also very useful for identifying foreign pathogens. Recently, success has been reported in imaging the rRNA of a fungal pathogen, *Aspergillus fumigatus*, in cell culture and mice with a ^{99m}Tc-labeled antisense MORPH probe, suggesting that the rRNA can be intercepted before being packaged into ribosomes [40].

tRNAs, used in translating mRNA into protein, are also highly expressed, and mutations in mitochondrial tRNAs have been linked to diseases [41]. tRNAs are heavily modified and compact, however, and not expected to be very accessible to antisense agents unless they are first intercepted during transcription.

Another very abundant class of RNA is small nuclear RNA (snRNA) involved in RNA processing, but this RNA becomes tied up in small nuclear ribonucleoproteins (snRNPs) and likewise not expected to be very accessible unless intercepted during transcription. A more promising RNA target is the class of noncoding RNAs known as miRNA [42–44]. miRNAs are produced from primary miRNA (pri-miRNA) or from pre-mRNA following splicing to yield hairpin pre-miRNA. The pre-miRNA is then cleaved by dicer to yield a short 22-mer duplex, one strand of which (the guide strand) becomes assembled into a ribonucleoprotein RISC complex and the other (the passenger strand) is discarded and presumably degraded. The genome may encode about 1000 of these miRNAs, and many diseases have been correlated with their disregulation, making them of great interest as disease markers [45–48]. It is not clear, however, how accessible miRNAs or their precursors would be to an antisense agent, though there are reports that miRNA can be imaged *in vivo* in animals with antisense probes [49].

The most diverse class of RNA is mRNA, which is translated into proteins. Most mRNAs are initially produced as pre-mRNAs in the nucleus, which are then spliced to remove introns to give the mature mRNA and exported to the cytoplasm [50, 51]. Some inducible mRNAs are produced without introns, such as RNAs produced and sequestered in the mitochondria. Disease-specific information resides in both the sequence of these mRNAs, in the case of mutations, deletions, and rearrangements, and in their relative expression levels. Unique disease-related mRNAs are also produced from transcription of retroviral DNA (such as HIV), viral DNA (such as adenovirus), or RNA viruses resident in the cytoplasm (such as hepatitis C, influenza) [52]. The most highly expressed mRNAs belong to the ribosomal proteins, in which mutations have been associated with disease [53]. mRNA is usually in much lower abundance than protein targets, with mRNAs ranging from 1 to 100,000 copies/cell corresponding to 1 pM to 100 nM. mRNAs also fold, though most of the highly structured and protein binding sections of the mRNA are in the noncoding regions [37], thereby offering the opportunity for antisense agents to binding in the coding sections.

13.3 ANTISENSE RNA IMAGING

The ability to noninvasively image RNA in a patient would enable one to monitor disease states in which the disease state is associated with the expression of a unique gene, such as an oncogene [54], or a viral gene [55, 56], or an overexpressed gene [57]. Antisense imaging of RNA could also be used to monitor the level and distribution of therapeutic genes in gene therapy or to monitor the effectiveness of therapeutic treatments that are based on suppressing gene expression. Once a disease-specific mRNA sequence for a patient is identified, it would be straightforward, at least in principle, to design and synthesize a patient-specific antisense nucleic acid probe within days that can recognize and bind the RNA with high affinity by using only Watson–Crick base-pairing rules. The problem at the moment is that currently available antisense oligonucleotides and analogs are membrane impermeable, and attempts to attach cell-targeting and cell-penetrating ligands to enable antisense imaging by PET have not been very successful [54, 58]. Thus, much of the development of antisense imaging agents has shifted to the design of the antisense delivery system and the use of optical turn-on probes that circumvent problems with membrane permeability. The universality and patient specificity of an antisense imaging approach, however, make it well worth the initial investment in developing generic, FDA-approved delivery systems for delivering RNA imaging probes.

To design an antisense imaging probe for a specific patient requires (i) identification of a suitable RNA for imaging and (ii) identification of sites on the RNA accessible to an antisense oligomer. To make any given antisense probe functional will require (i) a method for the efficient delivery of the antisense agent to the candidate cells, (ii) efficient intracellular delivery of the antisense agent to the cytoplasm or target organelle, (iii) efficient efflux of unbound antisense probe, or (iv) RNA target-specific probe activation. We will now briefly discuss these technical requirements individually in the following sections.

13.3.1 Selection of an RNA Target

The RNA target will depend on the particular disease state. In the case of cancer, there may be a specific mutation or rearrangement in the DNA that becomes transcribed into a mutated or rearranged mRNA sequence [59–61]. In principle, genetic changes could be detected by sequencing genomic DNA from a biopsy and compared to DNA sequence from normal tissue, as the cost for doing this has been rapidly declining with the advent of NexGen sequencing technologies [62–64]. In cancers that arise from retroviruses, such as HPV, a unique mRNA may be present that can be detected. Another target RNA would be one that is overexpressed and could be again detected for a specific cancer or a biopsy by DNA chip analysis in comparison to a sample of normal tissue. In some conditions, such as inflammation, there may be highly upregulated mRNAs. The RNA could also be an miRNA [60, 65], or other ncRNA, or an RNA, such as rRNA from a pathogen.

13.3.2 Identifying Antisense Accessible Sites on RNA Targets

Once candidate RNAs are identified, high-affinity antisense binding sites must be identified. Though RNA, unlike DNA, is single stranded, all parts of the sequence are not equally accessible to an antisense agent because RNA molecules will spontaneously fold into complicated structures, which may also bind proteins [37]. Thus, it would appear to be important to be able to identify sequences along the RNA that would be able to bind to an antisense agent with high affinity. In humans, to be unique, a sequence generally has to be about 18 nucleotides long, but 1 may not necessarily require that all 18 nucleotides of the sequence be accessible. It may be sufficient for an antisense agent to get a toehold on a single-strand section of the RNA, which can then displace an adjacent base-paired section of RNA by a strand displacement process that becomes favored by the base-pairing properties of the antisense agent used [66]. Even if the antisense agent is able to bind to the RNA before folding, thermodynamically favorable RNA folding could result in the displacement of the probe. Therefore, the challenge for selection of a suitable antisense agent is to find antisense accessible sites on the RNA inside a cell. Generally, candidate antisense agents have been selected through the use of RNA folding calculations and screening the candidates for activity [67].

There are a number of experimental methods for mapping antisense accessible sites on *in vitro* transcribed RNA involving hybridization to random oligonucleotide libraries [68, 69] or for initiating reverse transcriptase PCR [70]. This latter approach has been extended to determine the accessible sites on mRNAs produced by *in vitro* transcription [69] or isolated directly from cells (Fig. 13.4) [71]. This mapping method uses a library of random 10-mer sequences fused to a PCR tag to prime reverse transcriptase from sites accessible to the primer. The resulting cDNA is then PCR amplified using an mRNA-specific primer and the PCR tag to produce bands that can be aligned with the known mRNA sequence to map the antisense accessible sites. Antisense ODNs are then synthesized to these sites, and their affinity for the mRNA is determined by a binding assay. The affinity of chemical analogs of the highest-affinity antisense ODNs can then be determined directly or through a competition assay. These techniques have been successfully applied for the identification of high-affinity sites for the unr mRNA that is overexpressed in MCF-7 cells [69] and iNOS mRNA that is overexpressed in the mouse macrophage cell line in an inflammatory response to gIFN and LPS [71]. The problem with using *in vitro* transcribed RNA is that it lacks the 5′-capping sequence and may not fold in the same way as *in vivo* and would certainly lack any bound proteins that may alter the conformation of the mRNA

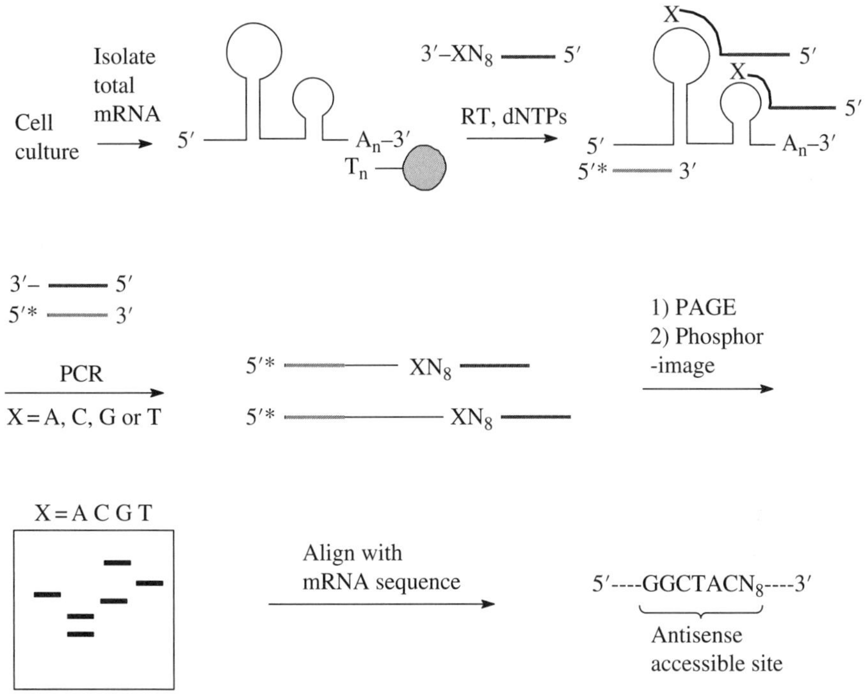

FIGURE 13.4 General method for determining antisense accessible sites on native mRNA. mRNA is first extracted from whole cells using a complementary oligoT bead. It is then incubated with a random 8 mer oligonucleotide terminating in a known base with a PCR tag (black) and extended with reverse transcriptase. It is then PCR amplified with the reverse primer (black) and forward gene-specific primer (gray) that is 5'-^{32}P-endlabeled (denoted with *) and run on a denaturing gel. From the length and pattern of bands, an antisense accessible site on the mRNA can be identified.

or block otherwise accessible sites. While mRNA directly isolated from cells would have the native sequence, it too would lack the native bound proteins unless methods were used to maintain their presence. More recently, methods have been reported for mapping mRNA structure *in vivo* through structure-specific chemical probes in conjunction with next-generation high-throughput sequencing technologies that might be able to identify antisense accessible sites [72].

13.3.3 Stable Antisense ODN Analogs for Imaging

The problem with DNA as an antisense agent is that the DNA•RNA hybrid that forms is a substrate for RNAseH that will cleave the RNA strand of the DNA•RNA hybrid and thereby destroy the binding affinity of the DNA for the target mRNA. In addition, DNA is susceptible to degradation in the serum and in cells. Likewise, RNA is even more susceptible to enzymatic degradation and can likewise induce mRNA strand cleavage through the RICS complex (RNAi). To circumvent these problems, nucleic acid analogs have been developed, which do not cause RNAseH activation and confer much greater stability to the antisense agent (Fig. 13.5). Nucleic acid analogs that have these properties are 2′-O-alkyl derivatives, thiophosphates, morpholidates (MORF), locked nucleic acids (LNAs), and PNAs [73–76]. Some of these analogs also show enhanced

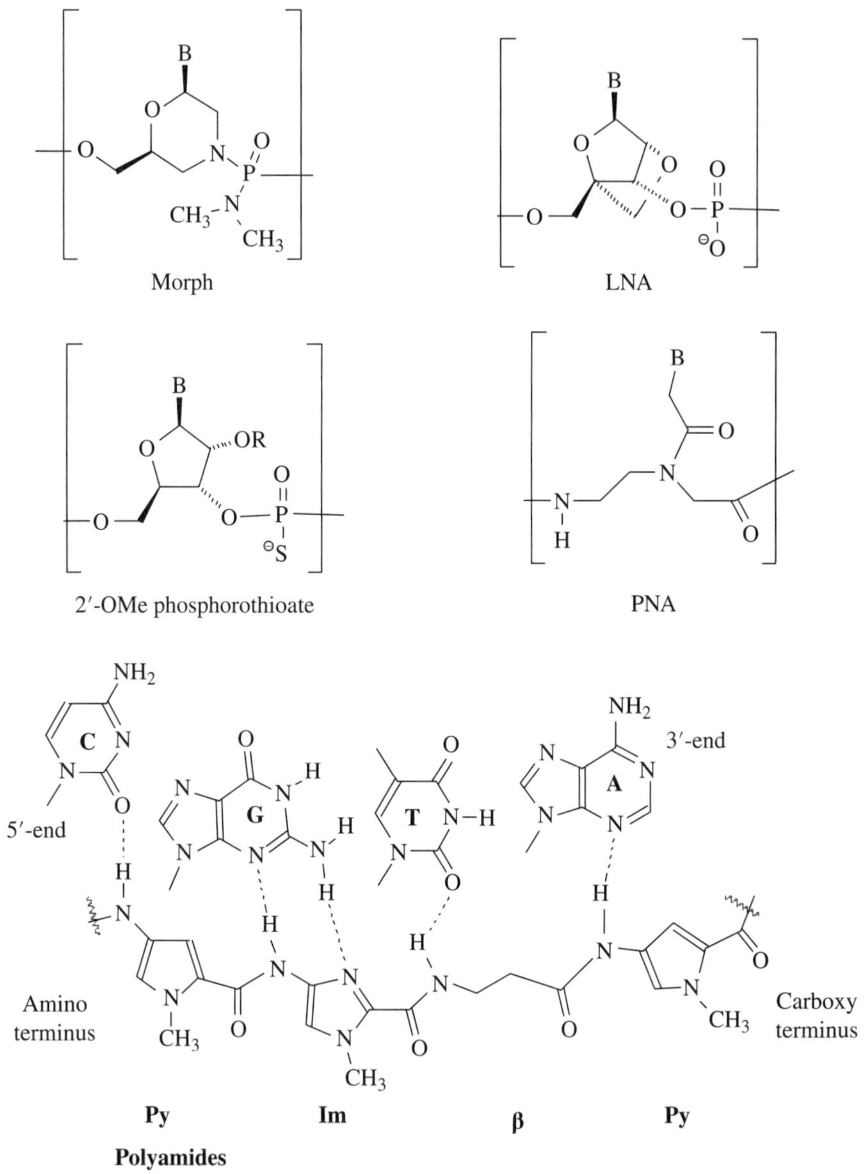

FIGURE 13.5 Structure of (a) degradation-resistant antisense nucleic acid analogs and (b) the antigene polyamides that recognize a double-stranded DNA sequence through hydrogen bonding interactions in the minor groove.

binding affinity for RNA, such as the LNAs and PNAs. Analogs based on the phosphodiester backbone can be conveniently synthesized by standard automated DNA synthesis, whereas PNAs can be synthesized by standard automated peptide synthesis. Various types of conjugation strategies have been developed for attaching reporter groups to the nucleotide analogs. Generally, these involve postsynthetic modification of the antisense agent by amide bond formation between an NHS ester of the reporter and amino group linked to the antisense agent or a maleimide of the reporter and a thiol on the antisense agent [77]. In the case of PNAs that are synthesized by solid-phase peptide methods, coupling can be done directly on column by amide bond-forming chemistry. More recently, click chemistry has been used to form bioconjugates [77, 78].

13.4 REPORTER SYSTEMS FOR RNA IMAGING

There are two general classes of reporter systems that can be used. Always ON and turn-on reporter systems. Always ON reporters are those reporters that always give a signal, whether or not bound to the target mRNA, such as a radioactive PET imaging agent, a fluorescent probe, or a relaxation agent. The problem with the always ON probes is that an RNA-specific signal will only arise if unbound probe can efficiently exit cells and the tissue environment, which is problematic for antisense agents due to their extremely low membrane permeability (Fig. 13.6).

This problem is greatly diminished with fluorescent turn -on probes, which only become significantly activated upon binding to the target mRNA through a variety of mechanisms (Fig. 13.6) [79–88]. A large number of turn -on probes are based on FRET, which is a distance-dependent singlet excitation energy transfer mechanism. The FRET mechanism has been utilized for the construction of molecular beacons, quenched strand displacement probes, and binary FRET imaging agents. In the molecular beacon design, the antisense sequence is made to form a hairpin with a short stem, one end of which is attached to a fluorescent reporter group and the other end is attached to a quencher (Fig. 13.7a) [89–91]. In the absence of the mRNA target, the fluorescence of the reporter is quenched. In the presence of the target, the hairpin opens and the quencher is then held at quite some distance from the reporter and fluorescence is restored.

The quenched strand displacement probe operates in a similar fashion to molecular beacons, only that the probe consists of a bimolecular hybrid between the antisense agent bearing the reporter and a shorter strand bearing the quencher (Fig. 13.7b) [92, 93]. In the presence of the target, the shorter quencher strand is displaced by the longer RNA target and fluorescence is restored. The potential problem with these two types of FRET-based agents is that the reporter and quenchers are susceptible to separation by other mechanisms than by target RNA binding, such as protein binding and degradation of the linkers and/or nucleic acid strands. Binary FRET probes do not suffer from this problem, as the signal only arises when the antisense strand bearing the donor fluorophore is bound to the mRNA next to the antisense strand bearing the acceptor fluorophore (Fig. 13.7c) [81]. The downside of the binary FRET system is that it requires a larger accessible site on the mRNA and requires separate delivery of the donor and acceptor probes to avoid FRET arising from close proximity

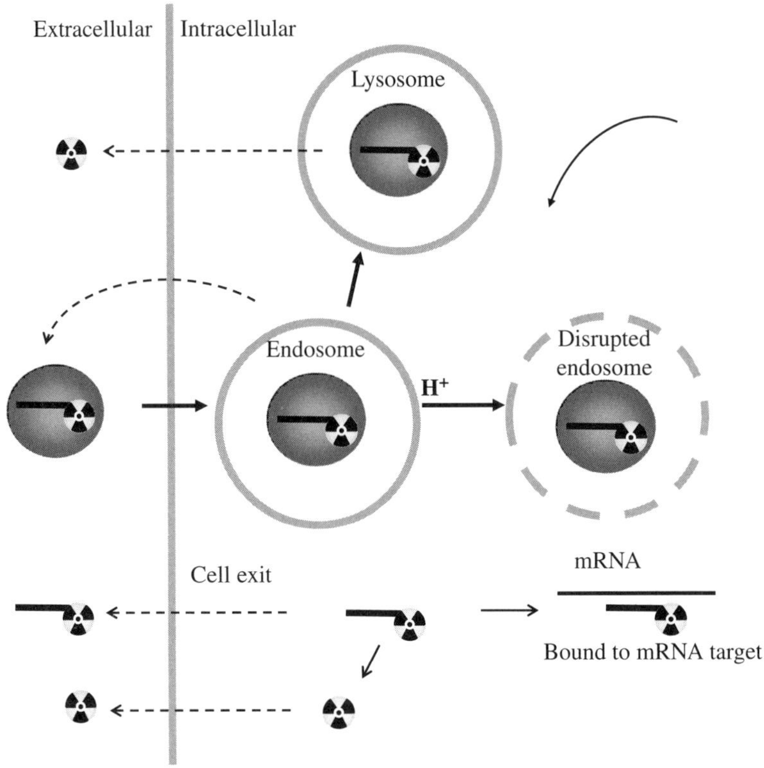

FIGURE 13.6 Problems attendant with always ON probes such as PET imaging agents. To obtain a nucleic acid target-specific signal, unbound probe or reporter must exit the cell. Any probe trapped in endosomes and lysosomes or unable to exit the cytoplasm will result in a background signal that can overwhelm the target-specific signal. Therefore, ligands promoting cell exit or environmentally sensitive linkers to the reporter that will allow it to efflux from the cell may need to be installed.

in the delivery vehicle. Other turn-on probes have also been developed that rely on a change in the fluorescence of a single fluorophore upon binding of an antisense agent to the target mRNA as exemplified by forced intercalation (FIT) probes (Fig. 13.7d) [94, 95]. The sensitivity and selectivity of the FIT probes can also be enhanced by FRET mechanisms in combination with quenchers and acceptor fluorophores.

Another class of turn-on probes relies on a chemical reaction to activate fluorescence (Fig. 13.7e) [84, 85]. Many of these are based on binary probes in which a reaction takes place between the two antisense strands. In one formulation of this approach, a catalyst is appended to one antisense strand and a fluorogenic substrate is attached to another, shorter antisense strand. The binding of the catalytic component converts the mRNA into a fluorescence-activating enzyme, which can then turnover multiple fluorogenic substrates. This would be especially important for imaging low-abundance mRNA probes. This type of system has been shown to work *in vitro*, but not yet *in vivo* [96–101]. The major hurdle is to develop a catalyst that can catalyze

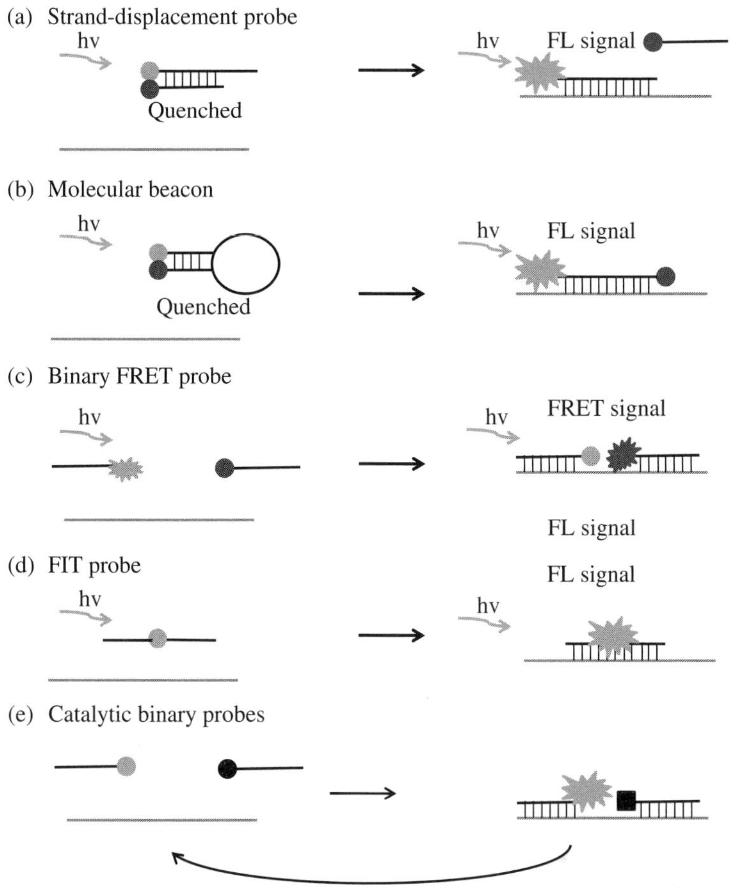

FIGURE 13.7 Optical antisense turn-on probes. To circumvent unwanted background signal from unbound probe, a number of clever systems labeled (a)-(e) have been developed that result in a large increase in fluorescence signal upon hybridizing to the target RNA. AQ3

a bio-orthogonal reaction in water, at pH 7 and at 37°C. Another approach has been to use the RNA itself as a catalyst to bring together two components that react stoichiometrically by a bio-orthogonal reaction to activate fluorescence as first shown using a Staudinger reaction [102] and then dissociate, thereby allowing for turnover. Since then, a number of catalytic Staudinger reaction-based turn-on systems for detecting nucleic acids have been developed, some of which have been shown to also work *in vivo* [103–105].

13.4.1 Antisense Imaging Agent Delivery

The problem with the individual antisense imaging probes for *in vivo* applications in animals is that their pharmacokinetics depend on their structure, with unmodified nucleic acids being rapidly degraded, while some modified nucleic acids are rapidly cleared from the blood and are excreted, whereas others are retained in tissues [106, 107]. Another problem relates to the delivery of the antisense probes to the cytoplasm of the cell. Many methods have been reported for imaging RNA in cell culture that rely on the addition of some cell permeabilizing agent or physical method for transient disruption of the cell membrane to allow the probes to enter the cells [108, 109]. Most of these agents, however, would be unsuitable for use in humans. This is where nanoparticles (NP) have a distinct advantage as a delivery agent [110–112]. NP can be engineered to have a size that is retained by the circulatory system and can be adorned with antisense probes along with a wide array of functionality to enable targeting to a specific tissue and to gain intracellular entry (Fig. 13.8).

Among the first types of agents for antisense delivery were cationic lipids, such as Lipofectamine [113–117]. These cationic agents can assemble into cationic micelles and liposomes that can bind negatively charged antisense imaging agents, or to hybrids formed between an uncharged ODN analog, such as a PNA, and a partially complementary ODN (Fig. 13.8a). The antisense agents can also be encapsulated within liposomes (Fig. 13.8b), and these complexes can either fuse directly with the outer cellular membranes or be endocytosed. More recently, naturally occurring exosomes have been investigated for the delivery of siRNA [118, 119]. Cationic delivery

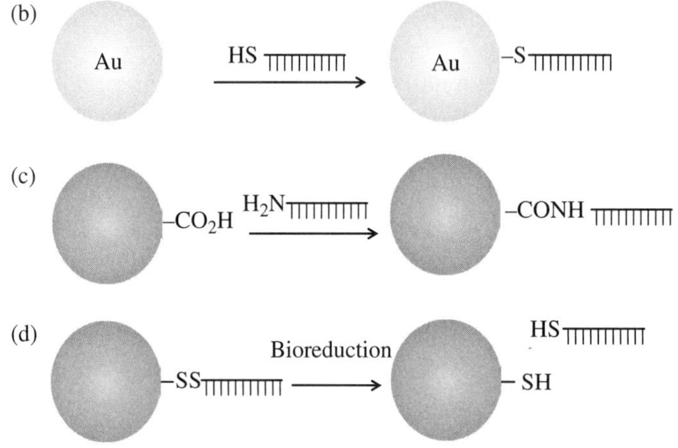

FIGURE 13.8 General methods for ligating or sequestering antisense agents to nanocarriers. From top to bottom: (a) direct electrostatic association with a cationic nanoparticle or lipid or via hybridization of a neutral antisense agent to a negatively charged complementary oligomer, (b) entrapment in a liposome or exosome, (c) thiol–gold bond formation, (d) amide bond formation, and (e) bioreductively cleavable disulfide bond formation.

agents based on polyethyleneimine (PEI), polylysine, or other polyamine-based NP have also been extensively studied as antisense delivery agents [120–122]. Because of their high positive charge, they interact strongly with the cell membrane and become endocytosed through one or more of many possible endocytotic mechanisms, such as pinocytosis, macropinocytosis, clathrin-mediated, and caveolin-mediated mechanisms [123, 124]. Both cationic lipid- and polymer-based systems show toxicity, however, and will require further development before they can be used in humans [120, 125]. Pretargeting and endocytosis of NP can also be achieved by receptor-mediated endocytosis [126]. To this end, the folate and ferritin have been widely used to deliver cargo into cancer cells [127, 128].

In addition to electrostatic binding, antisense agents can also be covalently attached to the NP through noncleavable linkers, such as to gold NP through thiol–gold bonds (Fig. 13.8c). Such gold-ODN NP are actively taken up by cells, presumably by binding to serum proteins that are then endocytosed [129]. Alternatively, antisense agents can be attached via other covalent reactions, such as amide bond formation (Fig. 13.8d) or through an acid-labile or bioreductively cleavable linker such as a disulfide bond (Fig. 13.8e). The specific linker used will depend on whether or not the NP would hinder binding of the antisense probe to the target mRNA. Having the antisense agent covalently bound to the NP runs the risk, however, that much of it may remain trapped in endosomes along with the NP. This appeared to be the case in a study with shell cross-linked NP linked to antisense PNAs via an amide linkage that did not show bioactivity, whereas the same PNAs linked via a bioreductively cleavable disulfide linker did [130]. Confocal studies showed that while the PNA was able to exit the endocytic compartments and enter the cytoplasm, the NP was not. Equally problematic with a noncleavable linker is that excess unbound or partially bound NP may not be able to exit the cell leading to a high background signal with an always -on probe. Even with turn-on NP probes, the NP could become occluded with target mRNA that could prevent the binding of other target mRNAs. Given these problems, it makes more sense to have an intracellularly releasable antisense agent, unless one wants to use an attached NP to restrict the probe from entering and accumulating in certain organelles. For example, tethering a molecular beacon probe to a large protein complex was found to prevent unwanted nonspecific accumulation of the probe in the nucleus [131]. In the case of binary probes, it would make most sense to bind them to separate NP to prevent any interaction that might lead to a false signal.

The general problem with endocytosis as a mechanism for intracellular delivery is that endocytosed NP generally become trapped in the endosomal/lysosomal pathway and do not enter the cytoplasm where the mRNA targets are. Thus, the NP need to incorporate additional functionality to enable the NP to escape the endosomes or at least facilitating the exit of the probes. Many amine-based NP appear to cause endosomal rupture by the pH sponge effect, which occurs as a result of proton pumps that acidify the interior of endosomes to a pH of 4–5 [132–134]. NP that contain a high concentration of amines have a number of amines that remain unprotonated at neutral pH but can be protonated at lower pH (a proton sponge). As a consequence, the NP function as a buffer that resists the lowering of the pH of the endosome and results in many more protons being pumped into the interior of the endosome along with counterions and water. The increase in solutes and water increases the internal pressure, leading to destabilization and rupture of the endosome causing its contents to be released. Another strategy for destabilizing the endosome is to use NP constructed from acid-labile polymers, such as dextran, that degrade to monomers inside the endosome causing an increase in osmotic pressure that likewise destabilizes and ruptures the endosome [135]. Other strategies make use of pH-sensitive peptides derived from virus particles that undergo conformational changes upon acidification that destabilize the membrane or form pores that facilitate release of the contents such as the GALA peptide [136, 137].

13.4.2 Efflux of Unbound Antisense Probes

Even if the antisense agent makes it efficiently to the cytoplasm, unbound antisense agent utilizing an always -on reporter must be able to exit the cell efficiently to reduce the nonspecific background signal (Fig. 13.6). This would be necessary for radioactive probes such as PET imaging agents, and in most cases, this aspect of the design of the imaging agent is neglected and not even discussed. While small molecule probes can easily diffuse out of cells because of their membrane permeability, the antisense agents cannot, and so some mechanism must be provided for the exit of the antisense agent or the reporter itself. It is possible that a cell-penetrating peptide might be able to facilitate exit of the antisense agent from the cell by exocytosis or that unbound probe is degraded to membrane-permeable fragments at a greater rate than RNA-bound probes. This aspect has received little to no attention but is critical to the success of the antisense imaging agent.

13.5 RECENT EXAMPLES OF NANOPARTICLE ANTISENSE IMAGING AGENTS

Most of the literature concerning genetically targeted NP deals with siRNA and gene therapy, and very little deals with antisense imaging. There are a few examples, however, of NP-based antisense imaging systems that have been developed that illustrate some of the design concepts discussed previously. These NP systems are based on turn-on optical probes, which avoid the requirement for always -on probes that excess unbound antisense probe be able to diffuse easily out of the cell.

13.5.1 Multiplexed Gold Nanoparticle-Based Strand Displacement Probes or Nanoflares

An mRNA turn-on antisense gold NP probe system was developed that was based on quenched strand displacement probe technology (Fig. 13.9) [138–140]. The idea was to make use of a well-known property of gold as a fluorescent quencher to create a gold NP with numerous antisense oligodeoxynucleotides that were covalently linked to the gold surface via a thiol group. These antisense strands were then hybridized to

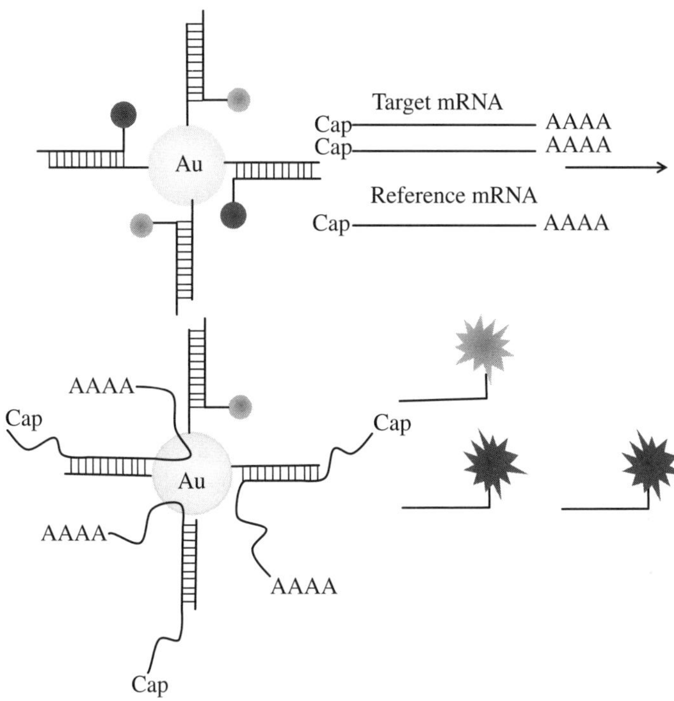

FIGURE 13.9 Multiplexed nanoflare. In this nanoparticle antisense imaging system, a gold nanoparticle is used as a carrier and fluorescence quencher. The amount of a target mRNA sequence is recorded by the ratio of the signal arising from the target mRNA to that from a reference mRNA sequence. Entry into the cell is thought to occur via endocytosis mediated by bound serum proteins, but how the nanoparticle escapes the endosomes is not known.

a complementary ODN strand bearing a fluorescent molecule. In the absence of target mRNA, the fluorescence of the fluorophore would be quenched, but upon binding of the antisense strand to the target mRNA, the strand bearing the fluorophore would be displaced and become fluorescent. Because the fluorescent strand is released, the fluorescence only reports on the existence of the mRNA and not its location. By having two antisense strands hybridized to different fluorophore-containing strands, the NP are able to report on the relative ratio of two RNAs, which would have much greater diagnostic utility than afforded by an individual signal. These particles appeared to enter cells, presumably by endocytosis via serum proteins bound to the NP [129]. It is not clear how the NP was able to exit the endosomes as there was no specific functionality present for that purpose. Because the antisense strand is covalently linked to the gold NP, the target mRNA becomes likewise bound to the NP and may sterically inhibit other mRNAs from triggering release of more fluorescent strands. The antisense strands may also become displaced from the gold in an exchange reaction with glutathione, which could result in a false signal. Another problem with the gold NP systems is that they are not biodegradable and often accumulate in the liver, which would not make them ideal for human use [141].

13.5.2 Bioreductively Releasable Strand Displacement Probe Nanoparticle

Another approach to a strand displacement-based detection system was reported that could detect miR-221, a miRNA (Fig. 13.10) [142]. In this system, the antisense strand displacement probe was designed to be bioreductively released from the NP and then transfer the fluorescent antisense strand to the target mRNA. The NP consisted of a silica-coated cobalt ferrite core with a carboxylated surface that was linked via a disulfide bond to a quenched strand displacement turn-on probe and also to a pretargeting aptamer ligand. The strand displacement probe consisted of an antisense strand linked to Cy 5.0 fluorophore that was hybridized to a shorter sense strand linked to a quencher, BHQ2 on the 5′-end. The DNA aptamer AS 1411 that targets nucleolin was used for pretargeting and to promote cellular internalization, presumably by endocytosis. There is some uncertainty about exactly how AS 1411 facilitates internalization and the role of nucleolin in the process [143, 144]. Nonetheless, miR-221-specific fluorescence enhancement was observed in cell culture and in mouse xenografts. For this NP to have worked, some feature must have facilitated the escape of the NP and/or the strand displacement probe from the endosomal pathway by which it is presumed to have entered. There were no specific auxiliaries attached to the NP for this purpose, though it is possible that the remaining surface carboxylates were functioning as a pH sponge. Because miRNAs are ultimately sequestered in a RISC complex and presumably inaccessible, the strand displacement probe must have been able to intercept the miRNA somewhere along the maturation pathway, or while in the RISC complex, but this remains to be elucidated. It is remarkable that this NP system was able to work *in vivo* in a mouse since there was no other functionality to impart stealth properties to the NP and prevent its uptake by the mononuclear phagocyte system (MPS). Apparently, some fraction of the NP must have been able to accumulate in the xenograft, presumably by the enhanced permeability and retention (EPR) effect.

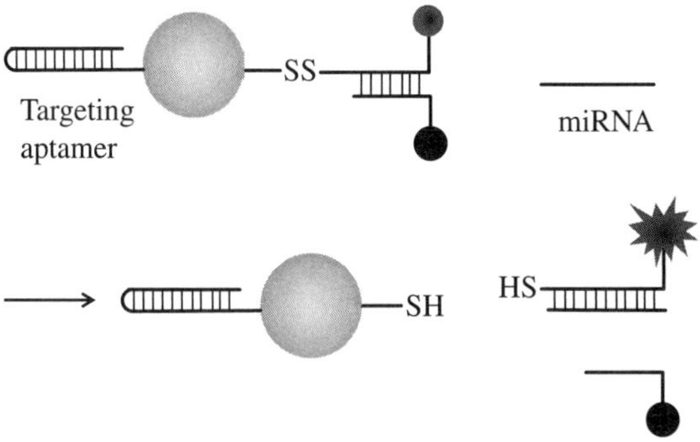

FIGURE 13.10 Targeted miRNA-sensing NP. This nanoparticle antisense imaging system makes use of an aptamer to pretarget an miRNA-sensing quenched strand displacement probe to a nucleolin-bearing cell, which then internalizes the particle into the cytoplasm. Because of the reducing environment, the disulfide bond is cleaved freeing the probe to bind to the miRNA target.

13.5.3 Noncovalent Delivery of Strand Displacement Probes by Cationic Nanoparticles

Another system based on quenched strand displacement probes was delivered by a cationic shell cross-linked NP (cSCK) (Fig. 13.11) [145]. This NP consisted of a polystyrene core and a poly-N-2-aminoethylacrylamide shell, which is largely protonated at pH 7, though a number of amines remain unprotonated [146]. The cSCK served three purposes. The first was to electrostatically bind to the negatively charged antisense turn-on probe. The second was to facilitate endocytosis through interaction of the positively charged shell with the cell membrane. The third was to facilitate endosomal destabilization by the pH sponge effect mediated by the remaining unprotonated amines within the shell, thereby neutralizing protons during endosomal acidification leading to an increase in osmotic pressure. The strand displacement probe was designed to target iNOS mRNA, which is upregulated over 100-fold in response to LPS/γ-IFN. The antisense binding site on the iNOS mRNA had been determined through the use of an RT-ROL assay on endogenous mRNA recovered from whole cell extracts of mouse macrophage cell line [71]. The strand displacement probe was constructed from a neutral, high-affinity, degradation-resistant PNA strand bearing a fluorescein fluorophore to a five-nucleotide shorter negatively charged partially complementary DNA strand bearing a DABCYL fluorescence quencher. The remaining five single-strand bases of the PNA strand serve as a "toehold" to facilitate binding to the target mRNA sequence, which then causes displacement of the quencher strand by branch migration. The higher affinity that the PNA strand has for the mRNA strand due to the additional five base pairs that can form drives the exchange reaction

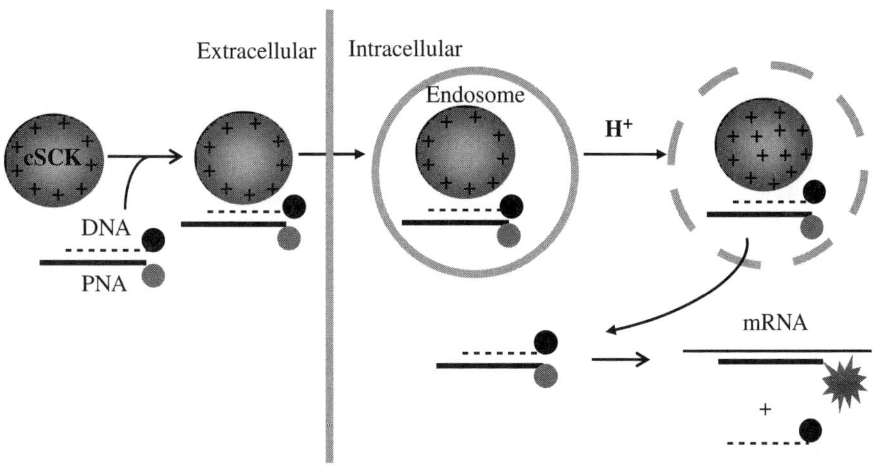

FIGURE 13.11 Cationic shell cross-linked nanoparticle-mediated delivery of a quenched strand displacement probe system. In this system, a cationic nanoparticle electrostatically binds to a quenched probe consisting of a neutral PNA hybridized to a partially complementary oligodeoxynucleotide. The NP then mediates entry of the probe into the cell via endocytosis facilitated by its positive shell and subsequent escape from the endosome via the proton sponge effect.

to favor the PNA•mRNA hybrid. The strand displacement cSCK system was able to detect a fourfold increase in iNOS expression induced by LPS/γ-IFN, though not the 100-fold level expected from the RT-PCR assays. The observed difference in expression level could be due to the fact that PCR was detecting whole cell RNA, while the probe might only be detecting cytosolic RNA. It is also possible that the DNA quenching strand or the quencher was being degraded in the cell. The cationic cSCK NP delivery system was also investigated for its ability to image iNOS mRNA in an injured lung mouse with a radioactively labeled PNA•ODN duplex [147]. In this case, the PNA was linked to a $TyrArg_9$ peptide, the tyrosine of which was used to introduce radioactive ^{123}I and the Arg_9 of which was a cell-penetrating peptide used to facilitate exit of unbound PNA from the cell. A statistically significant difference was observed in the retention of the targeted PNA probe compared to the untargeted probe in the injured lungs, but it was not sufficiently different that it could be used diagnostically.

13.6 CONCLUSIONS

The ability to image gene expression with antisense agents would be a very attractive and general way for noninvasively diagnosing and monitoring disease. Unfortunately, there are many hurdles yet to be surmounted before such methodology could become practical. While optical turn -on probes do not have to be able to exit the cytoplasm, they do not have the sensitivity or detection depth that can be achieved by PET imaging agents. What is therefore still needed to achieve clinically useful antisense imaging is a general, nontoxic method for efficient cytosolic delivery of antisense PET agents in humans and a system for enabling unbound probe or the attached radionuclide reporter to efficiently escape from the cell.

REFERENCES

[1] Juran BD, Lazaridis KN. Applying genomics to the study of complex disease. Semin Liver Dis 2007;**27**:3–12.

[2] Pflughoeft KJ, Versalovic J. Human microbiome in health and disease. Annu Rev Pathol 2012;**7**:99–122.

[3] Knight JC. Resolving the variable genome and epigenome in human disease. J Intern Med 2012;**271**:379–391.

[4] Teperino R, Lempradl A, Pospisilik JA. Bridging epigenomics and complex disease: the basics. Cell Mol Life Sci 2013;**70**:1609–1621.

[5] Ametamey SM, Honer M, Schubiger PA. Molecular imaging with PET. Chem Rev 2008;**108**:1501–1516.

[6] Ray P. Multimodality molecular imaging of disease progression in living subjects. J Biosci 2011;**36**:499–504.

[7] Pysz MA, Gambhir SS, Willmann JK. Molecular imaging: current status and emerging strategies. Clin Radiol 2010;**65**:500–516.

[8] Kaur S, Venktaraman G, Jain M, Senapati S, Garg PK, Batra SK. Recent trends in antibody-based oncologic imaging. Cancer Lett 2012;**315**:97–111.

[9] Soontornworajit B, Wang Y. Nucleic acid aptamers for clinical diagnosis: cell detection and molecular imaging. Anal Bioanal Chem 2011;**399**:1591–1599.

[10] Xing H, Wong NY, Xiang Y, Lu Y. DNA aptamer functionalized nanomaterials for intracellular analysis, cancer cell imaging and drug delivery. Curr Opin Chem Biol 2012;**16**: 429–435.

[11] Dunphy MP, Lewis JS. Radiopharmaceuticals in preclinical and clinical development for monitoring of therapy with PET. J. Nucl. Med 2009;**50 Suppl 1**:106S–121S.

[12] Skotland T. Molecular imaging: challenges of bringing imaging of intracellular targets into common clinical use. Contrast Media Mol Imaging 2012;**7**:1–6.

[13] Yang CJ, Jockusch S, Vicens M, Turro NJ, Tan W. Light-switching excimer probes for rapid protein monitoring in complex biological fluids. Proc Natl Acad Sci U S A 2005; **102**:17278–17283.

[14] Shi H, He X, Wang K, Wu X, Ye X, Guo Q, Tan W, Qing Z, Yang X, Zhou B. Activatable aptamer probe for contrast-enhanced *in vivo* cancer imaging based on cell membrane protein-triggered conformation alteration. Proc Natl Acad Sci U S A 2011;**108**:3900–3905.

[15] Lee S, Xie J, Chen X. Activatable molecular probes for cancer imaging. Curr Top Med Chem 2010;**10**:1135–1144.

[16] Ceulemans S, van der Ven K, Del-Favero J. Targeted screening and validation of copy number variations. Methods Mol Biol 2012;**838**:311–328.

[17] Deshpande A, White PS. Multiplexed nucleic acid-based assays for molecular diagnostics of human disease. Expert Rev Mol Diagn 2012;**12**:645–659.

[18] Gorreta F, Carbone W, Barzaghi D. Genomic profiling: cDNA arrays and oligoarrays. Methods Mol Biol 2012;**823**:89–105.

[19] Yu AC, Vatcher G, Yue X, Dong Y, Li MH, Tam PH, Tsang PY, Wong AK, Hui MH, Yang B, Tang H, Lau LT. Nucleic acid-based diagnostics for infectious diseases in public health affairs. Front Med 2012;**6**:173–186.

[20] Mirkin SM. DNA structures, repeat expansions and human hereditary disorders. Curr Opin Struct Biol 2006;**16**:351–358.

[21] Knauert MP, Glazer PM. Triplex forming oligonucleotides: sequence-specific tools for gene targeting. Hum Mol Genet 2001;**10**:2243–2251.

[22] Praseuth D, Guieysse AL, Helene C. Triple helix formation and the antigene strategy for sequence-specific control of gene expression. Biochim Biophys Acta 1999;**1489**:181–206.

[23] Izvolsky KI, Demidov VV, Nielsen PE, Frank-Kamenetskii MD. Sequence-specific protection of duplex DNA against restriction and methylation enzymes by pseudocomplementary PNAs. Biochemistry 2000;**39**:10908–10913.

[24] Nielsen PE. Sequence-selective targeting of duplex DNA by peptide nucleic acids. Curr Opin Mol Ther 2010;**12**:184–191.

[25] Dervan PB. Molecular recognition of DNA by small molecules. Bioorg Med Chem 2001;**9**:2215–2235.

[26] Ghosh I, Stains CI, Ooi AT, Segal DJ. Direct detection of double-stranded DNA: molecular methods and applications for DNA diagnostics. Mol Biosyst 2006;**2**:551–560.

[27] Klug A. The discovery of zinc fingers and their development for practical applications in gene regulation and genome manipulation. Q Rev Biophys 2010;**43**:1–21.

[28] Fujimoto J, Bando T, Minoshima M, Uchida S, Iwasaki M, Shinohara K-I, Sugiyama H. Detection of triplet repeat sequences in the double-stranded DNA using pyrene-functionalized pyrrole–imidazole polyamides with rigid linkers. Bioorg Med Chem 2008;**16**:5899–5907.

[29] Ooi AT, Stains CI, Ghosh I, Segal DJ. Sequence-enabled reassembly of β-lactamase (SEER-LAC): a sensitive method for the detection of double-stranded DNA. Biochemistry 2006;**45**:3620–3625.

[30] Costantini DL, Hu M, Reilly RM. Peptide motifs for insertion of radiolabeled biomolecules into cells and routing to the nucleus for cancer imaging or radiotherapeutic applications. Cancer Biother Radiopharm 2008;**23**:3–24.

[31] Chin JY, Schleifman EB, Glazer PM. Repair and recombination induced by triple helix DNA. Front Biosci 2007;**12**:4288–4297.

[32] Jain A, Wang G, Vasquez KM. DNA triple helices: biological consequences and therapeutic potential. Biochimie 2008;**90**:1117–1130.

[33] Vafai SB, Mootha VK. Mitochondrial disorders as windows into an ancient organelle. Nature 2012;**491**:374–383.

[34] Yamada Y, Akita H, Kogure K, Kamiya H, Harashima H. Mitochondrial drug delivery and mitochondrial disease therapy—an approach to liposome-based delivery targeted to mitochondria. Mitochondrion 2007;**7**:63–71.

[35] Yamada Y, Harashima H. Mitochondrial drug delivery systems for macromolecule and their therapeutic application to mitochondrial diseases. Adv Drug Deliv Rev 2008;**60**: 1439–1462.

[36] Wan Y, Kertesz M, Spitale RC, Segal E, Chang HY. Understanding the transcriptome through RNA structure. Nat Rev Genet 2011;**12**:641–655.

[37] Mauger DM, Siegfried NA, Weeks KM. The genetic code as expressed through relationships between mRNA structure and protein function. FEBS Lett 2013;**587**:1180–1188.

[38] Konig J, Zarnack K, Luscombe NM, Ule J. Protein-RNA interactions: new genomic technologies and perspectives. Nat Rev Genet 2011;**13**:77–83.

[39] Wilson DN, Doudna Cate JH. The structure and function of the eukaryotic ribosome. Cold Spring Harb Perspect Biol 2012;**4**:pii: a011536.

[40] Wang Y, Chen L, Liu X, Cheng D, Liu G, Liu Y, Dou S, Hnatowich DJ, Rusckowski M. Detection of Aspergillus fumigatus pulmonary fungal infections in mice with 99mTc-labeled MORF oligomers targeting ribosomal RNA. Nucl Med. Biol 2012;**40** (1):89–96.

[41] Yarham JW, Al-Dosary M, Blakely EL, Alston CL, Taylor RW, Elson JL, McFarland R. A comparative analysis approach to determining the pathogenicity of mitochondrial tRNA mutations. Hum Mutat 2011;**32**:1319–1325.

[42] Shruti K, Shrey K, Vibha R. Micro RNAs: tiny sequences with enormous potential. Biochem Biophys Res Commun 2011;**407**:445–449.

[43] Liu Q, Paroo Z. Biochemical principles of small RNA pathways. Annu Rev Biochem 2010;**79**:295–319.

[44] Shrey K, Suchit A, Nishant M, Vibha R. RNA interference: emerging diagnostics and therapeutics tool. Biochem Biophys Res Co 2009;**386**:273–277.

[45] Etheridge A, Lee I, Hood L, Galas D, Wang K. Extracellular microRNA: a new source of biomarkers. Mutat Res 2011;**717**:85–90.

[46] de Planell-Saguer M, Rodicio MC. Analytical aspects of microRNA in diagnostics: a review. Anal Chim Acta 2011;**699**:134–152.

[47] Jansson MD, Lund AH. MicroRNA and cancer. Mol Oncol 2012;**6**:590–610.

[48] Lundstrom K. Micro-RNA in disease and gene therapy. Curr Drug Discov Technol 2011;**8**:76–86.

[49] Wang F, Niu G, Chen X, Cao F. Molecular imaging of microRNAs. Eur J Nucl Med Mol I 2011;**38**:1572–1579.

[50] Darnell JE Jr. Reflections on the history of pre-mRNA processing and highlights of current knowledge: a unified picture. RNA 2013;**19**:443–460.

[51] Muller-McNicoll M, Neugebauer KM. How cells get the message: dynamic assembly and function of mRNA-protein complexes. Nat Rev Genet 2013;**14**:275–287.

[52] Wylie KM, Weinstock GM, Storch GA. Emerging view of the human virome. Transl Res 2012;**160**:283–290.

[53] Freed EF, Bleichert F, Dutca LM, Baserga SJ. When ribosomes go bad: diseases of ribosome biogenesis. Mol Biosyst 2010;**6**:481–493.

[54] Mukherjee A, Wickstrom E, Thakur ML. Imaging oncogene expression. Eur J Radiol 2009;**70**:265–273.

[55] Kummer S, Knoll A, Socher E, Bethge L, Herrmann A, Seitz O. Fluorescence imaging of influenza H1N1 mRNA in living infected cells using single-chromophore FIT-PNA. Angew Chem Int Ed Engl 2011;**50**:1931–1934.

[56] Kummer S, Knoll A, Socher E, Bethge L, Herrmann A, Seitz O. PNA FIT-probes for the dual color imaging of two viral mRNA targets in influenza H1N1 infected live cells. Bioconjug Chem 2012;**23**:2051–2060.

[57] Sun X, Fang H, Li X, Rossin R, Welch MJ, Taylor JS. MicroPET imaging of MCF-7 tumors in mice via unr mRNA-targeted peptide nucleic acids. Bioconjug Chem 2005;**16**: 294–305.

[58] Lendvai G, Estrada S, Bergstrom M. Radiolabelled oligonucleotides for imaging of gene expression with PET. Curr Med Chem 2009;**16**:4445–4461.

[59] Yamashita T, Honda M, Kaneko S. Application of serial analysis of gene expression in cancer research. Curr Pharm Biotechnol 2008;**9**:375–382.

[60] Nikitina EG, Urazova LN, Stegny VN. MicroRNAs and human cancer. Exp Oncol 2012;**34**:2–8.

[61] Omenn GS, Menon R, Zhang Y. Innovations in proteomic profiling of cancers: alternative splice variants as a new class of cancer biomarker candidates and bridging of proteomics with structural biology. J Proteomics 2013;**90**:28–37.

[62] Kwon SM, Cho H, Choi JH, Jee BA, Jo Y, Woo HG. Perspectives of integrative cancer genomics in next generation sequencing era. Genomics Inform 2012;**10**:69–73.

[63] Xuan, J., Yu, Y., Qing, T., Guo, L., and Shi, L., Next-generation sequencing in the clinic: promises and challenges, Cancer Lett. 2013, **340**(2):284-295.

[64] Gullapalli RR, Desai KV, Santana-Santos L, Kant JA, Becich MJ. Next generation sequencing in clinical medicine: challenges and lessons for pathology and biomedical informatics. J Pathol Inform 2012;**3**:40.

[65] Winter J, Diederichs S. MicroRNA biogenesis and cancer. Methods Mol Biol 2011; **676**:3–22.

[66] Zhang DY, Winfree E. Control of DNA strand displacement kinetics using toehold exchange. J Am Chem Soc 2009;**131**:17303–17314.

[67] Rhee WJ, Santangelo PJ, Jo H, Bao G. Target accessibility and signal specificity in live-cell detection of BMP-4 mRNA using molecular beacons. Nucleic Acids Res 2008;**36**:e30.

[68] Zhang HY, Mao J, Zhou D, Xu Y, Thonberg H, Liang Z, Wahlestedt C. mRNA accessible site tagging (MAST): a novel high throughput method for selecting effective antisense oligonucleotides. Nucleic Acids Res 2003;**31**:e72.

[69] Fang H, Yue X, Li X, Taylor JS. Identification and characterization of high affinity antisense PNAs for the human unr (upstream of N-ras) mRNA which is uniquely overexpressed in MCF-7 breast cancer cells. Nucleic Acids Res 2005;**33**:6700–6711.

[70] Allawi HT, Dong F, Ip HS, Neri BP, Lyamichev VI. Mapping of RNA accessible sites by extension of random oligonucleotide libraries with reverse transcriptase. RNA 2001;**7**: 314–327.

[71] Fang H, Shen Y, Taylor JS. Native mRNA antisense-accessible sites library for the selection of antisense oligonucleotides, PNAs, and siRNAs. RNA 2010;**16**:1429–1435.

[72] Spitale RC, Crisalli P, Flynn RA, Torre EA, Kool ET, Chang HY. RNA SHAPE analysis in living cells. Nat Chem Biol 2013;**9**:18–20.

[73] Karkare S, Bhatnagar D. Promising nucleic acid analogs and mimics: characteristic features and applications of PNA, LNA, and morpholino. Appl Microbiol Biotechnol 2006;**71**:575–586.

[74] Veedu RN, Wengel J. Locked nucleic acids: promising nucleic acid analogs for therapeutic applications. Chem Biodivers 2010;**7**:536–542.

[75] Prakash TP. An overview of sugar-modified oligonucleotides for antisense therapeutics. Chem Biodivers 2011;**8**:1616–1641.

[76] Deleavey GF, Damha MJ. Designing chemically modified oligonucleotides for targeted gene silencing. Chem Biol 2012;**19**:937–954.

[77] Weisbrod SH, Marx A. Novel strategies for the site-specific covalent labelling of nucleic acids. Chem Commun (Camb) 2008:5675–5685.

[78] Lallana E, Riguera R, Fernandez-Megia E. Reliable and efficient procedures for the conjugation of biomolecules through Huisgen azide-alkyne cycloadditions. Angew Chem 2011;**50**:8794–8804.

[79] Bao G, Rhee WJ, Tsourkas A. Fluorescent probes for live-cell RNA detection. Annu Rev Biomed Eng 2009;**11**:25–47.

[80] Tyagi S. Imaging intracellular RNA distribution and dynamics in living cells. Nat Methods 2009;**6**:331–338.

[81] Kolpashchikov DM. Binary probes for nucleic acid analysis. Chem Rev 2010;**110**: 4709–4723.

[82] Okamoto A. ECHO probes: a concept of fluorescence control for practical nucleic acid sensing. Chem Soc Rev 2011;**40**:5815–5828.

[83] Armitage BA. Imaging of RNA in live cells. Curr Opin Chem Biol 2011;**15**:806–812.

[84] Jun ME, Roy B, Ahn KH. "Turn-on" fluorescent sensing with "reactive" probes. Chem Commun (Camb) 2011;**47**:7583–7601.

[85] Shibata A, Abe H, Ito Y. Oligonucleotide-templated reactions for sensing nucleic acids. Molecules 2012;**17**:2446–2463.

[86] Guo J, Ju J, Turro NJ. Fluorescent hybridization probes for nucleic acid detection. Anal Bioanal Chem 2012;**402**:3115–3125.

[87] Su X, Xiao X, Zhang C, Zhao M. Nucleic acid fluorescent probes for biological sensing. Appl Spectrosc 2012;**66**:1249–1262.

[88] Wang K, Huang J, Yang X, He X, Liu J. Recent advances in fluorescent nucleic acid probes for living cell studies. Analyst 2013;**138**:62–71.

[89] Tan W, Fang X, Li J, Liu X. Molecular beacons: a novel DNA probe for nucleic acid and protein studies. Chemistry 2000;**6**:1107–1111.

[90] Santangelo P, Nitin N, Bao G. Nanostructured probes for RNA detection in living cells. Ann Biomed Eng 2006;**34**:39–50.

[91] Li Y, Zhou X, Ye D. Molecular beacons: an optimal multifunctional biological probe. Biochem Biophys Res Co 2008;**373**:457–461.

[92] Morrison LE, Halder TC, Stols LM. Solution-phase detection of polynucleotides using interacting fluorescent labels and competitive hybridization. Anal Biochem 1989;**183**: 231–244.

[93] Ellwood MS, Collins M, Fritsch EF, Williams JI, Diamond SE, Brewen JG. Strand displacement applied to assays with nucleic acid probes. Clin Chem 1986;**32**:1631–1636.

[94] Kohler O, Jarikote DV, Seitz O. Forced intercalation probes (FIT Probes): thiazole orange as a fluorescent base in peptide nucleic acids for homogeneous single-nucleotide-polymorphism detection. Chembiochem 2005;**6**:69–77.

[95] Socher E, Jarikote DV, Knoll A, Roglin L, Burmeister J, Seitz O. FIT probes: peptide nucleic acid probes with a fluorescent base surrogate enable real-time DNA quantification and single nucleotide polymorphism discovery. Anal Biochem 2008;**375**:318–330.

[96] Ma Z, Taylor JS. Nucleic acid-triggered catalytic drug release. Proc Natl Acad Sci U S A 2000;**97**:11159–11163.

[97] Ma Z, Taylor JS. Nucleic acid triggered catalytic drug and probe release: a new concept for the design of chemotherapeutic and diagnostic agents. Bioorg Med Chem 2001;**9**: 2501–2510.

[98] Ma Z, Taylor JS. PNA-based RNA-triggered drug-releasing system. Bioconjug Chem 2003;**14**:679–683.

[99] Brunner J, Mokhir A, Kraemer R. DNA-templated metal catalysis. J Am Chem Soc 2003;**125**:12410–12411.

[100] Zelder FH, Brunner J, Kramer R. DNA-templated catalysis using a metal-cleavable linker. *Chem Commun (Camb)* 2004:902–903.

[101] Boll I, Kramer R, Brunner J, Mokhir A. Templated metal catalysis for single nucleotide specific DNA sequence detection. J Am Chem Soc 2005;**127**:7849–7856.

[102] Cai J, Li X, Yue X, Taylor JS. Nucleic acid-triggered fluorescent probe activation by the Staudinger reaction. J Am Chem Soc 2004;**126**:16324–16325.

[103] Franzini RM, Kool ET. Efficient nucleic acid detection by templated reductive quencher release. J Am Chem Soc 2009;**131**:16021–16023.

[104] Furukawa K, Abe H, Hibino K, Sako Y, Tsuneda S, Ito Y. Reduction-triggered fluorescent amplification probe for the detection of endogenous RNAs in living human cells. Bioconjug Chem 2009;**20**:1026–1036.

[105] Pianowski Z, Gorska K, Oswald L, Merten CA, Winssinger N. Imaging of mRNA in live cells using nucleic acid-templated reduction of azidorhodamine probes. J Am Chem Soc 2009;**131**:6492–6497.

[106] Geary RS. Antisense oligonucleotide pharmacokinetics and metabolism. Expert Opin Drug Metab Toxicol 2009;**5**:381–391.

[107] Amantana A, Iversen PL. Pharmacokinetics and biodistribution of phosphorodiamidate morpholino antisense oligomers. Curr Opin Pharmacol 2005;**5**:550–555.

[108] Veldhoen S, Laufer SD, Restle T. Recent developments in peptide-based nucleic acid delivery. Int J Mol Sci 2008;**9**:1276–1320.

[109] Margus H, Padari K, Pooga M. Cell-penetrating peptides as versatile vehicles for oligonucleotide delivery. Mol Ther 2012;**20**:525–533.

[110] de Martimprey H, Vauthier C, Malvy C, Couvreur P. Polymer nanocarriers for the delivery of small fragments of nucleic acids: oligonucleotides and siRNA. Eur J Pharm Biopharm 2009;**71**:490–504.

[111] Fattal E, Barratt G. Nanotechnologies and controlled release systems for the delivery of antisense oligonucleotides and small interfering RNA. Br J Pharmacol 2009;**157**: 179–194.

[112] Kesharwani P, Gajbhiye V, Jain NK. A review of nanocarriers for the delivery of small interfering RNA. Biomaterials 2012;**33**:7138–7150.

[113] Rao NM. Cationic lipid-mediated nucleic acid delivery: beyond being cationic. Chem Phys Lipids 2010;**163**:245–252.

[114] Zhi D, Zhang S, Wang B, Zhao Y, Yang B, Yu S. Transfection efficiency of cationic lipids with different hydrophobic domains in gene delivery. Bioconjug Chem 2010;**21**: 563–577.

[115] Schroeder A, Levins CG, Cortez C, Langer R, Anderson DG. Lipid-based nanotherapeutics for siRNA delivery. J Intern Med 2010;**267**:9–21.

[116] Buyens K, De Smedt SC, Braeckmans K, Demeester J, Peeters L, van Grunsven LA, de Mollerat du Jeu X, Sawant R, Torchilin V, Farkasova K, Ogris M, Sanders N. Liposome based systems for systemic siRNA delivery: stability in blood sets the requirements for optimal carrier design. J Control Release 2012;**158**:362–370.

[117] Zhi D, Zhang S, Cui S, Zhao Y, Wang Y, Zhao D. The headgroup evolution of cationic lipids for gene delivery. Bioconjug Chem 2013;**24**:487–519.

[118] El-Andaloussi S, Lee Y, Lakhal-Littleton S, Li J, Seow Y, Gardiner C, Alvarez-Erviti L, Sargent IL, Wood MJA. Exosome-mediated delivery of siRNA in vitro and *in vivo*. Nat Protocols 2012;**7**:2112–2126.

[119] El Andaloussi S, Lakhal S, Mäger I, Wood MJA. Exosomes for targeted siRNA delivery across biological barriers. Adv Drug Deliv Rev 2013;**65**:391–7.

[120] Akhtar S. Cationic nanosystems for the delivery of small interfering ribonucleic acid therapeutics: a focus on toxicogenomics. Expert Opin Drug Metab Toxicol 2010;**6**: 1347–1362.

[121] Nimesh S, Gupta N, Chandra R. Cationic polymer based nanocarriers for delivery of therapeutic nucleic acids. J Biomed Nanotechnol 2011;**7**:504–520.

[122] Zhang K, Fang H, Shen G, Taylor JS, Wooley KL. Well-defined cationic shell crosslinked nanoparticles for efficient delivery of DNA or peptide nucleic acids. Proc Am Thorac Soc 2009;**6**:450–457.

[123] Juliano RL, Ming X, Nakagawa O. Cellular uptake and intracellular trafficking of antisense and siRNA oligonucleotides. Bioconjug Chem 2012;**23**:147–157.

[124] Canton I, Battaglia G. Endocytosis at the nanoscale. Chem Soc Rev 2012;**41**: 2718–2739.

[125] Lv H, Zhang S, Wang B, Cui S, Yan J. Toxicity of cationic lipids and cationic polymers in gene delivery. J Control Release 2006;**114**:100–109.

[126] Xu S, Olenyuk BZ, Okamoto CT, Hamm-Alvarez SF. Targeting receptor-mediated endocytotic pathways with nanoparticles: rationale and advances. Adv Drug Deliv Rev 2013;**65**:121–138.

[127] Li H, Qian ZM. Transferrin/transferrin receptor-mediated drug delivery. Med Res Rev 2002;**22**:225–250.

[128] Low PS, Kularatne SA. Folate-targeted therapeutic and imaging agents for cancer. Curr Opin Chem Biol 2009;**13**:256–262.

[129] Walkey CD, Olsen JB, Guo H, Emili A, Chan WC. Nanoparticle size and surface chemistry determine serum protein adsorption and macrophage uptake. J Am Chem Soc 2012;**134**:2139–2147.

[130] Fang H, Zhang K, Shen G, Wooley KL, Taylor JS. Cationic shell-cross-linked knedel-like (cSCK) nanoparticles for highly efficient PNA delivery. Mol Pharm 2009;**6**:615–626.

[131] Tyagi S, Alsmadi O. Imaging native β-actin mRNA in motile fibroblasts. Biophys J 2004;**87**:4153–4162.

[132] Behr JP. The proton sponge: a trick to enter cells the viruses did not exploit. Chimia 1997;**51**:34–36.

[133] Akinc A, Thomas M, Klibanov AM, Langer R. Exploring polyethylenimine-mediated DNA transfection and the proton sponge hypothesis. J Gene Med 2005;**7**:657–663.

[134] Cho YW, Kim JD, Park K. Polycation gene delivery systems: escape from endosomes to cytosol. J Pharm Pharmacol 2003;**55**:721–734.

[135] Cohen JL, Schubert S, Wich PR, Cui L, Cohen JA, Mynar JL, Frechet JM. Acid-degradable cationic dextran particles for the delivery of siRNA therapeutics. Bioconjug Chem 2011;**22**:1056–1065.

[136] Hatakeyama H, Ito E, Akita H, Oishi M, Nagasaki Y, Futaki S, Harashima H. A pH-sensitive fusogenic peptide facilitates endosomal escape and greatly enhances the gene silencing of siRNA-containing nanoparticles in vitro and *in vivo*. J Control Release 2009;**139**:127–132.

[137] Nakase I, Kobayashi S, Futaki S. Endosome-disruptive peptides for improving cytosolic delivery of bioactive macromolecules. Pept Sci 2010;**94**:763–770.

[138] Prigodich AE, Randeria PS, Briley WE, Kim NJ, Daniel WL, Giljohann DA, Mirkin CA. Multiplexed nanoflares: mRNA detection in live cells. Anal Methods 2012;**84**: 2062–2066.

[139] Prigodich AE, Seferos DS, Massich MD, Giljohann DA, Lane BC, Mirkin CA. Nanoflares for mRNA regulation and detection. ACS Nano 2009;**3**:2147–2152.

[140] Seferos DS, Giljohann DA, Hill HD, Prigodich AE, Mirkin CA. Nano-flares: probes for transfection and mRNA detection in living cells. J Am Chem Soc 2007;**129**: 15477–15479.

[141] Khlebtsov N, Dykman L. Biodistribution and toxicity of engineered gold nanoparticles: a review of in vitro and *in vivo* studies. Chem Soc Rev 2011;**40**:1647–1671.

[142] Kim JK, Choi KJ, Lee M, Jo MH, Kim S. Molecular imaging of a cancer-targeting theragnostics probe using a nucleolin aptamer- and microRNA-221 molecular beacon-conjugated nanoparticle. Biomaterials 2012;**33**:207–217.

[143] Kotula JW, Pratico ED, Ming X, Nakagawa O, Juliano RL, Sullenger BA. Aptamer-mediated delivery of splice-switching oligonucleotides to the nuclei of cancer cells. Nucleic Acid Ther 2012;**22**:187–195.

[144] Reyes-Reyes EM, Teng Y, Bates PJ. A new paradigm for aptamer therapeutic AS1411 action: uptake by macropinocytosis and its stimulation by a nucleolin-dependent mechanism. Cancer Res. 2010;**70**:8617–8629.

[145] Wang Z, Zhang K, Wooley KL, Taylor JS. Imaging mRNA expression in live cells via PNA. DNA strand displacement-activated probes. J Nucleic Acids 2012;**2012**:962652.

[146] Zhang K, Fang H, Wang Z, Taylor JS, Wooley KL. Cationic shell-crosslinked knedel-like nanoparticles for highly efficient gene and oligonucleotide transfection of mammalian cells. Biomaterials 2009;**30**:968–977.

[147] Shen Y, Shrestha R, Ibricevic A, Gunsten SP, Welch MJ, Wooley KL, Brody SL, Taylor J-SA, Liu Y. Antisense peptide nucleic acid-functionalized cationic nanocomplex for *in vivo* mRNA detection. Interface Focus 2013;**3**:20120059.

14

THE APPLICATION OF PLANT VIRAL NANOPARTICLES IN TISSUE-SPECIFIC IMAGING*

AMY M. WEN

Department of Biomedical Engineering, Case Western Reserve University, Cleveland, OH, USA

CHOI-FONG CHO

Harvey Cushing Neuro-Oncology Laboratories, Brigham and Women's Hospital, Harvard Medical School, Boston, MA, USA

JOHN D. LEWIS

Translational Prostate Cancer Research Group, Department of Oncology, University of Alberta, Edmonton, Alberta, Canada

NICOLE F. STEINMETZ

Departments of Biomedical Engineering, Radiology, Materials Science and Engineering, and Macromolecular Science and Engineering, Case Western Reserve University, Cleveland, OH, USA

Department of Macromolecular Science and Engineering, Case Western Reserve University, Cleveland, OH, USA

14.1 INTRODUCTION

The application of nanotechnology in medicine has led to advances in early detection, imaging, and treatment of disease. The ability to exploit nanoparticles to diagnose, treat, and monitor disease progression without surgery, biopsy, or other invasive methodologies has the potential to increase survival, improve quality of life, and reduce costs. Advancements in bioconjugation and self-assembly have led to the generation

*Amy M. Wen and Choi-Fong Cho contributed equally.

Nanotechnology for Biomedical Imaging and Diagnostics: From Nanoparticle Design to Clinical Applications, First Edition. Edited by Mikhail Y. Berezin.

of "smart" targeted nanoparticles, which combine targeting molecules with therapeutic and/or imaging agents all within one entity. These targeted nanoparticles hold promise for delivery of imaging agents and drugs to disease sites, resulting in the production of high contrast images with reduced noise levels as well as increased efficacy of therapy with reduced adverse side effects. Nanoparticles hold several key advantageous features: (i) multivalency, which allows for the integration of multiple targeting, imaging, and therapeutic agents within one platform; (ii) large surface area for functionalization with targeting ligands and extensive cargo volume for loading with imaging agents and therapeutic drugs; and (iii) suitable size and surface properties that enable them to have a prolonged blood circulation time and reduced uptake into the reticuloendothelial system (RES) (reviewed in Refs. [1–3]).

Several nanoparticles have been developed from various naturally occurring organic (lipids [4], proteins [5], polymers [6], etc.) and inorganic (gold [7, 8], silica [9], iron [10, 11], see chapter 3) building blocks. The inherent nature of the building blocks utilized can significantly (see also chapter 2) impact the physiochemical properties and function of the nanoparticles, offering investigators numerous choices to serve their specific purpose [12]. More than 25 nanoscale platforms have been approved for clinical use, including PEGylated liposomes containing doxorubicin (Doxil/Caelyx), albumin-bound nanoparticles of paclitaxel (Abraxane), and poly(lactic-co-glycolic acid) (PLGA) nanoparticles for drug delivery; superparamagnetic iron oxide nanoparticle contrast agents for magnetic resonance imaging (MRI) (GastroMARK); and virus-based vaccines (Gardasil) (reviewed in Refs. [13–15]). Although inorganic nanoparticles offer several advantages, such as wide availability, controlled shape and size, and easy surface manipulation, they generally have poor stability under aqueous conditions and low cellular transfer efficiency [12, 16]. Lipid-based nanoparticles such as liposomes and micelles can carry significant drug payloads, yet they suffer from poor *in vivo* stability [16, 17]. Nanomaterials comprised of plant viruses have qualities that address many of these limitations. Viral nanoparticles (VNPs) come in many shapes and sizes. Because the function of the viral capsid is to protect its genome under varying environmental conditions, they are naturally stable and monodisperse. VNPs combine the strengths of inorganic and lipid-based materials; essentially, they can be regarded as a fancy polymer assembled with atomic precision.

VNPs derived from plants or bacteria provide an ideal basis for the generation of targeted imaging agents and drug delivery vehicles (reviewed in Refs. [15, 18]). Viruslike particles (VLPs) are a subset of VNPs that can be produced in a heterologous expression systems at a large scale but lack any replicative genomic information. Thus, VLPs of eukaryotic viruses are safe, noninfectious, and nonhazardous for use in humans and other mammals (reviewed in Ref. [19]). VNPs and VLPs are self-assembling systems that are highly symmetrical, dynamic, polyvalent, and monodisperse, rendering them one of the most advanced nanomaterials produced in nature. They offer the advantages of biocompatibility and biodegradability over synthetic inorganic nanoparticles. In addition, they are extremely robust and well characterized and can be produced in large quantities within a short period of time [18]. The structure of VNPs can be modified in several ways to allow for the loading of drugs, imaging agents, and other nanoparticles within the internal cavity, as well as chemical conjugation of targeting ligands on the external surface for tissue-specific delivery [20].

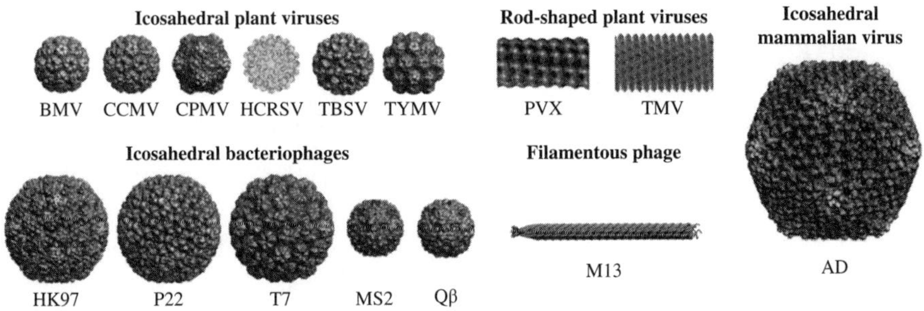

FIGURE 14.1 Viral nanoparticles (VNPs) currently developed for use in medicine. Icosahedral plant viruses: brome mosaic virus (BMV), cowpea chlorotic mottle virus (CCMV), cowpea mosaic virus (CPMV), Hibiscus chlorotic ringspot virus (HCRSV), tomato bushy stunt virus (TBSV), and turnip yellow mosaic virus (TYMV). Icosahedral bacteriophages: HK97, P22, T7, MS2, and Qβ. Icosahedral mammalian virus: adenovirus (Ad). Rod-shaped and filamentous viruses: potato virus X (PVX), tobacco mosaic virus (TMV), and bacteriophage M13. Images of BMV, CCMV, CPMV, TYMV, HK97, P22, T7, MS2, Qβ, and Ad were generated from the VIPER Database (URL: http://www.viperdb.scripps.edu/). (The structures of HCRSV, TMV, and M13 were reproduced with permission from Ref. [21]. © Elsevier. From Refs. [22] and [23]. © National Academy of Sciences).

Numerous VNPs have been developed and used in medicine (Fig. 14.1), and each platform can be tailored to serve distinct functions. Plant-based VNPs, such as brome mosaic virus (BMV), cowpea chlorotic mottle virus (CCMV), cowpea mosaic virus (CPMV), Hibiscus chlorotic ringspot virus (HCRSV), tomato bushy stunt virus (TBSV), and tobacco mosaic virus (TMV), can be produced and purified in large quantities from plants, while VNPs derived from bacteriophages, such as HK97, M13, MS2, and Qβ, can be produced in gram quantities from *Escherichia coli* bacterium cultures [19] (Fig. 14.2). VLPs derived from eukaryotic viruses can be produced from baculovirus in insect cells [24] or adenovirus in mammalian cells [25], and these particles maintain the morphology and cell-penetrating properties of the viruses from which they were derived [14].

The highly dynamic nature of VNPs allows them to undergo transitions that result in the formation of pores within the protein capsid, permitting access to their interior cavity for use as a constrained reaction vessel or storage unit [18]. Strategies to encapsulate guest molecules into the capsid's interior have been developed. For example, using a process called gating, the structure of the protein cages of CCMV and Red clover necrotic mosaic virus (RCNMV) can be changed by altering the pH and metal ion concentration to control the loading or release of cargo molecules from the interior cavity [26–29]. As VNPs are generated through the assembly of protein subunits, cargo encapsulation can also be achieved through a disassembly and reassembly process for particles such as CCMV and BMV. Self-assembly naturally occurs through electrostatic interactions between the protein subunits and RNA. Following nature's paradigm, negatively charged polymers and particles such as polystyrene sulfonate (PSS) for CCMV and citrate-coated nanoparticles for BMV can be used to initiate capsid reassembly around foreign cargo [30–32]. Some highly effective facilitators of reassembly include

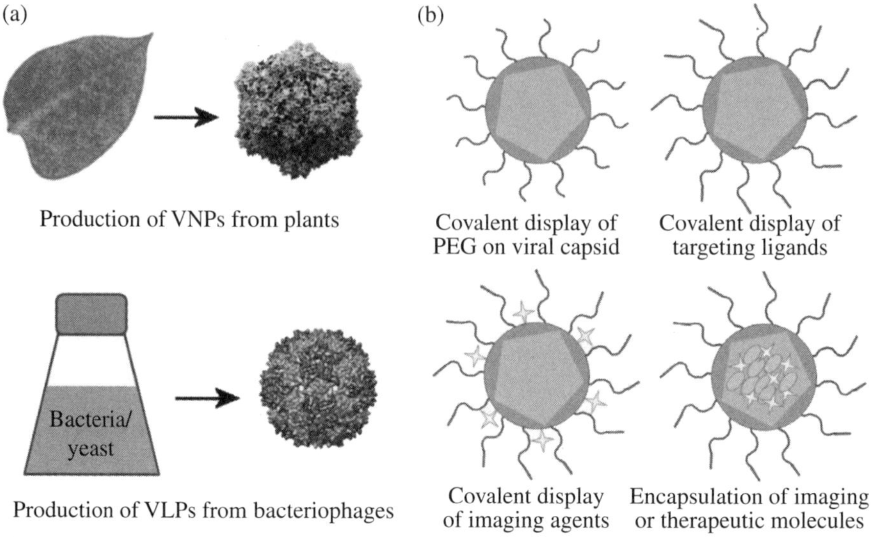

FIGURE 14.2 Production and chemical engineering of VNPs. (a) VNPs can be produced in their natural hosts, such as plants and bacteriophage. Viruslike particles (VLPs) can be produced in a heterologous expression system in bacteria and yeast. (b) After purification, VNPs can be chemically engineered and designed for conjugation or entrapment of targeting ligands, imaging moieties, and/or therapeutic molecules.

DNA and carboxylate-terminated polyethylene glycol (PEG) for the encapsulation of gold and magnetic iron oxide nanoparticles [33–35].

The structure and physicochemical properties of the protein subunits of VNPs allow them to be biochemically modified or genetically engineered for specific functionalization (Fig. 14.2). A wide variety of bio-orthogonal chemistries have been developed to allow loading with drug cargos, targeting ligands, and/or imaging moieties such as dyes and clinical contrast agents [15]. Examples of genetic modification include the insertion of amino acids onto the viral capsid to serve as ligation handles for bioconjugation, the introduction of peptide-based affinity tags for purification, and the incorporation of targeting ligands for cell-specific delivery [36, 37].

VNPs derived from plants and bacteriophages are safe for use in humans, as they are not the natural hosts for the viruses. Plant-derived CPMV, CCMV, and potato virus X (PVX) nanoparticles as well as Qβ and M13 bacteriophage particles localize in a wide variety of tissues throughout the body, with accumulation and clearance via the liver and spleen (organs of the RES) [38–42]. Despite broad biodistribution, no toxicity or other clinical symptoms have been observed after the administration of VNPs at doses of up to 100 mg/kg body weight in mice [39, 42]. A potential downside for the use of VNPs is immunogenicity of the protein coat. However, the likelihood of an immunogenic response can be reduced through the conjugation of PEG, an FDA-approved polymer that is nontoxic, neutrally charged, and strongly hydrophilic. This process, termed PEGylation, is a widely used strategy for enhancing solubility and stability of nanocarriers containing hydrophobic drugs, reducing biospecific interactions and nonspecific uptake, and increasing plasma circulation time [18, 43].

Several PEGylated VNPs, such as CPMV [44–46], PVX [40, 47], TMV [48, 49], and bacteriophage MS2 [50], have been generated. Minimal surface coverage (<1%) with PEG chains is sufficient to shield CPMV nanoparticles from interacting with cells, leaving a large surface area and numerous attachment sites available on the viral capsid for further modifications with targeting, imaging, and therapeutic agents [46]. Therapies comprised of PEGylated proteins are increasingly being utilized in the clinic, indicating that PEGylation is an accepted technology to overcome immunogenicity [51].

Besides toxicity and biodistribution, it is also essential to understand the pharmacokinetic properties of VNPs, as it is important to achieve the appropriate balance between nanoparticle accumulation at disease sites and system clearance. Although longer circulation times allow for the accumulation of drugs or imaging reagents into target tissues, the level of toxicity or background-to-noise ratio significantly increases as well. It is likely that the pharmacokinetics of VNPs tailored for drug delivery are different from those designed as imaging tools [18]. The pharmacokinetic properties of VNPs are dependent on their surface composition, such as surface charge and surface chemistry [52]. VNPs with a negative surface charge, such as CPMV and CCMV, tend to have short plasma half-lives (<15 min) [39, 42], while particles such as Qβ that have a positive surface charge have a much longer circulation time (more than 3 h) [38].

Chemical and genetic tuning resulting in altering the surface characteristics of a VNP will change its *in vivo* properties. For example, acetylating the surface lysine residues on Qβ and M13 capsid proteins neutralizes their positive charge and reduces the plasma half-lives of the particles [38, 41]. PEGylation of VNPs (as previously described) adds a hydration shell and has been shown to drastically increase circulation time [53, 54]. Other factors naturally have to be taken into account, and recent data show that circulation times cannot be generalized based on surface charge: positively charged, filamentous PVX has a charge comparable to Qβ but a circulation time similar to CPMV [40]. Model systems have been developed to rapidly determine the efficacy of VNP-based imaging agents *in vivo* [55, 56].

14.2 VNPs LABELED WITH FLUORESCENT DYES FOR OPTICAL IMAGING

Optical fluorescence imaging is one of the most widely applied imaging technologies with many applications in medicine and biological sciences. Fluorescence imaging relies on the property of certain molecules to absorb light (see chapter 9) at a specific wavelength and then emit light of a longer wavelength (reviewed in Ref. [57]). Studies to reconstruct light through tissues were facilitated by the discovery that near-infrared (NIR) photons penetrate tissue far more efficiently than those within visible range [58, 59]. Many targeted fluorescent imaging probes have since then been developed for *in vivo* imaging [60–62].

The incorporation of fluorescent dyes and proteins into modified VNPs has facilitated their characterization. For example, VNP drug formulations are typically colabeled with dyes and studied by fluorescence microscopy to evaluate cellular internalization and localization [63, 64]. Furthermore, for VNPs displaying targeted ligands, dye labeling and flow cytometry can be used together to make high-throughput measurements of specificity of uptake [37].While chemical conjugation to the protein coat is the most common technique for attaching dyes, several other methods can also be employed.

Encapsulation as described previously is one such method for forming optical constructs for intracellular imaging. It was found that dyes could be infused into RCNMV through gating by altering the pH and divalent ion concentrations to open or close the pores. The efficiency was charge dependent, with high loading of positively charged and neutral molecules but negligible loading of negatively charged dyes, likely due to their interaction with the RNA within the capsid interior [26]. Encapsulation through a reassembly process has also been demonstrated. Different types of molecular coatings, such as lipid micelles, streptavidin–biotin–DNA, dihydrolipoic acid (DHLA), and carboxylic PEG thiol, were investigated for encapsulating CdSe/ZnS quantum dots (QDs) in BMV. QDs functionalized with PEG were found to have the best reassembly efficiency and photostability [65]. Labeling BMV with fluorescent dyes through reassembly has also been demonstrated using an FDA-approved NIR fluorophore indocyanine green (ICG), which has an intrinsic negative charge to trigger its encapsulation without requiring a coating [66].

Noncovalent heterodimeric coiled-coil protein interactions can also be applied to couple fluorescent proteins with the capsid. By modifying the coiled-coil motif with glutamic acid (E-coil) or lysine (K-coil), heterodimerization from the association of the E-coil with the K-coil can be achieved. This was demonstrated using the CCMV capsid, introducing a K-coil at the N-terminus of its coat protein and an E-coil at the C-terminus of enhanced green fluorescent protein (EGFP) through genetic engineering. Complexes of the modified protein subunits and EGFP were formed *in vitro*. The EGFP–capsid protein complexes were then mixed with wild-type capsid proteins at various stoichiometric ratios to assemble CCMV capsids with good control over the amount of protein encapsulated, with up to 15 EGFP per particle obtained [67].

Preliminary dye-labeling studies such as these are crucial to the development of VNP molecular probes. Fluorescent VNPs modified with additional functionalities have been applied for visualizing biodistribution of the particles as well as for detecting and monitoring disease [15].

14.3 INTERACTION OF CPMV WITH CELLS AND ITS APPLICATION FOR INTRAVITAL IMAGING

CPMV nanoparticles have been studied extensively *in vivo* and applied for potential applications in biomedicine [18, 53, 68–70]. CPMV has been shown to recognize a variety of mammalian cells, such as murine fibroblast, PC-3, HeLa, and HT-29 cells, through their interaction with vimentin, a type III intermediate filament that is responsible for modulating the architecture and dynamics of cells [54, 71–73]. Vimentin is a cytoskeletal protein that is important for various cellular functions, including wound healing, cell adhesion, cell migration, proliferation, protein synthesis, gene expression, and signal transduction [74, 75].

Confocal microscopy revealed that the level of CPMV uptake by HeLa, HT-29, and MDA-MB-231 cancer cells correlated with the level of surface vimentin expression *in vitro* [54]. HeLa and HT-29 cells that are positive for surface vimentin staining showed detectable levels of CPMV internalization, while MDA-MB-231 cells with very low levels of surface vimentin displayed no detectable CPMV uptake (Fig. 14.3a). Similar CPMV targeting was observed in surface vimentin-positive prostate cancer PC-3 and LNCaP cell lines [73]. Surface vimentin is upregulated in many cancers and inflamed cells and has also been shown to mediate internalization of CPMV by endothelial cells. Immunohistochemical staining of rat aortic endothelium *ex vivo* revealed strong colocalization between surface vimentin expression and CPMV binding (Fig. 14.3b) [71]. Thus, CPMV is a particularly useful tool for imaging tumor neovasculature and vascular disease. The binding of CPMV to surface vimentin facilitates targeting of CPMV to areas of disease, which can be applied for tissue-specific drug delivery [64] and imaging [53].

Detection of atherosclerosis is one such application for CPMV. Compared to normal, healthy vasculature, there is an elevation in the expression of surface vimentin in atherosclerotic lesions. Using a LDL-receptor knockout mouse model of atherosclerosis, it was found that CPMV uptake increased at sites of inflammation within the lesion. Surface vimentin was detected via immunostaining of aorta tissue sections followed by confocal imaging. An increase of uptake was detected in lesion areas within 12 weeks of placing the mice on a high-fat diet. Correspondingly, fluorescence levels from dye-labeled CPMV injected intraorbitally increased by more than two-fold in these areas of lesion compared to nonlesion tissue. The confocal images were analyzed, and the areas within the endothelium that are positive for CPMV were determined. A 27-fold increase in the percent area where CPMV was detected was observed after 12 weeks, and a 32-fold increase was observed at 20 weeks. This data suggests that the interaction of CPMV with surface vimentin can allow it to be used as an imaging probe for early detection of atherosclerotic lesions where surface vimentin is upregulated [76].

The natural interaction of CPMV with vimentin can also be utilized for intravital imaging of vascular development in tumors. Fluorescently labeled CPMV nanoparticles have been used to image the vasculature of a chicken chorioallantoic membrane (CAM) tumor model and mouse embryos. In the CAM model, HT1080 fibrosarcoma tumor xenografts were implanted in the CAM of shell-less chicken embryos. Dyes were conjugated onto the surface of CPMV particles via their lysine residues, and the CPMV was then injected intravenously and imaged live under a fluorescence microscope. The particles were rapidly taken up by endothelial cells *in vivo* in both models (Fig. 14.3c–h). CPMV was observed to line the endothelial wall, leading to high-resolution intravital imaging of the vasculature for at least 72 h, enabling high-resolution vascular mapping studies to be conducted (Fig. 14.3d). Tumor-mediated angiogenesis can be examined by injecting CPMV labeled with a different dye after a period of 24 h, allowing one to image and distinguish the newly formed blood vessels. Visualization could be performed up to a depth of 500 μm for both microvasculature and larger vessels. Attempts to obtain the same resolution with QD nanoparticles were unsuccessful, as they tended to aggregate and obstruct

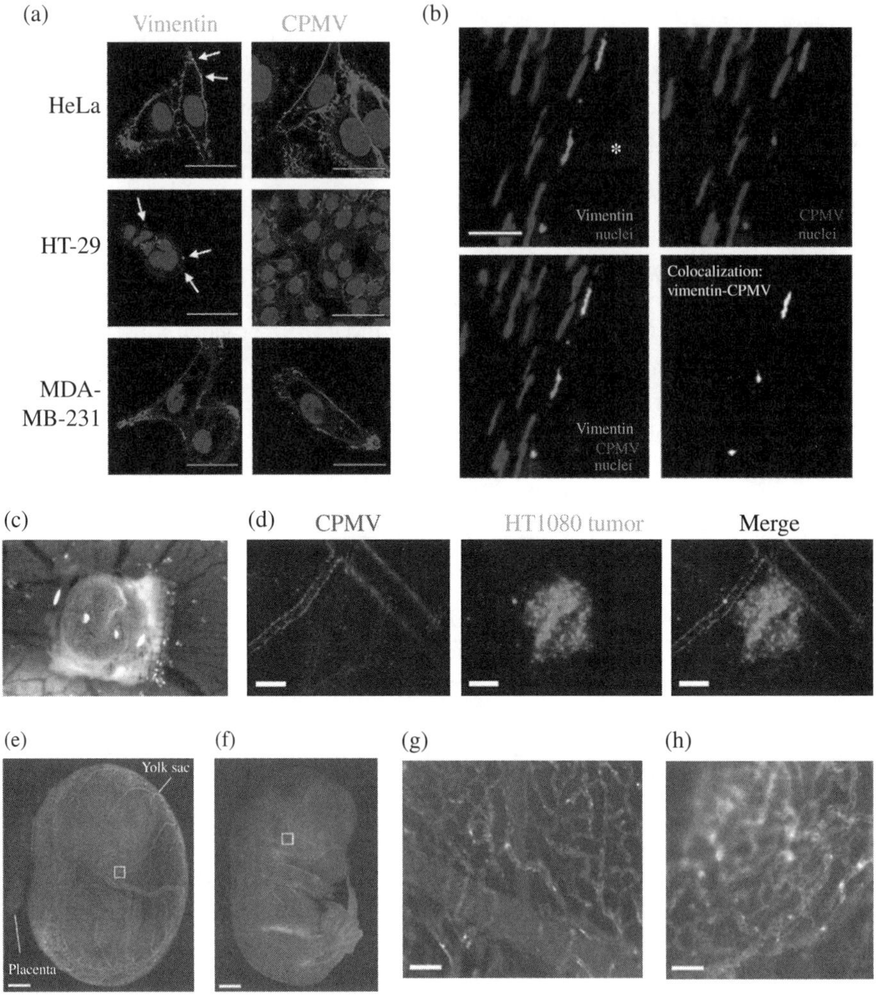

FIGURE 14.3 CPMV has natural affinity for cells through interaction with surface vimentin. (a) Left panel: surface vimentin expression of HeLa, HT-29, and MDA-MB-231 cells (indicated by arrows). Live cell staining was performed using the monoclonal antivimentin antibody v9. Right panel: CPMV-Oregon Green 488 (O488) nanoparticles uptake by HeLa, HT-29, and MDA-MB-231 cells. Scale bar is 30 mm. (Reproduced with permission from Ref. [54]. © Wiley.). (b) Colocalization of CPMV and surface vimentin on rat aortic endothelium. Scale bar is 30 mm, and * indicates aorta lumen. (Reproduced from Ref. [71].) (c) Bright-field image of HT1080 tumor CAM model after 7 days. Nylon mesh grid is used for angiogenesis quantification. (d) Image of CPMV-A555 lining the vasculature taken right after HT1080 tumor xenograft was performed (left), GFP-expressing HT1080 tumor (middle), and merged images (right). Scale bar is 100 μm. (e, f) Image of 11.5-day chick embryo injected with CPMV-A555 with yolk sac intact (e) and removed (f). White boxes are of magnified regions shown in (g) and (h). Scale bar is 1.1 mm. (g, h) Magnified images of yolk sac vasculature and capillaries in the head region, respectively. Scale bar is 25 μm. (Figure (c–h) were reprinted with permission from Ref. [53]. © Macmillan Publishers, Ltd.) (*See insert for color representation of the figure.*)

capillaries and arterioles [53]. Thus, CPMV is truly a valuable tool for intravital imaging of tumor neovasculature.

Utilizing its endogenous interactions and biocompatibility, CPMV holds promise as a tool for imaging vascular disease as well as tumors that express high levels of vimentin. Some recent data suggest that CPMV could also be naturally targeted to regions of inflammation in the central nervous system (CNS) [77]. This platform could be optimized through the use of NIR dyes and multiphoton fluorescence imaging to enhance the depth and sensitivity of visualization or paired with other imaging agents, such as radiotracers.

14.4 PASSIVE TARGETING OF TUMORS USING FLUORESCENT VNPs

Systemic nanoparticle drug delivery to solid tumors is largely governed by the enhanced permeability and retention (EPR) effect, where leaky vasculature and inefficient lymphatic drainage resulting from tumor hypervascularization lead to the passive diffusion of nanocarriers to the site of disease [78]. EPR-mediated tumor homing has been demonstrated using VNPs such as CPMV and PVX [40, 69, 79]. These VNP formulations were PEGylated to prevent nonspecific cell interactions and prolong plasma circulation time. In one study, CPMV nanoparticles were labeled either internally or externally with NIR fluorophores and compared side by side in a mouse model of colon cancer with HT-29 tumor xenografts. Twenty-four hours after intravenous injection, the tumors were excised and imaged *ex vivo*. Tumor homing was observed for both formulations, with higher fluorescence intensity observed for internally labeled CPMV particles [79]. The results show that CPMV is suitable for fluorescent imaging and can be further developed as a multifunctional formulation, as interior dye loading leaves the exterior surface free for additional modification with targeting ligands and other imaging agents.

Although most nanoparticulate systems currently under development are spherical in shape, there is growing interest in rod-shaped particles as recent data indicate that they have advantageous pharmacokinetic and tumor homing properties [80, 81]. It was recently shown that a similar trend also holds for VNPs. Tumor homing properties of flexible PVX filaments and icosahedral CPMV particles were compared in both the CAM and mouse tumor xenograft models. Particles tagged with red or NIR dyes were injected intravenously and observed using a fluorescence intravital imaging platform. There was no significant difference in tumor accumulation in the CAM model, but enhanced accumulation of PVX was found using a nude mouse HT-29 tumor xenograft model (Fig. 14.4). Although the CAM model may not accurately reflect the EPR effect as a result of differences in systemic clearance in the embryonic model, it can nevertheless provide insight on how differences in shape affect tumor penetration behavior. The filamentous PVX particles were able to penetrate into the tumor core, while there was no noticeable increase of CPMV in the tumor center (Fig. 14.4a). Furthermore, confocal images of tumor sections from both models showed PVX spread throughout the tumor and was detected in areas where CPMV was not apparent [40]. The superior tumor homing and tissue penetration properties of PVX and perhaps other rod-shaped VNPs can be used to advance cancer nanotechnology.

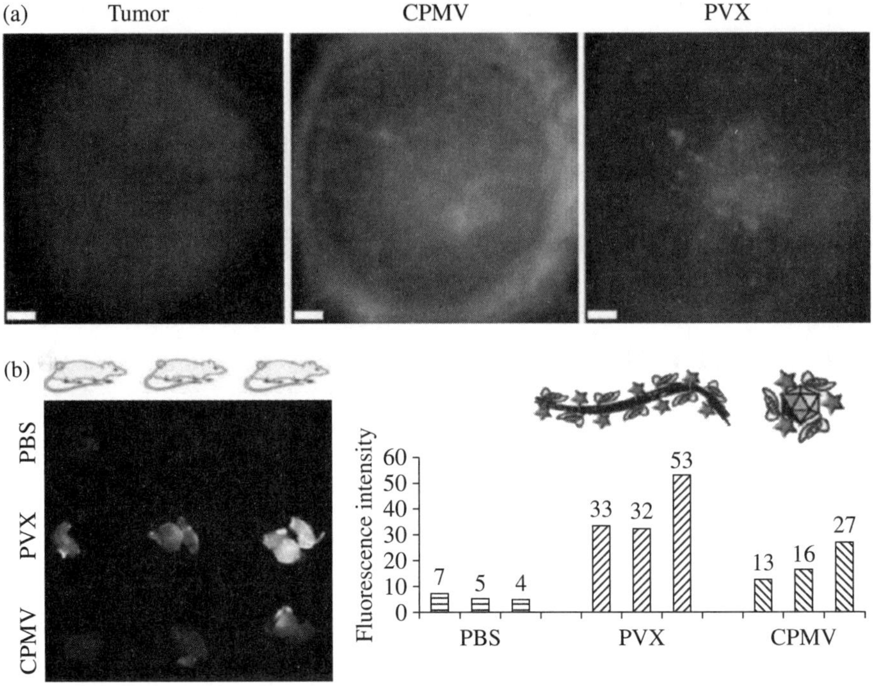

FIGURE 14.4 Images of tumor homing of PVX versus CPMV in human tumor xenograft models. (a) Intravital imaging of the CAM model. Chick embryos with human epithelial carcinoma HEp3 tumors (magenta) were coinjected with equal numbers of particles of Alexa Fluor 555-labeled PVX (PVX-A555) and Alexa Fluor 647-labeled CPMV (CPMV-A647). Imaging was performed 4 hours after injection. Scale bar is 190 μm. (b) Fluorescent images of HT-29 tumors excised from mouse tumor xenograft model. PVX-A647 and CPMV-A647 were injected by tail vein, and tumors were excised and imaged 24 h after injection (left). Fluorescence intensities of the tumors were then quantified (right). (Reprinted with permission from Ref. [40]. © American Chemical Society.)

14.5 MOLECULAR TARGETED FLUORESCENCE IMAGING

Several VNP formulations have been designed as active cancer-targeting optical imaging agents. A number of advances have been made to link targeting moieties to deliver VNPs containing imaging/contrast agents to specific disease sites while avoiding healthy tissues. Targeted imaging increases the signal-to-background ratio to allow more accurate diagnosis. VNPs provide a valuable platform for the development of imaging tools due to their high versatility for chemical bioconjugation. Combinatorial library approaches, including phage display and one-bead–one-compound (OBOC) libraries, have pioneered the discovery of tumor-specific ligands and vascular homing peptides to facilitate and advance the development of targeted VNPs [82–87].

The following sections highlight just a few of many examples of the utility of VNPs displaying specific ligands for targeted imaging. One method for cell targeting is the installation of nucleic acid aptamers on the viral capsid. Using the systematic evolution of ligands by exponential enrichment (SELEX) process, aptamer sequences can be developed that can bind to any chosen target with high specificity. Cell-specific imaging was demonstrated using MS2 mutants with *p*-aminophenylalanine (*p*aF) and cysteine amino acids inserted in their coat proteins on their exterior and interior surface, respectively. *p*aF allows for an efficient oxidative coupling strategy to be employed for aptamer attachment onto the capsid surface, while cysteine permits interior labeling through maleimide chemistry. As a preliminary proof of concept, a 41-nucleotide sequence that targets tyrosine kinase receptors on Jurkat T cells was substituted with phenylenediamine and coupled with the aniline groups of *p*aF in the presence of sodium periodate ($NaIO_4$), while a fluorescent dye was installed within the capsid. The particles were observed using confocal microscopy, and receptor-mediated internalization was confirmed, demonstrating that the installation of aptamers on VNPs is a promising method for the localization of specific cellular markers and cell imaging [88].

Other recent studies focused on engineering VNP formulations targeted to biomarkers that are preferentially expressed in tumors. The folate receptor (FR) is one such marker that is upregulated on cancer cells as a result of greater demand for folic acid (FA) during proliferation. FA is a vitamin essential for DNA replication, cell division, and growth, and FA uptake is mediated by binding to FR [89]. Conjugation of FA to VNPs was first investigated using CPMV and evaluated with KB and HeLa cancer cell lines. KB is a nasopharyngeal cancer cell line with high expression of FR, while HeLa is a cervical cancer cell line with reduced FR expression. Cells were incubated with the CPMV particles at 4°C to minimize endocytosis and monitored by flow cytometry for particle binding after permeabilization and treatment with anti-CPMV primary and fluorescent secondary antibodies. Direct conjugation of FA to surface lysines did not result in any improvement in cellular binding, which may be due to FA being inaccessible so close to the capsid or possibly native CPMV–cell interactions. Presentation of FA using a PEG linker resolved the difficulty and resulted in up to 94% binding to KB cells and a modest 22% binding to HeLa cells, while control formulations with just PEG showed negligible binding. A competition-binding assay using free FA demonstrated that binding of CPMV-FA was disrupted in the presence of free FA, indicating lack of a multivalent avidity effect. Regardless, conjugation of FA showed significant improvement in cell binding compared to FA alone, thus making CPMV-PEG-FA an attractive targeted design for cell-specific imaging [90].

Another possibility for a targeting ligand that could be displayed on VNPs is transferrin (Tfn), an 80 kDa iron-binding glycoprotein. Tfn receptors (TfnRs) are expressed at high levels on a variety of tumor cells and serve to transport Tfn loaded with iron into the cell through clathrin-mediated endocytosis [91]. Functionalization of Tfn for conjugation to VNPs can be accomplished through oxidation of its sialic acid residues. This method was used to make bacteriophage HK97 labeled with Tfn, and uptake of HK97-Tfn in tumor cells was compared with that of native HK97 using flow cytometry and anti-HK97 antibody staining. The nontargeted particles did not interact with the cells, possibly as a result of its strong negative surface charge. On the other hand, HK97-Tfn showed high levels of internalization in both cell lines tested, HeLa and HT-29. Internalization was shown to be TfnR-mediated using a competition-binding assay with free Tfn. Receptor-mediated endocytosis was further confirmed with visualization by confocal microscopy of HK97 dual-labeled with Tfn and fluorescent dye Oregon Green 488 (O488). Intracellular colocalization with LAMP-1 was observed, indicating that HK97-Tfn-O488 is transported into the endolysosomes following endocytosis [92]. This finding is in agreement with another investigation using a different Tfn-labeled bacteriophage, Qβ. Qβ-Tfn conjugates and native Qβ were labeled with a fluorescent dye, Alexa Fluor 568, and the particles were incubated with African green monkey kidney epithelial cells (BSC1) that express EGFP-labeled clathrin light chains (Clc) to study clathrin-mediated endocytosis. Incubation at 4°C followed by 37°C was used to determine binding of the particles to the cells and internalization, respectively. No binding or internalization was found for Qβ, while significant binding and uptake were found for Qβ-Tfn. The Qβ-Tfn colocalized with the clathrin-coated pits, and there was no binding or uptake observed when excess free Tfn was present, indicating that the internalization occurred through a TfnR-mediated endocytosis. This was further confirmed using total internal reflection fluorescence (TIRF) live cell microscopy, where the bottom surface of the cell membrane was imaged and the internalization of Qβ-Tfn through clathrin-coated pits was observed. Different ligand densities of Tfn were compared, and although the rate of uptake increased with more Tfn molecules loaded, there was not a significant increase in affinity on a per-Tfn basis. Nevertheless, on a per-particle basis, the Qβ-Tfn conjugates have greater affinities than free Tfn, which could be advantageous for targeted imaging of cells and delivery of therapeutics [93].

Although targeted platforms have been demonstrated mainly *in vitro*, they have also recently been validated *in vivo*. F56 is a synthetic peptide discovered through phage display that binds with high affinity to vascular endothelial growth factor receptor-1 (VEGFR-1) and as a result significantly hinders tumor angiogenesis and metastasis [94]. Multifunctional particles were made by attaching fluorescent PEG–fluorescein and F56–fluorescein to CPMV. Fibroblasts that do not express VEGFR-1 and human endothelial cells with high VEGFR-1 expression were treated with CPMV-PEG and CPMV-F56 conjugates and evaluated by confocal fluorescent imaging. The data demonstrated that the F56 peptide successfully targets the CPMV particles to endothelial cells, with no nonspecific binding to the fibroblast cells and no binding observed for CPMV-PEG. Additional studies *ex vivo* and *in vivo* were then performed with HT-29 colon cancer cells, which express high levels of VEGFR-1. Confocal microscopy was used to investigate binding of the particles to HT-29 tumor sections *ex vivo*, and there was significant binding to the tumor sections with F56 conjugates but not with PEG. Similar observations were found *in vivo* after intravenous injection of the particles in nude mice bearing HT-29 tumor xenografts. The tumors were sectioned and confocal microscopy was performed. No cellular uptake was found for CPMV-PEG, but accumulation throughout the tumor was observed for CPMV-F56 particles. Overall, confocal microscopy established that targeting with F56 is effective *in vitro*, *ex vivo*, and *in vivo*, although alternate dyes with better tissue penetration would be required for imaging *in vivo* [95].

Intravital imaging was also demonstrated using a targeted formulation incorporating the bombesin peptide. Bombesin targets the gastrin-releasing peptide receptor (GRPR), which is one of the most overexpressed receptors in human cancers, including prostate cancer [96]. CPMV nanoparticles functionalized with either PEG or PEG–bombesin and a NIR dye, Alexa Fluor 647, were utilized to analyze targeting specificity to PC-3 prostate cancer cells. *In vitro* studies using fluorescence confocal microscopy imaging showed that CPMV-PEG did not bind to the PC-3 cells at a noticeable level, while there was significant binding of CPMV–PEG–bombesin. A competition-binding assay using a 10-fold excess of free bombesin significantly decreased the uptake of CPMV–PEG–bombesin, demonstrating that internalization is due to bombesin–GRPR interactions. Targeting specificity was also evaluated using flow cytometry, and as expected, CPMV–PEG–bombesin had significantly higher uptake compared to CPMV-PEG. Nonspecific uptake of CPMV-PEG was somewhat elevated, most likely the result of insufficient PEG coverage to completely shield the CPMV–vimentin interaction. Targeting was also examined *in vivo* using the CAM model implanted with GFP-expressing PC-3 cells and intravital imaging. After intravenous injection, CPMV-PEG was quickly cleared from the tumor, while CPMV–PEG–bombesin accumulated over a period of 6h, with low background in the surrounding vasculature (Fig. 14.5). These observations indicate that CPMV–PEG–bombesin nanoparticles would be valuable for molecular imaging of GRPR-overexpressing tumors [69].

Targeted VNP formulations for molecular imaging have recently been described, and it is clear that there are many targeting ligands that can be incorporated for customized detection of a wide range of diseases. VNPs are fairly simple to modify, with several robust bio-orthogonal chemistries established for functionalization. As imaging techniques continue to improve, new fluorescent dyes developed, and targeting ligands discovered, targeted VNPs for *in vivo* imaging and diagnostics will continue to progress.

14.6 VNPs AS MRI CONTRAST AGENTS

MRI is used regularly as a noninvasive diagnostic tool, with the advantage of not requiring the use of ionizing radiation. MRI works by aligning protons of water molecules in a magnetic field and then exciting the protons with a pulse of radio-frequency (RF) waves applied perpendicular to the field. After the RF pulse, the protons "relax" and realign themselves with the magnetic field. The energy that is released during this relaxation process is detected and used to generate the MRI image. The return of the protons to equilibrium is typically characterized by two relaxation times, T_1 and T_2, where T_1 is the time constant describing the realignment of the protons with the magnetic field and T_2 is the time constant related to the loss of phase coherence between the protons that was instigated by the RF pulse (reviewed in Ref. [97]).

Contrast agents are typically used to shorten the relaxation time of nearby water protons in locations of disease, appearing bright T_1-weighted images and dark T_2-weighted images. In this manner, the difference in the area of disease is enhanced and

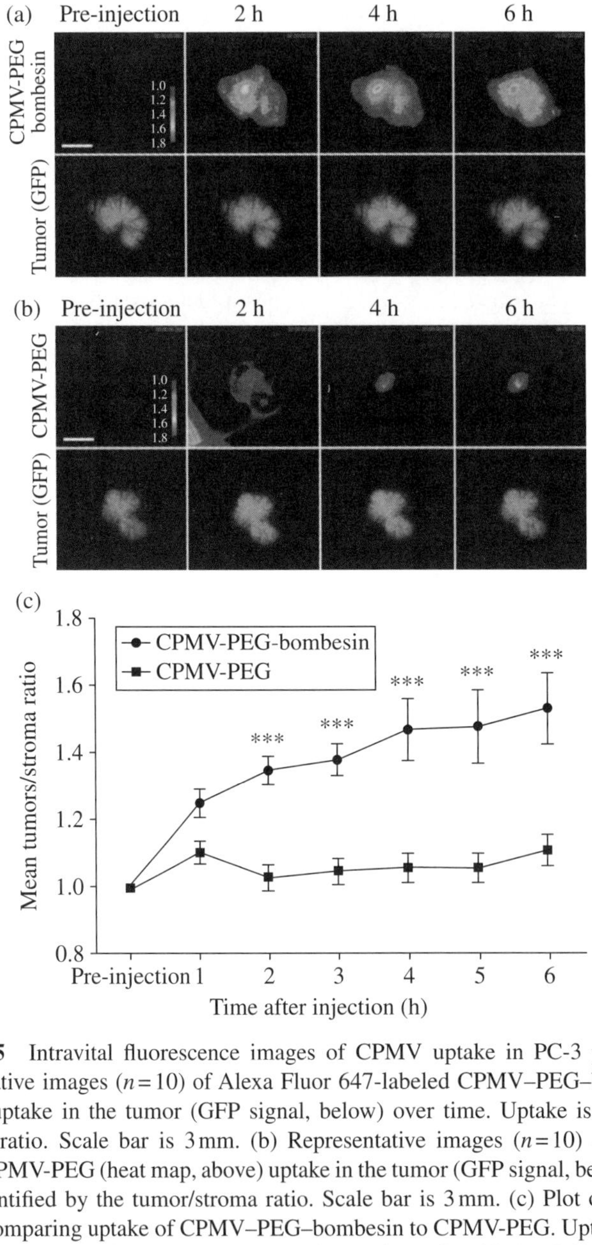

FIGURE 14.5 Intravital fluorescence images of CPMV uptake in PC-3 prostate tumor. (a) Representative images ($n = 10$) of Alexa Fluor 647-labeled CPMV–PEG–bombesin (heat map, above) uptake in the tumor (GFP signal, below) over time. Uptake is represented as tumor/stroma ratio. Scale bar is 3 mm. (b) Representative images ($n = 10$) of Alexa Fluor 647-labeled CPMV-PEG (heat map, above) uptake in the tumor (GFP signal, below) over time. Uptake is quantified by the tumor/stroma ratio. Scale bar is 3 mm. (c) Plot of mean tumor/stroma ratio comparing uptake of CPMV–PEG–bombesin to CPMV-PEG. Uptake of CPMV–PEG–bombesin was significantly higher at all time points after 2 h ($p < 0.0001$). (Reprinted with permission from Ref. [69]. © Elsevier.)

easier to distinguish. The ability to detect MRI contrast agents is extremely low, so there is interest in coordinating multiple Gd^{3+} ions to macromolecules in order to enhance the relaxivity and hence sensitivity. VNPs complexed with MRI contrast agents offer two advantages: (1) due to their multivalency (see chapter 8), a large payload of gadolinium in a small volume (high concentration) can be delivered in a tissue-specific manner; and (2) the size of a large macromolecular carrier slows down its tumbling rate, resulting in long correlation times and increased relaxivity.

Early work on VNP MRI contrast agents was based on CCMV. Interest in this platform came from earlier demonstration that CCMV has 180 metal binding sites in the interface between its subunits. These binding sites normally bind to calcium *in vivo* but were shown to have two orders of magnitude greater affinity for binding to terbium (Tb^{3+}) [27]. Since Gd^{3+} is also a lanthanide with a similar size and charge distribution as Tb^{3+}, it was hypothesized that CCMV could be used to coordinate Gd^{3+} and the Gd^{3+}-CCMV complex used as an MRI contrast agent. Indeed, the binding affinity to Gd^{3+} was found to be similar to Tb^{3+}, and the presence of 1.5 mM Ca^{2+} had very little effect, indicating that the metal binding is stable. The T_1 and T_2 relaxivities were measured to be around 250 and 350 mM^{-1} s^{-1} for Gd^{3+}-CCMV at a frequency of 20 MHz, much higher than the 20 mM^{-1} s^{-1} measured for the $GdCl_3$ control, 5 times higher than what was reported for Gd^{3+}–albumin [98], and 10 times higher than for Gd^{3+}–dendrimers [99, 100]. The properties of Gd^{3+}-CCMV still have to be explored *in vivo* to evaluate stability and possible immunogenicity. However, these initial results are very encouraging, showing that the size of these nanoparticles and large number of bound Gd^{3+} are ideal for greatly enhancing the relaxivity [101].

Since this initial study, there have been additional reports of paramagnetic VNP formulations that would also be promising candidates for applications in MRI. A popular platform that has been explored is the MS2 capsid. Although the direct binding of $Gd^{3+}$to CCMV demonstrated the power of VNP contrast agents, stronger chelates of Gd^{3+} is desired for reducing possible toxicity from free Gd^{3+}ions for *in vivo* applications. MS2 is a 27 nm icosahedral particle comprised of 180 identical coat proteins, with three lysine residues per coat protein available for modification on its exterior surface. The first account of MS2-based contrast agents utilized the 540 lysines to covalently attach an isothiocyanate (ITC) derivative of Magnevist; a Gd–diethylenetriaminepentaacetic acid (DTPA) MRI contrast agent is widely used in the clinic for the chelation of gadolinium. Using a 1.5 T clinical scanner (64 MHz, protons have a gyromagnetic ratio of 42.58 MHz/*T*), the resultant MS2(Gd-DTPA) contrast agents showed a three-fold increase in the T_1 relaxivity, with a relaxivity of 14.0–16.9 mM^{-1} s^{-1} per Gd^{3+}compared to 5.2 mM^{-1} s^{-1}for just Gd-DTPA (Fig. 14.6). Although not as high as what was found for direct attachment of Gd^{3+} to CCMV, it is expected for T_1 relaxivities to be lower at higher field strengths. Nevertheless, the relaxivity corresponds to up to 7200 mM^{-1} s^{-1} per MS2 particle, over 1000 times greater than any currently approved MRI contrast agent. In addition to demonstrating the potential use of MS2 for MRI, these particles could additionally be modified with fluorescein isothiocyanate (FITC) for optical imaging [102].

Further evaluation of MS2 looked at comparing the difference in relaxivity enhancement when Gd^{3+} is internally *versus* externally conjugated to an empty MS2 capsid. In addition to the three lysines on the exterior, MS2 also displays a tyrosine per coat protein on its interior surface and contains 1.8 nm pores that allow molecules to diffuse into the capsid. By means of these respective amino acid residues, orthogonal bioconjugation was used to functionalize MS2 with aldehyde

(a)

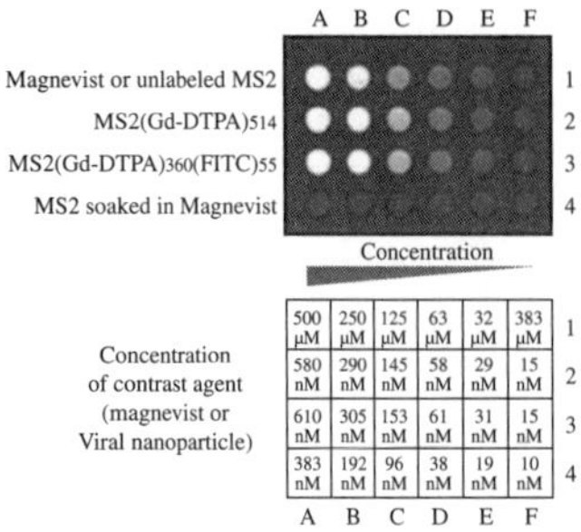

(b)

Contrast agent	Size (nm)	Ionic relaxivity ($mM^{-1}s^{-1}$) 1.5 T	9.4 T	Relaxivity per particle ($mM^{-1}s^{-1}$) 1.5 T	9.4 T
Magnevist	ND	5.2	5.2	5.2	5.2
MS2 (unlabeled)	26.9	None	None	None	None
MS2 soaked with Magnevist	28.4	None	None	None	None
MS2(Gd-DTPA-ITC)514	28.9	14.0	3.4	7200	1782
MS2(Gd-DTPA-ITC)360-(FITC)55	29.1	16.9	4.3	6100	1525

FIGURE 14.6 (a) T_1-weighted image of MS2-based contrast agents taken with a 1.5 T clinical scanner. Row 1, Magnevist as a positive control (A–E) and unlabeled MS2 as a negative control (F). Row 2, MS2(Gd-DTPA). Row 3, MS2(Gd-DTPA)(FITC). Row 4, MS2 soaked in Magnevist without covalent attachment. (b) Measurement of relaxivities. ND, not determined. None, no enhancement of relaxivity over buffer. (Reprinted with permission from Ref. [102]. © American Chemical Society.)

groups on either surface followed by oxime condensation to attach approximately 90 bis(hydroxypyridonate)terephthalamide (bis(HOPO)-TAM) ligands. These ligands were chosen for gadolinium chelation because contrast agents based on HOPO have been shown to have three-fold greater relaxivities compared to commercially available agents. It was found that internally modified MS2 conjugates had higher relaxivities than externally modified capsids, with relaxivities of 41.6 $mM^{-1}s^{-1}$ per Gd^{3+} ion or 3900 $mM^{-1}s^{-1}$ per particle compared to 30.7 $mM^{-1}s^{-1}$ per ion and 2500 $mM^{-1}s^{-1}$ per particle at 30 MHz. It was also found that internally functionalized particles was more soluble and had greater capsid stability [103]. Continuation of this work explored the parameters that affect relaxivity using nuclear magnetic resonance dispersion (NMRD). It was demonstrated that the higher relaxivity of the internally modified capsids was the result of easy accessibility of water through diffusion to the interior surface and the rigidity of the aromatic tyrosine linkers compared to the flexible lysines [104]. These results are highly encouraging, as internal attachment would diminish the potential interaction of gadolinium *in vivo* as well as allow the possibility of tissue-specific imaging through modification of exterior groups with targeting ligands.

For the intelligent design of more advanced MRI contrast agents, the effect of linker rigidity on relaxivity was further explored. HOPO-based gadolinium ligands were attached to genetically introduced cysteine residues on the interior of MS2 particles using flexible linear and rigid cyclohexyl linkers. The MS2 cysteine mutant was chosen due to a two-fold increase in bioconjugation efficiency over the previously used tyrosine residues. In addition, two enantiomers of the rigid *trans*-1,2-cyclohexyldiamine linker (*S,S* and *R,R*) were tested to investigate whether a difference in relaxivity would result from interaction with the chiral capsid. The design concept was confirmed, with some of the highest relaxivities reported for Gd^{3+} complexes observed. MS2-HOPO-*S,S* had an ionic relaxivity of 41.2 $mM^{-1}s^{-1}$ at 60 MHz or 7416 $mM^{-1}s^{-1}$ per particle, which was greater than that found for

MS2-HOPO-*R*,*R* of 29.6 $mM^{-1} s^{-1}$ and MS2-HOPO-Lin of 32.6 $mM^{-1} s^{-1}$ per Gd^{3+}. The relaxivity was thus improved by using a rigid linker, with linker chirality also playing a factor. The validation of the design concept guides the formulation of contrast agents with even greater sensitivity for future applications in targeting areas requiring greater enhancement [105].

Another contribution to the VNP contrast agent toolbox explored the use of CPMV and Qβ protein cages. These VNPs are similar to CCMV and MS2, also icosahedrons approximately 28 nm in diameter. The particles were modified using an azide linker conjugated to the lysines available on the exterior of both. Then, attachment of an alkyne derivative of Gd-DOTA (tetraazacyclododecane tetraacetic acid), a clinically approved gadolinium complex, was accomplished via Cu(I)-mediated azide–alkyne cycloaddition (CuAAC). Around 220 and 150 Gd-DOTA complexes were conjugated to CPMV and Qβ, respectively, using this method. In addition, the affinity of lanthanides for RNA was investigated, and it was found that approximately 80 Gd^{3+} ions could be loaded per CPMV, with similar values observed for Tb^{3+}. Dual-labeled CPMV particles were then made by covalent conjugation of Gd-DOTA followed by internal binding of Gd^{3+}, without any decrease in the efficiency Gd^{3+} loading found. The T_1 relaxivities were measured for the various CPMV and Qβ formulations. Relaxivities of up to 15.5 $mM^{-1} s^{-1}$ per Gd^{3+} and 4150 $mM^{-1} s^{-1}$ per particle were obtained for the dual-labeled particles at 64 MHz, values comparable to what was reported for the MS2(Gd-DTPA) contrast agents and reflecting a three-fold enhancement compared to Gd-DOTA alone [106]. The CPMV formulations were further evaluated in mice for their fate *in vivo*. CPMV with Gd-DOTA on the exterior and Gd^{3+} loaded in the interior showed similar plasma clearance kinetics, with very low levels of particles detected in the bloodstream after 30 min. At this time point, the biodistribution was analyzed, with over 90% found in the liver and a small amount in the spleen. Histological examination was also conducted and showed no evidence of pathology, indicating that the CPMV constructs are safely cleared from the system with no toxic effects [39].

A new class of nanoparticles for MRI that was previously unexplored is comprised of rod-shaped particles. Rods offer many advantages, such as greater tumor penetration and a larger surface area that offers more sites for functionalization [40]. Recently, TMV particles measuring 300 by 18 nm in size were investigated to test if the advantages of rod-shaped particles could be extended to applications in MRI. These particles were functionalized with Gd-DOTA on either their interior or exterior surface by targeting their solvent-exposed carboxylic acid and tyrosine residues, respectively. In addition, interior-modified TMV can undergo a thermal transition to form RNA-free spherical nanoparticles (SNPs) with a controlled size distribution. TMV with around 3,500 Gd-DOTA loaded within, TMV with 2,000 gadolinium chelates conjugated to its exterior surface, and 170 nm SNPs containing 25,000 Gd-DOTA were compared. T_1 relaxivities of 10.7, 18.4 $mM^{-1} s^{-1}$, and 15.2 $mM^{-1} s^{-1}$ per Gd^{3+} ion at 60 MHz, respectively, were obtained. This is a two- to three-fold increase compared to 4.9 $mM^{-1} s^{-1}$ measured for Gd-DOTA alone and similar to values found earlier for VNP Gd-DOTA and Gd-DTPA conjugates [102, 106]. In addition, there is a substantial increase in the relaxivity when the relaxivities per particle are considered, with a value of almost 400,000 $mM^{-1} s^{-1}$ for SNPs. Even greater relaxivities could be obtained with the more optimal HOPO ligand, paving the way for the use of TMV and other rod-shaped particles as supramolecular MRI contrast agents [107].

These VNP contrast agents have not yet been applied *in vivo*, with most evaluations carried out in test tubes in the laboratory setting. However, opportunities clearly exist for their translation into the clinic as these VNP constructs have so far been shown to be safe and biocompatible. There are many possibilities for Gd^{3+} incorporation into VNPs: direct binding at the interface between coat protein subunits, covalent attachment to the exterior and interior surface, and complexation with RNA molecules within the capsid. All these different approaches resulted in formulations with significant enhancements in relaxivity that are highly promising for future clinical use.

14.7 PET IMAGING

Recently, VNPs have also been explored as a carrier of radionuclides for positron emission tomography (PET) imaging, a noninvasive technique that uses radioisotopes that emit positrons when they undergo positive beta decay. PET imaging has been demonstrated as a valuable tool for evaluating biodistribution, tracking delivery, and monitoring therapeutic efficacy. The positrons travel a short distance after emission until they encounter electrons (see chapter 7), at which point the electron–positron pairs undergo annihilation and produce pairs of photons that travel in opposite directions. These photons are detected and the location of the annihilation events reconstructed, with the resulting image displaying where the molecular tracer has localized. A glucose analog, fluorodeoxyglucose (FDG) labeled with fluorine-18 (^{18}F), is the most common radiotracer used in PET imaging because it becomes localized in regions of high glucose metabolism, an attribute of most cancers. There are also other advantages of using ^{18}F, including its relatively long half-life of 110 min and the high-resolution images that can be obtained due to low-energy positron scattering (reviewed in Ref. [108]).

The preliminary exploration of using VNPs with PET imaging exploited these advantages of ^{18}F for biodistribution studies. Chemical conjugation to interior tyrosine residues of MS2 was used to sequester [^{18}F]fluorobenzaldehyde within the capsid so that the radionuclide would not affect the biodistribution of the native particle. PET imaging studies were performed to study the *in vivo* biodistribution of [^{18}F]-MS2 particles alongside [^{18}F]fluorobenzaldehyde after intravenous injection in Sprague-Dawley rats. The MS2 particles were able to avoid renal filtration due to their larger size, indicating that the capsids remained assembled. [^{18}F] Fluorobenzaldehyde was almost completely cleared through the kidneys within 5 min and collected in the urine, while [^{18}F]-MS2 circulated for at least 3 h and localized in the liver and spleen (Fig. 14.7a and b).The [^{18}F]-MS2 agents were further loaded inside with around 50 coumarin fluorophores per particle, and it was found that the distribution and clearance rate were not affected, verifying the hypothesis that interior cargos do not significantly alter the biodistribution [109].

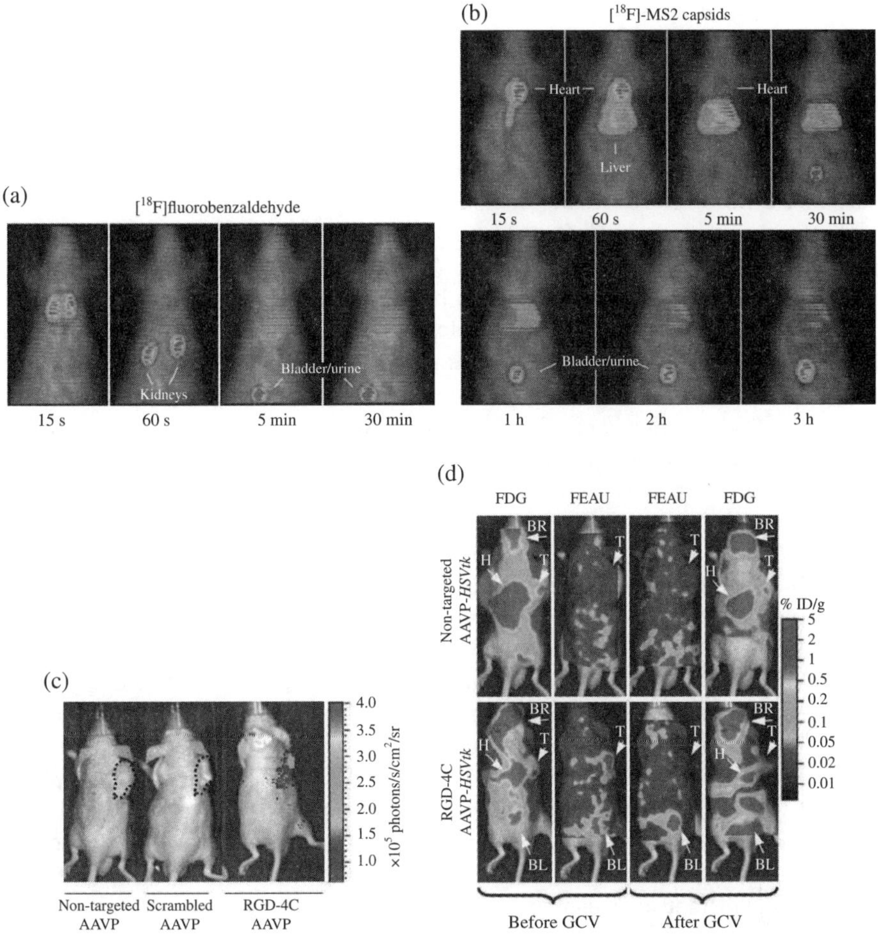

FIGURE 14.7 PET images of Sprague-Dawley rats injected with [^{18}F]fluorobenzaldehyde (a) and [^{18}F]-MS2 (b). Grayscale transmission scan renderings are shown in the background, and intensities were scaled to the maximum in each frame. (Figure (a, b) were reproduced with permission from Ref. [109]. © Springer Science and Business Media.) (c) Bioluminescent images of luciferase expression in nude mice bearing DU145 tumor xenografts ten days after delivery of AAVP. The targeted RGD formulation showed successful gene delivery, while the nontargeted and scrambled formulations did not. (d) PET images of [^{18}F]FEAU and [^{18}F]FDG accumulation in nude mice bearing DU145 tumor xenografts before and after GVC therapy. BL, bladder; BR, brain; H, heart; T, tumor. (Figure (c, d) were reprinted with permission from Ref. [110]. © Elsevier.) *(See insert for color representation of the figure.)*

The use of VNPs for PET imaging applications is very novel and work is still ongoing. Aside from MS2, hemagglutinating virus of Japan envelopes (HVJ-E) has also been employed as carriers of PET tracers. Superparamagnetic ferumoxide particles and ^{18}F-fluoride were loaded simultaneously within these envelopes, and it was found that coencapsulation of ^{18}F-fluoride with the cation poly-L-lysine (PLL) and ferumoxides resulted in more efficient packaging of the radiotracer. Innate biodistribution in rats was studied using PET imaging, and similar results were found as for MS2, with greater accumulation in the liver and spleen for animals injected intravenously with the HVJ-E formulation. Magnets placed externally on top of the head were able to alter the biodistribution, guiding the HVJ-E particles to accumulate in the head. These results demonstrate the use of combining viral imaging agents with external stimuli for region-specific delivery and with PET to study the fate of nanoparticles *in vivo* [111].

The power of PET imaging has also been harnessed in a different manner as a method to track the progress of gene delivery. Recombinant adeno-associated virus (AAV) was combined with a filamentous phage derived from M13, fd-tet, to form a chimeric AAV/phage (AAVP). In this approach, the fd-tet bacteriophage had incorporated within its genome *cis*-elements from AAV's genome to improve integration of the transgene after delivery to tumor cells. Targeted delivery was achieved using an RGD motif displayed on the phage's exterior, which binds to integrin receptors overexpressed in many tumor types. To confirm successful gene delivery to the tumor, PET imaging of a [^{18}F]2′-fluoro-5-ethyl-1-beta-D-arabinofuranosyluracil (FEAU) tracer for detection of herpes simplex virus thymidine kinase (HSVtk) reporter gene expression and bioluminescent imaging of a luciferase reporter gene were used (Fig. 14.7c and d). In addition, HSVtk served as a suicide gene, with cell death triggered by administration of ganciclovir (GCV). After GVC therapy, there was a marked decrease in [^{18}F]FEAU and [^{18}F]FDG accumulation at the tumor site, indicating cell death and successful suppression of tumor growth (Fig. 14.7d) [110].

Although this is a relatively young discipline, there is great potential for applications of VNPs in PET imaging. Multimodal VNP formulations have been established that combine PET tracers with targeting and other imaging modalities. Consequently, there are immense possibilities that exist for the development of innovative new imaging and therapeutic technologies.

14.8 FUTURE OUTLOOK

The use of VNPs for imaging is an emerging discipline, and several promising approaches have been taken that will lay the groundwork for future applications. For example, one can take advantage of the interaction between CPMV and vimentin for imaging vascular disease and tumor angiogenesis. VNPs that passively and actively target cancer cells can be designed as probes for the detection of solid tumors and sites of metastasis. There is also work underway for the application of VNPs for MRI and PET imaging. Several novel high relaxivity and sensitivity MRI contrast agents, as well as PET tracers, are being designed for biodistribution, disease metabolism, and therapeutic efficacy studies.

Going forward toward clinical translation, it will be important to assess the long-term fate of VNPs *in vivo*. Although numerous novel viral nanoparticulate systems have been developed, they have not yet been thoroughly investigated *in vivo*, with evaluations mainly providing evidence of the feasibility of these platforms on the benchtop or *in vitro*. Preliminary work has begun to evaluate the biodistribution and pharmacokinetics of these VNPs *in vivo*. Considering the advantages and versatility of the VNP platform compared to conventional nanoparticles, it is crucial that these nanoparticles are evaluated in more detail to accelerate translation from bench to bedside.

There is a great future for VNPs in imaging. We foresee that state-of-the-art multimodal formulations for simultaneous disease detection, imaging, and treatment will be developed by combining the various applications shown here for targeting, drug delivery, fluorescence, luminescence, and MRI and PET imaging. There are endless possibilities!

ACKNOWLEDGMENTS

This work was supported by NIH/NIBIB grants R00 EB009105 (to NFS) and P30 EB011317 (to NFS) and training grant T32 EB007509 (to AMW), Mt. Sinai Foundation (to NFS), and Prostate Cancer Canada grant 2011-742 (to JDL).

REFERENCES

[1] Longmire MR, Ogawa M, Choyke PL, Kobayashi H. Biologically optimized nanosized molecules and particles: more than just size. Bioconjug Chem 2011;**22**:993–1000.

[2] Sanvicens N, Marco MP. Multifunctional nanoparticles—properties and prospects for their use in human medicine. Trends Biotechnol 2008;**26**:425–433.

[3] Cai W, Chen X. Nanoplatforms for targeted molecular imaging in living subjects. Small 1840-1854;**2007**:3.

[4] Battaglia L, Gallarate M. Lipid nanoparticles: state of the art, new preparation methods and challenges in drug delivery. Expert Opin Drug Deliv 2012;**9**:497–508.

[5] Hawkins MJ, Soon-Shiong P, Desai N. Protein nanoparticles as drug carriers in clinical medicine. Adv Drug Deliv Rev 2008;**60**:876–885.

[6] Abeylath SC, Ganta S, Iyer AK, Amiji M. Combinatorial-designed multifunctional polymeric nanosystems for tumor-targeted therapeutic delivery. Acc Chem Res 2011;**44**:1009–1017.

[7] Boisselier E, Astruc D. Gold nanoparticles in nanomedicine: preparations, imaging, diagnostics, therapies and toxicity. Chem Soc Rev 2009;**38**:1759–1782.

[8] Xia Y, Li W, Cobley CM, Chen J, Xia X, Zhang Q, Yang M, Cho EC, Brown PK. Gold nanocages: from synthesis to theranostic applications. Acc Chem Res 2011;**44**:914–924.

[9] Lee JE, Lee N, Kim T, Kim J, Hyeon T. Multifunctional mesoporous silica nanocomposite nanoparticles for theranostic applications. Acc Chem Res 2011;**44**:893–902.

[10] Kievit FM, Zhang M. Surface engineering of iron oxide nanoparticles for targeted cancer therapy. Acc Chem Res 2011;**44**:853–862.

[11] Chen B, Wu W, Wang X. Magnetic iron oxide nanoparticles for tumor-targeted therapy. Curr Cancer Drug Targets 2011;**11**:184–189.

[12] Lee DE, Koo H, Sun IC, Ryu JH, Kim K, Kwon IC. Multifunctional nanoparticles for multimodal imaging and theragnosis. Chem Soc Rev 2012;**41**:2656–2672.

[13] Wagner V, Dullaart A, Bock AK, Zweck A. The emerging nanomedicine landscape. Nat Biotechnol 2006;**24**:1211–1217.

[14] Peek LJ, Middaugh CR, Berkland C. Nanotechnology in vaccine delivery. Adv Drug Deliv Rev 2008;**60**:915–928.

[15] Yildiz I, Shukla S, Steinmetz NF. Applications of viral nanoparticles in medicine. Curr Opin Biotechnol 2011;**22**:901–908.

[16] Xu ZP, Zeng QH, Lu GQ, Yu AB. Inorganic nanoparticles as carriers for efficient cellular delivery. Chem Eng Sci 2006;**61**:1027–1040.

[17] Lee SC, Huh KM, Lee J, Cho YW, Galinsky RE, Park K. Hydrotropic polymeric micelles for enhanced paclitaxel solubility: in vitro and *in vivo* characterization. Biomacromolecules 2007;**8**:202–208.

[18] Steinmetz NF. Viral nanoparticles as platforms for next-generation therapeutics and imaging devices. Nanomedicine 2010;**6**:634–641.

[19] Schneemann A, Young MJ. Viral assembly using heterologous expression systems and cell extracts. Adv Protein Chem 2003;**64**:1–36.

[20] Pokorski JK, Steinmetz NF. The art of engineering viral nanoparticles. Mol Pharm 2011;**8**:29–43.

[21] Doan DN, Lee KC, Laurinmaki P, Butcher S, Wong SM, Dokland T. Three-dimensional reconstruction of hibiscus chlorotic ringspot virus. J Struct Biol 2003;**144**:253–261.

[22] Ge P, Zhou ZH. Hydrogen-bonding networks and RNA bases revealed by cryo electron microscopy suggest a triggering mechanism for calcium switches. Proc Natl Acad Sci U S A 2011;**108**:9637–9642.

[23] Khalil AS, Ferrer JM, Brau RR, Kottmann ST, Noren CJ, Lang MJ, Belcher AM. Single M13 bacteriophage tethering and stretching. Proc Natl Acad Sci U S A 2007;**104**:4892–4897.

[24] Mäkelä AR, Ernst W, Grabherr R, Oker-Blom C. Baculovirus-based display and gene delivery systems. Cold Spring Harb Protoc 2010;**2010**:pdb.top72.

[25] Mok H, Park JW, Park TG. Enhanced intracellular delivery of quantum dot and adenovirus nanoparticles triggered by acidic pH via surface charge reversal. Bioconjug Chem 2008;**19**:797–801.

[26] Loo L, Guenther RH, Lommel SA, Franzen S. Infusion of dye molecules into Red clover necrotic mosaic virus. Chem. Commun. (Camb) 2008;**1**:88–90.

[27] Basu G, Allen M, Willits D, Young M, Douglas T. Metal binding to cowpea chlorotic mottle virus using terbium(III) fluorescence. J Biol Inorg Chem 2003;**8**:721–725.

[28] Speir JA, Munshi S, Wang G, Baker TS, Johnson JE. Structures of the native and swollen forms of cowpea chlorotic mottle virus determined by X-ray crystallography and cryo-electron microscopy. Structure 1995;**3**:63–78.

[29] Douglas T, Young M. Host-guest encapsulation of materials by assembled virus protein cages. Nature 1998;**393**:152–155.

[30] Dragnea B, Chen C, Kwak ES, Stein B, Kao CC. Gold nanoparticles as spectroscopic enhancers for in vitro studies on single viruses. J Am Chem Soc 2003;**125**:6374–6375.

[31] Hu Y, Zandi R, Anavitarte A, Knobler CM, Gelbart WM. Packaging of a polymer by a viral capsid: the interplay between polymer length and capsid size. Biophys J 2008;**94**:1428–1436.

[32] Bancroft JB, Hiebert E, Bracker CE. The effects of various polyanions on shell formation of some spherical viruses. Virology 1969;**39**:924–930.

[33] Chen C, Daniel MC, Quinkert ZT, De M, Stein B, Bowman VD, Chipman PR, Rotello VM, Kao CC, Dragnea B. Nanoparticle-templated assembly of viral protein cages. Nano Lett 2006;**6**:611–615.

[34] Huang X, Bronstein LM, Retrum J, Dufort C, Tsvetkova I, Aniagyei S, Stein B, Stucky G, McKenna B, Remmes N, Baxter D, Kao CC, Dragnea B. Self-assembled virus-like particles with magnetic cores. Nano Lett 2007;**7**:2407–2416.

[35] Sun J, DuFort C, Daniel MC, Murali A, Chen C, Gopinath K, Stein B, De M, Rotello VM, Holzenburg A, Kao CC, Dragnea B. Core-controlled polymorphism in virus-like particles. Proc Natl Acad Sci U S A 2007;**104**:1354–1359.

[36] Li F, Chen Y, Chen H, He W, Zhang ZP, Zhang XE, Wang Q. Monofunctionalization of protein nanocages. J Am Chem Soc 2011;**133**:20040–20043.

[37] Wu Z, Chen K, Yildiz I, Dirksen A, Fischer R, Dawson PE, Steinmetz NF. Development of viral nanoparticles for efficient intracellular delivery. Nanoscale 2012;**4**:3567–3576.

[38] Prasuhn DE Jr, Singh P, Strable E, Brown S, Manchester M, Finn MG. Plasma clearance of bacteriophage Qbeta particles as a function of surface charge. J Am Chem Soc 2008;**130**:1328–1334.

[39] Singh P, Prasuhn D, Yeh RM, Destito G, Rae CS, Osborn K, Finn MG, Manchester M. Bio-distribution, toxicity and pathology of cowpea mosaic virus nanoparticles *in vivo*. J Control Release 2007;**120**:41–50.

[40] Shukla S, Ablack AL, Wen AM, Lee KL, Lewis JD, Steinmetz NF. Increased tumor homing and tissue penetration of the filamentous plant viral nanoparticle Potato virus X. Mol Pharm 2013;**10**:33–42.

[41] Molenaar TJ, Michon I, de Haas SA, van Berkel TJ, Kuiper J, Biessen EA. Uptake and processing of modified bacteriophage M13 in mice: implications for phage display. Virology 2002;**293**:182–191.

[42] Kaiser CR, Flenniken ML, Gillitzer E, Harmsen AL, Harmsen AG, Jutila MA, Douglas T, Young MJ. Biodistribution studies of protein cage nanoparticles demonstrate broad tissue distribution and rapid clearance *in vivo*. Int J Nanomedicine 2007;**2**:715–733.

[43] van Vlerken LE, Vyas TK, Amiji MM. Poly(ethylene glycol)-modified nanocarriers for tumor-targeted and intracellular delivery. Pharm Res 2007;**24**:1405–1414.

[44] Chatterji A, Ochoa W, Shamieh L, Salakian SP, Wong SM, Clinton G, Ghosh P, Lin T, Johnson JE. Chemical conjugation of heterologous proteins on the surface of Cowpea mosaic virus. Bioconjug Chem 2004;**15**:807–813.

[45] Steinmetz NF, Lomonossoff GP, Evans DJ. Decoration of cowpea mosaic virus with multiple, redox-active, organometallic complexes. Small 2006;**2**:530–533.

[46] Steinmetz NF, Manchester M. PEGylated viral nanoparticles for biomedicine: the impact of PEG chain length on VNP cell interactions in vitro and ex vivo. Biomacromolecules 2009;**10**:784–792.

[47] Steinmetz NF, Mertens ME, Taurog RE, Johnson JE, Commandeur U, Fischer R, Manchester M. Potato virus X as a novel platform for potential biomedical applications. Nano Lett 2010;**10**:305–312.

[48] Schlick TL, Ding Z, Kovacs EW, Francis MB. Dual-surface modification of the tobacco mosaic virus. J Am Chem Soc 2005;**127**:3718–3723.

[49] Bruckman MA, Kaur G, Lee LA, Xie F, Sepulveda J, Breitenkamp R, Zhang X, Joralemon M, Russell TP, Emrick T, Wang Q. Surface modification of tobacco mosaic virus with "click" chemistry. Chembiochem 2008;**9**:519–523.

[50] Kovacs EW, Hooker JM, Romanini DW, Holder PG, Berry KE, Francis MB. Dual-surface-modified bacteriophage MS2 as an ideal scaffold for a viral capsid-based drug delivery system. Bioconjug Chem 2007;**18**:1140–1147.

[51] Leader B, Baca QJ, Golan DE. Protein therapeutics: a summary and pharmacological classification. Nat Rev Drug Discov 2008;**7**:21–39.

[52] Li SD, Huang L. Pharmacokinetics and biodistribution of nanoparticles. Mol Pharm 2008;**5**:496–504.

[53] Lewis JD, Destito G, Zijlstra A, Gonzalez MJ, Quigley JP, Manchester M, Stuhlmann H. Viral nanoparticles as tools for intravital vascular imaging. Nat Med 2006;**12**:354–360.

[54] Steinmetz NF, Cho CF, Ablack A, Lewis JD, Manchester M. Cowpea mosaic virus nanoparticles target surface vimentin on cancer cells. Nanomedicine (Lond) 2011;**6**:351–364.

[55] Cho CF, Ablack A, Leong HS, Zijlstra A, Lewis J. Evaluation of nanoparticle uptake in tumors in real time using intravital imaging. J Vis Exp 2011;**52**:e2808.

[56] Leong HS, Steinmetz NF, Ablack A, Destito G, Zijlstra A, Stuhlmann H, Manchester M, Lewis JD. Intravital imaging of embryonic and tumor neovasculature using viral nanoparticles. Nat Protoc 2010;**5**:1406–1417.

[57] Weissleder R, Pittet MJ. Imaging in the era of molecular oncology. Nature 2008;**452**: 580–589.

[58] Chance B. Optical method. Annu Rev Biophys Biophys Chem 1991;**20**:1–28.

[59] Jobsis FF. Noninvasive, infrared monitoring of cerebral and myocardial oxygen sufficiency and circulatory parameters. Science 1977;**198**:1264–1267.

[60] Folli S, Wagnieres G, Pelegrin A, Calmes JM, Braichotte D, Buchegger F, Chalandon Y, Hardman N, Heusser C, Givel JC, Chapuis G, Chatelain A, van Den Bergh H, Mach JP. Immunophotodiagnosis of colon carcinomas in patients injected with fluoresceinated chimeric antibodies against carcinoembryonic antigen. Proc Natl Acad Sci U S A 1992;**89**:7973–7977.

[61] Azad BB, Cho CF, Lewis JD, Luyt LG. Synthesis, radiometal labeling and in vitro evaluation of a targeted PPIX derivative. Appl Radiat Isot 2012;**70**:505–511.

[62] Whitney MA, Crisp JL, Nguyen LT, Friedman B, Gross LA, Steinbach P, Tsien RY, Nguyen QT. Fluorescent peptides highlight peripheral nerves during surgery in mice. Nat Biotechnol 2011;**29**:352–356.

[63] Wen AM, Ryan MJ, Yang AC, Breitenkamp K, Pokorski JK, Steinmetz NF. Photodynamic activity of viral nanoparticles conjugated with C(60). Chem Commun (Camb) 2012;**48**: 9044–9046.

[64] Aljabali AA, Shukla S, Lomonossoff GP, Steinmetz NF, Evans DJ. CPMV-DOX delivers. Mol Pharm 2013;**10**:3–10.

[65] Dixit SK, Goicochea NL, Daniel MC, Murali A, Bronstein L, De M, Stein B, Rotello VM, Kao CC, Dragnea B. Quantum dot encapsulation in viral capsids. Nano Lett 2006;**6**:1993–1999.

[66] Jung B, Rao AL, Anvari B. Optical nano-constructs composed of genome-depleted brome mosaic virus doped with a near infrared chromophore for potential biomedical applications. ACS Nano 2011;**5**:1243–1252.

[67] Minten IJ, Hendriks LJ, Nolte RJ, Cornelissen JJ. Controlled encapsulation of multiple proteins in virus capsids. J Am Chem Soc 2009;**131**:17771–17773.

[68] Rae CS, Khor IW, Wang Q, Destito G, Gonzalez MJ, Singh P, Thomas DM, Estrada MN, Powell E, Finn MG, Manchester M. Systemic trafficking of plant virus nanoparticles in mice via the oral route. Virology 2005;**343**:224–235.

[69] Steinmetz NF, Ablack AL, Hickey JL, Ablack J, Manocha B, Mymryk JS, Luyt LG, Lewis JD. Intravital imaging of human prostate cancer using viral nanoparticles targeted to gastrin-releasing Peptide receptors. Small 2011;**7**:1664–1672.

[70] Wen AM, Lee KL, Yildiz I, Bruckman MA, Shukla S, Steinmetz NF. Viral nanoparticles for *in vivo* tumor imaging. *J Vis Exp* 2012:e4352.

[71] Koudelka KJ, Destito G, Plummer EM, Trauger SA, Siuzdak G, Manchester M. Endothelial targeting of cowpea mosaic virus (CPMV) via surface vimentin. PLoS Pathog 2009;**5**:e1000417.

[72] Koudelka KJ, Rae CS, Gonzalez MJ, Manchester M. Interaction between a 54-kilodalton mammalian cell surface protein and cowpea mosaic virus. J Virol 2007;**81**:1632–1640.

[73] Steinmetz NF, Maurer J, Sheng H, Bensussan A, Maricic I, Kumar V, Braciak TA. Two domains of vimentin are expressed on the surface of lymph node, bone and brain metastatic prostate cancer lines along with the putative stem cell marker proteins CD44 and CD133. Cancers 2011;**3**:2870–2885.

[74] Kokkinos MI, Wafai R, Wong MK, Newgreen DF, Thompson EW, Waltham M. Vimentin and epithelial-mesenchymal transition in human breast cancer—observations in vitro and *in vivo*. Cells Tissues Organs 2007;**185**:191–203.

[75] Ivaska J, Pallari HM, Nevo J, Eriksson JE. Novel functions of vimentin in cell adhesion, migration, and signaling. Exp Cell Res 2007;**313**:2050–2062.

[76] Plummer EM, Thomas D, Destito G, Shriver LP, Manchester M. Interaction of cowpea mosaic virus nanoparticles with surface vimentin and inflammatory cells in atherosclerotic lesions. Nanomedicine (Lond) 2012;**7**:877–888.

[77] Shriver LP, Koudelka KJ, Manchester M. Viral nanoparticles associate with regions of inflammation and blood brain barrier disruption during CNS infection. J Neuroimmunol 2009;**211**:66–72.

[78] Maeda H, Wu J, Sawa T, Matsumura Y, Hori K. Tumor vascular permeability and the EPR effect in macromolecular therapeutics: a review. J Control Release 2000;**65**:271–284.

[79] Wen AM, Shukla S, Saxena P, Aljabali AA, Yildiz I, Dey S, Mealy JE, Yang AC, Evans DJ, Lomonossoff GP, Steinmetz NF. Interior engineering of a viral nanoparticle and its tumor homing properties. Biomacromolecules 2012;**13**:3990–4001.

[80] Chauhan VP, Popovic Z, Chen O, Cui J, Fukumura D, Bawendi MG, Jain RK. Fluorescent nanorods and nanospheres for real-time *in vivo* probing of nanoparticle shape-dependent tumor penetration. Angew Chem Int Ed Engl 2011;**50**:11417–11420.

[81] Geng Y, Dalhaimer P, Cai S, Tsai R, Tewari M, Minko T, Discher DE. Shape effects of filaments versus spherical particles in flow and drug delivery. Nat Nanotechnol 2007;**2**: 249–255.

[82] Aina OH, Liu R, Sutcliffe JL, Marik J, Pan CX, Lam KS. From combinatorial chemistry to cancer-targeting peptides. Mol Pharm 2007;**4**:631–651.

[83] Ladner RC, Sato AK, Gorzelany J, de Souza M. Phage display-derived peptides as therapeutic alternatives to antibodies. Drug Discov Today 2004;**9**:525–529.

[84] Devlin JJ, Panganiban LC, Devlin PE. Random peptide libraries: a source of specific protein binding molecules. Science 1990;**249**:404–406.

[85] Scott JK, Smith GP. Searching for peptide ligands with an epitope library. Science 1990;**249**:386–390.

[86] Houghten RA, Pinilla C, Blondelle SE, Appel JR, Dooley CT, Cuervo JH. Generation and use of synthetic peptide combinatorial libraries for basic research and drug discovery. Nature 1991;**354**:84–86.

[87] Lam KS, Lebl M, Krchnak V. The "one-bead-one-compound" combinatorial library method. Chem Rev 1997;**97**:411–448.

[88] Tong GJ, Hsiao SC, Carrico ZM, Francis MB. Viral capsid DNA aptamer conjugates as multivalent cell-targeting vehicles. J Am Chem Soc 2009;**131**:11174–11178.

[89] Salazar MD, Ratnam M. The folate receptor: what does it promise in tissue-targeted therapeutics? Cancer Metastasis Rev 2007;**26**:141–152.

[90] Destito G, Yeh R, Rae CS, Finn MG, Manchester M. Folic acid-mediated targeting of cowpea mosaic virus particles to tumor cells. Chem Biol 2007;**14**:1152–1162.

[91] Qian ZM, Li H, Sun H, Ho K. Targeted drug delivery via the transferrin receptor-mediated endocytosis pathway. Pharmacol Rev 2002;**54**:561–587.

[92] Huang RK, Steinmetz NF, Fu CY, Manchester M, Johnson JE. Transferrin-mediated targeting of bacteriophage HK97 nanoparticles into tumor cells. Nanomedicine (Lond) 2011;**6**:55–68.

[93] Banerjee D, Liu AP, Voss NR, Schmid SL, Finn MG. Multivalent display and receptor-mediated endocytosis of transferrin on virus-like particles. Chembiochem 2010;**11**:1273–1279.

[94] An P, Lei H, Zhang J, Song S, He L, Jin G, Liu X, Wu J, Meng L, Liu M, Shou C. Suppression of tumor growth and metastasis by a VEGFR-1 antagonizing peptide identified from a phage display library. Int J Cancer 2004;**111**:165–173.

[95] Brunel FM, Lewis JD, Destito G, Steinmetz NF, Manchester M, Stuhlmann H, Dawson PE. Hydrazone ligation strategy to assemble multifunctional viral nanoparticles for cell imaging and tumor targeting. Nano Lett 2010;**10**:1093–1097.

[96] Patel O, Shulkes A, Baldwin GS. Gastrin-releasing peptide and cancer. Biochim Biophys Acta 2006;**1766**:23–41.

[97] Caravan P. Strategies for increasing the sensitivity of gadolinium based MRI contrast agents. Chem Soc Rev 2006;**35**:512–523.

[98] Caravan P, Cloutier NJ, Greenfield MT, McDermid SA, Dunham SU, Bulte JW, Amedio JC Jr, Looby RJ, Supkowski RM, Horrocks WD Jr, McMurry TJ, Lauffer RB. The interaction of MS-325 with human serum albumin and its effect on proton relaxation rates. J Am Chem Soc 2002;**124**:3152–3162.

[99] Bryant LH Jr, Brechbiel MW, Wu C, Bulte JW, Herynek V, Frank JA. Synthesis and relaxometry of high-generation (G=5, 7, 9, and 10) PAMAM dendrimer-DOTA-gadolinium chelates. J Magn Reson Imaging 1999;**9**:348–352.

[100] Wiener EC, Brechbiel MW, Brothers H, Magin RL, Gansow OA, Tomalia DA, Lauterbur PC. Dendrimer-based metal chelates: a new class of magnetic resonance imaging contrast agents. Magn Reson Med 1994;**31**:1–8.

[101] Allen M, Bulte JW, Liepold L, Basu G, Zywicke HA, Frank JA, Young M, Douglas T. Paramagnetic viral nanoparticles as potential high-relaxivity magnetic resonance contrast agents. Magn Reson Med 2005;**54**:807–812.

[102] Anderson EA, Isaacman S, Peabody DS, Wang EY, Canary JW, Kirshenbaum K. Viral nanoparticles donning a paramagnetic coat: conjugation of MRI contrast agents to the MS2 capsid. Nano Lett 2006;**6**:1160–1164.

[103] Hooker JM, Datta A, Botta M, Raymond KN, Francis MB. Magnetic resonance contrast agents from viral capsid shells: a comparison of exterior and interior cargo strategies. Nano Lett 2007;**7**:2207–2210.

[104] Datta A, Hooker JM, Botta M, Francis MB, Aime S, Raymond KN. High relaxivity gadolinium hydroxypyridonate-viral capsid conjugates: nanosized MRI contrast agents. J Am Chem Soc 2008;**130**:2546–2552.

[105] Garimella PD, Datta A, Romanini DW, Raymond KN, Francis MB. Multivalent, high-relaxivity MRI contrast agents using rigid cysteine-reactive gadolinium complexes. J Am Chem Soc 2011;**133**:14704–14709.

[106] Prasuhn DE Jr, Yeh RM, Obenaus A, Manchester M, Finn MG. Viral MRI contrast agents: coordination of Gd by native virions and attachment of Gd complexes by azide-alkyne cycloaddition. *Chem. Commun. (Camb)* 2007:1269–1271.

[107] Bruckman MA, Hern S, Jiang K, Flask CA, Yu X, Steinmetz NF. Tobacco mosaic virus rods and spheres as supramolecular high-relaxivity MRI contrast agents. J Mater Chem B 2013;**1**:1482–1490.

[108] Fletcher JW, Djulbegovic B, Soares HP, Siegel BA, Lowe VJ, Lyman GH, Coleman RE, Wahl R, Paschold JC, Avril N, Einhorn LH, Suh WW, Samson D, Delbeke D, Gorman M, Shields AF. Recommendations on the use of 18F-FDG PET in oncology. J Nucl Med 2008;**49**:480–508.

[109] Hooker JM, O'Neil JP, Romanini DW, Taylor SE, Francis MB. Genome-free viral capsids as carriers for positron emission tomography radiolabels. Mol Imaging Biol 2008;**10**:182–191.

[110] Hajitou A, Trepel M, Lilley CE, Soghomonyan S, Alauddin MM, Marini FC III, Restel BH, Ozawa MG, Moya CA, Rangel R, Sun Y, Zaoui K, Schmidt M, von Kalle C, Weitzman MD, Gelovani JG, Pasqualini R, Arap W. A hybrid vector for ligand-directed tumor targeting and molecular imaging. Cell 2006;**125**:385–398.

[111] Flexman JA, Cross DJ, Lewellen BL, Miyoshi S, Kim Y, Minoshima S. Magnetically targeted viral envelopes: a PET investigation of initial biodistribution. IEEE Trans Nanobiosci 2008;**7**:223–232.

15

DESIGN AND DEVELOPMENT OF THERANOSTIC NANOMEDICINES*

JELENA M. JANJIC

Graduate School of Pharmaceutical Sciences, Mylan School of Pharmacy, Duquesne University, Pittsburgh, PA, USA

MINGFENG BAI

Molecular Imaging Lab, Department of Radiology, University of Pittsburgh, Pittsburgh, PA, USA

15.1 THERANOSTICS FOR PERSONALIZED NANOMEDICINE

Theranostics, or sometimes named "theragnostics," emerged in the past two decades as an "all-in-one" strategy that combines imaging and drug delivery. The number of publications grew dramatically since 2000 to date, while the diversity of fields involved in this area is very high. Figure 15.1 illustrates the emergence of this new nanotechnology area with reports published in various fields such as chemical engineering, biochemistry, and medicine. What this graph illustrates is the need for true multidisciplinary effort in order to bring personalized nanomedicine to clinical reality. Currently, we consider personalized medicine as an approach where the patient disease status information is combined with information of their genetic makeup relevant to the treatment. Fast metabolizers are given different doses for the same medicine as compared to slow metabolizers, for example.

Theranostics promises to go a step further and allow for adjustments and personalization during the actual treatment. The promise of theranostics to personalize

*Jelena M. Janjic and Mingfeng Bai are contributed equally.

Nanotechnology for Biomedical Imaging and Diagnostics: From Nanoparticle Design to Clinical Applications, First Edition. Edited by Mikhail Y. Berezin.

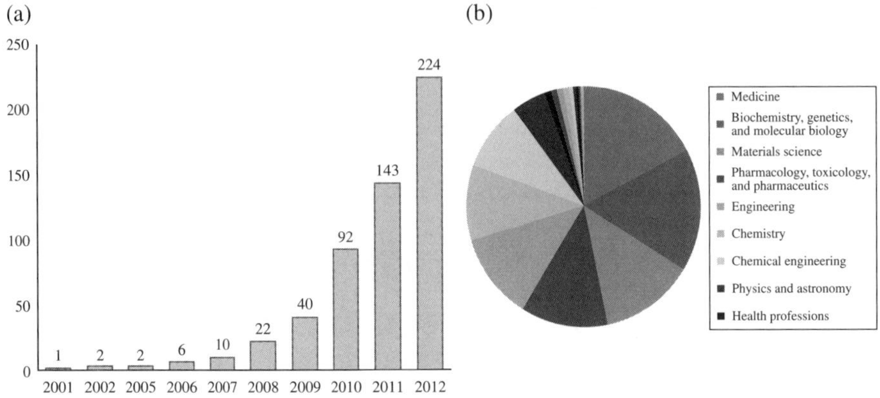

FIGURE 15.1 Theranostic nanotechnology publication frequency (a) and distribution of reports by field of study (b). Data obtained from Scopus (www.scopus.com) for years 2000–2012.

treatment became a very strong drive for basic and translational research around the world. Ideally, a physician would prescribe theranostic nanosystems to simultaneously deliver the drug of choice specifically to the diseased tissue or organ and monitor both the drug delivery and drug response in real time. This in turn would allow to further adjustments of drug dose based on feedback from the individual patient undergoing treatment. Theranostics as a concept promise to be that "smart." This ideal has not been reached yet, but much progress has been made to visualize drug delivery, image drug effects, and use nanosystems for biomarker detection *in vivo*. As theranostics continue to be further explored in a variety of disease models preclinically, some concerns emerged regarding their full translation to the clinic. One of the problems is the complexity of the system and its effects on future large-scale manufacturing. The second problem is how to evaluate the entire system considering its multiple components in respect to efficacy and safety. Specifically, some questions include: Are theranostic nanosystems too complicated to be produced on sufficient scale to be commercially viable? Do we need to revise our toxicity testing when both drug and imaging agents are codelivered as one formulation? How do we account for the unique effects of the diagnostic alone and drug alone and compare these to the whole system? It is easy to recognize that the toxicity profile of a nanosystem may present unique features and that it is not a simple sum of its parts. Choices of imaging agents, excipients, and drugs are critical contributing factors to theranostic nanosystem complexity and future manufacturing. Can we consider these issues early in the nanosystem design? Can existing examples in the literature currently in preclinical development provide enough input to explore the design space of theranostics and allow us to build predictable models for future manufacturing friendly formulations? If theranostic nanomedicines are to become future personalized medicines, these questions must be answered.

Theranostic nanosystem design seems to meet similar issues found in classical small-molecule drug discovery program: targeting, toxicity, efficacy, manufacturing, etc. Yet unique challenges exist simply due to their nanostructure nature that must be considered. Finally, if theranostics are to become true personalized medicines, should we contemplate manufacturing a tailored theranostic nanomedicine for each individual patient? Could we in the future become able to customize theranostic nanosystems at the bedside? And if this is the case, how would we assess the safety and efficacy of these systems? It is now easy to imagine the need for comprehensive assessment of this emerging field. We also argue that rational design will be critical to their future clinical utilization. Our hope is that theranostics manage to overcome these obstacles and live to become new medicines rather than interesting engineering exercise very successful in treating and imaging disease processes in animal models. However, we would also argue that nanotheranostics hold a great potential to help us elucidate *in vivo* disease mechanisms and this potential should be further explored.

The aim of this chapter is to survey some of the most prominent examples of theranostic nanomedicine design and offer arguments for future development of these agents. We aim to approach theranostics from a new medicine development point of view where we consider future pharmaceutical production considerations and offer ideas on how theranostics could be designed to meet these challenges.

It is important to note that theranostic nanomedicine belongs to incredibly active research discipline of nanomedicine, which itself is in early stages. Recent reviews attempted to both define nanomedicine and provide insights on how far this field has gone and how close it came to marketed products [1, 2]. Authors found that most of the nanomedicine formulations were designed for IV injection as the route of administration and more than 80% of nanosystems were delivered through passive targeting. These trends of course are repeating in the theranostic nanomedicine research, and we have to ask ourselves if we are maximizing their potential or simply doing what works and not fully exploring the vast design space of theranostics. Looking at the literature over the past 20 years, it is common to see progression from imaging nanoreagents and drug delivery nanoreagents to combined formulations, which we all now name theranostics. However, we should also ask, does a nanoparticle delivery system decorated with an imaging moiety necessarily qualify as a theranostic? Theranostic nanomedicine division of the National Institutes of Health defines theranostic nanomedicine as follows: "Theranostic Nanomedicine is the medical application of nanobiotechnology and refers to highly specific medical intervention at the nanoscale for diagnosing, curing or preventing diseases." This definition is broader than most currently published formulations, which typically are nanosystems carrying drugs or drug delivery nanosystems with added imaging entity. Therefore, we propose that designing theranostics should follow more specific criteria: (i) theranostics should be of small size and narrow size distribution, preferably monodisperse nanosystems; (ii) all excipients should be biocompatible and preferably biologically inert; (iii) the imaging capacity of the theranostic nanosystem should be at its highest level possible, unaffected by the remaining components of the system and the drug, and comparable to imaging nanoreagent alone; (iv) the drug loading and adequate drug release profile should be comparable of the theranostic to other drug delivery nanosystems without the imaging component; (v) the molecular imaging should provide insight into theranostics

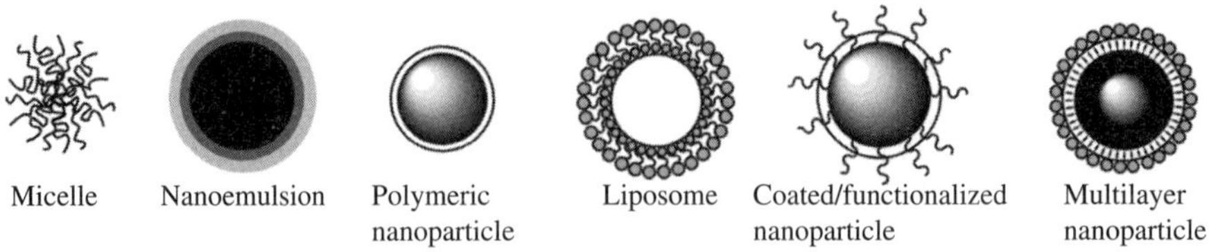

FIGURE 15.2 Theranostic nanosystems.

biodistribution, targeting efficacy, drug release, and/or disease response; (vi) the imaging component of the system should preferably not be released with the drug or have a different time course of release; and (vii) the imaging reagent should be nontoxic and pharmacologically inactive, so the drug effects alone are responsible for the theranostic pharmacological effects *in vivo*. Figure 15.2 shows some common examples of nanosystems used for theranostic formulation development. In all these formulations, the critical balance must be achieved: balance between imaging functionality, drug delivery, and targeting.

A long list of materials has been tested to prepare theranostic nanosystems: inorganic materials (e.g., iron oxide), polymers, and carbon materials. Here, we will discuss some major classes of materials explored for theranostics and their potential and challenges toward further development into personalized nanomedicines. Each theranostic nanomedicine class is illustrated by a few representative examples. We would like to point out that this chapter should not be considered as a comprehensive list. Rather, we aimed to provide examples, which can help understand the theranostic design. We have divided these examples based on the base material and nanoparticle structure rather than imaging modality since many contain more than one imaging agent. Finally, our discussion also includes concerns toward the future of theranostic nanosystems manufacturing and production. Pharmaceutical considerations on these agents are rarely discussed. Most reports provide little insight on the scale of their chemistry and formulation or even mention how these nanoparticles could be produced cost-effectively in the future. At this stage, theranostic nanomedicine is still fairly in the realm of exploratory pharmaceutical design and mostly without comprehensive discussion available in the literature on process development of theranostics or their manufacturing beyond a very small laboratory scale. Our group typically reports theranostic nanoemulsions produced on the scale of 20–30 g [3, 4]. Though these are sizable compared to many other reported theranostic nanoparticles, this is far from manufacturing pilot scale, which we would argue should be in the range of several hundred grams to kilograms. The examples in this chapter also serve as a framework for discussing potential challenges and opportunities in future manufacturing of these theranostic nanoparticles, if any were to advance further toward clinical scale production.

Though theranostic nanomedicines offer multifunctionality unlike any other formulation, they still need to go through development pipeline as any other medicine. Therefore, we propose that theranostic nanomedicines should be designed and developed with advanced development stages in mind (cost, scale, processing, shelf life, route of administration, etc.). The earlier we consider these issues, the more likely our successful theranostic nanomedicine candidate in the preclinical testing will reach the clinic.

15.2 SOLID NANOPARTICLE SYSTEMS

15.2.1 Iron Oxide Nanoparticles

Iron oxide nanoparticles (IONPs) are among the most commonly used nanomaterials for biomedical applications. In general, IONPs have two main components: a magnetic iron oxide core composed of magnetite (Fe_3O_4) and maghemite (γ-Fe_2O_3) and a hydrophilic surface coating with dextran, polyethylene glycol (PEG), or other polymers. IONPs have many favorable features as a theranostic platform, including:

1. Magnetic resonance imaging (MRI) signature for therapeutic monitoring. Small IONPs (typically <20 nm) become superparamagnetic when each particle has a single magnetic domain and thermal energy is high enough to overcome the energy barrier of magnetic flipping [5]. Such superparamagnetic IONPs are highly magnetic under external magnetic field but show zero magnetism without it. In MRI, superparamagnetic IONPs are effective in reducing T_2 relaxation time, therefore showing dark contrast in MRI images. This information allows for the imaging of IONPs uptake to monitor treatment efficacy [6].
2. Well-developed surface chemistry for bioconjugation. IONPs can be coated with a large number of functional groups for bioconjugation with therapeutic, signaling, and targeting molecules. It is therefore possible to use IONPs in target-specific drug delivery and multimodal imaging. For example (Fig. 15.3), a multifunctional polymeric matrix of poly(acrylic acid) (PAA)-coated IONPs were reported to have an encapsulated chemotherapeutic agent for cancer therapy, a surface folate receptor ligand for cancer targeting, an encapsulated near-infrared (NIR) dye for fluorescent imaging capabilities, and a superparamagnetic iron oxide core for MRI [7].
3. Biocompatibility and biodegradability. Unlike many other metallic compounds, iron oxide is considered suitable for *in vivo* applications as the iron cell homeostasis is well controlled by uptake, excretion, and storage [8]. Moreover, excessive iron can be efficiently cleared from the body [9]. In fact, several IONPs have been approved by the FDA for clinical usage, such as Feridex and Resovist. Overall, IONPs represent an outstanding class of nanomaterials for theranostic applications.

Although MRI provides outstanding anatomical information with good resolution, it suffers from poor sensitivity [10]. This limitation can be resolved by combining MRI with other sensitive imaging techniques, such as positron emission tomography (PET), single-photon emission computed tomography (SPECT), and optical imaging [10]. With large surface area and sophisticated surface chemistry, IONPs can serve

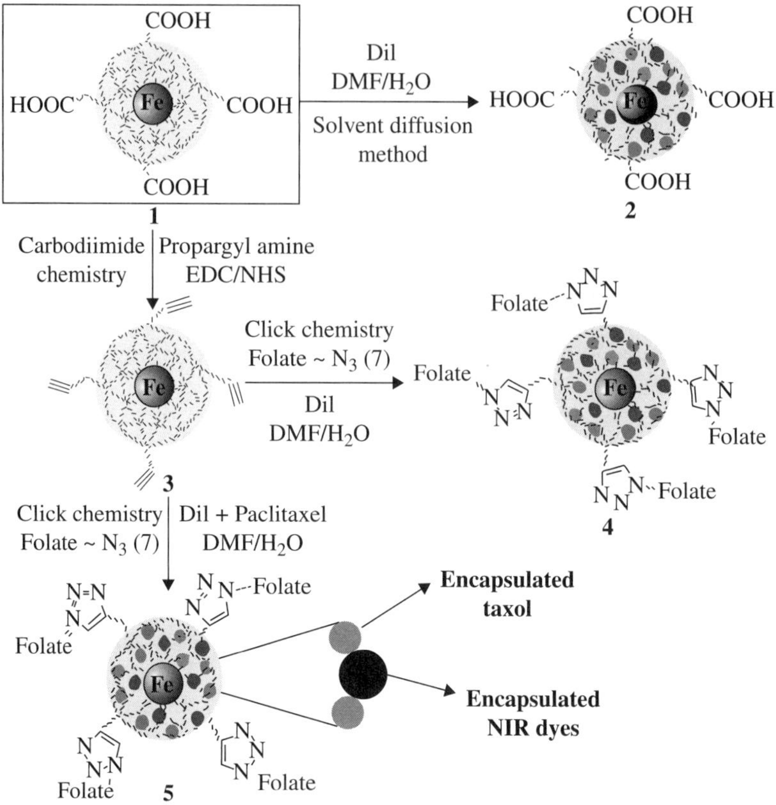

FIGURE 15.3 Schematic representation of the synthesis of theranostics and multimodal IONPs. Click chemistry and carbodiimide chemistry have been used for the synthesis of a library of functional IONPs. NIR dyes and paclitaxel-coencapsulated IONPs were prepared in water. (Reprinted from Ref. [7]. © Wiley.)

as a platform whose surface can be easily loaded with non-MRI signaling moieties. For example, it is possible to attach NIR fluorescent dyes onto IONPs for MRI and NIR fluorescence bimodal imaging [11]. Similarly, coupling ^{64}Cu- or ^{111}In-chelated DOTA to IONPs will allow for MRI/PET or MRI/SPECT imaging [12]. Attaching multiple types of imaging moieties to IONPs will also make multimodal imaging possible, such as MRI/PET/NIR fluorescence trimodal imaging [13].

Drug molecules can be loaded to IONP platforms through covalent coupling to the surface functional groups or physical absorption into porous iron oxide nanostructures [11, 14]. Similar to the approach of attaching signaling moieties to the surface of IONPs, drug molecules with appropriate functional groups can be covalently coupled to IONPs. The release of drug molecules is usually triggered by an acidic environment or enzymes. For example, Kohler *et al.* attached a cancer drug, methotrexate (MTX), to the surface of superparamagnetic IONP through amide coupling. The amide bond was cleaved in the lysosomal compartment with low pH and lysozymes after the drug nanoparticle conjugate was internalized in cancer cells [15]. For drug molecules without appropriate functional groups, they can be loaded into certain porous iron oxide nanostructures. For instance, commonly used cancer drugs, doxorubicin (DOX) and cisplatin, were found to be absorbed into hollow IONPs and released under physiologic conditions in a sustained manner [16, 17].

Although passive targeting mediated by enhanced permeability and retention (EPR) effect is commonly used to deliver IONPs to the target tumor region [16, 18], it is more desirable to introduce targeting moieties, such as antibodies, peptides, and ligands, to the surface of IONPs. Through specific interactions between the targeting moieties on IONPs and corresponding receptors on cancer cell membranes, IONPs can accumulate on the cell membranes or be internalized through receptor-mediated endocytosis, therefore enhancing imaging contrast and therapeutic efficacy. The success of targeted delivery of IONPs to cancer cells largely depends on the size and surface engineering. It is critical that IONPs have circulation time long enough to reach the target sites. IONPs with sizes smaller than 10 nm will be filtered out of the blood rather quickly because the basal lamina of the kidneys has pores of approximately 10 nm [5]. IONPs larger than 100 nm are prone to be uptaken by the reticuloendothelial system (RES) [11]. Therefore, IONPs with sizes between 10 and 100 nm are generally considered appropriate for targeted delivery. Other than the size, the other key consideration is surface engineering. It is important to graft IONP surface to achieve desired functionality, hydrophilicity, biostability, and specificity. A common approach to reduce nonspecific adsorption and enhance stability of IONPs in plasma is to introduce PEG to the surface. Some permeating molecules, such as amphipathic peptides and chitosan (a polysaccharide polymer), have been used to facilitate IONPs penetrate the blood–brain barrier (BBB) [19–21].

The classical synthetic method of IONPs involves coprecipitation of Fe(II) and Fe(III) with polymer under basic aqueous condition. Tangled with the nanocrystals, the polymer can protect the particles from aggregation and overgrowth [11]. A well-known example of IONPs synthesized from coprecipitation is cross-linked iron oxide (CLIO) nanoparticles, which use aminated or carboxylated dextran as the coating polymer [22, 23]. The functional groups (primary amino or carboxyl) distributed throughout the nanostructure allow for bioconjugation to various signaling, targeting, and therapeutic molecules.

Coprecipitation is simple and scalable; however, this method is challenged by limited control over the size, shape, and composition of IONPs. To address this limitation, the thermal decomposition method was later developed. This method involves high temperature reaction in organic solution and can produce highly monodispersed IONPs [24]. The major disadvantage is that the resulting IONPs are usually not soluble in water, thus limiting the extent of biological applications. To introduce hydrophilicity, biocompatibility, and functionality to IONPs developed from thermal decomposition method, surface modification is necessary. Several such modification techniques have been developed, such as ligand exchange [25] and ligand addition [26].

The superior magnetic properties and versatile surface chemistry have rendered IONPs a promising platform for *in vivo* therapeutic studies. For example, Yu *et al.*

developed DOX-loaded thermally cross-linked superparamagnetic IONPs (Dox@TCL-SPION) as a theranostic MRI agent and evaluated the agent in a lung cancer mouse model [27]. Dox@TCL-SPION has a hydrodynamic size of 21 nm and was able to accumulate in the tumor region by EPR effect. *In vivo*, Dox@TCL-SPION showed significantly larger antitumor effect than DOX alone, indicating enhanced therapeutic efficacy with passive targeting. At the same time, the therapeutic effect was monitored by T_2-MRI. In an effort to develop specifically targeted theranostic IONPs, prostate-specific membrane antigen (PSMA) aptamer was later on attached to the surface of Dox@TCL-SPION by the same group [28]. The resulting Dox@Apt-hybr-TCL-SPION showed selective drug delivery in a prostate cancer xenograft mouse model. Multimodal IONPs have also been applied in theranostic studies of cancers *in vivo*, such as MRI–optical imaging [29] and MRI/PET [30].

15.2.2 Gold Nanoparticles

Gold-based nanoparticles represent an attractive class of theranostic nanomaterials. This is largely due to the unique surface plasmon resonance (SPR) feature of gold. A gold nanostructure can create intense absorption and scattering peaks when illuminated with an electromagnetic wave [31]. Importantly, the absorption and scattering wavelengths can be tuned based on the size and shape of gold nanoparticles (GNPs), therefore allowing *in vivo* imaging applications in the favorable NIR optical region. GNPs have been developed in the shapes of spheres, cubes, rods, cages, and wires [14]. Although GNPs are poor fluorescence emitters, they are commonly used in photoacoustic imaging studies. In addition, the SPR property renders GNPs suitable for photothermal therapy (PTT) of tumors. Moreover, bioconjugation of targeting, imaging, and therapeutic entities to GNPs surfaces is well established. As such, GNPs have been extensively investigated in theranostic studies.

The intense absorption and scattering make GNPs suitable for photoacoustic and surface-enhanced Raman spectroscopy. Smaller GNPs also have good X-ray attenuation and can serve as contrast agent for X-ray and CT [32]. In addition, the surface functional groups of GNPs allow for multimodal imaging capabilities by incorporating other techniques, such as SPECT [33], PET [34, 35], CT [36], MRI [37], and optical [38].

The strong covalent interaction between gold and sulfur is of paramount importance in stabilizing nanostructures [39]. Such interaction is typically mediated through the sulfhydryl (–SH) group in thiols; therefore, surface attachment of drug molecules is often conducted via addition of thiolated molecules. For example, a modified paclitaxel (PTX) drug molecule with terminal carboxyl group was attached to surface sulfhydryl group on GNPs through a 4-mercaptophenol linker [40]. In another study, a colloidal GNP vector was developed by coupling tumor necrosis factor (TNF) with thiol-derivatized PEG on the surface [41].

As discussed earlier in this text, GNPs can also serve as a cancer therapy agent without the need of any additional drug molecules. This is because upon laser irradiation, GNPs are efficient in converting light to heat, which kills surrounding cancer cells via PTT. Compared to conventional chemotherapy drugs, which have notorious side effects, PTT is active within limited area and thus can reduce normal tissue damage. For instance, it has been demonstrated that PEG-coated gold nanocages increased the tumor surface temperature to 54°C within 2 min under illumination of NIR light [42].

Similar to many other nanomaterials, GNPs can be delivered to the tumor in a passive fashion, where circulating GNPs can penetrate the leaky tumor vasculature and accumulate in the tumor. For example, after being injected intravenously into tumor-bearing mice, NIR PEGylated nanoshells were shown to be heterogeneously located in tumors, mostly in the tumor cortex and absent from the tumor core [43]. For targeted delivery of the nanomaterial, targeting molecules can be conveniently attached to the surface of GNPs, typically through thiol-mediated conjugation. Some examples of targeting molecules that have been attached to GNPs include antibodies [44, 45], antibody fragments [46], receptor ligands [47, 48], and peptides [49, 50].

The most common method to synthesize GNPs is chemical reduction of gold salt to metallic gold with a capping ligand, which is typically a thiol [51]. One example is Brust–Schiffrin reaction, in which sodium borohydride is used to reduce gold chloride to gold with dodecanethiol as the capping ligand [52]. The GNPs produced have small sizes (1–3 nm) and are highly stable. However, ligand exchange or additional coating is needed for biological applications as these GNPs are not soluble in aqueous solutions. The most widely used coating molecule is PEG, which increases hydrophilicity and lengthens blood circulation time [53]. As previously mentioned, the thiol components involved in GNPs synthesis can be readily used to conjugate with other functional molecules, such as targeting, signaling, and therapeutic agents.

With stability, strong SPR, and ease of surface modification, GNPs have been exploited as an attractive platform for theranostic applications. For example, PEG-coated GNPs were used to treat human xenograft tumor in mice through PTT, and the therapy was guided by CT [54]. The PEG coating is important for GNPs to traverse the systemic circulation, deter protein binding and RES clearance, permeate through transendothelial pores in tumor blood vessels, and be retained in the tumor interstitium [54]. As a result, the circulation time of the PEGylated GNPs was as long as 17 h, thus allowing for cancer therapy through passive targeting. Compared to the controls, the PEGylated GNPs showed much higher potency.

The therapeutic efficacy of GNPs can be further improved by specific delivery and combining PTT and anticancer drug. In a study reported by You *et al.*, DOX and a targeting molecule for EphB4 (a receptor overexpressed by multiple tumors and angiogenic blood vessels) were loaded to gold hollow nanospheres [55]. The resulting multifunctional GNP, T-DOX@HAuNS, was evaluated in three mouse tumor models. In addition, ^{111}In was used to label the theranostic GNP to image the biodistribution and tumor uptake using SPECT. Treatment with T-DOX@HAuNS followed by NIR irradiation to release DOX resulted in significantly decreased tumor growth when compared with treatments with nontargeted DOX@HAuNS or HAuNS without DOX. In contrary to the group treated with DOX only, none of the treatment groups experienced a loss in body weight, suggesting lower side effects. These encouraging results indicate that combinatory therapy with chemo and PTT using targeted GNPs as the carrier is a promising approach for cancer therapy.

15.2.3 Carbon Nanomaterials

With rapid development in nanotechnology, carbon nanomaterials have become an attractive class of theranostic agents, owing to their unique physical and chemical properties. Several types of carbon nanomaterials have been applied in theranostic studies, including fullerene, carbon nanotubes (CNTs), graphene, carbon dots, and nanodiamonds. Among all theranostic carbon nanomaterials, CNTs are the most extensively studied. The large surface area of CNTs is available for efficient drug loading and bioconjugation with targeting and signaling molecules [56]. Moreover, CNTs can be developed with strong absorbance in the NIR region for photoacoustic and Raman imaging, as well as PTT of tumor.

CNTs can be categorized as single-walled carbon nanotubes (SWCNTs) and multiwalled carbon nanotubes (MWCNTs). SWCNTs are made up of a single rolled-up layer of a graphene sheet, and MWCNTs consist of multiple rolled-up layers of graphene [57]. Typical SWCNTs have diameters of less than 5 nm, whereas MWCNTs can have diameters as large as 100 nm. The lengths of CNTs can range from hundreds of nanometers to tens of micrometers.

Similar to GNPs, many CNTs have strong optical absorption in the NIR region, therefore suitable for photoacoustic or Raman imaging. Multimodal imaging approach is also readily achievable, as the surface of CNTs can be easily functionalized [57]. For example, SWCNTs can be treated to have a large number of carboxyl groups on the surface, which can be applied to conjugate various types of imaging moieties, such as NIR fluorescent dyes, radioisotopes, and magnetic nanoparticles [14].

The intense NIR absorption of CNTs has made them a promising PTT tool for cancer treatment. It has been shown that upon irradiation by NIR light, CNTs were internalized in cells and triggered endosomal rupture and cell death [58].

Anticancer drugs can also be attached to the functional surface of CNTs through covalent bonds. For instance, Pastorin *et al.* attached MTX to the amino functionalized surface of CNTs via amide coupling [59]. However, it was found that covalent amide bond was unfavorable to elicit intracellular drug release. To overcome this issue, Kam *et al.* coupled siRNA to CNTs via a disulfide bond, which can be cleaved by enzymes in the endolysosome [60]. In another study, cleavable ester bond was used to connect PTX and CNTs and conjugates showed higher efficiency in suppressing tumor growth than PTX alone [61]. The higher efficiency is largely due to the increased accumulation of conjugates in the tumor region, which was attributed to the prolonged circulation time [14].

Other than bioconjugation through covalent bonds, anticancer drugs can also be loaded to CNTs through aromatic stacking mechanism. For example, DOX could be efficiently loaded onto SWCNTs via supramolecular π–π stacking [62, 63]. The resulting SWNT-DOX complex demonstrated enhanced therapeutic efficacy and reduced toxicity compared to free DOX.

Although many types of nanoparticles without targeting molecules can be delivered to tumor through passive targeting, few have been reported to be uptaken by cells through endocytosis. With appropriate surface coatings, CNTs have been reported to be internalized by cells through endocytosis [64]. Even though the detailed mechanism behind the uptake remains unclear, this feature allows CNTs to deliver drug molecules to cells without the need of targeting molecules. Due to the ease of functional group introduction to CNTs' surface, targeting molecules can also be attached to CNTs through bioconjugation chemistry. For example, peptides and antibodies have been attached to CNTs and evaluated in tumor-targeted *in vivo* imaging with photoacoustic and Raman imaging [65, 66].

CNTs are developed based on graphene, which is inert and inhibitive to most conjugation chemistry. To introduce hydrophilicity and functionality to CNTs, extreme oxidative conditions can be applied to create defects on CNT surface where polar and functional groups can be mounted. One method developed by Dai's group involves refluxing SWCNTs in nitric acid for two 36-h periods separated by sonication for 30 min [67]. Multiple carboxyl groups are then generated on the surface, making SWCNTs water soluble and functional. PEGylated phospholipids are often introduced to the SWCNT surface to improve stability and biocompatibility as well as lengthen circulation time.

Although SWCNTs have intense NIR absorption, the molar extinction coefficients are lower than those of GNPs. In an effort to improve the absorption, Kim *et al.* developed gold-plated SWCNTs, which showed a 100-fold enhancement in photoacoustic signal [68]. Golden nanotubes were synthesized by coating a 4–8 nm-thick gold layer onto a SWCNT core with a diameter of 1.5–2 nm. These golden nanotubes are highly water soluble, well dispersed, relatively uniform in size, and rod shaped with average dimensions of 100 nm in length and 11 nm in diameter [68].

The strong NIR absorbance of CNTs has been utilized to treat tumor through PTT in cancer mouse models. It was demonstrated that NIR irradiation of PEGylated SWCNTs led to complete destruction of tumors in a human epidermoid mouth carcinoma mouse model [69]. No harmful side effects or recurrence of tumors were observed over 6 months after treatment, whereas tumors in control groups continued to grow. *In vivo* therapeutic studies using anticancer drug-loaded CNTs have also been reported. In a 4T1 breast cancer mouse model, SWCNT-PTX conjugate afforded much longer blood circulation time than PTX due to the PEGylated surface of SWCNTs [61]. This led to high tumor uptake and treatment efficacy. The PTX on SWCNT was likely released via cleavage of the ester bond between PTX and SWCNT by esterase. Bhirde *et al.* reported the first targeted *in vivo* therapeutic study using multifunctional SWCNT bioconjugates [70]. The two bioconjugates (SWNT–quantum dots (Qdot)–epidermal growth factor (EGF) and SWNT–cisplatin–EGF) have several components including EGF as the targeting molecule, cisplatin as the anticancer drug, Qdot as the fluorescence imaging moiety, and SWNT as the carrier. *In vivo* imaging showed that SWNT–Qdot–EGF injected into live mice was selectively taken up by tumors. Moreover, regression of tumor growth was rapid in mice treated with SWNT–cisplatin–EGF relative to nontargeted SWNT–cisplatin.

15.3 COLLOIDAL SYSTEMS

15.3.1 Liposomes

Liposomes for drug delivery have more than 50-year-long history of preclinical and clinical use. They represent, from a design point of view, the groundwork behind many nanomedicine systems we know today [71]. One of the first reports of targeted anticancer drug delivery was using liposomes [72]. Therefore, it is also easy to see them as the pioneering nanosystems for imaging and theranostic. By definition, liposomes are colloidal, thermodynamically stabilized systems comprised of a phospholipid bilayer. Based on their structure, liposomes are classified into several categories. Based on the size and number of bilayers, liposomes can be grouped into two categories: multilamellar vesicles (MLV), with onion-like structure of several phospholipid bilayers, and unilamellar vesicles, where single phospholipid bilayer surrounds a water-filled core. Unilamellar vesicles are then divided to large unilamellar vesicles (LUV) and small unilamellar vesicles (SUV) [71]. They are unique in their ability to carry both lipophilic and hydrophilic compounds (diagnostic or therapeutic), biologics, and macromolecules. Their surface can be modified to facilitate both passive and active drug targeting [73]. Liposomes are also well-explored platform for carrying imaging agents with magnetoliposomes the most well known [74, 75]. Here, we will highlight several examples where liposomes carrying both drugs and imaging agents for image-guided drug delivery.

Liposomes were explored as nanocarriers for a variety of imaging agents, such as radioisotopes, MRI, SPECT, PET, and optical imaging [76–81]. In these formulations, the imaging agents were conjugated to the liposomal surface, incorporated into the lipid bilayer, or added to the inner core of the liposome. Liposomes also showed great versatility for multimodal imaging. For example, Gd-decorated liposomes were used for target-selective imaging of tumor angiogenesis in rats [82]. In theranostic liposomes, radioisotopes, MRI, and optical imaging agents were used [83, 84]. The choice of imaging modality and how it is incorporated influences what role the imaging agent plays in the final theranostic formulation. If the agent is conjugated to the surface of the liposome while the drug is in its inner core, the imaging agent can provide insight into overall liposomal biodistribution and indicate whether the liposome reaches the target. Depending on the drug release mechanism involved, surface labeling of the liposome with imaging agent could support imaging of drug release. It is critical that drug and imaging agent should be chemically or physically compatible in the theranostic formulation. To facilitate chemical and physical compatibility, the drug and imaging agents reside in separate compartments of the liposome structure. Imaging liposomes offer insight into drug delivery effectiveness and potentially allow us to see if the drug is released at target site. These insights can in turn help us modify dose and treatment plan for a particular drug and therefore may decrease systemic side effects.

In recent years, MLs emerged as attractive formulations for not only imaging drug delivery but also controlling drug release using magnetic field. MLs is a versatile theranostic platform that offers treatment of cancers with molecular imaging monitoring of drug accumulation and even therapeutic response due to their biocompatibility, chemical functionality, and potential to combine drug delivery and hyperthermia [85]. The promising subclasses are thermosensitive MLs with tumor-targeting ligand, which release the drug payload by reaching critical temperature. Hyperthermia, achieved by alternating magnetic field, will trigger drug release and, synergistically with the drug, kill tumor cells [86]. MLs are a prime example of multipurposing a single theranostic formulation component for more than one role, which then simplifies the overall formulation while increasing its functionality. In a recent study, very small hydrophobic SPIONs, 5 nm maghemite (Fe_2O_3) capped with oleic acid (OA), were embedded within lipid bilayers of a liposome [87]. Particles in the magnetic field generate heat that triggers the drug release. The same particles also provide contrast for MRI.

Liposomes proved to be highly versatile drug delivery systems and were used to improve drug efficacy, bioavailability, targeting, and toxicity profile [88]. They found broad application in tumor therapy, immunology, dermatology, treatment of the eye disorders, brain targeting, and treating infection and as vaccine adjuvant formulations. Furthermore, liposomes were explored for delivery of poorly soluble small-molecule drugs to large macromolecules and biologics such as plasmid DNA, siRNA, and protein drugs. Therefore, liposomes as theranostic formulations provide a wide range of opportunities for personalized nanomedicine. For example, in a recent study, multimodal imaging liposome with targeted delivery and controlled release features was developed [84]. The liposomes were composed of DSPC/cholesterol/Gd-DOTA-DSPE/DOTA-DSPE with the molar ratio of 39:35:25:1 and ammonium sulfate/pH gradient. The presence of the gradient allowed them to postload the anticancer drug (DOX), the gamma imaging radionuclide technetium-99m (^{99m}Tc) ($T_{1/2}$ = 6.0 h, 141 keV γ-ray), and the therapeutic radionuclides rhenium-186/rhenium-188 (^{186}Re/^{188}Re). In addition, NIR lipid conjugate could also be introduced post liposome formation. The resulting formulation became a highly versatile theranostic platform where multiple imaging agents and therapeutic modalities would be combined during and postprocessing [84]. Therapeutic anticancer siRNA delivery with magnetic liposome was also reported [89]. These examples suggest that liposomes may represent the most versatile theranostic nanomedicine platform with high likelihood of reaching clinical success.

Liposomes have been investigated for targeted delivery since the late 1970s. Several liposomal formulations are currently approved for clinical use [90]. With the development of long-circulating STEALTH liposomes, passive targeting to tissues and organs was clinically achievable. The introduction of PEG to liposomal surface increased their steric stabilization *in vivo* and extended the circulation lifetime, which further facilitated the passive targeting via EPR effect [91]. Passive and active targeting was utilized in TPGS-coated theranostic liposomes [92]. However, passive targeting is typically combined with active targeting. Active targeting of liposomal formulations for drug delivery is well studied and extensively reviewed elsewhere. (In-depth discussion is available in the following References [93–97].) Folate is the most common cancer-targeting agent used for a large number of nanosystems and so far the only one applied to theranostic liposomes [92]. We argue that more ligands (e.g., cannabinoid receptor, sigma receptor, mannose) should be considered alone or in combination with folate to maximize targeting of theranostic liposomes and improve their drug delivery and imaging capacity.

There are several preparation methods for liposomes. In 1978, Szoka and Papahadjopoulos reported a systematic study for preparation of liposomes where they correlated the formulation composition to size [98]. This study also lays out characterization procedures, which involved size measurements, structural integrity testing, incorporation, and release of a fluorescent dye, as a surrogate for drug incorporation/release studies and microscopy for structural assessment. They used phosphatidylcholine (PtdCho), phosphatidylglycerol (PtdGro), and phosphatidylserine (PtdSer), purified from egg yolk and bovine brain, and synthetic phosphatidic acid and dipalmitoyl phosphatidylcholine (Pal2PtdCho). The lipids were dissolved with cholesterol in organic solvent first, and then solvent was removed under reduced pressure. Then the aqueous phase was added to the dried lipids, and the mixture was subjected to sonication, extrusion, and freeze-thaw cycles to obtain small-size vesicles. The challenges in these approaches were small scale and use of potentially toxic organic solvents. Since the late 1970s, liposomes were extremely well researched, and extensive literature exists on liposomal preparation [99]. Recent study evaluated the effects of liposomal composition to its physicochemical properties [100]. It was found that the greatest mean size and polydispersity values were obtained for liposomes composed of di-oleoyl-phosphatidylcholine (DOPC), which contains unsaturated phospholipids. Liposome size also increased with incorporation of unsaturated fatty acids in phospholipid chains due to augmentation of liposome membrane fluidity. Furthermore, regardless of the preparation methodology, the most important factors are the hydrophilic/hydrophobic interactions between lipid–lipid and lipid–water molecules. The input of energy (extrusion, sonication, etc.) will further lead to lipid molecule rearrangements and thermodynamic equilibrium necessary for overall liposomal stability and performance [101]. It is easy to appreciate that as we increase formulation complexity of liposomes by introducing one or more imaging agents and drugs, their effects on the liposome bilayer structure and fluidity cannot be ignored. However, for theranostic development, liposomes are highly advantageous due to well-established manufacturing on a large scale and long history as drug delivery systems in the clinic. Here, we focus on those liposomes prepared to carry both drugs and imaging agents. Depending on the drug chemistry, different drug incorporation approaches are applied. For example, for small molecules, there are two ways the drug is incorporated into the liposome: passive entrapment and active loading [94]. In a recent study, docetaxel was incorporated into TGPS-coated theranostic liposome [102]. D-alpha-tocopheryl polyethylene glycol 1000 succinate monoester (TPGS) is a PEGylated vitamin E commonly used in pharmaceutical and food industry as solubilizer and emulsifier. To incorporate the drug, ethanol injection/evaporation was followed by ultrasonication and filtration to obtain the final product. Briefly, the drug was passively incorporated by being codissolved with lipids in ethanol. The imaging agent, quantum dots, was added to ethanol solution in tetrahydrofuran. Both drug and imaging agent were introduced at the same time during liposomal formation. This method works well for lipophilic drugs, while the imaging agent is soluble in organic solvents [102].

TABLE 15.1 Representative Liposomal Hybrid Nanoparticles for Simultaneous Imaging and Therapy of Cancer[a]

Structural nanocomponent Type (incorporated therapeutic payload and other function)	Functional nanocomponent Type (diagnostic and other function)
Phospholipid liposome (no payload and membrane fusion)	Gold nanoparticle (optical imaging)
Phospholipid liposome (calcein (model drug))	Gold nanoparticle (photothermal heating-triggered drug release)
Phospholipid liposome (6-carboxyfluorescein (model drug))	Gold nanoparticle (photothermal heating-triggered drug release)
Phospholipid liposome (xylenol orange sodium salt (model drug))	Magnetic nanocrystal (MRI)
Phospholipid liposome (5,6-carboxyfluorescein (model drug))	Magnetic nanocrystal (electromagnetic heating-triggered drug release)
Phospholipid liposome (doxorubicin or cysteine protease inhibitor JPM-565)	Magnetic nanocrystal (MRI and magnetic targeting)

[a]Based on Ref. [103].

Though liposomes offer high level of manufacturing feasibility, adding multiple entities will certainly increase production complexity and the process modification to improve imaging agent and/or drug loading are needed. Radio-ligand-decorated liposomes as theranostic platform were extensively studied for PET or SPECT image-guided drug delivery, reviewed recently by Petersen *et al.* [83]. Liposomes are also an excellent platform for the development of hybrid nanosystems, which combine multiple imaging agents, carry high drug payload, and have one or more targeting moieties. Liposomes were shown to successfully carry nanocrystals, inorganic nanoparticles, and large macromolecules on their surface. Because of the presence of both hydrophobic and hydrophilic compartments, multiple chemical entities can be accommodated. Table 15.1 (based on Ref. [103]) summarizes several liposomal hybrid formulations.

Theranostic liposomes are at early stages of development, though they have promising potential for clinical translation due to their high versatility in respect of carrying diverse chemical entities and established preparation protocols. Limited information exists on their efficacy *in vivo*. MLs have been evaluated for antitumor effects *in vivo*. In one study, Yoshida *et al.* demonstrated that docetaxel-loaded MLs inhibited tumor growth in a mouse model by combination of magnetic field-induced hyperthermia and cytotoxic action of the drug [104]. In addition, they demonstrated that the same particles provide tumor-specific MR contrast *in vivo*. MRI-guided drug delivery was reported in a B16 melanoma mouse model using long-circulating liposomes coloaded with the glucocorticoid prednisolone phosphate (PLP) and the amphiphilic paramagnetic contrast agent Gd-DOTAMA(C(18)). In this example, the drug PLP was encapsulated into liposome hydrophilic core, while the imaging agent was introduced into the lipid bilayer. Biodistribution of the theranostic *in vivo* was demonstrated using MRI. It was also shown that the drug and imaging agent carried in the same liposome display differential release profile *in vivo* [105]. This example illustrates the need for careful selection of imaging agent and the drug in a theranostic formulation, and how the drug and imaging agent are incorporated will drive the theranostic performance.

15.3.2 Nanoemulsions

Nanoemulsions are kinetically stable emulsions with a droplet size typically between 100 and 500 nm, high oil content, and low amounts of surfactant [106]. Nanoemulsions have wide applications in pharmaceutical and food industry. Nanoemulsions are typically used to increase tissue penetration of poorly soluble drugs (transdermal), drug solubility, and bioavailability (injectable) [107–111]. Nanoemulsions are also easily incorporated into other dosage forms such as capsules and gels. They can be produced on an industrial scale [111–113]. Therefore, due to scalability and high drug carrying capacity, nanoemulsions also represent an attractive theranostic development platform with a high potential for clinical use.

In nanoemulsions, imaging modalities that have been explored are MRI, ultrasound, photoacoustic, and optical imaging. Perfluorocarbons (PFCs) are so far the most utilized materials for imaging nanoemulsions. PFCs for MRI *in vivo* were first reported by Longmaid *et al.* [114]. They used PFC nanoemulsions, originally designed as blood substitutes, and imaged the liver, tumor, and abscess in rats. Since this report, PFC nanoemulsions were explored extensively as MRI agents [115]. Recently, Ahrens and Zhong summarized the use of PFCs for cell tracking *in vivo* [116]. Figure 15.4 shows typical examples of ^{19}F MRI using PFC nanoemulsions.

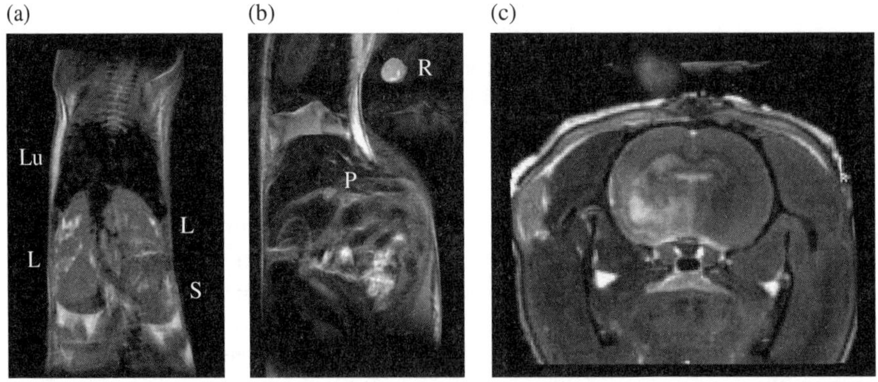

FIGURE 15.4 *In vivo* ^{19}F MRI using *ex vivo* perfluorocarbon (PFC)-labeled cells in rodent models. ^{19}F images of the labeled cells are displayed on a “hot-iron” intensity scale, and the anatomical (1H) images are shown in gray scale. (a) Composite image through the torso following intravenous inoculation with labeled dendritic cells (DCs). Cells are apparent in the liver (L) and spleen (S) and weakly in the lungs (Lu). (b) Adoptively transferred T cells selectively home to the pancreas in a prediabetic mouse model. The image shows T cells (pseudocolor) homing to the pancreas (P). R, reference capillary. (c) Neuronal stem cells injected into the infarct of a rat stroke model. A ^{19}F reference capillary is outside the brain. (Reprinted from Ref. [116]. © John Wiley & Sons, Ltd.) (*See insert for color representation of the figure.*)

Perfluorinated compounds are also contrast agents for ultrasound imaging. The FDA recently approved perflutren lipid microspheres (marketed as Optison™), composed of octafluoropropane encapsulated in an outer lipid shell and used for vascular imaging by medical ultrasound [117]. However, PFC microspheres are fairly large and have limited drug-targeting capacity. From an imaging point of view, PFC nanoemulsions are very attractive for theranostic development since they allow easy integration of multiple imaging modalities while at the same time they show significant drug loading capacity and potential for controlled drug release.

Janjic *et al.* [118] reported in 2008 perfluoropolyether (PFPE) conjugates to several fluorescent dyes. PFC nanoemulsions can also be labeled with NIR dyes by entrapment during nanoemulsification in the lipid or surfactant layer [3, 4, 119]. Additionally, PFC nanoemulsions designed as dual-mode imaging agents for optical and photoacoustic imaging were reported by Akers *et al.* [120]. PFC nanoemulsions were also investigated as ultrasound imaging contrast agents in preclinical models [121].

As a theranostic platform, PFC nanoemulsions have attracted increasing interest over the past 5 years. This is likely due to ease of preparation of PFC nanoemulsion on a reasonable scale, their ability to support multiple imaging modalities without added formulation complexity, and diverse drug loading options (entrapment, conjugation) with controlled release. Figure 15.5 shows PFC molecules used for theranostic nanoemulsion development to date. Recently, Janjic *et al.* reported theranostic nanoemulsions carrying a nonsteroidal anti-inflammatory drug (NSAID) [3, 4]. The lipophilic drug celecoxib was incorporated into the hydrocarbon core of the triphasic FC/HC/W nanoemulsion during processing. The nanoemulsion, once loaded into cells, provided ^{19}F MR signal, NIR signal, and suppressed COX-2-driven PgE2 release by releasing the drug in the cytoplasm.

Wickline *et al.* developed unique PFC nanoemulsions designed to deliver biologic cytolytic peptides specifically to tumors [122]. They named these particles "nanobees." This pioneering work laid out a foundation for PFC nanoemulsion theranostic development and also put these agents at the forefront of drug delivery vehicle design. Cytolytic peptides were identified as highly attractive and effective anticancer agents.

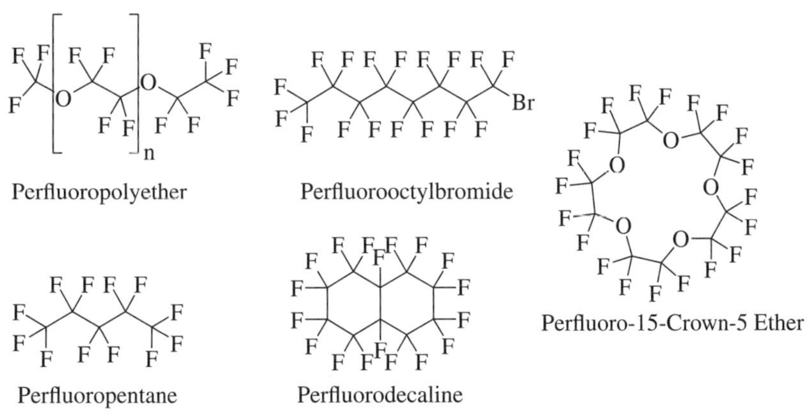

FIGURE 15.5 Perfluorocarbons used for theranostic nanoemulsions and nanoparticle development.

The bee venom-derived peptide melittin was incorporated into the lipid layer of the PFC nanoemulsion [123], and it was shown that melittin pharmacokinetics and tumor targeting were successfully achieved with dramatic tumor size reduction [124]. These nanoemulsions also provided tumor-specific ^{19}F MR images in the mouse model. Nanobees, therefore, represent a prime example of nanoemulsion theranostics. Rappaport *et al.* reported the delivery of PTX in a block-polymer-stabilized perfluoropentane nanoemulsion guided by ultrasound and MRI in a mouse tumor model [121]. In this formulation, the low-molecular-weight and low-boiling-point PFC perfluoropentane was used to promote ultrasound-triggered drug release. In this example, PFC is fulfilling two distinct roles, a dual-mode imaging agent (ultrasound and MR) and a drug release-controlling agent. Therefore, PFCs are again demonstrated to be very versatile materials for theranostic development. Beyond imaging, PFC nanodroplets can be used for controlled release using ultrasound [121, 125]. The ultrasound-responsive nanoemulsions are prepared with low-boiling-point PFCs, perfluorohexane and dodecafluoropentane (DDFP). Rappaport *et al.* describe DDFP nanoemulsions that grow from nanosized droplets to microbubbles under the effect of ultrasound [126]. This droplet increase in size can provide echogenic signal and at the same time induce drug release, as described in Figure 15.6.

Nanoemulsions are stabilized by a relatively low amount of surfactant. They can be stabilized by both ionic and nonionic surfactants. Low droplet size of nanoemulsions (150–200 nm) facilitates passive targeting to tumors and easy penetration to inflamed tissues via EPR effect. Passive targeting to circulating inflammatory cells by PFC nanoemulsions has been extensively investigated for the purpose of ^{19}F MRI of inflammation [119, 127–129] and potentially drug delivery to inflamed tissues [3]. A recent study used PFC nanoemulsions to deliver antigen to dendritic cells with the goal of boosting the immune response in DC-based vaccines [130]. For active targeting

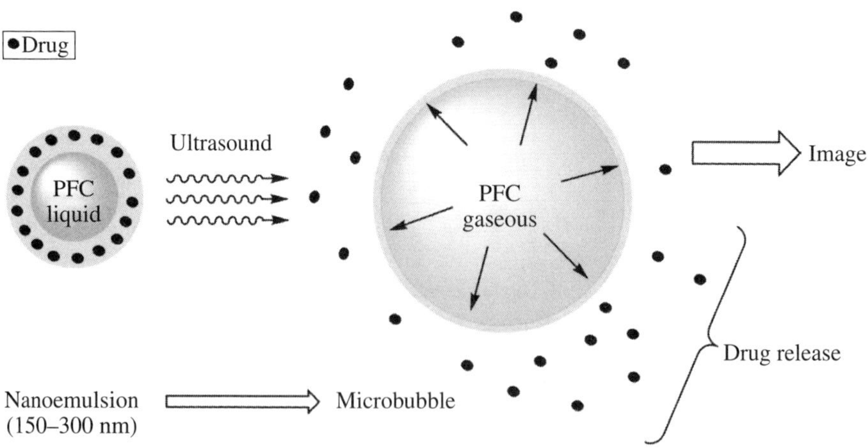

FIGURE 15.6 PFC nanoemulsion with drug loaded in the hydrophobic core of the surfactant layer when exposed to ultrasound expands in size as PFC shifts from liquid to gaseous phase. This change triggers the drug release and provides ultrasound imaging contrast.

strategies in nanoemulsion-based theranostics, the targeting agent is attached to the droplet surface by covalent conjugation to the surfactant molecule. For example, PFC nanoemulsions targeted to $\alpha_v\beta_3$ integrins for atherosclerosis imaging have been developed [131]. In an earlier study, Lanza *et al.* reported the targeted delivery of antiproliferative drugs to smooth muscle cells using perfluorooctyl bromide (PFOB) nanoemulsions [132]. The PFOB nanoemulsion lipid outer layer incorporated biotinylated phosphatidylethanolamine and tissue factor-specific antibody attached to the surface by using avidin. The hydrophobic drugs DOX and taxol were introduced to the formulation by adding them to the surfactant comixture prior emulsification.

Nanoemulsions are prepared by high-pressure homogenization, high energy processing (e.g., microfluidization and sonication), and low energy emulsification methods, such as phase inversion temperature (PIT) and concentration (PIC) methods [133]. Typical nanoemulsions are oil-in-water emulsions (O/W) with one or two surfactants to stabilize the dispersed phase. Recently, our group reported novel triphasic nanoemulsions (fluorocarbon/hydrocarbon/water, F/H/W), which contain two immiscible oils (F and H) as the dispersed phase stabilized by surfactants [3, 134]. Nanoemulsions show remarkable kinetic stability as demonstrated recently by combination of static and dynamic neutron scattering analyses [135]. Due to their small droplet size, nanoemulsions are translucent and typically resistant to creaming and sedimentation. They do not require high amounts of surfactant and can be stabilized by steric effects [106]. The most common degradation mechanism in nanoemulsions is Ostwald ripening, a gradual growth of the larger particles at the expense of smaller ones by molecular diffusion. Nanoemulsions can be produced using microfluidization or high-pressure homogenization on larger industrial scale reproducibly [136]. This last feature makes nanoemulsions very attractive systems for theranostic nanomedicine development. They offer high carrying capacity for imaging and therapeutic agents while being manufacturing friendly. For theranostic formulations, nanoemulsions are typically prepared using high energy methods, sonication, and microfluidization. Sonication is an attractive method for small-scale production in the laboratory due to its simplicity and ease of operation. However, microfluidization shows superior performance in droplet size reduction, small polydispersity, and potential for increased scale production. Therefore, microfluidization in our view is the preferred method for nanoemulsification.

There are two ways of preparing drug-loaded imaging-capable nanoemulsions: (1) direct entrapment and (2) conjugation. Patel *et al.* recently demonstrated that lipophilic drugs can be easily introduced into PFC nanoemulsion by dissolving them into a hydrocarbon-solubilizing oil first, mixing with PFC, and emulsifying the mixture by microfluidization [3]. Earlier studies by Wickline *et al.* added drugs to lipid surfactant commixture before the water phase was added and nanoemulsion was subsequently produced by sonication or microfluidization [123]. There are also reports of nonfluorinated theranostic nanoemulsions. In a recent study, O/W nanoemulsion loaded with iron oxide nanocrystals for MRI, the fluorescent dye Cy7 for NIRF imaging, and the hydrophobic glucocorticoid prednisolone acetate valerate (PAV) was evaluated in a mouse tumor model. The nanoemulsion surface was decorated with $\alpha_v\beta_3$-specific RGD peptides for angiogenesis targeting. In the study, both imaging modalities (MRI and NIRF) showed high specificity for the growing tumors. Though imaging was successfully used to visualize tumor growth and nanoemulsion distribution, confirmation on drug effects was obtained from histological data and molecular analysis of excised tissues [137].

15.4 SELECTED POLYMERIC NANOSYSTEMS

15.4.1 Dendrimers

Dendrimers are perfectly hyperbranched macromolecules with a large number of surface functional groups that define their properties and activity [138]. The architecture of dendrimers consists of a core surrounded by generations of covalently linked branched interiors [139]. This creates layered structure that grows as the dendrimer generation increases. Therefore, unlike most other types of nanoparticles, dendrimers have precisely controlled size, shape, and surface chemistry [140]. These unique properties allow dendrimers to be developed with high structural monodispersity, desired plasma circulation time, and biodistribution properties, as well as control over drug release. The abundant surface functional groups make dendrimers platforms for attachment of specific targeting groups, surface modifiers, imaging agents, and therapeutic drugs. These tunable properties, together with the precisely controlled structure, render dendrimers attractive theranostic agents. The most commonly studied dendrimeric structures are the polyamidoamine (PAMAM, Fig. 15.7) and polypropylenimine (PPI) dendrimers. Unlike IONP, CNTs, and GNPs, dendrimers do not possess any intrinsic signaling moieties. However, they can be used in a range of imaging modalities through their abundant surface functional groups. For example, NIR fluorescent dyes can be conjugated to dendrimers for optical imaging. It is also possible to use dendrimer in PET, SPECT, or MRI by attaching chelated $^{64}Cu^{2+}$, $^{111}In^{3+}$, or Gd^{3+} to the dendrimer surface. The emerging interest in multimodal imaging has also been met using dendrimers as the platforms. For instance, a bimodal MR/NIR fluorescence contrast agent was synthesized by attaching DTPA chelated Gd^{3+} and Cy5.5 dye to the surface of a generation 6 (G6) PAMAM dendrimer [142].

Drug molecules can be either encapsulated within the nanostructure or covalently conjugated to the surface of dendrimers. Noncovalent encapsulation is usually associated with hydrogen bonding or hydrophobic or ionic interactions [140]. Covalent conjugation between anticancer drugs and dendrimer surface is typically through a labile linker, such as a disulfide, peptide, or ester linker, that is tailored to release the drug in a controlled fashion. Although noncovalent encapsulation has the advantage of easy drug loading without synthesis and drug regulatory issues, it is challenged by the burst release of drug molecules in plasma, which often leads to very limited therapeutic benefit compared to free drug molecules [140]. Several drugs have been loaded into dendrimers through noncovalent encapsulation, such as MTX in a G2 polyester copolyether dendrimer [143], DOX in a G3 PAMAM dendrimer [144], PTX in a G5 polyglycerol dendrimer [145], and 5-fluorouracil (5-FU) in PAMAM dendrimers [146].

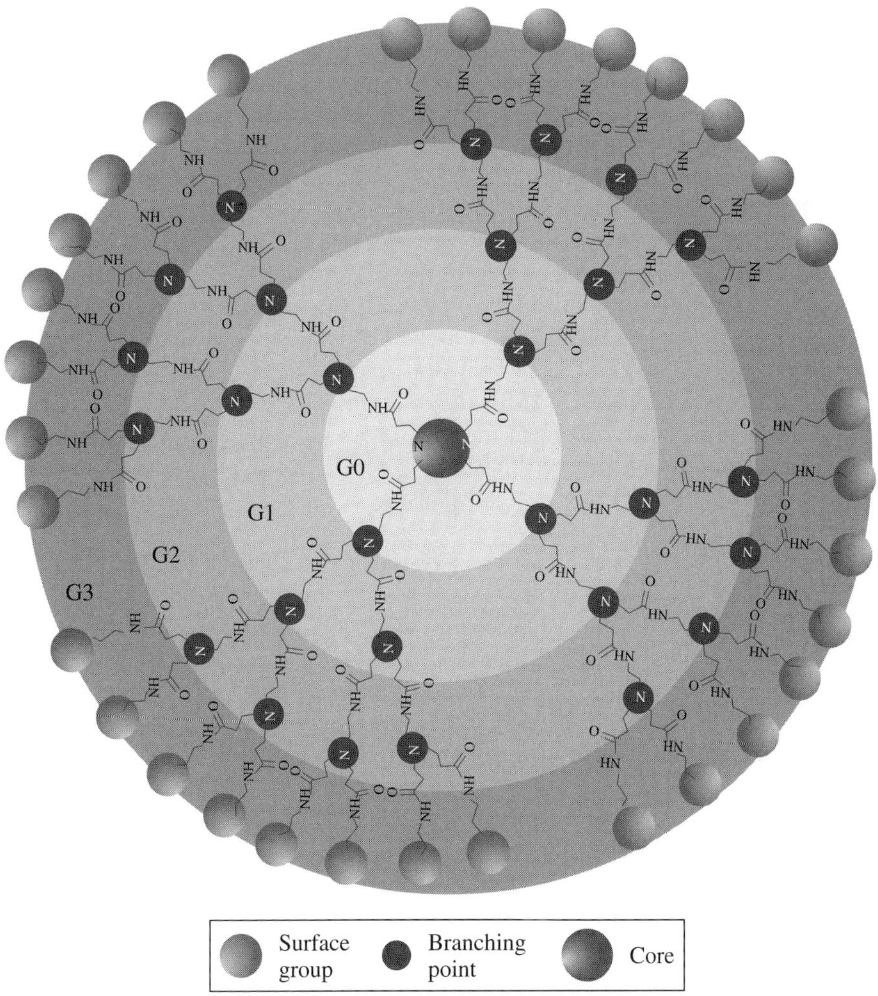

FIGURE 15.7 Structure of PAMAM dendrimers. (Reprinted from Ref. [141]. © Wiley.)

On the other hand, covalent conjugation of drugs with dendrimers enables better control over the drug release, as the labile linkers between drug and dendrimer can be specifically cleaved under certain tumor microenvironment. This will allow for reduced systematic toxicity and improved therapeutic efficacy. However, covalent conjugation involves additional synthetic steps and often requires modification of drug structure for conjugation reaction. As a result, it has the potential issues of releasing less active forms of the drug and drug regulations. There has been growing interest in loading drugs to dendrimers through covalent conjugation due to its significant advantage of controlled drug release. Some examples include MTX-G4 PAMAM with an amide linker [147], MTX-G5 polylysine dendrimer with a peptide linker [148], DOX-PAMAM with an acid-labile, *cis*-aconityl linker [149], and PTX–triazine dendrimer with an ester linker [150].

Like other types of theranostic nanoparticles, dendrimers can be delivered to a tumor region through both passive and active targeting mechanisms. For example, several drug-loaded dendrimers have been reported to be efficiently uptaken in prostate cancers through EPR effect [151, 152]. For active targeting, many dendrimers that specifically target receptors overexpressed by cancers have been developed through bioconjugation with targeting molecules on the surface. For example, PAMAM has been conjugated with antibodies [153], aptamers [154], and peptides [155] to target PSMA and $\alpha_v\beta_3$-integrin overexpressed by prostate cancer cells.

Two strategies have been reported for dendrimer synthesis, including divergent and convergent growth methods [156]. In the divergent method, synthesis begins with a desired functional core, and a branch component is constructed stepwise around the core to produce a core–shell structure [157]. Most commercially available dendrimers, such as PAMAM and PPI dendrimers, are synthesized using the divergent method. The convergent method starts with synthesis of the peripheral branches (dendrons) and proceeds inward to a reactive focal point at the root [157]. One example of dendrimers developed using convergent approach is dendritic poly(ether-imide)s [158].

The ability to carry various biologically active molecules in a well-defined and controlled manner makes dendrimers stand out as a promising theranostic platform. A recent study reported by Al-Jamal *et al.* uses a G6 polylysine dendrimer loaded with DOX (DOX-DM) as the theranostic agent [159]. The retention of DOX-DM was monitored by live whole-animal fluorescence imaging using fluorescence of DOX as the signal. In a lung cancer xenograft mouse model, DOX-DM showed enhanced therapeutic efficacy than free DOX.

15.4.2 Polymeric Nanoparticles

Some would refer to the end of twentieth and beginning of the twenty-first century as the era of polymers. Polymers are everywhere in our lives, from a coffee cup to a seat on an airplane to medicines and devices we use every day. No surprise, polymers became materials of choice for nanosystem development in drug delivery and imaging in the past several decades. In fact, polymers have been mentioned so far throughout this chapter but mostly in supporting roles.

Here, we'd discuss polymers in more depth and how they seem to be able to carry on the biggest weight when it comes to theranostic nanomedicine design. In their recent review on polymerasomes (or polymeric vesicles), De Oliveira *et al.* [160] argue that polymers are here to help us solve remaining problems in modern therapeutics such as efficacy, specificity, and controlled release of drugs to diseased tissues. There is yet much left to be done, and polymers may offer the most. Polymerasomes can be designed to respond to specific triggers (pH, temperature, magnetic field), carry drugs and targeting agents, and include multiple imaging modalities [160]. In that respect, polymerasomes are similar to liposomes but with one key difference—polymers allow for much more versatile chemistry and engineering over liposomes built primarily with lipids and also give us higher chemical stability and good biodegradable properties. So here, we discuss polymers in respect of their role as a

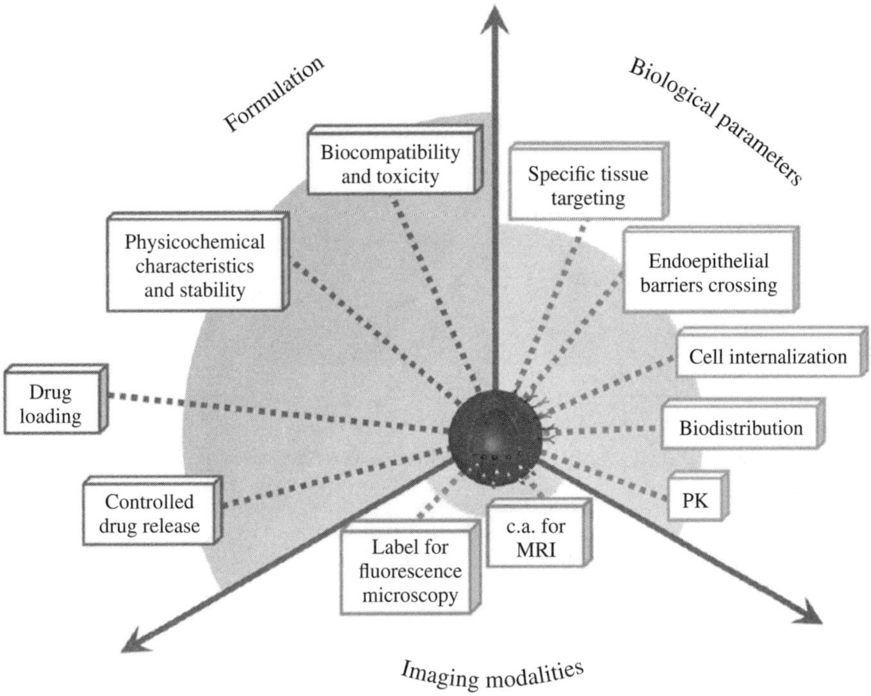

FIGURE 15.8 Challenges of polymersome development for theranostic applications. Size of the branches is proportional to the number of articles dealing with the subject in a logarithmic scale. (Reprinted from Ref. [160]. © Wiley.)

theranostic platform through specific examples. We also discuss how polymers may offer a solution to highly pressing problem of cost and manufacturing of theranostic nanomedicines. Challenges polymerasomes are to solve are presented in Figure 15.8, which also illustrates challenges ahead of us regardless of which nanosystem we are designing.

Polymeric particles have been formulated to serve as imaging agents alone for MRI, optical imaging, and ultrasound or in combination with varied drugs as potential theranostic systems [161]. It is important to ask a question here: if a polymeric nanoparticle carrying a drug carries a fluorescent dye that cannot be used for *in vivo* imaging, does this particle qualify as a theranostic? Though we do not wish to argue against much work done in this way, it is important to keep a perspective of what true theranostic is. Therefore, here we only discuss those polymeric particles that can be imaged *in vivo*. NIR *in vivo* imaging (discussed earlier in this text) became a cost-effective and highly versatile imaging method for preclinical and more recently clinical applications. NIR dyes can be directly conjugated to polymers [162] or incorporated into the nanoparticle by entrapment [154, 163]. For example, Zheng *et al.* doped indocyanine green (ICG), the only FDA-approved NIR dye, into poly(D,L-lactide-co-glycolide) (PLGA) lipid nanoparticles [163]. The resulting NIR nanoparticles exhibited good biocompatibility, monodispersity, excellent NIR

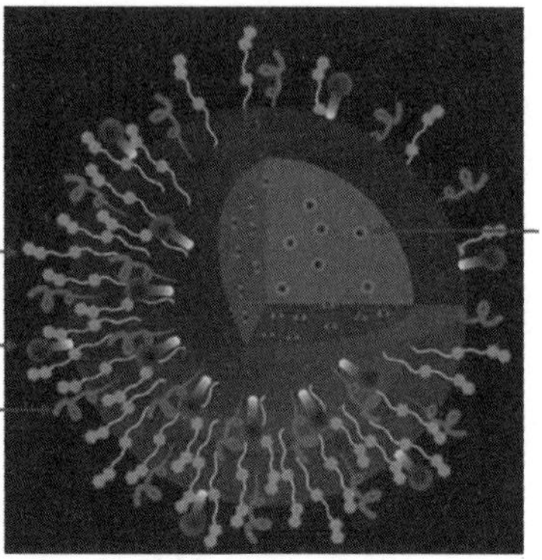

FIGURE 15.9 Schematic representation of the concept of a multifunctional core–shell nanoparticle system (PLGA/PFPL) composed of a PLGA core and a paramagnetic-folate-coated PEGylated liposome shell for MRI and drug delivery for targeted therapy. (Reprinted from Ref. [166]. © Wiley.)

penetration ability, significant stability against photobleaching, and long circulation time. Polymeric nanoparticles for theranostics can also be developed based on magnetic nanoparticles, such as paramagnetic gadolinium (Gd) nanoparticles [164] and SPIONs [165]. As an example, copolymers of poly-(N-isopropylacrylamide)-*co*-poly(N-acryloxysuccinimide)-*co*-poly(fluorescein *O*-methacrylate) were covalently attached to the surface of gadolinium metal–organic framework nanoparticles [164]. The gadolinium metal–organic core allows for imaging with T_1 MRI.

Similar to dendrimers, polymeric nanoparticles can be developed to carry therapeutic molecules either by encapsulation or bioconjugation on the surface. As such, a wide range of drugs can be incorporated into polymeric nanoparticles, regardless of the availability of functional groups on the drug molecules. For example, Liao *et al.* reported a multifunctional nanoscale platform with a hydrophobic poly (DL-lactide-co-glycolide) (PLGA) core and a hydrophilic paramagnetic-folate-coated PEGylated lipid shell (PFPL, PEG), which was designed for simultaneous MRI and targeted therapeutics [166]. Hydrophobic drug DOX was loaded inside the nanostructures by encapsulation with high loading percentage (~90%) (Fig. 15.9). The drug release study indicated that the modification of the PLGA core could significantly reduce the drug release rate. In another study, both DOX and gemcitabine (a well-known radiosensitiser) were covalently attached to *N*-(2-hydroxypropyl)methacrylamide (HPMA) copolymers via a lysosomally degradable spacer to study the synergistic effect of chemotherapy and radiotherapy [167]. *In vivo* therapeutic study indicated that the carrier-based radiochemotherapy was a promising approach for improving the efficacy of combined modality anticancer therapy.

Due to the well-developed polymer chemistry, a wide variety of polymeric nanoparticles have been developed with various sizes, surface charges, and functional groups, allowing for different targeting strategies. Many polymeric nanoparticles can be delivered through passive targeting, such as the DOX and gemcitabine-loaded HPMA copolymers mentioned earlier, whereas certain polymeric nanoparticles can be developed with targeting molecules on the surface, such as the folate-coated multifunctional PLGA/PFPL nanoparticle system shown in Figure 15.9. Theranostic polymers can be developed through various synthetic methods (direct polymerization, self-assembly, nanoprecipitation, emulsification/precipitation, etc.), which have been summarized in recent reviews [160, 168–170]. Polymeric nanoparticle formulation is best described through a recent nanoparticle example [171]. In this study, TPGS-functionalized PLGA particles were prepared for controlled release of PTX with a uniform size of 130–150 nm and a drug entrapment efficiency of 80%. The particles were synthesized by a modified inverse-phase nanoprecipitation method where the aqueous phase containing TPGS was slowly added into the water miscible organic phase. With the slow diffusion of the organic solvent into the aqueous solution, PLGA was precipitated to form nanoparticles and the TPGS coated as the monolayer around the PLGA polymeric core simultaneously. Simultaneous self-assembly and precipitation leads to small particle size with high uniformity and high drug loading. Summary of most common polymeric nanoparticle synthesis approaches is shown in Figure 15.10, and for detailed discussion, the reader is referred to Reference [172].

An early approach to develop theranostic polymers involves incorporation of photosensitizers into polymeric nanoparticles. For example, McCarthy *et al.* reported poly(lactic-co-glycolic acid) nanoparticles that encapsulate the photosensitizer meso-tetraphenylporpholactol [173]. Due to self-quenching, encapsulated meso-tetraphenylporpholactol did not produce significant fluorescence signal or cytotoxic singlet oxygen. However, when photosensitizer was released upon cellular uptake of the polymeric nanoparticles, considerable fluorescence signals and phototoxicity were recovered. *In vivo* phototherapy study in a prostate cancer mouse model showed complete eradication of tumors.

A major limitation of photosensitizer-incorporated polymeric nanoparticles is low tissue penetration of light, as most photosensitizers absorb in the visible region. To address this issue, NIR fluorescence dyes and chemotherapy agents were introduced

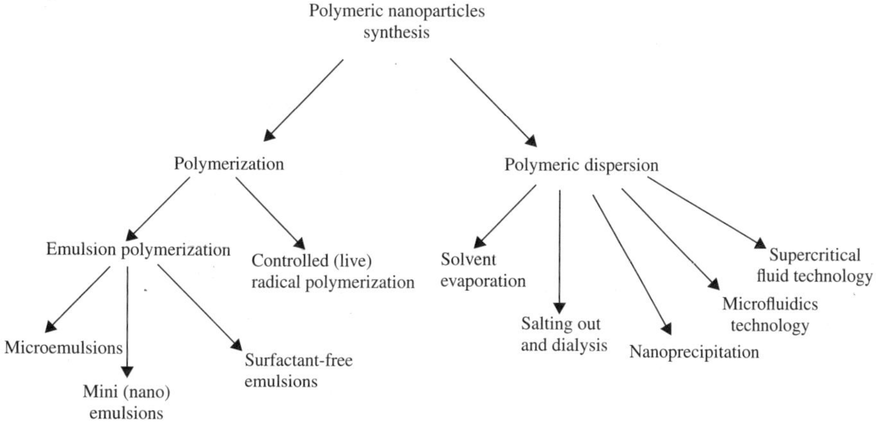

FIGURE 15.10 A brief summary of common methods for polymeric nanoparticle synthesis used in drug delivery and imaging.

to polymeric nanocarriers. For example, an NIR dye, Cy5.5, and an anticancer drug, PTX, were attached to glycol chitosan-based polymeric nanoparticles [162]. *In vivo*, these nanoparticles showed fourfold tumor/normal fluorescence signal contrast. Furthermore, they demonstrated significantly improved anticancer efficacy in a variety of tumor mouse models over free drugs, mainly due to the passive targeting effect.

Other than fluorescence imaging, MRI is also commonly used to image theranostic polymers. As mentioned earlier, many polymeric nanoparticles have been developed based on magnetic nanoparticles. For example, HPMA copolymers with DOX and gemcitabine were shown to circulate for prolonged periods of time and localize to tumors both effectively and selectively under MRI [167]. In rat prostate carcinoma tumor model, the copolymers were shown to interact synergistically with radiotherapy, with radiotherapy increasing the tumor accumulation of the copolymers, and with the copolymers increasing the therapeutic index of radiochemotherapy (both for DOX and for gemcitabine).

15.5 PHARMACEUTICAL ASPECTS OF THERANOSTICS

Pharmaceutical design and development approaches to theranostic nanomedicine have not been fully explored yet. Throughout this chapter, we argued for bringing pharmaceutical principles to early stages of development with hope that these would increase likelihood of success in the clinic. We talked about complexity of theranostic nanomedicines and shared a few examples where this was very evident. How this complexity of theranostic nanomedicines will play out in the real world when they are converted into medicines on the market is yet to be seen. As we continue to work on theranostic nanomedicine, we can learn from existing nanomedicine research and hopefully with the use of molecular imaging surpass many current challenges. Recently, Venditto and Szoka exclaimed in their comprehensive review on cancer nanomedicines entitled "So Many Papers and So Few Drugs!" [174]. In their critical overview of recent literature, they designated a timeline for nanomedicine development, which includes invention phase, innovation phase, and imitation phase. They point out to proliferation of nanomedicines with somewhat modest or moderate improvements over existing reports. Could it be that theranostic nanomedicine is riding the wave of existing innovation in nanomedicine? Or can theranostic nanomedicine stand alone as the solution for remaining challenges in nanomedicine such as lack of data supporting dosing in humans, direct measurements of efficacy, ability to personalize treatment to disease state, and whole-body state in an individual patient, mechanistic insights into nanomedicine effects on the body, and body effects on the nanomedicine? We would like to argue that if theranostic nanomedicines are designed rationally as future medicines, we could not only get closer to the goal of personalized medicine but also it can help us understand better the nanomedicine formulations in the body. Further, theranostic nanomedicines offer an extended value of becoming powerful probes for deducing disease mechanisms and mechanisms of drugs in live subjects, whether animals or people. For example, one can imagine a theranostic liposome, where the imaging agent is a biosensor for a particular change in the body that is expected only at a certain drug concentration, to be used to probe drug action in the subject as the drug is released. If the nanomedicine formulation is to become a tool for personalized medicine, more information is needed than just whether the drug delivery system reached its target or not. As we mentioned earlier, the question of cost and manufacturing feasibility is quite important to keep asking when theranostic nanomedicines are considered and designed. No patient would be able to afford a medicine that is too costly to produce no matter how much personalization it may offer. Many examples of multifunctional nanoparticles have its value for research, but may not always lead toward nanomedicine goal if manufacturing and safety are not fully considered. We do not wish to say that highly innovative nanomedicine formulations to date have no chance of reaching the clinic but caution that in their design pharmaceutical principles must be implemented early on. It is also wise to consider formulation platforms well tested as drug delivery systems or imaging agents with known clinical use for theranostic nanomedicine formulation. Those existing formulations come to theranostic field with already well-described manufacturing scale production know-how, patient safety, pharmacokinetics. We identified that for theranostic development liposomes, polymeric nanoparticles, and nanoemulsions may be pharmaceutically the most attractive platform. As nanodispersions are being developed to improve drug delivery, taking those into consideration for theranostic development is another wise approach. Finally, we also want to point out that theranostic principles could be applied to dosage forms beyond injectable nanosystems. Oral drug delivery or implants could benefit from theranostic design principles.

At the end, we would like to see theranostics as the personalized nanomedicines in the future and hope the efforts already existing in engineering, chemistry, material science, and pharmaceutics, when combined into joint forces, can lead us there faster.

REFERENCES

[1] Etheridge ML, Campbell SA, Erdman AG, Haynes CL, Wolf SM, McCullough J. The big picture on nanomedicine: the state of investigational and approved nanomedicine products. Nanomedicine 2013;**9**:1–14.

[2] Wang R, Billone PS, Mullett WM. Nanomedicine in action: an overview of cancer nanomedicine on the market and in clinical trials. J Nanomater 2013;**2013**:1–12.

[3] Patel SK, Zhang Y, Pollock JA, Janjic JM. Cyclooxgenase-2 inhibiting perfluoropoly (ethylene glycol) ether theranostic nanoemulsions-in vitro study. PLoS One 2013;**8**:e55802.

[4] O'Hanlon CE, Amede KG, O'Hear MR, Janjic JM. NIR-labeled perfluoropolyether nanoemulsions for drug delivery and imaging. J Fluor Chem 2012;**137**:27–33.

[5] Kievit FM, Zhang M. Surface engineering of iron oxide nanoparticles for targeted cancer therapy. Acc Chem Res 2011;**44**:853–862.

[6] Jain TK, Foy SP, Erokwu B, Dimitrijevic S, Flask CA, Labhasetwar V. Magnetic resonance imaging of multifunctional pluronic stabilized iron-oxide nanoparticles in tumor-bearing mice. Biomaterials 2009;**30**:6748–6756.

[7] Santra S, Kaittanis C, Grimm J, Perez JM. Drug/dye-loaded, multifunctional iron oxide nanoparticles for combined targeted cancer therapy and dual optical/magnetic resonance imaging. Small 2009;**5**:1862–1868.

[8] Tran PH, Tran TT, Vo TV, Lee BJ. Promising iron oxide-based magnetic nanoparticles in biomedical engineering. Arch Pharm Res 2012;**35**:2045–2061.

[9] Sanvicens N, Marco MP. Multifunctional nanoparticles—properties and prospects for their use in human medicine. Trends Biotechnol 2008;**26**:425–433.

[10] Cai W, Chen X. Multimodality molecular imaging of tumor angiogenesis. J Nucl Med 2008;**49** (Suppl 2):113S–128S.

[11] Chen K, Xie J, Xu H, Behera D, Michalski MH, Biswal S, Wang A, Chen X. Triblock copolymer coated iron oxide nanoparticle conjugate for tumor integrin targeting. Biomaterials 2009;**30**:6912–6919.

[12] Lee HY, Li Z, Chen K, Hsu AR, Xu C, Xie J, Sun S, Chen X. PET/MRI dual-modality tumor imaging using arginine-glycine-aspartic (RGD)-conjugated radiolabeled iron oxide nanoparticles. J Nucl Med 2008;**49**:1371–1379.

[13] Xie J, Chen K, Huang J, Lee S, Wang J, Gao J, Li X, Chen X. PET/NIRF/MRI triple functional iron oxide nanoparticles. Biomaterials 2010;**31**:3016–3022.

[14] Xie J, Lee S, Chen X. Nanoparticle-based theranostic agents. Adv Drug Deliv Rev 2010;**62**:1064–1079.

[15] Kohler N, Sun C, Wang J, Zhang M. Methotrexate-modified superparamagnetic nanoparticles and their intracellular uptake into human cancer cells. Langmuir 2005;**21**:8858–8864.

[16] Piao Y, Kim J, Na HB, Kim D, Baek JS, Ko MK, Lee JH, Shokouhimehr M, Hyeon T. Wrap-bake-peel process for nanostructural transformation from beta-FeOOH nanorods to biocompatible iron oxide nanocapsules. Nat Mater 2008;**7**:242–247.

[17] Cheng K, Peng S, Xu C, Sun S. Porous hollow Fe(3)O(4) nanoparticles for targeted delivery and controlled release of cisplatin. J Am Chem Soc 2009;**131**:10637–10644.

[18] Medarova Z, Pham W, Farrar C, Petkova V, Moore A. *In vivo* imaging of siRNA delivery and silencing in tumors. Nat Med 2007;**13**:372–377.

[19] Guerrero S, Araya E, Fiedler JL, Arias JI, Adura C, Albericio F, Giralt E, Arias JL, Fernandez MS, Kogan MJ. Improving the brain delivery of gold nanoparticles by conjugation with an amphipathic peptide. Nanomedicine 2010;**5**:897–913.

[20] Veiseh O, Sun C, Fang C, Bhattarai N, Gunn J, Kievit F, Du K, Pullar B, Lee D, Ellenbogen RG, Olson J, Zhang M. Specific targeting of brain tumors with an optical/magnetic resonance imaging nanoprobe across the blood-brain barrier. Cancer Res 2009;**69**:6200–6207.

[21] Hombach J, Bernkop-Schnurch A. Chitosan solutions and particles: evaluation of their permeation enhancing potential on MDCK cells used as blood brain barrier model. Int J Pharm 2009;**376**:104–109.

[22] Wunderbaldinger P, Josephson L, Weissleder R. Crosslinked iron oxides (CLIO): a new platform for the development of targeted MR contrast agents. Acad Radiol 2002;**9** (Suppl 2):S304–S306.

[23] Harisinghani M, Ross RW, Guimaraes AR, Weissleder R. Utility of a new bolus-injectable nanoparticle for clinical cancer staging. Neoplasia 2007;**9**:1160–1165.

[24] Sun S, Zeng H. Size-controlled synthesis of magnetite nanoparticles. J Am Chem Soc 2002;**124**:8204–8205.

[25] Xu CJ, Xu KM, Gu HW, Zheng RK, Liu H, Zhang XX, Guo ZH, Xu B. Dopamine as a robust anchor to immobilize functional molecules on the iron oxide shell of magnetic nanoparticles. J Am Chem Soc 2004;**126**:9938–9939.

[26] Hultman KL, Raffo AJ, Grzenda AL, Harris PE, Brown TR, O'Brien S. Magnetic resonance imaging of major histocompatibility class II expression in the renal medulla using immunotargeted superparamagnetic iron oxide nanoparticles. ACS Nano 2008;**2**:477–484.

[27] Yu MK, Jeong YY, Park J, Park S, Kim JW, Min JJ, Kim K, Jon S. Drug-loaded superparamagnetic iron oxide nanoparticles for combined cancer imaging and therapy *in vivo*. Angew Chem 2008;**47**:5362–5365.

[28] Yu MK, Kim D, Lee IH, So JS, Jeong YY, Jon S. Image-guided prostate cancer therapy using aptamer-functionalized thermally cross-linked superparamagnetic iron oxide nanoparticles. Small 2011;**7**:2241–2249.

[29] Kirui DK, Khalidov I, Wang Y, Batt CA. Targeted near-IR hybrid magnetic nanoparticles for *in vivo* cancer therapy and imaging. Nanomedicine 2012;**9**:702–711.

[30] Higuchi T, Anton M, Dumler K, Seidl S, Pelisek J, Saraste A, Welling A, Hofmann F, Oostendorp RA, Gansbacher B, Nekolla SG, Bengel FM, Botnar RM, Schwaiger M. Combined reporter gene PET and iron oxide MRI for monitoring survival and localization of transplanted cells in the rat heart. J Nucl Med 2009;**50**:1088–1094.

[31] Xia Y, Li W, Cobley CM, Chen J, Xia X, Zhang Q, Yang M, Cho EC, Brown PK. Gold nanocages: from synthesis to theranostic applications. Acc Chem Res 2011;**44**:914–924.

[32] Xu C, Tung GA, Sun S. Size and concentration effect of gold nanoparticles on X-ray attenuation as measured on computed tomography. Chem Mater 2008;**20**:4167–4169.

[33] Melancon MP, Lu W, Yang Z, Zhang R, Cheng Z, Elliot AM, Stafford J, Olson T, Zhang JZ, Li C. In vitro and *in vivo* targeting of hollow gold nanoshells directed at epidermal growth factor receptor for photothermal ablation therapy. Mol Cancer Ther 2008;**7**:1730–1739.

[34] Bhatnagar P, Li Z, Choi Y, Guo J, Li F, Lee DY, Figliola M, Huls H, Lee DA, Zal T, Li KC, Cooper LJ. Imaging of genetically engineered T cells by PET using gold nanoparticles complexed to Copper-64. Integr Biol (Camb) 2013;**5**:231–238.

[35] Xie H, Wang ZJ, Bao A, Goins B, Phillips WT. *In vivo* PET imaging and biodistribution of radiolabeled gold nanoshells in rats with tumor xenografts. Int J Pharm 2010;**395**: 324–330.

[36] van Schooneveld MM, Cormode DP, Koole R, van Wijngaarden JT, Calcagno C, Skajaa T, Hilhorst J, Hart DC, Fayad ZA, Mulder WJ, Meijerink A. A fluorescent, paramagnetic and PEGylated gold/silica nanoparticle for MRI, CT and fluorescence imaging. Contrast Media Mol Imaging 2010;**5**:231–236.

[37] Qin H, Zhou T, Yang S, Chen Q, Xing D. Gadolinium(III)-gold nanorods for MRI and photoacoustic imaging dual-modality detection of macrophages in atherosclerotic inflammation. Nanomedicine 2013;**8**:1611–1624.

[38] Wang J, Wheeler D, Zhang JZ, Achilefu S, Kang KA. NIR fluorophore-hollow gold nanosphere complex for cancer enzyme-triggered detection and hyperthermia. Adv Exp Med Biol 2013;**765**:323–328.

[39] Hakkinen H. The gold-sulfur interface at the nanoscale. Nat Chem 2012;**4**:443–455.

[40] Gibson JD, Khanal BP, Zubarev ER. Paclitaxel-functionalized gold nanoparticles. J Am Chem Soc 2007;**129**:11653–11661.

[41] Paciotti GF, Myer L, Weinreich D, Goia D, Pavel N, McLaughlin RE, Tamarkin L. Colloidal gold: a novel nanoparticle vector for tumor directed drug delivery. Drug Deliv 2004;**11**:169–183.

[42] Chen JY, Glaus C, Laforest R, Zhang Q, Yang MX, Gidding M, Welch MJ, Xia YN. Gold nanocages as photothermal transducers for cancer treatment. Small 2010;**6**: 811–817.

[43] Li ML, Wang JC, Schwartz JA, Gill-Sharp KL, Stoica G, Wang LV. In-vivo photoacoustic microscopy of nanoshell extravasation from solid tumor vasculature. J Biomed Opt 2009;**14**:010507.

[44] Lukianova-Hleb EY, Oginsky AO, Samaniego AP, Shenefelt DL, Wagner DS, Hafner JH, Farach-Carson MC, Lapotko DO. Tunable plasmonic nanoprobes for theranostics of prostate cancer. Theranostics 2011;**1**:3–17.

[45] Copland JA, Eghtedari M, Popov VL, Kotov N, Mamedova N, Motamedi M, Oraevsky AA. Bioconjugated gold nanoparticles as a molecular based contrast agent: implications for imaging of deep tumors using optoacoustic tomography. Mol Imaging Biol 2004;**6**:341–349.

[46] Qian X, Peng XH, Ansari DO, Yin-Goen Q, Chen GZ, Shin DM, Yang L, Young AN, Wang MD, Nie S. *In vivo* tumor targeting and spectroscopic detection with surface-enhanced Raman nanoparticle tags. Nat Biotechnol 2008;**26**:83–90.

[47] Chen H, Li S, Li B, Ren X, Mahounga DM, Cui S, Gu Y, Achilefu S. Folate-modified gold nanoclusters as near-infrared fluorescent probes for tumor imaging and therapy. Nanoscale 2012;**4**:6050–6064.

[48] Chen Q, Li K, Wen S, Liu H, Peng C, Cai H, Shen M, Zhang G, Shi X. Targeted CT/MR dual mode imaging of tumors using multifunctional dendrimer-entrapped gold nanoparticles. Biomaterials 2013;**34**:5200–5209.

[49] Kumar A, Ma H, Zhang X, Huang K, Jin S, Liu J, Wei T, Cao W, Zou G, Liang XJ. Gold nanoparticles functionalized with therapeutic and targeted peptides for cancer treatment. Biomaterials 2012;**33**:1180–1189.

[50] Kim YH, Jeon J, Hong SH, Rhim WK, Lee YS, Youn H, Chung JK, Lee MC, Lee DS, Kang KW, Nam JM. Tumor targeting and imaging using cyclic RGD-PEGylated gold nanoparticle probes with directly conjugated iodine-125. Small 2011;**7**: 2052–2060.

[51] Mieszawska AJ, Mulder WJ, Fayad ZA, Cormode DP. Multifunctional gold nanoparticles for diagnosis and therapy of disease. Mol Pharm 2013;**10**:831–847.

[52] Brust M, Walker M, Bethell D, Schiffrin DJ, Whyman R. Synthesis of thiol-derivatized gold nanoparticles in a 2-phase liquid-liquid system. *J Chem Soc Chem Commun* 1994:801–802.

[53] Niidome T, Yamagata M, Okamoto Y, Akiyama Y, Takahashi H, Kawano T, Katayama Y, Niidome Y. PEG-modified gold nanorods with a stealth character for *in vivo* applications. J Control Release 2006;**114**:343–347.

[54] von Maltzahn G, Park JH, Agrawal A, Bandaru NK, Das SK, Sailor MJ, Bhatia SN. Computationally guided photothermal tumor therapy using long-circulating gold nanorod antennas. Cancer Res 2009;**69**:3892–3900.

[55] You J, Zhang R, Xiong C, Zhong M, Melancon M, Gupta S, Nick AM, Sood AK, Li C. Effective photothermal chemotherapy using doxorubicin-loaded gold nanospheres that target EphB4 receptors in tumors. Cancer Res 2012;**72**:4777–4786.

[56] Liu Z, Liang XJ. Nano-carbons as theranostics. Theranostics 2012;**2**:235–237.

[57] Choi KY, Liu G, Lee S, Chen XY. Theranostic nanoplatforms for simultaneous cancer imaging and therapy: current approaches and future perspectives. Nanoscale 2012;**4**: 330–342.

[58] Kam NWS, O'Connell M, Wisdom JA, Dai HJ. Carbon nanotubes as multifunctional biological transporters and near-infrared agents for selective cancer cell destruction. Proc Natl Acad Sci U S A 2005;**102**:11600–11605.

[59] Pastorin G, Wu W, Wieckowski S, Briand JP, Kostarelos K, Prato M, Bianco A. Double functionalisation of carbon nanotubes for multimodal drug delivery. *Chem Commun (Camb)* 2006:1182–1184.

[60] Kam NW, Liu Z, Dai H. Functionalization of carbon nanotubes via cleavable disulfide bonds for efficient intracellular delivery of siRNA and potent gene silencing. J Am Chem Soc 2005;**127**:12492–12493.

[61] Liu Z, Chen K, Davis C, Sherlock S, Cao Q, Chen X, Dai H. Drug delivery with carbon nanotubes for *in vivo* cancer treatment. Cancer Res 2008;**68**:6652–6660.

[62] Liu Z, Fan AC, Rakhra K, Sherlock S, Goodwin A, Chen XY, Yang QW, Felsher DW, Dai HJ. Supramolecular stacking of doxorubicin on carbon nanotubes for *in vivo* cancer therapy. Angew Chem Int Ed Engl 2009;**48**:7668–7672.

[63] Liu Z, Sun X, Nakayama-Ratchford N, Dai H. Supramolecular chemistry on water-soluble carbon nanotubes for drug loading and delivery. ACS Nano 2007;**1**:50–56.

[64] Jin H, Heller DA, Strano MS. Single-particle tracking of endocytosis and exocytosis of single-walled carbon nanotubes in NIH-3T3 cells. Nano Lett 2008;**8**:1577–1585.

[65] De La Zerda A, Zavaleta C, Keren S, Vaithilingam S, Bodapati S, Liu Z, Levi J, Smith BR, Ma TJ, Oralkan O, Cheng Z, Chen XY, Dai HJ, Khuri-Yakub BT, Gambhir SS. Carbon nanotubes as photoacoustic molecular imaging agents in living mice. Nat Nanotechnol 2008;**3**:557–562.

[66] Xiang LZ, Yuan Y, Xing D, Ou ZM, Yang SH, Zhou FF. Photoacoustic molecular imaging with antibody-functionalized single-walled carbon nanotubes for early diagnosis of tumor. J Biomed Opt 2009;**14**:021008.

[67] Shi Kam NW, Jessop TC, Wender PA, Dai H. Nanotube molecular transporters: internalization of carbon nanotube-protein conjugates into Mammalian cells. J Am Chem Soc 2004;**126**:6850–6851.

[68] Kim JW, Galanzha EI, Shashkov EV, Moon HM, Zharov VP. Golden carbon nanotubes as multimodal photoacoustic and photothermal high-contrast molecular agents. Nat Nanotechnol 2009;**4**:688–694.

[69] Moon HK, Lee SH, Choi HC. *In vivo* near-infrared mediated tumor destruction by photothermal effect of carbon nanotubes. ACS Nano 2009;**3**:3707–3713.

[70] Bhirde AA, Patel V, Gavard J, Zhang G, Sousa AA, Masedunskas A, Leapman RD, Weigert R, Gutkind JS, Rusling JF. Targeted killing of cancer cells *in vivo* and in vitro with EGF-directed carbon nanotube-based drug delivery. ACS Nano 2009;**3**:307–316.

[71] Akbarzadeh A, Rezaei-Sadabady R, Davaran S, Joo SW, Zarghami N, Hanifehpour Y, Samiei M, Kouhi M, Nejati-Koshki K. Liposome: classification, preparation, and applications. Nanoscale Res Lett 2013;**8**:102–111.

[72] Urdal DL, Hakomori S. Tumor-associated ganglio-N-triosylceramide. Target for antibody-dependent, avidin-mediated drug killing of tumor cells. J Biol Chem 1980;**255**: 10509–10516.

[73] Elbayoumi TA, Torchilin VP. Current trends in liposome research. Methods Mol Biol 2010;**605**:1–27.

[74] Soenen SJ, Velde GV, Ketkar-Atre A, Himmelreich U, De Cuyper M. Magnetoliposomes as magnetic resonance imaging contrast agents. Wiley Interdiscip Rev Nanomed Nanobiotechnol 2011;**3**:197–211.

[75] Bothun GD, Lelis A, Chen Y, Scully K, Anderson LE, Stoner MA. Multicomponent folate-targeted magnetoliposomes: design, characterization, and cellular uptake. Nanomedicine 2011;**7**:797–805.

[76] Silindir M, Erdogan S, Ozer AY, Maia S. Liposomes and their applications in molecular imaging. J Drug Target 2012;**20**:401–415.

[77] Strijkers GJ, Kluza E, Van Tilborg GA, van der Schaft DW, Griffioen AW, Mulder WJ, Nicolay K. Paramagnetic and fluorescent liposomes for target-specific imaging and therapy of tumor angiogenesis. Angiogenesis 2010;**13**:161–173.

[78] Boerman OC, Laverman P, Oyen WJ, Corstens FH, Storm G. Radiolabeled liposomes for scintigraphic imaging. Prog Lipid Res 2000;**39**:461–475.

[79] Torchilin VP. Liposomes as delivery agents for medical imaging. Mol Med Today 1996;**2**:242–249.

[80] Gore JC, Sostman HD, Caride VJ. Liposomes for paramagnetic contrast enhancement in n.m.r. imaging. J Microencapsul 1986;**3**:251–264.

[81] Caride VJ. Liposomes as carriers of imaging agents. Crit Rev Ther Drug Carrier Syst 1985;**1**:121–153.

[82] Zhang D, Feng XY, Henning TD, Wen L, Lu WY, Pan H, Wu X, Zou LG. MR imaging of tumor angiogenesis using sterically stabilized Gd-DTPA liposomes targeted to CD105. Eur J Radiol 2009;**70**:180–189.

[83] Petersen AL, Hansen AE, Gabizon A, Andresen TL. Liposome imaging agents in personalized medicine. Adv Drug Deliv Rev 2012;**64**:1417–1435.

[84] Li S, Goins B, Zhang L, Bao A. Novel multifunctional theranostic liposome drug delivery system: construction, characterization, and multimodality MR, near-infrared fluorescent, and nuclear imaging. Bioconjug Chem 2012;**23**:1322–1332.

[85] Fattahi H, Laurent S, Liu F, Arsalani N, Elst LV, Muller RN. Magnetoliposomes as multimodal contrast agents for molecular imaging and cancer nanotheragnostics. Nanomedicine 2011;**6**:529–544.

[86] Pradhan P, Giri J, Rieken F, Koch C, Mykhaylyk O, Doblinger M, Banerjee R, Bahadur D, Plank C. Targeted temperature sensitive magnetic liposomes for thermo-chemotherapy. J Control Release 2010;**142**:108–121.

[87] Chen Y, Bose A, Bothun GD. Controlled release from bilayer-decorated magnetoliposomes via electromagnetic heating. ACS Nano 2010;**4**:3215–3221.

[88] Samad A, Sultana Y, Aqil M. Liposomal drug delivery systems: an update review. Curr Drug Deliv 2007;**4**:297–305.

[89] Kenny GD, Kamaly N, Kalber TL, Brody LP, Sahuri M, Shamsaei E, Miller AD, Bell JD. Novel multifunctional nanoparticle mediates siRNA tumour delivery, visualisation and therapeutic tumour reduction *in vivo*. J Control Release 2011;**149**:111–116.

[90] Immordino ML, Dosio F, Cattel L. Stealth liposomes: review of the basic science, rationale, and clinical applications, existing and potential. Int J Nanomedicine 2006;**1**:297–315.

[91] Maeda H, Sawa T, Konno T. Mechanism of tumor-targeted delivery of macromolecular drugs, including the EPR effect in solid tumor and clinical overview of the prototype polymeric drug SMANCS. J Control Release 2001;**74**:47–61.

[92] Muthu MS, Kulkarni SA, Raju A, Feng SS. Theranostic liposomes of TPGS coating for targeted co-delivery of docetaxel and quantum dots. Biomaterials 2012;**33**:3494–3501.

[93] Honda M, Asai T, Oku N, Araki Y, Tanaka M, Ebihara N. Liposomes and nanotechnology in drug development: focus on ocular targets. Int J Nanomedicine 2013;**8**:495–503.

[94] Allen TM, Cullis PR. Liposomal drug delivery systems: from concept to clinical applications. Adv Drug Deliv Rev 2013;**65**:36–48.

[95] Urbinati G, Marsaud V, Renoir JM. Anticancer drugs in liposomal nanodevices: a target delivery for a targeted therapy. Curr Top Med Chem 2012;**12**:1693–1712.

[96] Kumar P, Gulbake A, Jain SK. Liposomes a vesicular nanocarrier: potential advancements in cancer chemotherapy. Crit Rev Ther Drug Carrier Syst 2012;**29**:355–419.

[97] Fanciullino R, Ciccolini J. Liposome-encapsulated anticancer drugs: still waiting for the magic bullet? Curr Med Chem 2009;**16**:4361–4371.

[98] Szoka F Jr, Papahadjopoulos D. Comparative properties and methods of preparation of lipid vesicles (liposomes). Annu Rev Biophys Bioeng 1980;**9**:467–508.

[99] Szoka F Jr, Papahadjopoulos D. Procedure for preparation of liposomes with large internal aqueous space and high capture by reverse-phase evaporation. Proc Natl Acad Sci U S A 1978;**75**:4194–4198.

[100] Maherani B, Arab-tehrany E, Kheirolomoom A, Reshetov V, Stebe MJ, Linder M. Optimization and characterization of liposome formulation by mixture design. Analyst 2012;**137**:773–786.

[101] Mozafari MR. Liposomes: an overview of manufacturing techniques. Cell Mol Biol Lett 2005;**10**:711–719.

[102] Muthu MS, Kulkarni SA, Xiong J, Feng SS. Vitamin E TPGS coated liposomes enhanced cellular uptake and cytotoxicity of docetaxel in brain cancer cells. Int J Pharm 2011;**421**:332–340.

[103] Sailor MJ, Park JH. Hybrid nanoparticles for detection and treatment of cancer. Adv Mater 2012;**24**:3779–3802.

[104] Yoshida M, Watanabe Y, Sato M, Maehara T, Aono H, Naohara T, Hirazawa H, Horiuchi A, Yukumi S, Sato K, Nakagawa H, Yamamoto Y, Sugishita H, Kawachi K. Feasibility of chemohyperthermia with docetaxel-embedded magnetoliposomes as minimally invasive local treatment for cancer. Int J Cancer 2010;**126**:1955–1965.

[105] Cittadino E, Ferraretto M, Torres E, Maiocchi A, Crielaard BJ, Lammers T, Storm G, Aime S, Terreno E. MRI evaluation of the antitumor activity of paramagnetic liposomes loaded with prednisolone phosphate. Eur J Pharm Sci 2012;**45**:436–441.

[106] McClements DJ. Nanoemulsions versus microemulsions: terminology, differences, and similarities. Soft Matter 2012;**8**:1719–1729.

[107] Shakeel F, Shafiq S, Haq N, Alanazi FK, Alsarra IA. Nanoemulsions as potential vehicles for transdermal and dermal delivery of hydrophobic compounds: an overview. Expert Opin Drug Deliv 2012;**9**:953–974.

[108] Kotta S, Khan AW, Pramod K, Ansari SH, Sharma RK, Ali J. Exploring oral nanoemulsions for bioavailability enhancement of poorly water-soluble drugs. Expert Opin Drug Deliv 2012;**9**:585–598.

[109] McClements DJ, Rao J. Food-grade nanoemulsions: formulation, fabrication, properties, performance, biological fate, and potential toxicity. Crit Rev Food Sci Nutr 2011;**51**:285–330.

[110] Rajpoot P, Pathak K, Bali V. Therapeutic applications of nanoemulsion based drug delivery systems: a review of patents in last two decades. Recent Pat Drug Deliv Formul 2011;**5**:163–172.

[111] Sarker DK. Engineering of nanoemulsions for drug delivery. Curr Drug Deliv 2005;**2**:297–310.

[112] Mitri K, Vauthier C, Huang N, Menas A, Ringard-Lefebvre C, Anselmi C, Stambouli M, Rosilio V, Vachon JJ, Bouchemal K. Scale-up of nanoemulsion produced by emulsification and solvent diffusion. J Pharm Sci 2012;**101**:4240–4247.

[113] Muller RH, Harden D, Keck CM. Development of industrially feasible concentrated 30% and 40% nanoemulsions for intravenous drug delivery. Drug Dev Ind Pharm 2012;**38**:420–430.

[114] Longmaid HE III, Adams DF, Neirinckx RD, Harrison CG, Brunner P, Seltzer SE, Davis MA, Neuringer L, Geyer RP. *In vivo* 19F NMR imaging of liver, tumor, and abscess in rats. Preliminary results. Invest Radiol 1985;**20**:141–145.

[115] Janjic JM, Ahrens ET. Fluorine-containing nanoemulsions for MRI cell tracking. Wiley Interdiscip Rev Nanomed Nanobiotechnol 2009;**1**:492–501.

[116] Ahrens ET, Zhong J. *In vivo* MRI cell tracking using perfluorocarbon probes and fluorine-19 detection. NMR Biomed 2013;**26**:860–871.

[117] Main ML, Goldman JH, Grayburn PA. Ultrasound contrast agents: balancing safety versus efficacy. Expert Opin Drug Saf 2009;**8**:49–56.

[118] Janjic JM, Srinivas M, Kadayakkara DK, Ahrens ET. Self-delivering nanoemulsions for dual fluorine-19 MRI and fluorescence detection. J Am Chem Soc 2008;**130**: 2832–2841.

[119] Balducci A, Wen Y, Zhang Y, Helfer BM, Hitchens TK, Meng WS, Wesa AK, Janjic JM. A novel probe for the non-invasive detection of tumor-associated inflammation. Oncoimmunology 2013;**2**:e23034.

[120] Akers WJ, Kim C, Berezin M, Guo K, Fuhrhop R, Lanza GM, Fischer GM, Daltrozzo E, Zumbusch A, Cai X, Wang LV, Achilefu S. Noninvasive photoacoustic and fluorescence sentinel lymph node identification using dye-loaded perfluorocarbon nanoparticles. ACS Nano 2011;**5**:173–182.

[121] Rapoport N, Nam KH, Gupta R, Gao Z, Mohan P, Payne A, Todd N, Liu X, Kim T, Shea J, Scaife C, Parker DL, Jeong EK, Kennedy AM. Ultrasound-mediated tumor imaging and nanotherapy using drug loaded, block copolymer stabilized perfluorocarbon nanoemulsions. J Control Release 2011;**153**:4–15.

[122] Pan H, Soman NR, Schlesinger PH, Lanza GM, Wickline SA. Cytolytic peptide nanoparticles ("NanoBees") for cancer therapy. Wiley Interdiscip Rev Nanomed Nanobiotechnol 2011;**3**:318–327.

[123] Soman NR, Lanza GM, Heuser JM, Schlesinger PH, Wickline SA. Synthesis and characterization of stable fluorocarbon nanostructures as drug delivery vehicles for cytolytic peptides. Nano Lett 2008;**8**:1131–1136.

[124] Soman NR, Baldwin SL, Hu G, Marsh JN, Lanza GM, Heuser JE, Arbeit JM, Wickline SA, Schlesinger PH. Molecularly targeted nanocarriers deliver the cytolytic peptide melittin specifically to tumor cells in mice, reducing tumor growth. J Clin Invest 2009;**119**:2830–2842.

[125] Rapoport NY, Christensen DA, Fain HD, Barrows L, Gao Z. Ultrasound-triggered drug targeting of tumors in vitro and *in vivo*. Ultrasonics 2004;**42**:943–950.

[126] Rapoport N. Phase-shift, stimuli-responsive perfluorocarbon nanodroplets for drug delivery to cancer. Wiley Interdiscip Rev Nanomed Nanobiotechnol 2012;**4**:492–510.

[127] Stoll G, Basse-Lusebrink T, Weise G, Jakob P. Visualization of inflammation using (19) F-magnetic resonance imaging and perfluorocarbons. Wiley Interdiscip Rev Nanomed Nanobiotechnol 2012;**4**:438–447.

[128] Kadayakkara DK, Beatty PL, Turner MS, Janjic JM, Ahrens ET, Finn OJ. Inflammation driven by overexpression of the hypoglycosylated abnormal mucin 1 (MUC1) links inflammatory bowel disease and pancreatitis. Pancreas 2010;**39**:510–515.

[129] Weise G, Basse-Luesebrink TC, Wessig C, Jakob PM, Stoll G. *In vivo* imaging of inflammation in the peripheral nervous system by (19)F MRI. Exp Neurol 2011;**229**:494–501.

[130] Dewitte H, Geers B, Liang S, Himmelreich U, Demeester J, De Smedt SC, Lentacker I. Design and evaluation of theranostic perfluorocarbon particles for simultaneous antigen-loading and (19)F-MRI tracking of dendritic cells. J Control Release 2013;**169**:141–149.

[131] Caruthers SD, Cyrus T, Winter PM, Wickline SA, Lanza GM. Anti-angiogenic perfluorocarbon nanoparticles for diagnosis and treatment of atherosclerosis. Wiley Interdiscip Rev Nanomed Nanobiotechnol 2009;**1**:311–323.

[132] Lanza GM. Targeted Antiproliferative drug delivery to vascular smooth muscle cells with a magnetic resonance imaging nanoparticle contrast agent: implications for rational therapy of restenosis. Circulation 2002;**106**:2842–2847.

[133] Maali A, Mosavian MTH. Preparation application of nanoemulsions in the last decade (2000-2010). J Dispers Sci Tech 2013;**34**:92–105.

[134] Patel SK, Patrick MJ, Pollock JA, Janjic JM. Two-color fluorescent (NIR and visible) triphasic perfluorocarbon nanoemuslions. J Biomed Opt 2013;**18**:101312.

[135] Hoffmann I, Heunemann P, Farago B, Grillo I, Holderer O, Päch M, Gradzielski M. Structure and dynamics of nanoemulsions: insights from combining dynamic and static neutron scattering. Phys Rev E 2012;**86**:061407.

[136] Constantinides PP, Chaubal MV, Shorr R. Advances in lipid nanodispersions for parenteral drug delivery and targeting. Adv Drug Deliv Rev 2008;**60**:757–767.

[137] Gianella A, Jarzyna PA, Mani V, Ramachandran S, Calcagno C, Tang J, Kann B, Dijk WJ, Thijssen VL, Griffioen AW, Storm G, Fayad ZA, Mulder WJ. Multifunctional nanoemulsion platform for imaging guided therapy evaluated in experimental cancer. ACS Nano 2011;**5**:4422–4433.

[138] Ornelas C, Pennell R, Liebes LF, Weck M. Construction of a well-defined multifunctional dendrimer for theranostics. Org Lett 2011;**13**:976–979.

[139] Longmire M, Choyke PL, Kobayashi H. Dendrimer-based contrast agents for molecular imaging. Curr Top Med Chem 2008;**8**:1180–1186.

[140] Kaminskas LM, McLeod VM, Porter CJ, Boyd BJ. Association of chemotherapeutic drugs with dendrimer nanocarriers: an assessment of the merits of covalent conjugation compared to noncovalent encapsulation. Mol Pharm 2012;**9**:355–373.

[141] Lazniewska J, Milowska K, Gabryelak T. Dendrimers—revolutionary drugs for infectious diseases. Wiley Interdiscip Rev Nanomed Nanobiotechnol 2012;**4**:469–491.

[142] Koyama Y, Talanov VS, Bernardo M, Hama Y, Regino CA, Brechbiel MW, Choyke PL, Kobayashi H. A dendrimer-based nanosized contrast agent dual-labeled for magnetic resonance and optical fluorescence imaging to localize the sentinel lymph node in mice. J Magn Reson Imaging 2007;**25**:866–871.

[143] Dhanikula RS, Hildgen P. Influence of molecular architecture of polyether-co-polyester dendrimers on the encapsulation and release of methotrexate. Biomaterials 2007;**28**:3140–3152.

[144] Davis SC, Pogue BW, Springett R, Leussler C, Mazurkewitz P, Tuttle SB, Gibbs-Strauss SL, Jiang SS, Dehghani H, Paulsen KD. Magnetic resonance-coupled fluorescence tomography scanner for molecular imaging of tissue. Rev Sci Instrum 2008;**79**:064302.

[145] Ooya T, Lee J, Park K. Hydrotropic dendrimers of generations 4 and 5: synthesis, characterization, and hydrotropic solubilization of paclitaxel. Bioconjug Chem 2004;**15**:1221–1229.

[146] Ren Y, Kang CS, Yuan XB, Zhou X, Xu P, Han L, Wang GX, Jia Z, Zhong Y, Yu S, Sheng J, Pu PY. Co-delivery of as-miR-21 and 5-FU by poly(amidoamine) dendrimer attenuates human glioma cell growth in vitro. J Biomater Sci Polym Ed 2010;**21**:303–314.

[147] Jiang YY, Tang GT, Zhang LH, Kong SY, Zhu SJ, Pei YY. PEGylated PAMAM dendrimers as a potential drug delivery carrier: in vitro and *in vivo* comparative evaluation of covalently conjugated drug and noncovalent drug inclusion complex. J Drug Target 2010;**18**:389–403.

[148] Kaminskas LM, Kelly BD, McLeod VM, Sberna G, Boyd BJ, Owen DJ, Porter CJ. Capping methotrexate alpha-carboxyl groups enhances systemic exposure and retains the cytotoxicity of drug conjugated PEGylated polylysine dendrimers. Mol Pharm 2011;**8**:338–349.

[149] Zhu S, Hong M, Tang G, Qian L, Lin J, Jiang Y, Pei Y. Partly PEGylated polyamidoamine dendrimer for tumor-selective targeting of doxorubicin: the effects of PEGylation degree and drug conjugation style. Biomaterials 2010;**31**:1360–1371.

[150] Lim J, Chouai A, Lo ST, Liu W, Sun X, Simanek EE. Design, synthesis, characterization, and biological evaluation of triazine dendrimers bearing paclitaxel using ester and ester/disulfide linkages. Bioconjug Chem 2009;**20**:2154–2161.

[151] Lim J, Lo ST, Hill S, Pavan GM, Sun X, Simanek EE. Antitumor activity and molecular dynamics simulations of paclitaxel-laden triazine dendrimers. Mol Pharm 2012;**9**: 404–412.

[152] Lo ST, Stern S, Clogston JD, Zheng J, Adiseshaiah PP, Dobrovolskaia M, Lim J, Patri AK, Sun X, Simanek EE. Biological assessment of triazine dendrimer: toxicological profiles, solution behavior, biodistribution, drug release and efficacy in a PEGylated, paclitaxel construct. Mol Pharm 2010;**7**:993–1006.

[153] Patri AK, Myc A, Beals J, Thomas TP, Bander NH, Baker JR Jr. Synthesis and in vitro testing of J591 antibody-dendrimer conjugates for targeted prostate cancer therapy. Bioconjug Chem 2004;**15**:1174–1181.

[154] Wu X, Ding B, Gao J, Wang H, Fan W, Wang X, Zhang W, Ye L, Zhang M, Ding X, Liu J, Zhu Q, Gao S. Second-generation aptamer-conjugated PSMA-targeted delivery system for prostate cancer therapy. Int J Nanomed 2011;**6**:1747–1756.

[155] Shukla R, Thomas TP, Peters J, Kotlyar A, Myc A, Baker JR Jr. Tumor angiogenic vasculature targeting with PAMAM dendrimer-RGD conjugates. Chem Commun (Camb) 2005:5739–5741.

[156] Oliveira JM, Salgado AJ, Sousa N, Mano JF, Reis RL. Dendrimers and derivatives as a potential therapeutic tool in regenerative medicine strategies: a review. Prog Polym Sci 2010;**35**:1163–1194.

[157] Esfand R, Tomalia DA. Poly(amidoamine) (PAMAM) dendrimers: from biomimicry to drug delivery and biomedical applications. Drug Discov Today 2001;**6**:427–436.

[158] Leu CM, Shu CF, Teng CF, Shiea J. Dendritic poly(ether-imide)s: synthesis, characterization, and modification. Polymer 2001;**42**:2339–2348.

[159] Al-Jamal KT, Al-Jamal WT, Wang JT, Rubio N, Buddle J, Gathercole D, Zloh M, Kostarelos K. Cationic poly-l-lysine dendrimer complexes doxorubicin and delays tumor growth in vitro and *in vivo*. ACS Nano 2013;**7**:1905–1917.

[160] De Oliveira H, Thevenot J, Lecommandoux S. Smart polymersomes for therapy and diagnosis: fast progress toward multifunctional biomimetic nanomedicines. Wiley Interdiscip Rev Nanomed Nanobiotechnol 2012;**4**:525–546.

[161] Krasia-Christoforou T, Georgiou TK. Polymeric theranostics: using polymer-based systems for simultaneous imaging and therapy. J Mater Chem B 2013;**1**:3002.

[162] Kim K, Kim JH, Park H, Kim YS, Park K, Nam H, Lee S, Park JH, Park RW, Kim IS, Choi K, Kim SY, Kwon IC. Tumor-homing multifunctional nanoparticles for cancer theragnosis: simultaneous diagnosis, drug delivery, and therapeutic monitoring. J Control Release 2010;**146**:219–227.

[163] Zheng C, Zheng M, Gong P, Jia D, Zhang P, Shi B, Sheng Z, Ma Y, Cai L. Indocyanine green-loaded biodegradable tumor targeting nanoprobes for in vitro and *in vivo* imaging. Biomaterials 2012;**33**:5603–5609.

[164] Rowe MD, Thamm DH, Kraft SL, Boyes SG. Polymer-modified gadolinium metal-organic framework nanoparticles used as multifunctional nanomedicines for the targeted imaging and treatment of cancer. Biomacromolecules 2009;**10**:983–993.

[165] Mahmoudi M, Sant S, Wang B, Laurent S, Sen T. Superparamagnetic iron oxide nanoparticles (SPIONs): development, surface modification and applications in chemotherapy. Adv Drug Deliv Rev 2011;**63**:24–46.

[166] Liao ZY, Wang HJ, Wang XD, Zhao PQ, Wang S, Su WY, Chang J. Multifunctional nanoparticles composed of a poly(dl-lactide-coglycolide) core and a paramagnetic liposome shell for simultaneous magnetic resonance imaging and targeted therapeutics. Adv Funct Mater 2011;**21**:1179–1186.

[167] Lammers T, Subr V, Peschke P, Kuhnlein R, Hennink WE, Ulbrich K, Kiessling F, Heilmann M, Debus J, Huber PE, Storm G. Image-guided and passively tumour-targeted polymeric nanomedicines for radiochemotherapy. Br J Cancer 2008;**99**:900–910.

[168] Liao J, Wang C, Wang Y, Luo F, Qian Z. Recent advances in formation, properties, and applications of polymersomes. Curr Pharm Des 2012;**18**:3432–3441.

[169] Wang Z, Niu G, Chen X. Polymeric materials for theranostic applications. Pharm Res 2014;**31**:1358–1376.

[170] Zhu C, Liu L, Yang Q, Lv F, Wang S. Water-soluble conjugated polymers for imaging, diagnosis, and therapy. Chem Rev 2012;**112**:4687–4735.

[171] Wang G, Yu B, Wu Y, Huang B, Yuan Y, Liu CS. Controlled preparation and antitumor efficacy of vitamin E TPGS-functionalized PLGA nanoparticles for delivery of paclitaxel. Int J Pharm 2013;**446**:24–33.

[172] Vauthier C, Bouchemal K. Methods for the preparation and manufacture of polymeric nanoparticles. Pharm Res 2009;**26**:1025–1058.

[173] McCarthy JR, Perez JM, Bruckner C, Weissleder R. Polymeric nanoparticle preparation that eradicates tumors. Nano Lett 2005;**5**:2552–2556.

[174] Venditto VJ, Szoka FC Jr. Cancer nanomedicines: so many papers and so few drugs!. Adv Drug Deliv Rev 2013;**65**:80–88.

16

ANIMAL MODELS FOR PRECLINICAL IMAGING

Grayson Talcott and Walter J. Akers

Department of Radiology, Washington University School of Medicine, St. Louis, MO, USA

16.1 INTRODUCTION

The purpose of biomedical imaging is to noninvasively detect the presence and investigate the progression of pathologic processes (disease). The requirements for imaging are dependent on the level of detail needed for adequate characterization of the disease at hand. Diagnosis of a broken bone may only require planar X-ray, while diagnosis of metastatic cancer requires molecular imaging with ^{18}F-fluorodeoxyglucose (^{18}F-FDG) positron emission tomography (PET).

Biomedical imaging has been driven by the need to see below the surface into the intricate workings of the body in health and disease. Interrogation of anatomical structures is accomplished via X-ray, X-ray CT, and MRI, which have become higher in quality over the last 30 years. Contrast enhancement has further improved the capability of these modalities to evaluate physical structures of the cardiovascular system and central nervous system (CNS). Physiologic imaging began with nuclear medicine using radiotracers to assess renal clearance and blood flow. We are now in the age of molecular imaging in which expression and activity of disease-specific molecules and genes can be assessed using targeted contrast agents.

The goals of *preclinical* imaging are not the same as clinical biomedical imaging. In general, animal models are selected based on predictable and reproducible outcomes with known anatomy, physiology, and even molecular and genetic expressions. Imaging modalities are generally utilized in basic research for noninvasive or minimally invasive

---

Nanotechnology for Biomedical Imaging and Diagnostics: From Nanoparticle Design to Clinical Applications, First Edition. Edited by Mikhail Y. Berezin.

investigation of biological processes relevant to human disease. They can be separated into (i) development of novel imaging techniques, (ii) development of therapeutic and/or diagnostic agents, and (iii) reporting of protein and gene expression.

As the theme of this book is nanomaterials in biomedical imaging, this chapter will focus on relevant animal models that require only basic animal handling and surgical skills used in most research institutions. It is not in the scope of this single chapter to elaborate on all animal models available at this time, but to present a survey of animal models that broadly reflect the biology of human disease sufficiently for molecular imaging with nanomaterial contrast agents. Rodent models are most prevalent in biomedical research for practical and economic reasons and will encompass the most attention.

16.2 ETHICS IN ANIMAL RESEARCH

The use of animals in research has provided immeasurable benefit to our understanding and ability to treat human disease. While significant progress has been made in the development of alternatives to animals in research, the progression of disease, delivery of diagnostic and therapeutic agents, and response to therapies cannot be adequately evaluated by *in vitro* assays. Regardless of scientific impact, preclinical imaging studies should always be designed with respect for animal life and to minimize any pain and suffering.

Local Institutional Animal Care and Use Committees (IACUC) govern the appropriate use of animals in research at each institution and must approve all animal research procedures prior to initiation. The IACUC must include veterinarians, institutional scientists, nonscientists, and members of the public not otherwise associated with the institution to provide a balanced view of the value of biomedical research with the humane treatment of animals [1]. While specific guidelines for appropriate use of animals in research are provided in the *Guide for the Care and Use of Laboratory Animals* [1], these rules are interpreted by each IACUC for each instance, leading to minor differences in experimental standards between institutions. Scientific justification must also be supported by evidence that studies cannot be adequately completed using nonanimal methods.

The goal of preclinical imaging with nanomaterials is to assess the biodistribution, pharmacokinetics, and toxicity of imaging agents in healthy animals and/or disease models prior to translational studies and human clinical trials. Secondary goals may include assessment of drug delivery and efficacy of therapeutic and diagnostic (theranostic) multifunctional nanomaterials or other biological effects not directly related to imaging, but are not addressed here. To accomplish this goal, nanomaterials are administered to animals, animals are anesthetized for imaging for one or more sessions, and then the animals are euthanized to measure organ-specific biodistribution. Minimal pain and distress occur for the animals, including anesthesia and injections, qualifying these studies as *Category C* for pain and distress according to the USDA Animal Welfare Act. Invasive imaging procedures that require deep anesthesia and analgesics to manage pain would go under *Category D*. Classification of pain and distress for model-specific procedures should be determined separately, with advice from the local IACUC if needed.

16.3 CONSIDERATIONS IN ANIMAL CARE

Proper maintenance of animal health and welfare is also necessary for producing reliable experimental results. Inadequate care can adversely affect imaging results due to infection, inflammation, and alterations of blood flow.

16.3.1 Drug Formulation

Current regulations on animal use in research compel investigators to use pharmaceutical grade materials (FDA approved) in all animal studies, including acute, nonsurvival studies when available [1]. Scientific justification is needed for use of unapproved agents, which include almost all nanomaterials. The local IACUC may require evaluation of grade, purity, sterility, pyrogenicity, stability, and other factors prior to approval for use in animal studies. While the primary aim of these regulations is animal safety, these factors are also important for the validity of experimental results and conclusions. Investigators are therefore encouraged to evaluate these factors for each formulation prior to conducting animal studies.

Synthesis of nanomaterial imaging agents is performed in a chemical laboratory, often utilizing organic solvents and nonbiocompatible constituents. Nanomaterials must be transitioned to biocompatible, pharmaceutical grade formulations prior to administration. In general, nanomaterials should be formulated in sterile, pH-buffered aqueous solution such as PBS for intravenous injection. Analysis of nanomaterials in the final formulation should be performed to ensure that the desired characteristics are maintained, including solubility and stability. For example, gold nanoparticles (NPs) were 50 nm in diameter in ultrapure water but increased to 200–250 nm when dispersed in PBS [2]. Aggregation of NPs will significantly alter the pharmacokinetics and biodistribution relative to well-dispersed solutions and adversely affect experimental conclusions.

16.3.2 Dose Calculations

The dose of material administered to laboratory animals for imaging studies should be based on several factors. Animals vary significantly in size (Fig. 16.1) as determined by body weight (BW), body surface area (BSA), or body mass index (BMI) [3, 4]. Metabolic rates also increase with decreasing size, which affects pharmacokinetics and organ-specific uptake rates. The administered dose is of particular concern for molecular imaging using contrast agents that target saturable receptors or enzymes. Doses close to or above that required for saturation of the relevant binding sites can adversely affect conclusions from an imaging study by reducing the target-specific uptake and contrast [4]. Mass effect must be considered when choosing a dose for molecular imaging alongside detection sensitivity and resolution of the imaging modality used.

16.3.3 Anesthesia

Imaging of animals is generally performed under heavy sedation or full anesthesia. Anesthesia can be delivered by systemic injection (e.g., ketamine/xylazine) or by inhalation (e.g., isoflurane). Injectable anesthesia is beneficial for procedures less

Mouse (15–30 g) Rat (50–350 g) Rabbit (1–4 kg) Human (65–90 kg)

FIGURE 16.1 Comparison of relative body weights for laboratory animals and humans for dose estimation for preclinical imaging.

than 30 min as it produces heavy sedation to surgical plane of anesthesia and easy positioning and manipulation without connection to hoses. Most injectable anesthetics include controlled substances (ketamine, phenobarbital, etc.) requiring DEA licensing and detailed record keeping. Injectable anesthesia is relatively nonreversible, has a lower safety threshold, and is slower acting relative to isoflurane. On the other hand, it is economical, requires no special equipment, and is relatively easy to use. Formularies for injectable anesthesia regimens can be found in [5].

Inhalation anesthesia is used for many imaging procedures, particularly those requiring more than 30 min. Isoflurane is administered via precision vaporizer in 100% oxygen. Anesthesia for small animals can be induced in an airtight chamber prior to being placed in the imaging system equipped with nose cone for anesthesia maintenance (Fig. 16.2). While isoflurane is not toxic, care must be taken to minimize leakage of isoflurane vapor into the environment to protect personnel from anesthetic effects.

16.3.4 Temperature

It is important to maintain body temperature both for the health of the animal and the quality of imaging results. It has been shown in many studies that body temperature affects normal physiology and delivery of drugs and imaging agents within the body of small animals. Due to the large surface area relative to weight of mice and rats, they lose normal body temperature quickly. Anesthesia decreases the body's ability to maintain normal body temperature. Thus, if external heat is not supplied, animals can quickly become hypothermic, negatively affecting animal health and reliability of experimental results. For example, hypothermia results in constricted vasculature, decreased cellular metabolism, and poor lymphatic drainage. These factors can lead to decreased perfusion and nonspecific accumulation relative to that seen in normothermic animals. Maintenance of body temperature can be accomplished by two different methods: surface warming or heated air. The method of choice is dependent on the imaging modality and the animal containment

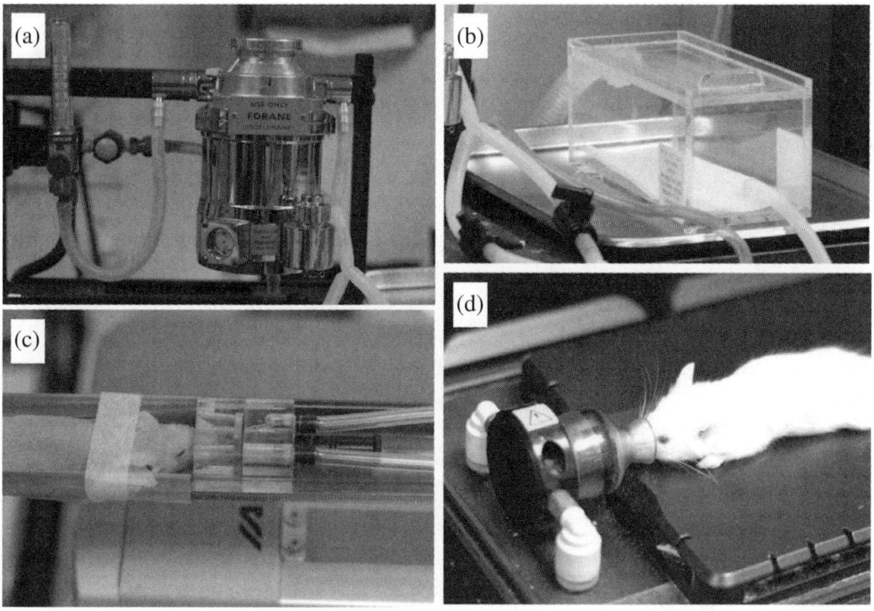

FIGURE 16.2 Setup for inhalation anesthesia of laboratory animals for imaging. A precision vaporizer (a) ensures delivery of desired concentration of anesthetic regardless of gas flow rate. Animals are initially induced in anesthetic chamber (b) before being transferred to imaging bed (c, d). Imaging platforms for nuclear imaging (c, nanoSPECT/CT, Bioscan) and optical imaging (d, Pearl, LI-COR Biosciences) utilize nose cones for direct delivery to mice with minimal leaking to the environment.

used. Electrical heat plates can be used with caution as superficial burns can occur if not carefully controlled. These plates may interfere with imaging in the case of MRI, CT, and nuclear imaging. Heated-water circulation is a safer method and is compatible with most imaging modalities. Heated, forced air is another option when the animal is tightly contained or air can be constantly blown over the subject. Many commercial preclinical imaging systems feature methods for maintaining normothermic conditions.

16.3.5 Positioning and Motion

Adequate maintenance of position and reduction of motion are important factors in acquiring reproducible and comparable imaging datasets. This is especially true for planar imaging modalities including X-ray, scintigraphy, fluorescence reflectance imaging (FRI), and bioluminescence imaging (BLI). While three-dimensional (3D) tomographic imaging modalities are not as dependent on position, reliable determination of anatomical features is best when animals are positioned correctly.

Imaging of the skull and brain is generally accomplished by holding of the head in a modified stereotactic frame. This accomplishes several purposes including standardized positioning between subjects and a reduced motion artifact. The position of internal organs within the body varies with position. For comparable images, it is important to standardize the position of each subject.

Breathing motion can introduce significant reduction in resolution for imaging, particularly in small animals. This becomes most important for regions near the lungs or abdomen. Gating of image acquisition based on the heart rate is necessary for cardiac imaging and requires fast data acquisition. Breath and cardiac gating of acquisition can be automated by inputting physiologic signals from the monitoring systems to trigger data acquisition. An alternative to gated image acquisition is postprocessing of image data. Automated processing can best be accomplished by programmatic selection based on physiologic monitoring acquired simultaneously with the imaging data. Otherwise, a manual selection process can be effective, but time-consuming.

16.3.6 Other Considerations

For optical, ultrasound, and photoacoustic (optoacoustic) imaging experiments, the hair covering the region of interest must be removed. Hair significantly attenuates signal in these modalities. Hair removal must be done carefully to prevent damage to the skin that may cause artifacts in imaging. Gentle clipping can be followed with chemical depilatory such as Nair® to remove hair over a large section of the body. It is not practical to remove all of the hair. Nude mice that are naturally hairless are often used in optical imaging for this reason.

16.4 CHOICE OF ANIMAL MODEL

The primary considerations in choosing an animal model should be based on the biological question at hand. In development of diagnostic agents, the first question is usually: "can we detect the target?" In this case, the decisions on the choice of animal model should include the imaging modality for sensitivity and resolution. Selections of animal models include the following criteria: (i) similarity in biology to human disease, (ii) quality of model (ease of producing, reproducibility, and short time of onset), and (iii) translation of results to human medicine.

In many studies, the question is, can we detect a biological response? Imaging is a very powerful method for noninvasive detection of anatomical and molecular responses. MRI, CT, and ultrasound have been used for many years to measure the size of tumors in preclinical chemotherapy development. As with the clinical analog, response to chemotherapy is determined by the reduction or lack of reduction in tumor size. In the current age of molecular imaging, responses can now be measured by glucose utilization, incorporation of radiolabeled nucleotides, or alterations in other molecular markers of disease progression. Again here, the choice of animal model depends heavily on the imaging modality for resolution and sensitivity. In general, well-established models of response that have traditionally been used for survival studies are good choices for these studies. For example, traditional methods use survival curves to assess response to therapy or external measurement of subcutaneous xenograft size by caliper measurements. Imaging enables noninvasive measurements in orthotopic cancer studies and internal pathologies.

16.4.1 Animal Models for Imaging Arthritis and Inflammation

Arthritis affects nearly all people at some point in their lives due to the natural aging process. Other forms of arthritis develop in younger people due to injuries, autoimmune disease (rheumatoid arthritis (RA)), and genetic disorders affecting cartilage growth or joint conformation (e.g., hip dysplasia). Diagnosis of arthritis must include determining the etiology of the joint pain to select the appropriate treatment. While these diseases have been known for many years, the complex mechanisms that result in damage have not been elucidated fully, and more investigation is needed.

Arthritis describes a broad range of diseases that cause inflammation and pain of the joints. The most common models for inflammation requires direct injection of irritants.

Arthritis was induced in rabbits by intra-articular injection of albumin and scanned with BOLD MRI periodically for 28 days [6]. Inflammation of the joint (synovitis) was specifically detected compared with control (untreated and saline injected). Animal models have been used in the development of pharmaceuticals prior to use in humans. Arthritis may be induced by chemical or other injury directly to the joints, systemic arthritis-inducing drugs, or genetic engineering of animals to mimic human disease.

Inflammation results in dilation of blood vessels, vascular leakiness, and infiltration of tissue with white blood cells. Thus, inflammation is an obvious choice for optical imaging with fluorescent contrast agents that accumulate nonspecifically due to enhanced permeability and retention (EPR). This can give a better measurement of inflammation than traditional measures of swelling, particularly at early time points. It also allows more sensitive measurement with less inflammation needed to see a response. Fluorescence imaging can be used to measure the relative reduction in inflammation due to pharmaceutical intervention. It can also be used to measure molecular markers of inflammation associated with diseases such as cancer, cardiovascular disease and others.

Animal models have also been used extensively in basic research to study the genetic predispositions to arthritis and immunology of RA. RA is an autoimmune disease that affects all the joints of the body and can occur at any time of life. To date, there has been limited success in treatment due to poor efficacy over time or adverse side effects at effective doses. Animal models of RA have been predictive of outcomes in humans. Adjuvant-induced arthritis models are produced by intradermal injection of immune-stimulating compounds such as Freund's complete adjuvant (FCA) including mycobacterium, synthetic adjuvant, or type II collagen (typically bovine).

Osteoarthritis (OA) is a degenerative disease of the joints that affects the cartilage and bone and is diagnosed by symptoms, X-ray radiography, and CT. Generally, X-ray and CT cannot detect pathology of the cartilage or synovia. Soft tissue (cartilage, ligaments, tendons) defects are best observed with MRI; however, contrast enhancement, such as microCT arthrography, and other special techniques can elucidate more information about the status of disease detected by MRI. OA develops from increased friction and wear of the joint as a result of joint incongruity, laxity, or injury or age-related wear. OA can cause significant pain and generally gets worse with time unless cause can be eliminated through joint resurfacing or replacement. Drug development for OA has been very successful in managing OA-related pain. However, at this time, there are no drugs that have successfully demonstrated repair of cartilage defects and reduction of OA. Further investigation into the pathophysiology of OA and mechanism for regenerating synovial cartilage is needed to improve outcomes in patients with OA.

Certain strains of guinea pig develop spontaneous OA of the knee that can be detected by CT at 11 months [7]. Intra-articular injection of mono-iodoacetate (MIA) induces OA-like pathology by interfering with cartilage metabolism [8, 9]. While commonly used as a model of inflammation, injection of FCA or *N,N*-dioctyldecyl-*N'*,*N*-bis(2-hydroxyethyl)propanediamine (LA) suspended in oil in the hind paw of rats results in OA of the joints of that foot [10]. Glucose-6-phosphate isomerase-induced arthritis was imaged with ^{18}F-FDG PET/CT (Fig. 16.3) [11].

Paw edema can be induced by intraplantar injection of 10 μl of 10 mg/ml β-zymosan in sterile 0.9% NaCl solution and followed by molecular imaging of cathepsin activity using activatable and affinity-based fluorescent contrast agents [12]. Molecular imaging methods are also used to detect cells and molecular markers of inflammation associated with specific diseases. Thus, surgical models of OA have been developed to mimic injury-related OA. These include direct cartilage injury [9] and destabilization of the medial meniscus [13].

16.4.2 Animal Models for Cancer Imaging

Cancer, in particular, is a disease amenable to imaging for diagnosis as it usually does not produce definitive signs at an early stage. Breast cancer screening through mammography has significantly improved outcomes in women through early detection and therefore better prognosis. Cancer diagnosis is generally determined by imaging through anatomical abnormalities induced at the primary site and sometimes in distant locations at late stages. This imaging guides surgical therapy and is used to monitor the progress of neoadjuvant, adjuvant, and radiation therapy. The majority of preclinical cancer research models are murine due to the variety of strains available and repeatability. Some common types of mouse models used in biomedical research are shown in Figure 16.4.

16.4.2.1 Spontaneous Cancer Models Cancer occurs naturally in many animals, with the wide variety seen in the human counterpart. Like the human disease, spontaneous cancer occurrence is generally unpredictable and with long latency. Certain lines of mice and rats more predictably develop tumors.

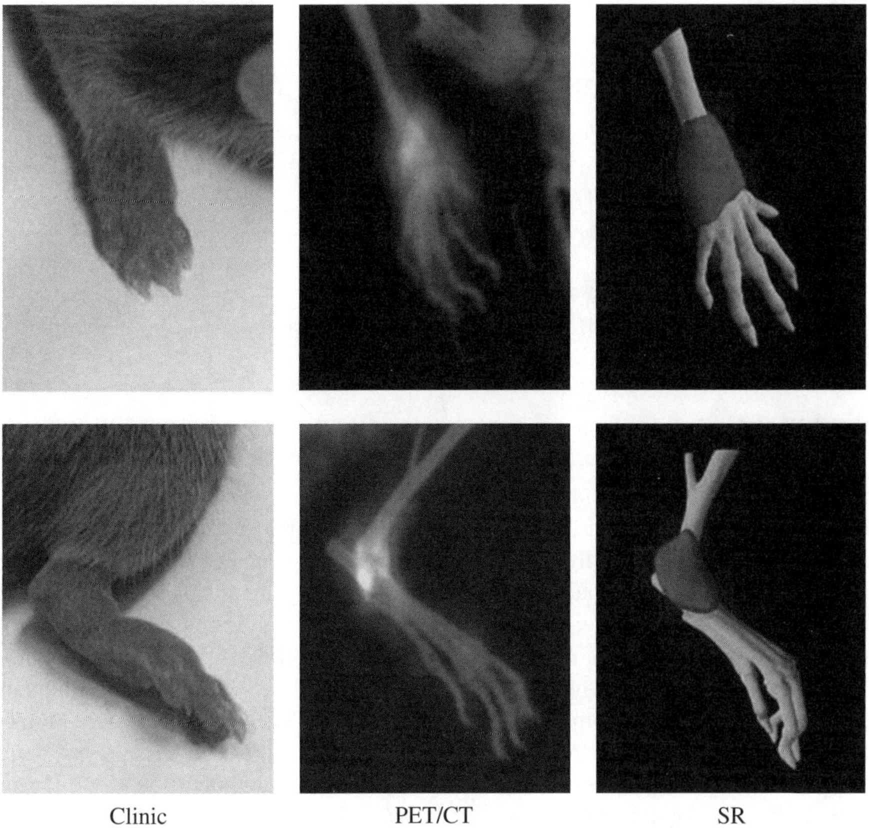

FIGURE 16.3 Demonstration of arthritis imaging in mice using ^{18}F-FDG PET/CT. Clinically evident swelling of the hind paw after intradermal injection of FCA correlates with high uptake of ^{18}F-FDG in the ankle of the mouse as seen with PET/CT. (With permission from Ref. [11].) (*See insert for color representation of the figure.*)

16.4.2.2 Chemical-Induced Cancer Models Cancers can be induced in rodent models by administration of chemical irritants. The advantage of these tumor models is the repeatable development of tumors in adequate study numbers for statistics. These tumors are similar to some types of human cancers, but are not generally useful compared with other models.

The carcinogen *N*-methyl-*N*'-nitro-*N*-nitrosoguanidine (MNNG), when added to the drinking water, leads to stomach cancer in rats [15], hamsters [16], and dogs [17] after 25–32 weeks. Intrarectal administration of MNNG causes formation of aberrant crypt foci followed by adenocarcinoma development by 30 weeks [18]. Intrauterine administration also induces uterine cancer in rats [19]. The combination of azoxymethane and dextran sulfate sodium (AOM-DSS) treatment of mice and rats reliably induces colorectal cancer with histopathologic and molecular characteristics of human colorectal cancer [20, 21].

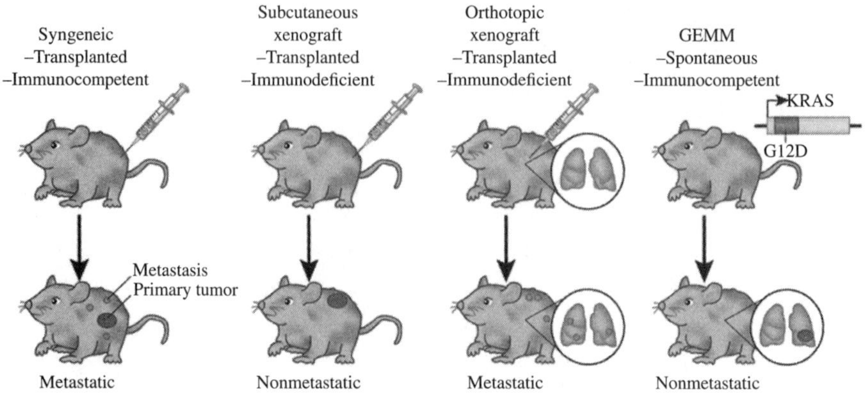

FIGURE 16.4 Cartoon of common mouse used in biomedical research. Source: Katie Vicari. (Reprinted with permission from Ref. [14]. © Macmillan Publishers Ltd.)

16.4.2.3 Xenografts Early research in cancer required models in which cancer development could be easily assessed, such as the subcutaneous xenografts. The most commonly used animal models of cancer are the human cancer xenografts in immunocompromised mice. For optical imaging, the nude mouse is particularly well suited as these strains are relatively hairless, which benefits optical imaging. They are also sufficiently immunocompromised to allow growth of many human cancer cell lines. Severe combined immunodeficiency (SCID) mice are more immunocompromised and allow growth of more cell lines and human tissues but have hair. In these models, tumors grow after implantation of human cancer cells from immortalized cell cultures or transplanted tissues. Implantations can be done ectopically such as underneath the skin (subcutaneous) or in the tissue of origin (orthotopic). Subcutaneous xenografts with human cancer cell lines are relatively simple to perform and enable large cohorts for experiments. Subcutaneous xenografts have been called "walking petri dishes" because of the relatively homogenous growth and lack of stromal tissues and appropriate architecture compared with natural human disease. Also, of particular importance to molecular imaging, vasculature tends to be more chaotic and leakier than in human disease, leading to EPR effects and higher nonspecific accumulation.

Bilateral tumor xenografts can provide additional sample numbers for tumors if the same cell lines are implanted. Different cell lines can be implanted for direct comparison in the same animal. For example, the M21 melanoma cell line expresses high levels of alpha-v-beta-3 (avb3) integrin receptors, while the M21L expresses very low levels of avb3. These cell lines have been implanted in opposite flanks of nude mice for molecular imaging studies of molecular probes targeting avb3. The difficulty with this approach is that the tumors grow at different rates and implantation time must be varied accordingly. It is also likely that other aspects of tumor biology may be different in these tumors as avb3 is associated with angiogenesis. Vascularization and vascular permeability may be altered. These and other differences must be accounted for in imaging studies.

16.4.2.4 Orthotopic Xenografts These models can be produced similarly by injection of cells or implantation of tumor tissues in the equivalent location in the mouse. Implantation of breast cancer cells in the mammary tissue of mice results in faster growth and histology more like human breast cancer. It is believed that the cancer cells are better suited to the environment from which they originated, contributing to. the "seed and soil" hypothesis.

16.4.2.5 Metastatic Models Cancer metastasis can also be modeled through injection of cells into locations for delivery to other places. For example, a well-studied model of breast cancer metastasis to the bone marrow is by intracardiac injection of human breast cancer cells. Another example is ovarian cancer that frequently forms abdominal metastases. This disease process has been modeled by intraperitoneal injection of ovarian carcinoma cells. Metastases to the brain, lungs, lymph nodes, and liver have also been modeled in a similar manner. These models have contributed significantly to the "seed and soil" hypothesis as well as the reseeding hypothesis.

16.4.2.6 Human Tissue Explants Implantation of intact tissues has demonstrated even greater advantage for replicating the biology of human disease. Intact tissues freshly collected from human cancer patients during surgery are implanted in immunocompromised mice. In the original human-in-mouse breast cancer model, the mammary fat pad is cleared and replaced with human fibroblasts before insertion of breast cancer tissue removed from human patients. The goal of this model is to better replicate the biology of human breast cancer. Researchers have compared the histology, genetics, and molecular expression patterns of primary and metastatic human cancer to the transplanted tumors. In general, the transplanted tumors maintained similar biology and gene expression to the original tumor, but in some cases, gene expression better correlated with that of metastases.

16.4.2.7 Spontaneous and Transgenic Models Genetic predisposition to cancer development has been found in certain lines of mice, rats, and other animals. These models replicate natural cancer development in the respective organ. Such models are relatively specific in type and have long latency, which is a limitation for most imaging studies.

Transgenic models have been developed by genetic modification of host DNA to introduce oncogenes or eliminate tumor suppressor genes. These animal models were initially developed to investigate causative factors in cancer development. Many transgenic cancer models develop tumors in relatively short time and in predictable manner. For example, the mouse mammary tumor virus (MMTV) gene selectively affects the mammary tissues and can be altered to include oncogenes such as HER2/neu and PyMT. Hemizygous MMTV-PyMT model predictably develops multifocal tumors throughout the mammary tissues at 8–12 weeks.

16.4.2.8 Viral-Induced Cancer Viral-induced cancers are known in animal models. Viruses can be genetically engineered to locally introduce desired DNA leading to cancer development. This reverse of gene therapy can be beneficial for investigation of cancer-inducing genes and targeted therapy. It has been known for years that viral infections can result in tumor development. Recently, immunization against HPV has been recommended for women to prevent HPV-induced cervical cancer. Some preclinical cancer models have since been developed, which involve viral infections that result in cancer formation. These models do not include the potential for genetic engineering through viral vectors to introduce oncogenes. In general, viral-induced cancer development is used to investigate early stages of cancer development as the progression from dysplasia to neoplasia and metastasis can be monitored in the presence of a fully functioning immune system. These models are fully nonhuman at this time, though humanized variants could be developed. Introduction of imaging reporter genes could enable detection of molecular events during carcinogenesis.

16.4.3 Animal Models for Cardiovascular Disease Imaging

16.4.3.1 GEM/Diet and Surgical Modification/Injury Imaging of cardiovascular disease primarily involves ischemic injury, atherosclerosis, cholesterol plaques, and stroke. Stroke will be discussed in the neuroimaging Section "Imaging of the nervous system" as it is more concerned with neurologic dysfunction.

In humans, atherosclerosis is commonly associated with poor diet and lack of exercise, but often includes a genetic predisposition. Plaques develop presumably as a result of inflammation of the arterial walls followed by cholesterol deposition and further infiltration by white blood cells. Cholesterol deposition can constrict the lumen of blood vessels and effectively block blood flow resulting in a "heart attack" when coronary blood vessels are affected. Prolonged obstruction causes tissue ischemia and loss of heart function.

Another sequela to cholesterol deposition is development of "unstable plaque" that can release small fragments into the bloodstream or completely rupture with large pieces breaking free. These fragments may lodge in smaller blood vessels with disastrous results, particularly if they block vessels in the brain (stroke). Thus, detection of cholesterol plaques within vessels is an important goal in diagnosis of atherosclerosis. Detection of unstable plaque is even more crucial in the diagnosis, therapy, and periodic monitoring to prevent stroke. Unstable plaque can be modeled in animals by a two-step injury process, which causes damage to the smooth muscle as well as endothelial lining of a large blood vessel. Rabbits are most commonly used for this model, and the injury is induced intravascularly, which can be difficult to perform in smaller animals.

This small size of blood vessels in animals makes imaging of plaques difficult as well. MRI is used to diagnose atherosclerosis and to visualize cholesterol deposition in humans and is therefore used predominantly in preclinical research. The resolution of PET/CT is sufficiently low as to reduce the utility of this modality. Intravascular ultrasound and optical imaging techniques are being developed to aid in diagnosis and intervention as well and also requires large vessels. Due to the depth of lesions and the relatively poor spatial resolution of optical imaging, most atherosclerotic plaque imaging studies have utilized invasive, postmortem optical imaging or have focused on peripheral models. Recently, atherosclerosis of the aortic arch has been imaged using noninvasive diffuse optical tomography.

Arterial injuries in mice and rats can be implemented as models of angioplasty injury and healing. This type of injury requires moderate technical skill for isolation and cannulation of the carotid or femoral artery. For carotid artery injury, the bifurcation of the common carotid artery into the internal and external branches is isolated. A small wire or bulbous probe is inserted through the external carotid artery and into the common carotid and to denude the endothelial layer by repeated abrasion. The external carotid is then ligated. Care must be taken to prevent debris from entering the internal carotid artery and causing stroke. Care must also be taken in imaging with this model as damage is done to surrounding tissues, which may result in nonspecific accumulation of imaging agents, obscuring intravascular signal.

Ischemic injuries, such as myocardial infarction, have been investigated using the murine hindlimb ischemia model. In this model, the femoral artery is removed, significantly reducing the blood delivery in the lower limb. Blood delivery is regenerated over time through expansion of collateral circulation by arteriogenesis and angiogenesis. Angiogenesis is a desired method for tissue healing, and drugs to improve angiogenesis are needed.

16.4.3.2 Thrombosis/Clot Models Thrombosis occurs when blood vessels are occluded by blood clots or intravascular debris and can result in injury and pain in peripheral tissues or stroke when intracranial blood vessels are occluded. Diagnosis of thrombotic disease is relatively straightforward clinically, but treatment can be a challenge when the chronology is unknown. Animal models of thrombosis are needed for development of therapies to ameliorate the disease and to better understand sequelae.

16.4.3.3 Aneurysm Aneurysm is a silent killer that develops over time. A pocket forms in the wall of an artery after a tiny rupture in the smooth muscle lining. The pocket enlarges and may rupture at any time resulting in immediate death when affecting the aorta or large artery in the brain. Early detection of aortic aneurysm is inhibited by lack of clinical signs or indications of predisposition. A diagnostic method for the detection of an aneurysm prior to rupture would save many lives.

Animal models of aortic aneurysm require surgical isolation of the vessel followed by local injection of elastase or other agents to damage the vessel wall. Mouse models have been used for many years and are well characterized. The difficulty in imaging is the relatively small size of the abdominal aorta in mice and the difficulty in reducing motion influence on data collection. Gating of image acquisition can significantly improve the resolution of these studies.

16.4.4 Infectious Disease Models for Imaging

16.4.4.1 Bacterial Infection Advancements in imaging of bacterial infections will help to improve therapies and treatments and aid in the prevention of progression of the disease within the host. Currently, the most common method for imaging bacterial infection is by using bacteria transfected with fluorescent proteins and/or bioluminescent constructs (luciferase).

Postarthroplasty infection models using bioluminescent *Staphylococcus aureus* have shown success in determining bacterial load within the joint. *S. aureus* is an advantageous bacterium, which given the opportunity will spread and cause a potentially serious infection. During orthopedic surgery, complications can arise with postarthroplasty infections caused by *S. aureus* during both acute and chronic recovery periods. Acute postarthroplasty infections are less severe and are more amenable to treatment. Treatment of chronic infections is much more difficult and often requires reoperation, which can lead to prolonged disability and require extended antibiotics [22]. In cases where biomaterials or devices are implanted, the infection can adhere to the implant, creating a biofilm that prevents immune cells or antibiotics to penetrate. Using bioluminescent bacterial agents, the pathogenesis of the infection can be tracked and the bacterial burden can be determined over time.

In research, the advantage of *in vivo* bioluminescence is the ability to have the animal survive and allow for longitudinal quantification of bacterial burden as opposed to performing multiple histologic analyses. This allows for reduction of overall animal numbers as well as for valuable animals to be evaluated over time.

Optical imaging of bacterial infection in animal models also has been demonstrated using fluorescent contrast agents that specifically target the bacteria. *Yersinia pestis* (black plague) infection has been reoccurring for centuries and has never been fully eradicated. Recently, bioluminescence has been used to monitor microbial dissemination in infected animals, *in vivo*, without the need to sacrifice animals to determine infection [23]. The ability to monitor bacterial dissemination with bioluminescents creates a successful noninvasive procedure that is comparable to viable plate counts.

Combining both bioluminescent and fluorescent optical imaging using microCT and X-ray has shown the ability to provide noninvasive and longitudinal measurements of the dynamic changes in bacterial burden, neutrophil recruitment, and bone damage in a mouse orthopedic implant infection model [24]. Combining information from both optical and structural imaging allows for the observation of biological process in a 3D model. With these dual modalities, the accuracy of the images increases, as well as the precision of the measurements for pathological changes.

16.4.4.2 Fungal Infections Diagnoses of fungal infections associated with immunocompromised patients has increased over the last several decades, but the number of effective antifungal drugs remains relatively unchanged and limited. Resistance to these drugs is becoming an increasing concern, and there is an increasing demand for new antifungal treatments. In addition, little progress has been made toward finding new modalities to detect fungal infections and identify regression of the disease.

Candida albicans is the most common fungal infection in people and can progress to disseminated candidiasis. Fifty percent of these cases progress to the CNS leading to a 90% mortality rate. Disseminated candidiasis is a fungal infection that can be modeled in mice and typically targets the kidneys and brain, but can also affect other viscera such as the liver, spleen, and heart [25]. The infection has shown the ability to deteriorate the blood–brain barrier (BBB), which can lead to fungal meningoencephalitis. An MRI following the injection of gadolinium–diethylenetriaminepentaacetic acid (Gd-DTPA) demonstrated that the integrity of the BBB was lost during the fungal invasion [25]. The ability to follow the infection and visualize its ability to weaken the BBB will help to understand the pathogenesis of the infection and lead to faster and more efficient means of treating *C. albicans* infections.

Aspergillus fumigatus is also reported to be a significant pathogen in immunocompromised patients causing high mortality. Infections are difficult to identify due to the lack of rapid screening tests, and early diagnoses go undetected. Ribosomal RNA (rRNA) targeting using oligonucleotide probes has been used to identify bacterial pathogens *in vitro* and a ^{99m}Tc-labeled phosphorodiamidate morpholino (MORF) oligomers (a DNA analog) targeting fungal rRNA [26]. A recent report used this targeting method for imaging detection of fungal infections in mice with *A. fumigatus* pulmonary infections.

16.4.4.3 Viral Infection The ability to image progression or regression of viral infections would greatly benefit the development of vaccines and antiviral therapy and would contribute to the understanding of the pathogenesis of the viruses.

Some viral infections can be imaged by the manipulation of viral DNA to include fluorescent or bioluminescent reporters (poxviruses) or by loading viral particles with other contrast agents. The recombinant virus is expected to replicate and encode the bioluminescent reporter into oncoming generations. This process could help visualize viral reproduction and host response simultaneously and provide insight into the pathogenesis of the virus along with the process of its replication [27].

In addition, there are reports of a well-known PET radiotracer, ^{18}F-FDG, showing the ability to visualize inflammatory response processes [28]. The radiotracer is taken up by cells, phosphorylated and selectively trapped within cells with high rates of glycolysis. The inflammatory response of both infectious and noninfectious disease is visualized as the radiotracer is taken up nonspecifically by both lymphocytes and neutrophils. Theoretically, FDG PET can be used to measure the inflammatory response of viral infections and provide an understanding of whether or not anti-inflammatory or antiviral therapies are beneficial.

Bioluminescence imaging using luciferase has recently been an effective imaging tool following the transmission of viral infections. Using luciferase, murine gammaherpesvirus-68 (MVH-68) can be traced from nasal inoculation to vaginal excretion in a female mouse; however, it was dependent on the presence of estrogen [29]. The virus is not transmitted to the litter, but direct transmission to a naïve male mouse is possible. The sexual transmission of MHV-68 to the male mouse has been documented in the laboratory using bioluminescence imaging, which could lead to a greater depth of understanding of the disease as well as development of strategies for preventing spread in a population.

Influenza A is a viral respiratory disease that can cause serious disease worldwide. The virus's ability to have a segmented genome allows for reassortment between human, avian, and swine strains and can lead to new pandemic strains [30]. Luciferase can be transgenically implemented into influenza A so that it will encode it into insertion sites of host cells. This will allow for real-time observations of the virus as it spreads, as well as give insight into how monoclonal antibodies interact for therapeutic analysis.

16.4.5 Imaging of the Nervous System

As the science community entered the twenty-first century, neurological studies have become a prominent mainstay within all disciplines and modalities. CNS disorders and injuries have been under intensive study since the beginning of the so-called (brain age) of medical studies. Studying the CNS has proven difficult however due to its complexity and depth. Also, due to the relatively nonregenerative properties of cells within the CNS, finding means to repair the CNS has become the main goal of those within the field of research.

Alzheimer's disease is a growing health concern for older generations of humans. It is a degenerative disease that slowly erodes the neuronal structure and pathways within the brain. How and why the disease develops is still relatively unknown; however, advances in imaging techniques are providing an in-depth look into the formation and progression of the disease.

The challenge in neurological disease has always been developing methods to image across the BBB. The BBB is a semipermeable membrane that is glucose soluble. This creates difficulties in crossing the barrier with bioluminescence and other fluorescent materials that are required for imaging the brain. As nanomaterials are generally excluded from crossing the BBB, nanomaterial contrast agents are limited to imaging of intravascular disease such as stroke or cancer in which the BBB is broken. In the case of a compromised BBB, tissue uptake is primarily nonspecific. Nanomaterials have shown great promise in the delivery of small-molecule therapeutic agents to the brain and facilitate crossing of the BBB. This strategy is beginning to show promising results for imaging purposes.

A biodegradable nanocarrier system made up of poly-n-butyl cyanoacrylate (PBCA) dextran polymers coated with polysorbate 80 (PBCA NPs) has shown great promise in having the ability to cross the BBB [31]. PBCA has great potential because it is able to carry BBB-impermeable materials across the barrier and allows for advancement of diagnostic neuroimaging. The process begins with a Nissl stain for cellular imaging; then moves to visualization of amyloid plaques, *in vivo*, using non-BBB-permeable reagents that detect plaques; and finally finishes with the delivery of a gadolinium-based contrast agent into the brain of the mouse for a whole-brain MRI *in vivo* [31]. Diagnostic neuroimaging of the mouse will provide key information to the mechanisms of Alzheimer's disease, as well as insight into how to repair the damaged tissue and how to prevent the disease from progressing.

One recent report indicates that nanomaterials have demonstrated drug delivery to the brain by incorporating mechanisms for translocation of agents across/through the BBB. It was shown that the PBCA NPs encapsulated with different fluorophores were successful at crossing the BBB. The NPs were administered IV into mice, and then the first several hundred micrometers of cortex were imaged *in vivo* using intravital multiphoton microscopy through a craniotomy sealed with a glass coverslip [31] (Fig. 16.5).

Stem cell transplantation has shown promise in therapeutic treatment strategies for CNS disorders and injuries; however, there is no definitive model for evaluating stem cell graft survival and interaction with the CNS environment postimplantation

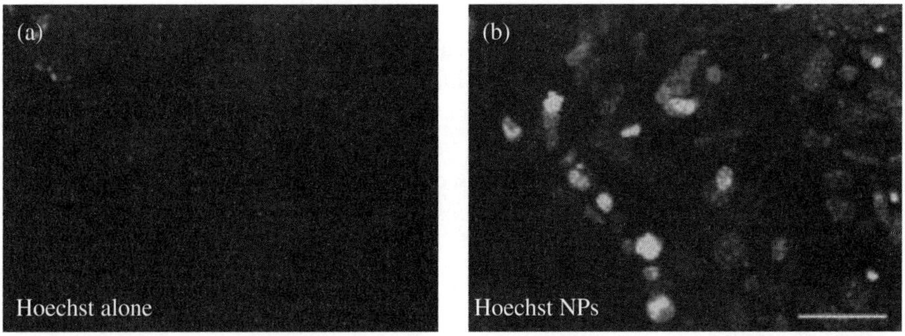

FIGURE 16.5 Hoechst alone does not cross the BBB (a), but PBCA NPs (coated with polysorbate 80) inoculated with Hoechst dye will cross the BBB and stain the nuclei in the brain (b). (Reprinted with permission from Ref. [31]. © National Academy of Sciences.)

[32]. Using bioluminescents, stem cell survival and follow-up studies of interactions with the environment of the CNS can be evaluated in ways never before possible. Using neural stem cells (NCS), which differentiate toward neurons, astrocytes, and oligodendrocytes, and bone marrow-derived stromal cells (BMSC), which have immune-modulating properties, studies have shown that they can monitor stem cell survival rates, as well as their behavior with the CNS environment [32].

The ability to visualize neuronal injuries with a simple MRI down to a glial cell level would drastically improve treatments and facilitate new modalities for imaging the CNS. Manganese (Mn^{+2}) can enter into neurons through voltage-gated calcium channels, and from there, it can provide insight into the brain's physiology much more precisely than other imaging agents such as gadolinium or iron oxide can [33].

16.5 COMPANION ANIMAL DISEASE MODELS

The healthcare for companion animals follows as closely as possible to the current state of human healthcare as allowed by the difference in economics. Pet dogs, cats and other animals are generally well cared for and live considerably long lives. Many of the diseases common in humans are represented in the companion animal population as well, including cancer, diabetes, and neurologic injuries and disorders. In many cases, the presentation and biology of these diseases is very similar to that of human disease, much more so than induced or even genetically modified rodent models. Pets with these ailments are often brought to veterinary specialists at regional teaching hospitals for care. The presentation of pets with naturally occurring disease negates the necessity of waiting for natural disease to occur in research animals.

Clinical trials with companion animals are thus relatively easy to arrange. Benefits of companion animal clinical trials for preclinical trials are significant. In fact, the NIH has initiated the Companion Animal Clinical Trials Consortium for the purpose of aiding companies and researchers in the design, implementation, and analysis of clinical trials through collaborations of institutions throughout the nation.

Treatment of pets is not regulated by the FDA, but is left up to the owners and attending veterinarians. Use of experimental agents, whether therapeutic or diagnostic, is therefore not as restrictive as in humans. Compounds need not be produced in GMP facilities or be approved for investigational use by the FDA. This does not mean that agents should not be tested first in rodent models, but that promising agents may be tested in closer-to-clinic companion animal models to confirm preclinical test results prior to initiating the very expensive processes required for investigational new drug (IND).

REFERENCES

[1] National Research Council. *Guide for the Care and Use of Laboratory Animals*. 8th ed. Washington, DC: The National Academies Press; 2011.

[2] Gosens I, Post JA, de la Fonteyne LJ, Jansen EH, Geus JW, Cassee FR, de Jong WH. Impact of agglomeration state of nano- and submicron sized gold particles on pulmonary inflammation. Part Fibre Toxicol 2010;**7**:37.

[3] Pai MP. Drug dosing based on weight and body surface area: mathematical assumptions and limitations in obese adults. Pharmacotherapy 2012;**32**:856–868.

[4] Kung MP, Kung HF. Mass effect of injected dose in small rodent imaging by SPECT and PET. Nucl Med Biol 2005;**32**:673–678.

[5] Hawk CT, Leary SL, Morris TH, American College of Laboratory Animal Medicine., and European College of Laboratory Animal Medicine. *Formulary for Laboratory Animals*. 3rd ed. Ames, IA: Blackwell Publishing; 2005.

[6] Doria AS, Crawley A, Gahunia H, Moineddin R, Rayner T, Tassos V, Zhong A, Pritzker K, Mendes M, Jong R, Salter RB. Correlative BOLD MR imaging of stages of synovitis in a rabbit model of antigen-induced arthritis. Pediatr Radiol 2012;**42**:63–75.

[7] Wang SX, Arsenault L, Hunziker EB. Stereologic analysis of tibial-plateau cartilage and femoral cancellous bone in guinea pigs with spontaneous osteoarthritis. Clin Orthop Relat Res 2011;**469**:2796–2805.

[8] Guingamp C, Gegout-Pottie P, Philippe L, Terlain B, Netter P, Gillet P. Mono-iodoacetate-induced experimental osteoarthritis: a dose-response study of loss of mobility, morphology, and biochemistry. Arthritis Rheum 1997;**40**:1670–1679.

[9] Siebelt M, Waarsing JH, Kops N, Piscaer TM, Verhaar JA, Oei EH, Weinans H. Quantifying osteoarthritic cartilage changes accurately using *in vivo* microCT arthrography in three etiologically distinct rat models. J Orthop Res 2011;**29**:1788–1794.

[10] Almarestani L, Fitzcharles MA, Bennett GJ, Ribeiro-da-Silva A. Imaging studies in Freund's complete adjuvant model of regional polyarthritis, a model suitable for the study of pain mechanisms, in the rat. Arthritis Rheum 2011;**63**:1573–1581.

[11] Irmler IM, Opfermann T, Gebhardt P, Gajda M, Brauer R, Saluz HP, Kamradt T. *In vivo* molecular imaging of experimental joint inflammation by combined (18)F-FDG positron emission tomography and computed tomography. Arthritis Res Ther 2010;**12**:R203.

[12] Caglic D, Globisch A, Kindermann M, Lim NH, Jeske V, Juretschke HP, Bartnik E, Weithmann KU, Nagase H, Turk B, Wendt KU. Functional *in vivo* imaging of cysteine cathepsin activity in murine model of inflammation. Bioorg Med Chem 2011;**19**: 1055–1061.

[13] Moodie JP, Stok KS, Muller R, Vincent TL, Shefelbine SJ. Multimodal imaging demonstrates concomitant changes in bone and cartilage after destabilisation of the medial meniscus and increased joint laxity. Osteoarthr Cartil 2011;**19**:163–170.

[14] Francia G, Kerbel RS. Raising the bar for cancer therapy models. Nat Biotechnol 2010;**28**:561–562.

[15] Sugimura T, Fujimura S. Tumour production in glandular stomach of rat by *N*-methyl-*N*'-nitro-*N*-nitrosoguanidine. Nature 1967;**216**:943–944.

[16] Kogure K, Sasadaira H, Kawachi T, Shimosato Y, Tokunaga A. Further studies on induction of stomach cancer in hamsters by *N*-methyl-*N*'-nitro-*N*-nitrosoguanidine. Br J Cancer 1974;**29**:132–142.

[17] Sugimura T, Tanaka N, Kawachi T, Kogure K, Fujimura S. Production of stomach cancer in dogs by *N*-methyl-*N*'-nitro-*N*-nitrosoguanidine. Gann 1971;**62**:67.

[18] Maurin N, Forgue-Lafitte ME, Levy P, Zimber A, Bara J. Progression of tumors arising from large ACF is associated with the MUC5AC expression during rat colon MNNG carcinogenis. Int J Cancer 2007;**120**:477–483.

[19] Tanaka T, Mori H. Experimental induction of uterine cancer in rats by *N*-methyl-*N*'-nitro-*N*-nitrosoguanidine. Pathol Res Pract 1983;**178**:20–26.

[20] Thaker AI, Shaker A, Rao MS, Ciorba MA. Modeling colitis-associated cancer with azoxymethane (AOM) and dextran sulfate sodium (DSS). J Vis Exp 2012;**67**:pii: 4100.

[21] De Robertis M, Massi E, Poeta ML, Carotti S, Morini S, Cecchetelli L, Signori E, Fazio VM. The AOM/DSS murine model for the study of colon carcinogenesis: from pathways to diagnosis and therapy studies. J Carcinog 2011;**10**:9.

[22] Pribaz JR, Bernthal NM, Billi F, Cho JS, Ramos RI, Guo Y, Cheung AL, Francis KP, Miller LS. Mouse model of chronic post-arthroplasty infection: noninvasive *in vivo* bioluminescence imaging to monitor bacterial burden for long-term study. J Orthop Res 2012;**30**:335–340.

[23] Sha J, Rosenzweig JA, Kirtley ML, van Lier CJ, Fitts EC, Kozlova EV, Erova TE, Tiner BL, Chopra AK. A non-invasive *in vivo* imaging system to study dissemination of bioluminescent Yersinia pestis CO92 in a mouse model of pneumonic plague. Microb Pathog 2013;**55**:39–50.

[24] Niska JA, Meganck JA, Pribaz JR, Shahbazian JH, Lim E, Zhang N, Rice BW, Akin A, Ramos RI, Bernthal NM, Francis KP, Miller LS. Monitoring bacterial burden, inflammation and bone damage longitudinally using optical and muCT imaging in an orthopaedic implant infection in mice. PLoS One 2012;**7**:e47397.

[25] Navarathna DH, Munasinghe J, Lizak MJ, Nayak D, McGavern DB, Roberts DD. MRI confirms loss of blood-brain barrier integrity in a mouse model of disseminated candidiasis. NMR Biomed 2013;**26**:1125–1134.

[26] Wang Y, Chen L, Liu X, Cheng D, Liu G, Liu Y, Dou S, Hnatowich DJ, Rusckowski M. Detection of Aspergillus fumigatus pulmonary fungal infections in mice with (99m) Tc-labeled MORF oligomers targeting ribosomal RNA. Nucl Med Biol 2013;**40**:89–96.

[27] Bray M, Lawler J, Paragas J, Jahrling PB, Mollura DJ. Molecular imaging of influenza and other emerging respiratory viral infections. J Infect Dis 2011;**203**:1348–1359.

[28] Jonsson CB, Camp JV, Wu A, Zheng H, Kraenzle JL, Biller AE, Vanover CD, Chu YK, Ng CK, Proctor M, Sherwood L, Steffen MC, Mollura DJ. Molecular imaging reveals a progressive pulmonary inflammation in lower airways in ferrets infected with 2009 H1N1 pandemic influenza virus. PLoS One 2012;**7**:e40094.

[29] Francois S, Vidick S, Sarlet M, Desmecht D, Drion P, Stevenson PG, Vanderplasschen A, Gillet L. Illumination of murine gammaherpesvirus-68 cycle reveals a sexual transmission route from females to males in laboratory mice. PLoS Pathog 2013;**9**:e1003292.

[30] Heaton NS, Leyva-Grado VH, Tan GS, Eggink D, Hai R, Palese P. *In vivo* bioluminescent imaging of influenza a virus infection and characterization of novel cross-protective monoclonal antibodies. J Virol 2013;**87**:8272–8281.

[31] Koffie RM, Farrar CT, Saidi LJ, William CM, Hyman BT, Spires-Jones TL. Nanoparticles enhance brain delivery of blood-brain barrier-impermeable probes for *in vivo* optical and magnetic resonance imaging. Proc Natl Acad Sci U S A 2011;**108**:18837–18842.

[32] Reekmans K, De Vocht N, Praet J, Le Blon D, Hoornaert C, Daans J, Van der Linden A, Berneman Z, Ponsaerts P. Quantitative evaluation of stem cell grafting in the central nervous system of mice by *in vivo* bioluminescence imaging and postmortem multicolor histological analysis. Methods Mol Biol 2013;**1052**:1–17.

[33] Bade AN, Zhou B, Epstein AA, Gorantla S, Poluektova LY, Luo J, Gendelman HE, Boska MD, Liu Y. Improved visualization of neuronal injury following glial activation by manganese enhanced MRI. J Neuroimmune Pharmacol 2013;**8**:1027–1036.

INDEX

absorption, 112–13, 119, 121, 126–8, 134, 151, 173–5, 265–7, 269–70, 281, 287, 300–301, 353
absorption coefficients, optical, 112–13, 126, 294, 303, 352
activity, antitumor, 65, 360
affibody, 283
albumin, 8, 100, 172, 228, 232
Alexa Fluor, 208, 267, 410, 412–14
Amersham Inc., 12–13
anesthesia, inhalation, 469–70
angiograms, 13
animal models, 113, 130, 366, 431, 467–84
animals, 3, 102, 114, 223, 227, 350, 351, 378, 385, 454, 468–71, 473, 474, 478–80, 483
anti-angiogenic perfluorocarbon nanoparticles, 187, 463
antisense agents, 374, 376, 378–84, 386–8
antisense imaging agents, 377, 379, 385–92
antisense probes, 374, 378, 379, 386–8
antisense RNA imaging, 379–83
antisense strands, 383–4, 388–90
applications of viral nanoparticles, 401–21

barriers, blood–brain (BBB), 435, 480–483
Bell Labs, 13
bilayers, 53, 85–7, 92, 197, 440
biodegradable nanoparticles, 207, 249
biodegradable polymeric nanoparticles, 207
bioimaging applications, 14, 245, 326, 329, 330
biomarkers, 19, 171–2, 264–5, 282, 285, 374–9, 430
biomedical applications of polymer nanocapsules, 96–103
blood pool agents, 228, 230–231, 233, 238, 244, 247
blood pool imaging, 66, 69
blood vessels, 5, 13, 115–16, 121, 130, 134–6, 142, 228, 298–9, 303, 314, 478–9
body weight (BW), 62, 205, 359, 366–7, 437, 469–70
brain tumor imaging, 50, 72, 339

Nanotechnology for Biomedical Imaging and Diagnostics: From Nanoparticle Design to Clinical Applications, First Edition. Edited by Mikhail Y. Berezin.

brain tumors, 205, 243, 246, 304, 305, 310–311, 339, 341
BRIGHTs (nanoparticles), 337–8
brome mosaic virus (BMV), 122, 403
BW *see* body weight

CA *see* contrast agent (CA)
cancer
  development, 350, 476–8
  drugs, 434–5
  imaging, 71, 288, 474–8
  therapy, 100, 102, 349–69, 433, 436, 437, 452
  tumors, 358, 359, 367–8
capsid, 245, 406, 411, 415, 416, 418, 419
capsules, 84, 86–7, 90–91, 93–4, 96–7, 99–100, 103–4, 444
carbon nanotubes (CNTs), 13, 122–6, 201–2, 244, 438–9
cationic nanoparticles, 386, 391–2
CCMV *see* cowpea chlorotic mottle virus (CCMV)
cell-penetrating peptide (CPP), 208–9, 308, 388, 392
cells, living, 96, 100, 232
central nervous system (CNS), 231, 467, 480, 482–3
characterization, 8, 83, 139, 163, 166, 326, 405, 442, 467
  of nanocapsules, 85–92
chelates, 58, 227, 230–231, 234, 237, 241–5, 247
chemotherapy, 296, 350, 367–8, 452, 472
choice of animal model, 472–83
chromophores, 112–13, 173, 269–70, 298, 309
circulating tumor cells (CTCs), 123, 310
circulation, 15–16, 52, 59–62, 66, 67, 208, 239, 471
CLIO *see* cross-linked iron oxide (CLIO)
CNS *see* central nervous system (CNS)
CNTs *see* carbon nanotubes (CNTs)
colloidal systems, 440–448
companion animals, 349–51, 359, 367, 483–4
complexes, 10, 17, 58, 224, 226–8, 240, 244, 247–8, 386
confocal microscopy, 407, 411–13
continuous-wave (CW), 270, 273, 275, 286, 368
contrast agents (CA)
  exogenous nonplasmonic, 120–126
  liposomal, 50, 65
  macromolecular MR, 248
  macromolecular MR imaging, 252
  nanoparticle-based OA, 113, 139
  optical, 268
copolymers, 55, 69, 232, 452, 454
core, 39, 43, 51, 69, 89–90, 196, 206, 207, 234, 236–8, 240–242, 245–7, 336, 448–50, 453
  micelle, 54, 55
core-shell NPs, 38–40, 42, 180
cowpea chlorotic mottle virus (CCMV), 245, 403–5, 415, 417
cowpea mosaic virus (CpMV), 406–9, 411–13, 417, 420
  binding, 407
  uptake, 407, 414
CpMV-A647, 410
CpMV-F56 particles, 412
CpMV nanoparticles, 405–7, 409, 413
  labeled, 409
  shield, 405
CpMV particles, 407, 411, 412
  icosahedral, 409
  labeled, 409
CpMV-PEG, 412, 413
CpMV-PEG–bombesin, 413, 414
  uptake of, 412–13
CpMV-PEG–bombesin nanoparticles, 413
CpMV-vimentin interaction, 413
CPP *see* cell-penetrating peptide (CPP)
cross-linked iron oxide (CLIO), 35–6, 39, 435
cross-linkers, 85–8, 91, 207
CTCs *see* circulating tumor cells (CTCs)
CT imaging, 5, 50, 51, 59, 63–5, 69, 77
CuS nanoparticles, 210
CW *see* continuous-wave (CW)
Cy7, 12–13, 447
cyclic RGD-PEGylated gold nanoparticle probes, using, 203
cytoplasm, 202, 374–5, 378, 379, 384, 386–8, 390, 392, 445

delivery systems, liposomal drug, 441, 442
dendrimers, 171–2, 178, 196, 200–201, 230, 234–40, 249, 448–50, 452
depth, 112, 119, 121, 274–5, 281–2, 287, 313, 329, 331, 339, 352, 450, 478, 481, 482
detectors, 12, 112, 159, 161, 165, 173, 175–7, 179, 181, 270–275, 277, 284–7, 327–8
development of theranostic nanomedicines, 429–55
development of viral nanoparticles, 403
diagnostic imaging, 63, 196, 200–202, 212
diagnostic neuroimaging, 482
Diaphanography, 12
diethylene triamine pentaacetic acid (DTPA), 10, 11, 53, 55–6, 68, 198, 202, 205, 230–234, 240–241, 243–7, 415
diffuse fluorescence optical imaging instruments, 265, 273–7
diffuse optical tomography (DOT), 287, 478
diffusion equations, 278–80, 286
2, 3-dimercaptosuccinic acid (DMSA), 31, 34, 35, 38–9, 43
DLS *see* dynamic light scattering (DLS)
DNA, 103, 182, 183, 245, 333, 334, 350, 372, 374–87, 380–383, 391, 392, 477, 481
dogs, 11, 17, 67, 349–50, 360, 362–4, 366, 367, 475, 483
DOT *see* diffuse optical tomography (DOT)
DOTA, 165, 206, 207, 228, 230–232, 236, 239, 245, 434
doxorubicin (DOX), 65, 100, 199, 364, 435–8, 443, 447, 448, 450, 452, 454
drug(s), 60–61, 67–8, 71, 90, 92, 97, 263–5, 364, 430–432, 440–442, 454–5, 479, 480
  anticancer, 100, 102, 103, 437–9, 440, 448, 454
  carriers, 59–61, 103
  delivery, 125, 212, 245, 405, 421, 429–32, 440–442, 450, 453, 455, 468, 482
  molecules, 40, 93, 97, 100, 434–6, 439, 448, 452
  release, 100, 432, 440–441, 446, 448–9, 452
drug delivery systems, multifunctional theranostic liposome, 441
DTPA *see* diethylene triamine pentaacetic acid (DTPA)
dual-mode nanoparticle contrast agent (DMCA), 43, 44
dyes, 12–13, 90, 92, 93, 97, 121–3, 178, 181–3, 274, 282, 308, 313, 434
dynamic light scattering (DLS), 86, 153, 156, 161–3, 184

Eastman Kodak laboratories, 12
EB *see* Evans blue (EB)
ECF *see* extracellular fluid (ECF)
echo time (TE), 168–70
  short, 168–9
EGFR *see* epidermal growth factor receptor (EGFR)
electron capture (EC), 197, 205
ELISA, 264, 283–4
emission wavelengths, 265–6, 268, 274, 279–80, 286
endocytosis, 202, 387, 389–91, 412, 439
endogenous chromophores, 12, 113–20
endosomes, 207, 208, 374, 384, 387–9, 391
engineered gold nanoparticles, 221, 399
enhanced tumor MR imaging, 65
entrapment, 58, 61, 83, 94, 386, 404, 445, 451
epidermal growth factor receptor (EGFR), 40, 126, 205
Evans blue (EB), 291, 312–13
evidence of tumor cell death, 363
excitation, 96, 122, 175, 180, 182, 266–9, 274, 279–80, 286, 327, 331, 336, 338, 342
  light, 173, 182, 266, 274, 275, 284, 287
  source, 175, 275, 276, 279, 282, 327, 330, 337
exogenous, 302–3, 305, 307, 309, 311, 313, 315
exogenous contrast agents, 112–13, 120–121, 264, 295, 300, 303, 326, 328, 333
experiments, 68, 87–8, 91, 92, 243, 244, 282, 356, 476
extinction coefficient, 116, 267, 269–70, 286, 308

extracellular fluid (ECF), 228–9
extratumoral vasculatures, 299–300

FA *see* folic acid (FA)
FCS *see* fluorescence correlation spectroscopy (FCS)
FD *see* frequency domain (FD)
FeAU, 419–20
FePt nanoparticles/NPs, 40
ferrite, 34–5, 37–9, 43–4
FIT probes *see* forced intercalation probes (FIT probes)
FL signal, 385
fluid, extracellular *see* extracellular fluid (ECF)
fluorescein isothiocyanate (FITC), 415–16
fluorescence, 175, 180, 181, 265–9, 279, 283, 285, 287–8, 293, 294, 384–5, 389, 390
  signals, 274–7, 279, 281, 282, 284, 385, 453, 454
  targets, 274–5, 281, 286
fluorescence anisotropy, 153, 181–2
fluorescence correlation spectroscopy (FCS), 153, 183–4
fluorescence imaging, 18, 41, 122, 263–5, 275, 281, 287, 454, 473
fluorescence intensity, 93, 102, 177, 181, 183, 185, 267, 273–6, 281–3, 287
fluorescence lifetime (FLT), 102, 180–181, 267, 273, 284–5
  tumor, 285
fluorescence reflectance imaging (FRI), 471
fluorescence resonance energy transfer (FRET), 268, 383, 384
fluorescent probes, 264–5, 267–8, 282, 286, 383
fluorescent proteins, 118–20, 183, 479
fluorobenzaldehyde, 418, 419
fluorophore(s), 99, 178, 180–182, 200, 265, 267, 268, 274, 275, 279, 281–6, 384, 482
  concentrations, 181, 274–5, 282, 286
folic acid (FA), 103, 411
forced intercalation probes (FIT probes), 384–5
formulations, liposomal, 63, 71, 441, 442
frequency, 65, 180, 248, 276, 296, 328, 338, 365
frequency domain (FD), 180, 271, 276–8, 280, 286
FRET *see* fluorescence resonance energy transfer (FRET)
FRI *see* fluorescence reflectance imaging (FRI)
function, photon density, 279, 286
functional groups, 39, 85, 91–2, 203, 243, 433–6, 439, 448, 452

gadolinium, 10, 152, 171, 223–30, 246–9, 416, 417, 452, 483
gadolinium-based nanoparticles, 223–49
gadolinium-labeled liposomes, 21, 77, 257
gadolinium oxide nanoparticles, 245–6
gamma-imaging, 57, 63–4, 67, 71–2, 198–9
Gd-DOTA, 417, 441
Gd-DTPA, 43–4, 54, 63, 170–171, 226, 229, 231, 415–17, 480
gene delivery, 419–20
GNCs *see* gold nanocages (GNCs)
GNPs *see* paramagnetic gold nanoparticles (GNPs)
gold, 128, 130, 132, 137, 152, 154, 158–60, 335, 354, 356, 358, 367, 388, 389, 436–7
  colloids, 5, 183, 340
gold nanobeacons (GNBs), 126, 136, 301, 303–4
gold nanocages (GNCs), 127, 130–134, 210, 301–3, 308, 309, 437
gold nanoparticles, 126–7, 129, 134, 136, 244, 334–6, 366–7, 436–7, 443
  hollow, 130, 132
  synthesis, 437
  toxicity of engineered, 221, 399
gold nanorods, 125–30, 137–8, 186, 301, 309–12, 349–51, 353, 354, 359, 361
  using, 130
gold nanoshells (NS), 126–9, 137, 210, 301

half-life, 13
$HAuCl_4$ (auric acid), 40, 130, 131, 244–5, 301
HAuNS *see* hollow gold nanospheres/nanoshells
HeLa, 201, 406–8, 411, 412

hemoglobin, 12, 113–16, 116, 117, 120, 122, 298–9
HER2 receptors, 283–5
hollow gold nanospheres/nanoshells (HAuNs), 134, 136, 138, 437
hollow nanocapsules, 83–104
human serum albumin (HSA), 306
hybrid nanoparticle probe, 47, 218
hyperthermia, 103, 351–2, 441

ICG *see* indocyanine green (ICG)
ICP-MS, 151–5, 204
ICP-OES, 152–6
ID *see* injected dose (ID)
image contrast, 30, 121, 169–70, 247
image particles, 158
image reconstruction, 7, 281, 285, 287
imaging
  algorithms, 265, 275, 281–7
  infarct, 67
  mmPA, 309
  nanoparticles, 2, 14, 151–2, 178
  nanoreagents, 431
  nuclear, 5–6, 196, 203, 311–12, 471
  planar, 165, 274, 281–2
  scintigraphic, 65
  small animal, 276
  targeted, 402, 410, 412
  voxel, 286, 301, 303
  whole-body, 114, 296
imaging agents
  dual-mode, 445–6
  genetic, 375–7
imaging modalities, 2–3, 50, 59, 63, 65, 66, 69, 71, 112, 115, 120, 199, 440, 444, 445, 448, 467, 470–472
  multiple, 445, 450
imaging probes, molecular, 264
imaging systems, nanoparticle antisense, 388–92
imaging techniques, 2, 6, 16, 139, 156, 264, 339, 482
  nuclear, 5–6, 196
immunoliposomes, 17–18, 59, 67
indocyanine green (ICG), 3, 12–13, 15, 100, 102, 113, 121–2, 125, 173–4, 181, 303–5, 307, 312–13, 451
infection, 67, 232, 441, 469, 478, 479
injected dose (ID), 62, 200, 204, 359, 367
injection
  intradermal, 302–4, 312–13, 475
  intratumoral, 199
injection of X-gal, 120
intensity, 88, 97, 102, 114, 152, 159, 163, 178, 181, 248, 274–6, 281–4, 307, 328, 336
in vivo biodistribution, 195, 200–201, 205, 210
in vivo bioimaging, 329, 334, 336, 338
in vivo bioluminescence imaging, 480
in vivo cancer detection, 338
in vivo imaging, 14, 63, 119, 198, 202, 439, 451
  applications, 335, 436
  systems, 264
in vivo measurements of nanoparticle concentrations, 140, 144
in vivo molecular fluorescence imaging, 263–88
in vivo OA imaging, 116, 131
in vivo PET imaging, 219, 447
in vivo pharmacokinetics, 203, 206
in vivo photoacoustic imaging, 302, 304, 305
in vivo Raman-based bioimaging, 333–8
in vivo stability, 18, 50, 52, 53, 57, 199, 200, 208, 211, 227
in vivo tests, 40–41, 43
in vivo tumor imaging, 338, 340
iron oxide, 35, 37, 199–200, 206, 432, 433, 483
iron oxide-based magnetic nanoparticles (IOMNPs), 27–8, 30–45
iron oxide nanoparticles (IONPs), 17, 34, 70–71, 171, 195, 205–6, 404, 433–6

Kavli Prize in Nanoscience, 13
Kodak, 12

labeling, 8, 13, 52, 57, 165, 182, 196, 202, 206
laboratory animals, 468–71
laser light, 161, 163, 352
LCLs *see* long-circulating liposomes (LCLs)
LI-COR Biosciences, 471
lifetime, 17, 102, 180–181, 267, 274–6, 280, 284–7

ligands, 402, 404, 409–11, 416, 418
light
  intensity, 161–2, 269
  polarized, 181
  propagation, 265, 270, 274–5, 277–81, 285
linear polymers, 230–234, 236, 237
lipid-based pharmaceutical nanocarriers, 49–72
lipid nanoparticles, 196, 197, 242, 451
liposomal contrast, 63
liposomal imaging agents, 63–8
liposome(s), 2–3, 6–10, 16–18, 49–61, 63–8, 71–2, 87–8, 171–2, 196–9, 230, 240–242, 440–444
  circulation time of, 61
  diagnostic, 71, 72
  disposition, 76
  folate-modified, 65
  formulations, 63, 71, 441
  indium-111-labelled, 77
  membrane, 55, 172, 241, 442
  multilamellar, 7–8, 10
  peptide-conjugated, 242
  phospholipid, 443
  radiolabeled, 9, 64, 197–9
  systems, 198
  theranostic, 71, 440–443, 454
  tumor-targeted, 78, 258
liposome particles, lamellar phase of, 10, 11
live cells, 103, 408, 412
liver, 5, 8, 10, 11, 16, 17, 42, 58–60, 62–3, 66, 102, 125, 197, 198, 200–202, 206, 210, 228, 229, 234, 238–40, 242, 243, 245–7, 299, 303, 340, 359, 366, 367, 389, 404, 417, 418, 420, 444, 477, 480
living mice, 123–5, 304, 305, 310, 311, 341
LNAs *see* locked nucleic acids (LNAs)
loading liposomes, 51–61
localized surface plasmon resonance (LSPR), 301, 308, 335
locked nucleic acids (LNAs), 382, 383
long-circulating drugs, 60–61
long-circulating liposomes (LCLs), 64, 197, 443
LSPR *see* localized surface plasmon resonance (LSPR)
lysosomes, 384
macrophages, 16, 65, 67, 69, 97, 98, 205, 232, 244, 380, 391
magnetic imaging, 10–11
magnetic nanoparticles, 27–45, 65, 161, 182, 205, 310, 438, 452, 454
  iron oxide shell of, 46, 456
  multifunctional, 45, 218
magnetic properties, 28, 35, 37, 43, 199
  of nanoparticles, 152, 167–72
magnetic resonance imaging (MRI), 2, 10, 11, 18, 27–30, 39, 40, 43, 44, 50, 51, 54, 55, 57, 59, 63–5, 68–72, 102, 103, 152, 167, 170–172, 199, 223–49, 287, 310, 339, 341, 415, 417, 420, 421, 433, 434, 436, 440, 441, 443, 444, 446–8, 451, 452, 454, 467, 471–4, 478, 480, 482, 483
  contrast agents, 10, 11, 28, 32–9, 43, 54, 99, 169, 171, 172, 309, 413–18, 420
  scanner, 10, 30, 34, 35, 168
  signal, 55, 169, 170
magnetite nanoparticles (MNPs), 11, 17, 27–34, 182, 309, 310, 438, 452, 454
magnetization, 27–30, 32–9, 43, 44, 167, 168, 248
magnetoliposomes, 50, 440
Magnevist, 228, 229, 415, 416
Mallinckrodt Inc., 13, 15, 16
MB *see* methylene blue (MB)
melamine formaldehyde (MF), 89, 129
melanin, 12, 113, 114, 116–18, 120, 268, 269, 298
  production, 116–18
melanomas, 64, 116, 134, 200, 211, 298, 301, 302, 443, 476
membrane, 89, 90, 103, 240–242, 374, 379, 386, 388
  liposomal, 54, 55, 91
metabolic rate of oxygen consumption (MRO2), 295, 299
methylene blue (MB), 300, 312, 313
MF *see* melamine formaldehyde (MF)
micellar imaging agents, 68–72
micelles, 40, 49–61, 68–72, 104, 171, 172, 184, 241, 242, 303, 402
microbubbles, 14–17, 65, 305, 306, 446
MicroCT image, 311–12
microparticles, 208
miRNAs, 375, 378, 390

mitochondria, 103, 375, 376, 378
mmPA imaging modes, 309
MMTV see mouse mammary tumor virus (MMTV)
MNP-Au nanoparticles, 309
MNPs see magnetite nanoparticles (MNPs)
models, rabbit liver VX-2 tumor, 63
molecular beacon, 383, 387
monomers, 31, 85–8, 91, 178, 200, 231, 303, 388
mouse, 13, 67, 70, 102, 115–18, 123–6, 130–131, 136, 198, 199, 205–7, 211, 232, 281–3, 298–300, 302, 311–15, 351, 359, 380, 390–392, 407, 409, 410, 436, 439, 443, 444, 446, 450, 453, 454, 474–7, 479–82
mouse brain, 115, 134, 135
mouse mammary tumor virus (MMTV), 477
mouse tumor models, 437, 446, 447
MPR nanoparticles, 310–311
injection of, 310–311
MR angiography, 228, 230, 231
MRI see magnetic resonance imaging (MRI)
MR imaging, 109, 187, 252, 255–8
liposomal Gd-DTPA-enhanced, 77, 257
mRNA, 373–5, 378–81, 383, 384, 387–92
MRO2 see metabolic rate of oxygen consumption (MRO2)
  particles, 295
multifunctional nanoparticles, 455
multimodality imaging agents, 39–45, 196
multimodal silica nanoparticles, 191, 221

NA *see* numerical aperture (NA)
Nanobubbles (NBs), 2, 15, 306
nanocages, 113, 127, 128, 130–134, 158, 159, 301–3
nanocapsules (NCs), 83–104, 178
nanocarrier(s), 49–72, 103, 199, 386, 404, 409, 440
  labeling, 52, 57
  membrane, 54, 55, 58, 59
nanoemulsions, 432, 444–8, 455
nanomaterials, 210, 242–4, 263, 352, 402, 433, 436–9, 468, 469, 482
nanomedicine, 136, 195, 202, 212, 353, 368, 429–55
nanoparticle(s)
  absorbing, 101, 112, 113
  active, 122
  carriers, 171, 172
  characterization, 152, 173–7
  component, 18, 247
  concentrations, 125, 126, 185
  cross-linked, 2, 391
  design, 2, 18, 249
  development, 3, 445
  ellipsoidal, 22, 146
  fluorescent, 183, 264
  gold-based, 125, 436
  hybrid, 443
  hydrocarbonized porous silicon, 210
  inorganic, 195, 199, 402, 443
  labeling, 165
  new triple-modality MRI-photoacoustic-Raman, 323, 343
  optical, 14, 173, 178, 184
  organic, 166, 303
  polymer, 195, 200
  research, 2, 12, 175, 176, 180, 181, 185
  size of, 14, 159
  supermagnetic, 10–11
  surface, 121, 137, 139, 333
  surface modifications, 16–18
  using, 263
Nanopartz, 352–4, 357–9, 361, 368
  gold nanorods, 349, 351, 353, 361
nanorods, 125, 127, 129, 130, 138, 184, 202, 335, 353
nanoSPECT/CT, 471
nanostructures, magnetic nanoparticle-gold core–shell, 309, 310
nanovesicles, 52, 53, 55, 58–63, 65, 103
NCs *see* nanocapsules (NCs)
near-infrared (NIR), 12, 70, 71, 97, 100, 101, 111, 113, 114, 116, 117, 121, 122, 127–9, 132, 134, 135, 138, 139, 153, 173–9, 182, 202, 268–70, 287, 298, 300, 301, 303, 307, 308, 313, 326, 327, 334–6, 338, 352, 360, 363, 364, 366, 368, 405, 406, 409, 433, 434, 436–9, 445, 448, 451, 453, 454
near-infrared cyanine dyes, 12–13, 178
nephrogenic systemic fibrosis (NSF), 246, 249
night vision scopes, 13

NIR *see* near-infrared (NIR)
normalized rate of accumulation (NRA), 282–4
NP *see* nuclear pore (NP)
NPs, dumbbell, 28, 40, 41
NRs, 353–6, 362, 364, 368
NSF *see* nephrogenic systemic fibrosis (NSF)
nuclear pore (NP), 31–4, 37, 43, 44, 332, 376, 377, 386–92
nucleation, 14, 30, 31, 38, 40, 354–6
nucleic acids, 335, 374–9, 381–5, 411
numerical aperture (NA), 327, 329

OA *see* oleic acid (OA)
objects, 3, 121, 242, 247, 249, 274, 294, 295, 306, 331
OCT *see* optical coherence tomography (OCT)
OEF *see* oxygen extraction fraction (OEF)
oleic acid (OA), 31, 32, 34, 40, 441
oleylamine, 31, 32, 38–40
optical absorption, 112–15, 117, 122, 125, 127, 129, 131, 132, 134, 138, 139, 269, 294, 301, 303–5, 307, 308, 313, 314, 352, 438
optical coherence tomography (OCT), 264, 293, 295
optical imaging, 12–14, 71, 112, 264, 288, 311, 405–6, 415, 433, 440, 443, 444, 448, 451, 471–3, 476, 478, 480
optical techniques, 152, 173–84
optical tomography, 18, 285–8, 478
optoacoustic (OA)
  contrast agents, 111–39
  imaging, 111–39, 472
  imaging techniques, 111–39
  nanoparticles, 113–20
optoacoustic tomography (OAT), 112, 114–16, 118, 120–122, 124–6, 130, 131, 134–6
organic solution phase syntheses of IOMNPs, 30–32
oxide nanoparticles
  functional iron, 47, 456
  sized iron, 45, 186
  ultra-small gadolinium, 261
oxygen extraction fraction (OEF), 299, 300

PAC *see* perturbed angular correlation (PAC)
paclitaxel, 100, 402, 434
PAMAM dendrimers, 201, 236, 237, 239
Pandia, 349–68
PARACEST agents, 248, 249
paramagnetic gold nanoparticles (GNPs), 244, 367, 436–9, 448
paramagnetic liposomes, antitumor activity of, 78, 461
paramagnetic species, 224, 225, 240, 245, 247, 249
particle(s)
  organic, 152, 158, 162
  samples, 155–7, 164
  size, 13, 61, 69, 156, 157, 159, 162, 171, 172, 453
passive, 102, 201, 333, 409, 431, 435–8, 440–442, 446, 450, 452, 454
PAT *see* photoacoustic tomography (PAT)
PA waves, 294, 295, 297
PEG *see* polyethyleneglycol (PEG)
PEG-HAuNS, 134–6
PEG-NRs, 358–9
PEGylated liposomes, 18, 60, 65, 66, 102, 198, 402, 452
PEGylated nanoparticles, 18
PE micelles, 68
peptide nucleic acids (PNAs), 118, 376, 382, 383, 387
perfluorocarbons (PFCs), 444–6
perfluorocarbon nanoparticles, 240, 241, 307
  using dye-loaded, 190, 323, 462
permeability, 15, 43, 60, 62, 85, 87, 90, 91, 94, 103, 104, 172, 203, 228, 230, 240, 242, 358, 379, 383, 388, 390, 409, 435, 473, 476
perturbed angular correlation (PAC), 9
PET *see* positron emission tomography (PET)
PET/CT, 474, 475, 478
  images, 210, 211
PET Imaging, 66, 71, 165, 203–12, 383, 384, 388, 392, 418–21
PFC nanoemulsions, 444–7
PFC nanoparticles, 171, 181, 308
  dye-loaded, 308
PFCs *see* perfluorocarbons (PFCs)
phagocytic cells, 16, 57, 206

photoacoustic imaging, 311, 339, 341, 436, 445
 noninvasive, 317
photoacoustics (PA), 2, 14, 18, 111, 293–315, 339, 341, 436, 438, 439, 444
 contrast agents, 14, 309–12
 images, 124, 294, 295, 298, 300, 302–6, 308–11, 313, 314, 341
 signals, 14, 122, 123, 125, 129, 133, 138, 294, 295, 300, 301, 303–11, 313, 314, 341, 439
photoacoustic tomography (PAT), 114, 264, 293–315
photomultiplier tube (PMT), 6, 175, 177, 273, 275, 276
photons
 absorbed, 112, 266, 267
 ballistic, 271
 snake, 271, 331
photosensitizers, 70, 100, 453
photothermal imaging, 349, 358–64
photothermal therapy (PTT), 202, 358, 436–9
plants, 401–21
plasmonic nanoparticles, 126–38, 331, 333
PMT *see* photomultiplier tube (PMT)
PNAs *see* peptide nucleic acids (PNAs)
polyethyleneglycol (PEG), 18, 32, 57, 60, 61, 64, 65, 68, 71, 72, 125, 130, 197, 201, 207, 230, 231, 242, 243, 246, 364–6, 404–6, 411–13, 433, 435–7, 441, 452
polymeric nanoparticles, 207, 450–455
polymeric nanosystems, 448–54
polymerization, 85–7, 90, 91, 158, 159, 206, 453
polymer nanocapsules/NCs, 83, 84, 96–103
 synthesis of, 106, 317
polymers, 18, 28, 60, 85, 87, 103, 137, 230–234, 237, 247, 248, 365, 366, 402, 403, 432, 433, 435, 450
polymer shell, 84, 86, 90
polymersomes, 87, 89, 90, 97–100, 102–4, 303, 451
positron emission tomography (PET), 2, 6, 7, 18, 39, 50, 52, 66, 71, 165, 196, 203–12, 379, 383, 384, 388, 392, 418–21, 433, 436, 440, 443, 448, 467
postinjection, 58, 62, 98, 129, 134, 211, 238, 242, 303, 308, 312, 341
potato virus X (PVX), 403–5, 409, 410
PPCy-C8-loaded PFC nanoparticles, 308
preclinical, 1, 11, 12, 152, 196, 203, 293, 301, 304, 308, 312, 314, 329, 342, 351, 366, 430, 433, 440, 445, 451, 483
 imaging, 467–84
preformed nanocarriers, 52–9
preparation, nanocarrier, 52–3, 55, 58
proton signals, 167, 168
PVX *see* potato virus X (PVX)

quantum dots (QDs), 2, 13–14, 70, 99, 121–2, 180, 182–4, 186, 195, 326, 327, 406, 442
quantum yield, 13, 71, 177–80, 264, 267, 286, 307
quencher, 267, 268, 383, 384, 388–92

rabbits, 63, 67–70, 230, 231, 473, 478
radioactivity, 9, 62, 165, 202, 208
Radioactivity measurement of nanoparticles, 152, 165–6
radio-frequency (RF), 29, 30, 49, 103, 167, 168, 224, 248, 352, 413
radioisotopes, 5, 6, 9, 52, 53, 57, 196, 206, 418, 438, 440
radiolabeled nanoparticles, 165, 166, 196–212
radiolabeling, 58, 165, 166, 196, 198, 200, 202, 203, 205–7, 210
radionuclide, 9, 51–4, 58, 61, 165, 166, 196–7, 200, 202, 203, 205, 239, 392, 418, 441
Raman-based bioimaging, 333–8, 342
Raman imaging, 326, 339, 438, 439
Raman reporters, 326, 333, 335–42
Raman scattering, surface-enhanced, 325–42
Raman signals, 311, 328, 330, 331, 333, 336, 338, 341
rate of accumulation, normalized *see* normalized rate of accumulation (NRA)
reduction, 5, 9, 11, 40, 51, 62, 100, 234, 244, 355, 365, 437, 446, 447, 471–4, 480
region of interest (ROI), 71, 129, 265, 299, 472

relaxation, 10, 29, 30, 43, 55, 167–72, 223–6, 243, 248, 266, 294, 383, 413, 433
  rate, 10, 30, 171, 224–6, 248
  times, 10–11, 29, 30, 168–71, 224, 413
relaxivity, 30, 32–4, 36–9, 44, 54, 55, 57, 170–172, 199, 205, 224–8, 230–234, 236, 237, 239–47, 249, 414, 416–18, 420
  high, 171, 228, 231, 232, 234, 243, 249, 420
  higher, 32, 226, 230, 232, 236, 237, 240, 416
reporter, 338–42, 374, 383–8, 392
reticuloendothelial system (RES), 16, 18, 50, 54, 59, 61, 63, 197, 200, 201, 205, 207, 208, 211, 338, 358, 402, 404, 435, 437
RF *see* radio-frequency (RF)
RNA, 373–5, 377–81, 383, 385, 386, 392, 403, 406, 417, 418
RNA imaging, 379–88
ROI *see* region of interest (ROI)

scanning electron microscopy (SEM), 87, 89, 102, 127, 151, 153, 159–61, 337
scattering, 12, 112, 116, 127, 128, 135, 136, 173, 174, 182, 185, 186, 264, 269–75, 277–9, 281, 284, 285, 287, 300, 301, 436
scintigraphy, 55, 69, 471
SCK-NPs, 206
seed-mediated synthesis, 354–6
SEM *see* scanning electron microscopy (SEM)
sentinel lymph nodes (SLN), 122, 123, 133, 134, 200, 296, 302, 303, 307, 310, 312, 313
sequences, 169, 375–8, 380, 381
SERRS *see* surface-enhanced resonance Raman scattering (SERRS)
SERS *see* surface-enhanced Raman scattering
SERS-based bioimaging, 326, 328, 330, 333–8
silica nanoparticles, 210–212
  modified, 191, 221
silver nanoparticles, 131, 132, 137, 334–5
single nanoparticles, 121, 155
single-photon emission computed tomography (SPECT), 2, 6, 10, 18, 55, 71, 165, 196–203, 311, 312, 433, 436, 437, 440, 443, 448
  imaging, 10, 71, 165, 196–203
single-walled carbon nanotubes (SWCNTs), 304–5, 438, 439
single-walled nanotubes (SWNTs), 122–6, 202
size analysis, 152, 156–63
size-controlled synthesis of magnetite nanoparticles, 45, 456
SLN *see* sentinel lymph nodes (SLN)
small animals, 113, 129, 130, 180, 204, 275, 276, 295, 296, 298, 301–4, 307, 311–13, 329, 470, 472
small unilamellar liposomes, 73, 76
small unilamellar vesicles (SUVs), 8, 62, 63, 207, 209, 211, 440
solid nanoparticle systems, 433–9
solution, seed, 355, 356
source, 62, 179, 269, 272–8, 284–7, 328, 476
source–detector pair, 274, 278, 286, 287
SPECT *see* single-photon emission computed tomography (SPECT)
spectrophotometer, conventional, 173–5
spheres, 112, 174, 175, 177, 179, 180, 225, 228, 335, 356, 436
SPIONs *see* super paramagnetic iron oxide nanoparticles (SPIONs)
spleen, 5, 9, 16, 17, 42, 58, 62, 63, 125, 131, 197, 198, 200, 201, 210, 242, 243, 245, 299, 359, 366, 367, 404, 417, 418, 420, 444, 480
SPR *see* surface plasmon resonance (SPR)
SteP2-labeled PEG-liposomes, 205
stereotactic frame, 471
strand displacement probe, 383, 388–92
subtraction images, 123, 124
super paramagnetic iron oxide nanoparticles (SPIONs), 102, 200, 441, 452
surface
  dendrimer, 239, 448
  liposomal, 440, 441
  nanocarrier, 52, 55, 60
  surface-enhanced Raman, 325–42
surface-enhanced Raman nanoparticle tags, 343, 369, 458

surface-enhanced Raman scattering (SERS), 325–42, 368
labels, 339
probes, 326–8, 331, 333–6, 338, 339, 342
surface-enhanced resonance Raman scattering (SERRS), 331, 338
surface modification, 2, 11, 16–18, 40–42, 137, 202, 313, 435, 437
surface plasmon resonance (SPR), 113, 121, 126, 128, 129, 132, 353, 436, 437
surfactants, 31, 32, 38, 85–7, 90, 96, 137, 210, 337, 354, 444–7
SUVs *see* small unilamellar vesicles (SUVs)
SWCNTs *see* single-walled carbon nanotubes (SWCNTs)
SWNT-RGD, 123, 125
SWNTs *see* single-walled nanotubes (SWNTs)

T2-weighted, 30, 39, 168, 229, 246
targeted liposomes, 17, 50
targeted nanoparticles, 402
targeting molecules, 185, 240, 402, 433, 437–9, 450, 453
target mRNA, 381, 383, 384, 387, 389–91
TBSV *see* tomato bushy stunt virus (TBSV)
TCSPC *see* time-correlated single-photon counter (TCSPC)
TD systems, 273, 275, 276
TE *see* echo time (TE)
TEM *see* transmission electron microscopy (TEM)
TEM images of silver nanoparticles, 131, 132
Texas Instruments, 13
theranostic(s), 203, 429–34, 436, 440, 442, 444–6, 451, 452, 454–5, 468
nanomedicines, 50, 429–55
nanoparticles, 432, 450
nanosystems, 71, 430–432
thermal ablation of tumors, 350
throat, 12
time-correlated single-photon counter (TCSPC), 275, 276
tissue(s)
imaging, 12, 67–8, 118, 274, 287, 313
mammary, 477
molecular imaging of, 292, 463
normal, 50, 62, 264, 265, 268, 282, 283, 352, 380, 437
samples, 158
TNF *see* tumor necrosis factor (TNF)
tobacco mosaic virus (TMV), 403, 405, 417, 418
tomato bushy stunt virus (TBSV), 403
tomography, optical coherence, 264, 293, 295
translation, clinical, 295, 296, 314–15, 420, 443
transmission electron microscopy (TEM), 34, 38, 41, 86, 87, 98, 127, 151, 153, 157–9, 203, 356
images, 33, 36, 43, 88, 89, 127, 131, 132, 203, 204, 301, 302, 304, 306, 310, 336
transverse resolutions, 296–7
tri-n-octylphosphine oxide (TOPO), 14
tumor(s), 42, 43, 62–7, 69–71, 116–18, 123–5, 197–9, 210–212, 281–5, 313–15, 339–41, 350–352, 366–8, 436–41, 472, 474–8
high HER2 expression, 283
imaging, 63–6, 366
margins, 13, 315, 329, 352
mouse mammary gland, 313–14
neck, 198, 199, 284
region, 204, 282, 300, 435, 436, 438, 450
site, 130, 197, 198, 239, 282, 340, 420
size, 446, 472
solid, 102, 198, 199, 295, 360, 368, 409, 420
tissues, 58, 97, 102, 352, 477
transplanted, 477
uptake, 200, 201, 205, 206, 208, 210, 212, 437
tumor accumulation, 58, 62, 198, 200, 203, 409, 454
higher, 62, 202, 206
tumor angiogenesis, 116, 125, 412, 420, 440
molecular imaging of, 142, 456
tumor-bearing mice, 64, 201, 203, 208, 310, 358, 437
tumor cells, 103, 116, 117, 120, 121, 198, 247, 301, 363, 411, 441
tumor necrosis factor (TNF), 311, 436

tumor-to-muscle ratio, high, 203, 206, 210, 212
tumor-to-muscle SUV ratios, 211
turbid media, 175, 280–281, 331, 332

ultrasound imaging, 15, 52, 111, 112, 293–315, 445, 446
ultrasound scanner, 15
US and PA images, 295, 306

vesicles, 52–4, 58, 59, 61–3, 86, 90, 93, 102, 197
vimentin, 406, 407, 409, 420
  surface, 407, 408
  surface expression, 407, 408
viral nanoparticle potato virus, 423
viral nanoparticles, 184, 402, 403
  engineering, 260, 422
  PEGylated, 423
  using, 424, 425
virus, cowpea chlorotic mottle, 245, 403
virus-based nanoparticles (VNPs), 402–6, 409–11, 413–21
  pharmacokinetic properties of, 405
  using fluorescence, 413
viruslike particles (VLPs), 402–4
visualization, 4, 8–10, 54, 61–2, 65–71, 116, 164, 198, 199, 224, 228–30, 232–4, 238, 239, 242, 243, 246, 247, 339, 341, 407, 409, 412, 482
  aid, 169, 170

wavelength of light, 272, 273

x-ray imaging, 4–5, 244

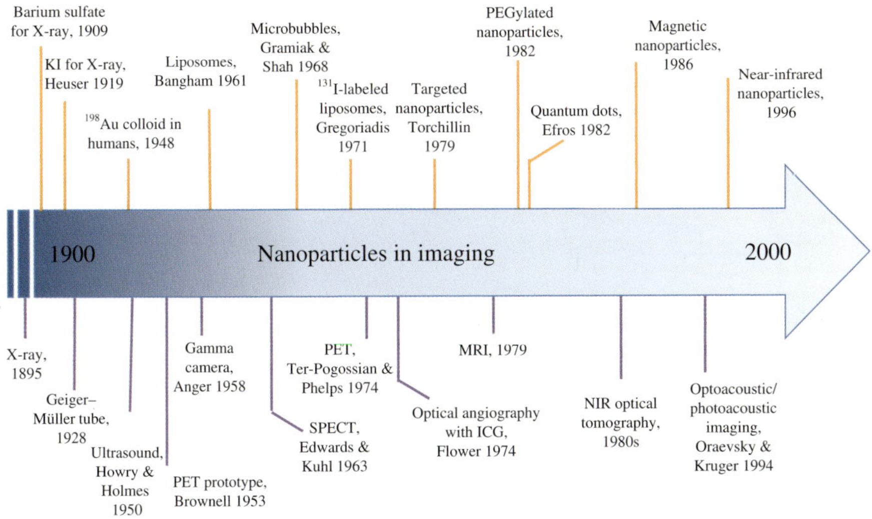

FIGURE 1.2 Timeline of the most important events in the development of nanoparticles for imaging and diagnostics covering the period from the twentieth century. The upper part corresponds to nanoparticles, and the lower part to the development of imaging modalities.

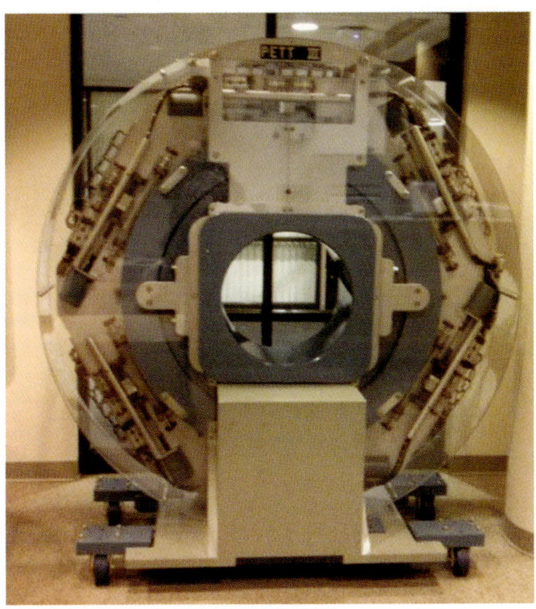

FIGURE 1.4 First human PEN scanner PETT III (1974) located in the hall of the Department of Radiology Washington University School of Medicine in St. Louis, where this scanner had been invented. The inventors had given the name "positron emission transaxial tomography" (PETT). The name was reduced to PET because transaxial was no longer the only plane used for image reconstruction.

Nanotechnology for Biomedical Imaging and Diagnostics: From Nanoparticle Design to Clinical Applications, First Edition. Edited by Mikhail Y. Berezin.

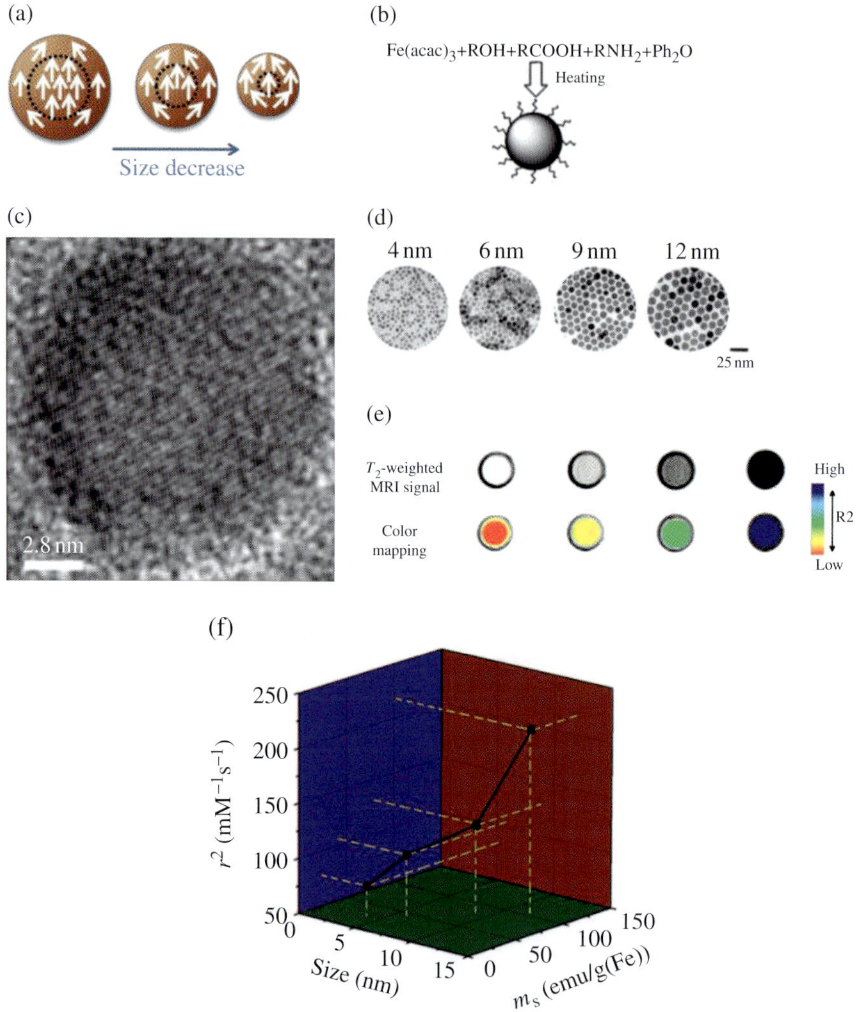

FIGURE 2.4 (a) Schematic illustration of size-induced surface spin-canting effect in MNPs. (b) The synthesis of Fe_3O_4 NPs *via* thermal decomposition of $Fe(acac)_3$ and (c) the HRTEM image of a 16 nm Fe_3O_4 NP. (d)–(f) Size effect on MRI contrast: (d) TEM images of Fe_3O_4 NPs with different sizes, (e) the relevant T_2-weighted MR images and color-coded maps, and (f) relation of size, saturation magnetization, and relaxivity r_2. (Reprinted with permission from Refs. [32] and [33]. © American Chemical Society).

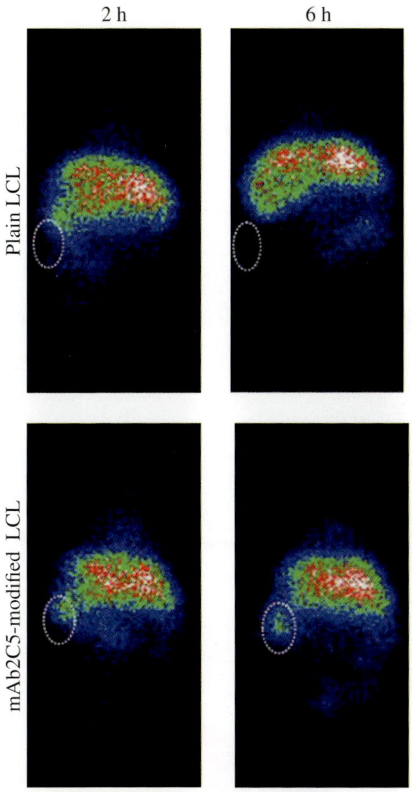

FIGURE 3.3 Whole-body gamma-images of LLC tumor-bearing mice receiving various ^{111}In-labeled liposomal preparations (nontargeted long-circulating liposomes (LCLs), plain LCL; tumor-targeted LCLs modified with tumor-specific monoclonal antibody 2C5 (mAb2C5), mAb2C5-modified LCL), captured at different times post IV tail vein injection. Tumor is encircled. (Reprinted with kind permission from Ref. [101]. © Springer Science and Business Media.)

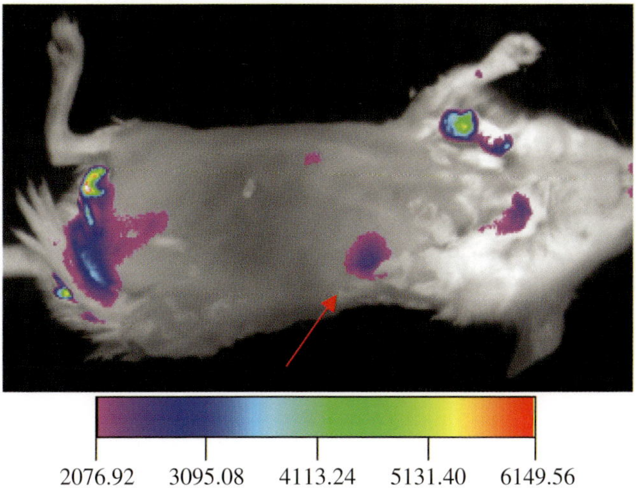

FIGURE 3.5 NIR imaging. Side view of mouse injected with Alexa-labeled micelles, at 2 h after the injection. Arrow points the tumor. (With permission from Ref. [153]. © DovePress.)

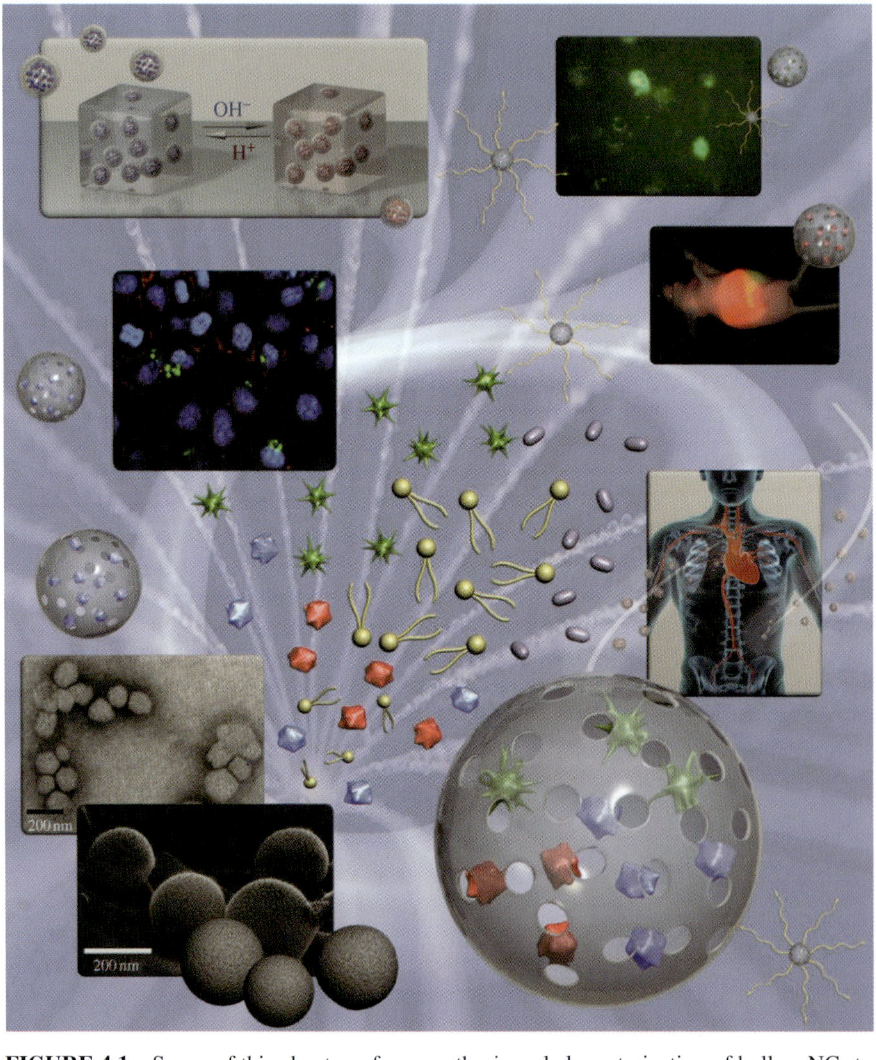

FIGURE 4.1 Scope of this chapter—from synthesis and characterization of hollow NCs to building nanodevices and biomedical applications.

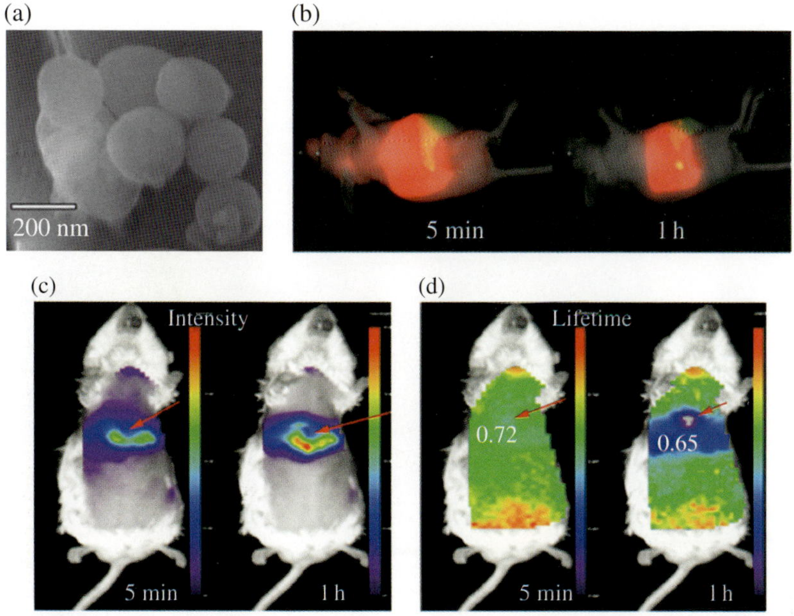

FIGURE 4.20 (a) SEM images of NCs with ICG after treatment with 4% w/v BSA. (b) Fluorescence intensity after administration of ICG-NCs. Red (ICG) channel emission (800 nm) and green channel emission (700 nm) correspond to endogenous fluorescence (from the diet). Red fluorescence is primarily localized in the liver. (c) Intravenous (IV) injection of ICG-NC shows an accumulation of the fluorescent dye in the liver. Left, fluorescence intensity; right, fluorescence lifetime, ex/em = 775/820 nm.

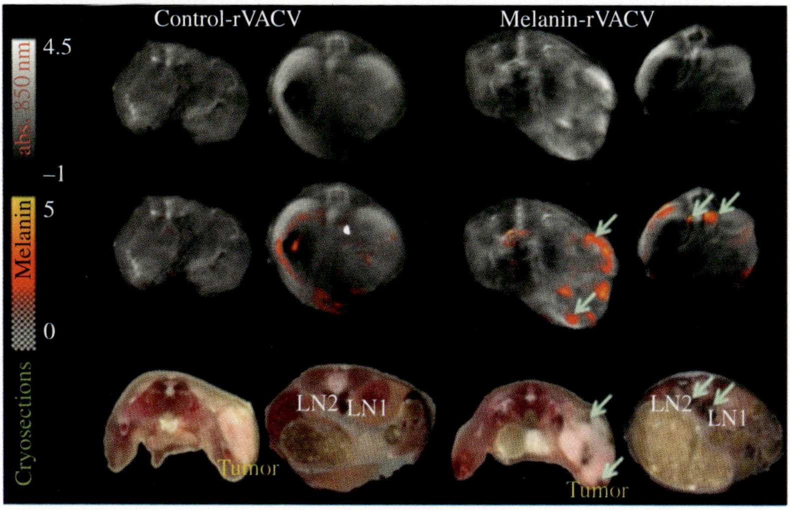

FIGURE 5.4 Optoacoustic imaging of melanin-rVACV in live mice. Multispectral optoacoustic tomography (MSOT) of PC-3-RFP tumor and lymph node metastasis (LN1 and LN2) bearing live mice 14 days after rVACV injection. (Top) Background optoacoustic images taken at 850 nm excitation wavelength. (Middle) Melanin spatial distribution (pseudocolor) is overlaid on the background images using a variable transparencies function. (Bottom) As indicated by the *arrows*. (With permission from Ref. [54]. © PNAS.)

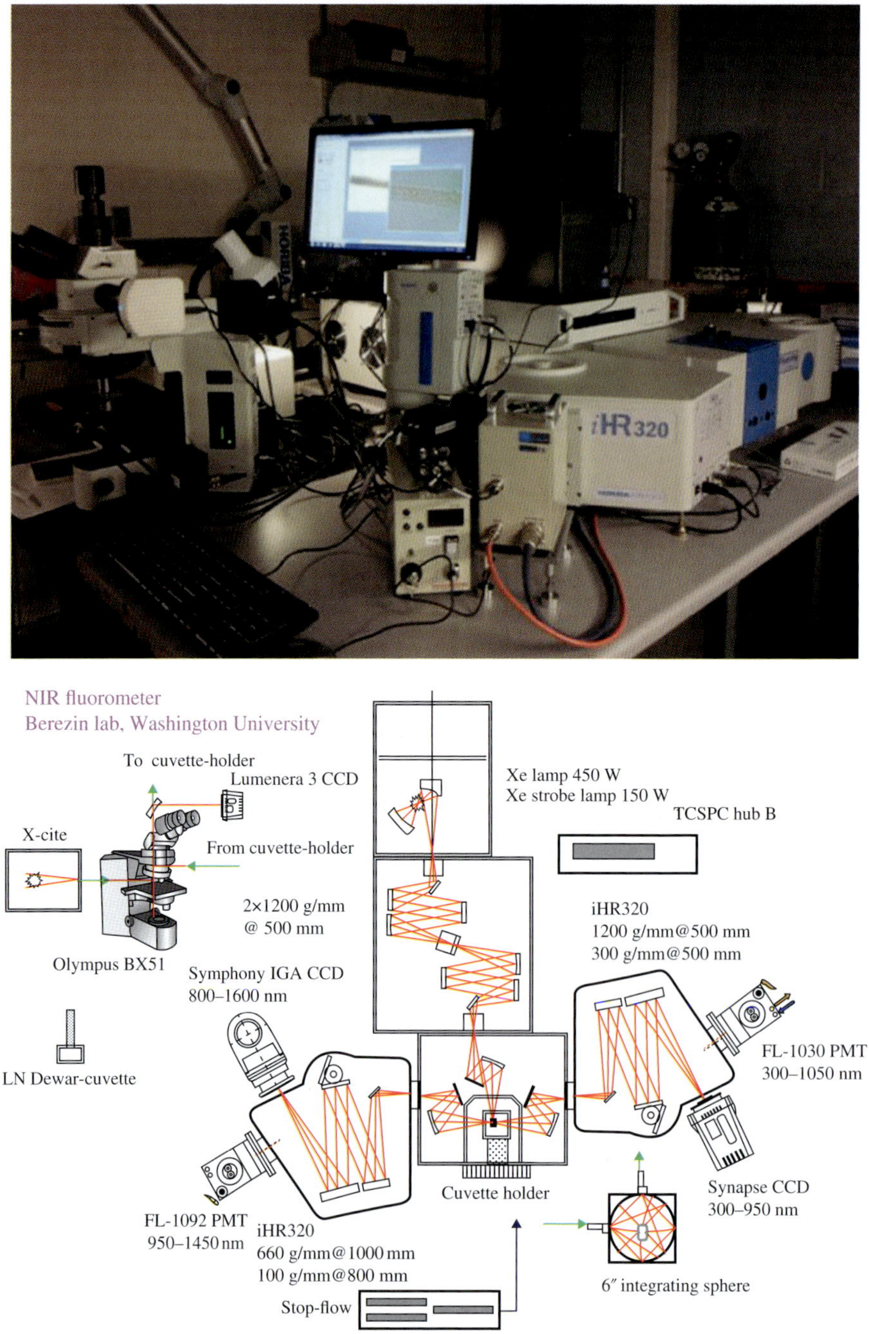

FIGURE 6.19 NIR spectrophotometer "NanoLog" for nanoparticle research located in the Optical Spectroscopy Core, Washington University, St. Louis. The system features a T-shaped geometry; Xe lamps; a double grating excitation monochromator for highest stray-light rejection; cuvette holder with 90° angle and front face detection for measuring fluorescence in solution, solid, and opaque systems; and imaging spectrographs with a kinematic turret for multiple gratings and multiple NIR detectors. (The schematics were kindly provided by Horiba, Inc.)

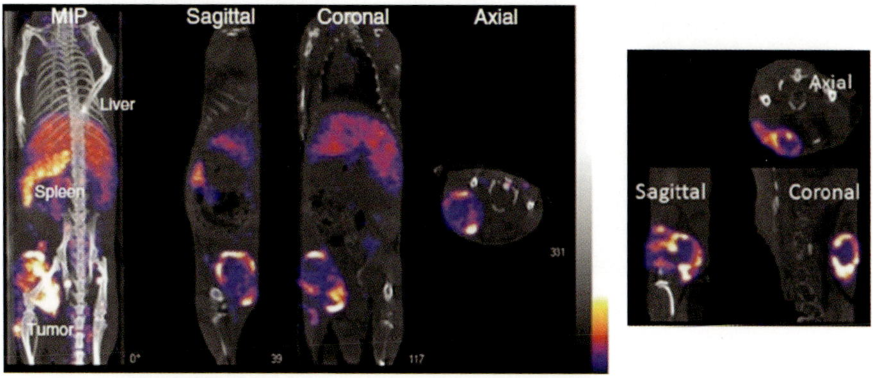

FIGURE 7.1 *In vivo* SPECT/CT images of sarcoma 180-bearing mice 24 h after administration of ^{111}In-labeled liposomes. Case #1 (left panel): whole-body images containing maximum intensity projection (MIP) and representative sagittal, coronal, and axial slice images containing the tumor. Case #2 (right panel): representative sagittal, coronal, and axial slice images of the tumor site in another mouse. The SPECT projection data were acquired under the optimized conditions for *in vivo* imaging. The acquisition time was 150–300 s per projection with 24 (6×4) directions, total = 1–2 h. (With permission from Ref. [50]. © Springer.)

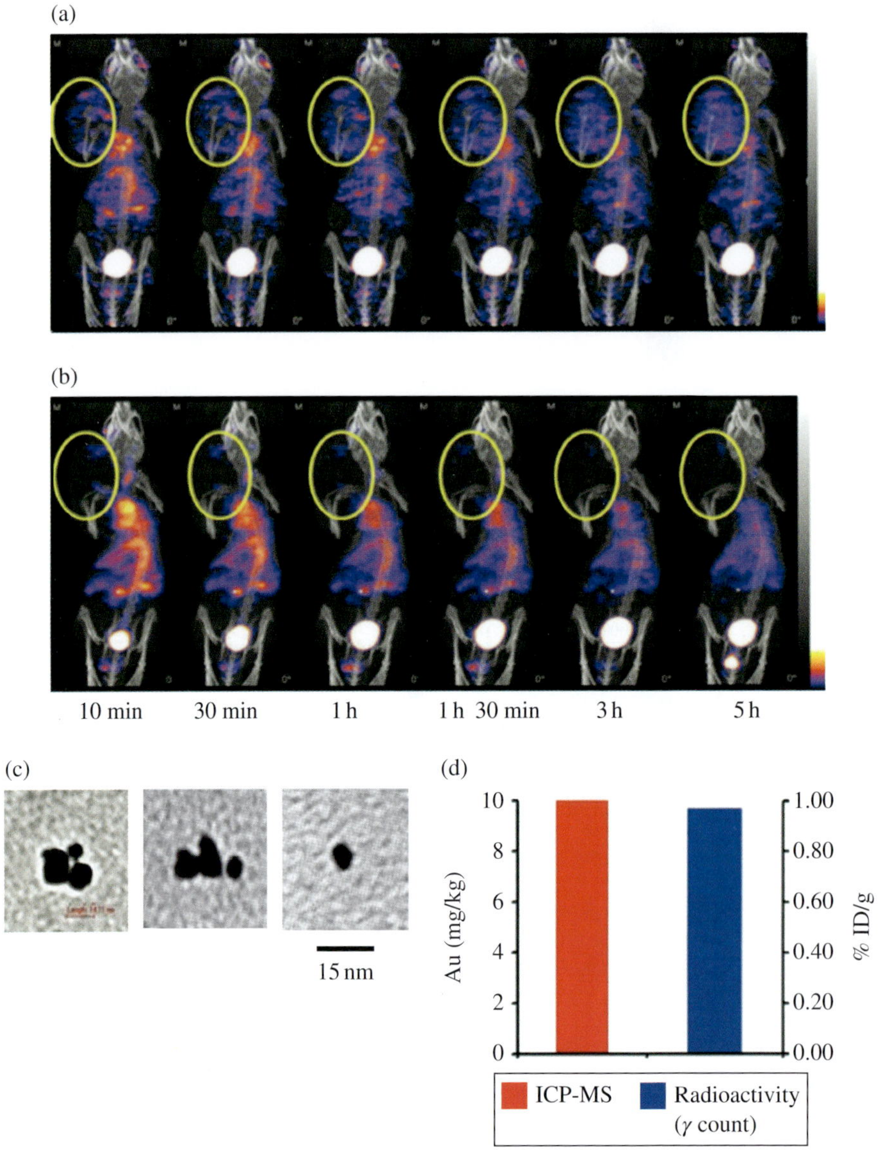

FIGURE 7.2 (a) Small animal SPECT/CT of ^{125}I-cRP-AuNPs and (b) cold-form-blocked ^{125}I-cRGD-AuNPs in nude mice bearing U87MG tumors xenografts after injection through tail vein. The SPECT/CT studies were performed serially from 0 to 5 h after injection of 11.1 MBq radiolabeled AuNPs. Yellow circles indicates tumor region. (c) TEM images of gold nanoparticles obtained from urine samples. (d) Biodistribution comparison of cRPAuNPs (inductively coupled plasma mass spectrometry [ICP-MS]) and iodine-125 (γ count). Data obtained from tumor 1 h after injection. %ID = injection dose. (With permission from Ref. [87]. © Wiley.)

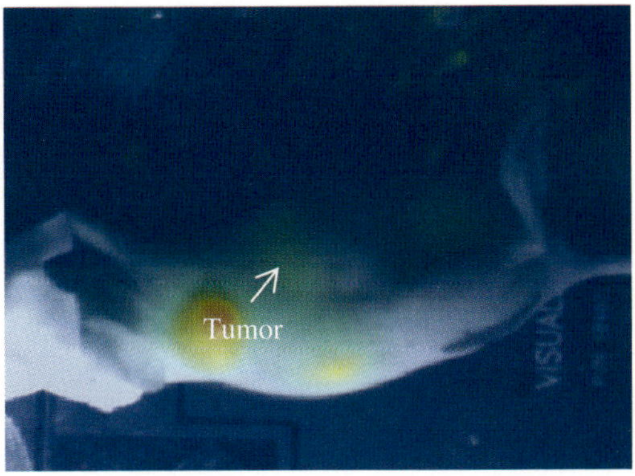

FIGURE 9.12 Superimpose of planar white-field and fluorescence images after injection of HER2-targeted fluorescence probe in a mouse with high HER2-expressed tumor.

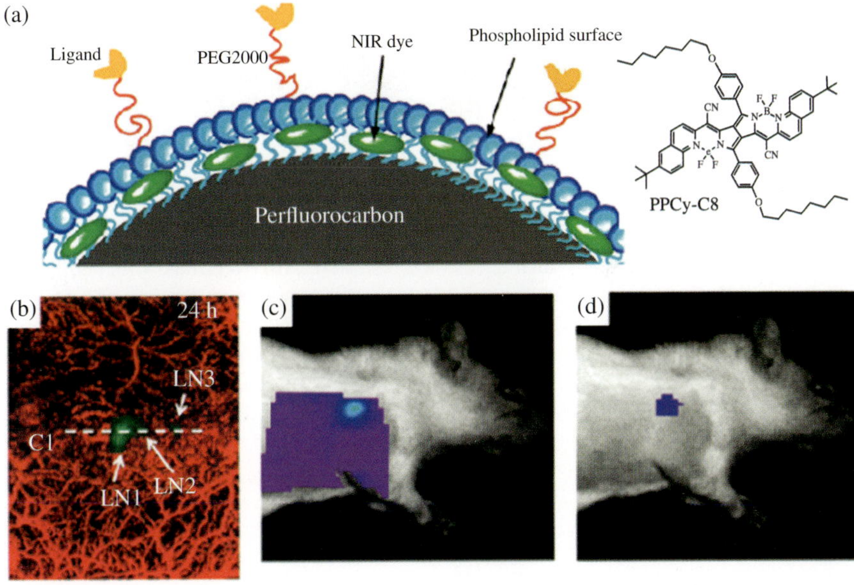

FIGURE 10.11 Dual-modality photoacoustic (PA) and fluorescence (FL) contrast agents. (a) Schematic of dye-loaded perfluorocarbon nanoparticles and its chemical structure. (b) *In vivo* noninvasive PA mapping of sentinel lymph node in small animals. *In vivo* (c) intensity- and (d) lifetime-based FL mapping of the sentinel lymph node in a rat. (Reprinted with permission from Ref. [144]. © American Chemical Society.)

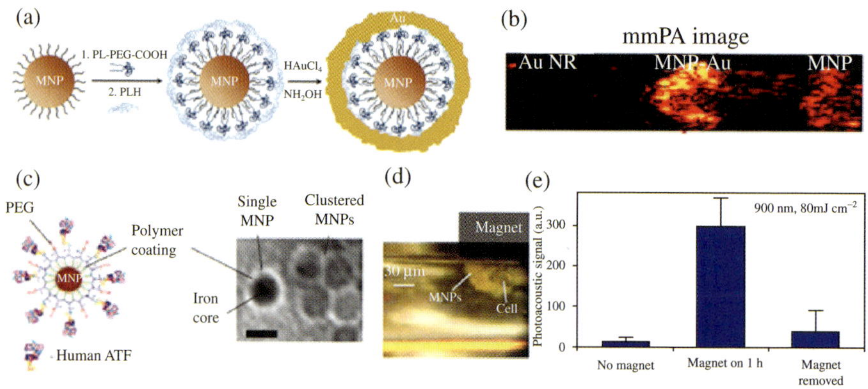

FIGURE 10.13 Magnetomotive photoacoustic (PA) contrast agents. (a) Schematic of magnetic nanoparticle-gold core–shell nanostructures (MNP-Au) (Reprinted with permission from Ref. [147]. © Macmillan Publishers Ltd.). (b) Magnetomotive PA imaging of gold nanorods (AuNR), MNP-Au, and magnetic nanoparticles (MNPs). (c) Schematic (left) and TEM image (right) of MNPs. (d) *In vitro* magnetic trapping of MNP-labeled cancer cells at a flow velocity of 5 cm/s. (e) *In vivo* PA detection of magnetic enrichment of circulating tumor cells in tumor-bearing mice. (Reprinted with permission from Ref. [148]. © Macmillan Publishers Ltd.)

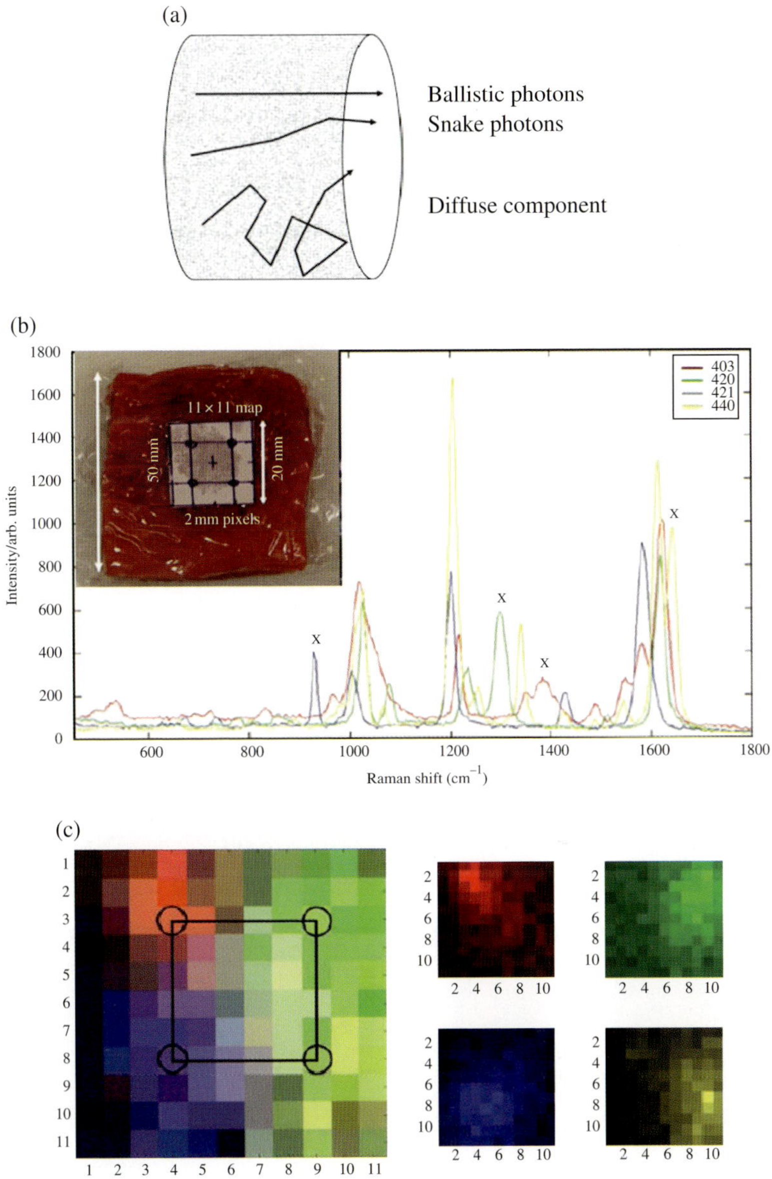

FIGURE 11.3 (a) Three main light components within turbid media. (b) SERS spectra of Cabot nanoparticles (NP) with different labels. The Xs signify the identity peaks used to measure the NP flavor signal in the SESORS spectrum. Inset image is a 20 × 50 × 50 mm porcine block with grid used to align injection points of NP. (Reproduced with permission from Ref. [25]. © Royal Society of Chemistry.) (c) SERS mapping measured in a 11 × 11 grid, pixel size 2 mm. Left image shows all different signals plotted together with the injection points marked, and right image shows the SERS signal of each labels. (Reproduced with permission from Ref. [26]. © Royal Society of Chemistry.)

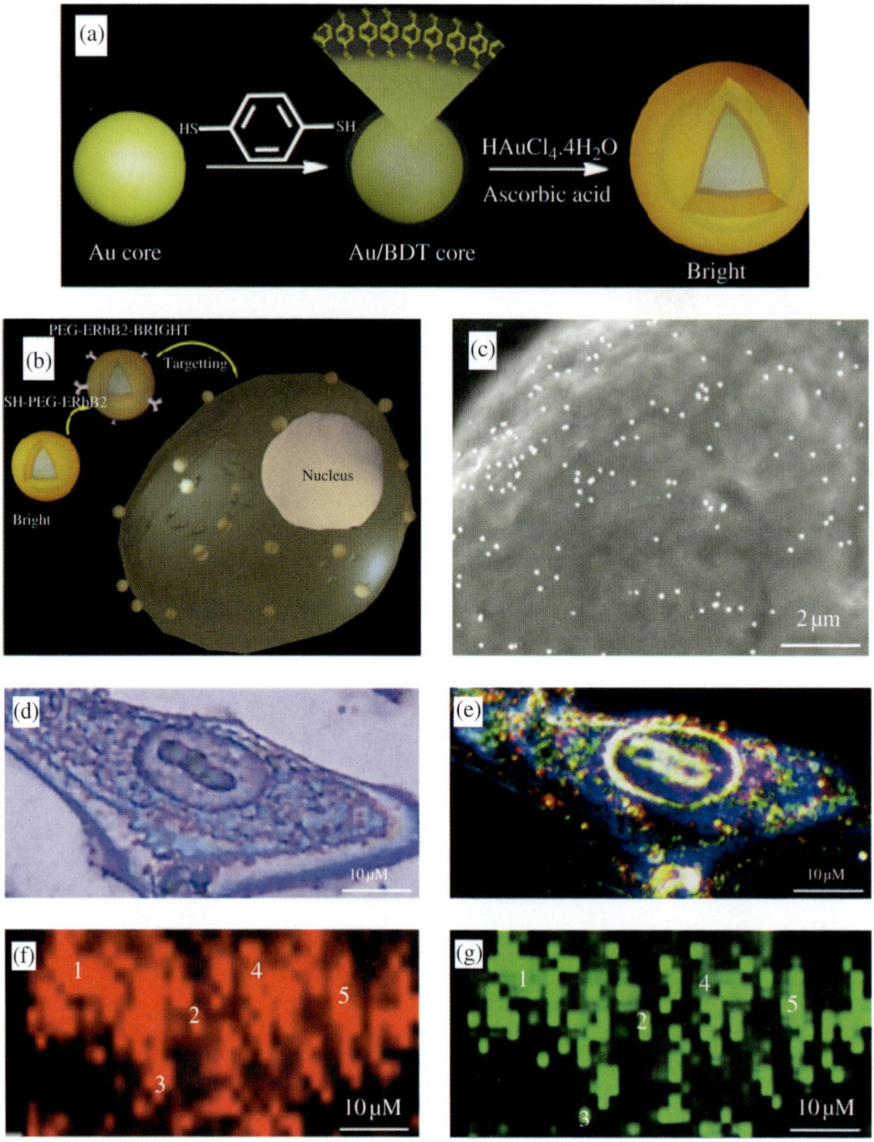

FIGURE 11.6 (a) Synthesis of BRIGHTs involves the exchange of gold nanoparticle surfactant with a Raman reporter. (b) Schematic diagram illustrating the biofunctionalization *in vitro* targeting of BRIGHTs to SKBR3 breast cancer cells. (c) SEM image of BRIGHT-targeted fixed SKBR3 cell reveals that the nanostructures largely remain as individual particles. (d) A bright-field optical micrograph of BRIGHT-targeted SKBR3 cells and (e) corresponding dark-field micrograph showing the intense Rayleigh scattering from BRIGHTs. Intensity map of 1078 cm^{-1} Raman band from BRIGHTs targeted to SKBR3 cell collected using (f) 785 nm and (g) 632.8 nm as excitation sources, respectively. (Reprinted with permission from Ref. [61]. © Wiley.)

FIGURE 12.4 Commercial samples of gold nanorods from Nanopartz, Inc.

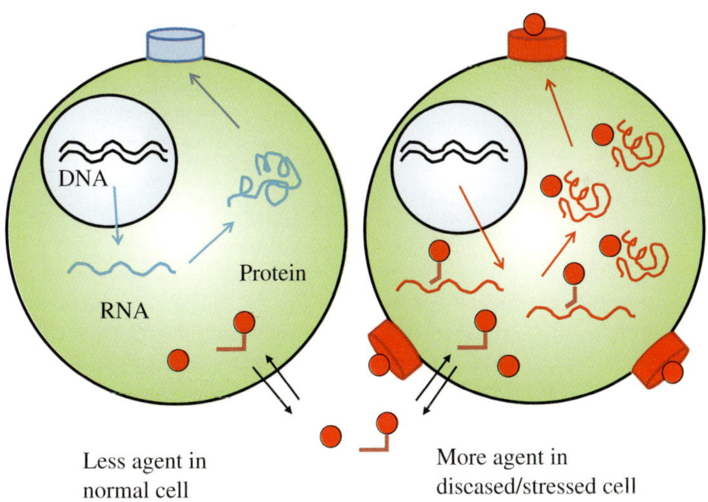

FIGURE 13.1 Idealized picture of gene expression imaging in which antisense probes accumulate in a cell expressing a unique or overexpressed mRNA sequence. If an always-on probe is used, such as a radiolabeled antisense agent, the probe must be able to enter the cytoplasm, and any unbound probe or reporter must be able to efficiently exit the cytoplasm. Turn-up probes only need to enter efficiently. The problem is that antisense agents themselves are membrane impermeable and thus must be attached to cell-penetrating carriers. Another problem is that mRNA is folded and bound to proteins, making many sites inaccessible to a probe in its folded state.

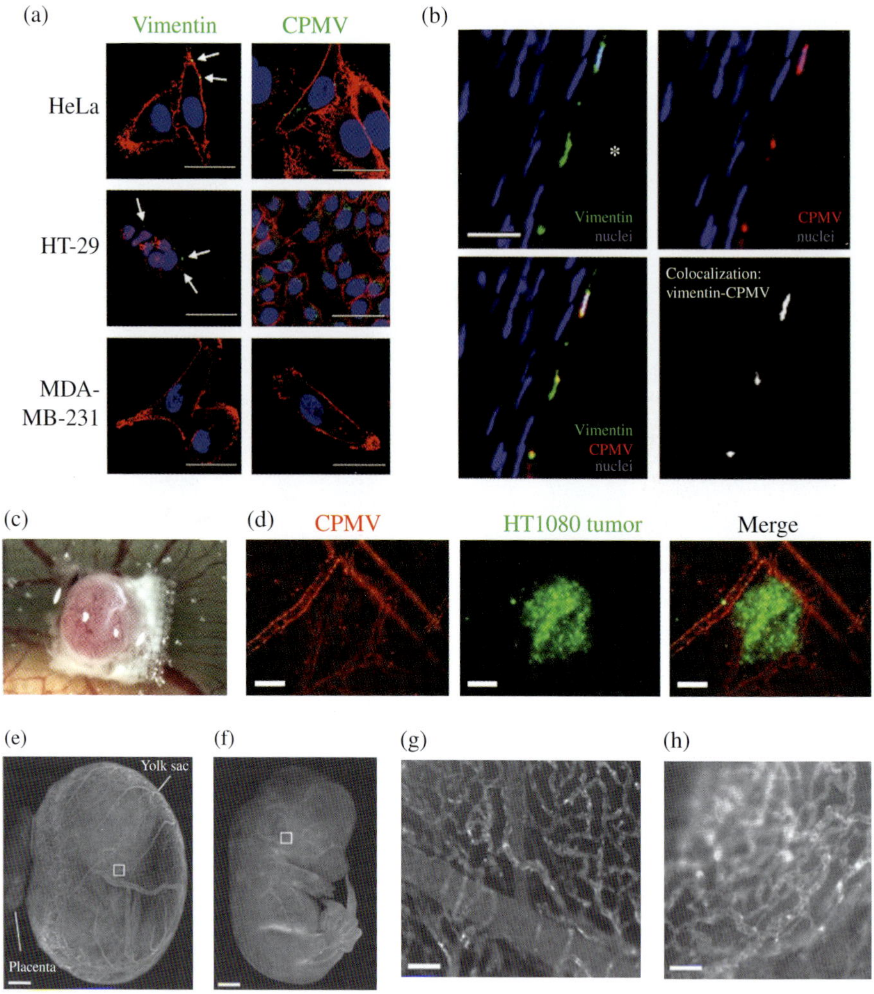

FIGURE 14.3 CPMV has natural affinity for cells through interaction with surface vimentin. (a) Left panel: surface vimentin expression of HeLa, HT-29, and MDA-MB-231 cells (indicated by arrows). Live cell staining was performed using the monoclonal antivimentin antibody v9. Red, wheat germ agglutinin-Alexa Fluor 555 (WGA-A555) (cell membrane); green, surface vimentin; blue, DAPI (nuclei). Right panel: CPMV-Oregon Green 488 (O488) nanoparticles uptake by HeLa, HT-29, and MDA-MB-231 cells. Red, WGA-A555 (cell membrane); green, CPMV, blue: DAPI (nuclei). Scale bar is 30 mm. (Reproduced with permission from Ref. [54]. © Wiley.). (b) Colocalization of CPMV and surface vimentin on rat aortic endothelium. Green, vimentin; red, CPMV; blue, DAPI (nuclei); white, vimentin–CPMV colocalized signal. Scale bar is 30 mm, and * indicates aorta lumen. (Reproduced from Ref. [71].) (c) Bright-field image of HT1080 tumor CAM model after 7 days. Nylon mesh grid is used for angiogenesis quantification. (d) Image of CPMV-A555 lining the vasculature taken right after HT1080 tumor xenograft was performed (left), GFP-expressing HT1080 tumor (middle), and merged images (right). Scale bar is 100 μm. (e, f) Image of 11.5-day chick embryo injected with CPMV-A555 with yolk sac intact (e) and removed (f). White boxes are of magnified regions shown in (g) and (h). Scale bar is 1.1 mm. (g, h) Magnified images of yolk sac vasculature and capillaries in the head region, respectively. Scale bar is 25 μm. (Figure (c–h) were reprinted with permission from Ref. [53]. © Macmillan Publishers, Ltd.)

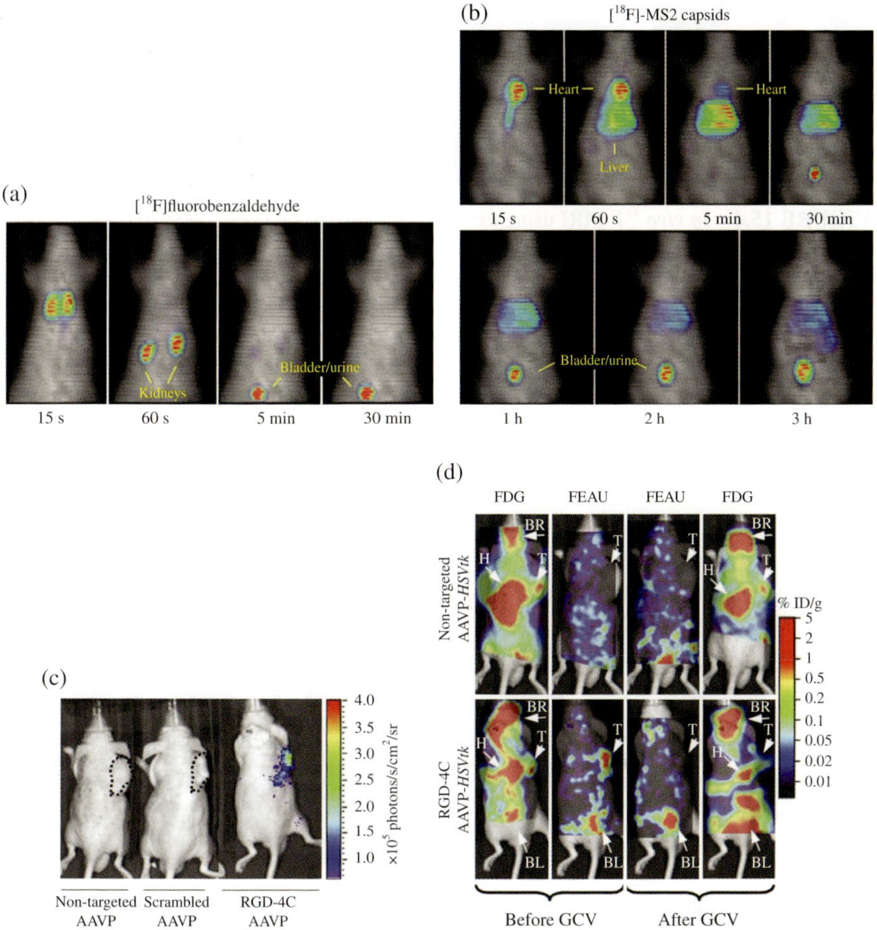

FIGURE 14.7 PET images of Sprague-Dawley rats injected with [^{18}F]fluorobenzaldehyde (a) and [^{18}F]-MS2 (b). Grayscale transmission scan renderings are shown in the background, and intensities were scaled to the maximum in each frame. (Figure (a, b) were reproduced with permission from Ref. [109]. © Springer Science and Business Media.) (c) Bioluminescent images of luciferase expression in nude mice bearing DU145 tumor xenografts ten days after delivery of AAVP. The targeted RGD formulation showed successful gene delivery, while the nontargeted and scrambled formulations did not. (d) PET images of [^{18}F]FEAU and [^{18}F]FDG accumulation in nude mice bearing DU145 tumor xenografts before and after GVC therapy. BL, bladder; BR, brain; H, heart; T, tumor. (Figure (c, d) were reprinted with permission from Ref. [110]. © Elsevier.)